Singer's

Engineering Mechanics

Statics and Dynamics

Third Edition (SI Units)

Singer's

Engineering Mechanics
Statics and Dynamics

Third Edition (SI Units)

Ferdinand L. Singer

SI Edition prepared by:

K. Vijaya Kumar Reddy
Professor of Mechanical Engineering,
JNT University, Hyderabad.

J. Suresh Kumar
Associate Professor in Mechanical Engineering,
JNT University, Hyderabad.

BSP BS Publications
An Imprint of **BSP Books Pvt. Ltd.**
4-4-309/316, Giriraj Lane,
Sultan Bazar, Hyderabad - 500 095

Singer's Engineering Mechanics Statics and Dynamics, Third Edition (SI Units)
2010 – SI edition prepared
***by* K. Vijaya Kumar Reddy and J. Suresh Kumar**

23rd *Reprint* 2026

Published by:

BSP BS Publications
An Imprint of **BSP Books Pvt. Ltd.**
4-4-309/316, Giriraj Lane, Sultan Bazar,
Hyderabad - 500 095
Phone: 040 – 23445688
e-mail: info@bspbooks.net
website: www.bspbooks.net

Printed at

Aditya Offset Process (I) Pvt. Ltd.
Hyderabad.

Price: Rs. 995.00

ISBN: 978-81-7800-219-4 (Paperback)

Preface to Adapted Edition

This book has its identity throughout the world among the engineers, in particular civil and mechanical engineering as Singer's Engineering Mechanics: Statics and Dynamics.

Now, it has been adapted into SI units for the convenience of students with the same vision as the previous edition.

K. Vijaya Kumar Reddy
J. Suresh Kumar

Preface to Third Edition

The previous editions endeavored to show how a few basic concepts - the relations between a force and its components, the principle of moments, and Newton's laws of motion - may be combined and applied to a wide variety of practical situations that are encountered by engineers. Another purpose was to help the student develop the logical, orderly processes of thinking that characterize an engineer. Both of these objectives have been emphasized to an even greater extent in this third edition.

This book has been completely rewritten and expanded to include the suggestions received from many users of the previous editions. There are two very significant changes: a reorganization that coordinates usually disparate topics into unified entities in both statics and dynamics and an integrated treatment of scalar-geometric analysis and vector analysis.

No previous experience in vector analysis is necessary because each application of vector notation is preceded or accompanied by detailed explanations that emphasize its geometric significance. The use of force multipliers permits vectors to be expressed in a simple manner rather than the decimalized notation so common in other books. In addition, force multipliers also simplify the transition from scalar-geometric notation to vector notation and vice versa.

Neither vector notation nor the scalar-geometric method is used to the exclusion of the other; instead each is employed where it is most appropriate. For two-dimensional analysis, the scalar-geometric method is retained as the simplest and most direct approach. For three-dimensional analysis and the development of general concepts, vector notation provides the most direct approach, particularly for explaining the effects of a change in direction of a body's motion. But primarily the book emphasizes how to analyze and comprehend the various areas of mechanics irrespective of the mathematical method.

Other noteworthy changes and additions in Part I, Statics, include a unified and coordinated treatment of plane and space statics; an expansion of the analysis of structures to incorporate the criteria for stability; a more organized approach to frames containing multi-force members; an optional section on space trusses; and additional material on friction, shear and bending moment in beams, and virtual work.

Part II, Dynamics, has been reorganized and rewritten to take full advantage of vector notation. Especially noteworthy is the chapter on the general principles of dynamics that is preceded and followed, respectively, by the kinematics and kinetics of a particle. The immediate application of kinematics to the kinetics of a particle contributes to understanding their interaction and reinforces each. Similarly, the kinematics of plane motion immediately precedes its application to kinetics. These fundamental approaches to the force-acceleration method are followed by chapters on the work-energy and impulse-momentum methods. Much thought was given to adding the particle motions of these chapters to particle kinetics but was discarded in favor of concentrating on basic dynamics instead of a multiplicity of approaches. However, the book is written so that those portions of energy and momentum methods that are pertinent to particle motion can be combined with particle dynamics, if so desired. After a chapter on spatial dynamics, the concluding chapter is, an introduction to mechanical vibrations, which seems to be the natural place to discuss simple harmonic motion and various types of pendulums without interrupting the continuity of other basic developments.

Each portion of the book contains more material than can reasonably be covered in a semester. Sections on advanced or specialized topics are identified by an asterisk. These sections are independent of the rest of the book so that flexibility of course outlines or objectives may be attained.

It is hoped that these topics, as well as the others in the text, are presented in a manner that will relieve instructors of the burden of detailed explanations, thereby leaving them free to expand and develop those details that they wish to emphasize. Not only are the theoretical aspects rigorously presented, but their practical applications are also emphasized. Numerous illustrative problems (most of them new) show in detail how principles are applied. The explanations are complete-nothing is taken for granted. As each equation or principle is applied, it is stated in brackets at the left-hand side of the page. Adjacent to it, values are substituted in the respective order in which the symbols appear in the equation. This procedure enables students to readily follow the various steps of the solution without continually referring to the body of the text.

But primarily, the students' points of view and their special problems have been kept in mind. The prose style is almost conversational, and, without being too brief or lengthy, should be conducive to self-study. The physical significance of fundamental concepts and the assumptions and limitations made in developing them are carefully discussed so that understanding is achieved and memorization reduced to a minimum. The summaries appended to each chapter are intended to give the student a concise statement of key elements, which should be useful in review and post-college work.

There is a completely revised set of more than 1200 problems that have been carefully prepared to give wide variety in type and scope. It is anticipated that an instructor can make an adequate selection without repetition for four or five years. The problems have been arranged approximately in their order of difficulty and answers to two-thirds of them have been given; the others may on occasion be used for quizzes or to demonstrate mastery of concepts. No problem is presented that is not preceded by an adequate text discussion of the concepts involved, and usually numerical values that simplify the arithmetical computations are selected.

The numbering plan used in this revision enables one to locate quickly any cross reference. With this plan, all figures, equations, tables, and problems are preceded by the chapter and section numeral in which they appear and are numbered consecutively throughout each section. Figures for assigned problems are given the same number of the problem to which they refer in order to simplify correlation of a problem figure with the corresponding problem data.

The author wishes to acknowledge his indebtedness to his colleagues all over the nation for the many valuable suggestions that they so generously offered him for this revision. To identify them individually would make too lengthy a list (with possibly an inadvertent omission), but each has received my thanks. Particular mention, however, must go to the publisher's reviewers who painstakingly noted inconsistencies and offered suggestions for changes that were very helpful, especially those of Professor Andrew Pytel of The Pennsylvania State University and Professor William C. Plumtree of California State University, Los Angeles. Although great care was taken to eliminate errors, it is inevitable that some will still be found. The author will appreciate being informed about them and will welcome any comments that readers may care to offer.

Ferdinand L. Singer

List of Symbols and Abbreviations*

A	area, ampltidue of vibration
a, a	acceleration
$\bar{\mathbf{a}}$, $\bar{a}$	acceleration of mass center
$\mathbf{a}_n$, a_n	normal acceleration
$\mathbf{a}_t$, a_t	tangential acceleration
b	breadth, width
C	Couple
C	centroid, instant center
d	depth, diameter, distance, moment arm
E	modulus of elasticity in tension or compression
$\hat{\mathbf{e}}$	unit vector along axes whose orientation changes
e	coefficient of restitution, natural base of logarithms
e_{st}	static elongation
F	frictional resistance
f	coefficient of friction, frequency
G	modulus of elasticity in shear, gravity or mass center
g	gravitational acceleration (9.81 m/sec^2)
h	height
I	moment of inertia
$\bar{I}$	centroidal moment of inertia
$\hat{\mathbf{i}}, \hat{\mathbf{j}}, \hat{\mathbf{k}}$	unit vectors along coordinate axes
J	polar moment of inertia
$\bar{J}$	centroidal polar moment of inertia
KE	kinetic energy

* With very few exceptions, these symbols and abbreviations agree with those approved by the American Standards Association.

k	radius of gyration, spring constant
L	length
M	moment of force
M	bending moment
m	mass (W/g)
$\hat{\mathbf{n}}$	general unit vector
n	revolutions per minute, subscript denoting normal direction
P, Q, F	forces or loads, concentrated
P_{uv}, P_{xy}	products of inertia
p	linear momentum
R	reaction
R_F	resultant force
r	absolute position vector
r, R	radii
RW	resultant work
s	magnitude of displacement, arc length
t	thickness, time, subscript denoting tangential direction
U	work
u, υ, *w*	rectangular coordinates
V	volume, transverse shearing force
v, v	velocity
$\bar{\mathbf{v}}, \bar{v}$	velocity of mass center
W, W	weight, load
w	load per unit length
X Y Z	inertial reference
xyz	moving reference frame
x, y, z	rectangular coordinates, moving or fixed
$\bar{x}, \bar{y}, \bar{z}$	coordinates of centroid, center of gravity, or mass center
$\boldsymbol{\alpha}$, α	angular acceleration
α, β, γ. . .	angles
β	phase angles, angle of contact for belt friction
γ	weight per unit volume (specific weight)

δ r, $\delta\theta$	virtual displacement (linear or angular)
δ s, $\delta\theta$	magnitude of virtual displacement (liner or angular)
δU	virtual work
θ	angular coordinate, second Eulerian angle
ρ	relatives position vector
ρ	mass density, radius of curvature, variable radius
ρ, θ	polar coordinates
τ	period, periodic time
ϕ	angle of friction
φ	first Eulerian angle
ψ	third Eluerin angle
$\boldsymbol{\omega}$, ω	angular velocity
$\boldsymbol{\omega}$ or ω_n	natural circular frequency ($2\pi f$)
$\boldsymbol{\Omega}$	angular velocity of moving reference
Ω	angular velocity of precession
^	circumflex denoting unit vector
deg	degrees
FBD	free-body diagram
cm	centimeter
mph	meter per hour
W	Watt
m	metre
N	newtons
kw	kilowatt
rad	radians
rev	revolutions
rpm	revolutions per minute
rps	revolutions per second
sec	seconds
shm	simple harmonic motion

Contents

Preface to Adapted Edition .... (v)
Preface to Third Edition .... (vii)
List of Symbols and Abbreviations .... (xi)

PART - I
STATICS

Chapter 1
Introductory Concepts

1-1 Scope of Mechanics 1
1-2 Preview if Statics 1
1.3 Fundamental Concepts and Axioms 2
1-4 Scalar and Vector Quantities 5
1-5 Introduction to Free-body Diagrams 6
1-6 Guide to Solving Problems 7
1-7 Dimensional Checks 8
1-8 Conversion of Units 9
1-9 Numerical Computations Slide Rules and Calculators 10
1-10 Useful Techniques and Information 12

Chapter 2
Resultants of Force System

2-1 Introduction 16
2-2 Parallelogram Law 16
2-3 Forces and Components 18
2-4 Resultant of Coplanar Concurrent Forces 25
2-5 Components of Forces in Space 30
2-6 Vector Notation 37
2-7 Vector Algebra, Dot Product 40
2-8 Vector Algebra, Cross Product 46
2-9 Moment of Force Principle of Moments 54

2-10 Coplanar Applications 57
2-11 Spatial Applications 63
2-12 Couples 73
2-13 Resultant of any Force System 74

Chapter 3

Equilibrium of Force Systems

3-1 Introduction 91
3-2 Free-body Diagrams 92
3-3 Equations of Equilibrium 93
3-4 Equilibrium of Planar Systems 94
3-5 Further Discussion of Planar Equilibrium 104
3-6 Equilibrium of Spatial Systems 118

Chapter 4

Analysis of Structures

4-1 Introduction 137
4-2 Construction of Trusses 138
4-3 Method of Joints 143
4-4 Methods of Sections 154
4-5 Redundant Members; Counter Diagonals 167
4-6 Method of Members Frames 169
4-7 Space Trusses 184

Chapter 5

Friction

5-1 Introduction 195
5-2 Theory of Friction 195
5-3 Angle of Friction 198
5-4 Laws of Friction 200
5-5 Further Applications of Friction 210
5-6 Square-threaded Screws 223
5-7 Belt Friction 226
5-8 Disk Friction 233
5-9 Rolling Resistance 238

Chapter 6

Centroids and Centers of Gravity

6-1 Introduction 241
6-2 Centre of Gravity of Flat Plate 242
6-3 Centroids of Areas and Lines 243
6-4 Importance of Centroids and Moments of Area 244
6-5 Centroids Determined by Integration 245
6-6 Centroids of Composite Figures 252
6-7 Theorem of Pappus 264
6-8 Center of Gravity of Bodies Centroids of Volumes 270
Summary 276

Chapter 7

Moments of Inertia

7-1 Definition of Moment of Inertia 278
7-2 Polar Moment of Inertia 280
7-3 Radius of Gyration 281
7-4 Transfer Formula for Moment of Inertia 281
7-5 Moments of Inertia by Integration 283
7-6 Moments of Inertia for Composite Areas 289
7-7 Product of Inertia 297
7-8 Product of Inertia is Zero with Respect to Axes of Symmetry . 298
7-9 Transfer Formula for Product of Inertia 299
7-10 Moment of Inertia with Respect to Inclined Axes 306
7-11 Mohr's Circle for Moments of Inertia 309
7-12 Maximum and Minimum Moments of Inertia Principal Axes .. 312
7-13 Moments of Inertia of Masses 316
7-14 Radius of Gyration 317
7-15 Transfer Formula for Mass Moments of Inertia 317
7-16 Mass Moments of Inertia by Integration 319
7-17 Moments of Inertia of Composite Bodies 327
Summary 337

Chapter 8
Special Topics

8-1 Introduction 339
8-2 Shear and Moment Equations and Diagrams 341
8-3 Relations between Load, Shear and Bending Moment 354
8-4 Definition and Meaning of Work 373
8-5 Introductory Concepts of Virtual Work Method 375
8-6 Principle and Application of Virtual Work 376
8-7 Introduction 394
8-8 Application of Parallelogram Law 394
8-9 Resultant of Coplanar Force Systems 395
8-10 String Polygon 396
8-11 Equilibrium of Force Systems by Graphic Methods 401
8-12 Graphic Analysis of Trusses Maxwell Diagrams 405

PART - II
DYNAMICS

Chapter 9
Kinematics of a Particle

9-1 Introduction 415
9-2 Motion of a Particle 417
9-3 Rectilinear Motion 422
9-4 Motion Curves 431
9-5 Introduction to Vector Calculus 443
9-6 Rectangular Components of Curvilinear Motion 444
9-7 Normal and Tangential Components of Acceleration 454
9-8 Radial and Transverse Components Cylindrical Coordinates ... 463
Summary 474

Chapter 10

General Principles of Dynamics

10-1 Introduction 477
10-2 Newton's Laws of Motion for a Particle 478
10-3 Fundamental Equation of Kinetics for a Particle 480
10-4 Gravitational and Absolute System of Units 481
10-5 D'Alembert's Principle – Motion of the Mass Center 482
10-6 Moment Effect of External Forces 486
Summary 489

Chapter 11

Kinetics of Particles

11-1 Introduction 490
11-2 Translation – Analysis as a Particle 491
11-3 Further Discussion of Particle Kinetics 504
11-4 Translation – Analysis as a Rigid Body 518
Summary 530

Chapter 12

Kinematics of Rigid Bodies

12-1 Introduction. Types of Rigid-Body Motion 531
12-2 Angular Motion – Fixed-Axis Rotation 532
12-3 Definition and Analysis of Plane Motion 542
12-4 Application of Kinematic Equations 547
12-5 Instant Center and Instantaneous Axis of Rotation 565
12-6 The Omega Theorem 581

Chapter 13

Kinetics of Rigid Bodies

13-1 Introduction 585
13-2 Equations of Plane Motion 585
13-3 Fixed-Axis Rotation 590
13-4 Rolling Bodies 607
13-5 General Plane Motion 621
Summary 637

Chapter 14

Work-Energy Method

14-1 Introduction 639
14-2 Work-Energy Equation for Translation 640
14-3 Interpretation and Computation of Work 641
14-4 Work-Energy Applied To Particle Motion 646
14-5 Power Efficiency 656
14-6 Work-Energy Applied to Connected Systems 659
14-7 Work-Energy Method Applied to Fixed-Axis Rotation 668
14-8 Work-Energy Applied to Plane Motion 685
Summary 701

Chapter 15

Impulse and Momentum

15-1 Introduction 704
15-2 Linear Impulse-Momentum 704
15-3 Dynamic Action of Jet Streams 711
15-4 Conservation of Linear Momentum 719
15-5 Elastic Impact 726
15-6 Impulse Momentum in Plane Motion 734
15-7 Satellite Motion 750
15-8 Introduction to Gyroscopic Action 759
Summary 764

Chapter 16

Mechanical Vibrations

16-1 Introduction, Definitions and Concepts 768
16-2 Simple Harmonic Motion. Free Vibrations 769
16-3 Simple Pendulum 774
16-4 Compound Pendulum 776
16-5 Torsion Pendulum 779

16-6 Graphical Representation of Simple Harmonic Motion............ 782
16-7 Free Vibrations Without Damping : General Case 784
16-8 Free Vibrations Analyzed by Work-Energy Method 793
16-9 Forced Vibrations 798
Summary 806

Index 809

PART - I
STATICS

CHAPTER 1

INTRODUCTORY CONCEPTS

1-1 SCOPE OF MECHANICS

The subject of mechanics occupies a unique position in the physical sciences because it is fundamental to so many fields of study. In its broadest sense, mechanics may be defined as the science which describes and predicts the conditions of rest or motion of bodies under the action of forces. In a narrower sense, mechanics may be divided into areas which, with some overlap, are of separate interest to physicists and to engineers.

The topics of mechanics which interest physicists and astronomers (and, to some extent, electrical engineers) are particle mechanics, celestial mechanics, quantum mechanics, and relativistic mechanics. To engineers, the important areas of mechanics are the mechanics of solids and of fluids and their concomitant, continuum mechanics. A visual outline of how these topics are inter-related is shown in Fig. 1-1.1. In this book, we consider only the mechanics of rigid bodies which is divided into two parts: statics and dynamics. In statics, we consider the effects and distribution of forces on rigid bodies which are and remain at rest. In dynamics, we consider the various motions of rigid bodies and the correlation of these motions with the forces causing them.

1-2 PREVIEW OF STATICS

Throughout mechanics will recur the concepts of finding the components of a force and the moment of a force, plus the concept of a free-body diagram. In a very real sense, the components of a force and its moment effect are the building blocks of statics. You will learn these concepts in your first week of study. Thereafter with the aid of a free-body diagram, (to be introduced shortly and discussed in detail in Chapter 3) you will learn how to apply these concepts to solve a broad variety of situations. In particular, the solution of many practical problems will teach you a method of reasoning, called engineering methodology, which is the hallmark of the engineer.

- Engineering Mechanics
  - Mechanics of solids
    - Rigid bodies
      - Statics
      - Dynamics
        - Kinematics
        - Kinetics
    - Deformable bodies
      - Strength of materials
      - Theory of elasticity
      - Theory of plasticity
  - Mechanics of fluids
    - Ideal fluids
    - Viscous fluids
    - Compressible fluids

Figure 1-1.1 Abbreviated outline of engineering mechanics.

In Chapter 2 the concepts of the components of a force and its moment effect are applied to determine the resultant of any force system. In Chapter 3, we discuss the conditions which reduce the resultant of a force system to zero and thereby obtain the two basic equations of static equilibrium; namely, $\Sigma F = 0$ and $\Sigma M = 0$. In succeeding chapters, these two basic equations are applied to the analysis of structures and other situations.

1.3 FUNDAMENTAL CONCEPTS AND AXIOMS

RIGID BODY

A rigid body is defined as a definite amount of matter the parts of which are fixed in position relative to one another. Actually, solid bodies are never rigid; they deform under the action of applied forces. In many cases, this deformation is negligible compared to the size of the body and the body may be *assumed* rigid. Bodies made of steel or cast iron, for example, are of this type. The study of *strength of materials*, however, is based on the deformation (however small) of such bodies.

MASS

Mass is that invariant property of a body which measures its resistance to a change of motion.

FORCE

Force is the action exerted by one body upon another. Its *external* effect upon a body is manifested by a change in, or a tendency to change, the state of motion of the body upon which it acts. The *internal* effect of a force is to produce stress and deformation in the body. The *characteristics* of a force are:

1. its magnitude,
2. the position of its line of action, and
3. the direction (or sense) in which the force acts along its line of action.

The *unit* of force commonly used in India is Newton (N) or multiples of the Newton such as the kN (i.e., 1 kN = 1000 N).

The *principle of transmissibility* of a force states that the *external* effect of a force on a rigid body is the same for all points of application along its line of action; i.e., it is independent of the point of application. Its *internal* effect, however, definitely is associated with the point of application of the force. Thus, in Fig. 1-3,1(a), the motion of the block will be the same whether it is pushed at *A* or pulled at *B*. The local internal effects at *A* and *B*, however, will be quite different. Note that the principle of transmissibility applies only to the external effect of a force on the *same* rigid body. Observe that the system of two bars in Fig. 1-3.1(b), hinged together at *D*, is not inherently rigid; it is the hinge supports at *C* and *E* that restrict relative movement of the bars. In this case,

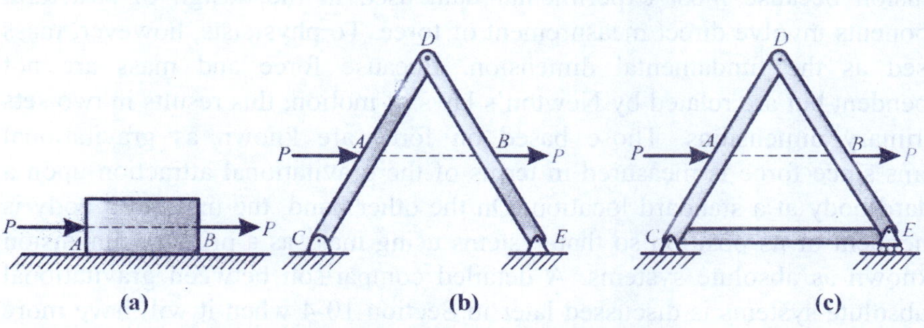

Figure 1-3.1 Principle of transmissibility is valid for (a) and (c), but not for (b).

therefore, since we do not have a single rigid body, both the external reactions and the internal effects will be different if *P* is applied first at *A* and then at *B*. However, if the hinge support at *E* is replaced by a roller support and another bar *CE* added as in Fig. 1-3.1(c), we shall have a triangular truss equivalent to a single rigid body. Here the external reactions will be the same whether P is applied at A or at B, although the internal effects will differ since P applied at A tends to bend bar CD while P acting at B tends to bend bar DE.

AXIOMS OF MECHANICS

The principles of mechanics are postulated upon several more or less self-evident facts which cannot be proved mathematically but can only be demonstrated to be true. We shall call these facts the axioms of mechanics. They will be discussed at length in subsequent sections as they are used. At this time, we merely collate them for reference and state them in the following form:

1. The parallelogram law: The resultant of two forces is the diagonal formed on the vectors of these forces.
2. Two forces are in equilibrium only when equal in magnitude, opposite in direction, and collinear in action.
3. A set of forces in equilibrium may be added to any system of forces without changing the effect of the original system.
4. Action and reaction forces are equal but oppositely directed.

UNITS AND DIMENSIONS

We define *unit* as the numerical standard used to measure the qualitative *dimension* of a physical quantity. The three primary dimensions basic to mechanics are length, time, and either force or mass[1]. Combinations of these determine the secondary dimensions in which all other quantities can be expressed. Engineers everywhere prefer force instead of mass as the basic dimension because most experimental data used in the design of structural components involve direct measurement of force. To physicists, however, mass is used as the fundamental dimension. Because force and mass are not independent but are related by Newton's laws of motion, this results in two sets of primary dimensions. Those based on force are known as gravitational systems since force is measured in terms of the gravitational attraction upon a standard body at a standard location. On the other hand, the mass of a body is independent of its position so that systems using mass as a primary dimension are known as absolute systems. A detailed comparison between gravitational and absolute systems is discussed later in Section 10-4 when it will have more significance. For the present, Table 1-3.1 summarizes various dimensional systems and the units in which these dimensions are measured. Table 1-3.2 lists various equivalents between these systems of units.

[1]In addition, there are two other primary dimensions, namely temperature and electric charge, but these are not used in this book.

Table 1-3.1 Systems of Units.

System	Length	Time	Force	Mass	Used by
Gravitational English (FPS)	foot (ft)	sec	pound (lb)	slug[a] (1b-sec^2/ft)	Engineers in English-speaking countries
Gravitational Metric (MKS)	meter (m)	sec	kilogram (kg)	metric slug[a] (kg.sec^2/m)	Engineers elsewhere
Absolute Metric (CGS)	centimeter (cm)	sec	dyne[a] (g-cm/sec^2)	gram (g)	Physicists Rarely
Absolute Metric (MKS)	meter (m)	sec	newton[a] (kg-m/sec^2)	kilogram (kg)	Physcists generally

[a]Derived unit

Table 1-3.2 Conversion Equivalents.

Lengths:	1 m = 3.281 ft = 39.37 inches
	1 km = 0.6214 mile
Force:	1 N = 0.2248 lb
Velocity:	1 km/hr = 0.9115 ft/sec
Acceleration:	g = 9.81 m/sec^2 = 32.2 fps^2
Power:	1 kw = 1.341 hp
	1 hp = 746 N-m/sec
	1 w = 1 N-m/sec
Pressure:	1 atm = 760 mm Hg
	1 mm Hg = 2.7852 psi
Volume:	1 litre = 0.03531 ft^3
	= 0.2642 gal

1-4 SCALAR AND VECTOR QUANTITIES

The quantities encountered in mechanics are of three types scalars, vectors and tensors. Quantities which possess magnitude only and can be combined arithmetically are defined a scalars. Common examples in mechanics are area, energy, mass, and time.

Vectors are defined as quantities having *both* magnitude and direction which combine *geometrically* according to the parallelogram law (Section 2-2). The result must be commutative; i.e., independent of the order of geometric addition. A vector of a quantity is represented geometrically (i.e., graphically) by drawing a line acting in the direction of the quantity, the length of this line representing to some scale the magnitude of the quantity. An arrowhead is placed on the line, usually at the end, to denote the sense of the direction of the quantity. Also in this book, vectors are generally distinguished from scalars by a notation which is described in Section 2-6. Common examples of vectors are force, moment, displacement, velocity, acceleration, and momentum.

Quantities which possess magnitude but require two or more directional aspects to describe them completely are known as tensors. Other examples are the stress tensor and the strain tensor used in deformable body mechanics. The number of directions needed to fully describe a tensor is called the order of the tensor. Accordingly, observe that scalars and vectors are special cases of tensors, being respectively of zero order and first order, but we shall not treat them in this context. Incidentally, the study of mechanics rarely, if ever, involves higher than second-order tensors.

1-5 INTRODUCTION TO FREE-BODY DIAGRAMS

One of the most important concepts in mechanics is that of the free-body diagram. This concept is discussed in detail in Chapter 3 where we first really use it. It is introduced here to help you distinguish between action and reaction forces. To do so, it is necessary to *isolate* the body being considered by removing all elements that act upon the isolated body. A sketch of the isolated body which shows only the forces acting *upon it* by the removed elements is defined as a *free-body diagram*. The forces acting on the isolated or free body are the action forces, also called the applied forces. The reaction forces are those exerted by the free body upon other bodies. One exception to this concept concerns ground supports whose actions on the free body are commonly called reactions.

The free body may consist of an entire assembled structure or an isolated part of it, depending on which quantities are to be determined. For example, consider the derrick shown in Fig. 1-5.1(a). To determine the force in the boom AC, we would isolate pin C whose free-body diagram (Fig. 1-5.1(b)) exposes this force as one of those acting upon it. These forces consist of the gravitational attraction of the weight W, the pull T exerted on the pin by the cable, and the force P exerted by the boom. If the effect of the supports at D and E were desired, a free-body diagram of the entire derrick, showing it isolated from these supports as in Fig 1-5.1(c), would expose their action upon the derrick. Observe that the force P in the boom is not shown here since the boom is part of the entire derrick. Remember that a free-body diagram shows only forces exerted by the removed elements that are necessary to isolate the body. Therefore, since P acts in an element that was not removed, its action is internal and is not exposed.

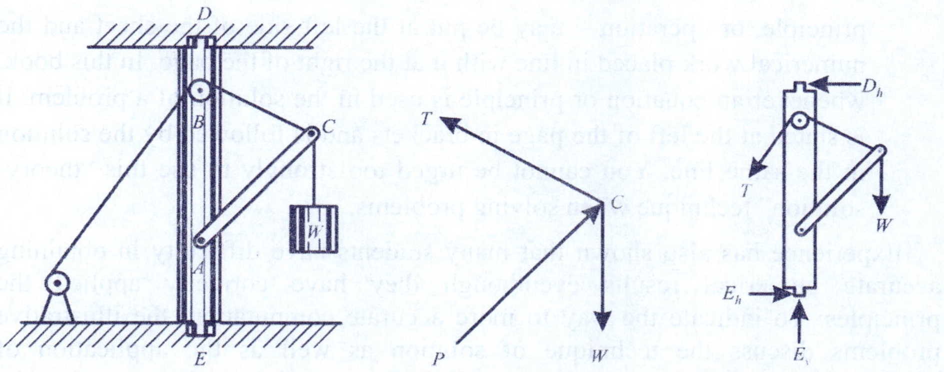

(a) Derrick (b) Free-body diagram of pin C (c) Free-body diagram of derrick

Figure 1-5.1 Free-body diagrams.

1-6 GUIDE TO SOLVING PROBLEMS

One of the first things a student should acquire is the ability to organize his work in a neat and orderly fashion. Properly arranged work not only helps to eliminate personal errors but also permits easy checking by another person – a frequent occurrence in engineering offices. To aid the student to achieve orderly work habits the following suggestions are offered:

1. After identifying the problem, start by constructing a neat diagram of the quantities involved. This diagram should be of sufficient size so that pertinent data and dimensions may be added without affecting its legibility. A freehand sketch is usually acceptable, although the use of a straight edge is preferred and will take little additional time. Some students use different-colored pencils to distinguish between known and unknown quantities.
2. State as concisely as possible what data are given and what information is required. Students who fail to realize what is required, often find themselves in the unenviable position of obtaining the right answer to the wrong question.
3. Errors are frequently caused by mental substitution in equations and subsequent failure to include the term in the equation. For this reason, write out the equation you intend to use before substituting in it. This will also make the process apparent to any person who may check it. If an equation is not used, write a short note indicating the principle used or the operation performed. This short statement of theory – be it equation,

principle, or operation – may be put at the left side of the sheet and the numerical work placed in line with it at the right of the page. In this book, whenever an equation or principle is used in the solution of a problem, it is stated at the left of the page in brackets and is followed by the solution in the same line. You cannot be urged too strongly to use this "theory-solution" technique when solving problems.

Experience has also shown that many students have difficulty in obtaining accurate numerical results eventhough they have correctly applied the principles. To indicate the way to more accurate computation, the illustrative problems discuss the technique of solution as well as the application of principles. The following sections offer several additional items for the student's guidance.

1-7 DIMENSIONAL CHECKS

The equations used in engineering computations must be dimensionally homogeneous; that is, the units on each side must be of the same dimensional form. An easy way to check the dimensions in an equation is to substitute the dimensional equivalents of each term and then multiply or divide these equivalents as though they were algebraic quantities. This process determines the dimensional unit of each term.

For example, consider the equation $v^2 = v_0^2 + 2as$ where v and v_0 are in meter per second, a in meter per second squared, and s in meter. The numeral 2 is a dimensionless number.[2] Substituting dimensionally in the equation, we get

$$\frac{\text{m}^2}{\text{sec}^2} = \frac{\text{m}^2}{\text{sec}^2} + \frac{\text{m}}{\text{sec}^2} \times \text{m}$$

which checks the equation since each term is in the same dimensional units.

A similar process may be used to determine the unit of an expression. For example, determine the dimensional unit of kinetic energy if it is expressed by the relation $\text{KE} = \frac{W}{2g}\text{v}^2$, where W is in Newtons, v in meter per second, and g in meter per second squared. Substituting dimensionally, we have

$$\text{KE} = \frac{\text{N}}{\text{m per sec}^2} \times \frac{\text{m}^2}{\text{sec}^2} = \text{N-m} \qquad \textbf{\textit{Ans.}}$$

Even the definition of acceleration, $\text{a} = \frac{d^2s}{dt^2}$ (Section 9-3), can be checked dimensionally. Here d^2s is a second differential of length, whereas dt^2 is a

[2] In section 9-3, the numeral 2 is shown to be a factor of integration.

square of a differential of time. With units of length in meter and time in seconds, dimensional substitution in the definition for *a* yields,

$$a = \frac{\text{m}}{\text{sec}^2} \qquad \textit{Ans.}$$

1-8 CONVERSION OF UNITS

Occasionally it is necessary to convert a term from one system of units to another to make an equation dimensionally correct. The conversion is accomplished by multiplying the given term by unity where unity is a ratio of units (of the same kind but different in size) containing the required units and those given.

For example, convert a velocity of 60 miles per hour to units of meters per second. Begin by writing

$$v = 60\frac{\text{miles}}{\text{hour}}$$

To express v in feet per second, multiply the right side by the following ratios, each of which has the value of unity.

$$v = 60\frac{\text{miles}}{\text{hour}} \times \frac{1609\text{ m}}{\text{mile}} \times \frac{\text{hour}}{3600\text{ sec}}$$

Canceling out like units, we obtain

$$v = 60 \times \frac{1609}{3600} = 26.816\text{ m/sec}$$

The ratio $\frac{1609}{3600}$ is the conversion factor by which miles per hour must be multiplied to yield meter per second. It is not necessary to remember conversion factors once the method is mastered.

As another example, consider the equation $\theta = \omega_o t + \frac{1}{2}\alpha t^2$ where ω_o is in revolutions per minute, t in seconds, and α in revolutions per minute per second. It is required to express θ in radians. Substituting dimensionally, we have

$$\theta(\text{rad}) = \omega_o\left(\frac{\text{rev}}{\text{min}}\right) \times t(\text{sec}) + \frac{1}{2} \times \alpha\left(\frac{\text{rev}}{\text{min} \times \text{sec}}\right) \times t^2(\text{sec}^2)$$

Converting by multiplying by ratios having the value unity gives

$$\theta(\text{rad}) = \omega_o \left(\frac{\text{rev}}{\text{min}} \times \frac{2\pi \text{ rad}}{\text{rev}} \times \frac{\text{min}}{60 \text{ sec}} \right) \times t(\text{sec})$$

$$+ \frac{1}{2} \times \alpha \left(\frac{\text{rev}}{\text{min} \times \text{sec}} \times \frac{2\pi \text{ rad}}{\text{rev}} \times \frac{\text{min}}{60 \text{ sec}} \right) \times t^2 (\text{sec}^2)$$

$$\theta(rad) = \omega_o t \times \frac{2\pi}{60}(rad) + \frac{1}{2}\alpha t^2 \times \frac{2\pi}{60}(rad) \qquad \textbf{\textit{Ans.}}$$

Each term is now expressed in units of radians. In this example, θ was obtained in radians by multiplying each right-hand term by the factor $\frac{2\pi}{60}$.

1-9 NUMERICAL COMPUTATIONS SLIDE RULES AND CALCULATORS

In engineering computations, the numerical data are rarely accurate to more than three significant figures. No increase in accuracy is attained by carrying more figures through a computation than those in the least accurate data. As in this book, slide rule computations will generally be sufficiently accurate since the known data are usually less precise than the percentage error in misreading the last three figures on a slide rule. Recently, however, pocket calculators at reasonable cost have become available to provide the engineer with an alternate means of computation. Indeed, pocket calculators have begun to replace the slide rule, but a suitable choice may be confusing since they are available in such a variety of types and advertising claims. We therefore present a short discussion of the major differences between calculators currently available. It is also pertinent to present a short comparison between the slide rule and calculator together with their relative advantages and disadvantages.

The advantages of the calculator over the slide rule, aside from its greater precision, are its ability to add and subtract, to automatically locate the decimal point, and the comparative ease of learning to use it. The following analogies should explain the major differences between calculators. In terms of the slide rule scales, most calculators possess only the equivalent of C and D scales and some have a "memory" which is a desirable feature. A slightly more advanced type of calculator has the equivalent of A, B, C, D, and CI scales. These are advertised as "slide rule" calculators since they automatically can square numbers, extract square roots, and take reciprocals. Finally, some calculators are available that have, in addition, the logarithmic and trigonometric scales common to all log–log slide rules and can truly be called "electronic slide rules."

With the foregoing explanation, the choice of a calculator suitable for the engineer should be simplified. However, calculators vary widely in price according to their capability and numerical capacity so some additional information may be useful. At a minimum, the calculator should have square root, square, and reciprocal keys plus a separate memory and be able to perform mixed and chain numerical computations. Frequently, supplementary logarithmic and trigonometric tables must be available to augument their use. The necessity for supplemental tables, however, is offset by their relative low cost. Of this type, the Melchor 400 seems to offer the best value.

The few pocket calculators which have the functional capability of the log–log slide rule are two or three times as expensive as the Melchor 400. In a class by themselves are the Hewlett Packard models HP 35 and HP 45 which have direct entry keys for logarithmic and trigonometric functions. Although the HP 35 has fewer memory registers than the companion HP 45, it is completely satisfactory for practically all engineering computations. Its capability and versatility are truly remarkable because it uses a four-register operational stack and reverse "Polish" (Lukasiewicz) notation which enables it to solve equations directly without the need for writing down intermediate steps. Presently, the HP 35 is so much superior to its competitors that its higher cost is justified.

Other pocket calculators[3] achieve logarithmic and trigonometric capability by using double function entry keys. They usually require more key board entries than the HP 35 and have a smaller numerical range. However, their lower cost tends to offset these and several other minor deficiencies.

The use of a slide rule instead of a pocket calculator is still a viable choice for the student because of its low cost, versatility, and long life. As indicated before, the fact that a slide rule is not as accurate as a pocket calculator is not a

[3]Typical of this group are the Rockwell International Unicorn 202/SR, the Sharp PC 1801, and the Bowmar MX-100. While all of these have essentially the same features, the Bowmar MX-100 appears to be best. Generally, the more direct entry keys a calculator has, the more useful it will be. Also available, but much more expensive, is the extremely versatile Hewlett Packard HP 65 which, in addition to being a pocket calculator of the same size and features of the HP 45, is also a programmable mini-computer especially useful for repetitive computations. It accepts programs up to 100 steps long which may be fed in either by direct keyboard entry or by using tiny magnetic program cards. Similar in many respects to the HP 35 is Texas Instrument's model SR 50.

serious deficiency in engineering calculations. Indeed, almost all the answers to problems in this book are computed only to slide rule accuracy. In some instances, moreover, such as computing proportions or applying the sine law, the slide rule will be more convenient and faster than a calculator although admittedly not as precise. These and other useful slide rule techniques are explained at pertinent places in the text. Also note that with a slide rule, the several parts of a computation are visually evident whereas a calculator usually shows only the last entry.

1-10 USEFUL TECHNIQUES AND INFORMATION

SIGNIFICANT FIGURES

It frequently happens that numerical work is computed to a greater degree of accuracy than is warranted by the given data. The accuracy of the final result depends on the least accurate figure used in the computation. For example, if the value of the gravitational constant g is taken as 9.81, the answer should be computed to three significant figures only. No increase in accuracy is attained by carrying more figures through a computation than those in the least accurate data. If for some reason four figures are carried through a computation where the data in the example are accurate to only three figures, the answer should be rounded off to three figures.

If a result is found to be 24.2, for example, the number indicates that the result is greater than 24.15 but less than 24.25; the last 2 of the result is doubtful. On the other hand, 24.20 means that the result is greater than 24.195 but less than 24.205; obviously, the last zero should not be added unless it is a significant figure.

LOCATION OF THE DECIMAL POINT

In slide rule work difficulty is sometimes experienced in locating the decimal point. It is not advisable to rely upon systems that locate this point by counting the number of times the slide extends to the left or right, or by counting the significant figures, etc. A better method is to compute an approximate answer for the problem by rounding off the figures and canceling as far as possible. This method not only will locate the decimal point but will check the slide rule work.

For example, compute the value of

$$\frac{2.196 \times 48.2 \times 289}{3420 \times 68.2}$$

Rounding off figures gives approximately

$$\frac{2\times 50\times 300}{3500\times 70}=\frac{6}{49}\approx\frac{6}{50}=0.12$$

By the slide rule, the numerals in the result are 131; hence, to resemble the approximate value; the answer must be 0.131.

SOLUTION OF QUADRATIC EQUATIONS

Quadratic equations may be solved by formula, slide rule, completing the square, etc. Completing the square is probably as fast and accurate as any other method. It is obtained as follows:

Consider a quadratic equation of the form $x^2 + ax - b = 0$. Rearranging this, we get

$$x^2+ax+\left(\frac{a}{2}\right)^2=\left(\frac{a}{2}\right)^2+b$$

hence

$$\left(x+\frac{a}{2}\right)^2=\left(\frac{a}{2}\right)^2+b$$

$$x+\frac{a}{2}=\pm\sqrt{\left(\frac{a}{2}\right)^2+b}$$

from which the value of x is easily found.

For example, solve for the roots of

$$x^2 - 22.8x - 31.3 = 0$$

From the above discussion, we obtain

$$x^2 - 22.8x + (11.4)^2 = (11.4)^2 + 31.3$$

$$(x - 11.4)^2 = 130 + 31.3 = 161.3$$

$$x - 11.4 = \pm 12.7$$

$$x = 24.1 \quad \text{and} \quad x = -1.3$$ ***Ans.***

TRIGONOMETRIC FUNCTIONS

The slide rule gives the functions of angles ranging from slightly below 0.6° to 84.3°. It is occasionally useful to know how to use the slide rule to find the functions of angles outside these limits.

For example, to find the value of sin 86°, proceed as follows:

Since

$$\sin 2\theta = 2 \sin \theta \cos \theta$$

we have

$$\sin 86^\circ = 2 \sin 43^\circ \cos 43^\circ$$
$$= 2 \times 0.682 \times 0.731 = 0.998$$

For angles less than 0.6° a similar technique is available. Since $\sin \theta \approx \theta$ for small angles (where θ is in radians), the value of $\sin \dfrac{1}{2}^\circ$, for example, is given by the equation

$$\frac{1}{2}^\circ = \frac{1}{2}\deg \times \frac{2\pi \text{ rad}}{360 \deg} = \frac{\pi}{360} \text{ rad}$$

Therefore

$$\sin \frac{1}{2}^\circ = \frac{\pi}{360} = 0.00873 \qquad \textbf{\textit{Ans.}}$$

Likewise for cosine functions of small angles, we note that $\cos \theta \approx 1 - \dfrac{\theta^2}{2}$.

Hence, in computing the value of $\cos \dfrac{1}{2}^\circ$ we get

$$\cos \frac{1}{2}^\circ = 1 - \frac{1}{2}\left(\frac{\pi}{360}\right)^2 = 1 - 0.0000381 = 0.9999619$$

Trigonometric Functions

Angle	15°	30°	45°	60°	75°
sin	0.259	$\frac{1}{2} = 0.500$	$\frac{1}{2}\sqrt{2} = 0.707$	$\frac{1}{2}\sqrt{3} = 0.866$	0.966
cos	0.966	$\frac{1}{2}\sqrt{3} = 0.866$	$\frac{1}{2}\sqrt{2} = 0.707$	$\frac{1}{2} = 0.500$	0.259
tan	0.268	$\frac{1}{3}\sqrt{3} = 0.577$	1.000	$\sqrt{3} = 1.732$	3.732

$$\sin(-\theta) = -\sin\theta;\ \cos(-\theta) = \cos\theta;\ \tan(-\theta) = \tan\theta$$
$$\sin^2\theta + \cos^2\theta = 1;\ 1 + \tan^2\theta = \sec^2\theta;\ 1 + \text{ctn}^2\theta = \csc^2\theta$$

$$\sin^2\theta = \frac{1}{2} - \frac{1}{2}\cos^2\theta;\ \cos^2\theta = \frac{1}{2} - \frac{1}{2}\sin^2\theta$$

$$\sin 2\theta = 2\sin\theta\cos\theta$$

$$\cos 2\theta = \cos^2\theta - \sin^2\theta = 1 - 2\sin^2\theta = 2\cos^2\theta - 1$$

$$\sin(\theta \pm \phi) = \sin\theta\cos\phi \pm \cos\theta\sin\phi$$

$$\cos(\theta \pm \phi) = \cos\theta\cos\phi \mp \sin\theta\sin\phi$$

$$\tan(\theta \pm \phi) = \frac{\tan\theta \pm \tan\phi}{1 \mp \tan\theta\tan\phi}$$

$$\sin\theta = \theta - \frac{\theta^3}{3!} + \frac{\theta^5}{5!} - \frac{\theta^7}{7!} + \ldots.$$

$$\cos\theta = 1 - \frac{\theta^2}{2!} + \frac{\theta^4}{5!} - \frac{\theta^6}{6!} + \ldots.$$

Differentials and Integrals

$$\frac{dx^n}{dx} = nx^{n-1};\quad \frac{d(uv)}{dx} = u\frac{dv}{dx} + v\frac{du}{dx};\quad \frac{d\left(\frac{u}{v}\right)}{dx} = \frac{v\frac{du}{dx} - u\frac{dv}{dx}}{v^2}$$

$$\int x^n\,dx = \frac{x^{n+1}}{n+1} + C \qquad \int \frac{dx}{x} = \log_e x + C$$

$$\int \sin x\,dx = -\cos x + C \qquad \int \cos x\,dx = \sin x + C$$

$$\int \sin^2 x\,dx = \frac{x}{2} - \frac{\sin 2x}{4} + C \qquad \int \cos^2 x\,dx = \frac{x}{2} + \frac{\sin 2x}{4} + C$$

$$\int u\,dv = uv - \int v\,du$$

CHAPTER 2

RESULTANTS OF FORCE SYSTEMS

2-1 INTRODUCTION

The effect of a system of forces on a body is usually expressed in terms of its resultant, since the value of this resultant determines the motion of the body. As we shall see, if the resultant is zero, the body will be in equilibrium and will not change its original state of motion. This is the province of statics. If the resultant of a force system is not zero, the body will have a varying state of motion, thereby creating a problem in dynamics.

Insofar as forces are concerned, this chapter is the foundation of mechanics. Especially important are the concepts of finding the relation between a force and its components, and that of determining the moment of a force and the related principle of moments. These concepts and their associated applications should be thoroughly mastered since they will recur throughout the subsequent engineering theory and applications. At appropriate intervals in this chapter especially, we shall discuss techniques of computation which help to increase accuracy and shorten the time to obtain an answer.

2-2 PARALLELOGRAM LAW

The method of vector addition is based on the parallelogram law which is one of the fundamental axioms of mechanics. The parallelogram law cannot be proved; it can only be demonstrated by experiment. One method of demonstrating the law is by means of the apparatus shown in Fig. 2-2.1. Tie three cords together and fasten the weights P, Q, and W to the free ends. (The sum of P and Q must be greater than W.) Place the cords to which P and Q are attached over the smooth pegs as shown and allow the system to reach a position of equilibrium.

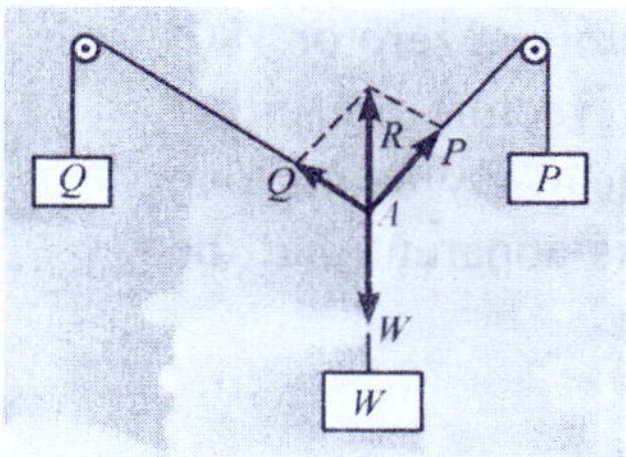

Figure 2-1.1 Parallelogram law.

The tensions in these cords will then be equal to the weights P and Q. Draw vectors P and Q to scale from point A where the cords are tied together and construct a parallelogram with these vectors as the initial sides. It will be found that the diagonal R of the parallelogram scales exactly to the value of W and is in line with the vector representing W.

Since it is axiomatic that two equal, opposite, collinear forces are in equilibrium, we conclude that weight W will be perfectly supported by the force R. In other words, the net effect of the forces P and Q may be replaced by the single force R. Such a force is called a resultant. The parallelogram law may now be stated as follows: *The resultant of two forces is the diagonal of the parallelogram whose initial sides are the vectors of these forces.* The diagonal to be used is that which emanates from the intersection of the initial sides.

The parallelogram formed by two vectors P and Q is divided by their resultant R into two congruent triangles as is evident in Fig. 2-2.2(b). If the triangle ABC were drawn alone as in part (c), the vector joining the tail of P to the tip of Q would have the same magnitude and direction as the resultant R defined by the parallelogram law. In this instance, however, force Q has been represented by the free vector BC. A *free vector* is defined as one which may be freely moved in space, as distinguished from a *localized vector* which is fixed or bound to a specific point of application. It is also evident that the triangle ADC in part (a) may also be used to determine R. In this case, P is taken as the free, vector whereas Q is the localized vector.

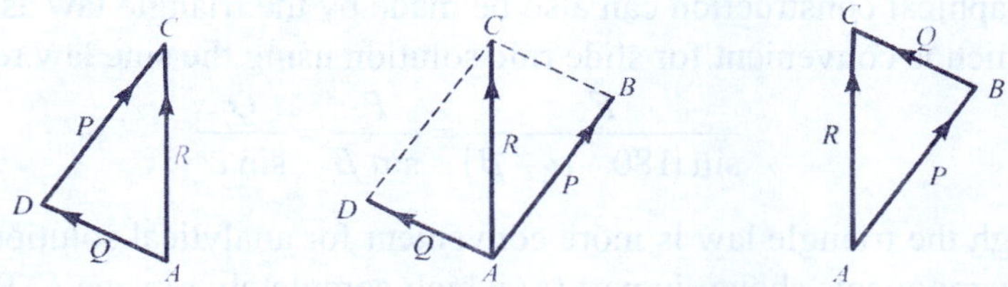

Figure 2-2.2 Triangle law.

We may now state the *triangle law* as a convenient corollary of the parallelogram law: *If two forces are represented by their free vectors placed tip-to-tail, their resultant is the vector directed from the tail of the first vector to the tip of the second vector.*

The special case in which the angle between two free vectors is zero or 180° is extensively used in analytical solutions. Evidently such vectors are either parallel or they act along the same line; i.e., they are collinear. By taking one direction as positive and the other direction as negative, it is apparent that the resultant of two collinear vectors is their algebraic sum.

2-3 FORCES AND COMPONENTS

The parallelogram law shows how to combine two forces into a resultant force. Of equal importance is the inverse operation, called resolution, in which a given force is replaced by two components which are equivalent to the given force. The method is demonstrated in Fig. 2-3.l(a) in which we are to replace force F by components directed along the lines OA and OB radiating from the tail O of F. We need to merely draw lines from the tip of F parallel to the specified directions to form the parallelogram shown. The initial sides P and Q of this parallelogram are the desired components. Obviously for the parallelogram law to be applied the components P and Q must intersect on F.

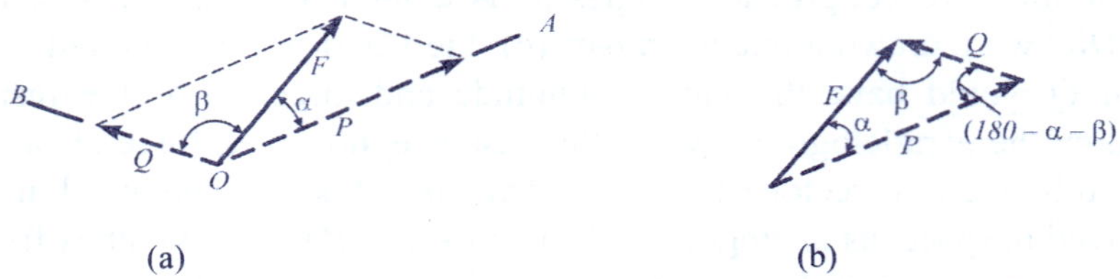

Figure 2-3.1 Oblique components.

This graphical construction can also be made by the triangle law as shown in part (b) which is convenient for slide rule solution using the sine law relation:[1]

$$\frac{F}{\sin(180-\alpha-\beta)}=\frac{P}{\sin\beta}=\frac{Q}{\sin\alpha}$$

Although the triangle law is more convenient for analytical solution, it is the localized components shown in part (a) which completely replace F. Recall that the basic demonstration of the parallelogram law required that the components of the resultant (F in this instance) must intersect on the line of action of the resultant force.

---

[1]To solve these proportions on the slide rule, merely set the numerators on scale D opposite the corresponding angles on scale S (for sine) and read on scale D the values of P and Q.

Depending on the directions specified, there are an infinite number of pairs of oblique components of F that may be found. Such nonorthogonal components, however, are of limited use. Analytically, it is much more convenient to resolve a force into a pair of perpendicular components. Such rectangular components are then readily combined with similarly oriented rectangular components of other forces by adding these components algebraically as we shall show in the next section.

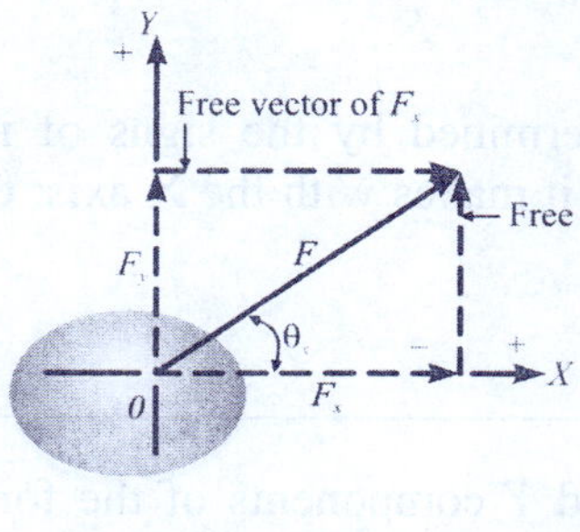

Figure 2-3.2 Rectangular components.

Consider now Fig. 2-3.2 in which force F acts upon the given body. The effect of the force is to move the body rightward and upward. Choosing these directions as the positive directions of perpendicular X and Y reference axes, we project the force F upon them to obtain the perpendicular components F_x and F_y. More precisely, we should draw parallels to the X and Y axes to obtain the basic parallelogram, but when the reference axes are perpendicular, the projected length of the force yields the same components. The relations between these components and F is determined by the basic definitions of sine and cosine of the angle θ_x between F and the X axis; i.e., sin $\theta_x = F_y/F$ and cos $\theta_x = F_x/F$ which are usually rewritten in the following form.

$$F_x = F\cos\theta_x$$
$$F_y = F\sin\theta_x \qquad (2\text{-}3.1)$$

The components F_x and F_y are considered positive if they act in the positive directions of the X and Y axes and negative if directed oppositely. Usually the X and Y axes are horizontal and vertical, and their positive directions are those of the common Cartesian coordinate axes. However, the orientation of the X and Y axes is arbitrary; their directions are adapted to the situation. The relations given above are independent of the orientation of the X axis. If desired, the angle θ_y between F and the Y axis may also be used; the components are then given by $F_x = F \sin \theta_y$, and $F_y = F \cos \theta_y$.

It is, obvious that when the rectangular components of a force are known, they completely specify the magnitude, inclination, and direction of the force. For example, assuming values of F_x and F_y to be known, we obtain from Fig. 2-3.2 the following equations:

$$\left.\begin{aligned} F &= \sqrt{(F_x)^2 + (F_y)^2} \\ \tan\theta_x &= \frac{F_y}{F_x} \end{aligned}\right\} \qquad (2\text{-}3.2)$$

The direction of F is determined by the signs of its components and its inclination by the acute angle it makes with the X axis; this is further explained in the illustrative problems.

ILLUSTRATIVE PROBLEMS

2-3.1 Determine the X and Y components of the forces P and F shown in Fig. 2-3.3.

Solution

Since P is directed upward to the left, its components act in these directions as shown. With respect to the positive directions of the reference axes, the sign of P_y is plus and of P_x is minus. Applying Eq. (2-3.1). we obtain

$$[F_x = F\cos\theta_x\} \qquad P_x = -200\cos 60^\circ = -200(0.5)$$

$$= -100\ \text{N} = 100\ \text{N}\leftarrow \qquad \textbf{\textit{Ans.}}$$

$$[F_y = F\sin\theta_x\} \qquad P_y = 200\sin 60^\circ = 200(0.866)$$

$$= 173.2\ \text{N} = 173.2\ \text{N}\uparrow \qquad \textbf{\textit{Ans.}}$$

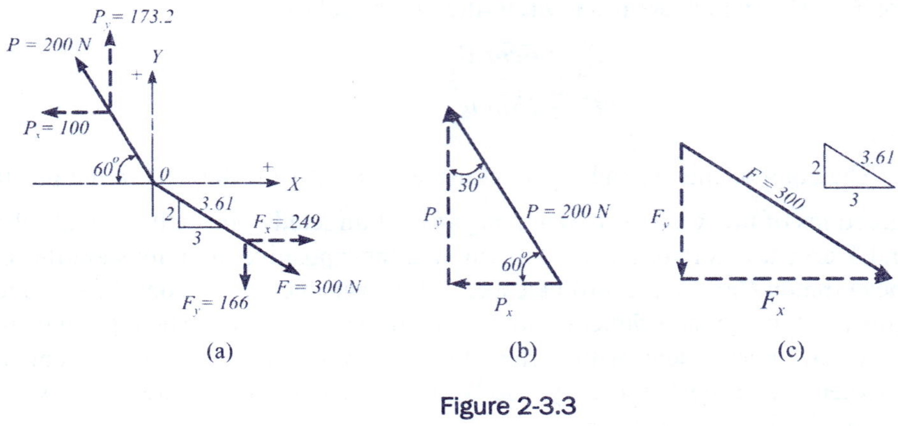

Figure 2-3.3

Although the signs of the components are sufficient to denote their direction, it is good practice to interpret these signs by directional arrows following the absolute values. The components are also drawn intersecting on the force P as shown in part (a) with the absolute values enclosed in parentheses to denote that they are computed values. By using dashed vectors to denote the components and solid vectors for the given forces, the effects of a force are clearly evident.

The components of P could also be found by mentally visualizing the triangle formed by P and its components as shown in part (b) and applying the sine law to set 'the following proportions directly on the slide rule:

$$\frac{200}{\sin 90} = \frac{P_x}{\sin 30} = \frac{P_y}{\sin 60}$$

Then using the slide rule technique described previously of setting numerical values on scale D Opposite the corresponding angles on scale S, we read directly the values $P_x = 100$ N and $P_y = 173.2$ N.

A similar procedure may be used to find the components of F by taking the preliminary step of computing θ_x as defined by the tangent of its slope. However, when the direction of a force is defined by its slope, it is simpler and more direct to note the similarity between the slope triangle and the force triangle as shown in part (c). The hypotenuse of the slope triangle is $\sqrt{(2)^2 + (3)^2} = \sqrt{13} = 3.61$. From, the proportionality between the corresponding sides of the force triangle and the slope triangle, we obtain

$$\left[\frac{F_x}{x} = \frac{F_y}{y} = \frac{F}{d}\right] \qquad \frac{F_x}{3} = \frac{F_y}{2} = \frac{300}{3.61}$$

whence

$$F_x = 249 \text{ N} \rightarrow \qquad \text{and} \qquad F_y = 166 \text{ N} \downarrow \qquad \textit{Ans.}$$

The directional arrows following these results again denote the direction of the absolute value of each component.

2-3.2 The components of a certain force are defined by $F_x = 240$ N and $F_y = -160$ N. Determine the magnitude, inclination with the X axis, and pointing of the force.

Solution

The determination of a force from its components is a frequent occurrence in mechanics. Two analytical techniques are available; one using Eq. (2-3.2) and the other a direct slide rule method.[2]

Using Eq. (2-3.2), the magnitude of the force is

$$\left[F=\sqrt{(F_x)^2+(F_y)^2}\right] F=\sqrt{(240)^2+(160)^2}\quad F=289\text{ N} \qquad \textbf{\textit{Ans.}}$$

while the inclination with the X axis is

$$\left[\tan\theta_x = \frac{F_y}{F_x}\right] \quad \tan\theta_x = \frac{160}{240} = 0.667 \quad \theta_x = 33.7^\circ \qquad \textbf{\textit{Ans.}}$$

Note particularly that, by neglecting the given signs of the components, the angle found is the acute angle between the force and the *X* axis. The direction of the force is found by sketching a tip-to-tail summation of the components as shown in Fig. 2-3.4 or by visualizing it mentally. Note that the minus sign of F_y means it is directed downward. Hence the force *F* points down to the right. If desired, the angle can also be specified using the mathematical convention of defining a positive angle as measured counterclockwise from the positive *X* axis. Accordingly, in the given example, θ_x might be defined as –33.7° or + 326.3°.

---

[2] If you use a slide rule method, start with $\tan\theta = \frac{F_y}{F_x}$ but think of it in the form of the following proportion with settings to be made on the indicated scales:

$$\frac{F_y}{\tan\theta} = \frac{F_x}{1} \begin{matrix}\leftarrow \text{set on scale D}\\ \leftarrow \text{set on scale T}\end{matrix}$$

The practical manner of doing this is to set the hairline over the *smaller* component (F_y = 160 here) on scale D and then move the slide so its index is opposite the larger component (F_x = 240 in this example) on scale D. The angle θ = 33.7° is then read at the hairline on the T scale. The reason for making the initial setting of the hairline on scale D over the smaller component (whether it be F_y, or F_x) is to obtain a value of θ less than 45° which can be read directly on scale T.

Without moving the hairline, refer now to Fig. 2-3.4 to visualize the sine law in the form

$$\frac{F_y = 160}{\sin 33.7^\circ} = \frac{F}{\sin 90^\circ = 1} \begin{matrix}\leftarrow \text{set on scale D}\\ \leftarrow \text{set on scale S}\end{matrix}$$

The hairline being already set on scale D at F_y = 160, now move the slide until 33.7° on the sine scale appears under the hairline. The slide rule is thus set for the sine law proportion above, hence opposite the index of the slide we read *F* = 289 N on scale D.

One precaution must be observed. If F_y should be larger than F_x, the angle read will be the angle the force makes with the *Y* axis, but the magnitude of *F* will be correct.

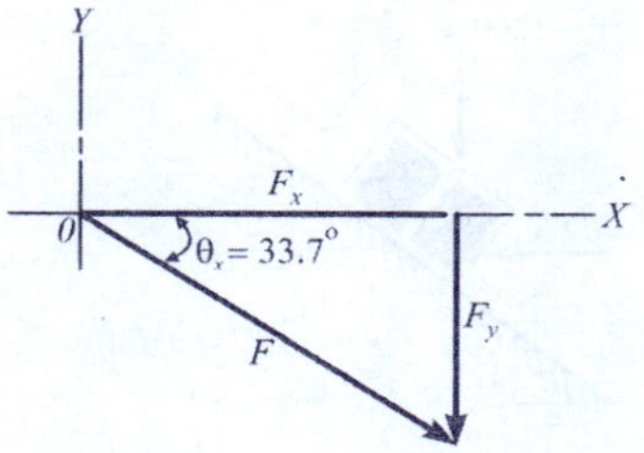

Figure 2-3.4

PROBLEMS

2-3.3 Determine the X and Y components of each of the forces shown in Fig. P-2-3.3.

For T: $T_x = -307$ N; $T_y = -257$ N ***Ans.***

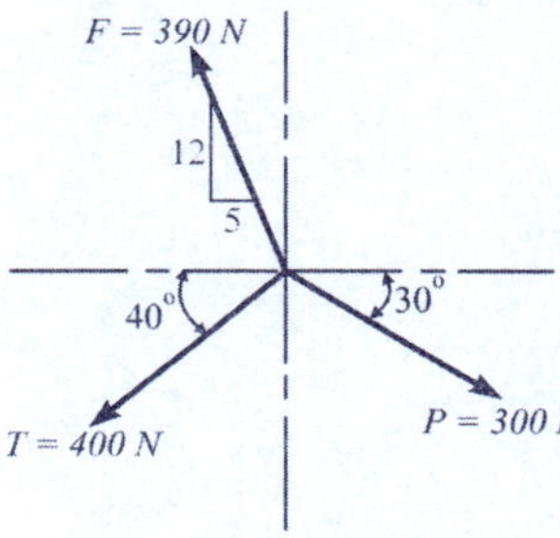

Figure P-2-3.3

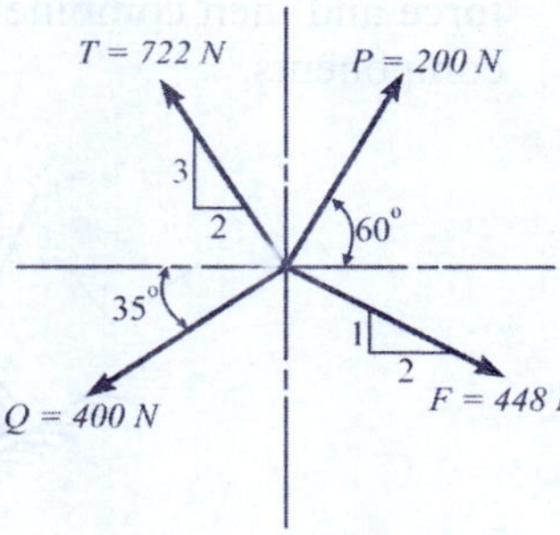

Figure P-2-3.4

2-3.4 Compute the X and Y components of each of the four forces shown in Fig. P-2-3.4.

For Q: $Q_x = -326$ N; $Q_y = -230$ N ***Ans.***

2-3.5 The body on the incline in Fig. P-2-3.5 is subjected to the vertical and horizontal forces shown. Find the components of each force along X-Y axes oriented parallel and perpendicular to the incline.

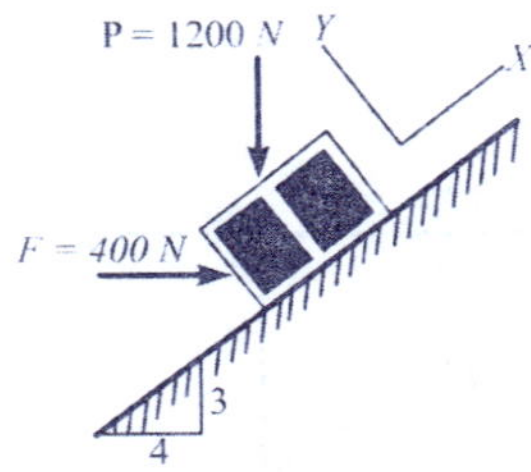

Figure P-2-3.5

$P_x = -720$ N; $F_y = -240$ N **Ans.**

2-3.6 Rework Prob. 2-3.5 if the slope of the incline is changed to become 5 vertical to 12 horizontal.

2-3.7 Referring to Fig. P-2-3.7, determine the components of force P along the X-Y axes which are parallel and perpendicular to the incline.

Hint: First determine the horizontal and vertical components of the force and then combine the parallel and perpendicular effects of these components.

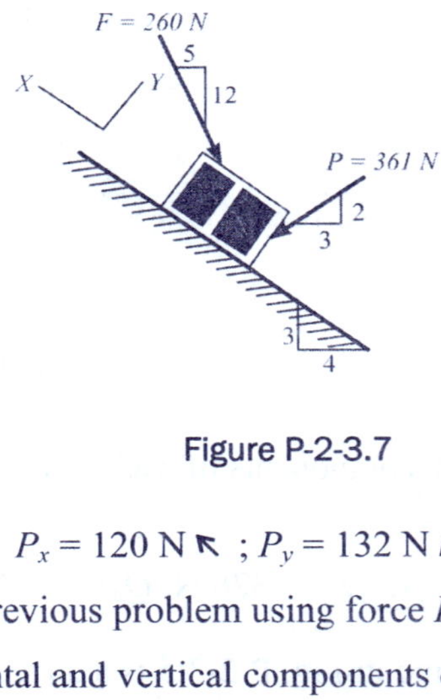

Figure P-2-3.7

$P_x = 120$ N ↖ ; $P_y = 132$ N ↙ **Ans.**

2-3.8 Solve the previous problem using force F instead of force P.

2-3.9 The horizontal and vertical components of several forces are:

(a) $P_h = -100$ N and $P_v = 200$ N; (b) $F_h = 300$ N and $F_v = -200$ N;

(c) $T_h = -50$ N and $T_v = -70$ N. Determine each force.

(a) $P = 224$ N up to left at $\theta_h = 53.45°$ **Ans.**

2-3.10 Repeat Prob. 2-3.9 if (a) $P_h = -240$ N and $P_v = -150$ N; (b) $F_h = -210$ N and $F_v = 270$ N; (c) $T_h = -500$ N and $T_v = -400$ N.

(c) $T = 639$ N down to left at $\theta_h = 38.7°$ ***Ans.***

2-3.11 The body on the 30° incline in Fig. P-2-3.11 is acted upon by a force P inclined at 20° with the horizontal. If P is resolved into components parallel and perpendicular to the incline and the value of the parallel component is 300 N, compute the value of the perpendicular component and of P.

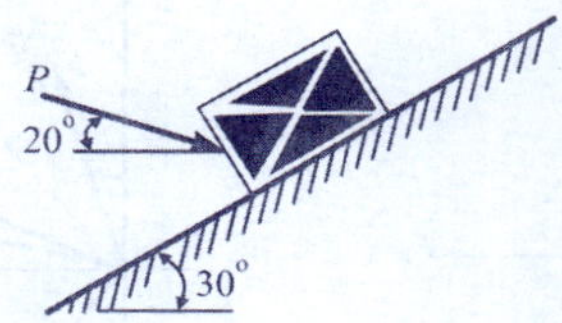

Figure P-2-3.11

2-3.12 In Fig. P-2-3.12, the X component of P is 893 N. Determine P and its Y component.

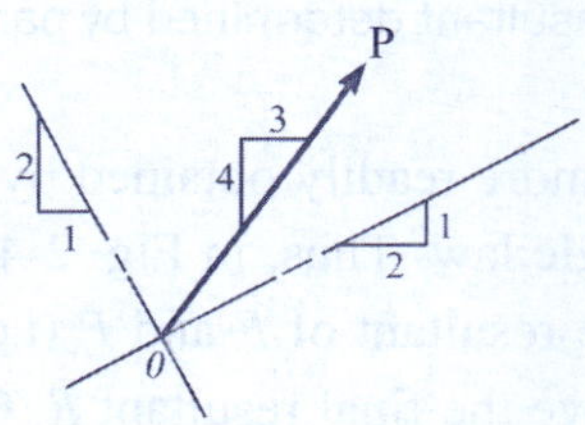

Figure P-2-3.12

2-4 RESULTANT OF COPLANAR CONCURRENT FORCES

The determination of the resultant of three or more concurrent forces that are not collinear requires determining the sum of three or more vectors. There are two ways of accomplishing the addition of three or more vectors: graphically and analytically.

Graphically

Two vectors can be added to give a resultant; this resultant in turn can be added to a third vector, etc., until all the vectors have been added together to give an overall resultant. These vectors can be added in any order.

Consider the system of three concurrent forces shown in Fig. 2-4.1. If the parallelogram method of vector addition is used, forces F and P may be combined to give a resultant R_1 as shown in Fig. 2-4.1(b). Since R_1 is equivalent to and replaces F and P, the original system of three forces now consists of only two: R_1 and Q. These may also be combined by the parallelogram method to give the final resultant R. If the original system consists of more than three forces, this same technique can be extended to include the additional forces.

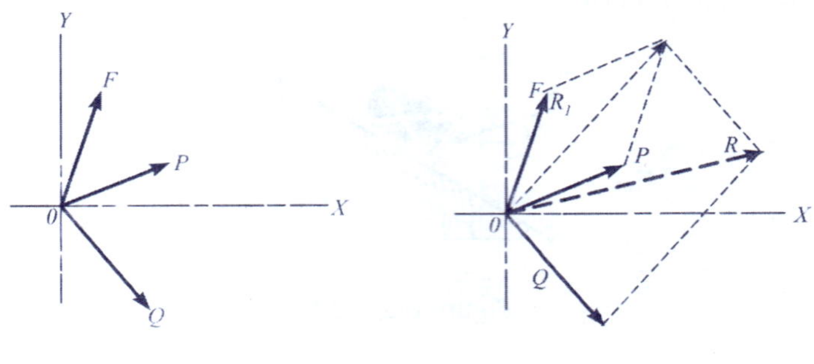

(a) (b)

Figure 2-4.1 Resultant determined by parallelogram law.

The same result can be more readily obtained by the use of free vectors and the application of the triangle law. Thus, in Fig. 2-4.2, by using the free vector P, one can easily obtain the resultant of F and P (i.e., R_1). To this resultant the free vector Q is added to give the final resultant R. Observe that R_1 need not be drawn at all, the total resultant of the system being obtained by joining the tail of the first vector (F) with the tip of the last vector (Q). The same result would be obtained if the order of addition had been P, F, and Q. In fact, any convenient order of tip-to-tail vector addition may be used.

Analytically

The vectors can be resolved into components that coincide with arbitrarily chosen axes. The components of each vector with respect to these axes can be added algebraically, and the resulting additions will be the components of the overall resultant vector.

Figure 2-4.2 can be redrawn as in Fig. 2-4.3 to show the X and Y components of each force by projection upon the reference axes. It is apparent that R_x, the X

component of R, is equivalent to the algebraic sum of the X components of F, P, and Q; also that R_y is equivalent to the algebraic sum of the Y components of F, P and Q. Denoting such algebraic summations of the components of the forces by ΣX and ΣY respectively, we have

$$R_x = \Sigma X \text{ and } R_y = \Sigma Y$$

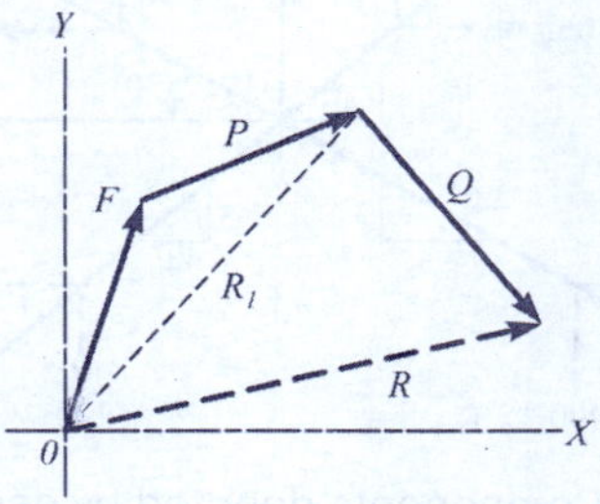

Figure 2-4.2 Resultant determined by triangle law.

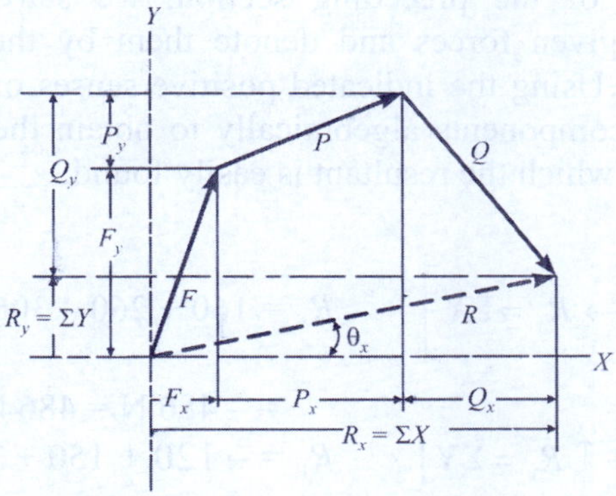

Figure 2-4.3 Rectangular components of resultant.

Having thereby computed the components of the resultant R, we can now determine its magnitude and inclination by using the method discussed in Section 2-3. Doing this transforms Eq. (2-3.2) into

$$R = \sqrt{(\Sigma X)^2 + (\Sigma Y)^2} \tag{2-4.1}$$

$$\tan \theta_x = \frac{\Sigma Y}{\Sigma X} \tag{2-4.2}$$

The pointing of R is determined by the signs of its rectangular components.

ILLUSTRATIVE PROBLEMS

2-4.1 Determine completely the resultant of the coplanar concurrent force system shown in Fig. 2-4.4.

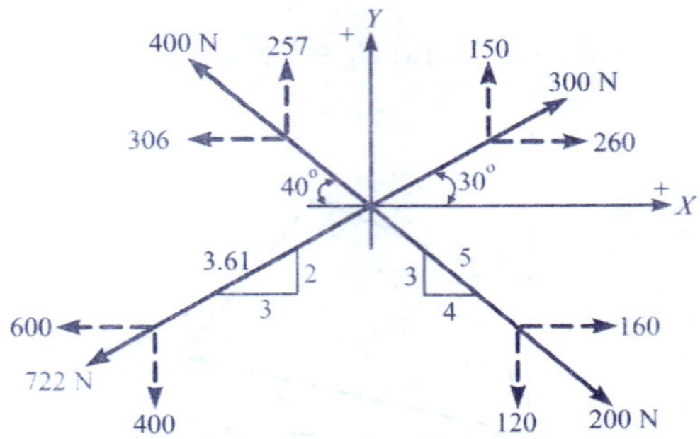

Figure 2-4.4 Components denoted by dashed vectors.

Solution

Using the methods of the preceding section, we start by computing the components of the given forces and denote them by the dashed vectors as shown in Fig. 2-4.4. Using the indicated positive senses of the reference axes, we next sum these components algebraically to obtain the components of the resultant force, from which the resultant is easily found.

We thus obtain

$$\left[\overset{+}{\rightarrow} R_x = \Sigma X\right] \qquad R_x = 160 + 260 - 306 - 600$$

$$= -486 \text{ N} = 486 \text{ N} \leftarrow$$

$$\left[+\uparrow R_y = \Sigma Y\right] \qquad R_y = -120 + 150 + 257 - 400 = -113 \text{ N}$$

$$= -113 \text{ N} = 113 \text{ N} \downarrow$$

Combining these components of R, we obtain

$$\text{R} = 498 \text{ N} \swarrow; \; \theta_x = 13.1^\circ \qquad \textit{Ans.}$$

If desired, a check may be made graphically by plotting the given forces to scale in tip-to-tail fashion.

PROBLEMS

2-4.2 Determine the resultant of the four forces acting on the body shown in Fig. P-2-4.2.

R = 214 N down to the right at $\theta_x = 59^\circ$ ***Ans.***

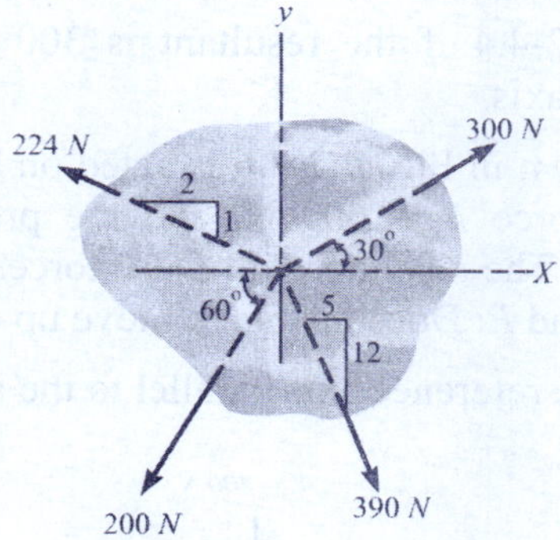

Figure P-2-4.2

2-4.3 The body shown in Fig. P-2-4.3 is acted on by four forces. Determine the resultant.

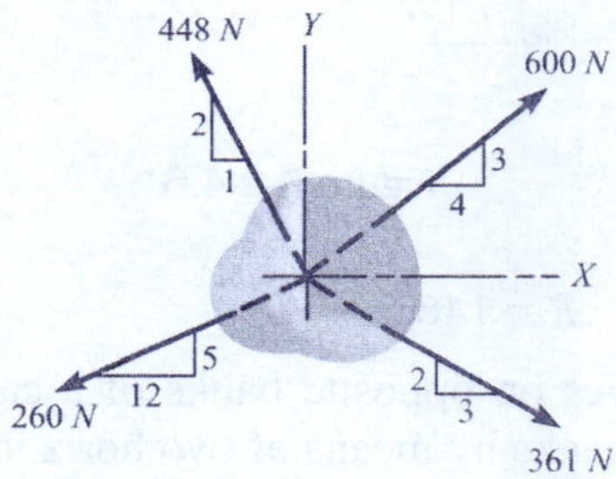

Figure P-2-4.3

R = 572 N up to the right at $\theta_x = 53.5^\circ$ ***Ans.***

2-4.4 The force system shown in Fig. P-2-4.4 has a resultant of 200 N pointing up along the Y axis. Compute the values of F and θ required to give this resultant.

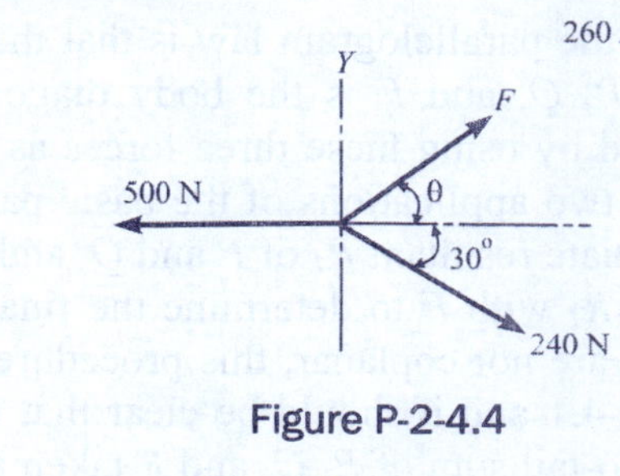

Figure P-2-4.4

F = 369 N at $\theta_x = 47.6^\circ$ ***Ans.***

2-4.5 Repeat Prob. 2-4.4 if the resultant is 300 N down to the right at 60° with the X axis.

2-4.6 The block shown in Fig, P-2-4.6 is acted on by its weight $W = 400$ N. a horizontal force $F = 600$ N, and the pressure P exerted by the inclined plane. The resultant R of these forces is parallel to the incline. Determine P and R. Does the block move up or down the incline?

Hint: Take one reference axis parallel to the incline.

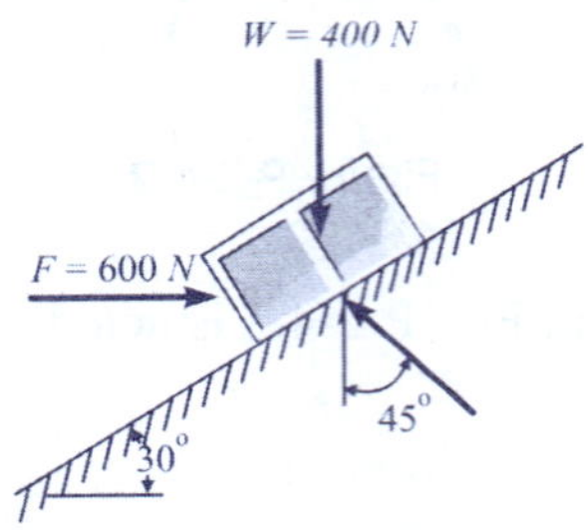

Figure P-2-4.6

$R = 146.5$ N ***Ans.***

2-4.7 Two locomotives on opposite banks of a canal pull a vessel moving parallel to the banks by means of two horizontal ropes. The tensions in these ropes are 2000 N and 2400 N while the angle between them is 60°. Find the resultant pull on the vessel and the angle between each of the ropes and the sides of the canal.

$R = 3820$ N; 33° and 27° ***Ans.***

2-4.8 Rework Prob. 2-4.7 if the cable tensions are 2000 N and 1500 N.

2-5 COMPONENTS OF FORCES IN SPACE

The spatial counterpart of the parallelogram law is that the resultant R of three concurrent spatial forces P, Q, and F is the body diagonal of the rhomboid shown in Fig. 2-5.1 formed by using these three forces as its initial sides. This conclusion is the result of two applications of the basic parallelogram law; one applied to find an intermediate resultant R_I of P and Q, and the other combining this intermediate resultant R_I with F to determine the final resultant R. Except for the fact that the forces are not coplanar, this procedure is the exact parallel of that described in Fig. 2-4.1 and it should be clear that the resultant R could also be formed by the tip-to-tail sum of P, Q, and F taken as free vectors in any order.

Although this construction is straightforward, actually doing it on paper is difficult to perform. Practically, we resort to the concept of finding the components of each force along mutually perpendicular directions, whence combining these components algebraically, we obtain three mutually perpendicular components of the resultant force which is then readily found. This procedure, of course, is the analytical spatial counterpart of Section 2-4.

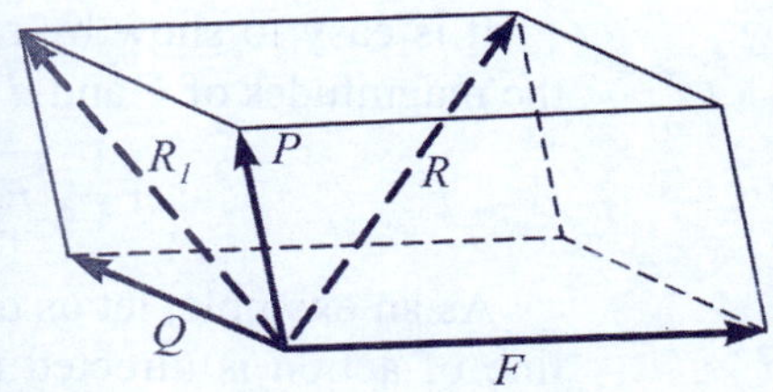

Figure 2-5.1 R is body diagonal of rhomboid having P, Q, and F as initial sides.

To determine the rectangular components of a space force, consider Fig. 2-5.2 showing a force F whose direction is specified by two points along its line of action. The rectangular components of the force are directly proportional to the rectangular components of the distance d separating the two points. This proportionality is an extension of the relations between the sides of a force triangle and its corresponding slope triangle that was discussed in Illus. Prob. 2-3.1. It is expressed by

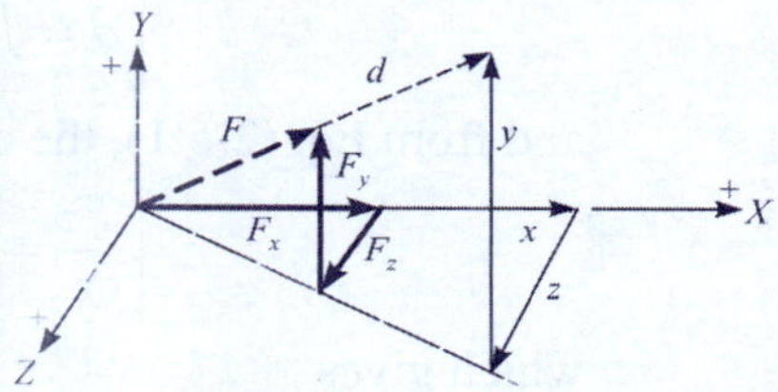

Figure 2-5.2 Proportionality of force components to distance components.

$$\frac{F_x}{x}=\frac{F_y}{y}=\frac{F_z}{z}=\frac{F}{d}=F_m \tag{2-5.1}$$

where F_m is the value of the equal ratios and is known as the multiplier of the force F (or more briefly, as the force multiplier) expressed in units of force per unit of length.

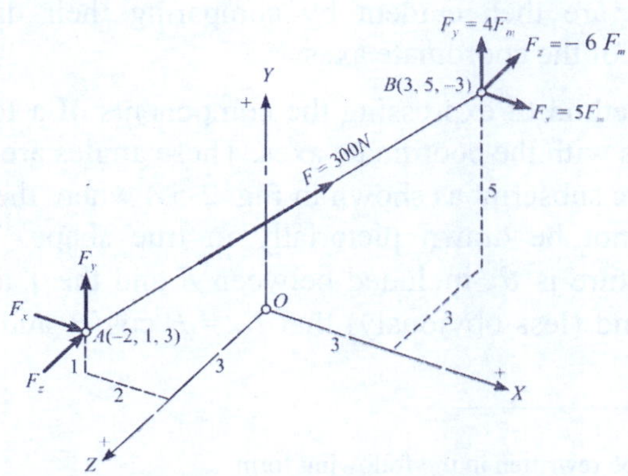

Figure 2-5.3 The components of a force as determined by the coordinates of two points on its line of action.

It is easy to show by repeated applications of the Pythagorean theorem that the magnitudes of *F* and *d* are given by

$$F = \sqrt{F_x^2 + F_y^2 + F_z^2} \text{ and } d = \sqrt{x^2 + y^2 + z^2} \qquad (2\text{-}5.2)$$

As an example, let us determine the components of a force F = 300 N whose line of action is directed from point *A* (–2, 1, 3) toward point *B* (3, 5, –3) as shown in Fig. 2-5.3. From this figure, or by taking the difference in the coordinates of *B* and *A*, the components of the distance *d* between *A* and *B* are $x = 5$, $y = 4$, and $z = -6$; applying Eq. (2-5.2), the length of *d* is found to be

$$\left[d = \sqrt{x^2 + y^2 + z^2}\right] \quad d = \sqrt{(5)^2 + (4)^2 + (6)^2} \quad d = 8.78$$

and from Eq. (2-5.1), the components of *F* are

$$\left[\frac{F_x}{x} = \frac{F_y}{y} = \frac{F_z}{z} = \frac{F}{d}\right] \qquad \frac{F_x}{5} = \frac{F_y}{4} = \frac{F_z}{-6} = \frac{300}{8.78}$$

which gives

$$F_x = 171\text{ N};\ F_y = 137\text{ N};\ F_z = -205\text{ N}$$

Note that the components of *F*, which here is directed from *A* toward *B*, will point in the directions corresponding to the paths taken in moving from *A* to *B* along directions parallel to the coordinate axes. The components, which must intersect on the action line of the force, may be shown as acting at *A* or *B*, or at any other convenient point on the line of action. At *B*, the components have been expressed as the product of the force multiplier and the absolute values of the rectangular components of the distance between *A* and *B*. This is usually the simplest way of writing the components of a force. The correct signs of the force components are then evident by comparing their directions with the positive directions of the coordinate axes.

An alternate method of expressing the components of a force is in terms of the angles it makes with the coordinate axes. These angles are generally denoted by θ with a suitable subscript as shown in Fig. 2-5.4 where they appear distorted because they cannot be drawn pictorially in true shape. The angle shown clearest in the picture is θ_y, included between *F* and the *Y* axis. It is apparent that $F_y = \cos\theta_y$ and (less obviously) that $F_x = F\cos\theta_x$ and $= F_z = F\cos\theta_z$.[3]

---

[3]These relations may be rewritten in the following form

$$\frac{F_x}{\cos\theta_x} = \frac{F_y}{\cos\theta_y} = \frac{F_z}{\cos\theta_z} = \frac{F}{1} \qquad (2\text{-}5.3)$$

which is convenient for slide rule computation by setting the components on the D scale opposite the corresponding angles on the cosine scale.

The cosines of the angles θ_x, θ_y and θ_z are defined as the direction cosines of the force.

It should be noted that the direction and the angles are not independent. From the fact that the sum of the squares of the components of a force is equal to the square of its magnitude, the relations of Eq. (2-5.3) reduce to

$$\cos^2 \theta_x + \cos^2 \theta_y + \cos^2 \theta_z = 1 \tag{2-5.4}$$

which enables us to find the third angle once any two have been specified.

Figure 2-5.4 also verifies the extension to spatial systems regarding the conclusion obtained in Section 2-4 that the resultant force can be found from any sequence of tip-to-tail addition of its components. The resultant of a system of concurrent space forces may now be found in a fashion similar to that used for coplanar concurrent forces. The X component of the resultant is equal to the algebraic summation of the X components of the forces comprising the system, and the same holds true for the Y and Z components as well, i.e.,

$$R_x = \Sigma X \qquad R_y = \Sigma Y \qquad R_z = \Sigma Z$$

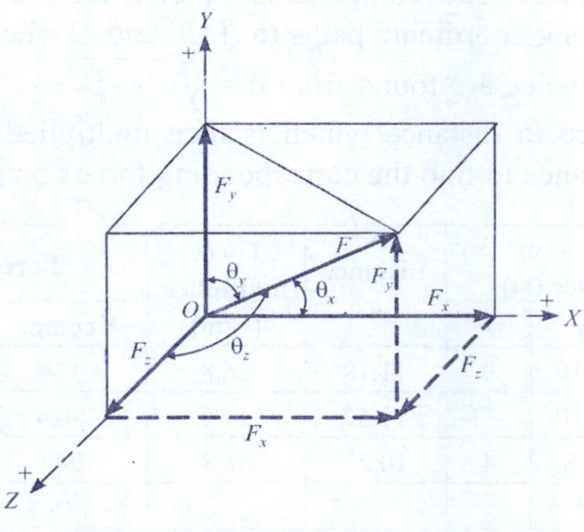

Figure 2-5.4 Rectangular components.

when R is substituted for F, Eq. (2-5.2) then determines the magnitude of the resultant while Eq. (2-5.3) determines the direction cosines and direction angles.

ILLUSTRATIVE PROBLEM

2-5.1 Find the resultant of the concurrent force system shown in Fig. 2-5.5 which consists of the forces T = 300 N, P = 200 N, and F = 500 N directed from D toward A, B, and C, respectively.

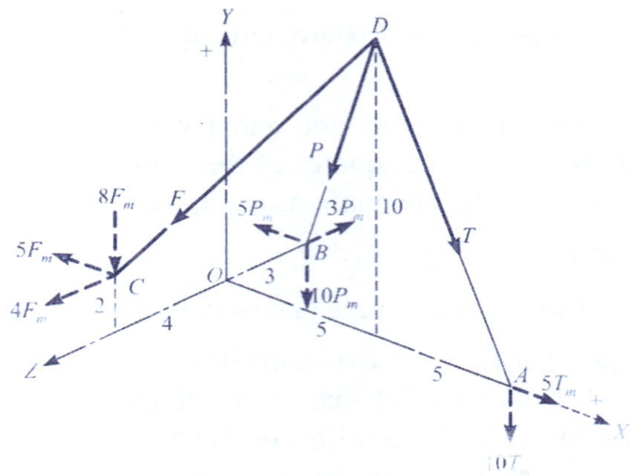

Figure 2-5.5

Solution

It is convenient to tabulate the computations as shown below. The table is almost self-explanatory The components of distance are the signed lengths traveled from D along coordinate paths to A, B, and C whence the terms in the column headed Distance are found from d = $\sqrt{x^2 + y^2 + z^2}$. The force multiplier is the ratio of force to distance which is then multiplied into the respective components of distance to find the corresponding force components.

Force (N)	Comps. of distance (m)			Distance *d*	Force multiplier (N/m)	Force components (N)		
	x	y	z			*X* comp.	*Y* comp.	*Z* comp.
T = 300	5	– 10	0	11.18	26.8	+ 134	– 268	0
P = 200	– 5	– 10	– 3	11.57	17.3	– 86.4	– 173	– 51.9
F = 500	– 5	– 8	4	10.25	48.8	– 244	– 390	+ 195
Totals						– 196.4	– 831	+ 143.1

An alternate plan is to draw the force components on the space diagram as shown by the dashed vectors, the direction of each component corresponding with the direction taken in moving along paths parallel to the reference axes from D to A, B, and C. Each component is shown as the product of the force multiplier by the absolute length of the coordinate paths, and the sign of each component is evident by comparison with the positive senses of the reference axes. The coefficient of each component then corresponds to the components of distance in the table above, whence the distance and force multiplier are evaluated as before and used to find the force components.

From the algebraic summation of the force components, we obtain

$$R_x = \Sigma X = -\ 196.4\ \text{N}$$

$$R_y = \Sigma Y = -\ 831\ \text{N}$$

$$R_z = \Sigma Z = +\ 143.1\ \text{N}$$

By visualizing a vector addition of these terms, we see that the resultant points to the left, down, and forward. The magnitude of the resultant and its direction cosines and direction angles are determined as follows;

$$\left[R = \sqrt{R_x^2 + R_y^2 + R_z^2}\right] \qquad \text{R} = \sqrt{(196.4)^2 + (831)^2 + (143.1)^2}$$

$$= 865\ \text{N} \qquad \textit{Ans.}$$

$$\left[\frac{R}{1} = \frac{R_x}{\cos\theta_x} = \frac{R_y}{\cos\theta_y} = \frac{R_z}{\cos\theta_z}\right]$$

$$\frac{865}{1} = \frac{196.4}{\cos\theta_x} = \frac{831}{\cos\theta_y} = \frac{143.1}{\cos\theta_z}$$

whence

$$\theta_x = 76.88^\circ; \qquad \theta_y = 16.1^\circ; \qquad \theta_z = 80.48^\circ \qquad \textit{Ans.}$$

PROBLEMS

2-5.2 Determine the resultant of the system of concurrent forces having the following magnitudes and passing through the origin and the indicated points: P = 280 N (+ 12, + 6, – 4); T = 520 N (– 3, – 4, + 12); F = 270 N (+ 6, – 3, – 6).

R = 394 N pointing forward, down, and to the right *Ans.*

2-5.3 Solve Prob. 2-5.2 using the following data; P = 140 N (+ 3, – 6, + 2); T = 260 N (– 12, + 4, – 3); F = 90 N (+ 3, + 6, – 6). Also find the direction angles of the resultant.

R = 171 N pointing backward, up, and to the left;
$\theta_x = 28.7^\circ$; $\theta_y = 83.29^\circ$; $\theta_z = 62.1^\circ$ *Ans.*

2-5.4 Find the resultant of the force system shown in Fig. P-2-5.4 in which P = 280 N, T = 260 N, and F = 210 N.

R = 674 N pointing forward, down, and to the left *Ans.*

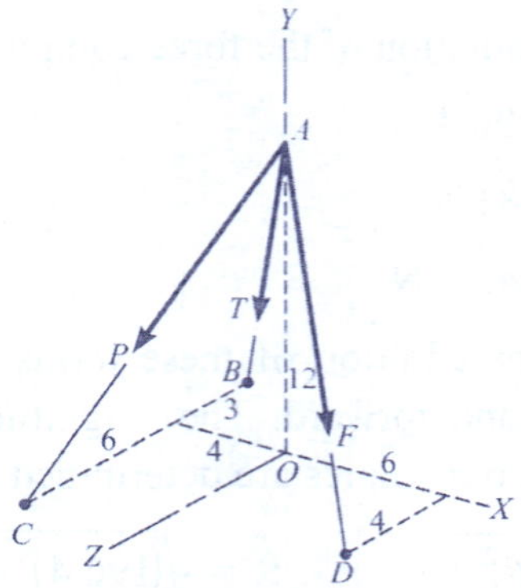

Figure P-2-5.4

2-5.5 In Fig. P-2-5.5, a vertical boom *AE* is supported by guy wires from *A* to *B*, *C*, and *D*. If the tensile load in *AD* = 252 N, find the forces in *AC* and *AB* so that the resultant force on *A* will be vertical.

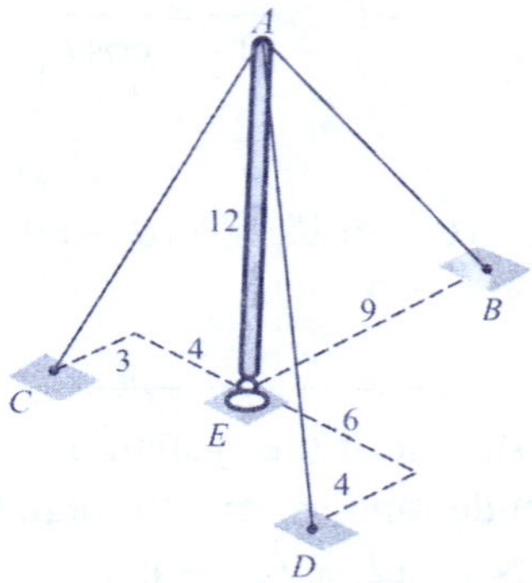

Figure P-2-5.5

AB = 255 N; *AC* = 351 N **Ans.**

2-5.6 Three concurrent forces *P*, *T*, and *F* have a resultant of 10 N directed forward and up to the right at $\theta_x = 60°$, $\theta_y = 60°$, $\theta_z = 45°$. *P* equals 21 N and passes from the origin through point (3, 2, 6). The value of *T* is 18 N and is directed from the origin toward point (–6, 6, –3). Determine the magnitude of the third force *F* and the angles it makes with the reference axes.

F = 16.06 N backward and down to the right;

$\theta_x = 60.1°$; $\theta_y = 25.9^0$; $\theta_z = 72.1°$ **Ans.**

2-6 VECTOR NOTATION

In the preceding sections, we learned how to resolve a vector into its rectangular components. This process is a basic step in all practical solutions of problems, and we have seen how to apply it to determine analytically the resultant of concurrent force systems.

For a general development of the theoretical aspects of mechanics, however, a more rigorous treatment is possible by using vector analysis. This is especially true in Dynamics where the principles of vector analysis are invaluable for analyzing the time rate of change of both the magnitude and direction of the vector quantities encountered. As we shall see, the distinction between magnitude and direction is achieved by a notation which expresses a vector as the product of its magnitude and a unit vector which defines its direction.

Recall that vectors were defined as mathematical quantities possessing magnitude and direction which add according to the parallelogram law. As we saw in Section 2-2, the parallelogram law applied to forces is restricted to *localized* vectors having a definite point of application. It is this restriction that requires the components of a force to intersect at its point of application. Somewhat more liberty is permitted when considering the *external effect* of a force on a rigid body; then the principle of transmissibility permits the components of a force to be applied so that they intersect anywhere on its line of action. In this case the forces may be called *sliding* vectors.

A *free* vector is one which is independent of the point of application of the vector or which represents a quantity such as a couple which, as shown in Section 2-12, has no particular point of application and hence may be freely moved in space. Unless otherwise specified, *vector analysis generally deals with free vectors* so that two vectors are said to be equal when they have the same *magnitude and direction*, whether or not they have the same point of application. Being equal, they are represented by the same symbol, although their physical effect may be different if they are not free vectors. In particular, free vectors may always be added in tip-to-tail fashion (the polygon rule) to determine the magnitude and direction of their resultant.

When using vector analysis, we denote a vector by printing it in boldface type to distinguish it from its magnitude printed in italic type; for example, **F** represents a vector having a magnitude F and a specified direction. The magnitude of **F** may also be denoted by $|\mathbf{F}|$. In longhand writing, a vector may be denoted by drawing a short arrow above the letter used to represent it ($\vec{F}$) or by underlining the letter ($\underline{F}$). We recommend the method of underlining because it can also be used on a typewriter. This method of indicating a quantity is a vector permits us to use the symbology of ordinary algebra instead of a special symbol (like $\nrightarrow$) to denote vector (or geometric) addition. Thus we may write

$$\mathbf{F} = \mathbf{F}_x + \mathbf{F}_y + \mathbf{F}_z \quad \text{for} \quad F = F_x \nrightarrow F_y \nrightarrow F_z \qquad (a)$$

or

$$\mathbf{R} = \mathbf{P} + \mathbf{F} + \mathbf{T} \quad \text{for} \quad \mathrm{R} = \mathrm{P} \nrightarrow \mathrm{F} \nrightarrow \mathrm{T} \qquad (b)$$

Equations written in boldface type are always to be understood to be vector equations and each term in such an equation .nust always be a dimensionally homogeneous vector. This brings us to the concept of a unit vector which is a special symbol used to specify the direction of the scalar magnitude of a vector.

A unit vector is defined as a vector of unit magnitude in a specified direction. We shall denote them by placing a circumflex (^) over them as well as printing them in boldface. For longhand writing, the circumflex is sufficient to denote a unit vector; it need not also be underlined. Multiplying a unit vector by a scalar denotes a vector having the direction of the unit vector and a magnitude equal to that of the scalar. In this connection, a negative scalar denotes a vector direction opposite to that of the unit vector.

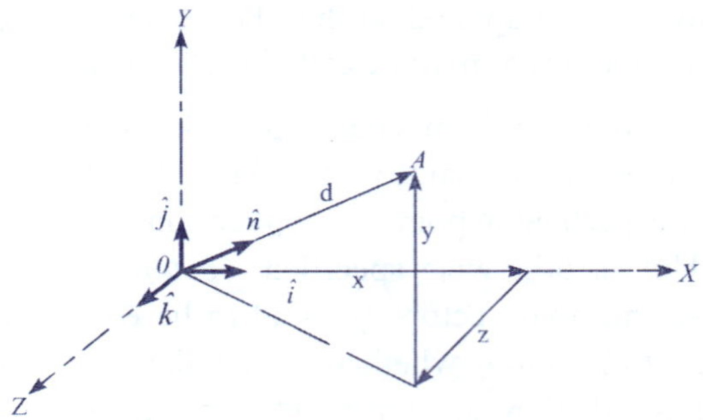

Figure 2-6.1

It is conventional to let $\hat{\mathbf{i}}$, $\hat{\mathbf{j}}$, and $\hat{\mathbf{k}}$ represent vectors of unit length directed along the positive senses of the X, Y, Z coordinate axes respectively as shown in Fig. 2-6.1. Applying this convention, the vector component $\mathbf{F}_x$ may be written as $\mathbf{F}_x\hat{\mathbf{i}}$ (meaning that the vector $\mathbf{F}_x$ has the scalar magnitude F_x in the + X direction). Thus an alternate form of writing Eq. (a) in this new notation is

$$\mathbf{F} = \mathbf{F}_x\hat{\mathbf{i}} + \mathbf{F}_y\hat{\mathbf{j}} + \mathbf{F}_z\hat{\mathbf{k}} \qquad \text{(c)}$$

This is known as the standard Cartesian form of representing a vector. It does not appear to be much of an improvement over Eq. (*a*) but its value will be apparent after we learn how to correlate a unit vector $\hat{\mathbf{n}}$ in any direction with the $\hat{\mathbf{i}}, \hat{\mathbf{j}}, \hat{\mathbf{k}}$ unit vectors.

Consider now a unit vector $\hat{\mathbf{n}}$ in an arbitrary direction such as $\overrightarrow{OA}$ in Fig. 2-6.1. By definition, the unit vector $\hat{\mathbf{n}}$ in the direction from O to A is the vector **d** divided by the magnitude d of the distance from O to A. Thus,

$$\hat{\mathbf{n}} = \frac{\mathbf{d}}{d} = \frac{x\hat{\mathbf{i}} + y\hat{\mathbf{j}} + z\hat{\mathbf{k}}}{d} = \frac{1}{d}\left(x\hat{\mathbf{i}} + y\hat{\mathbf{j}} + z\hat{\mathbf{k}}\right) \tag{2-6.1}$$

where $\quad \mathrm{d} = \sqrt{x^2 + y^2 + z^2}$

In particular, the proportionality ratios between a force **F** parallel or coincident with **d** discussed in Section 2-5 which are,

$$\frac{F_x}{x} = \frac{F_y}{y} = \frac{F_z}{z} = \frac{F}{d} = F_m \tag{2-5.1}$$

permit the force **F** to be expressed in any of the following variations:

$$\mathbf{F} = \mathbf{F}\hat{n} = \frac{\mathbf{F}}{\mathbf{d}}\left(x\hat{i} + y\hat{j} + z\hat{k}\right) = \mathbf{F}_{\mathbf{m}}\left(x\hat{i} + y\hat{j} + z\hat{k}\right) \tag{2-6.2}$$

In many cases, it will be especially convenient to use the third of these variations where we combine the force multiplier with the rectangular components of the distance between any two points on the action line of **F**. The use of a force multiplier instead of a force magnitude and its associated unit vector (i.e., $F\hat{\boldsymbol{n}}$) has two advantages:

First, the vector representation of the force (and its corresponding components) are expressed directly in terms of the coordinate component distances separating any two points on its line of action, and

Second, computations are easier since unit vectors in terms of decimals or fractions are avoided.

For consistency, vector methods generally use a right-handed coordinate system in which a rotation of the X axis toward the Y axis about the Z axis would advance a right-handed screw along the positive direction of the Z axis. The several different arrangements of right-handed coordinate systems are shown in Fig. 2-6.2. They are obtained by cyclic interchange of the coordinate directions. Which to use is a maker of individual preference.

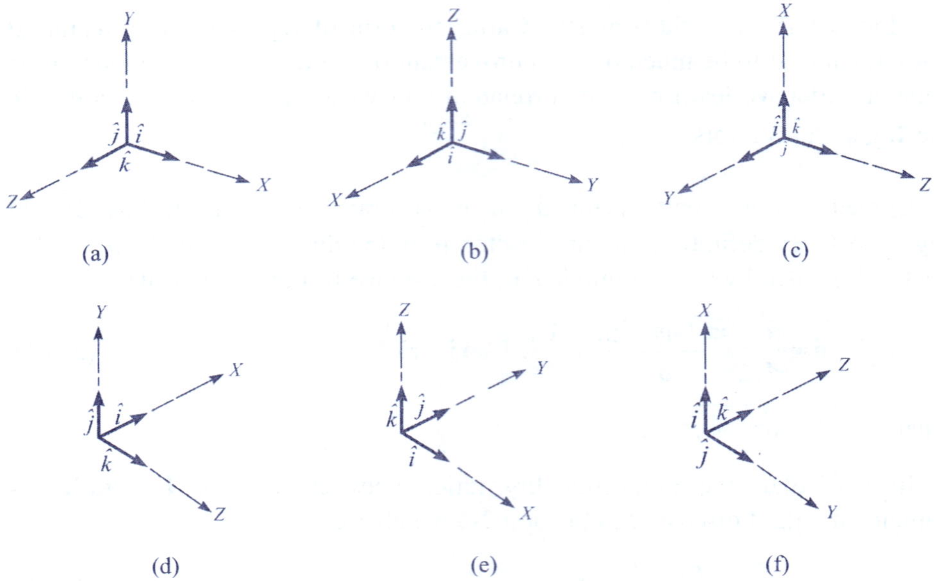

Figure 2-6.2 Various arrangements of right-handed coordinate system obtained by cyclic interchange of the coordinate directions.

2-7 VECTOR ALGEBRA, DOT PRODUCT

There are two types of vector multiplication which seem almost to have been invented to find the component of a vector along any spatial direction, or the moment of the vector about any center[4]. These types are known as the scalar (or dot) product and the vector (or cross) product. In each case, the result will be the product of the magnitudes of the vectors times respectively the cosine or the sine of their included angle.

Let us consider first the *dot product.* It is defined to be a scalar quantity determined by multiplying the magnitudes of two vectors by the cosine of their included angle. The name comes from using a dot between two vectors to denote their multiplication. Thus, the dot product of multiplying **a** by **b** is

$$\mathbf{a} \cdot \mathbf{b} = ab \cos \theta \tag{2-7.1}$$

This product is interpreted geometrically in Fig. 2-7.1. It shows how to obtain the component of one vector along the direction of another vector. The angle θ is the angle smaller than 180° included between the vectors (or their

[4]There are other uses also, but those mentioned are especially advantageous in mechanics.

directions if they do not have a common origin). As shown in parts (a) and (b), the dot product may be used to find either the orthogonal component of **b** upon **a** or the orthogonal of **a** upon **b**. Either component is obtained by dividing the dot product by the magnitude of the vector in whose direction the component is desired. Observe that **a**/a and **b**/b are actually unit vectors $\hat{n}_a$ and $\hat{n}_b$ in the respective directions of **a** and **b**. This observation is formalized by this rule: *The component of a vector in any direction is the dot product of the vector with a unit vector in the desired direction.* Its application is discussed in Illus. Prob. 2-7.1.

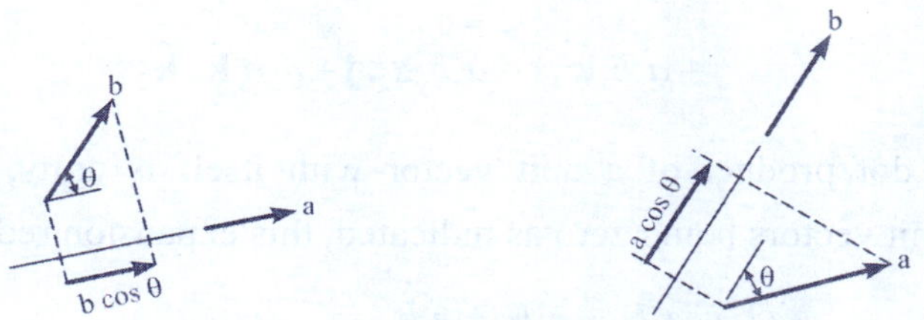

(a) $b\cos\theta = \dfrac{\text{a.b}}{\text{a}} = \hat{\text{n}}_\text{a}.\text{b}$ (b) $a\cos\theta = \dfrac{\text{a.b}}{\text{b}} = \text{a}.\hat{\text{n}}_\text{b}$

Figure 2-7.1 Interpretation of dot product a. b = $ab\cos\theta$.

As a preliminary step to applying the dot product, consider the dot products of various combinations of the orthogonal unit vectors $\mathbf{i}, \mathbf{j}$ and $\hat{\mathbf{k}}$. From its definition, the dot product of a unit vector with itself is unity since it is the product of unit magnitudes by the cosine of the zero angle between their directions; however, all the other combinations of dot products of the unit vectors will be zero because their included angle is 90° whose cosine is zero. Thus, we obtain

$$\hat{\mathbf{i}}.\hat{\mathbf{i}} = \hat{\mathbf{j}}.\hat{\mathbf{j}} = \hat{\mathbf{k}}.\hat{\mathbf{k}} = 1$$

$$\hat{\mathbf{i}}.\hat{\mathbf{j}} = \hat{\mathbf{j}}.\hat{\mathbf{k}} = \hat{\mathbf{k}}.\hat{\mathbf{i}} = 0 \tag{2-7.2}$$

Some other properties of the dot product, as a consequence of its definition, are the following:

Commutative: $\mathbf{A}.\mathbf{B} = \mathbf{B}.\mathbf{A}$

Associate: $m\mathbf{A}.n\mathbf{B} = mn\mathbf{A}.\mathbf{B}$

Distributive $\mathbf{C}.(\mathbf{A}+\mathbf{B}) = \mathbf{C}.\mathbf{A} + \mathbf{C}.\mathbf{B}$

These properties of the dot product enable us to evaluate the expansion of **a . b** in terms of their scalar components as follows:

$$\mathbf{a}\cdot\mathbf{b} = \left(a_x\hat{i} + a_y\hat{j} + a_z\hat{k}\right)\cdot\left(b_x\hat{i} + b_y\hat{j} + b_z\hat{k}\right)$$

$$= a_xb_x\hat{\mathbf{i}}\cdot\hat{\mathbf{i}} + a_xb_y\underbrace{\hat{\mathbf{i}}\cdot\hat{\mathbf{j}}}_{=0} + a_xb_z\underbrace{\hat{\mathbf{i}}\cdot\hat{\mathbf{k}}}_{=0}$$

$$+ a_yb_x\underbrace{\hat{\mathbf{j}}\cdot\hat{\mathbf{i}}}_{=0} + a_yb_y\hat{\mathbf{j}}\cdot\hat{\mathbf{j}} + a_yb_z\underbrace{\hat{\mathbf{j}}\cdot\hat{\mathbf{k}}}_{=0}$$

$$+ a_zb_x\underbrace{\hat{\mathbf{k}}\cdot\hat{\mathbf{i}}}_{=0} + a_zb_y\underbrace{\hat{\mathbf{k}}\cdot\hat{\mathbf{j}}}_{=0} + a_zb_z\hat{\mathbf{k}}\cdot\hat{\mathbf{k}}$$

Since the dot product of a unit vector with itself is unity, all other dot products of unit vectors being zero as indicated, this expansion reduces to

$$\mathbf{a}\cdot\mathbf{b} = a_xb_x + a_yb_y + a_zb_z \qquad (2\text{-}7.3)$$

which is summarized by the rule: *The dot product of two vectors is equal to the sum of the products of their respective scalar components*. In particular,

$$\mathbf{a}\cdot\mathbf{a} = a_x^2 + a_y^2 + a_z^2 = a^2$$

Further, if **a . b** = 0 and it is known that neither **a** nor **b** are zero, then the cosine of their included angle must be zero; i.e., θ must be 90°, so that two vectors are perpendicular when their dot product is zero. Conversely, if two nonzero vectors are known to be perpendicular, their dot product must be zero.

The following illustrative problem discusses the application of the dot product.

ILLUSTRATIVE PROBLEM

2-7.1 In the tripod shown in Fig. 2-7.2, forces **F** and **P** act as shown along the legs *DC* and *DB*. In terms of its force multiplier, determine the component of **F** along the direction of **P** and also the angle between **F** and **P**. Note that points *B* and *C* are in the same vertical plane and that *C* is located 2 m above the horizontal plane containing *B* and *A*.

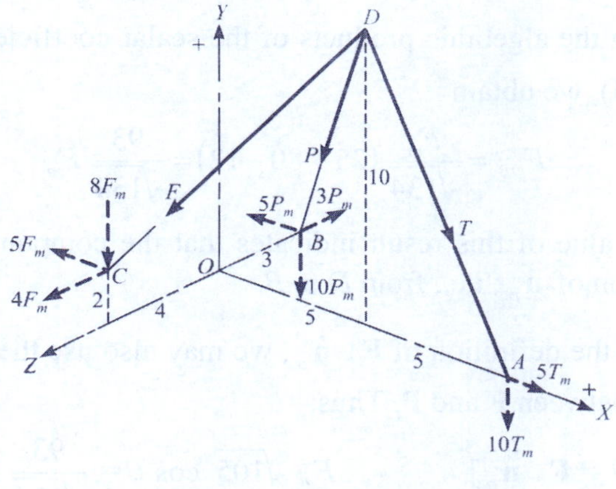

Figure 2-7.2

Solution

Following the method discussed on p. 39, the components of **F** and **P**, acting as shown, have been expressed as the product of their respective force multipliers and the components of the distances traveled in moving along coordinate directions from D to C and from D to B. On comparing the directions of these components with the positive directions of the coordinate axes, we obtain

$$\mathbf{F} = F_m(-5\,\hat{\mathbf{i}} - 8\,\hat{\mathbf{j}} + 4\,\hat{\mathbf{k}}); \qquad F = F_m\sqrt{105}$$

and $$\mathbf{P} = P_m(-5\,\hat{\mathbf{i}} - 10\,\hat{\mathbf{j}} + 3\,\hat{\mathbf{k}}); \qquad P = P_m\sqrt{134}$$

The component F_{DB} of **F** along the direction of DB (i.e., along **P**) may now be found by applying the rule deduced on p. 40 that the component of a vector in any direction is the dot product of the vector with a unit vector in the desired direction. The unit vector $\hat{\mathbf{n}}_p$ along **P** being

$$\hat{\mathbf{n}}_p = \frac{\mathbf{P}}{P} = \frac{1}{\sqrt{134}}\left(-5\hat{\mathbf{i}} - 10\hat{\mathbf{j}} - 3\hat{\mathbf{k}}\right)$$

the component F_{DB} of **F** along **P** is

$[F_{DB} = \mathbf{F} \cdot \hat{\mathbf{n}}_p]$

$$F_{DB} = F_m\left(-5\hat{\mathbf{i}} - 8\hat{\mathbf{j}} + 4\hat{\mathbf{k}}\right) \cdot \frac{1}{\sqrt{134}}\left(-5\hat{\mathbf{i}} - 10\hat{\mathbf{j}} - 3\hat{\mathbf{k}}\right)$$

whence, summing the algebraic products of the scalar coefficients of $\hat{\mathbf{i}}$, $\hat{\mathbf{j}}$, and $\hat{\mathbf{k}}$ as in Eq. (2-7.3), we obtain

$$F_{DB} = \frac{F_m}{\sqrt{134}}(25 + 80 - 12) = \frac{93}{\sqrt{134}} F_m \qquad \textit{Ans.}$$

The positive value of this result indicates that the component of **F** upon **P** acts in the direction of $\hat{\mathbf{n}}_p$; i.e., from *D* to *B*.

By reapplying the definition of $\mathbf{F} \cdot \hat{\mathbf{n}}_p$, we may also use the preceding result to find the angle between **F** and **P**. Thus;

$$[F \cos\theta = \mathbf{F} \cdot \hat{\mathbf{n}}_p] \qquad F_m \sqrt{105} \cos\theta = \frac{93}{\sqrt{134}} F_m$$

whence

$$\cos\theta = \frac{93}{\sqrt{(105)(134)}} = 0.785; \quad \theta = 38.2^\circ \qquad \textit{Ans.}$$

A general interpretation of this procedure is that the cosine of the angle between two directions equals the dot product of the unit vectors along the specified directions.

PROBLEMS

2.7.2 In Fig. P-2-7.2, a boom *AC* is supported by a ball-and-socket joint at C and by the cables *BE* and *AD*. If the force multiplier of a force **F** acting from *B* to *E* is F_m = 10 N/m and that of a force **P** acting from *A* to *D* is P_m = 20 N/m, find the component of each force along *AC*.

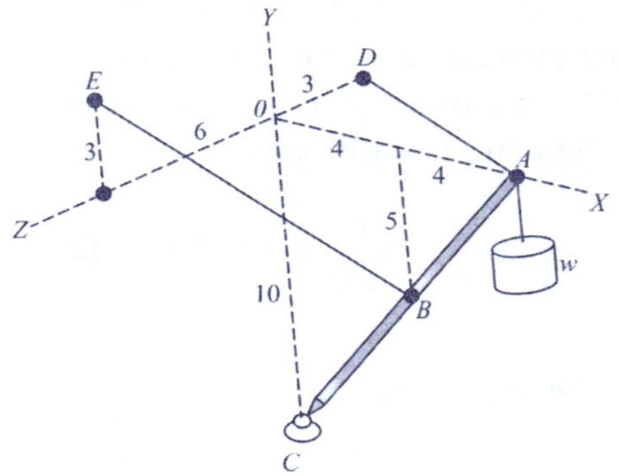

Figure P-2-7.2

F_{AC} = –37.5 N; P_{AC} = 100 N *Ans.*

2.7.3 Using the data of the previous problem, find the components of **F** and **P** along the direction *DE*.

2.7.4 In the system shown in Fig. P.2-74, it is found that the force multiplier of force F acting from B to D is F_{m}, = 150 N/m and that of force **P** acting from *A* to *E* is P_m 100 N/m. Find the component of force along AC. What angle does each force make with *AC*?

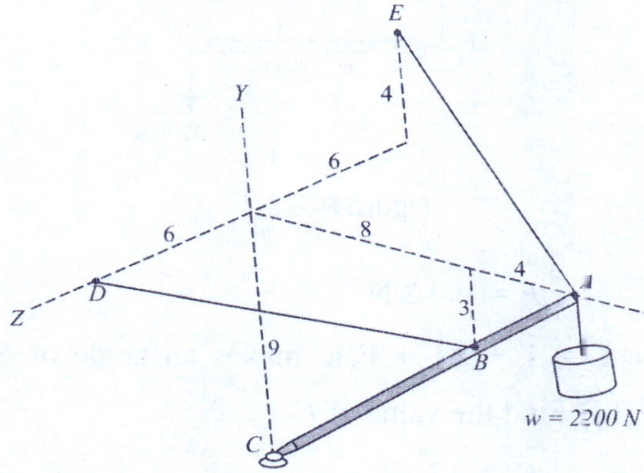

Figure P-2-7.4

F_{AC} = 690 N; P_{AC} = 720 N; $\theta_F = 63.85°$; $\theta_P = 59.1°$ ***Ans.***

2.7.5 Using the data of the preceding problem find the components of forces **F** and **P** along the direction *DE*. What angle does each make with *DE*?

2-7.6 In the unsymmetrical cantilever framework shown in Fig. P-2-76 points *C* and *D* are in the same vertical plane while *B* is 3 m in front of this plane. Assume a force **F** to act from *A* to *C* and another force **P** to act from *A* to *D*. In terms of their force multipliers, find the components of these forces along the direction from *A* to *B*. What is the angle between **F** and **P**?

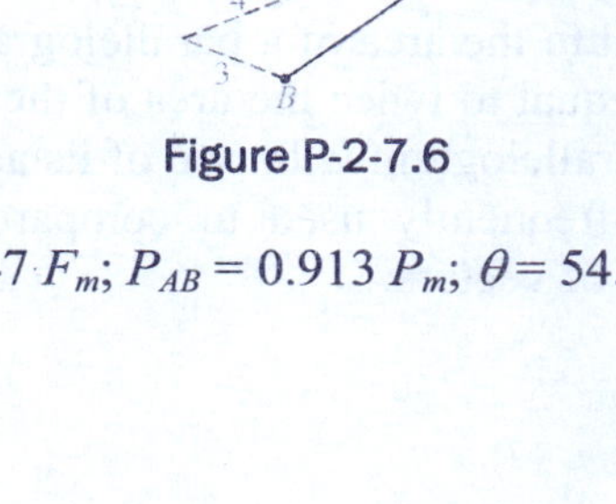

Figure P-2-7.6

$F_{AB} = 5.47\,F_m$; $P_{AB} = 0.913\,P_m$; $\theta = 54.2°$ ***Ans.***

2.7.7 Referring to Fig. P-2.7.7, find the magnitude of the force **F** acting from D toward A that will have a component of 100 N along DB.

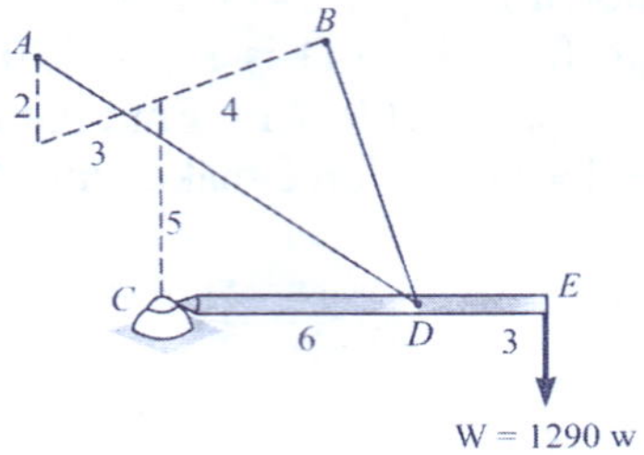

Figure P-2-7.7

$F = 144.3$ N **Ans.**

2.7.8 The force $\mathbf{F} = \hat{\mathbf{i}} + 2\hat{\mathbf{j}} + F_z\hat{\mathbf{k}}$ makes an angle of 60° with the line $\mathbf{L} = 4\hat{\mathbf{i}} + 3\hat{\mathbf{j}}$. Find the value of F_z.

$F_z = \pm 3.32$ **Ans.**

2-8 VECTOR ALGEBRA, CROSS PRODUCT

The second type of vector multiplication is known as the cross product because of the symbol × used to denote this type of multiplication. The cross product of two vectors **a** and **b** (written as **a** × **b**) is defined as the product of their magnitudes by the sine of their included angle. In addition, this result is a new vector (hence the alternate name of vector product) acting perpendicular to the plane of **a** and **b** in the direction a right-hand screw would advance as **a** is turned toward **b**. An alternate to the right-hand screw rule is the three-finger right-hand rule: Place the thumb in the direction of **a**, the first finger in the direction of **b**; then the second finger points in the direction of **a** × **b**. Perhaps best of all is the right-hand rule in which the fingers of the right-hand curl in the direction of **a** rotating toward **b**; then the extended thumb points in the positive direction of the cross product. Denoting $\hat{\mathbf{n}}$ as the unit vector of this positive direction, the explicit form of the cross product is

$$\mathbf{a} \times \mathbf{b} = ab \sin \theta\, \hat{\mathbf{n}} \tag{2-8.1}$$

which is interpreted geometrically in Fig. 2-8.1. We see that the magnitude of the cross product is equal to the area of a parallelogram constructed with initial sides **a** and **b**. It is also equal to twice the area of the triangle formed by taking either diagonal of this parallelogram with two of its adjacent sides. These rather obvious comments are frequently used to compare geometrically the cross products of different sets of vectors.

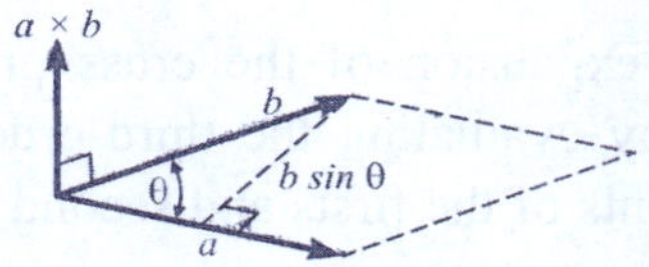

Figure 2-8.1 a × b is perpendicular to plane of a and b.

Due to the right-hand rule, note that **b** × **a** acts in the opposite sense to **a** × **b**, i.e.,

$$\mathbf{b}\times\mathbf{a} = -\,\mathbf{a}\times\mathbf{b} \qquad (2\text{-}8.2)$$

Therefore, the vector product is *not* commutative; the order of the terms may not be interchanged without also changing the sign of the cross product.

Consider now the cross products of various combinations of the orthogonal unit vectors. A cross product of a unit vector with itself (or of any vector with itself) is zero because the sine of the included angle is zero. A convenient scheme to remember other combinations is that if any two follow each other alphabetically (i.e., in the clockwise cyclic order shown below), their cross product is the positive value of the third. Thus, we obtain

$$\begin{aligned}
&\hat{\mathbf{i}}\times\hat{\mathbf{i}}=\hat{\mathbf{j}}\times\hat{\mathbf{j}}=\hat{\mathbf{k}}\times\hat{\mathbf{k}}=0\\
&\hat{\mathbf{i}}\times\hat{\mathbf{j}}=\hat{\mathbf{k}}\,; \qquad \hat{\mathbf{j}}\times\hat{\mathbf{k}}=\hat{\mathbf{i}}\,; \qquad \hat{\mathbf{k}}\times\hat{\mathbf{i}}=\hat{\mathbf{j}}\\
&\hat{\mathbf{j}}\times\hat{\mathbf{i}}=-\,\hat{\mathbf{k}}\,; \qquad \hat{\mathbf{k}}\times\hat{\mathbf{j}}=\hat{\mathbf{i}}\,; \qquad \hat{\mathbf{i}}\times\hat{\mathbf{k}}=-\,\hat{\mathbf{j}}
\end{aligned} \qquad (2\text{-}8.3)$$

These properties of the cross product of the orthogonal unit vectors enable us to evaluate the expansion of **a** × **b** in terms of their scalar components as follows:

$$\mathbf{a}\times\mathbf{b} = \left(a_x\hat{\mathbf{i}}+a_y\hat{\mathbf{j}}+a_z\hat{\mathbf{k}}\right)\times\left(b_x\hat{\mathbf{i}}+b_y\hat{\mathbf{j}}+b_z\hat{\mathbf{k}}\right)$$

$$= a_xb_x\,\overset{=0}{(\hat{\mathbf{i}}\times\hat{\mathbf{i}})} + a_xb_y\,\overset{=\hat{\mathbf{k}}}{(\hat{\mathbf{i}}\times\hat{\mathbf{j}})} + a_xb_z\,\overset{=-\hat{\mathbf{j}}}{(\hat{\mathbf{i}}\times\hat{\mathbf{k}})}$$

$$+ a_yb_x\,\overset{=-\hat{\mathbf{k}}}{(\hat{\mathbf{j}}\times\hat{\mathbf{i}})} + a_yb_y\,\overset{=0}{(\hat{\mathbf{j}}\times\hat{\mathbf{j}})} + a_yb_z\,\overset{=\hat{\mathbf{i}}}{(\hat{\mathbf{j}}\times\hat{\mathbf{k}})}$$

$$+ a_zb_x\,\overset{=\hat{\mathbf{j}}}{(\hat{\mathbf{k}}\times\hat{\mathbf{i}})} + a_zb_y\,\overset{=-\hat{\mathbf{i}}}{(\hat{\mathbf{k}}\times\hat{\mathbf{j}})} + a_zb_z\,\overset{=0}{(\hat{\mathbf{k}}\times\hat{\mathbf{k}})}$$

which reduces to

$$\mathbf{a}\times\mathbf{b} = \hat{\mathbf{i}}\left(a_yb_z-a_zb_y\right)+\hat{\mathbf{j}}\left(a_zb_x-a_xb_z\right)+\hat{\mathbf{k}}\left(a_xb_y-a_yb_x\right) \qquad (2\text{-}8.4)$$

Instead of this direct expansion of the cross product, the same result is usually easier to obtain by evaluating the third-order determinant formed by taking the scalar components of the firsts and second terms of the cross product as respectively the elements the first and second rows and using $\hat{\mathbf{i}}$, $\hat{\mathbf{j}}$, and $\hat{\mathbf{k}}$ as the elements the third row. Thus, we get for a × b,

$$\mathbf{a} \times \mathbf{b} = \begin{vmatrix} a_x & a_y & a_z \\ b_x & b_y & b_z \\ \hat{\mathbf{i}} & \hat{\mathbf{j}} & \hat{\mathbf{k}} \end{vmatrix} = \hat{\mathbf{i}} \begin{vmatrix} a_y & a_z \\ by & b_z \end{vmatrix} + \hat{\mathbf{j}} \begin{vmatrix} a_z & a_x \\ b_z & bx \end{vmatrix} + \hat{\mathbf{k}} \begin{vmatrix} ax & a_y \\ bx & b_y \end{vmatrix}$$

$$= \hat{\mathbf{i}}\,(a_y b_z - a_z b_y) + \hat{\mathbf{j}}\,(a_z b_x - a_x b_z) + \hat{\mathbf{k}}\,(a_x b_y - a_y b_x)$$

Here we expanded the third-order determinant by first reducing it to minors of the third row in the following way: Visualize (or actually write out) the first column of the determinant repeated after the third column as shown below. Then the minors of the third row are those elements of the second-order determinants which are enclosed by the dashed boxes. The arrows indicate which elements the third row are associated with each second-order determinant.

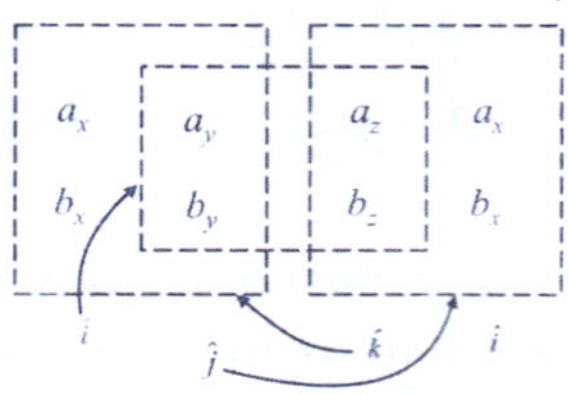

Remember that in expanding the second-order determinants, the products of terms lying on a diagonal downward to the right are positive, whereas the products of terms on a diagonal upward to the right are negative. With a little practice, you will be able to write out directly the expansion of the third-order determinant by eliminating the step of explicitly stating the second-order determinants since these are implied in the dashed boxes shown here.

Some further properties of the cross product are that it is associative when multiplied by a scalar, namely

$$m\,(\mathrm{a} \times \mathrm{b}) = m\mathrm{a} \times \mathrm{b} = \mathrm{a} \times m\mathrm{b} \tag{2-8.5}$$

but it is *not* associative when cross-multiplied by a vector; i.e.,

$$\mathrm{a} \times (\mathrm{b} \times \mathrm{c}) \neq (\mathrm{a} \times \mathrm{b}) \times \mathrm{c}$$

The inequality of these vector triple products is evident by comparing the following reduction formulas for a vector triple product:

$$a \times (b \times c) = (a \,.\, c)b - (a \,.\, b)c$$
$$(a \times b) \times c = (c \,.\, a)b - (c \,.\, b)a \qquad (2\text{-}8.6)$$

In these reduction formulas, the vector triple product is equal to the difference of the vectors enclosed in the parentheses; each of these vectors being multiplied by the scalar dot product of the remaining two others. In writing the difference of these terms, give the positive sign to that vector enclosed in the parentheses which next to the vector outside these parentheses.

We have stated these reduction formulas to demonstrate the nonassociative property mentioned above. Although we have no immediate need for these reduction formulas at this introductory stage, they can be so useful in later work that we present them here for future reference.

Another property of the cross product is that it is distributive over vector addition; thus

$$\mathrm{c} \times (\mathrm{a} + \mathrm{b}) = \mathrm{c} \times \mathrm{a} + \mathrm{c} \times \mathrm{b} \qquad (2\text{-}8.7)$$

but we must be careful to maintain the original order of the vectors.

For some applications of the cross product, consider the following illustrative problems.

ILLUSTRATIVE PROBLEM

2-8.1 Find the shortest distance from the origin to the line passing through points A (– 2, 1, 3) and B (4, 5, 0).

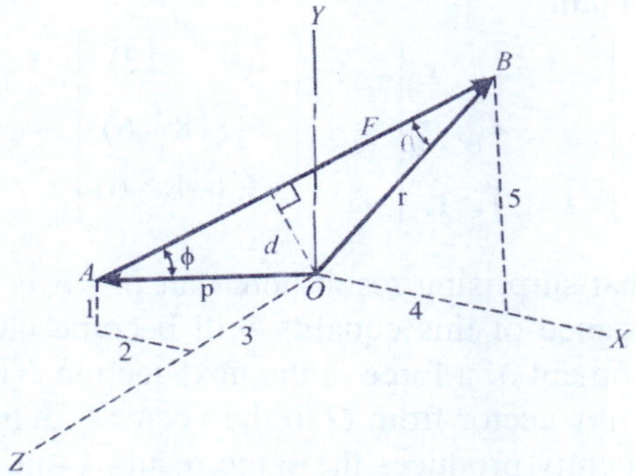

Figure 2-8.2

Solution

Figure 2-8.2 represents the given data and will aid the analysis. Let the directed line segments *AB*, *OB*, and *OA* be denoted by **F**, **r**, and **p**, respectively. Then.

$$\mathbf{F} = 6\hat{\mathbf{i}} + 4\hat{\mathbf{j}} - 3\hat{\mathbf{k}} \qquad F = \sqrt{36+16+9} = \sqrt{61}$$

$$\mathbf{r} = 4\hat{\mathbf{i}} + 5\hat{\mathbf{j}} \qquad r = \sqrt{16+25} = \sqrt{41}$$

$$\mathbf{p} = 2\hat{\mathbf{i}} + \hat{\mathbf{j}} + 3\hat{\mathbf{k}} \qquad p = \sqrt{4+1+9} = \sqrt{14}$$

As shown in the figure, the required distance is $d = r \sin \theta$. If we recall that $r \sin \theta$ is part of the magnitude of $r \times$ F, i.e., $|\mathbf{r} \times \mathbf{F}| = rF \sin \theta$, our procedure is clear. First, we evaluate r × F as explained on p. 48 to obtain

$$\mathbf{r} \times \mathbf{F} = \begin{vmatrix} 4 & 5 & 0 \\ 6 & 4 & -3 \\ \hat{\mathrm{i}} & \hat{\mathrm{j}} & \hat{\mathrm{k}} \end{vmatrix} = \hat{\mathbf{i}} \begin{vmatrix} 5 & 0 \\ 4 & -3 \end{vmatrix} + \hat{\mathbf{j}} \begin{vmatrix} 0 & 4 \\ -3 & 6 \end{vmatrix} + \hat{\mathbf{k}} \begin{vmatrix} 4 & 5 \\ 6 & 4 \end{vmatrix}$$

$$= \hat{\mathbf{i}}(-15) + \hat{\mathbf{j}}(12) + \hat{\mathbf{k}}(16-30) = -15\hat{\mathbf{i}} + 12\hat{\mathbf{j}} - 14\hat{\mathbf{k}}$$

Computing its magnitude as the square root of the sum of the squares of its orthogonal components, we have

$$|\mathbf{r} \times \mathbf{F}| = rF \sin \theta = \sqrt{225 + 144 + 196} = \sqrt{565}$$

whence

$$d = r \sin \theta = \frac{|\mathbf{r} \times \mathbf{F}|}{F} = \frac{\sqrt{565}}{\sqrt{61}} = 3.04 \qquad \textit{Ans.}$$

The, distance d could also be found from $d = \mathrm{p} \sin \phi$ which suggests the use of **p** × **F**. Thus, we obtain

$$\mathbf{p} \times \mathbf{F} = \begin{vmatrix} -2 & 1 & 3 \\ 6 & 4 & -3 \\ \hat{\mathbf{i}} & \hat{\mathbf{j}} & \hat{\mathbf{k}} \end{vmatrix} = \hat{\mathbf{i}} \begin{bmatrix} \hat{\mathbf{i}}(-3-12) \\ +\hat{\mathbf{j}}(18-6) \\ +\mathbf{k}(-8-6) \end{bmatrix} = -15\hat{\mathbf{i}} + 12\hat{\mathbf{j}} - 14\hat{\mathbf{k}}$$

Observe the somewhat surprising result here that **p** × **F** is identical with **r** × **F.** The physical significance of this equality will become clear when we discuss the concept of the moment of a force in the next section. There we shall see that the cross product of *any* vector from *O* to the vector *F* defines the moment of **F** about *O* and consequently produces the same result. Continuing the solution, it is evident that the magnitude of **p** × **F** is

$$|\mathbf{p} \times \mathbf{F}| = pF \sin \phi = \sqrt{565}$$

Whence as before

$$d = p \sin \phi = \frac{|\mathbf{p} \times \mathbf{F}|}{F} = \frac{\sqrt{565}}{\sqrt{61}} = 3.04 \qquad \textit{Check}$$

2-8.2 Find the length of the common perpendicular between two nonintersecting vectors **F** and **P**.

Solution

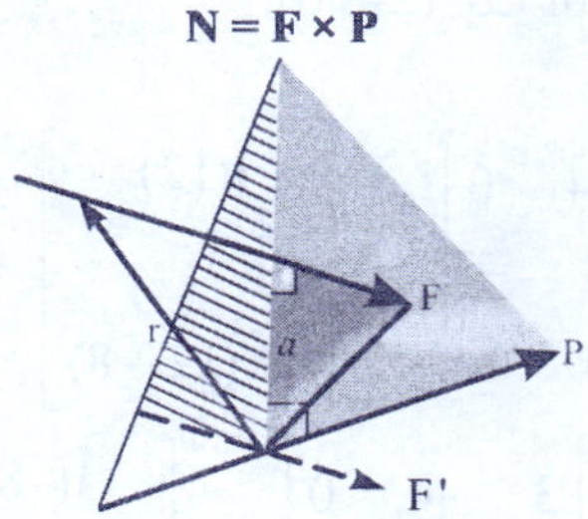

Figure 2-8.3

In Fig, 2-8.3 the vector **N = F x P** represents the direction of their common perpendicular. The projection upon **N** of any vector **r** joining **F** and **P** will be the required length of the common perpendicular. This projection is obtained by taking the dot product of **r** with the unit vector **n̂** in the direction of **N.** This procedure is expressed by

$$d = \mathbf{r} \cdot \hat{\mathbf{n}} = \frac{\mathbf{r} \cdot \mathbf{N}}{N} = \frac{\mathbf{r} \cdot (\mathbf{F} \times \mathbf{P})}{|\mathbf{F} \times \mathbf{P}|} \qquad \textit{Ans.}$$

The numerator of the last term in this result is an example of what is known as a scalar triple product. The scalar triple product may be evaluated in a straightforward manner by first evaluating the cross product and then taking the dot product of this result with **r.** An alternative and preferred equivalent to this is to evaluate the third-order determinant in which the rows are the respective scalar components of the vectors, viz.,

$$\mathbf{r} \cdot (\mathbf{F} \times \mathbf{P}) = \begin{vmatrix} r_x & r_y & r_z \\ F_x & F_y & F_z \\ P_x & P_y & Pz \end{vmatrix}$$

It is interesting to observe that if the dot and cross in the scalar triple product be interchanged, we get the same determinant. Hence, it is only the *order* of the factors that matters, i.e.,

$$\mathbf{r} \cdot (\mathbf{F} \times \mathbf{P}) = (\mathbf{r} \times \mathbf{F}) \cdot \mathbf{P} = \mathbf{P} \cdot (\mathbf{r} \times \mathbf{F}) = (\mathbf{P} \times \mathbf{r}) \cdot \mathbf{F}$$

These identities may be verified by interchanging the rows in the determinant above,

2-8.3 If $\boldsymbol{\omega} = 3\hat{\mathbf{i}} - 4\hat{\mathbf{j}}$ and $\mathbf{r} = 2\hat{\mathbf{i}} + 4\hat{\mathbf{j}} - 3\hat{\mathbf{k}}$, evaluate the vector triple product $\boldsymbol{\omega} \times (\boldsymbol{\omega} \times \mathbf{r})$.

Solution

We will solve this problem in two ways. First we will evaluate ω × r and then cross ω into this result. As a check, we will use the reduction formula for the vector triple product given in Eq. (2-8.6).

***Method* 1:**

$$\boldsymbol{\omega} \times \mathbf{r} = \begin{vmatrix} 3 & -4 & 0 \\ 2 & 4 & -3 \\ \hat{\mathbf{i}} & \hat{\mathbf{j}} & \hat{\mathbf{k}} \end{vmatrix} = \begin{bmatrix} \hat{\mathbf{i}}(12) \\ +\hat{\mathbf{j}}(9) \\ +\hat{\mathbf{k}}(12+8) \end{bmatrix} = 12\hat{\mathbf{i}} + 9\hat{\mathbf{j}} + 20\hat{\mathbf{k}}$$

$$\boldsymbol{\omega} \times (\boldsymbol{\omega} \times \mathbf{r}) = \begin{vmatrix} 3 & -4 & 0 \\ 12 & 9 & 20 \\ \hat{\mathbf{i}} & \hat{\mathbf{j}} & \hat{\mathbf{k}} \end{vmatrix} = \begin{bmatrix} \hat{\mathbf{i}}(-80) \\ +\hat{\mathbf{j}}(-60) \\ +\hat{\mathbf{k}}(27+48) \end{bmatrix}$$

$$= 80\hat{\mathbf{i}} - 60\hat{\mathbf{j}} + 75\hat{\mathbf{k}} \qquad \textit{Ans.}$$

***Method* 2**: Applying the rule given on p. 49 for the reduction of a vector triple product, we obtain

$$\boldsymbol{\omega} \times (\boldsymbol{\omega} \times \mathbf{r}) = (\boldsymbol{\omega} \,.\, \mathbf{r})\mathbf{w} - (\boldsymbol{\omega} \,.\, \boldsymbol{\omega})\mathbf{r}$$
$$= (6-16)\,(3\hat{\mathbf{i}} - 4\hat{\mathbf{j}}) - (9+16)\,(2\hat{\mathbf{i}} - 4\hat{\mathbf{j}} - 3\hat{\mathbf{k}})$$
$$= -30\hat{\mathbf{i}} + 40\hat{\mathbf{j}} - 50\hat{\mathbf{i}} - 100\hat{\mathbf{j}} + 75\hat{\mathbf{k}}$$
$$= -\ 80\hat{\mathbf{i}} - 60\hat{\mathbf{j}} + 75\hat{\mathbf{k}} \qquad \textit{Check}$$

PROBLEMS

2-8.4 If two vectors a and b have the values a $2\hat{\mathbf{i}} + 3\hat{\mathbf{j}} - 4\hat{\mathbf{k}}$ and b = $3\hat{\mathbf{i}} - 4\hat{\mathbf{j}} + 2\hat{\mathbf{k}}$, find the result of a × b; a × (a × b); b × (a × b).

$-10\hat{\mathbf{i}} - 16\hat{\mathbf{j}} - 17\hat{\mathbf{k}}$; $-115\hat{\mathbf{i}} + 74\hat{\mathbf{j}} - 2\hat{\mathbf{k}}$; $100\hat{\mathbf{i}} + 31\hat{\mathbf{j}} - 88\hat{\mathbf{k}}$ ***Ans.***

2-8.5 Given the vectors: a = $2\hat{\mathbf{i}} - 3\hat{\mathbf{j}} - 4\hat{\mathbf{k}}$; b = $4\hat{\mathbf{i}} + 2\hat{\mathbf{j}} + \hat{\mathbf{k}}$, c = $3\hat{\mathbf{i}} - \hat{\mathbf{j}} - 2\hat{\mathbf{k}}$, evaluate a × b; c × a; (a × b) × (a × c).

$5\hat{\mathbf{i}} - 18\hat{\mathbf{j}} + 16\hat{\mathbf{k}}$; $-2\hat{\mathbf{i}} + 8\hat{\mathbf{j}} - 7\hat{\mathbf{k}}$; $2\hat{\mathbf{i}} - 3\hat{\mathbf{j}} - 4\hat{\mathbf{k}}$

2-8.6 Given the four vectors: $a = 2\hat{i} + 3\hat{j} - 4\hat{k}$, $b = 4\hat{i} + 2\hat{j} + \hat{k}$, $c = \hat{i} + 2\hat{j} - 3\hat{k}$, $d = 3\hat{i} + 4\hat{j} + 5\hat{k}$, find the result of $a \times (b \times c)$ and of $(b \times c) \times d$.

2-8.7 using the vectors specified in Prob. 2-8.6, evaluate $(a - b) \cdot (c \times d)$ and $(a \times b) \cdot (c - d)$.

124; – 246 ***Ans.***

2-8.8 Refer to the cantilever framework shown in Fig. P-2-8.8 and find the shortest distance from point *B* to line *AC* and to line *AD*.

d to line *AC* = 7.82 m; *d* to line *AD* = 8.73 m

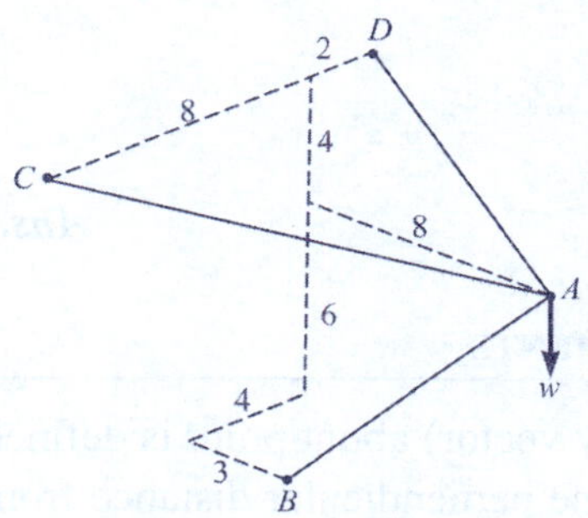

Figure P-2-8.8

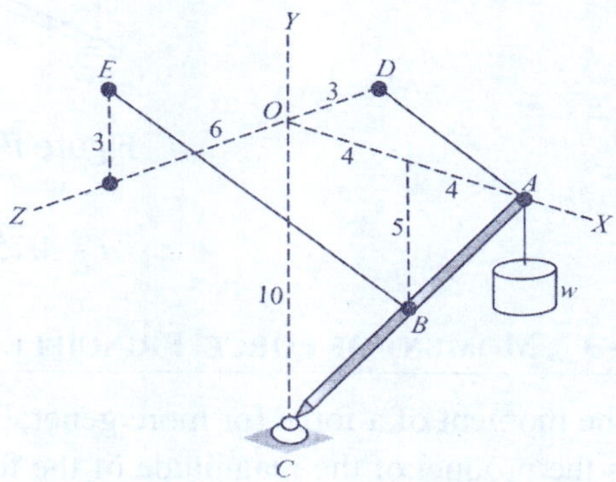

Figure P-2-8.9

2-8.9 In Fig. P 2-8.9, a boom *AC* is supported by a ball-and-socket joint at *C* and by the cables *BE* and *AD*. The force multiplier of a force **F** acting from *B* to *E* is F_m = 10 N/m, find the component of **F** that is perpendicular to the plane *DAC*.

89.4 N ***Ans.***

2.8.10 In Fig. P-2-8.9, if the force multiplier of a force **P** acting from *A* to *D* is P_m = 20 N/m, determine the component of **P** that is perpendicular to the plane defined by points *E*, *A*, and *C*

122.3 N ***Ans.***

2-8.11 Refer to the system shown in Fig. P-2-8.9 and find the length of the common perpendicular between lines *BE* and *AD*.

d = 4.34 m ***Ans.***

2-8.12 In the system shown in Fig. P-2-8.12, determine the length of the common perpendicular between lines *AE* and *BD*.

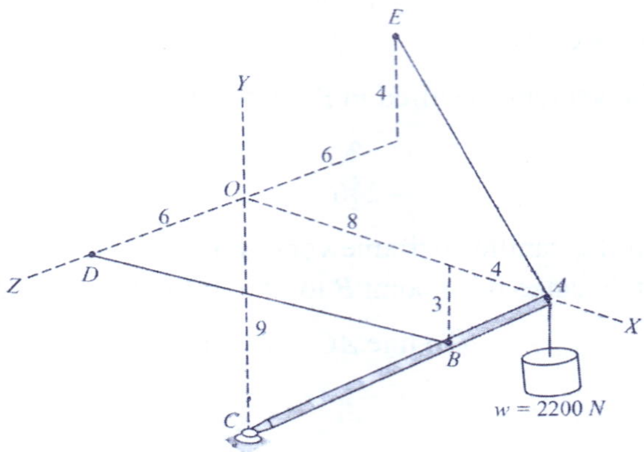

Figure P-2-8.12

$d = 4.15$ m ***Ans.***

2-9 MOMENT OF FORCE. PRINCIPLE OF MOMENTS

The moment of a force (or more generally, of any vector) about point is defined as the product of the magnitude of the force by the perpendicular distance from the point to the action line of the force. In Fig. 2-9.1, this perpendicular distance *d* is called the moment arm whence the magnitude of the moment of a force *F* about a center *O* is expressed by

$$M_O^F = Fd \qquad (a)$$

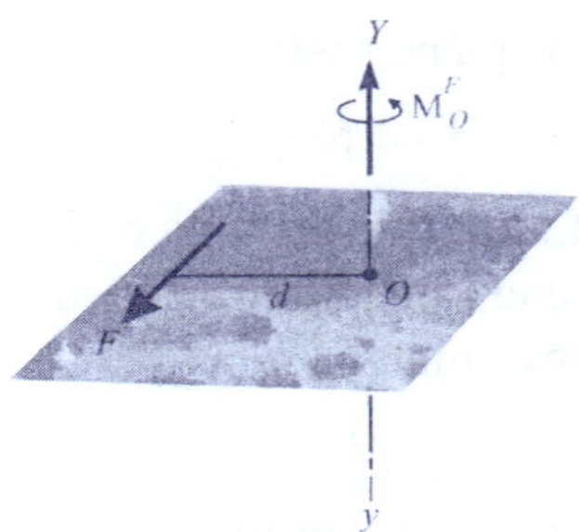

Figure 2-9.1 Moment of a force.

The moment of a force about a point represents the tendency to rotate the moment arm (or the body on which the force acts) about an axis which is perpendicular to the plane defined by the force and its moment arm. The moment axis passes through the moment center. By curling the fingers of the right hand about the moment axis in the sense of this rotation, the extended thumb represents the direction of the moment vector. The moment vector may be considered to be a sliding vector whose line of action coincides with the moment axis. Because a moment is the product of a force and a distance, the unit of moment is their dimension product e.g., Newton-meter (N-m), gram-centimeters (gm-cm), etc[5]. The moment vector is usually denoted by a double-line arrow to distinguish it from a force vector, or by a curved arrow about the moment axis.

The vector nature of moment can be deduced from Fig. 2-9.2 where we have drawn position vectors **r** and **p** from the moment center O to any two points on the action line of **F**. Since **r** and **p** lie in the plane defined by **F** and O; it is evident that the moment arm d is given by either $d = r \sin \theta$ or $d = p \sin \phi$. Applying the definition of moment, we then have

$$M_O^F = Fd = Fr \sin \theta = Fp \sin \phi \qquad (b)$$

where, from the definition of a cross product, the last two terms represent the magnitudes of **r x F** and **p x F**. (This explains the observation made in Prob. 2-8.1). Either cross product represents a vector perpendicular to the plane defined by **F** and d, the direction of which is consistent with the moment vector of ***F*** about O described by the right-hand rule of the preceding paragraph. We conclude that the moment of **F** about O is given by

$$\mathbf{M}_O^F = \mathbf{r} \times \mathbf{F} \qquad (2\text{-}9.1)$$

where **r** is *any* position vector extending from the moment center to any point on the action line of the force.

---

[5]It makes no difference whether the unit of force or the unit of length is stated first: it would be equally correct to use meter-Newton (m-N) as the unit of moment if desired.

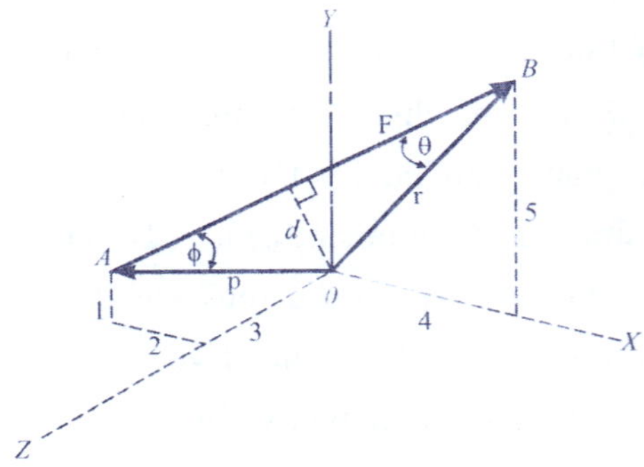

Figure 2-9.2

Note that the moment cross product must be written as $\mathbf{r} \times \mathbf{F}$ and *not* as $\mathbf{F} \times \mathbf{r}$ so that consistency with the right-hand rule for a moment vector can be maintained. This statement is so important that we restate it explicitly thus: The moment of a force about a moment center is the product of the position vector from the moment center to *any* point on the line of action of the force crossed with the force vector.

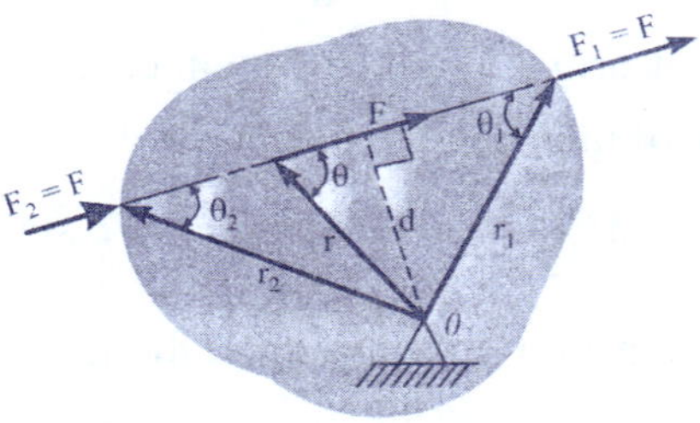

Figure 2-9.3 The moments of F, F_1, and F_2 about O are all equal since $d = r \sin \theta = r_1 \sin \theta_1 = r_2 \sin \theta_2$.

This general definition of moment as a cross product can be used to demonstrate the *principle of transmissibility*, namely that the external effect of a force on a rigid body is independent of where it is applied along its line of action. Thus, in Fig. 2-9.3, the rotational effect of force $\mathbf{F}$ upon a body free to rotate about an axle at O is equivalent to that of a numerically equal force $\mathbf{F}_1$ applied along the action line of $\mathbf{F}$ at the right edge of the body or to that of $\mathbf{F}_2$ applied at the left edge inasmuch as

$$M_O^F = |\mathbf{r} \times \mathbf{F}| = |\mathbf{r}_1 \times \mathbf{F}_1| = |\mathbf{r}_2 \times \mathbf{F}_2| = Fd \qquad (c)$$

PRINCIPLE OF MOMENTS

We now consider a very important concept known as the principle of moments which states that the moment of a force is equal to the moment sum of its components. This almost self-evident statement is known as Varignon's theorem[6] and is demonstrated as follows:

In Fig. 2-9.4, let **R** be the resultant of the concurrent, forces **P**, **F**, and **T**. The force system may be either coplanar or spatial, *but it must be concurrent.*

We have seen that **R** = **P** + **F** + **T**. About any point *O* as moment center, the moments of these forces is

$$\mathbf{r} \times \mathbf{R} = \mathbf{r} \times (\mathbf{P} + \mathbf{F} + \mathbf{T}) = \mathbf{r} \times \mathbf{P} + \mathbf{r} \times \mathbf{F} + \mathbf{r} \times \mathbf{T}$$

which proves the theorem. Note carefully that the resultant and its components must be concurrent or it will be impossible to draw a common position vector to them from the common moment center

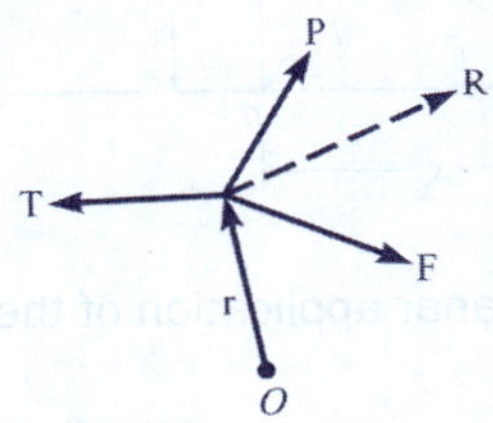

Figure 2-9.4

A general symbolic statement of the theorem is

$$M_o^F = \Sigma M_o = \Sigma(r \times F) \quad (2\text{-}9.2)$$

2-10 COPLANAR APPLICATIONS

Frequently, it is more convenient to determine the moment of a force from the moment sum of its rectangular components rather than from the force itself. For example, in Fig. 2-10.1, suppose a force **F**, making an angle θ_x with the *X axis,* passes through the point A having the known coordinates (x, y). In this case, it is clumsy to determine the moment of **F** about *O* as the product *Fd* since it is

[6]This concept was developed by the French mathematician Varignon (1654-1722) using geometric methods. Vector algebra permits a proof which is both simpler and more general.

too inconvenient to calculate the moment arm d. Instead, we note that the position vector to A (but not shown) is $\mathbf{r}_A = x\hat{\mathbf{i}} + y\hat{\mathbf{j}}$ and that $\mathbf{F} = F_x\hat{\mathbf{i}} + F_y\hat{\mathbf{j}}$. Applying moment as a cross product, we obtain

$$\mathbf{M}_O^F = \mathbf{r}_A \times \mathbf{F} = (x\hat{\mathbf{i}} + y\hat{\mathbf{j}}) \times (F_x\hat{\mathbf{i}} + F_y\hat{\mathbf{j}})$$

which, after applying the algebra of unit vectors reduces to

$$M_O^F\,\hat{\mathbf{k}} = (xF_y)\,\hat{\mathbf{k}} - (yF_x)\,\hat{\mathbf{k}} \qquad (d)$$

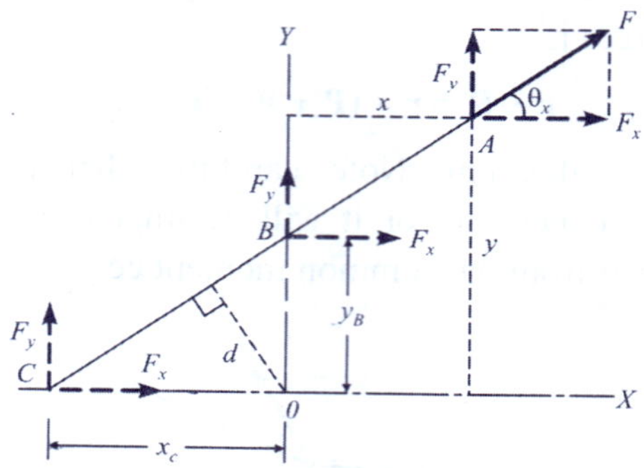

Figure 2-10.1 Coplanar application of the principle of moments.

In coplanar applications, it is obvious that the moment vectors are perpendicular to the XY plane so that the unit vector $\hat{\mathbf{k}}$ really need not be written. Most engineers, therefore, prefer a scalar method for computing moments in coplanar force systems. The positive rotational effect is simply denoted by a curved arrow about a plus sign. If we choose counterclockwise rotation as positive (denoted by $+\circlearrowleft$) to achieve consistency with the cross product, we replace $\mathbf{F}$ by its components F_x an F_y, acting at A in Fig. 2-10.1 and we may note that the moment arm of F_y, about O is the coordinate distance x and the moment arm of F_x about O is the coordinate distance y. Then using the principle that the moment of a force is equal to the moment sum of its components, we obtain

$$+\circlearrowleft\; M_O^F = xF_y - yF_x \qquad (e)$$

which is identical with Eq. (d), but avoids the complicated "bookkeeping" of the vector method.

Some engineers prefer clockwise rotation as the positive sense moment instead of counterclockwise. In coplanar work, it is immaterial which convention is adopted provided we are consistent throughout any particular problem, and use a curved arrow to indicate the positive sense of moment. It maybe convenient, for example, take clockwise moments as positive in Fig. 2-10.1 since it is obvious that F has a clockwise moment about O.

We may also determine the intercepts of **F** with the X and Y axes by placing the components of **F** at B and C. as shown. The principle of moments then gives

$$\circlearrowleft + \; M_O^F = yF_x - xF_y = Fd = F_x(y_B) = F_y(x_C) \qquad (f)$$

from which the values of the moment arm d and the intercepts x_C and y_B are readily found.

Note that F_y, at B and F_x at C both have zero moment about O since they pass through O and therefore have zero moment arms. It is often useful to apply the converse of this observation. This is illustrated in Fig. 2-10.2 where we seek the moment about A of the force **P** acting on the roof truss. By resolving **P** into the components P_x and P_y acting on its line of action at B, we eliminate the moment of P_x about A and obtain

$$\circlearrowleft + \; M_A^P = P_y(\overline{AB}) \qquad (g)$$

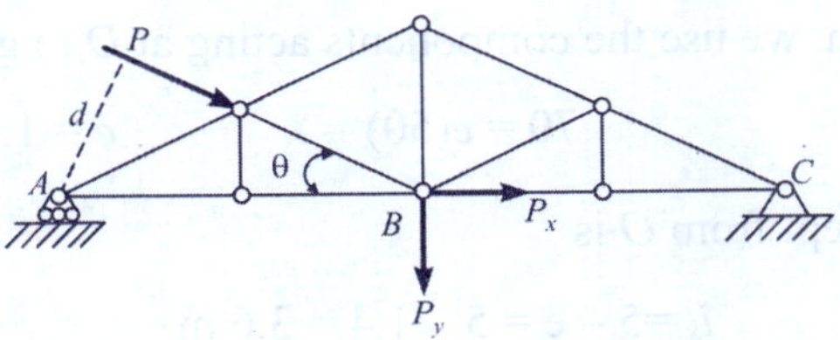

Figure 2-10.2 Moment of a force in terms of its components.

ILLUSTRATIVE PROBLEMS

2-10.1 In Fig. 2-10.3, a force **F** passing through C causes a clockwise moment of 120 N-m about A and a clockwise moment of 70 N-m about B. Determine the force and its x intercept.

Solution

This problem demonstrates how to interpret and apply the physical meaning of moment. By resolving the force into its components at C, we observe that since F_y passes through A, the moment of **F** about A is due only to F_x which must act leftward as shown in order to cause the specified clockwise moment about A. Next, in order to create the specified clockwise moment about B, the action line of **F** must intersect the *X axis* to the left of *B* at *D* with F_y acting upward as shown.

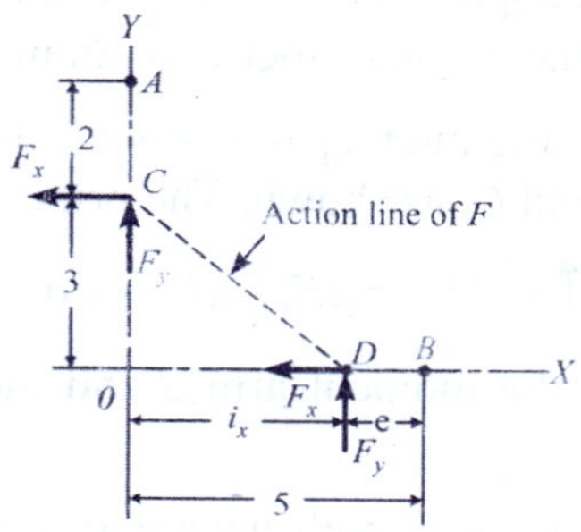

Figure 2-10.3

We can now apply the principle of moments to obtain

$[\circlearrowleft + M_A^P = yFx]$ $\qquad 120 = 2F_x \qquad F_x = 60\text{ N} \leftarrow$

$[\circlearrowleft + M_B^F = xF_y - yF_x]$ $\qquad 70 = 5F_y - 3(60) \qquad F_y = 50\text{ N} \uparrow$

With F_y now known, we use the components acting at *D* to get

$[\circlearrowleft + M_B^F = eF_y]$ $\qquad 70 = e(50) \qquad e = 1.4\text{ m}$

whence the *x* intercept from *O* is

$$i_x = 5 - e = 5 - 1.4 = 3.6\text{ m}$$ ***Ans.***

This procedure illustrates the application of Varignon's theorem, but it may be simpler in this instance to determine i_x directly, using the slope of the action line of **F** as specified by its components. Doing this yields

$\left[\frac{3}{i_x} = \frac{F_y}{F_x}\right]$ $\qquad \frac{3}{i_x} = \frac{50}{60} \qquad i_x = 3.6\text{ m}$ ***check***

Query

How can the *x* intercept be found directly from the moment of *A* about *O* ?

The positive values of F_x and F_y obtained confirm our analysis. We combine these components to obtain $F = 78.2$ N up to the left θ_x, = 39.8° with the *X axis*.

Ans.

If we had assumed arbitrary directions of F_x and F_y instead of deducing them as used earlier, a positive sign would confirm our assumption of direction while a negative sign would indicate the opposite direction.

PROBLEMS

2-10.2 In Fig. P-2-10.2 a force **F** passing through C causes a clockwise moments as positive, compute the moment of force $F = 450$ N and of force $P = 361$ N about points A, B, C and D.

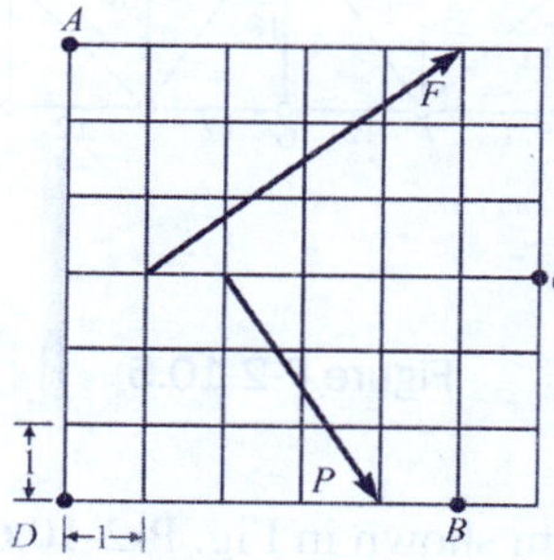

Figure P-2-10.2

For F: $M_D^F = 810$ N-m; for P: $M_C^P = -1200$ N-m **Ans.**

2-10.3 In Fig. P-2-10.2, if the moment of P is 480 N-m about D, find the moment of P about B. Also, if the moment of F about C is 750 N-m, find its moment about D.

2.10.4 In Fig. P-2-10.4, find the y coordinate of point A so that the 361N force will have a clockwise moment of 400 N-m about O. Also determine the Y intercept of the action line of the force.

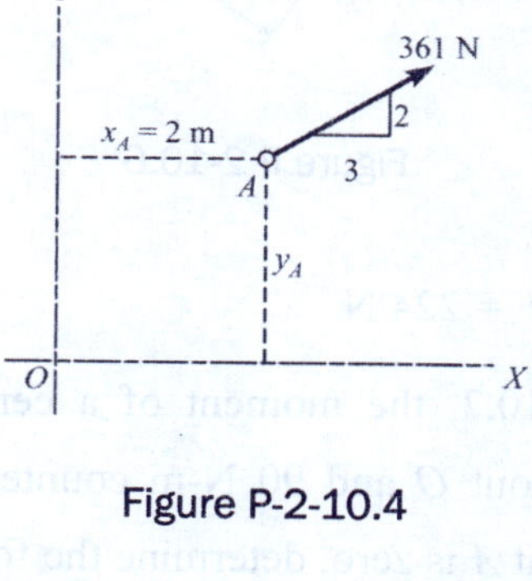

Figure P-2-10.4

$Y_A = 2.67$ m; $i_y = 1.33$ m above O **Ans.**

2.10.5 For the truss shown in Fig. P-2-10.5, compute the perpendicular distance from E and from G to the line BD.

Hint: Imagine a force P directed along BD. The required distances are the moment arms of P about E and G.

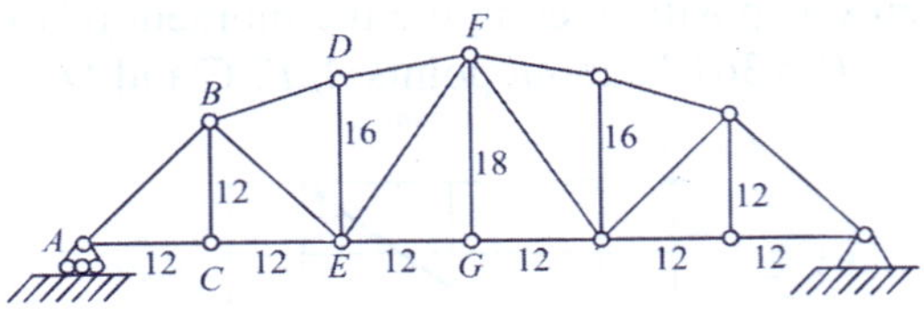

Figure P-2-10.5

2-10.6 In the rocker arm shown in Fig. P-2-10.6, the moment of F about O balances that of P about O. Find F.

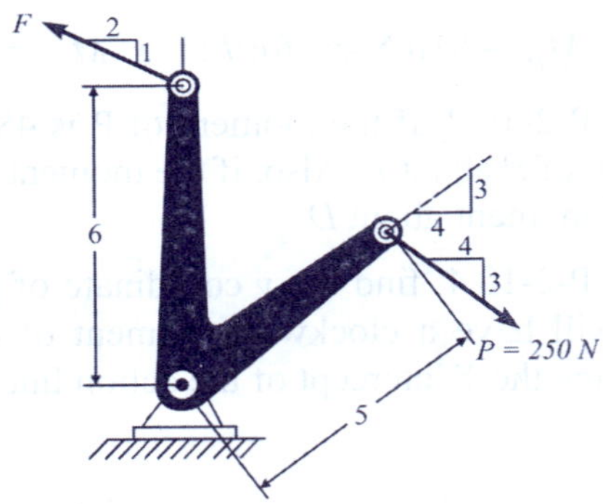

Figure P-2-10.6

$F = 224$ N **_Ans._**

2-10.7 In Fig. P-2-10.7, the moment of a certain force F is 180 N-m clockwise about O and 90 N-m counterclockwise about B. If its moment about A is zero, determine the force.

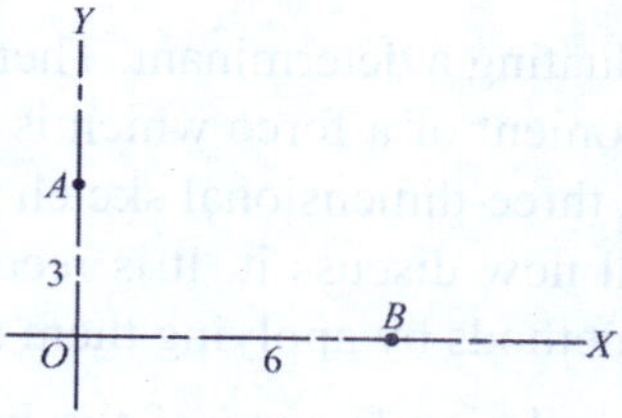

Figure P-2-10.7

F = 75 N down to the right at $\theta_x = 36.9°$ ***Ans.***

2-10.8 In Fig. P-2-10.7, a force *P* intersects the *X* axis at 4 m to the right of *O*. If its moment about *A* is 170 N-m counterclockwise and its moment about *B* is 40 N-m clockwise, determine its *y* intercept.

2-10.9 In Fig. P-2-10.9, a force *F* passing through *C* produces a clockwise moment of 600 N-m about *A* and a counterclockwise moment of 300 N-m about *B*. Determine the moment of *F* about *O*.

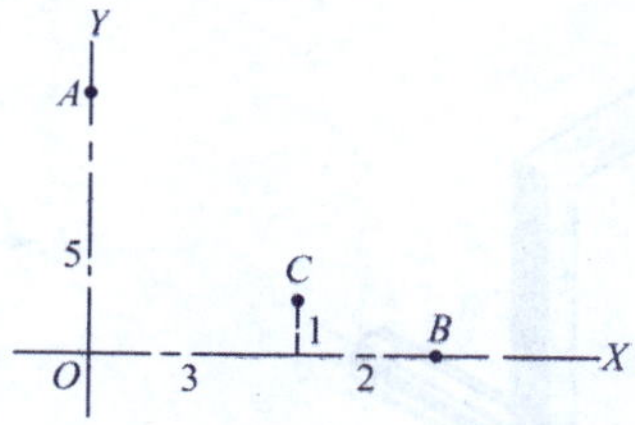

Figure P-2-10.9

M_C^F = 300 N-m clockwise ***Ans.***

2-10.10 In Fig. P-2-10.9, a force *F*, having a slope of 2 vertical to 3 horizontal, produces a clockwise moment of 330 N-m about *A* and a counter clockwise moment of 420 N-m about *B*. Compute the moment of *F* about *C*.

M_C^F = 210 N-m ***Ans.***

2-11 SPATIAL APPLICATIONS

In most cases, the moment of force is found by expanding the determinant resulting from **M** = **r x F**. We shall show the details in the illustrative problems. It is a very concise approach, but obscures the physical concept of moment by relying on the rules for evaluating a determinant. There is an alternate geometric approach for finding the moment of a force which is also useful. However, this alternate method requires a three-dimensional sketch and the ability to visualize its spatial aspects. We shall now discuss it. It is recommended that the student gain familiarity with both methods by applying them alternately in assignments.

To understand the physical significance of the moment of any force about any axis, let us imagine a device like that shown in Fig. 2-11.1. Assume the axis (the X axis in this instance) to be a taut wire on which is mounted a structure free to slide and rotate about the axis. A force **F**, represented by its rectangular components, is applied at point A. The component F_x which is parallel to the axis (i.e., the wire) can only slide the structure along the wire; it is evidently impossible for F_x to rotate the structure about the axis. From another point of view, we could say that F_x intersects the X *axis* at infinity and therefore has no moment about it. In vector terms, we would note that $\boldsymbol{x} \times \boldsymbol{F_x} = 0$. In any event, we conclude that a force has no moment about a parallel axis.

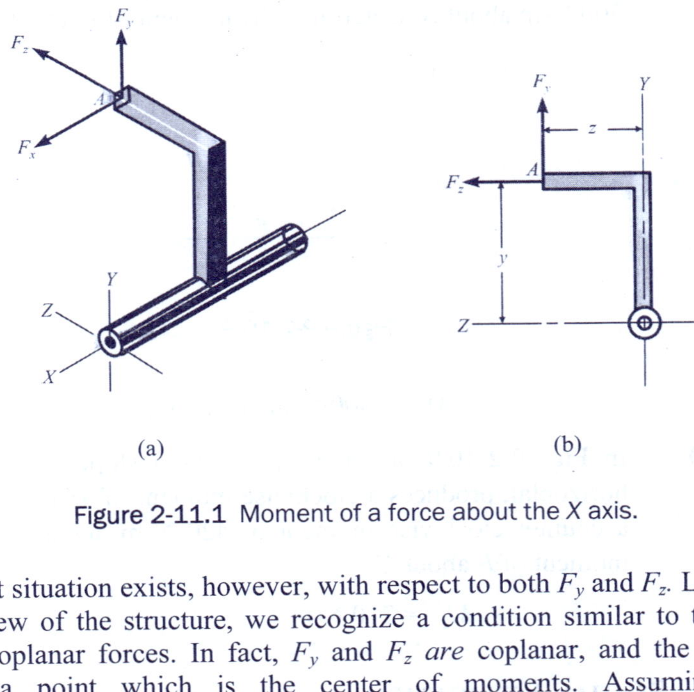

Figure 2-11.1 Moment of a force about the X axis.

A different situation exists, however, with respect to both F_y and F_z. Looking at the end view of the structure, we recognize a condition similar to that for moment of coplanar forces. In fact, F_y and F_z *are* coplanar, and the X axis appears as a point which is the center of moments. Assuming the counterclockwise sense of rotation to be positive, the moment of **F** about the X axis is given by

$$+\uparrow M_x = yF_z - zF_y$$

A similar analysis may be made to determine the moment M_y of **F** about the *Y* axis or the moment M_z of **F** about the **Z** axis. *In general, the moment of a force about any axis is due to the components of the force lying in the plane perpendicular to the axis of moments.* Thus, as developed above, the moment of a force about the *X* axis is due solely to the *Y* and *Z* components of the force. It is helpful to observe that there is always one *X* term, one *Y* term, and one *Z* term involved in the moment of a force component about a coordinate axis; for example, the moment arm of the *Y* component of a force about the *X* axis is its *Z* coordinate.

To conclude this discussion, consider Fig. 2-11.2 which shows the components of a force **F** acting at a point *A* whose position vector from a moment center at *O* is **r**. The vector form of the moment of **F** about *O* is

$$\mathbf{M}_O^F = \mathbf{r} \times \mathbf{F} = \begin{vmatrix} x & y & z \\ F_x & F_y & F_z \\ \mathbf{i} & \mathbf{j} & \mathbf{k} \end{vmatrix}$$

which, on expanding the determinant, becomes

$$\mathbf{M}_O^F = \hat{\mathrm{i}}\,(yF_z - zF_y) + \hat{\mathrm{j}}(zF_x - xF_z) + \hat{\mathrm{k}}\,(xF_y - yF_x)$$

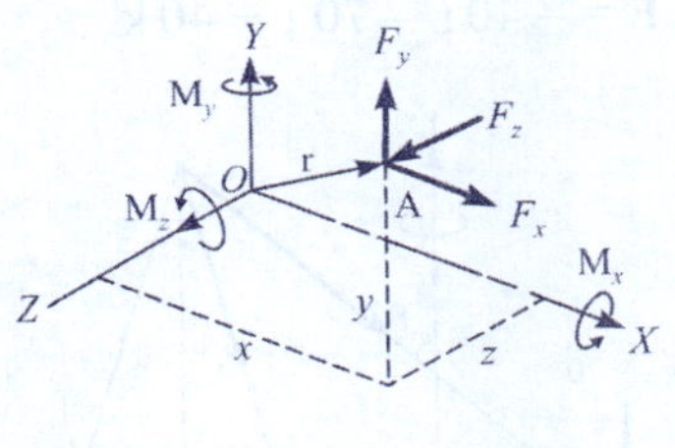

Figure 2-11.2 $M_O^F = \mathrm{r} \times \mathrm{F}$

The coefficients of the unit vectors are respectively the moment of **F** about the coordinate axes, or

$$M_x = yF_z - zF_y$$

$$M_y = zF_x - xF_z$$

$$M_z = xF_y - yF_x$$

thereby giving results which correspond exactly with the geometric approach discussed above.

Since moment is a vector, its component (i.e., rotational effect) about any arbitrary axis through the moment center can be found as the dot product of the moment **M** with a unit vector $\hat{\mathbf{n}}$ which specifies the direction of this arbitrary axis. This concept will be applied in Illus. Prob. 2-11.1.

ILLUSTRATIVE PROBLEMS

2-11.1 A tripod consists of three bars joined at *D* as shown schematically in Fig. 2-11.3. If 10 N/m is the value of the force multiplier of **F** acting along bar *DA* from *D* to *A*, find the moment of **F** about point *C* and about the line *CB*. Also find the value of **P** acting in bar *DB* from *D* to *B* that will cause the moment of **P** about the axis *AC* to be 2000 N-m.

Preliminary

Draw the components of force **F** at any convenient point on its line of action. These components point in the directions taken when moving along coordinate paths from *D* to *A* (i.e., the direction of **F**) and have the value of the force multiplier times the distance traveled along these coordinate paths. With a force multiplier of 10 N/m. the values in Newtons of the components of **F** are as shown. The vector form of **F**, consistent with the positive senses of the indicated reference axes, is then

$$\mathbf{F} = -40\hat{\imath} - 70\hat{\jmath} + 40\hat{k}$$

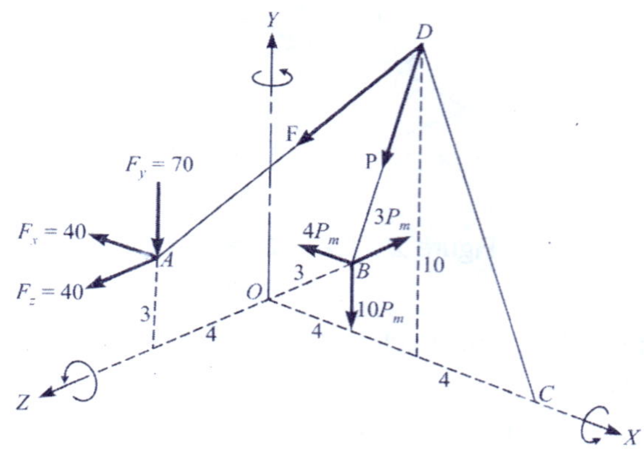

Figure 2-11.3

In a similar fashion, the components of **P** in terms of its as yet unknown force multiplier are as shown; whence its vector form becomes

$$\mathbf{P} = P_m(-4\hat{\imath} - 10\hat{\jmath} - 3\hat{k})$$

Solution

The moment of **F** about C is $\mathbf{M}_C^F = \mathbf{r}_{CD} \times \mathbf{F}$ where $\mathbf{r}_{CD} = -4\hat{\mathbf{i}} + 10\hat{\mathbf{j}}$ is the most convenient position vector from C to a point D on **F**. Express this cross product in determinant form as shown below, noting that the first row is the xyz order of the components of $\mathbf{r}_{CD}$; the second row is the xyz order of the components of **F**; and the third row consists of the unit vectors in the same xyz order. By following this scheme, the vector form of $\mathbf{r}_{CD}$ and **F** need not be written out; the respective components with their signs being obtained directly from the space diagram. Thus, we obtain

$$\mathbf{M}_C^F = \mathbf{r}_{CD} \times \mathbf{F} = \begin{vmatrix} -4 & 10 & 0 \\ -40 & -70 & 40 \\ \hat{\mathbf{i}} & \hat{\mathbf{j}} & \hat{\mathbf{k}} \end{vmatrix} = \begin{vmatrix} & \hat{\mathbf{i}}(400) \\ + & \hat{\mathbf{j}}(160) \\ + & \hat{\mathbf{k}}(280 + 400) \end{vmatrix}$$

$$\mathbf{M}_C^F = 400\,\hat{\mathbf{i}} + 160\,\hat{\mathbf{j}} + 680\,\hat{\mathbf{k}} \text{ N-m} \qquad \textit{Ans.}$$

The coefficients of these unit vectors represent respectively the magnitude of the moment of **F** about the X, Y, and Z coordinate directions through C. Combining these components, the magnitude of the moment is $\mathbf{M}_C^F = 805$ N-m.

The moment component of $\mathbf{M}_C^F$ about the direction CB is expressed by $\mathbf{M}_{CB}^F = \mathbf{M}_C^F \cdot \hat{\mathbf{n}}_{CB}$ where the unit vector along CB is

$$\hat{\mathbf{n}}_{CB} = \frac{\overrightarrow{CB}}{CB} = \frac{-8\hat{\mathbf{i}} - 3\hat{\mathbf{k}}}{\sqrt{73}}$$

Recalling that the dot product is the summation of the products of the respective coefficients of the unit vectors, we obtain

$$M_{CB}^F = \mathbf{M}_C^F \cdot \hat{\mathbf{n}}_{CB} = (400\,\hat{\mathbf{i}} + 160\,\hat{\mathbf{j}} + 680\,\hat{\mathbf{k}}) \cdot \left(\frac{-8\hat{\mathbf{i}} - 3\hat{\mathbf{k}}}{\sqrt{73}}\right)$$

$$= \frac{1}{\sqrt{73}}(-3200 - 2040) = -614 \text{ N-m} \qquad \textit{Ans.}$$

The negative sign of this result denotes that the vector corresponding to this moment component is directed opposite to the direction CB, i.e., $\mathbf{M}_{CB}^F = -614\,\hat{\mathbf{n}}_{CB}$. Of course, a positive component would be obtained if the unit vector had been $\hat{\mathbf{n}}_{BC}$ directed from B to C.

Now to find the value of **P** to cause its moment about the axis AC to be 2000 N-m, refer again to Fig. 2-11.3 and, by using the scheme explained previously, first find the moment of **P** about a center on AC as follows:

$$\mathbf{M}_C^P = \mathbf{r}_{CD} \times \mathbf{P} = P_m \begin{vmatrix} -4 & +10 & 0 \\ -4 & -10 & -3 \\ \hat{\mathbf{i}} & \hat{\mathbf{j}} & \hat{\mathbf{k}} \end{vmatrix}$$

$$= P_m\,[\hat{\mathbf{i}}(-30) + \hat{\mathbf{j}}(-12) + \hat{\mathbf{k}}(40 + 40)] \text{ N-m}$$

The data imply that the moment component of **P** about AC is 2000 N-m in the direction AC, so we use a unit vector $\hat{\mathbf{n}}_{AC}$ in forming the following dot product:

$$[M_{AC}^P = \mathbf{M}_C^P \cdot \hat{\mathbf{n}}_{AC}]$$

$$2000 = P_m(-30\hat{\mathbf{i}} - 12\hat{\mathbf{j}} + 80\hat{\mathbf{k}}).\left(\frac{8\hat{\mathbf{i}} - 3\hat{\mathbf{j}} - 4\hat{\mathbf{k}}}{\sqrt{89}}\right)$$

$$= \frac{P_m}{\sqrt{89}}(-240 + 36 - 320)$$

$$= -\frac{524}{9.43}P_m, \quad P_m = -36$$

Finally, multiplying this value of P_m into the second row of the determinant written above (since this row represents the scalar components of **P**), we obtain the vector form of **P** to be

$$\mathbf{P} = 144\hat{\mathbf{i}} + 360\hat{\mathbf{j}} + 108\hat{\mathbf{k}} \text{ and } P = 403 \text{ N}$$ ***Ans.***

An alternate procedure, preferred by some, is to write $M_{AC}^P = \mathbf{M}_C^P \cdot \hat{\mathbf{n}}_{AC}$ as $(\mathbf{r}_{CD} \times \mathbf{P}) \cdot \hat{\mathbf{n}}_{AC}$ which is a scalar triple product. The value of this scalar triple product is the expansion of the third-order determinant as written below, in which the first, second, and third rows are the xyz order of the scalar components of the respective order of the vectors in the triple product. Doing this gives

$$M_{AC}^P = \frac{P_m}{\sqrt{89}} \begin{vmatrix} -4 & -10 & 0 \\ -4 & -10 & -3 \\ 8 & -3 & -4 \end{vmatrix}$$

$$= \frac{P_m}{\sqrt{89}}[8(-30) - 3(-12) - 4(40 + 40)]$$

$$= \frac{P_m}{\sqrt{89}}[-240 + 36 - 320]$$

as before.

2.11.2 Solve the previous problem by using the alternate geometric approach to find the moments of **F** and **P** about coordinate directions passing through point C.

Solution

Using Fig. 2-11.3, we may find the moment of **F** about an X axis through C as the moment sum of its components acting at A. The positive sense of moments is indicated by the curved arrows about the coordinate axes. These curved arrows correspond to the curling of the fingers of the right hand when the extended thumb points in the positive direction of these axes.

Observe that F_x is parallel to the X axis and hence has no moment about it. The moment arm of F_y about the X axis through C (i.e., X_C) is the z coordinate of 4 m while the moment arm of F_z is the y coordinate of 3 m from this axis. Corresponding with the sign convention of the right-hand rule, we then obtain

$$[+M_x = +\curvearrowleft] \qquad M^F_{X_C} = 70(4) + 40(3) = 400 \text{ N-m}$$

Similarly, the moment sum of the force components of **F** acting at A about Y and Z axes through C is found to be

$$[+M_y = \curvearrowleft] \qquad M^F_{Y_C} = 40(8) - 40(4) = 160 \text{ N-m}$$

$$[+M_z = +\curvearrowleft] \qquad M^F_{Z_C} = 40(3) + 70(8) = 680 \text{ N-m}$$

These results are the same as the coefficients of the unit vectors defining $\mathbf{M}^F_C$ in the previous problem. Note that if we had visualized the components of **F** to be acting at D, instead of at A, the above approach would give results that are identical term for term with the expansion of the determinant for $\mathbf{M}^F_C$.

Similarly, the moment of **P** about coordinate axes through are found to be

$$[\Sigma M_x = \curvearrowleft] \qquad M^P_{X_C} = -3(10P_m) = -30P_m$$

$$[\Sigma M_y = \curvearrowleft] \qquad M^P_{Y_C} = 3(4P_m) - 8(3P_m) = -12P_m$$

$$[\Sigma M_z = \curvearrowleft] \qquad M^P_{Z_C} = 8(10P_m) = -80P_m$$

These results also agree with the coefficients of the vector form of $\mathbf{M}^P_C$ obtained in the preceding problem. However, even if this geometric approach is used to find the vector form of a moment, we would still need the dot product approach of vector algebra to find the moment component in any desired direction.

It should now be apparent that the geometric approach as stated emphasizes the physical meaning of moment which is somewhat obscured by the vector approach, but it does require a space sketch and the ability to visualize its three-dimensional aspects.

PROBLEMS

2-11.3 In Fig. P-2-11.3, find the moment of force P about the diagonal *AB*. Use this result to show that the length of the common perpendicular between *P* and *AB* is 4 m.

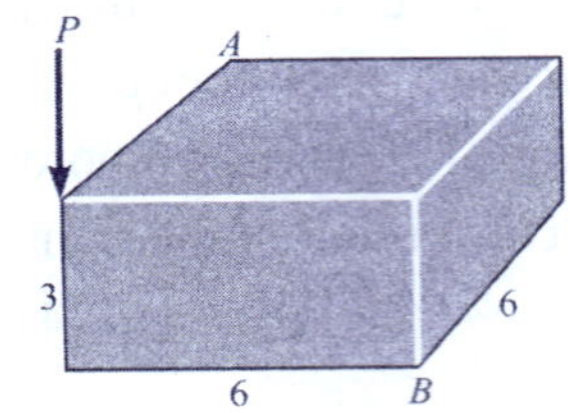

Figure P-2-11.3

2-11.4 **F** is a 400-m force directed from *A* (2, 1, –4) to *B* (4, 4, 1). (a) Find the moment of **F** about the origin and the magnitude of this moment. (b) Find the moment of **F** about the point *C* (3, 4, 2) and its moment arm from *C*. (c) Find the moment arm of **F** about the axis passing through the origi n and point *C*.

(a) $\mathbf{M}_O^F = (17\hat{\mathbf{i}} - 18\hat{\mathbf{j}} + 4\hat{\mathbf{k}})400/\sqrt{38}$; $M_O^F = 1628$ N-m

(b) $\mathbf{M}_O^F$ $(3\hat{\mathbf{i}} - 7\hat{\mathbf{j}} + 3\hat{\mathbf{k}})400/\sqrt{38}$; $d = 1.328$ m

(c) $d = 0.392$ m ***Ans.***

2-11.5 In Fig. P-2-11.5, a boom *AC* is supported by a ball-and-socket joint at *C* and by the cables *BE* and *AD*. If the force multiplier of force **F** acting from *B* to *E* is F_m = 10 N/m, find (a) the moment of **F** about point *C*, (b) the moment of **F** about point *D*, (c) the moment of **F** about a line directed from *C* to *D*.

(a) $\mathbf{M}_C^F = 300\hat{\mathbf{i}} - 240\hat{\mathbf{j}} + 520\hat{\mathbf{k}}$ N-m;

(b) $\mathbf{M}_D^F = -540\hat{\mathbf{i}} - 360\hat{\mathbf{j}} + 120\hat{\mathbf{k}}$ N-m;

(c) $\mathbf{M}_{CD}^F = -379$ N-m. (What does the minus sign mean?) ***Ans.***

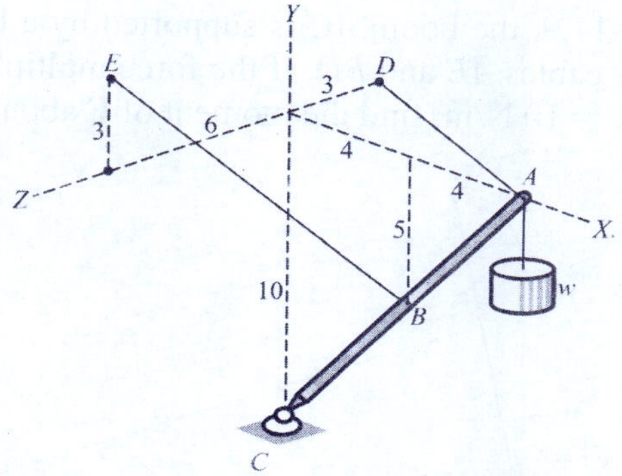

Figure P-2-11.5

2-11.6. In Fig. P-2-11.5, a force **P** acts from *A* toward *D*. Determine the magnitude of **P** to cause a moment of 2000 N-m about the line directed from *B* toward *E*.

$P = 465$ N ***Ans.***

2-11.7 For the system shown in Fig. P-2-11.7, the force multiplier of **P** acting from *A* to *E* is $P_m = 100$ N/m. Find the moment of **P** about *C* and about *D*. What is the moment of **P** about an axis directed from *C* to *D*?

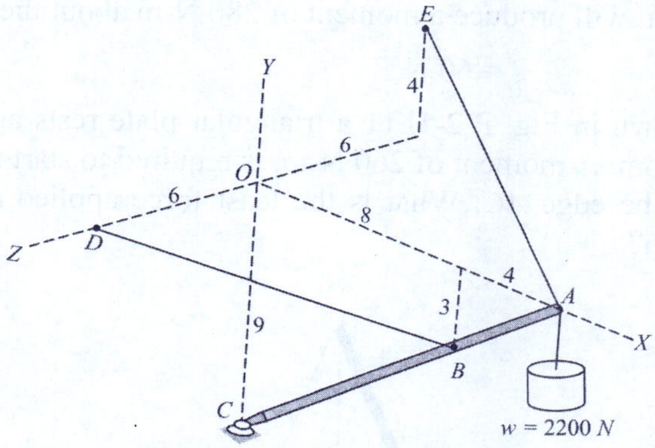

Figure P-2-11.7

2.11.8 Referring to Fig. P-2-117, compute the magnitude of the force **F** acting from *B* to *D* that will cause a moment of 960 N-m about an axis directed from *A* to *E*.

2-11.9 In Fig. P-2-11.9, the boom *AC* is supported by a ball-and-socket joint at *C* and by cables *AE* and *BD*. If the force multiplier of **F** acting from *B* to *D* is $F_m = 10$ N/m, find the moment of **F** about the axis *AE*.

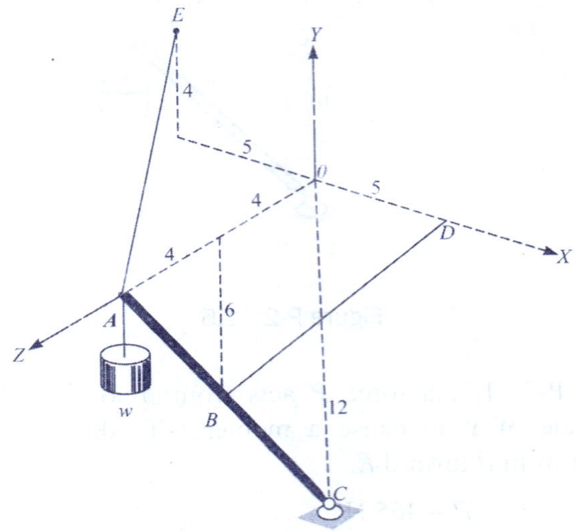

Figure P-2-11.9

2-11.10 Referring to Fig. P-2-11.9, determine the value of **P** directed along *A* to *E* that will produce a moment of 280 N-m about the axis *DB*.

P = 60 N *Ans.*

2-11.11 As shown in Fig. P-2-11.11 a triangular plate rests against the corner of a room. A moment of 260 N-m, is required to start rotating the plate about the edge *AC*. What is the least force applied at *B* to start this rotation?

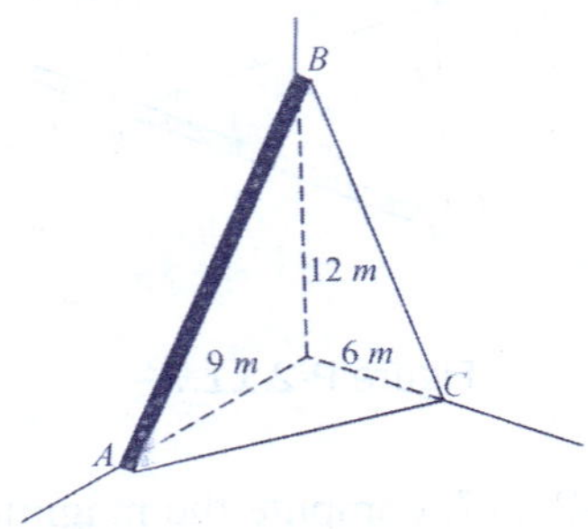

Figure P-2-11.11

20 N *Ans.*

2-12 COUPLES

A couple is defined as a pair of equal, parallel, oppositely directed forces. As shown in Fig. 2-12.1, the perpendicular distanced d between the action lines of the forces is called the moment arm of the couple. The vector sum of these two forces is zero, but their moment sum is not zero. The only effect of a couple on a body is a tendency to rotate the body about an axis perpendicular to the plane of the couple. *A unique property of a couple is that the moment sum of its forces is constant and independent of any moment center.* This is proved by selecting moment centers at A and B to give respectively

$$\circlearrowleft + \Sigma M_A = Fd$$

$$\circlearrowleft + \Sigma M_B = F(d + a) - Fa = Fd$$

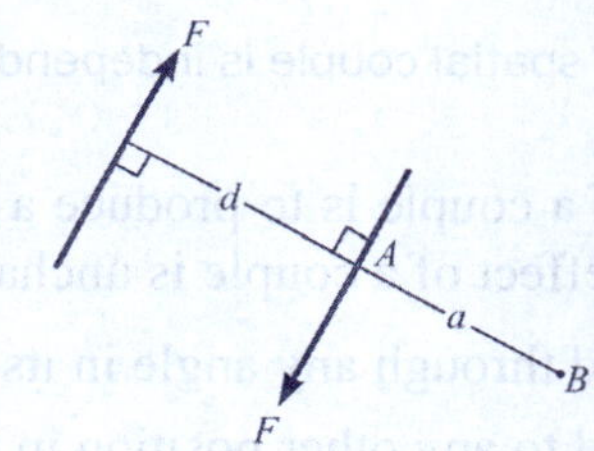

Figure 2-12.1 Moment of a couple is independent of moment center.

We conclude that the moment of a couple $\mathbf{C}$ is equal to the product of one of the forces composing the couple multiplied by the perpendicular distance between their action lines. This relation is expressed by the equation

$$|\mathbf{C}| = Fd$$

For the general case of two parallel, equal, oppositely directed spatial forces $\mathbf{F}$ and $-\mathbf{F}$ shown in Fig. 2-12.2, their moment sum also is constant and independent of any moment center. Thus, with respect to any origin O, draw position vectors $\mathbf{r}_A$ to any point A on $\mathbf{F}$ and draw $\mathbf{r}_B$ to any point B on $-\mathbf{F}$. Adding the moment of each force about O, we have

$$\mathbf{M} = \mathbf{r}_A \times \mathbf{F} + \mathbf{r}_B \times (-\mathbf{F}) = (\mathbf{r}_A - \mathbf{r}_B) \times \mathbf{F} = \boldsymbol{\rho} \times \mathbf{F}$$

from which we conclude that the moment vector of a couple is perpendicular to the plane of the couple, and independent of any moment center since the magnitude of $\boldsymbol{\rho} \times \mathbf{F}$ is always equal to Fd where d is the perpendicular distance between the action lines of the forces. Furthermore, since the origin and points A and B may be chosen arbitrarily, it follows that the moment vector of a couple is a free vector; i.e., it may be placed any where parallel to itself. The position vector ρ may be drawn from either of the forces to the other and the resultant cross product used to determine the couple vector. Usually the sense of rotation of the couple is evident by inspection.

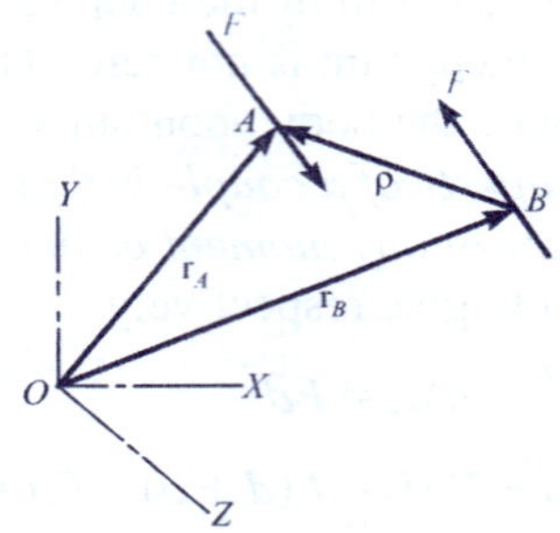

Figure 2-12.2 Moment of a spatial couple is independent of any movement center.

Since the only effect of a couple is to produce a moment that is independent of the moment center, the effect of a couple is unchanged if

(a) the couple is rotated through any angle in its plane

(b) the couple is shifted to any other position in its plane

(c) the couple is shifted to a parallel plane

(d) the couple is replaced by another pair of forces in its plane whose product *Fd* and sense of rotation is unchanged.

For the somewhat special case in which a given force system is composed entirely of couples in the same or parallel planes, the resultant couple will consist of another couple equal to the algebraic summation of the moment sum of the original couples. For all other cases, the resultant couple will be the geometric summation of the moment vectors of the original couples.

2-13 RESULTANT OF ANY FORCE SYSTEM

In this section, we shall show how to replace any force system by single resultant force acting through any arbitrary point plus a single resultant couple which is the moment sum of the original force system about that arbitrary point. This reduction of a given force system to an equivalent force-couple system determines either the criteria for equilibrium or else the type of motion it imparts to a rigid body.

The transformation of a given force system into an equivalent force-couple system is accomplished as shown in Fig. 2-13.1 where the original spatial force system in (a) is modified to that in (b) by adding balanced pairs of forces through the chosen reference point O. Each force of these balanced pairs has a magnitude equal to one of the original forces. Obviously, the addition of these balanced pairs of forces does not alter the effect of the original force system on the state of motion of the body. System (b) is now separated into that shown in (c) where we see that each of the original forces is equivalent to the same force now acting through O plus a couple equal to the moment of that force about O. The resultant of the concurrent system of the left portion of (c) is the resultant force **R** acting through O in part (d) whereas the system of couples acting on the right portion of (c) produces the resultant couple shown as $C = \Sigma M_O$, in (d).

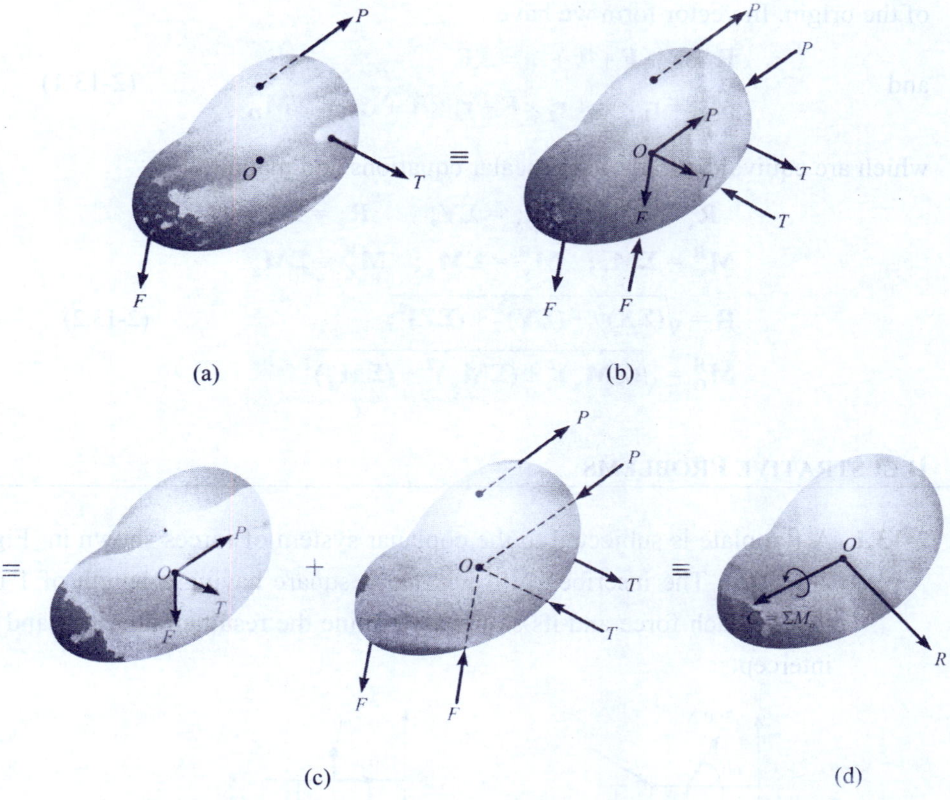

Figure 2-13.1 Reduction of a nonconcurrent system to a force-couple system.

This discussion is valid for either coplanar or spatial systems. For coplanar systems, if the resultant is not merely a couple, the reduction may be extended one more step to consist of a single force so located that it causes the same rotational effect about O as the couple $\mathbf{C} = \Sigma \mathbf{M}_O$, but for spatial systems the reduction usually ends with the force-couple system. It is possible to reduce spatial systems one step further to a *wrench* consisting of the resultant force and a parallel couple vector, but this has limited interest. Refer to the hint in Prob. 2-13.24 for the procedure of determining a wrench.

Summarizing this discussion, we compute the resultant force as though the original forces were concurrent using the methods of Sections 2-4 and 2-5. The value of the resultant couple is equal to the moment sum of the original forces about the selected origin. Note that the resultant couple depends on the position of the selected origin, but that the resultant force is independent of the position of the origin. In vector form we have

$$\mathbf{R} = \mathbf{P} + \mathbf{F} + \mathbf{T} + \ldots = \Sigma\mathbf{F}$$

and

$$\mathbf{M}_O^R = \mathbf{r}_1 \times \mathbf{P} + \mathbf{r}_2 \times \mathbf{F} + \mathbf{r}_3 \times \mathbf{T} + \ldots = \Sigma\mathbf{M}_O \qquad (2\text{-}13.1)$$

which are equivalent to these six scalar equations and magnitudes:

$$\mathbf{R}_x = \Sigma\mathbf{X}; \quad \mathbf{R}_y = \Sigma\mathbf{Y}; \quad \mathbf{R}_z = \Sigma\mathbf{Z}$$
$$\mathbf{M}_x^R = \Sigma\mathbf{M}_x; \quad \mathbf{M}_y^R = \Sigma\mathbf{M}_y; \quad \mathbf{M}_Z^R = \Sigma\mathbf{M}_z$$
$$\mathbf{R} = \sqrt{(\Sigma\mathbf{X})^2 + (\Sigma\mathbf{Y})^2 + (\Sigma\mathbf{Z})^2}; \qquad (2\text{-}13.2)$$
$$\mathbf{M}_O^R = \sqrt{(\Sigma\mathbf{M}_x)^2 + (\Sigma\mathbf{M}_y)^2 + (\Sigma\mathbf{M}_z)^2}$$

ILLUSTRATIVE PROBLEMS

2-13.1 A flat plate is subjected to the coplanar system of forces shown in Fig. 2-13.2(a). The inscribed grid with each square having a length of 1 m locates each force and its slope. Determine the resultant and its x and y intercepts.

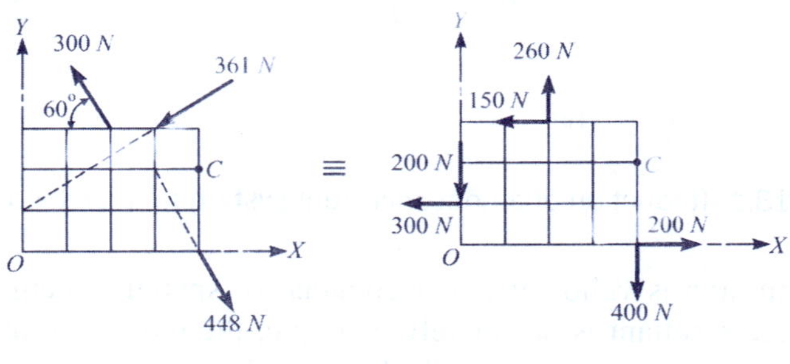

Figure 2-13.2

Solution

We start by resolving each force into its X and Y components. These components may be drawn as dashed vectors on the original diagram, but they are shown here on a duplicate diagram in (b) to help follow the computations. Each set of components must intersect on the action line of the force they replace (see Sections 2-9 and 2-10). By judiciously locating these components as shown, we can simplify the moment summation about O.

The effects of the resultant are now found to be

$[\overset{+}{\rightarrow} R_x = \Sigma X]$ $\quad R_x = 200 - 150 - 300 = -250 \text{ N} \leftarrow$

$[+\uparrow R_y = \Sigma Y]$ $\quad R_y = 260 - 200 - 400 = -340 \text{ N} \downarrow$

$[\circlearrowleft + M_O^R = \Sigma M_O]$ $\quad M_O^R = 400(4) - 260(2) - 150(3) - 300(1)$

$= 330 \text{ N-m}$

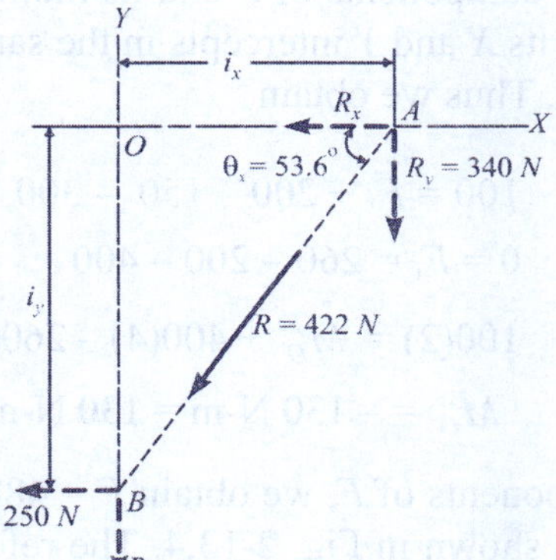

Figure 2-13.3 Location of resultant.

Combining these effects, the resultant has a magnitude of 422 N directed down to the left at $\theta_x = 53.6°$ with the X axis as shown in Fig. 2-13.3. The position of the reference origin must be above R as shown in order for R to have the clockwise moment effect about O found above. Using this figure, we apply the principle that the moment of a force equals the moment sum of its parts by resolving R into its components at A and B. From the components located at A, we find the x intercept to be

$[M_O^R = (R_y)\, i_x]$ $\quad 330 = 340\, i_x \quad i_x = 0.97$ m right of O

and similarly from the components located at B, the y intercept is

$[M_O^R = (R_x)\, i_y]$ $\quad 330 = 250\, i_y \quad i_y = 1.32$ m below O

Note that at point A, R_x passes through O and hence has no moment about it; likewise at B, R_y has no moment about O.

In the case of coplanar forces, it is possible to reduce a force-couple system to a single resultant force as we have done above because the vector of the resultant couple is perpendicular to the resultant force. This usually does not occur in a spatial force system except for parallel spatial forces.

2-13.2 Find a fourth force F to be added to the coplanar system of Prob. 2-13.1 that will cause F and the original three forces shown in Fig. 2-13.2a to have a horizontal resultant of 100 N acting rightward through point C.

Solution

By equating the effects of the specified resultant to the effects of the force system, we can find the components of F and its moment effect. These are then combined to find F and its X and Y intercepts in the same manner as we found R in the previous problem. Thus we obtain

$[\overset{+}{\rightarrow} R_x = \Sigma X]$ $\quad 100 = F_x + 200 - 150 - 300, \quad F_x = 350 \text{ N} \rightarrow$

$[+\uparrow R_y = \Sigma Y]$ $\quad 0 = F_y = 260 - 200 - 400 \quad F_y = 340 \text{ N} \uparrow$

$[\uparrow\!+\; M_O^R = \Sigma M_O]$ $\quad 100(2) = M_O^F + 400(4) - 260(2) - 150(3) - 300(1)$

$$M_O = -130 \text{ N-m} = 130 \text{ N-m}$$

Combining the components of F, we obtain $F = 488$ N up to the right at $\theta_x = 44.2^\circ$ with the X axis as shown in Fig. 2-13.4. The reference origin O is located as shown so that F will produce the counterclockwise moment about it as determined above. Applying the components of F at its intersection with the X axis, we find the x intercept of F to be

$[M_O^F = (F_y) i_x]$ $\quad 130 = 340 i_x \quad i_x = 0.382$ m right of O ***Ans.***

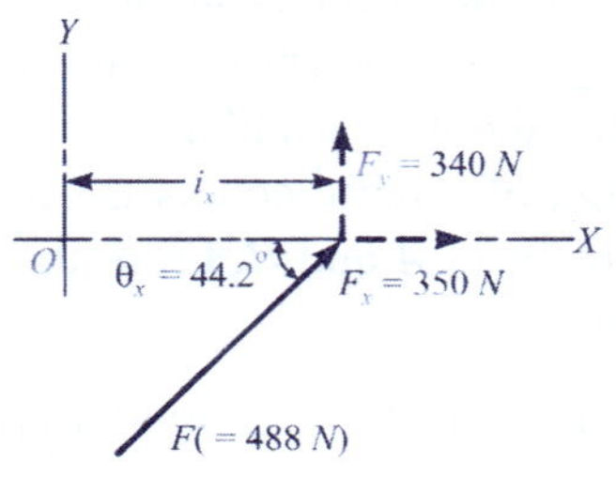

Figure 2-13.4

2-13.3 The rectangular block shown in Fig. 2-13.5 is acted upon by the following forces:

$P = 538$ N directed from B to C

$Q = 490$ N directed from C to E

$T = 269$ N directed from F to G

Replace the specified set of forces by a single resultant force acting at O and a couple.

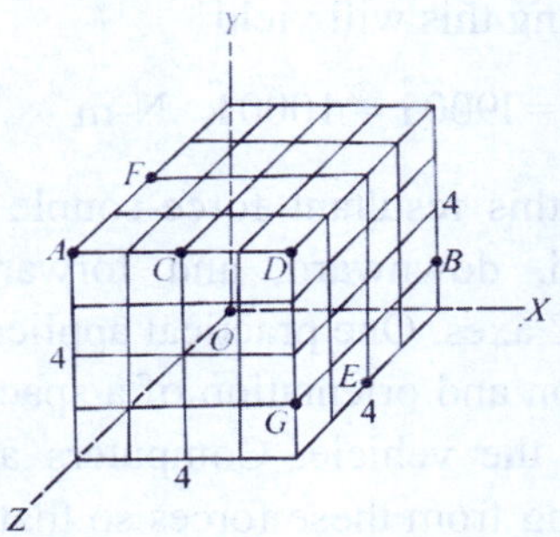

Figure 2-13.5

Solution

Using vector notation, the components of the various forces are

$$\mathbf{P} = \frac{538}{\sqrt{29}}(-2\hat{\mathbf{i}} + 3\hat{\mathbf{j}} + 4\hat{\mathbf{k}}) = -200\hat{\mathbf{i}} + 300\hat{\mathbf{j}} + 400\hat{\mathbf{k}}$$

$$\mathbf{Q} = \frac{490}{\sqrt{24}}(2\hat{\mathbf{i}} - 4\hat{\mathbf{j}} - 2\hat{\mathbf{k}}) = 200\hat{\mathbf{i}} - 400\hat{\mathbf{j}} - 200\hat{\mathbf{k}}$$

$$\mathbf{T} = \frac{269}{\sqrt{29}}(4\hat{\mathbf{i}} - 3\hat{\mathbf{j}} + 2\hat{\mathbf{k}}) = 200\hat{\mathbf{i}} - 150\hat{\mathbf{j}} + 100\hat{\mathbf{k}}$$

Since $\mathbf{R} = \Sigma\mathbf{F}$, this arrangement is especially convenient to find $\mathbf{R}$ by summing the coefficients of the unit vectors. This gives

$$\mathbf{R} = 200\hat{\mathbf{i}} - 250\hat{\mathbf{j}} + 300\hat{\mathbf{k}} \text{ N} \qquad \textit{Ans.}$$

The resultant couple is found by summing the moments of each of the forces about O. The position vectors from O to convenient points on P, Q, and T respectively being

$$\mathbf{r}_{OB} = 4\hat{\mathbf{i}} + \hat{\mathbf{j}}$$

$$\mathbf{r}_{OE} = 4\hat{\mathbf{i}} + 2\hat{\mathbf{k}}$$

$$\mathbf{r}_{OF} = 4\hat{\mathbf{j}} + 2\hat{\mathbf{k}}$$

the corresponding forms of $\mathbf{M}_O^R = \Sigma \mathbf{r} \times \mathbf{F}$ in determinant form are

$$\mathbf{M}_O^R = \begin{vmatrix} 4 & 1 & 0 \\ -200 & 300 & 400 \\ \hat{\mathbf{i}} & \hat{\mathbf{j}} & \hat{\mathbf{k}} \end{vmatrix} + \begin{vmatrix} 4 & 0 & 2 \\ 200 & -400 & -200 \\ \hat{\mathbf{i}} & \hat{\mathbf{j}} & \hat{\mathbf{k}} \end{vmatrix} + \begin{vmatrix} 0 & 4 & 2 \\ 200 & -150 & 100 \\ \hat{\mathbf{i}} & \hat{\mathbf{j}} & \hat{\mathbf{k}} \end{vmatrix}$$

As a suggested technique, perform this, evaluation by placing the computed coefficients of the unit vectors from each determinant in vertical columns to facilitate their addition. Doing this will yield

$$\mathrm{C} = \mathbf{M}_O^R = 1900\,\hat{\mathbf{i}} - 1000\,\hat{\mathbf{k}} \text{ N-m} \qquad \textit{Ans.}$$

The physical effect of this resultant force-couple system on a body is to move the body rightward, downward, and forward while simultaneously rotating it about the X and Z axes. One practical application of these results is in the prediction of the position and orientation of a space vehicle as produced by the action of thrusters on the vehicle. Computers are used to integrate the equations of motion resulting from these forces so that predictions can be made of its position and orientation. Or conversely, we could determine the amount and direction of thrust necessary to accomplish a desired position and orientation. Such force-acceleration-velocity-position relations represent problems in dynamics that are considered there.

2-13.4 Express the resultant of the previous problem in terms of a single resultant force acting at A in Fig. 2-13.5 and a couple.

Solution

The resultant force **R** will obviously remain the same, but the couple will change since we have a different reference point. Instead of repeating the procedure of the previous problem, we may transform the resultant force-couple system acting at O into an equivalent force-couple system acting at A. The new couple will be $\mathbf{M}_A^R = \mathbf{M}_O^R + \mathbf{r}_{AO} \times \mathbf{R}$. Evaluating this latter term, we have

$$\mathbf{r}_{AO} \times \mathbf{R} = \begin{vmatrix} 0 & -4 & -4 \\ 200 & -250 & 300 \\ \hat{\mathbf{i}} & \hat{\mathbf{j}} & \hat{\mathbf{k}} \end{vmatrix} = -2200\,\hat{\mathbf{i}} - 800\,\hat{\mathbf{j}} + 800\,\hat{\mathbf{k}}$$

which is added to the previously found value of $\mathbf{M}_O^R$ to yield

$$\mathrm{C} = \mathbf{M}_A^R = -300\,\hat{\mathbf{i}} - 800\,\hat{\mathbf{j}} - 200\,\hat{\mathbf{k}} \text{ N-m} \qquad \textit{Ans.}$$

PROBLEMS

2-13.5 Replace the system of forces acting on the frame in Fig. P-2-13.5 by a resultant force R through A and a couple acting horizontally through B and C.

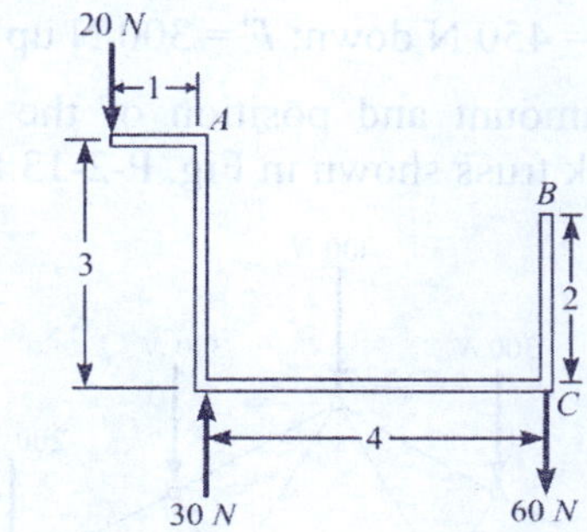

Figure P-2-13.5

$R = 50$ N down; $B = 110$ N right; $C = 110$ N left ***Ans.***

2-13.6 Fig. P-2.13.6 shows a plate subjected to loads. Replace the loads by an equivalent force through C and a couple acting through A and B. Solve if the forces of the couple are (a) horizontal and (b) vertical.

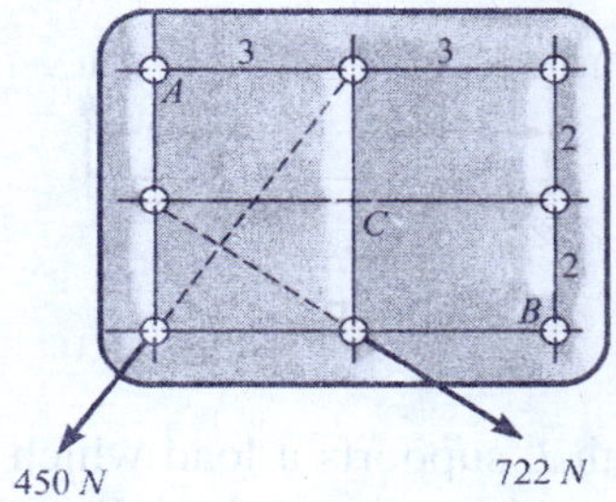

Figure P-2-13.6

2-13.7 A vertical force P at A and another vertical force F at B which act on the bar shown in Fig. P-2-13.7 produce a resultant force of 150 N down at D and a counterclockwise couple $C = 300$ N-m. Find the magnitude and direction of forces P and F.

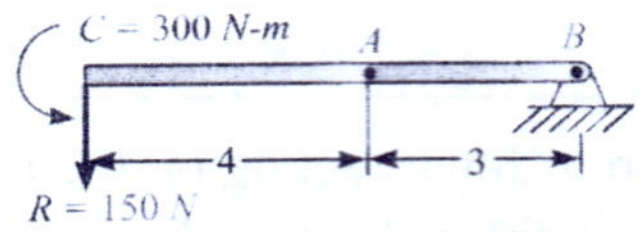

Figure P-2-13.7

$P = 450$ N down; $F = 300$ N up ***Ans.***

2-13.8 Determine the amount and position of the resultant of the loads acting on the Fink truss shown in Fig. P-2-13.8.

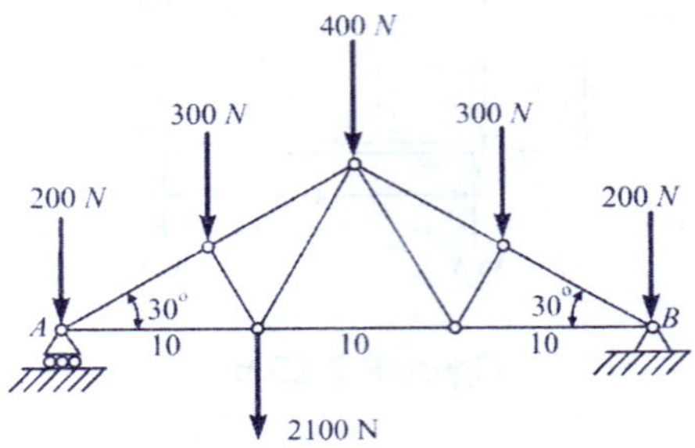

Figure P-2-13.8

R = 3500 N down at 12 m to right of *A* ***Ans.***

2-13.9 Find the values of *P* and *F* so that the four forces shown in Fig. P-2-13.9 produce an upward resultant of 300 N acting at 4 m from the left end of the bar.

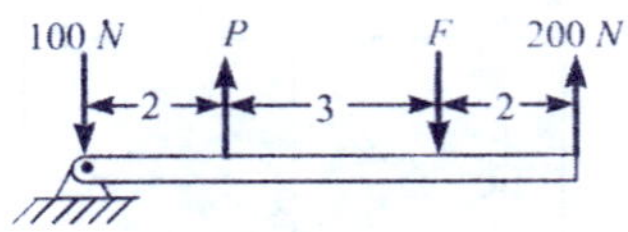

Figure P-2-13.9

2-13.10 A beam of length *L* supports a load which varies uniformly from *w* N/m at the right end to zero at the left end. Show that the resultant load is $W = wL/2$ acting at $L/3$ from the right end.

2-13.11 The beam *AB* in Fig. P-2-13.11 supports a load which varies uniformly from an intensity of 60 N/m to 180 N/m. Calculate the magnitude and position of the resultant load.

Hint: Replace the given loading by two equivalent triangular loadings each similar to the loading of Prob. 2-13.10.

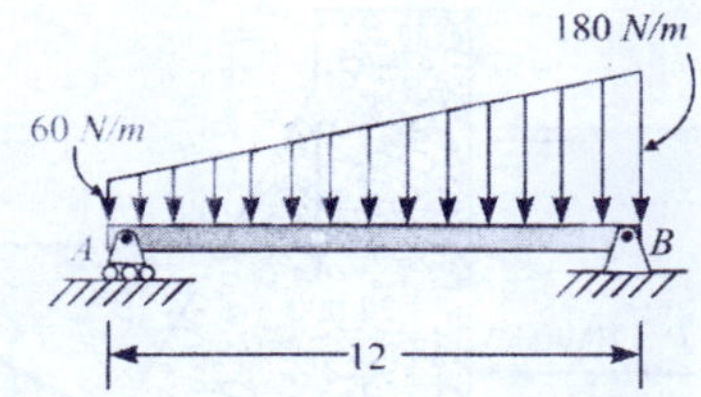

Figure P-2-13.11

2-13.12 The 16-m wing of a small airplane is subjected to a lift which varies from zero at the tip to 360 N/m at the fuselage according to $w = 90x^{1/2}$ where x is measured in feet from the tip. Compute the resultant lift and its location from the wing tip.

R = 3840 N at 9.60 m ***Ans.***

2-13.13 Compute the resultant of the three forces acting on the plate shown in Fig. P-2-13.13. Locate its intersection with *AB* and *BC*.

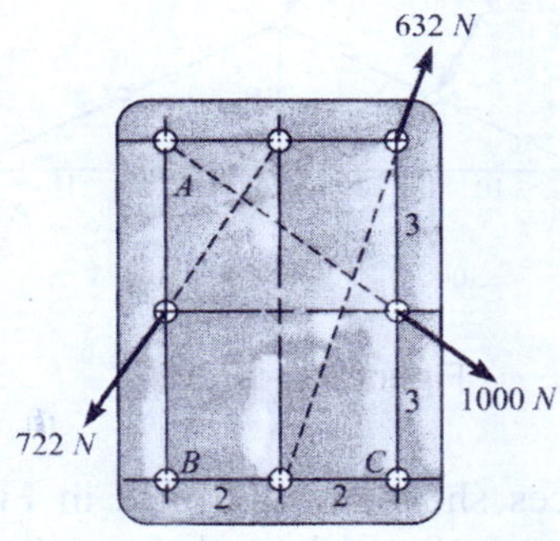

Figure P-2-13.13

R = 776 N down to the right at $\theta_h = 40.6°$
acting 3.43 m. above B and 4 m. right of B ***Ans.***

2-13.14 Determine the resultant of the three forces acting on the dam shown in Fig. P-2-13.14 and locate its intersection with the base *AB*. For good design, this intersection should occur within the middle third of the base. Does it?

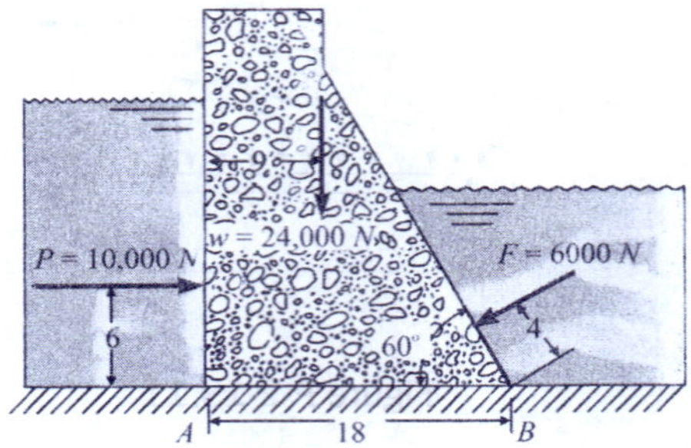

Figure P-2-13.14

Yes ***Ans.***

2-13.15 The Howe roof truss shown in Fig. P-2-13.15 carries the given loads. The wind loads are perpendicular to the inclined members. Determine the resultant and its intersection with AB.

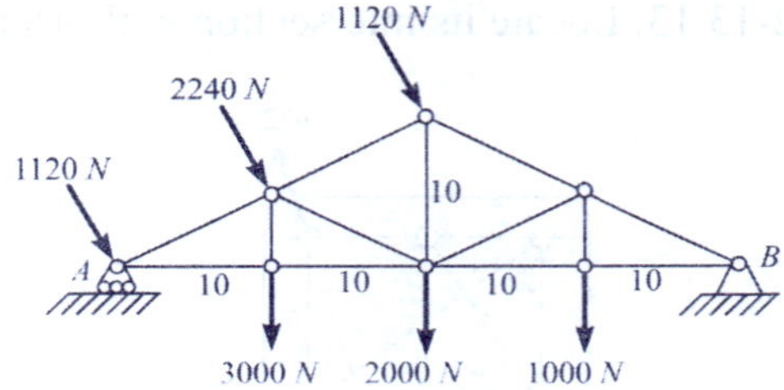

Figure P-2-13.15

2-13.16 The three forces shown on the grid in Fig. P-2-13.16 produce a horizontal resultant force through point A. Find the magnitude and sense of P and F.

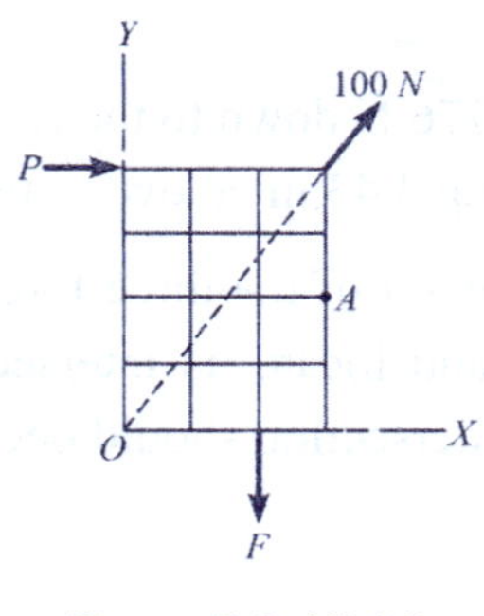

Figure P-2-13.16

2-13.17 The resultant of four forces, of which three are shown in Fig. P-2-13.16, is 390 N acting down to the right with a slope of 5 to 12 through point A. If P = 150 N and F = 130 N, determine the missing force T and its x intercept.

T = 180.5 N acting down to the right at $\theta_h = 33.7^o$ located 3.1 m to right of O ***Ans.***

2.13.18 The three forces shown in Fig. P-2-13.18 are required to cause a horizontal resultant acting through point A. If T = 316 N, determine the values of P and F.

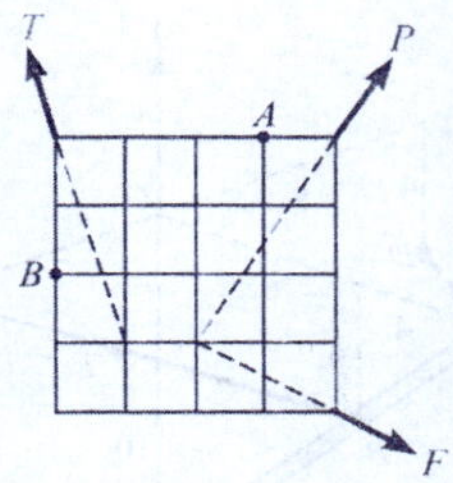

Figure P-2-13.18

P = – 180.5 N; F = 336 N ***Ans.***

2.13.19 The three forces in Fig. P-2-13.18 create a vertical resultant acting through point B. If P = 361 N, compute the values of T and F.

2-13.20 The rectangular block of Fig. P-2-13.20 is acted upon by a set of forces whose force multipliers and directions are as follows: P_m = 20 N/m directed from A to E, Q_m = 10 N/m directed from B to C, T_m = 30 N/m directed from G to F. Reduce these forces to a single resultant force acting at O and a couple.

$\mathbf{R} = -60\,\hat{\mathbf{i}} + 40\,\hat{\mathbf{j}} - 60\,\hat{\mathbf{k}}$ N; $C = -220\,\hat{\mathbf{i}} - 80\,\hat{\mathbf{j}} + 300\,\hat{\mathbf{k}}$ N-m ***Ans.***

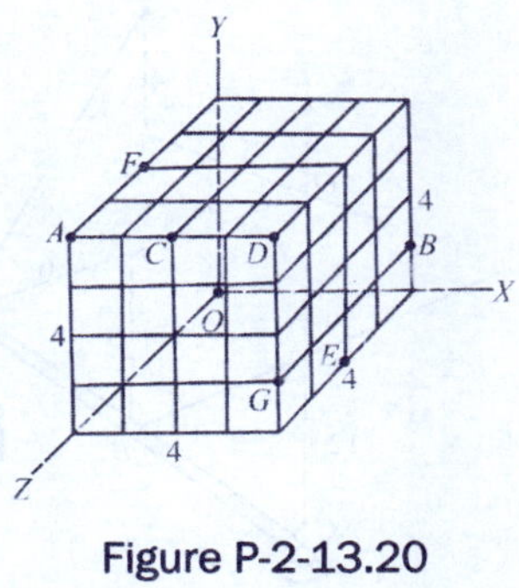

Figure P-2-13.20

2-13.21 Reduce the following set of forces to a single resultant force acting at the origin of the coordinate axes and a couple:

E = 140 N acting from A(3, 7,–1) toward B (5, 1,2)

P = 260 N acting from C(–6, 4, 1) toward D (6, 7,5)

T = 270 N acting from E(–2, 4, 1) toward G (4, –2, 4)

2-13.22 In Fig. P-2-13.22, a force **T** acts along BE and a force **P** acts along AD. Assuming their force multipliers to be T_m = 20 N/m and P_m = 10 N/m, find the force **F** to be applied at C to reduce their resultant to a couple. What is this resultant couple?

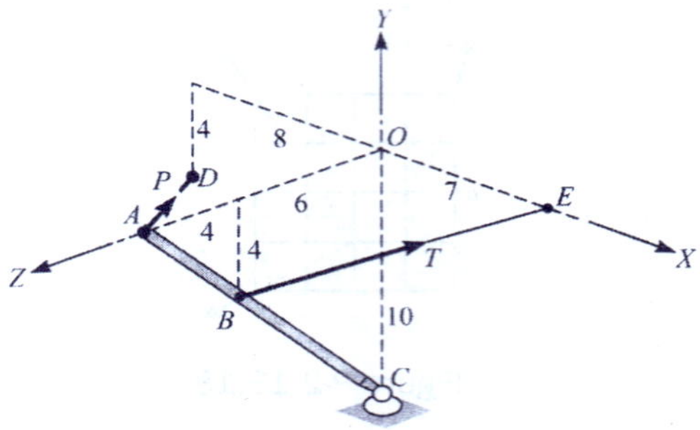

Figure P-2-13.22

$$\mathrm{F} = -60\,\hat{\mathbf{i}} - 40\,\hat{\mathbf{j}} + 220\,\hat{\mathbf{k}} \ \mathrm{N};$$

$$\mathrm{C} = -\,1800\,\hat{\mathbf{i}} + 40\,\hat{\mathbf{j}} - 40\,\hat{\mathbf{k}} \ \mathrm{N\text{-}m}$$ ***Ans.***

2.13.23 In Fig. P-2-13.23, the boom AC is acted upon by a vertical load W = 600 N, a force T = 440 N directed from B to D, and a force P = 280 N directed from A to E. Find the force **F** to be applied at C to reduce the resultant on the boom to a couple. What is this couple?

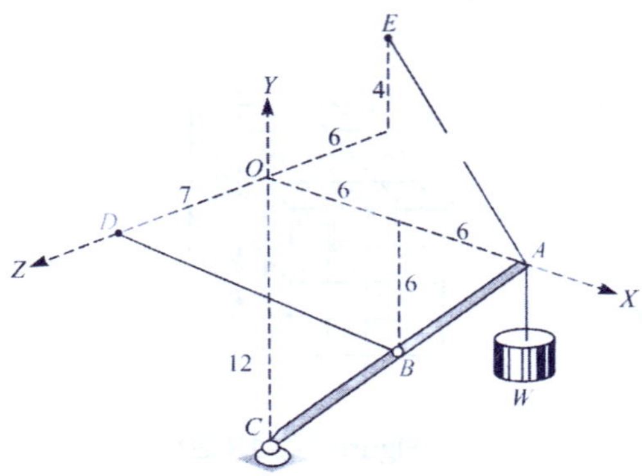

Figure P-2-13.23

2.13.24 The resultant of a certain force system is $\mathbf{F} = 20\hat{\mathbf{i}} - 30\hat{\mathbf{j}} + 10\hat{\mathbf{k}}$ N at the origin of coordinate axes and a couple $\mathbf{C} = 56\hat{\mathbf{i}} + 42\hat{\mathbf{j}} - 28\hat{\mathbf{k}}$ N-m. Replace this force and couple by an equivalent system (known as a wrench) consisting of the force **F** and a collinear couple $\mathbf{C}_F$.

Hint: The collinear couple $\mathbf{C_F}$ is the projection of **C** upon **F**; i.e., $\mathbf{C_F} = (\mathbf{C} \cdot \hat{\mathbf{n}}_F)\hat{\mathbf{n}}_F$. Also the wrench, acting through a point A, must produce the same moment about O as does the original system. Hence, $\mathbf{C} = \mathbf{r}_{OA} \times \mathbf{F} + \mathbf{C}_F$ which may be solved for the coordinates of point A. It is convenient to let point A denote the intersection of the line of action of the wrench with one of the coordinate planes. For instance, if here we assume A to be at (x, y, 0) in the XY plane, the moment relation rewritten as $\mathbf{C} - \mathbf{C_F} = \mathbf{r}_{OA} \times \mathbf{F}$ is easily solved for the x and y coordinates of A. Observe that $\mathbf{C} - \mathbf{C_F}$ represents that component of **C**, denoted by $\mathbf{C}_p$, which is perpendicular to **F**.

$F = 20\hat{\mathbf{i}} - 30\hat{\mathbf{j}} + 10\hat{\mathbf{k}}$ N; passing through A (–3.3, 6.2, 0) m;

$C_F = -6\hat{\mathbf{i}} + 9\hat{\mathbf{j}} - 3\hat{\mathbf{k}}$ N ***Ans.***

2.13.25 Replace the force and couple shown in Fig. P-2-13.25 by a wrench passing through point E in the XZ plane. Refer to the hint in the previous problem.

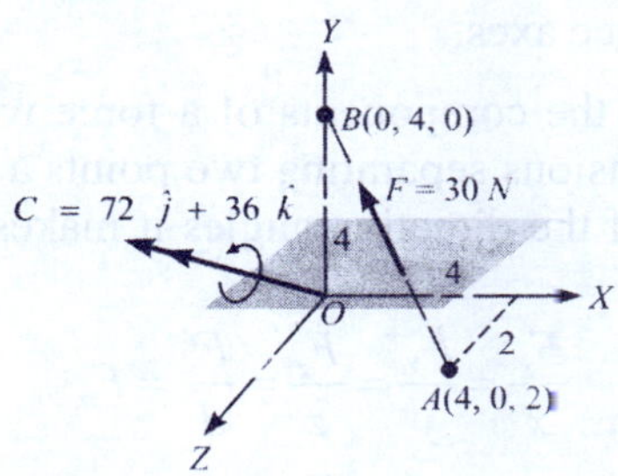

Figure P-2-13.25

2-13.26 Replace the force system shown in Fig. P-2-13.26 by a wrench passing through point A in the YZ plane. Apply the hint given in Prob. 2-13.24.

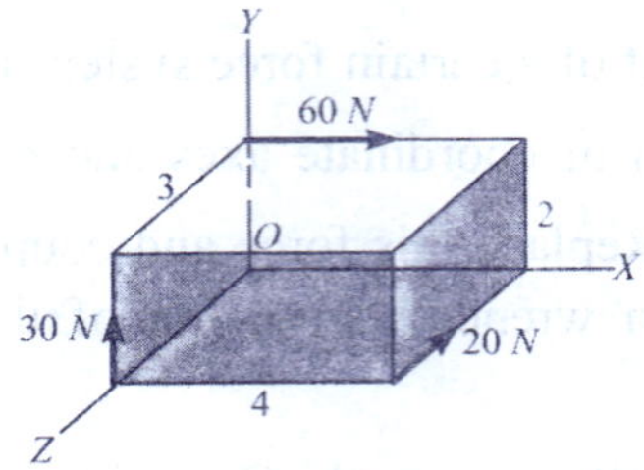

Figure P-2-13.26

$F = 60\hat{\mathbf{i}} - 30\hat{\mathbf{j}} + 20\hat{\mathbf{k}}$ N; acting through A (0, 2.08, 1.454) m;

$C_F = -14.7\hat{\mathbf{i}} - 7.35\hat{\mathbf{j}} + 4.9\hat{\mathbf{k}}$ N-m **Ans.**

SUMMARY

In this chapter, we develop the basic concepts of how to determine the components of a force and its moment effect about any point. These concepts are then extended, to determine the reduction of any system of forces to an equivalent resultant force-couple system.

Starting with the parallelogram law and its spatial extension, the components of each force in a coplanar system is expressed by

$$F_x = F\cos\theta_x \qquad F_y = F\sin\theta_x \qquad (2\text{-}3.1)$$

These components are considered positive when directed along the positive directions of the reference axes.

For spatial systems, the components of a force whose direction is specified by the coordinate dimensions separating two points a distance d apart on its line of action, or in terms of the direction angles it makes with the coordinate axes, are respectively

$$\frac{F_x}{x} = \frac{F_y}{y} = \frac{F_z}{z} = \frac{F}{d} = F_m \qquad (2\text{-}5.1)$$

$$\frac{F_x}{\cos\theta_x} = \frac{F_y}{\cos\theta_y} = \frac{F_z}{\cos\theta_x} = \frac{F}{1} \qquad (2\text{-}5.3)$$

The moment of a force (or more generally, of any vector) about a point is defined as the product of the force by a moment arm which is the perpendicular distance from the moment center to the action line of the force. The moment axis through the moment center is perpendicular to the plane defined by the force and its moment arm. The principle of moments permits the moment of a force to be expressed as the moment sum of its components. This principle is especially useful in all cases in which the moment arm of the force is difficult to compute but the moment arms of its components are readily available.

Sections 2-6, 2-7, and 2-8 discuss vector notation and the basic elements of vector algebra. For spatial systems, the scalar dot product and the vector cross product are ideal for determining respectively the magnitude of the component of a force (or any vector) along a desired direction and the moment of a force (or any vector) about any point; Thus, the component of a force **F** along a direction AB defined by the unit vector $\hat{\mathbf{n}}$ is

$$F_{AB} = \mathrm{F} \cdot \hat{\mathbf{n}}$$

whereas the moment of **F** is

$$\mathbf{M}_O^F = \boldsymbol{r} \times \boldsymbol{F} \tag{2-9.1}$$

where **r** is *any* position vector extending from the moment center O to the action line of the force. The vector form of the principle of moments is

$$\mathbf{M}_O^F = \Sigma \boldsymbol{M}_O = \Sigma(\boldsymbol{r} \times \boldsymbol{F}) \tag{2-9.2}$$

Detailed explanations (both vector and geometric) of coplanar and spatial applications of the principle of moments are given in Sections 2-10 and 2-11.

The resultant of a concurrent force system passes through the point of concurrency. Its components are the algebraic summations of the rectangular components of each force, whence its magnitude is given by

$$R = \sqrt{(\Sigma X)^2 + (\Sigma Y)^2 + (\Sigma Z)^2}$$

and its direction cosines and direction angles are found from **Eq.** (2-5.3) by substituting R for F.

Using the concept of a couple (Section 2-12), we can express the effect of other than concurrent systems of force in terms of a resultant force-couple system referred to any arbitrary origin. The resultant force-couple system in vector form is

$$\mathbf{R} = \Sigma\mathbf{F} \text{ and } \mathbf{M}_O^R = \Sigma\mathbf{M}_O \tag{2-13.1}$$

equivalent to the following six scalar equations and magnitudes:

$$\mathbf{R}_x = \Sigma\mathbf{X}; \quad \mathbf{R}_y = \Sigma\mathbf{Y}; \quad \mathbf{R}_z = \Sigma\mathbf{Z}$$
$$\mathbf{M}_x^R = \Sigma\mathbf{M}_x; \quad \mathbf{M}_y^R = \Sigma\mathbf{M}_y; \quad \mathbf{M}_Z^R = \Sigma\mathbf{M}_z$$
$$\mathbf{R} = \sqrt{(\Sigma\mathbf{X})^2 + (\Sigma\mathbf{Y})^2 + (\Sigma\mathbf{Z})^2};$$
$$\mathbf{M}_O^R = \sqrt{(\Sigma\mathbf{M}_x)^2 + (\Sigma\mathbf{M}_y)^2 + (\Sigma\mathbf{M}_z)^2}$$

Observe that the resultant force is independent of the choice of a reference origin, but that the resultant moment is not.

For coplanar systems, the effect of the force system may be extended one more step to consist of a single resultant force **R** so located that it causes the same rotational effect about O as the couple $\mathbf{C} = \Sigma \mathbf{M}_O$, but for spatial systems, the reduction usually ends with the resultant force-couple system. Occasionally, the magnitude of the resultant force may be zero although the system may produce a moment effect. In such cases, the resultant is a couple **C** consisting of two equal, parallel, oppositely directed forces having a moment equal to the product of one of the forces and the perpendicular distance between them. The moment of the couple is then equal to the moment sum of the given system about any center since the moment of a couple is independent of a moment center.

CHAPTER 3

EQUILIBRIUM OF FORCE SYSTEMS

3-1 INTRODUCTION

Equilibrium is the term used to designate the condition where the resultant of a force system is zero. The physical meaning of equilibrium, as applied to a body, is that the body is either at rest or is moving along a straight line path with constant velocity. Both these statements are implied in Newton's first law of motion; namely, a particle acted upon by a balanced force system has no acceleration. Balanced force systems compose the subject matter of statics which, as its name implies, deals essentially with the action of forces on bodies which are at rest. Unbalanced force systems cause bodies to accelerate; the resulting motion and the correlation between force and acceleration is the area of dynamics which we study in Dynamics.

In this chapter we shall determine and apply the conditions necessary to produce equilibrium for both coplanar and spatial force systems. The principles and techniques developed in this chapter are the basic fundamentals of statics; the student must master them. Actually what is studied is a method of reasoning–of learning how to apply the basic concepts of the components of a force and its moment effect in the most efficient manner.

Perhaps the most significant difference between the previous chapter and this one is that in Chapter 2, the force systems to be analyzed were specified explicitly, whereas here, the student must originate the force system. This is always the initial step in solving a problem, often the most difficult step, and is achieved by applying the free-body concept which we consider next.

3-2 FREE-BODY DIAGRAMS

Problems in mechanics always involve the interaction of bodies upon one another. Successful solution of these problems generally requires that the bodies be isolated from one another so that the forces involved may be analyzed and unknown forces determined. An isolated view of a body which shows only the external forces exerted on the body is called a free-body diagram (abbreviated as FBD). The external forces are caused either by direct bodily contact or by gravitational or magnetic attraction.

Table 3.2.1 describes some of the more common types of bodily contact or support and shows how to represent their action upon the isolated free body. The earth pull is always shown as a downward vertical force equal to the weight W of the free body and acting through its center of gravity. The action of a flexible cord, rope, or cable is represented by a tensile pull T directed along the cord away from the body. The action of a smooth surface is shown by a force S acting perpendicular to the smooth surface because there can be no resistance to sliding along such a surface. Similarly a roller support exerts a reaction R that is perpendicular to the surface along which the roller can move. At a smooth pin or hinge, however, the supporting force F can be exerted in any direction perpendicular to the pin axis; such a force is usually represented by two perpendicular components which, when known, can be combined to determine the supporting force and its direction. If the body is not free to rotate about the pin because of too tight a fit, for example, then a couple or moment will act in addition to the force components shown. A smooth ball-and-socket joint can support a force in any direction which is usually represented by three orthogonal components.

Notice that although forces always occur in pairs (i.e., action and reaction forces which are collinear, equal, but oppositely directed), a free-body diagram shows only the forces acting *upon* the isolated body. It does not show the forces exerted by the free-body upon other bodies or those within the body.

The steps involved in drawing a free-body diagram are:

1. Draw a diagram of the body completely isolated from all other bodies. The free body may consist of an entire assembled structure or any combination or part of it.
2. Represent the action of each body or support that has been removed by a force (or its components) as shown in Table 3-2.1.
3. Label each force by its magnitude if known, or by a symbol if unknown. If the sense of an unknown force along its line of action should be incorrectly assumed, the solution will give a negative sign, but its magnitude will nevertheless be correct.

4. The free-body diagram should be legibly and neatly drawn, and of sufficient size that all pertinent data can be clearly depicted.

Table 3-2.1 Action of Forces

Type of body removed	Sketch of reacting bodies	Action of body removed on the free body
Earth		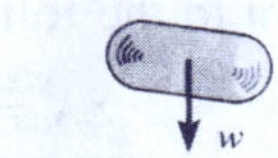
Flexible cord, rope, or cable (weight neglected)	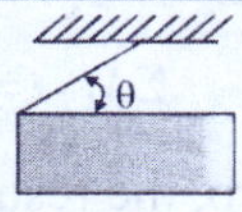	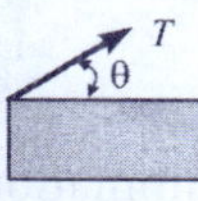
Smooth surface	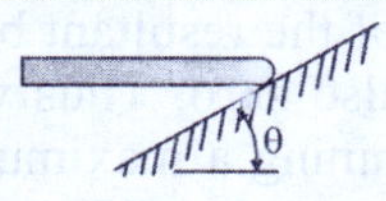	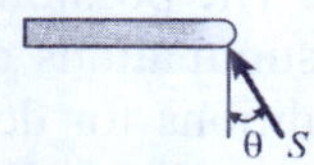
Roller	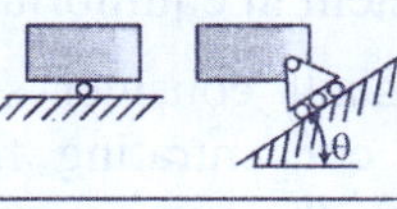	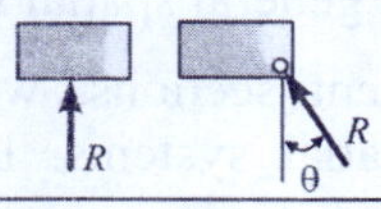
Smooth pin or hinge	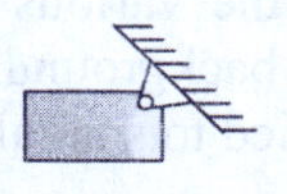	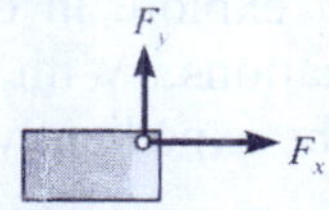
Ball-and-socket joint		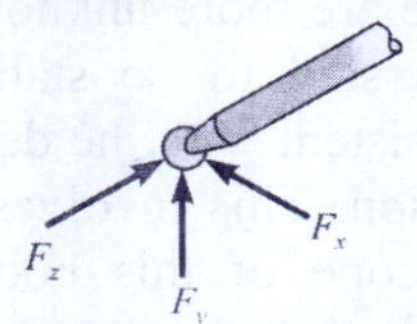

The technique of constructing a correct free-body diagram consists of applying the preceding steps in conjunction with the equations of equilibrium. A detailed discussion accompanies the illustrative problems of the next few sections.

3-3 EQUATIONS OF EQUILIBRIUM

As stated in Section 3-1, equilibrium is defined as that condition when the resultant of a force system is equal to zero. The equations of equilibrium therefore will be obtained from those conditions necessary and sufficient to cause a zero resultant.

It was shown in Section 2-13 that the most general force system may be reduced to a concurrent force system at any arbitrary point plus a couple which equals the moment sum of the original system about that point. When the force-couple system is equal to zero, the basic vector equations of equilibrium therefore are

$$\sum \mathrm{F} = 0 \text{ and } \sum \mathrm{M} = 0 \tag{3-3.1}$$

which are equivalent to the following six scalar equations:

$$\begin{aligned} \Sigma X &= 0 & \qquad \Sigma M_x &= 0 \\ \Sigma Y &= 0 \quad \text{and} & \Sigma M_y &= 0 \\ \Sigma Z &= 0 & \Sigma M_z &= 0 \end{aligned} \tag{3-3.2}$$

If all the force summations are zero, the resultant cannot be a single force but may be a couple. The possibility of the resultant being a couple is eliminated if all the moment summations are also zero. Thus we obtain a maximum of six independent conditions for determining a maximum of six unknown quantities that maintain a general spatial system in equilibrium.

In subsequent sections, we shall consider separately the equilibrium of planar and spatial systems. By concentrating first upon the simpler planar systems, we can explore in depth the various methods of applying the basic equilibrium equations. With this background, we shall discuss how these techniques may be expediently applied to spatial systems.

When there are more unknowns than independent equations of equilibrium, the system is said to be statically indeterminate. When this occurs, other equations consistent with the deformations of the body must be used in order to obtain a solution. This involves the mechanics of deformable bodies which is beyond the scope of this book.[1] In general, for a system to be statically determinate, there must be no more unknowns than there are independent equations of equilibrium. Occasionally, there may be fewer unknowns than independent equations of equilibrium; this usually indicates incomplete constraint against motion or else that the system is unstable. In later sections where it is more appropriate, we shall discuss the conditions of constraint or stability.

3-4 EQUILIBRIUM OF PLANAR SYSTEMS

For equilibrium of planar force systems, three of the equations listed as Eq. (3-3.2) become trivial because they are automatically satisfied.

---

[1]See, for example, the author's *Strength of Materials*, Harper & Row, 1962.

For example, if the force system lies wholly in the *XY* plane, it is evident that $\Sigma Z = 0$, $\Sigma M_x = 0$, $\Sigma M_y = 0$. For this reason, a maximum of only three unknowns can be found to maintain equilibrium in a statically determinate planar system since only three equations of equilibrium are available for planar systems

$$\begin{aligned} \Sigma X &= 0 \\ \Sigma Y &= 0 \\ \Sigma M &= 0 \end{aligned} \tag{3-4.1}$$

In the moment equation, the subscript is omitted because it is obvious that moment summations will always be with respect to a *Z* axis through any convenient point in the *XY* plane.

In applying these equations, the orientation of the *XY* axes is completely arbitrary, and vary with the situation to be solved. Sometimes it is convenient to select them as horizontal and vertical; at other times as perpendicular axes rotated to a more convenient orientation; and sometimes along two oblique directions.

ILLUSTRATIVE PROBLEM

3-4.1 In Fig. 3-4.1, two cylinders, *A* of weight 400 N and *B* of weight 200 N, rest on smooth inclines. They are connected by a bar of negligible weight hinged to each cylinder at its geometric center by smooth pins. Find the force *P* acting as shown that will hold the system in the given position.

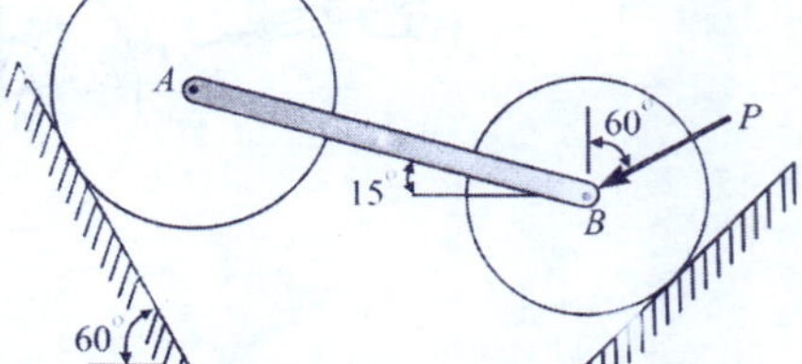

Figure 3-4.1

Preliminary

The first step in solving any problem is to draw the free-body diagrams which show the force system to be analyzed. Since this step is so important, we discuss in detail how the free-body diagrams are obtained.

In Fig. 3-4.2 is shown exploded views of the separate bodies composing the original system of Fig. 3-4.1. On each isolated body, show the action of the removed body or support, referring to Table 3-2.1 if necessary. The directions of these forces must be consistent with the equations of equilibrium.

Considering first bar *AB*, the forces exerted by the smooth pins *A* and *B* upon the bar may theoretically act in any direction. However, since the bar, has a fixed spatial position and the weight of the bar is neglected, the equation of equilibrium $\Sigma M_B = 0$ will only be satisfied if the force *F* at *A* passes through *B*. Similarly, $\Sigma M_A = 0$ requires that the force *Q* at *B* pass through A. Finally, a force summation along the direction *AB* will be balanced only if *Q* is equal and opposite to *F*. We conclude that a member fixed in space and loaded by only two forces requires the forces to be equal and oppositely directed along a line joining their points of application. This conclusion is of great importance in the analysis of pin-connected trusses since it determines the action lines of the forces existing in the bars of such trusses. However, this conclusion does not specify whether the forces are directed toward or away from each other; this is determined by solving the equilibrium equations when a positive result will confirm the assumed sense of an unknown force. For the assumed directions of *F* and *Q*, the bar *AB* tends to shorten and is said to be in compression; the opposite action is called tension.

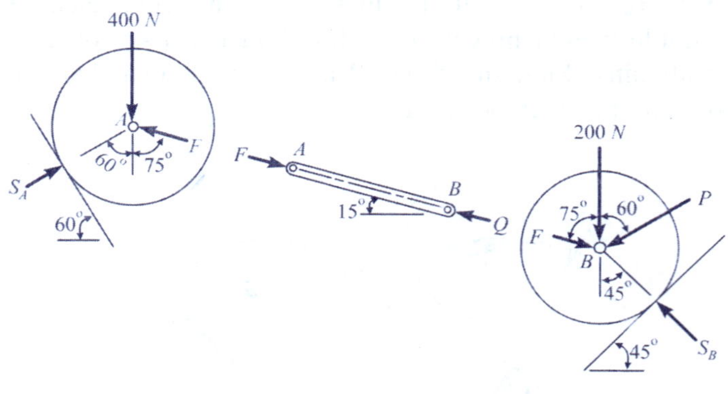

Figure 3-4.2 Free-body diagrams.

Consider next the FBD of cylinder *A*. In addition to its weight of 400 N, the cylinder is acted upon by the equal but opposite force *F* that is exerted by pin *A* on bar *AB*, plus the contact force S_A exerted by the inclined surface. Because the incline is smooth, S_A is perpendicular to it and passes through point *A*. Hence the force system on cylinder *A* is concurrent. Another approach which determines the direction of S_A is to note that S_A must pass through *A* in order to satisfy $\Sigma M_A = 0$; it therefore is normal to the incline whence the name normal force. In this approach we have used one of the three available equations of equilibrium to determine the direction of S_A; the other two equations will be used to find the unknown magnitudes of F and S_A by methods we discuss shortly.

Similarly, the FBD of cylinder B appears as shown where we have replaced Q by its equal value F to avoid denoting unessential unknowns, and where $\Sigma M_B = 0$ has been used to show that the contact force S_B is normal to the 45° incline. Although the FBD of cylinder B apparently has three unknowns and only two equations of equilibrium remaining available to solve for them, it becomes statically determinate after we determine the value of F from the FBD of cylinder A.

Observe that the FBD of the assembled cylinders and bar shown in Fig. 3-4.3 does not show the force F because here F is internal to the system and its action and reaction effects balance out. This FBD contains only three unknowns P, S_A, and S_B and may be solved directly by using three equations of equilibrium now that we have established the directions of S_A and S_B. For example, we might apply $\Sigma M_B = 0$ to eliminate P and S_B to find S_A, whence two force summations would determine P and S_B. Or P might be found directly from a moment summation about the intersection of S_A and S_B. Which of the various alternative approaches to use is a matter of judgment that is acquired only by experience. In this example, we believe the most efficient approach is to consider the FBD of each cylinder even though it introduces the additional unknown force F acting in the bar. We shall see later in Section 4-6 that the procedure of drawing the free-body diagrams of the various bodies in a system which is apparently statically indeterminate in its assembled state will supply additional equations of equilibrium from each separate FBD to enable a complete solution to be obtained.

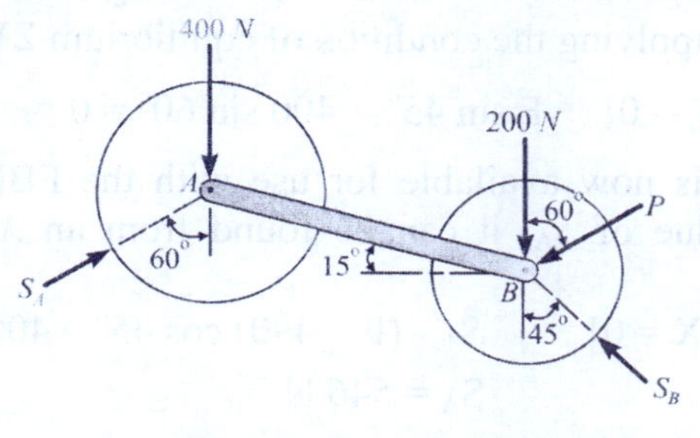

Figure 3-4.3 Free-body diagram of assembled cylinders and bar.

Solution

Several methods are available for the solution of concurrent force systems. Let us discuss each of these methods so that you can judge which is the most efficient and rapid method to use when force summations of equilibrium are involved.

***Method* 1: Using Horizontal and Vertical Axes.** The use of horizontal and vertical reference axes is recommended whenever one of the unknown forces is horizontal or vertical. Separate horizontal and vertical force summations can be obtained having one unknown in each which is readily found .This method may also be routinely applied if both are inclined to the horizontal, but horizontal and vertical summations will then each include both unknowns, necessitating solving a set of simultaneous equations. This unnecessary complication can be avoided in such cases by a rotation of the reference axes as described next.

***Method* 2: Using Rotated Axes.** Since the reference axes arbitrarily selected in the first place, a better choice of them eliminate simultaneous equations; this usually simplifies the numerical work and reduces the chance for error As applied to the FBD of cylinder *A*, redrawn here as Fig. 3-4.4, let the *X* reference be coincident with the unknown force S_A. Then S_A will have component along a *Y* axis assumed to be perpendicular to this axis and will be automatically eliminated from such a *Y* summation

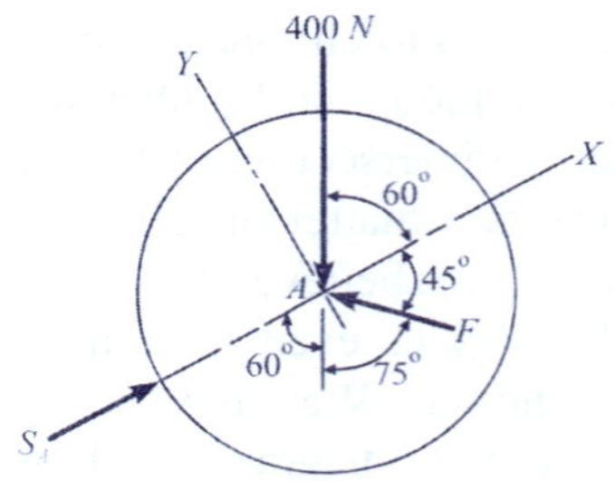

Figure 3.4.4 Method 2; Using rotated axes.

Using elementary geometry, we determine the angle between each force and this *X* axis, whence applying the condition of equilibrium ΣY = 0, we obtain

$$[+\nwarrow \Sigma Y = 0] \quad F \sin 45^\circ - 400 \sin 60^\circ = 0 \qquad F = 490 \text{ N}$$

The value of *F* is now available for use with the FBD of cylinder if we should need the value of S_A, it can be found from an *X* summation forces, namely

$$[+\nearrow \Sigma X = 0] \quad S_A - (F = 490) \cos 45^\circ - 400 \cos 60^\circ = 0$$

$$S_A = 546 \text{ N}$$

If only the value S_A had been desired, it could have been found directly and independently of *F* by passing a different *X* reference axis through *F*. As an exercise, do this to check the value S_A.

***Method* 3: Using Force Triangle.** When three concurrent forces are in equilibrium as is the case with cylinder *A*, the easiest solution generally obtained by applying the sine law to the triangle representing the polygon of forces. Since forces in equilibrium have a zero resultant, the tip of the last vector must touch the tail of the first vector. Starting with the known force of 400 N, this tip-to-tail addition gives the force triangle shown in Fig. 3-4.5 to which we apply of sines to obtain the equation[2]

$$\frac{400}{\sin 45^\circ} = \frac{F}{\sin 60^\circ} = \frac{S_A}{\sin 75^\circ}$$

whence as before,

$$F = 490 \text{ N} \qquad \text{and} \qquad S_A = 546 \text{ N}$$

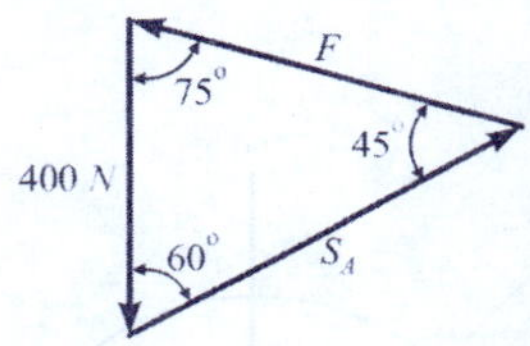

Figure 3.4.5 Method 3; Using forces triangle.

Comparing this solution from a force triangle with Method 2 will be seen to be equivalent to using two sets of rotated axes, one set drawn with its X axis through one unknown force and the other set drawn so that its X axis passes through the other unknown force.

We are now ready to determine the forces P and S_B holding cylinder B in equilibrium. Of the three methods discussed, Method 2 is the most practical here. A closed polygon of forces for this system forms a quadrilateral which may be used graphically, but is impractical analytically because the sine law does not apply. Although a diagonal of this quadrilateral can be drawn that will subdivide it into two triangles to which the sine law can be applied, this procedure will prove to be more cumbersome than using Method 2.

---

[2]Using a slide rule in which the **S** (sine) and **D** scales are aligned with each other, these proportions are rapidly solved by setting 45° on **S** opposite 400 on **D**, whence, as shown in the following table, the values of F and S_A are found opposite 60° and 75°, respectively.

On scale S	Set 45°	Opposite 60°	Opposite 75°
On scale D	Opposite 400	Read F = 490	Read S_A = 546

Applying the method of rotated axes to the FBD of cylinder B, we draw the X axis coinciding with S_B as in Fig. 3-4.6, thereby eliminating S_B from a Y summation. Hence, we obtain P from

$$[+\nearrow \Sigma Y = 0] \qquad 490 \sin 30° - 200 \sin 45° - P \sin 75° = 0$$

$$P = 107.7 \text{ N} \qquad \textbf{\textit{Ans.}}$$

If desired, the remaining unknown is now found from

$$[\searrow + \Sigma X = 0] \qquad 490 \cos 30° + 200 \cos 45° - (P = 107.7) \cos 75° - S_B = 0$$

$$S_B = 538 \text{ N} \qquad \textbf{\textit{Ans.}}$$

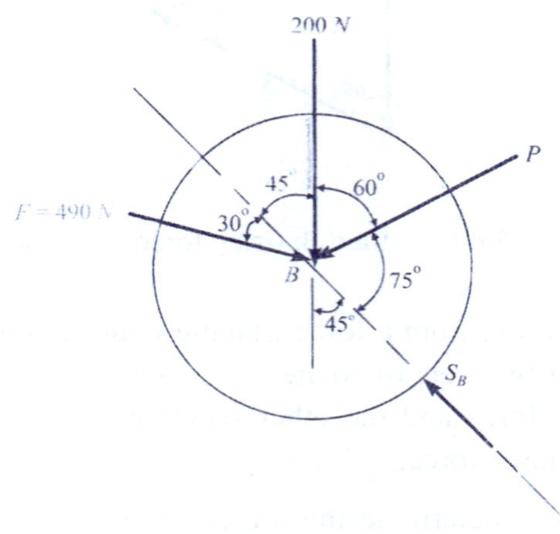

Figure 3.4.6 Method of rotated axes applied to FBD of cylinder B.

One final observation is pertinent. Method 2 is equivalent to taking the dot product of the force system with a unit vector perpendicular to the force to be eliminated. However, this use of the dot product offers no advantage over the use of rotated axes; indeed, it is more cumbersome except when all the forces are defined by their slopes, and even then alternate methods are more efficient as the next section shows.

PROBLEMS

3.4.2 A 300-N box is held at rest on a smooth incline by a force P making an angle θ with the incline as shown in Fig. P.3-4.2. If $\theta = 45°$, determine the value of P.

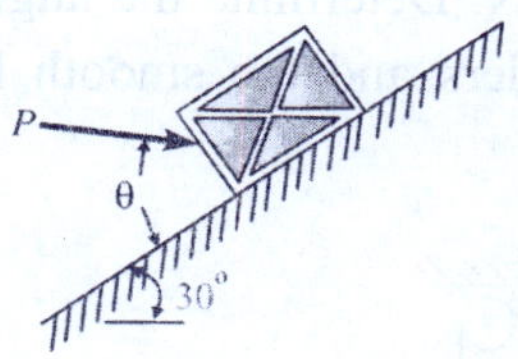

Figure P-3-4.2

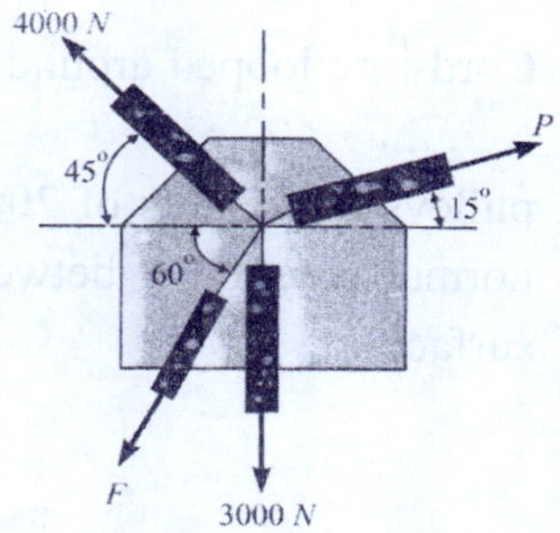

Figure P.3-4.4

P = 212 N ***Ans.***

3-4.3 If the value of P in Fig. P-3-4.2 is 180 N, determine the angle θ it must make with the smooth incline to hold the 300 N box at rest.

$\theta = 33.55°$ N ***Ans.***

3-4.4 The forces on the gusset plate of a joint in a bridge truss act as shown in Fig. P-3-4.4, Determine the values of P and F to maintain equilibrium of the joint.

$F = 800$ N ***Ans.***

3-4.5 The 300 N sphere in Fig. P-3-4.5 is supported by the pull P and a 200 N weight passing over a frictionless pulley. If $\alpha = 30°$, compute the values of P and θ.

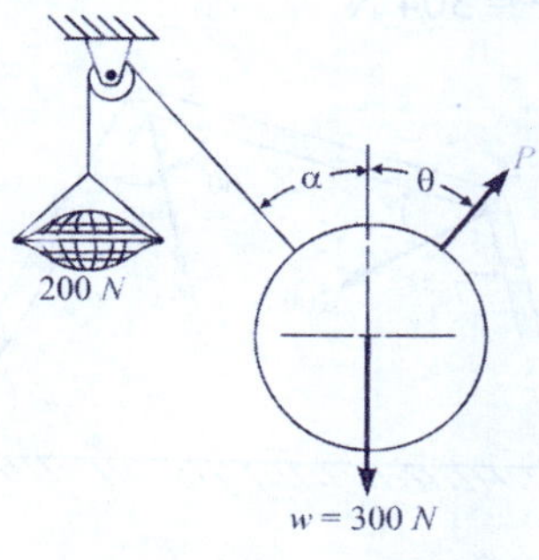

Figure P-3-4.5

3-4.6 In Fig. P-3-4.5, if P is known to be 160 N, determine the values of α and θ that maintain equilibrium.

$\alpha = 29.5°$; $\theta = 38°$ ***Ans.***

3-4.7 Cords are looped around a small spacer separating two cylinders each weighing 400 N and pass, as shown in Fig. P-3-4.7, over frictionless pulleys to weights of 200 N and 600 N. Determine the angle and the normal reaction S between the cylinders and the smooth horizontal surface.

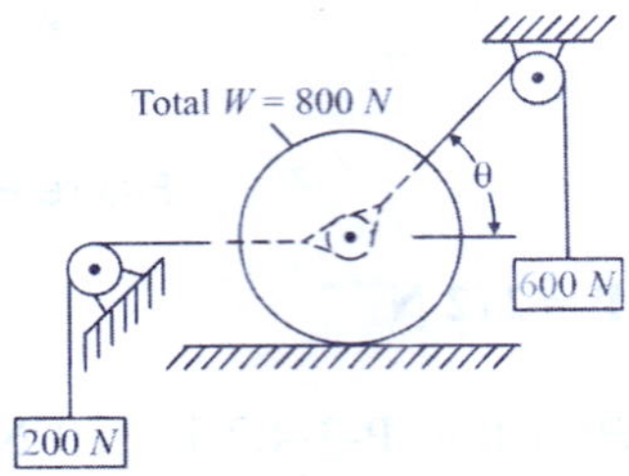

Figure P-3-4.7

3-4.8 Solve the previous problem if the cylinders rest on a smooth surface inclined up to the right at 15° with the horizontal.

$\theta = 63.2°$; N = 273 N ***Ans.***

3-4.9 Three bars, pinned together at B and C and supported by hinges at A and D as shown in Fig. P-3-4.9 form a four-link mechanism. Determine the value of P that will prevent motion.

P = 304 N ***Ans.***

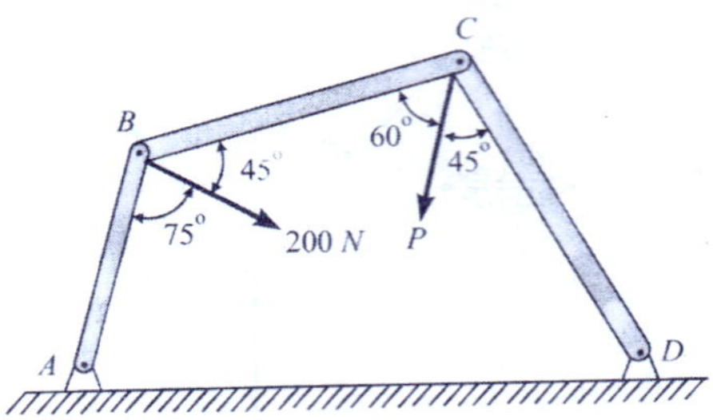

Figure P-3-4.9

3-4.10 Determine the amount and direction of the smallest force P required to start the wheel in Fig. P-3-4.10 over the block. What is the reaction at the block?

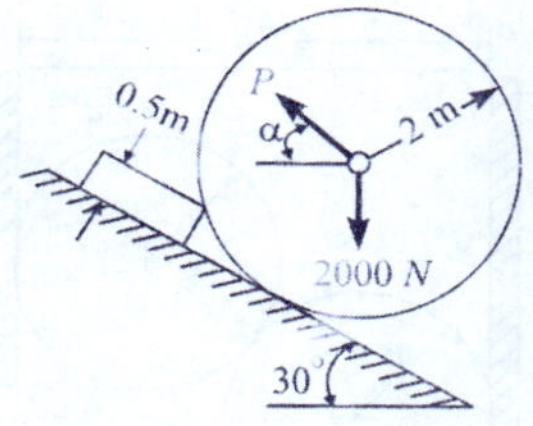

Figure P-3-4.10

$P = 1893$ N at 71.3° with horizontal; R = 642 N ***Ans.***

3-4.11 Three cylinders are piled in a rectangular ditch as shown in Fig. P-3-4.11. Neglecting friction, determine the reaction between cylinder A and the vertical wall.

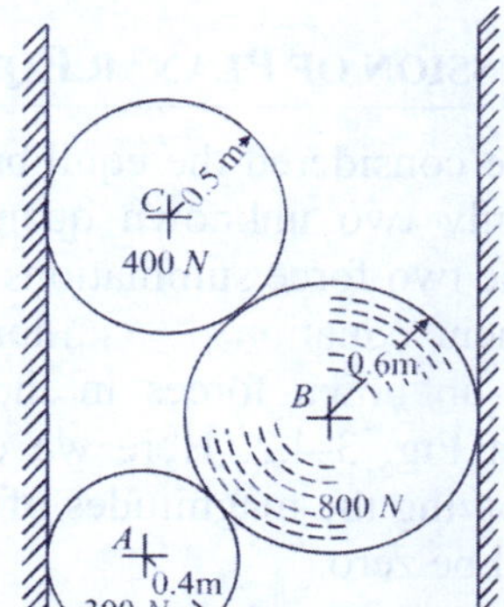

Figure P-3-4.11

1600 N ***Ans.***

3-4.12 Two smooth ball bearings, each of weight w and radius r, are placed inside a cylinder open at both ends. The assembly rests on a horizontal surface as shown in Fig. P-3-4.12. If the cylinder is of weight W and radius R < 2r, find (a) the force exerted by either ball bearing on the cylinder, (b) the smallest value of W that will prevent the cylinder from tipping over. (c) Could the cylinder possibly tip if it were closed at the bottom?

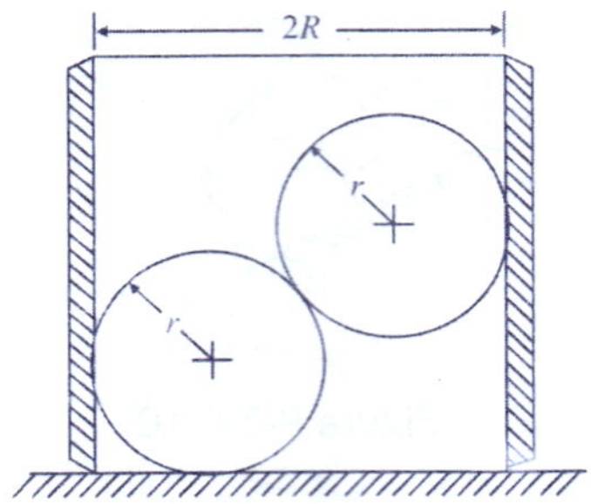

Figure P-3-4.12

(a) $F = \dfrac{\omega(R-r)}{\sqrt{2Rr - R^2}}$; (b) $W \geq \dfrac{2\omega(R-r)}{R}$; (c) No ***Ans.***

3-5 FURTHER DISCUSSION OF PLANAR EQUILIBRIUM

In the preceding section, we considered the equilibrium of coplanar concurrent force systems involving only two unknown quantities. We concentrated on various methods of using the two force summations of equilibrium to determine these unknowns. The moment equation of equilibrium was used primarily to establish the directions of unknown forces in the construction of free-body diagrams as for bar AB in Fig. 3-4.2. Here we consider the equilibrium of general planar systems involving the magnitudes of three unknown quantities to ensure that the resultant will be zero.

Frequently it is expedient to express force summations of equilibrium in terms of equivalent moment summations. The advantage of replacing a force summation by a moment summation is that any particular force can be eliminated from equilibrium consideration by taking moments about a center on its line of action. This method is especially useful for concurrent systems when the force to be eliminated is specified in direction by its slope. (See the second part of Illus. Prob. 3-5.1.) In nonconcurrent systems, the moment center is selected at the intersection of two unknown forces which thereby eliminates them from the moment summation and permits the third unknown to be found independently of the other two. Consequently, moment summations of equilibrium are frequently used instead of force summations. Actually, as we shall now show, moment summations of equilibrium are convenient alternate variations of force summations of equilibriums.

Recall that for a planar force system, the resultant reduces to a single force or a couple. For the body shown in Fig.3-5.1, if $\Sigma M_A = 0$, the resultant cannot be a couple, but may be a force R passing through A. If, in addition, $\Sigma M_B = 0$, the resultant (if it exists) must pass through both A and B. Finally, if about a third point C which is not collinear with A and B, we also have $\Sigma M_C = 0$, the resultant R cannot exist because this would mean that the resultant simultaneously has two different directions such as AB and AC which is impossible. The only possibility consistent with zero moment sums about three noncollinear points is that the resultant is zero. Hence three alternate equations of equilibrium are

$$\Sigma M_A = 0 \qquad (3\text{-}5.1)$$

$$\Sigma M_B = 0$$

$$\Sigma M_C = 0$$

It should not be assumed that Eq. (3-5.1) presents three new or additional equations of equilibrium. Two of them are actually equivalent to force summations of equilibrium as we can show by referring again to Fig. 3-5.1. Choosing reference axes as shown, we can express the moment of R about C as the moment sum of its components ΣX and ΣY acting at A. As the moment center C is on the Y axis, we have $\Sigma M_C = (\Sigma X)d$. Hence, if equilibrium exists, $(\Sigma X)d = 0$ from which d may be cancelled to yield $\Sigma X = 0$. A similar procedure may be used to show that $\Sigma M_B = 0$ is equivalent to a force summation of equilibrium perpendicular to the direction AB.

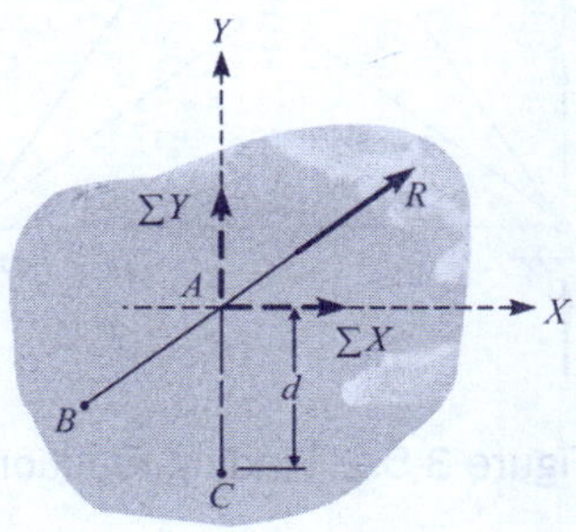

Figure 3-5.1

Summarizing this discussion, alternate sets of equilibrium equations are

$$\begin{array}{ccccc} \Sigma X = 0 & & \Sigma M_x = 0 & & \Sigma M_A = 0 \\ \Sigma Y = 0 & \text{or} & \Sigma M_y = 0 & \text{or} & \Sigma M_B = 0 \\ \Sigma M = 0 & & \Sigma M_z = 0 & & \Sigma M_C = 0 \end{array} \qquad (3\text{-}5.2)$$

The second and third sets of these equations are obtained by replacing a force summation by an equivalent moment summation.

Observe that there are only three independent equations of equilibrium. If occasionally we obtain equations that are not independent, it will be because we selected axes or moment centers that violate the restrictions discussed previously. Do not be disturbed if this occurs, but merely choose other axes or moment centers until you obtain three independent equations. In applying the moment summations, it is best (whenever possible) to select the moment center at the intersection of two of the unknowns, thereby eliminating these unknowns from the moment summation, This technique, as well as several others, is illustrated in the following sample problems.

ILLUSTRATIVE PROBLEMS

3-5.1. A truss, loaded as shown in Fig. 3-5.2, is supported by a hinge at *A* and a roller support at *D*. Find the reactions at the supports. Assuming the truss is composed of members connected by smooth pins, then find the forces transmitted by members *AB* and *AE*. Members length are in meters.

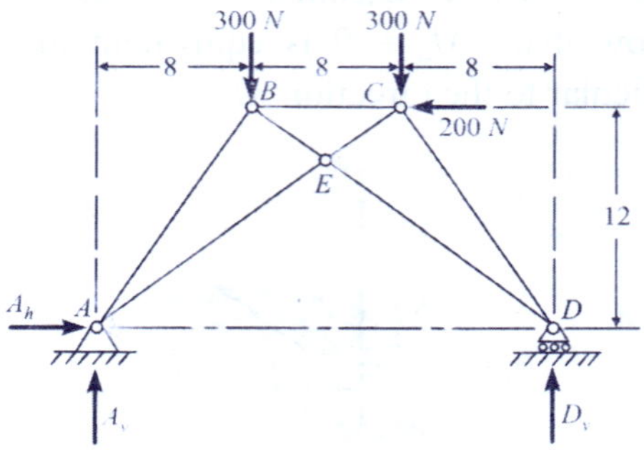

Figure 3-5.2 Finding reactions.

Solution

The roller at D constrains the reaction to be vertical. The reaction at the hinge *A* is resolved into its components A_h and A_v. These three unknown quantities on the FBD of the truss are determined now by a straight forward application of the three basic equations of equilibrium. To simplify computations, we choose equations of equilibrium which will each contain only one unknown.

Using a horizontal summation which thereby eliminates the vertical forces, we obtain A_h from

$$[\Sigma H = 0] \qquad A_h - 200 = 0 \qquad A_h = 200 \text{ N} \qquad \textbf{\textit{Ans.}}$$

A moment summation about A eliminates A_v and A_h and solves directly for D_v as follows:

$$[\,+\circlearrowleft \Sigma M_A = 0] \qquad 24\,D_v + 200(12) - 300(8) - 300(16) = 0$$

$$D_v = 200 \text{ N} \qquad \textbf{\textit{Ans.}}$$

Finally, A_v is determined from a moment summation about D which, by eliminating D_v and A_h, gives a result which is independent of the other unknowns.

$$[\circlearrowleft +\Sigma M_D = 0] \qquad 24\,A_v - 300(16) - 300(8) - 200(12) = 0$$

$$A_v = 400 \text{ N} \qquad \textbf{\textit{Ans.}}$$

A vertical summation may be used to check A_v and D_v but a more reliable check involving all the unknowns is obtained from a summation of moments about C. If desired, the components acting at A can be combined to give a total reaction at A of 448 N.

With the reactions known, a FBD of pin A will show the action of the unknown forces transmitted by members AB and AE. If we consider the weights of these members to be negligible compared with the applied loads, they are loaded only at their end pins by forces which must act along AB and AE, calling these forces P and F, and assuming them to act as shown on the FBD of pin A in Fig. 3-5.3, we consider which set of equilibrium equations will determine them most efficiently.

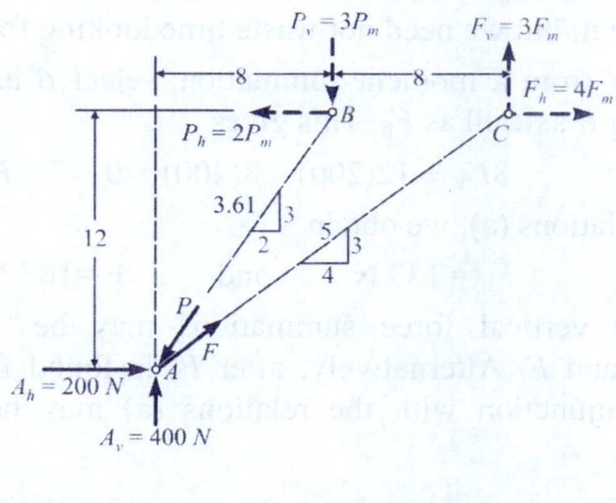

Figure 3-5.3 $\Sigma M_C = 0$ eliminates F while $\Sigma M_B = 0$ eliminates P.

We may use $\Sigma H = 0$ and $\Sigma V = 0$ which will yield a set of simultaneous equations which are quite simple to solve in this instance, but if we use equivalent moment summations as an alternate approach, each unknown can be found independently of the other. To eliminate an unknown force from an equilibrium equation, all we need do is to take a moment sum about a point on its line of action.

The dimensions of the truss specify the slopes of P and F to be as shown so that the relations among their components are

$$\frac{P_h}{2} = \frac{P_v}{3} = \frac{P}{3.61} = P_m \quad \text{and} \quad \frac{F_h}{4} = \frac{F_v}{3} = \frac{F}{5} = F_m \qquad (a)$$

As we shall soon see, it is convenient to draw the components of P at B and those of F at C.

To determine P, take a moment summation about any point on the action line of F thereby eliminates F from the moment equation. In this case, C is a convenient moment center since it not only eliminates F but also the component P_h which passes through C. Thus we obtain

$$[+\uparrow \Sigma M_C = 0] \qquad 8P_v + 12(200) - 16(400) = 0 \qquad P_v = 500 \text{ N}$$

whence, using the relations (a), we have

$$P_h = 333 \text{ N} \qquad \text{and} \qquad P = 601 \text{ N} \qquad \textbf{\textit{Ans.}}$$

Observe that any moment center on the action line of F may be selected to determine P, but both components of P may then appear in the moment summation. For example, this would occur if E in Fig. 3-5.2 were taken as the moment center and then it would be preferable to express each component in terms of its force multiplier. Solving for the force multiplier then determines the components and the force. Of course, if a moment center like C is available, we would prefer to use it, but we need not waste time looking for it.

To determine F from a moment summation, select B as a moment center, thereby eliminating P as well as F_h. This gives

$$[+\uparrow \Sigma M_B = 0] \qquad 8F_V + 12(200) - 8(400) = 0 \qquad F_v = 100 \text{ N}$$

whence from relations (a), we obtain

$$F_h = 133 \text{ N} \qquad \text{and} \qquad F = 167 \text{ N} \qquad \textbf{\textit{Ans.}}$$

Horizontal and vertical force summations may be used to check the components of P and F. Alternatively, after P_v is found from $M_C = 0$, these summations in conjunction with the relations (a) may be used to find the components of F.

3-5.2 Two basins, supported as shown in Fig. 3-5.4a and separated by a roller at C, jointly carry the indicated loads. On CD, the load varies uniformly from an intensity of 100 N/m to an intensity of 200 N/m, while AB supports, a concentrated load of 500 N. Determine the values of the reactions at A, B, and D.

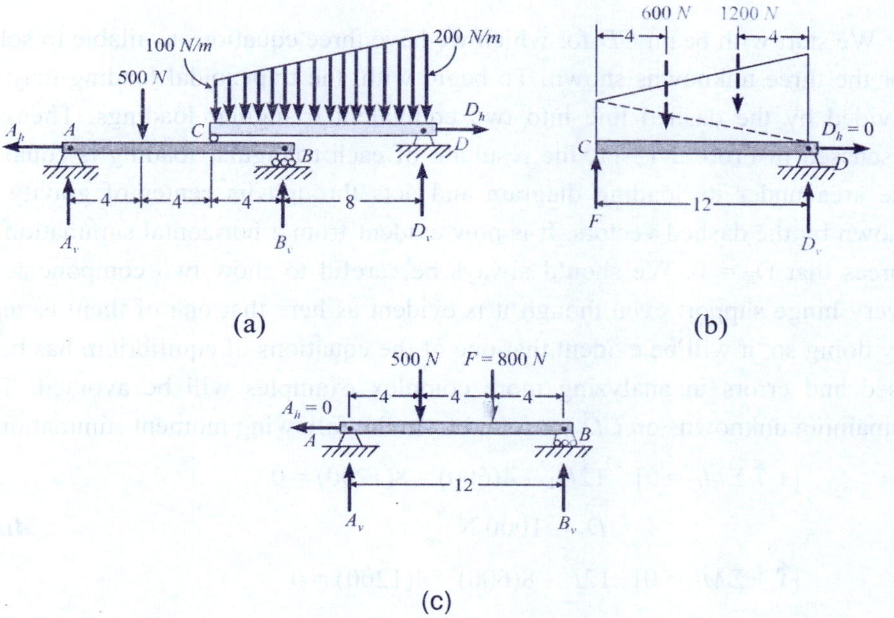

Figure 3-5.4

Solution

With two unknown components at each of the hinges A and D, the FBD of the assembled beams is statically indeterminate to the second degree since there are only three independent equations of equilibrium available to determine the five unknown supporting forces. By taking the assembly apart and drawing a separate FBD of each beam as shown in (b) and (c), we expose the contact force F exerted by the roller C as an additional unknown. As compensation for this additional unknown force, however, we may write three independent equations for each FBD to obtain a total of six independent equations of equilibrium.

This procedure of drawing the free-body diagrams of various bodies in a system which is apparently statically indeterminate in its assembled state will often supply sufficient additional equations of equilibrium from each separate FBD to enable a complete solution to be obtained. The criterion for this method to be successful is that the degree of statical indeterminacy of the original assembly (second degree in this problem), should equal the difference between the three additional equations of equilibrium and the number of additional unknowns exposed by each separate FBD. Do not concern yourself about this criterion now, however, since we shall apply it in a more codified fashion later in Chapter 4.

We start with beam *CD* for which we have three equations available to solve for the three unknowns shown. To begin with, the trapezoidal loading may be divided by the dashed line into two equivalent triangular loadings. Then, as discussed in Prob. 2-13.11, the resultant of each triangular loading is equal to the area under its loading diagram and acts through its center of gravity as shown by the dashed vectors. It is now evident from a horizontal summation of forces that $D_h = 0$. We should always be careful to show two components at every hinge support even though it is evident as here that one of them is zero. By doing so, it will be evident that one of the equations of equilibrium has been used and errors in analyzing more complex examples will be avoided. The remaining unknowns on *CD* are found from the following moment summations:

$$[+\circlearrowleft \Sigma M_C = 0] \quad 12D_V - 4(600) - 8(1200) = 0$$

$$D_V = 1000 \text{ N} \qquad \textbf{\textit{Ans.}}$$

$$[\circlearrowright + \Sigma M_D = 0] \quad 12F - 8(600) - 4(1200) = 0$$

$$F = 800 \text{ N} \qquad \textbf{\textit{Ans.}}$$

The value of *F* on *AB* acting equal and opposite to its action on *CD* is now used in Fig. 3-5.4c to obtain

$$[+\circlearrowleft \Sigma M_A = 0] \quad 12B_V - 4(500) - 8(800) = 0$$

$$B_V = 700 \text{ N} \qquad \textbf{\textit{Ans.}}$$

$$[+\circlearrowright \Sigma M_B = 0] \quad 12A_V - 8(500) - 4(800) = 0$$

$$A_V = 600 \text{ N} \qquad \textbf{\textit{Ans.}}$$

Note that a horizontal summation gives $A_h = 0$.

A check on these results is available from a vertical summation applied to the FBD of the original assembly in (a). Doing this gives

$$[\Sigma V = 0] \quad 600 + 700 + 1000 - 500 - 600 - 1200 = 0 \qquad \textbf{Check}$$

3-5.3 A 12 m bar of negligible weight is acted upon by a vertical load $P = 200$ N and a horizontal load $T = 100$ N applied at the positions shown in Fig. 3-5.5(a). The ends of the bar are in contact with a smooth vertical wall and a smooth incline. Determine the equilibrium position of the bar as defined by the angle θ it makes with the horizontal.

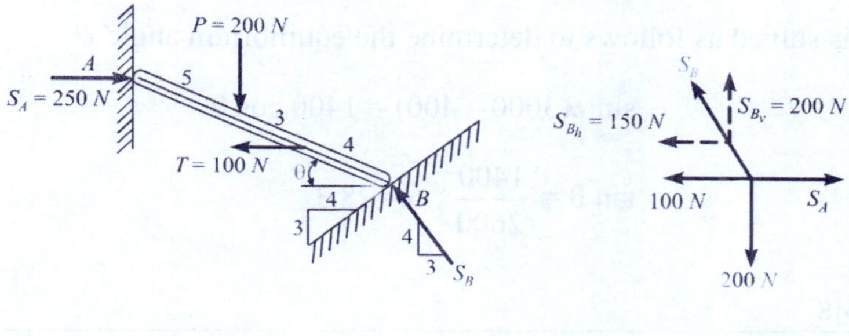

(a) (b)

Figure 3-5.5

Solution

We complete the FBD by drawing the reaction forces S_A and S_B at A and B acting respectively perpendicular to the smooth wall and incline as shown. The slope of S_B is the reciprocal of the slope of the incline. This FBD involves two unknown forces and an unknown angle θ which may be determined from two force summations and a moment summation.

To clarify the force summations, it is convenient to draw the forces at a common point as shown in Fig. 3-5.5(b), using the concept that any force system is equivalent to a concurrent system at any arbitrary point plus a couple. Since we have equilibrium, the resultant force is zero as is the resultant couple.

Applying first a vertical summation to the point diagram (b), which eliminates S_A, we have

$$[+\uparrow \Sigma V = 0] \qquad S_{B_v} - 200 = 0 \qquad S_{B_v} = 200 \text{ N} \qquad \textit{Ans.}$$

from which, using the proportionality relations between a force and its slope, we obtain

$$S_{Bh} = 150 \text{ m} \qquad \text{and} \qquad S_B = 250 \text{ N}$$

The value of S_A is found from a horizontal summation, namely,

$$\left[\overset{+}{\rightarrow} \Sigma H = 0\right] \qquad S_A - 100 - 150 = 0 \qquad S_A = 250 \text{ N}$$

For reference, this value of S_A is now inserted on the FBD of (a), using parentheses to denote its computed value.

Applying a moment summation about B, we now obtain

$$[\circlearrowleft + \Sigma M_B = 0] \qquad 250(12 \sin\theta) - 200(7\cos\theta) - 100(4 \sin\theta) = 0$$

which is solved as follows to determine the equilibrium angle θ.

$$\sin \theta(3000 - 400) = 1400 \cos \theta$$

$$\tan \theta = \frac{1400}{2600}\ ; \theta = 28.3^{\circ} \qquad \textbf{\textit{Ans.}}$$

PROBLEMS

3-5.4 Determine the value of T if the force system shown in Fig. P-3-5.4 is in equilibrium.

Suggestion: Use a moment summation.

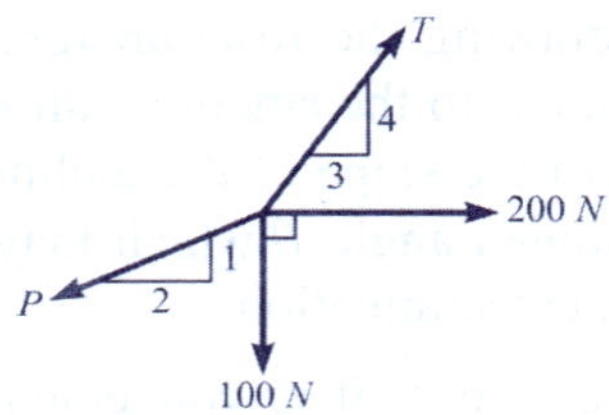

Figure P-3-5.4

3-5.5 Determine the forces P, F, and T required to keep the triangular frame ABC shown in Fig. P-3-5.5 in equilibrium

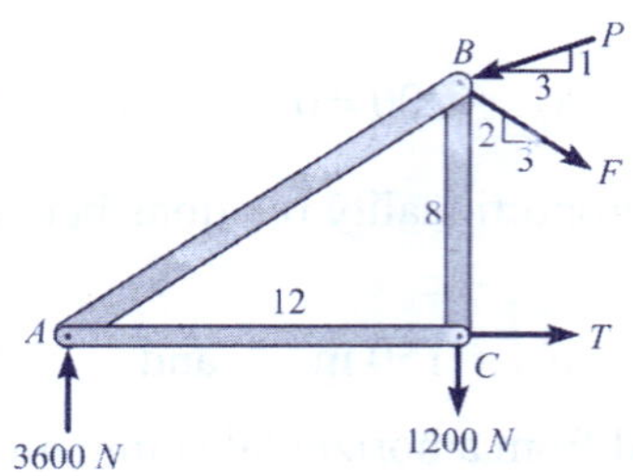

Figure P-3-5.5

$P = 6330$ N; $F = 722$ N; $T = 5400$ N ***Ans.***

3-5.6 The weight of the trapezoidal block is 7200 N acting shown in Fig. P-3-5.6. The ground reaction varies uniformly from an intensity of P_A N/m at A to P_B N/m at B. Determine P_A and P_B.

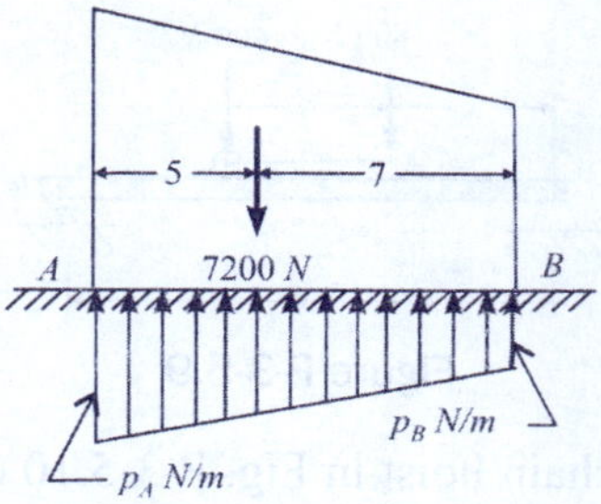

Figure P-3-5.6

$P_A = 900$ N/m; $P_B = 300$ N/m **Ans.**

3-5.7 A beam supports a load varying uniformly from an intensity of w N/m at the left end to p N/m at the right end. Find the values of w and p to cause the reactions shown in Fig. P-3-5.7.

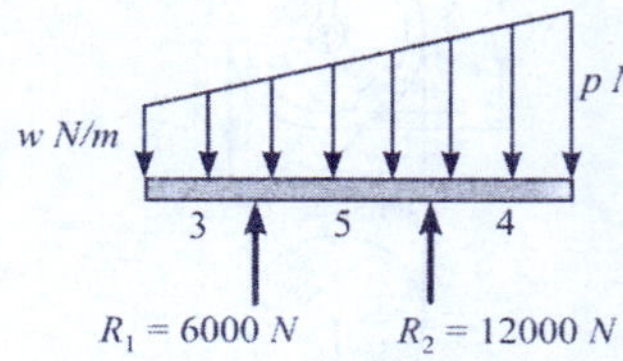

Figure P-3-5.7

$w = 1250$ N/m; $p = 1750$ N/m **Ans.**

3-5.8 As shown in Fig. P-3-5.8, the intensity of loading on a simply supported beam 10 m long is given by $y = 10x^3$ where y is in N/m and x is in m measured front A. Find the reactions at A and B.

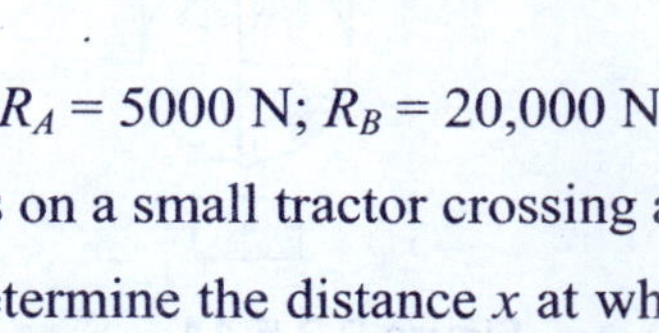

Figure P-3-5.8

$R_A = 5000$ N; $R_B = 20,000$ N **Ans.**

3-5.9 The wheel loads on a small tractor crossing an 18 m span are given in Fig. P-3-5.9. Determine the distance x at which the support reaction at A is twice that at B.

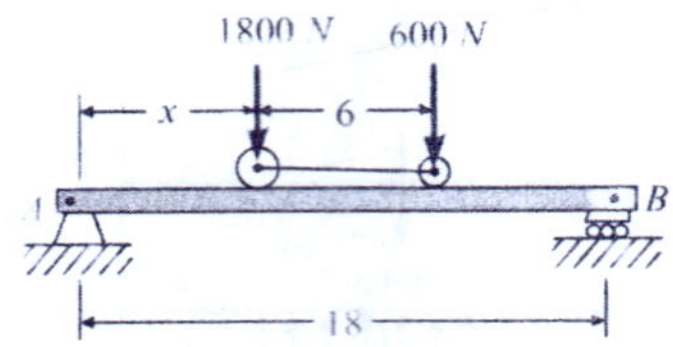

Figure P-3-5.9

3-5.10 The differential chain hoist in Fig. P-3-5.10 consists of two concentric pulleys rigidly fastened together. The pulleys form two sprockets for an endless chain looped over them in two loops. In one loop is mounted a movable pulley supporting a load W. Neglecting friction, determine the maximum load W that can just be raised by a pull P applied as shown. How is this result changed if P is not vertical?

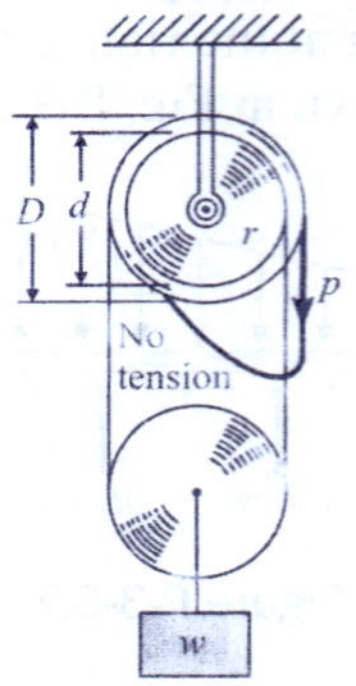

Figure P-3-5.10

$$W = \frac{2PD}{D - d}$$ ***Ans.***

3-5.11 If each pulley shown in Fig. P-3-5.11 weighs 36 N and $W = 720$ N find P to maintain equilibrium.

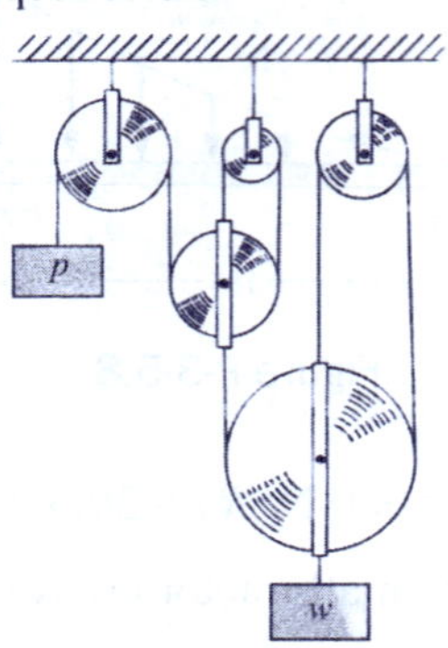

Figure P-3-5.11

3.5.12 A weight W rests on the bar AB as in Fig. P-3-5.12. The cable connecting W and B passes over frictionless pulleys. If bar AB has negligible weight, show that the reaction at A is $W(L-a)/(L+a)$.

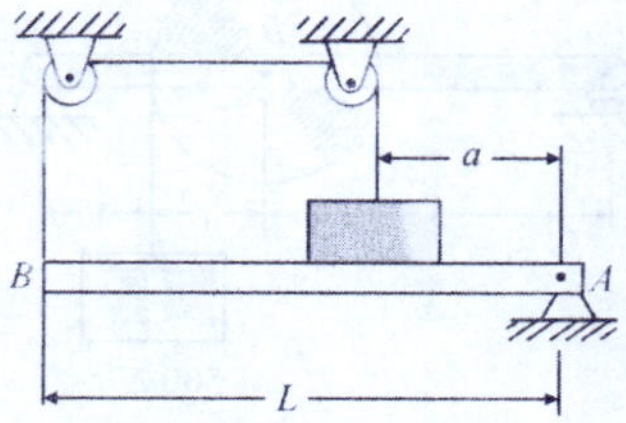

Figure P-3-5.12

3-5.13 A boom AB is supported in a horizontal position by a hinge A and a cable which runs from C over a small pulley at D as shown in Fig. P-3-5.13. Compute the tension T in the cable and the horizontal and vertical components of the reaction at A. Neglect the weight of the boom and the size of the pulley at D.

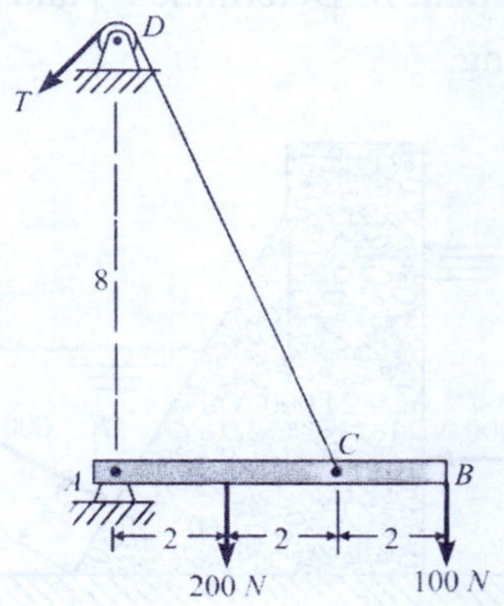

Figure P-3-5.13

3-5.14 Repeat Prob. 3-5.13 if the cable pulls the boom AB into a position at which it is inclined upward at a slope of 4 vertical to 3 horizontal. The loads remain vertical.

$T = 168$ N; $A_h = 75$ N; $A_v = 150$ N **Ans.**

3.5.15 A pulley of 1m radius, supporting a load of 500 N, is mounted at B on a horizontal beam as shown in Fig. P-3-5.15. If the beam weighs 200 N and the pulley weighs 50 N, find the hinge force at C.

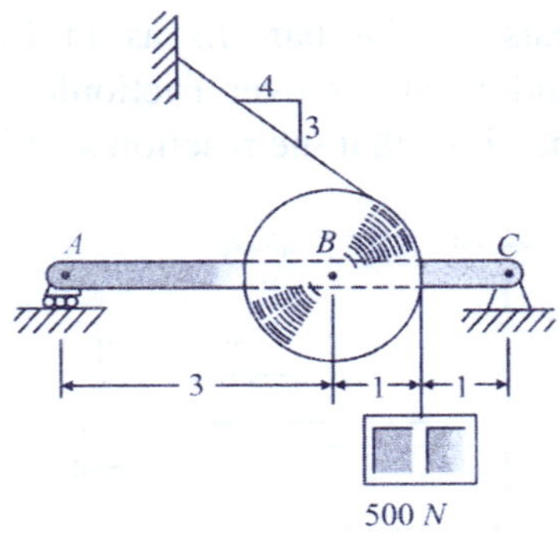

Figure P-3-5.15

$C = 472$ N up to the right at $\theta_h = 32°$ ***Ans.***

3-5.16 The forces acting on a 1 m length of a dam are shown in Fig. P-3-5.16. The upward ground reaction varies uniformly from an intensity of P_A N/m at A to P_B N/m at B. Determine P_A and P_B and also the horizontal resistance to sliding.

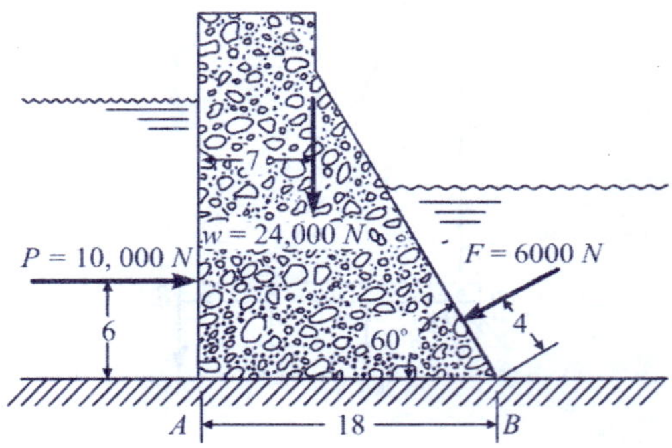

Figure P-3-5.16

3.5.17 The Fink truss in Fig. P-3-5.17 is supported by a hinge at A and by rollers on a 1 to 2 slope at B. The bracing bars are normal to the inclined members at their midpoints. The wind loads are perpendicular to AC. Determine the horizontal and vertical components of the reactions at A and B.

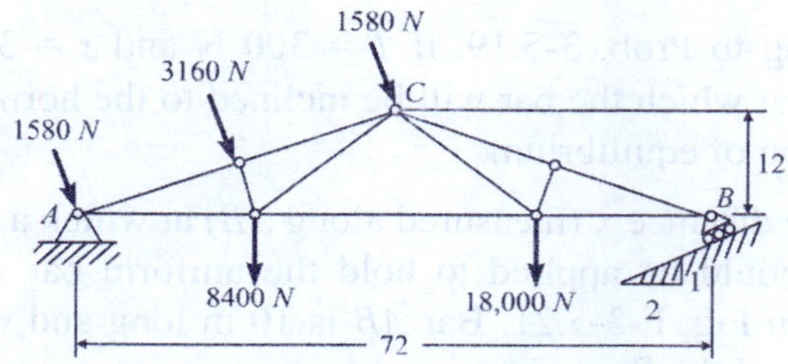

Figure P-3-5.17

$A_h = 6500$ N; $B_v = 17{,}000$ N **Ans.**

3.5.18 The roof truss in Fig. P-3-5.18 is supported by a hinge at *A* and a horizontal roller support at *B*. The wind loads are perpendicular to *BC*. Compute the total reactions at *A* and *B*.

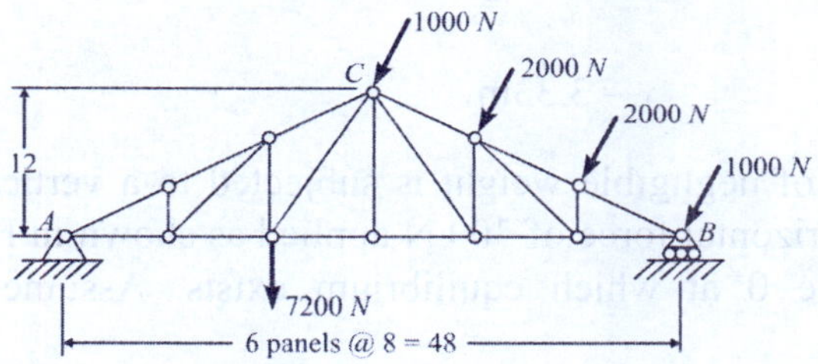

Figure P-3-5.18

3.5.19 A 12 m bar of negligible weight rests in a horizontal position on the smooth inclines in Fig. P-3-5.19. Compute the distance *x* at which load *T* = 100 N should be placed from point *B* to keep the bar horizontal.

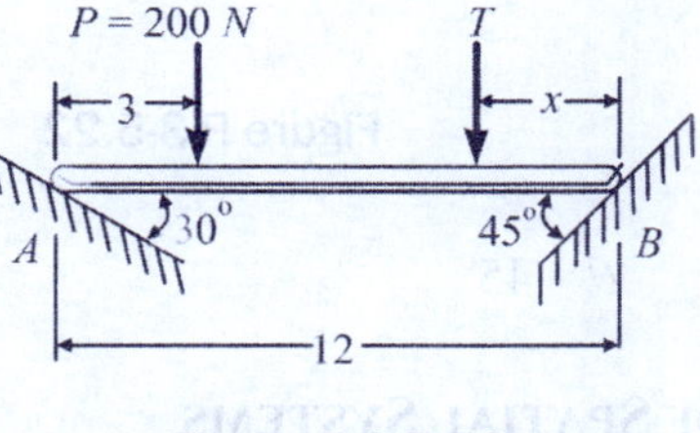

Figure P-3-5.19

$x = 4.83$ m **Ans.**

3-5.20 Referring to Prob. 3-5.19, if $T = 300$ N and $x = 3$ m, determine the angle θ at which the bar will be inclined to the horizontal when it is in a position of equilibrium.

3-5.21 Find the distance x (measured along AB) at which a horizontal force of 60 N should be applied to hold the uniform bar AB in the position shown in Fig. F-3-5.21. Bar AB is 10 m long and weighs 140 N. The incline and the floor are smooth.

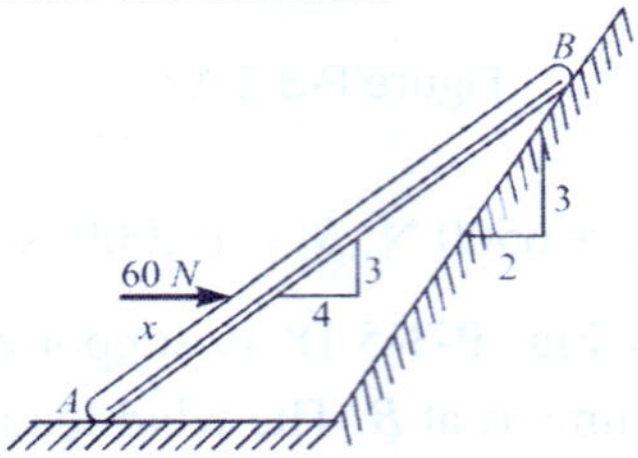

Figure P-3-5.21

$x = 3.33$m ***Ans.***

3-5.22 Bar AB of negligible weight is subjected to a vertical force of 600 N and a horizontal force of 300 N applied as shown in Fig. P-3-5.22. Find the angle θ at which equilibrium exists. Assume smooth inclined surfaces.

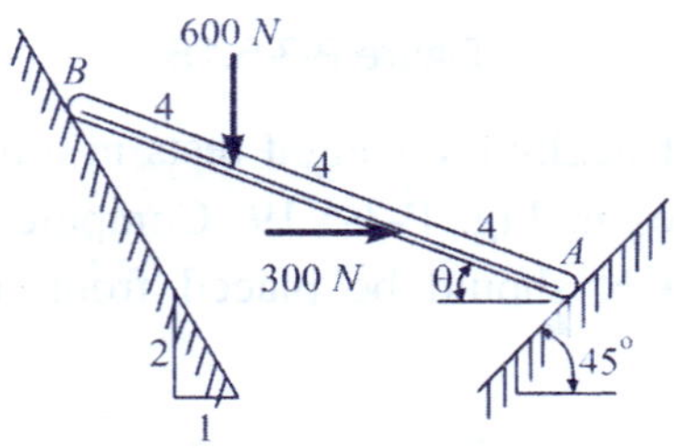

Figure P-3-5.22

$\theta = 45°$ ***Ans.***

3-6 EQUILIBRIUM OF SPATIAL SYSTEMS

As developed in Section 3-3, the general vector equations of equilibrium are

$$\Sigma F = 0 \text{ and } \quad \Sigma M = 0$$

which are equivalent to the following six scalar equations:

$$\Sigma X = 0 \qquad \Sigma M_x = 0 \tag{3.3.1}$$

$$\Sigma Y = 0 \text{ and} \qquad \Sigma M_y = 0$$

$$\Sigma Z = 0 \qquad \Sigma M_z = 0 \tag{3.3.2}$$

which may be used to solve for a maximum of six unknown quantities that maintain a general spatial system in equilibrium.

The general procedure of solving spatial systems in equilibrium follows the same pattern as developed for planar systems. After drawing the free-body diagram of the assembled system, or separate parts of it, we apply force and moment summations to determine the unknown quantities. Usually the most effective procedure is to start with a moment summation about a center or an axis which eliminates as many unknowns as possible. It is a matter of individual preference whether the equilibrium equations are applied directly in vector form or in scalar form; both methods are discussed in the illustrative problems. For moment summations about non coordinate axes, however, the vector form will be preferable

ILLUSTRATIVE PROBLEMS

3-6.1 The vertical mast shown in Fig. 3-6.1 is mounted in a ball-and-socket joint at A and supported by two guy wires extending from D to anchorages at B and C. At the midpoint of the mast, a force $P = 700$ N acts parallel to the Z axis and a force F = 1400 N acts parallel to the X axis. Find the tensions in the guy wires and the reaction at A.

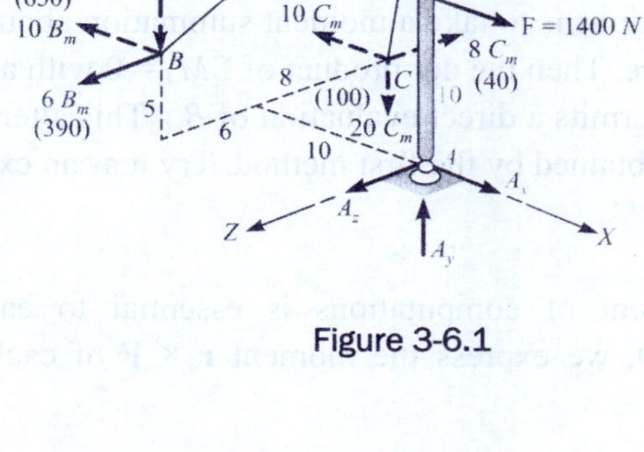

Figure 3-6.1

Analysis

We begin with the usual first step – creating the force system to be solved by drawing the free-body diagram of the mast. Here it is apparent that *BD* and *CD*, being loaded only at their ends, exert pulls on the mast that are directed along their lengths. By moving in these directions along coordinate paths, we express their components in terms of their force multipliers by dashed vectors applied at any point on their lines of action. Calling these forces **B** and **C**, we show their components for clarity at *B* and *C*. Moment balance about *D* of the mast eliminates these pulls but also requires that the moments of loads **P** and **F** about *D* be balanced by the moment of the reaction force at *A*; consequently, this reaction cannot pass through *D*. Because the direction of the reaction at *A* is unknown, we cannot represent it in terms of a common force multiplier, but must denote it by three unknown components A_x, A_y, and A_z assumed to act in the directions shown. We conclude that the unknowns acting on the mast consist of the force multipliers B_m and C_m, plus the unknown values of the components A_x, A_y, and A_z.

To determine these five unknowns, we have six scalar equations of equilibrium. Therefore, the system either is unstable or is not completely constrained. Actually, as we shall see, one of the six equations of equilibrium is automatically satisfied, implying a stable system, but constraint is incomplete since nothing prevents the mast from rotating about its vertical axis.

Consider now the choice of equations that will lead to the simplest solution. The guiding principle is to use a moment equation that will eliminate as many unknowns as possible. Here it is apparent that the three independent components A_x, A_y, and A_z, are eliminated by a moment summation about their common point *A*. Thus, the vector equation $\Sigma M_A = 0$ will determine the force multipliers B_m and C_m, after which the force summations $\Sigma X = 0$, $\Sigma Y = 0$, and $\Sigma Z = 0$ determine separately the components A_x, A_y, and A_z. This procedure is usually the most straight forward. An alternate method which avoids simultaneous equations is to take a moment summation about the noncoordinate axis *AC* for instance. Then the dot product of $\Sigma M_A = 0$ with a unit vector passing through *A* and *C* permits a direct evaluation of B_m. This alternate method can be used to check B_m obtained by the first method. Try it as an exercise.

Solution

Orderly arrangement of computations is essential to ease of solution. In applying $\Sigma \mathbf{M}_A = 0$, we express the moment $\mathbf{r} \times \mathbf{F}$ of each force about A in determinant form, using the procedure discussed in Illus. Prob. 2-11.1. Using the vertical distance from A to each force as the simplest position vector, we obtain

$$[\Sigma M_A = 0] \quad B_m \begin{vmatrix} 0 & 20 & 0 \\ -10 & -15 & 6 \\ \hat{\mathbf{i}} & \hat{\mathbf{j}} & \hat{\mathbf{k}} \end{vmatrix} + C_m \begin{vmatrix} 0 & 20 & 0 \\ -10 & -20 & -8 \\ \hat{\mathbf{i}} & \hat{\mathbf{j}} & \hat{\mathbf{k}} \end{vmatrix} + \begin{vmatrix} 0 & 10 & 0 \\ 1400 & 0 & -700 \\ \hat{\mathbf{i}} & \hat{\mathbf{j}} & \hat{\mathbf{k}} \end{vmatrix} = 0$$

Instead of writing out the complete expansion of these determinants, we now write separately a summation of the coefficients of the $\hat{\mathbf{i}}, \hat{\mathbf{j}}, \hat{\mathbf{k}}$ unit vectors, each coefficient being the expansion of their respective minor in each determinant. Since each coefficient of the unit vectors represents the moment of each force about the corresponding coordinate axis, this procedure gives the following scalar equations:

$$[\Sigma M_{x_A} = 0] \qquad B_m(120) + C_m(-160) - 7000 = 0$$

$$[\Sigma M_{z_A} = 0] \qquad B_m(200) + C_m(200) - 14{,}000 = 0$$

which are solved to yield

$$B_m = 65 \text{ and } C_m = 5$$

Observe that the equation $\Sigma M_{Y_A} = 0$ gives nothing since all the forces intersect a vertical Y axis through A and are automatically eliminated. As mentioned above, this indicates that rotation of the mast about this vertical axis is unconstrained.

To complete the solution for A_x, A_y, and A_z, a good method is to write the various forces in the following "quasi-vector" form, using the known values of B_m and C_m to find the components of B and C:

$\lvert A\rvert$	=	A_x	A_y	A_z
$\lvert B\rvert$	=	−650	−975	+390
$\lvert C\rvert$	=	−50	−100	−40
$\lvert P\rvert$	=	0	0	−700
$\lvert F\rvert$	=	1400	0	0
$\lvert \Sigma F\rvert$	=	0	0	0

Whence to satisfy force balance, we obtain

$$[\Sigma X = 0] \qquad A_x + 1400 - 650 - 50 = 0 \qquad A_x = -700 \text{ N} \qquad \textit{Ans.}$$

$$[\Sigma Y = 0] \qquad A_y - 975 - 100 = 0 \qquad A_y = 1075 \text{ N} \qquad \textit{Ans.}$$

$$[\Sigma Z = 0] \qquad A_z + 390 - 40 - 700 = 0 \qquad A_x = 350 \text{ N} \qquad \textit{Ans.}$$

Alternatively, we could write these force summations directly from the free-body diagram of Fig. 3-6.1 on which the components of B and C are shown in parentheses, but the procedure outlined has the advantage of not inadvertently overlooking a force component. Moreover, the indicated force summations need not be written out; the quasi-vector form implies them.

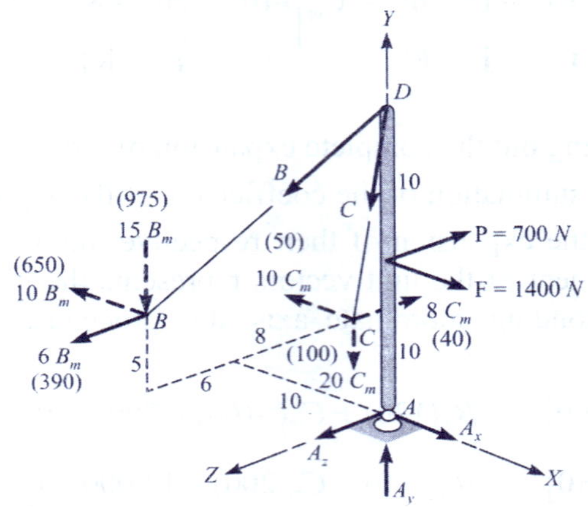

Figure 3-6.1

Finally, we combine the components of the various forces to give the following magnitudes:

$$A = 1332 \text{ N}; B = 1235 \text{ N}; C = 118.7 \text{ N} \qquad \textbf{\textit{Ans.}}$$

Alternate Solution

The geometric approach discussed in Illus. Prob. 2-11.2 may also be used if desired. If we apply it to Fig. 3-6.1, repeated here for convenient reference, it is evident[3] that moment summations about coordinate axes through A are

$$[\Sigma M_{x_A} = 0] \qquad 15B_m(6) + 6B_m(5) - 20C_m(8) - 700(10) = 0$$

reducing to

$$120\, B_m - 160C_m - 7000 = 0$$

$$[\Sigma M_{z_A} = 0] \qquad 10B_m(5) + 15B_m(10) + 20C_m(10) - 1400(10) = 0$$

reducing to

$$200B_m + 200C_m - 14{,}000 = 0$$

---

[3]If not readily evident, it will become so after you draw the projections of the spatial figure upon the coordinate planes.

These equations are identical with those obtained using vector algebra. Note that the moment effect of **B** might be simpler to obtain if we were to visualize its components applied at *D*.

The solution is completed as before by solving for B_m and C_m and using these force multipliers to compute the components of *B* and *C* which are enclosed in parentheses on Fig. 3-6.1. This figure is now used to obtain force summations along the *X*, *Y*, and *Z* coordinate directions which determine separately the components of the reaction at *A*.

3.6.2 The rigid body shown in Fig. 3-6.2 is supported by a ball-and-socket joint at *A* and by a hinge at *B* which provides only horizontal support. A weight W = 480 N is applied at corner *C* and equilibrium is maintained by the cable *DE*. Determine the components of the reactions at *A* and *B* and the pull **P** in the cable.

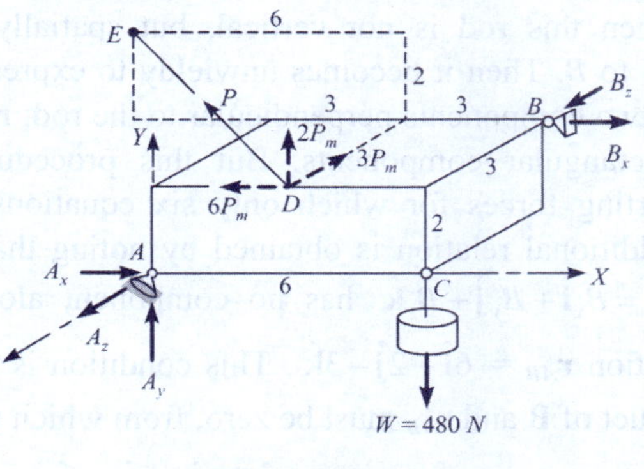

Figure 3-6.2

Solution

In the FBD of the body, the components at *A* and those at *B* are assumed to act as shown while the cable pull **P** is expressed in terms of its force multiplier P_m. To determine these six unknowns, there are available six scalar equations of equilibrium. The three components at A are eliminated by taking a moment summation about point A whence, using the geometric approach, we obtain

$$[+ \; \Sigma M_{xA} = 0] \qquad 2B_z - 2(3P_m) = 0 \qquad \text{(a)}$$

$$[+ \; \Sigma M_{yA} = 0] \qquad 3(3P_m) - 6B_z - 3B_z = 0 \qquad \text{(b)}$$

$$[+ \; \Sigma M_{zA} = 0] \qquad 2(6p_m) + 3(2P_m) - 2B_x - 6(480) = 0 \qquad \text{(c)}$$

Solving these equations, we get

$$P_m = 120 \text{ N/m};\ B_z = 360 \text{ N};\ B_x = -\,360 \text{ N} \qquad \textit{Ans.}$$

The magnitude of **P** is $P_m \sqrt{(6)^2 + (2)^2 + (3)^2} = 840$ N.

Force summations now give

$[\Sigma X = 0)$	$A_x - 6(120) + (-360) = 0$	$A_x = 1080$ N	***Ans.***
$[\Sigma Y = 0]$	$A_y - 480 + 2(120) = 0$	$A_y = 240$ N	***Ans.***
$[\Sigma Z = 0]$	$A_z - 3(120) + 360 = 0$	$A_z = 0$	

Instead of the above procedure, another approach might be to start with a moment summation about the noncoordinate axis AB. Since this eliminates the unknown forces at A and B, we could obtain **P** and its components directly as explained at the end of the next illustrative problem. Then moment summations about coordinate axes through A would determine the components of **B**.

Supplementary Note: It was implied that the hinge at B provided horizontal support by acting upon a vertical rod extending from the body at B. Consider the situation when this rod is not vertical, but spatially inclined along the direction from A to B. Then it becomes unwieldy to express the hinge reaction at B in terms of two components perpendicular to the rod; rather, it is preferable to use three rectangular components. But this procedure results in seven unknown supporting forces for which only six equations of equilibrium are available. An additional relation is obtained by noting that the hinge reaction expressed as $\mathbf{B} = B_x\hat{\mathbf{i}} + B_y\hat{\mathbf{j}} + B_z\hat{\mathbf{k}}$ has no component along the inclined rod having the direction $\mathbf{r}_{AB} = 6\hat{\mathbf{i}} + 2\hat{\mathbf{j}} - 3\hat{\mathbf{k}}$. This condition is equivalent to stating that the dot product of **B** and $\mathbf{r}_{AB}$ must be zero, from which we obtain

$$[\mathbf{B}.\mathbf{r}_{AB} = 0] \qquad (B_x\hat{\mathbf{i}} + B_y\hat{\mathbf{j}} + B_z\hat{\mathbf{k}}).(6\hat{\mathbf{i}} + 2\hat{\mathbf{j}} - 3\hat{\mathbf{k}}) = 0$$

$$6B_x + 2B_y - 3B_z = 0 \tag{d}$$

The solution is completed by considering Eq. (d) together with three moment summations about the coordinate axes through A, remembering to include the effect of B_y in these summations. Finally, force summations determine the components at A.

Also note that when the rod at B is vertical, as implied in the original data, the condition **B . r** = 0 results in $B_y = 0$, which is consistent with only horizontal support at B.

3-6.3 In the previous problem, the position of point E in the XY plane containing point B is to be changed so that the pull P to maintain equilibrium will be a minimum. Find the direction and magnitude of P under these conditions.

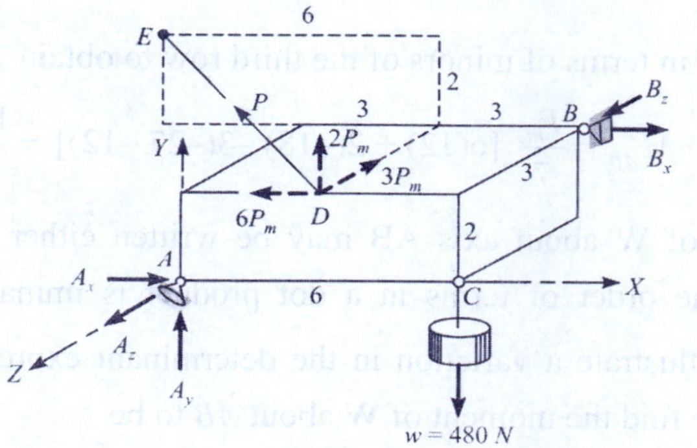

Figure 3-6.2

Solution

To compute the independent effect of **P** on the system, take a moment sum about the axis AB in Fig. 3-6.2, thereby eliminating the reactions at A and B. In order for **P** to be a minimum and yet produce a maximum moment effect about the axis AB, it must be perpendicular to the moment arm from D to this axis; i.e., **P** must be perpendicular to the plane ABD. The direction of **P**, therefore, must be parallel to that of the vector product $\overrightarrow{AB} \times \overrightarrow{AD}$ which, taking values from Fig. 3-6.2, is found to be

$$\overrightarrow{AB} \times \overrightarrow{AD} = \begin{vmatrix} 6 & 2 & -3 \\ 3 & 2 & 0 \\ \hat{i} & \hat{j} & \hat{k} \end{vmatrix} = 6\hat{\mathbf{i}} - 9\hat{\mathbf{j}} + 6\hat{\mathbf{k}}$$

The force P parallel to this direction, therefore, in terms of its multiplier, is

$$\mathbf{P} = P_m(6\hat{\mathbf{i}} - 9\hat{\mathbf{j}} + 6\hat{\mathbf{k}})$$

The moment of **P** about axis AB is $M_{AB}^{P} = M_{A}^{P}.\hat{n}_{AB} = (r_{AD} \times P).\hat{n}_{AB}$

where $\hat{n}_{AB} = \frac{1}{7}(6\hat{\mathbf{i}} + 2\hat{\mathbf{j}} - 3\hat{\mathbf{k}})$ is the unit vector along AB. This scalar triple product may be written as the following determinant in which the rows are the respective components of r_{AD}, P, and $\hat{n}_{AB}$ in the order they appear in the triple product:

$$M_{AB}^{P} = \frac{P_m}{7}\begin{vmatrix} 3 & 2 & 0 \\ 6 & -9 & 6 \\ 6 & 2 & -3 \end{vmatrix}$$

which we expand in terms of minors of the third row to obtain

$$M_{AB}^{P} = \frac{P_m}{7}\,[6(12) + 2(-18) - 3(-27 - 12)] = \frac{153}{7}P_m$$

The moment of W about axis AB may be written either as $\boldsymbol{M}_{A}^{W}.\hat{\boldsymbol{n}}_{AB}$ or as $\hat{\mathbf{n}}_{AB}.\boldsymbol{M}_{A}^{W}$ since the order of terms in a dot product is immaterial. Using the second form to illustrate a variation in the determinant expression of a scalar triple product, we find the moment of **W** about *AB* to be

$$M_{AB}^{W} = \hat{\mathbf{n}}_{AB}\cdot(\mathbf{r}_{AC}\times\mathbf{W}) = \frac{1}{7}\begin{vmatrix} 6 & 2 & -3 \\ 6 & 0 & 0 \\ 0 & -480 & 0 \end{vmatrix} = \frac{1}{7}(480)(18)$$

Finally, we now obtain

$$[\Sigma M_{AB} = 0] \qquad \frac{153}{7}P_m + \frac{(480)(18)}{7} = 0 \qquad P_m = -56.5\text{ N/m}$$

whence

$$\mathbf{P} = P_m(6\hat{\mathbf{i}} - 9\hat{\mathbf{j}} + 6\hat{\mathbf{k}}) = -338\hat{\mathbf{i}} + 507\hat{\mathbf{j}} - 338\hat{\mathbf{k}}$$

and

$$P = 698\text{ N}$$

As a supplement to this discussion, it should be noted that **P** in Prob. 3-6.2 could be found directly from $\Sigma M_{AB} = 0$. In that problem, the direction of **P** was specified so that we could write directly

$$\Sigma M_{AB} = \hat{\text{n}}_{AB}.\Sigma(\text{r}\times\mathbf{F}) = \frac{P_m}{7}\begin{vmatrix} 6 & 2 & -3 \\ 3 & 2 & 0 \\ -6 & 2 & -3 \end{vmatrix} + \frac{1}{7}\begin{vmatrix} 6 & 2 & -3 \\ 6 & 0 & 0 \\ 0 & -480 & 0 \end{vmatrix} = 0$$

Expanding in terms of elements of the third rows gives

$$P_m\,[-6(6) + 2(-9) - 3(12 - 6)] - 480(-18) = 0$$

from which

$$P_m = 120\text{ N/m} \qquad \textbf{Check}$$

Although this procedure checks the value of P_m of Prob. 3-6.2, it is considered preferable to use the method of the previous problem when all the reaction components are also required.

PROBLEM

3-6.4 A concurrent spatial system consists of three forces $\mathbf{P} = P_m(3\hat{\mathbf{i}} - 2\hat{\mathbf{j}} + 4\hat{\mathbf{k}})$, $\mathbf{Q} = Q_m(2\hat{\mathbf{i}} + \hat{\mathbf{j}} + 3\hat{\mathbf{k}})$, and $\mathbf{T} = T_m(4\hat{\mathbf{i}} + \hat{\mathbf{j}} - 2\hat{\mathbf{k}})$ held in equilibrium by a fourth force $\mathbf{R} = -400\,\hat{\mathbf{j}}$ N. Interpret and evaluate the expression $(\Sigma\mathbf{F}) \cdot (\mathbf{P} \times \mathbf{Q}) = 0$.

3-6.5 Solve for the value of Q_m in the previous problem by taking a summation of forces perpendicular to the plane defined by **T** and **P**.

$Q_m = 160$ ***Ans.***

3-6.6 The shear-leg derrick shown in Fig. P-3-6.6 supports a vertical load of 3000 N applied at *A*. Points *B*, *C*, and *D* are in the same horizontal plane and *A*, *O*, and *D* are in the *XY* plane. Determine the force in each member of the derrick.

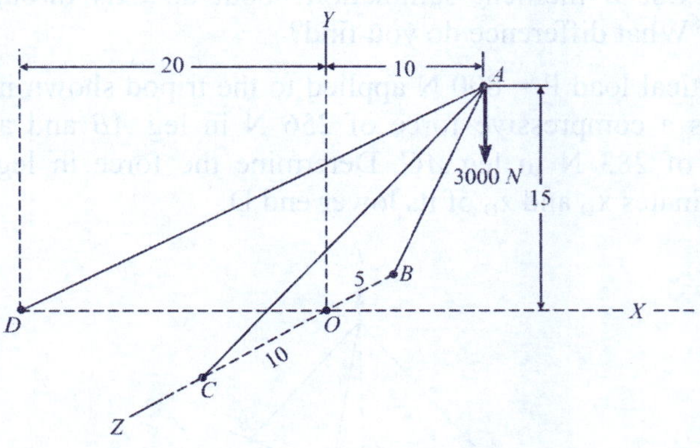

Figure P-3-6.6

3-6.7 A vertical load of 1100 N is supported by the three bars shown in Fig. P-3-6.7. Find the force in each bar. Points *C*, *O*, and *D* are in the *XZ* plane while B is 5 m above this plane.

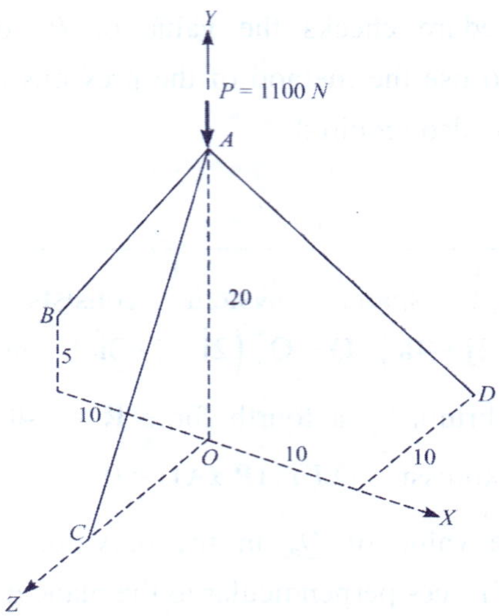

Figure P-3-6.7

$AB = 361$ m; $AC = 448$ N; $AD = 490$ N ***Ans.***

3.6.8 Referring to Prob. 3-6.7, compute the force in bar *AB* by each of the following methods:

(a) Use a force summation perpendicular to plane *ACD*.

(b) Use a moment summation about an axis through *C* and *D*. What difference do you find?

3.6.9 A vertical load P = 800 N applied to the tripod shown in Fig. P-3-6.9 causes a compressive force of 256 N in leg *AB* and a compressive force of 283 N in leg *AC*. Determine the force in leg *AD* and the coordinates x_D and z_D of its lower end D.

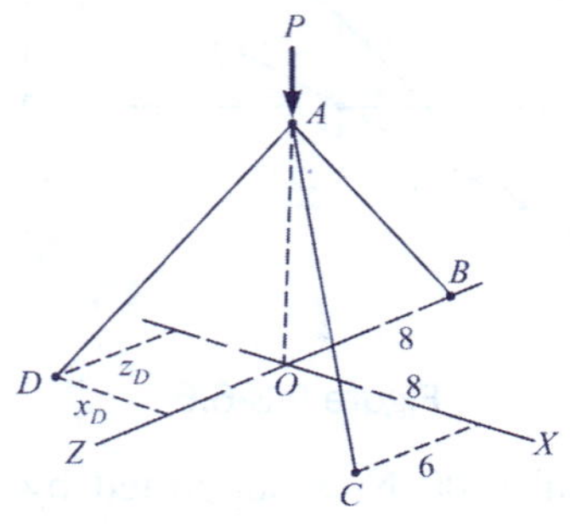

Figure P-3-6.9

AD = 433 N; $x_D = 4$ m; $z_D = 1$ m ***Ans.***

3-6.10 Determine the maximum safe vertical load W that can be supported by the tripod shown in Fig. P-3-6.10 without exceeding a compressive load of 2400 N in any member.

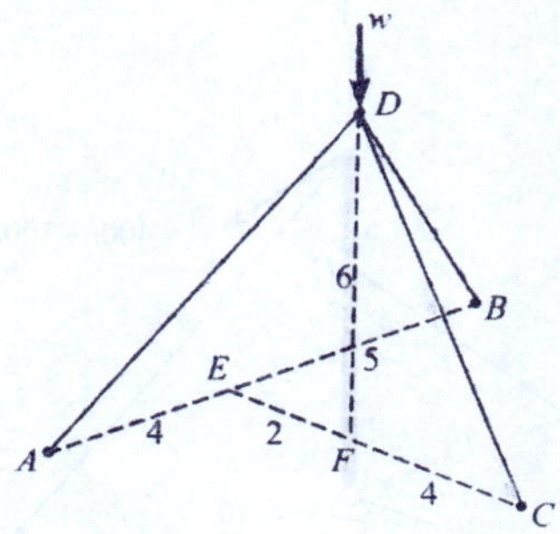

Figure P-3-6.10

$W = 5200$ N **Ans.**

3.6.11 The plate shown in Fig. P-3-6.11 carries a load of 1000 N applied at E and is supported in a horizontal position by three vertical cables attached at A, B, and C. Compute the tension in each cable.

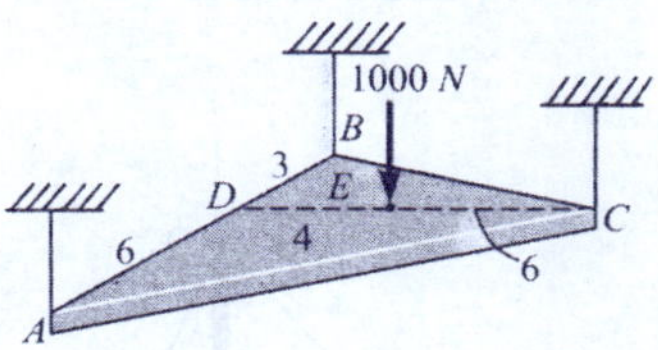

Figure P-3-6.11

3-6.12 An airplane with a total wingspread of 40 m is equipped with a tricycle landing gear. The two rear wheels of the landing gear have a 10 m tread and are 12 m behind the forward wheel. The airplane's weight of 5700 N acts 1 m in front of the rear wheels.

(a) Determine the downward force applied at one wingtip along a line 2 m ahead of the rear wheels that will tilt the airplane when it is standing on the runway.

(b) What upward force at the same place on the wingtip will tilt the airplane?

(a) 1650 N **Ans.**

3.6.13 A vertical mast *AE* is supported in a ball-and-socket joint at *A* and by cables *CE* and *BD* as shown in Fig. P-3-6.13. Find the components of the reaction at *A*.

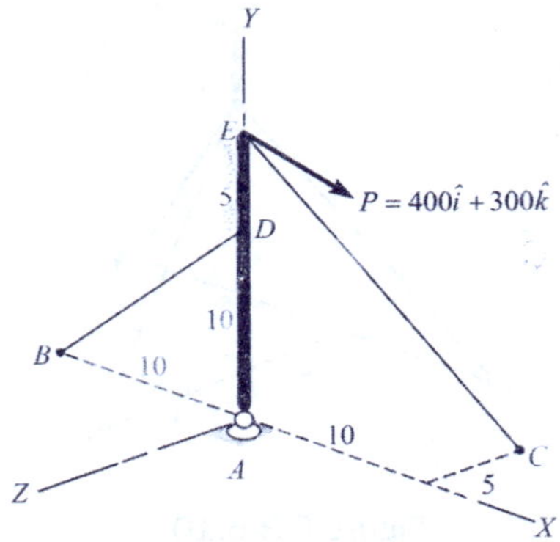

Figure P-3-6.13

3-6.14 A mast *AD* is supported in a ball-and-socket joint at *A* and by two guy wires *BD* and *CD* as shown in Fig. P-3-6.14. Find the magnitude of the force *A* if $\mathbf{F} = 600\,\hat{\mathbf{i}}$ N and P $400\,\hat{\mathbf{k}}$ N.

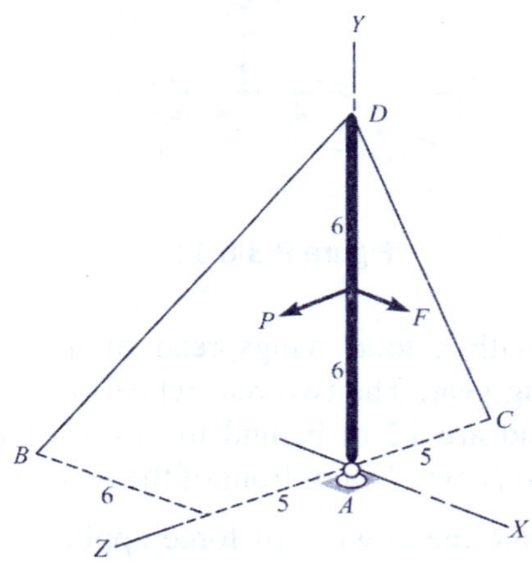

Figure P-3-6.14

A = 1718 N **Ans.**

3-6.15 The boom of the derrick in Fig. P-3-6.15 is rotated back 30° from the *XY* plane. Determine the forces in *BC* and *BD* and the components of the bearing reaction at the socket of the mast *AB*.

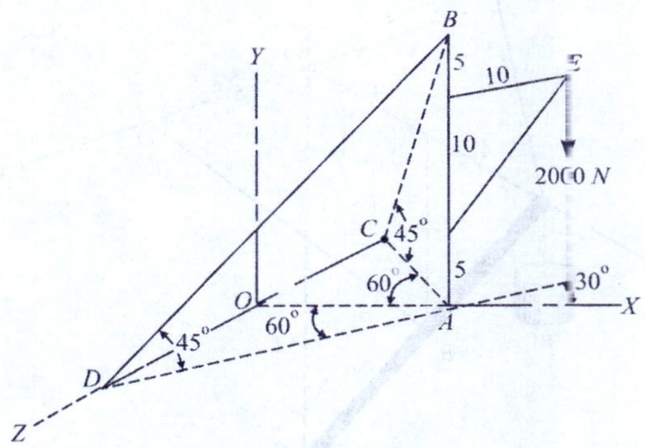

Figure P-3-6.15

3-6.16 In Prob. 3-6.15, determine the position of the boom which will cause a maximum force in *BC*. What is this maximum force?

The boom must be rotated 30° forward
from the *XY* plane; *BC* = 1630 N ***Ans.***

3-6.17 In Fig. P-3-6.17, the boom *AC* is supported in a ball-and-socket joint at *C* and by the cables *AD* and *BE*. If the weight *W* = 3300 N, find the force in the cables and the components of the reaction at *C*.

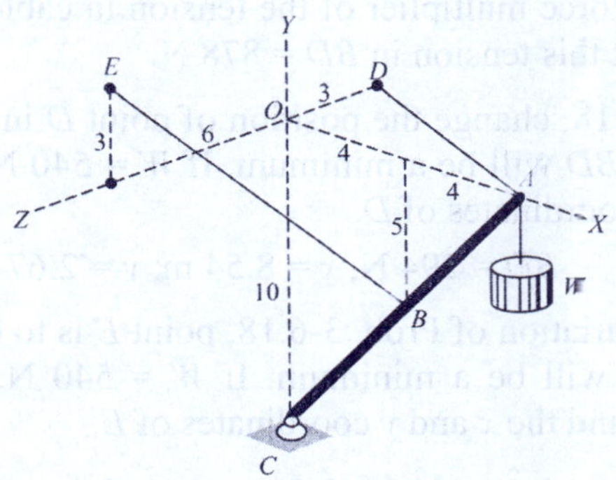

Figure P-3-6.17

3-6.18 The boom in Fig. P-3-6.18 supports a load W = 1400 N. Compute the forces in the cables *AE* and *BD* and also the components exerted by the ball-and-socket joint *C*.

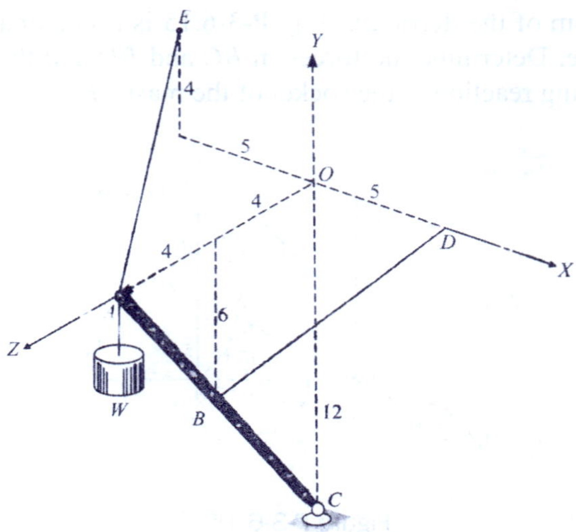

Figure P-3-6.18

$$AE = 512 \text{ N};\ BD = 878 \text{ N};$$
$$C_x = -250 \text{ N};\ C_y = 600 \text{ N};\ C_z = 800 \text{ N}$$ ***Ans.***

3-6.19 In Prob. 3-6.17, determine the force multiplier of the tension in cable *BE* directly by using only one equation of equilibrium. Then compute the force in the cable.

3-6.20 In Prob. 3-6.18, use only one equation of equilibrium to compute directly the force multiplier of the tension in cable *BD*. Use this result to verify that this tension in *BD* = 878 N.

3-6.21 In Prob. 3-6.18, change the position of point *D* in the *XY* plane so that the force in *BD* will be a minimum. If $W = 540$ N, find this force and the *x* and *y* coordinates of *D*.

$$BD = 294 \text{ N};\ x = 8.54 \text{ m};\ y = 2.67 \text{ m}$$ ***Ans.***

3-6.22 In another variation of Prob. 3-6.18, point *E* is to be shifted so that the force in *AE* will be a minimum. If $W = 540$ N, find this minimum force in *AE* and the *x* and *y* coordinates of *E*.

$$\overrightarrow{AE} = 144\hat{\mathbf{i}} + 60\hat{\mathbf{j}} - 90\hat{\mathbf{k}} \text{ N};\quad x = -12.8 \text{ m};\ y = 5.33\text{m}$$ ***Ans.***

3-6.23 The boom *AC* in Fig. P-36.23 rests in a ball-and-socket joint at *A* and is held in equilibrium by the cables *BE* and *CD*. The load applied at *C* is $\mathbf{P} = 900\hat{\mathbf{j}} - 300\hat{\mathbf{k}}$ N. Determine the coordinates of *D* in the *YZ* plane so that the force in *CD* will be a minimum. Neglect the weight of the boom and cables.

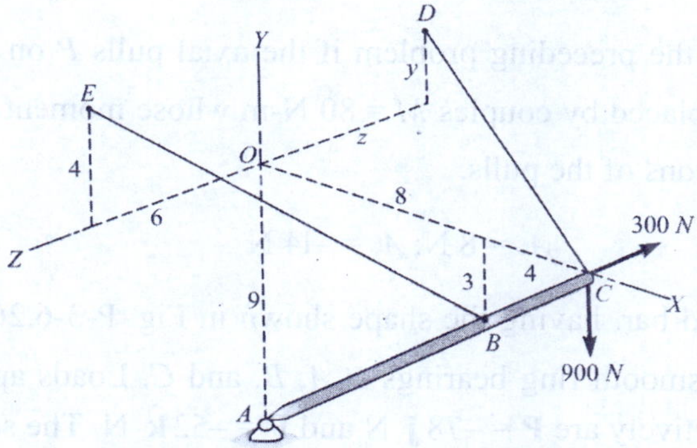

Figure P-3-6.23

$y = 16$ m; $z = -34.7$ m **Ans.**

3-6.24 Into a rigid hollow tube with two right angle bends, there is inserted a close-fitting flexible shaft to which axial pulls of $P = 40$ N are applied as shown in Fig. P-3-6.24. The assembly is supported by three smooth ring bearings at *A, B*, and *C* whose reactions are normal to the tube. Determine the reactions at the ring bearings.

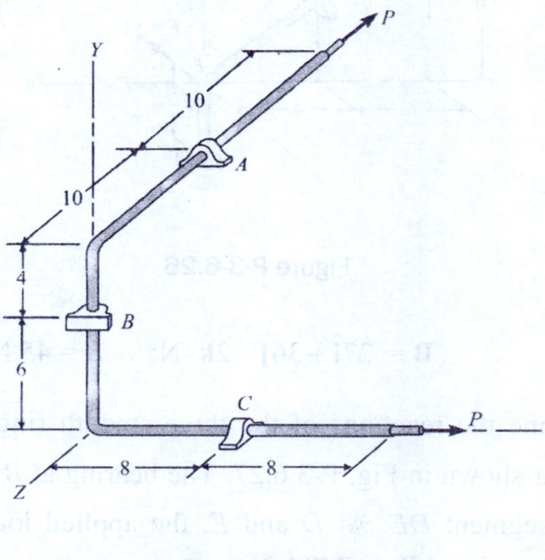

Figure P-3-6.24

$A_x = -56$ N; $A_y = 58$ N **Ans.**

3-6.25 Solve the preceding problem if the axial pulls P on the flexible shaft are replaced by couples $M = 80$ N-m whose moment vectors act in the directions of the pulls.

$A_x = 8$ N; $A_y = -14$ N **Ans.**

3-6.26 A rigid bar, having the shape shown in Fig. P-3-6.26, is supported by three smooth ring bearings at A, B, and C. Loads applied at D and E respectively are $\mathbf{P} = -78\,\hat{\mathbf{j}}$ N and $\mathbf{F} = -52\,\hat{\mathbf{k}}$ N. The segment DEC is in the same vertical plane. Determine the reaction components of the bearing at B upon the bar.

Hint: Since the total reaction at B must be normal to the bar, $\mathbf{B} \cdot \mathbf{r}_{DE} = 0$ Refer also to the supplementary note of problem 3.6.2.

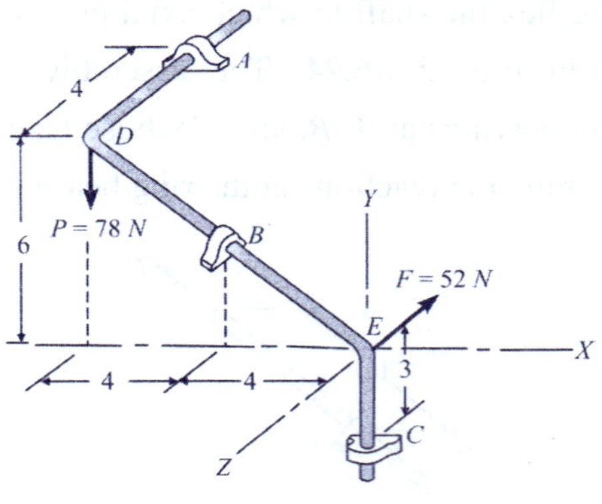

Figure P-3-6.26

$\mathbf{B} = 27\hat{\mathbf{i}} + 36\hat{\mathbf{j}} - 2\hat{\mathbf{k}}$ N; $B = 45$ N **Ans.**

3-6.27 Determine the reactions of the three smooth ring bearings upon the rigid bar shown in Fig. P-3.6.27. The bearing at B is at the mid length of the segment DE. At D and E, the applied loads are respectively $\mathbf{F} = 170\,\hat{\mathbf{k}}$ N and $\mathbf{P} = -340\,\hat{\mathbf{j}}$ N.

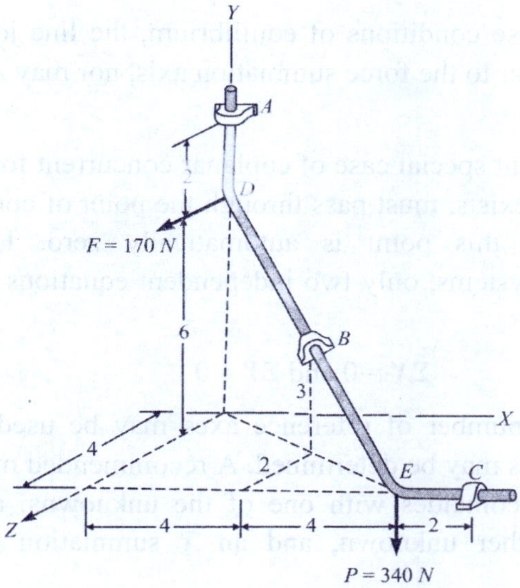

Figure P-3-6.27

$\mathbf{A} = 296\hat{\mathbf{i}} - 198\hat{\mathbf{k}}$ N; $\mathbf{B} = 296\hat{\mathbf{i}} + 360\hat{\mathbf{j}} - 52\hat{\mathbf{k}}$ N; $\mathbf{C} = -20\hat{\mathbf{j}} + 80\hat{\mathbf{k}}$ N ***Ans.***

SUMMARY

Equilibrium is the term used to express the condition existing when the force system acting on a body has a resultant force-couple system equal to zero. For planar force systems confined to the *XY* plane, the resultant force-couple system becomes zero if

$$\Sigma X = 0; \quad \Sigma Y = 0; \quad \Sigma M = 0 \tag{3-4.1}$$

where the moment summation is understood to be about a *Z* axis through any convenient point in the *XY* plane. These equations represent three independent equations of equilibrium from which no more than three unknowns can be determined.

Rather than use these equations routinely, it is preferable to replace force summations by moment summations about the intersection of two of the unknowns, thereby eliminating them from the equation and permitting a direct determination of the third unknown quantity. Any one of the following sets of equilibrium equations may be used; the choice depends on the specific problem to be solved.

$$\left.\begin{matrix}\Sigma X = 0\\ \Sigma Y = 0\\ \Sigma M = 0\end{matrix}\right\} \text{ or } \left.\begin{matrix}\Sigma X = 0\\ \Sigma M_A = 0\\ \Sigma M_B = 0\end{matrix}\right\} \text{ or } \left.\begin{matrix}\Sigma M_A = 0\\ \Sigma M_B = 0\\ \Sigma M_C = 0\end{matrix}\right\} \tag{3-5.2}$$

In applying these conditions of equilibrium, the line joining *A* and *B* must not be perpendicular to the force summation axis, nor may *A*, *B*, and *C* lie on the same straight line.

For the important special case of coplanar concurrent force systems in which the resultant, if it exists, must pass through the point of concurrency, a moment summation about this point is automatically zero. Hence, for coplanar concurrent force systems, only two independent equations of equilibrium exist, namely,

$$\Sigma X = 0 \text{ and } \Sigma Y = 0$$

Although any number of reference axes may be used, no more than two unknown quantities may be determined. A recommended method is to select the *X* axis so that it coincides with one of the unknowns; a *Y* summation then determines the other unknown, and an *X* summation determines the first unknown.

When the coplanar concurrent forces are specified by their slopes instead of their direction angles, a convenient variation of force summations is to use

$$\Sigma M_A = 0 \text{ and } \Sigma M_B = 0$$

Select the moment center *A* on the action line of one of the unknown forces, thereby permitting a direct evaluation of the other unknown force. Similarly, the first unknown is determined from a moment summation about a center *B* located on the action line of the second unknown. Note carefully that the line joining *A* and *B* must not pass through the point of concurrency of the system. Force summations may be used to check the results.

Equilibrium of concurrent spatial systems of forces is determined by applying either set of the following equations or a combination of them. However, since there are only three independent conditions of equilibrium, no more than three unknown quantities may be determined. (3-3.2)

$$\left.\begin{matrix}\Sigma X = 0\\ \Sigma Y = 0\\ \Sigma Z = 0\end{matrix}\right\} \text{ and } \left.\begin{matrix}\Sigma M_x = 0\\ \Sigma M_y = 0\\ \Sigma M_z = 0\end{matrix}\right\}$$

To create equilibrium of nonconcurrent spatial systems of forces, both sets of the above equations must be satisfied. Hence, a general spatial system in equilibrium may be solved for as many as, but no more than, six unknown quantities.

CHAPTER 4

ANALYSIS OF STRUCTURES

4-1 INTRODUCTION

The preceding chapter on equilibrium has discussed the basic principles and techniques of statical analysis. In this chapter, we extend these concepts to the determination of how the loads applied to a structure are distributed throughout the structure. In subsequent courses we shall learn how to determine the dimensions of the various parts of a structure so they can safely resist these loads.

Although there are many kinds of structures, we limit the discussion here to pin-connected types, which are assemblages of bars and beams fastened together by smooth bolts that act as hinges. As a preliminary to their design, our objective is to determine the forces acting in the bars and upon the pins or hinges of the beams. To do so, we must compose supplementary problems involving these forces by means of properly constructed free-body diagrams. The construction of these free-body diagrams requires an understanding of how the structure was assembled so that we can intelligently apply assumptions (when necessary) that simplify the analysis of the structure. We feel it is proper to remind you that engineering is the art of obtaining answers to problems that frequently cannot be solved exactly; what we must do is to create a model of the real situation that we can solve, recognizing that the closer the model resembles the real situation, the more reliable the answers will be.

After we have learned how structures are constructed, we shall see that the forces acting in or upon their various members create concurrent or nonconcurrent systems of forces in equilibrium. Depending on the situation, sometimes we consider the free-body diagrams of the joints that connect the members together; at other times we consider the free-body diagrams of the members. The unknown forces in these free-body diagrams are readily determined by applying the methods and techniques studied in the preceding chapter.

4-2 CONSTRUCTION OF TRUSSES

A truss is a structure composed of members fastened together in such a way as to resist change in shape; it is a rigid structure.[1] The purpose of a truss is to support a larger load or span a greater distance than any individual member from which the truss may be built. To make a structure rigid, its members must be fastened together in such a way as to prevent any movement between them. A pin-connected structure meeting this condition consists of three bars forming the triangular truss shown in Fig. 4-2.1. A pin-connected structure composed of four bars is not rigid but is a mechanism (see Fig. 4-2.2) in which a specified motion of *AB* produces a predictable controlled motion of *BC* and *CD*; it is the basis of all machines.

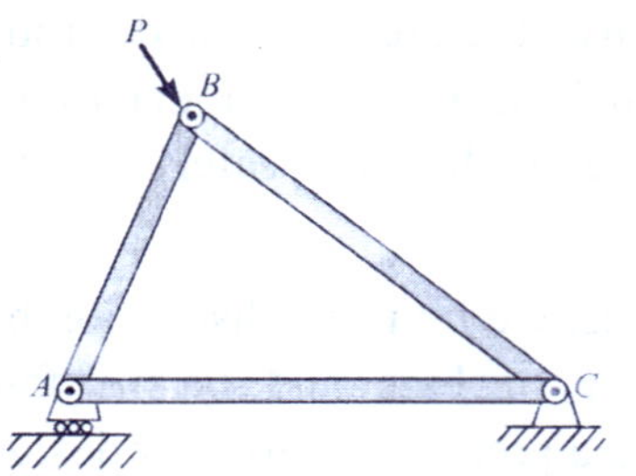

Figure 4-2.1 Rigid truss.

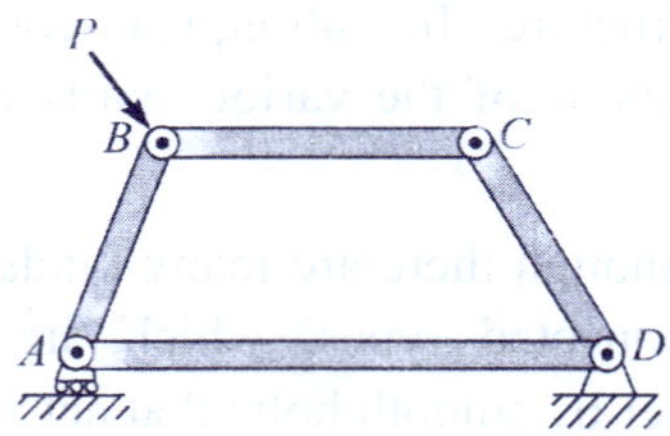

Figure 4-2.2 Mechanism.

In Fig. 4-2.3, adding a bar *BD* to the four-link mechanism of Fig. 4-2.2 restores rigidity, while adding two bars *BD* and *AC* as in Fig. 4-2.4 produces a structure which is over-rigid and statically indeterminate. However, removing

---

[1]The term *rigid* is used in the sense of having no deformation. Actual members are, of course, subject to deformation which can be neglected if very small compared with the dimensions of the truss. There are, however, some types of trusses that are statically indeterminate in which the deformations must be taken into account; they must be solved by considering the deformation of the members themselves. These types are beyond the range of this book; for their analysis, the student is referred to Timoshenko and Young, *Theory of Structures,* McGraw-Hill; J. S. Kinney, *Indeterminate Structural Analysis*, Addison-Wesley; or J. R. Benjamin, *Statically Indeterminate Structures,* McGraw-Hill.

one of the bars from this over- rigid structure produces the trusses shown in (a) and (b) of Fig. 4-2.5, while in part (c) two bars are removed but stability restored by a hinge at A instead of a roller support. After we discuss the formation of simple trusses and the assumptions used in solving them, we shall develop a simple criterion to determine whether a truss is stable, unstable, or over-rigid.

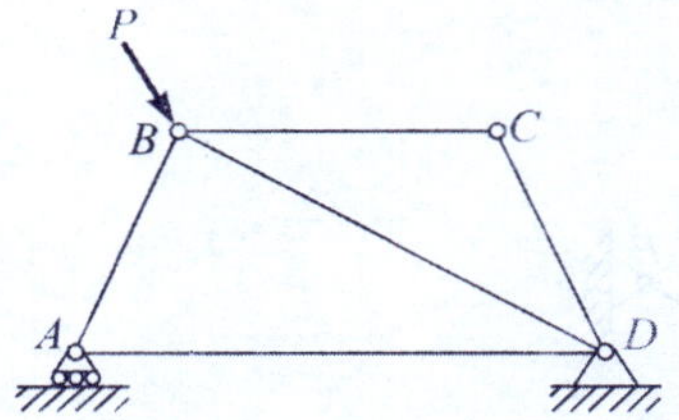

Figure 4-2.3 Just rigid truss.

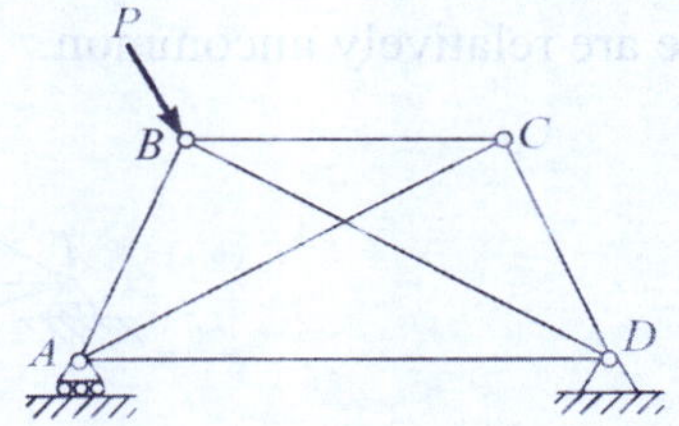

Figure 4-2.4 Over-rigid truss.

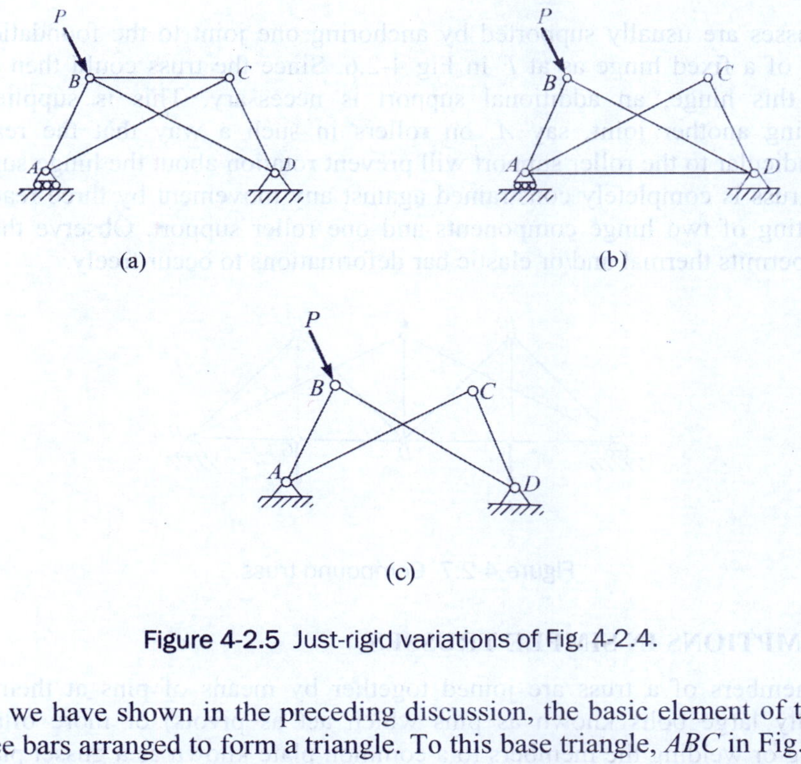

Figure 4-2.5 Just-rigid variations of Fig. 4-2.4.

As we have shown in the preceding discussion, the basic element of trusses is three bars arranged to form a triangle. To this base triangle, ABC in Fig. 4-2.6 two more bars may be added to locate a joint D which is rigid relative to the other joints. Since A and D are now fixed relative to each other, two more bars may be added at these joints to fasten together at E, after which two additional bars determine joint *F*. Thus, by continuing to add bars at joints rigid to one another, additional joints determine and a truss consisting of many members may be constructed. Trusses built up in this manner are known as *simple trusses*. Trusses consisting of two simple trusses joined by three nonintersecting bars or by a common pin and a bar are known as *compound trusses*. See Fig. 4-2.7 consisting of two simple trusses *ABCDE* and *DFGH* joined by a common pin *D* and the bar *EH*. Any other combination produces a *complex truss*, but these are relatively uncommon.

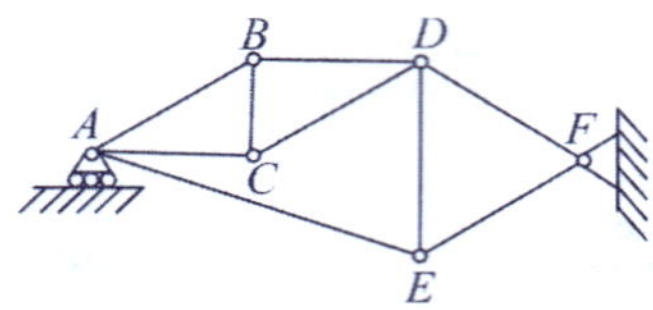

Figure 4-2.6 Formation of a sample truss.

Trusses are usually supported by anchoring one joint to the foundation by means of a fixed hinge as at *F* in Fig 4-2.6. Since the truss could then rotate about this hinge, an additional support is necessary. This is supplied by mounting another joint, say *A*, on rollers in such a way that the reaction perpendicular to the roller support will prevent rotation about the hinge support. Then truss is completely constrained against any movement by three reactions consisting of two hinge components and one roller support. Observe that the roller permits thermal and/or elastic bar deformations to occur freely.

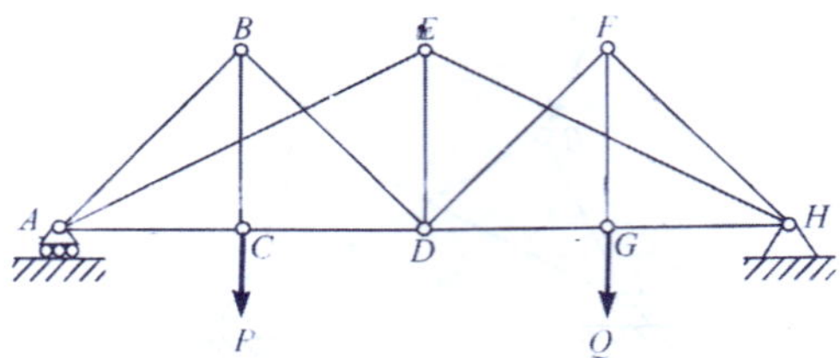

Figure 4-2.7 Compound truss.

ASSUMPTIONS IN SIMPLE TRUSSES

The members of a truss are joined together by means of pins at their ends (actually large bolts known as pins which act as pivots) or more often by riveting or welding the members to a common plate known as a gusset plate. If due care is taken to assemble the bars so that their centerlines intersect in a common point at each joint, experiments have shown that the forces in the bars may be calculated as if they were pin-connected, even though their ends are actually riveted or welded to a gusset plate.

Furthermore, trusses are designed so that all applied loads act at the ends of the members. This construction is illustrated in Fig. 4-2.8 for a simple bridge. Loads are carried by the flooring which is supported by stringers. The stringers in turn are supported by crossbeams which are attached to the joints. The net effect is that all loads are applied at the joints of the truss. Further, it is usually assumed that the weight of the truss members is negligible in comparison with the applied loads. In case their weight is not negligible, calculations are made on the assumption that the weight of each member is divided equally into two forces which act vertically downward on the end pins of the member.

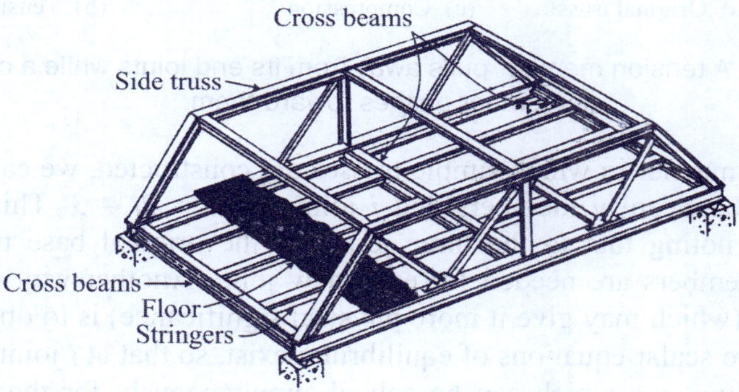

Figure 4-2.8 Simple bridge.

Since the space position of a truss member is fixed by the rigid construction of the truss and the loads are applied only at the joints, its free-body diagram would show it acted upon by only two equal, oppositely directed forces exerted by the pins at its ends. To satisfy equilibrium, these forces must be axially directed along the member as we learned in the discussion concerning bar *AB* in Fig. 3-4.2. Such members held in equilibrium by only two forces are called *two-force members*. The fact that the internal force in a two-force member is axial is important because it determines the *action line* of the force although its magnitude may be unknown.

Members which are stretched are said to be in tension while those that are shortened are said to be in compression. In a typical truss, Fig. 4-2.9a, if the member *CE* were in tension, isolating it from the truss would give the free-body diagram shown in part (b), and its effect on the joints of the truss would appear as forces T which pull away from these joints. Another member, *BD*, assumed to be in compression, would be isolated as shown so that its effect on its end joints would be represented by forces C which push toward the joints. From these diagrams, we deduce the following rule: A member in tension causes forces which *pull* away from its end joints whereas a member in compression causes forces which *push* toward its end joints.

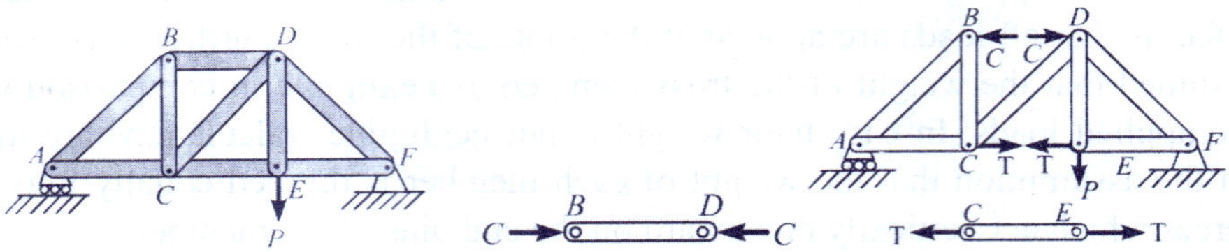

(a) Original truss (c) Compression (b) Tension

Figure 4-2.9 A tension member pulls away from its end joints while a compression member pushes toward them.

From the manner in which simple trusses are constructed, we can deuce that the relation between m members and j joints is $m = 2j - 3$. This relation is obtained by noting that to the three joints of the original base triangle, two additional members are needed for each new joint. Another way of obtaining this relation (which may give it more physical significance) is to observe that at each joint two scalar equations of equilibrium exist, so that at j joints there are a total of $2j$ equations which can be solved simultaneously for the forces in m members plus three external reactions; i.e.,

$$2j = m + 3 \tag{4-2.1}$$

Usually the three external reactions are found by applying three equations of equilibrium to a free-body diagram of the entire truss. As we shall see, solving first for the external reactions greatly simplifies the solution, since then there is no need to solve $2j$ sets of equations simultaneously.

If $2j < m + 3$, there are less equations available than there are unknowns; i.e., there are more members than there are independent equations and the truss is statically indeterminate. If $2j > m + 3$, there are too few members; i.e., the truss is unstable. The student should be warned that although $2j = m + 3$ is a necessary condition for stability, it is not a sufficient condition since the requisite number of members could be connected in a manner that would not produce a stable structure. See Fig. 4-2.10. However, this will not occur with simple trusses built up with triangular elements, or with compound trusses consisting of two simple trusses joined either by three nonparallel bars, or by a common hinge and a bar.

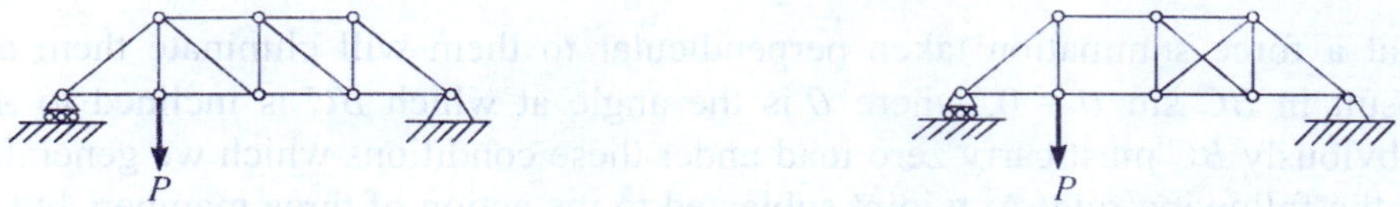

(a) Unstable (b) Stable

Figure 4-2.10

A more general form of Eq. (4-2.1) applicable to mechanisms and frames as well as trusses is

$$2j = m + c \tag{4-2.2}$$

where c is the number of constraint supports. Apply this equation to Fig. 4-2.5c to confirm stability.

4-3 METHOD OF JOINTS

The assumption that all members of a truss are two-force members transmitting axial loads means that the free-body diagram of any joint is a concurrent force system in equilibrium. Since only two independent equations of equilibrium can be written for a plane concurrent system, we start the analysis (after determining the reactions) with a joint acted on by only two members. The unknown force in these members are determined by using any of the methods previously discussed Illus. Probs. 3-4.1or 3-5.1.

Once the forces at a starting joint have been determined, their effects on adjacent joints are known. Continue by considering the next joint at which only two unknown forces act, and so on successively until the unknown force in each member has been found. This procedure is explicitly applied in the illustrative problems.

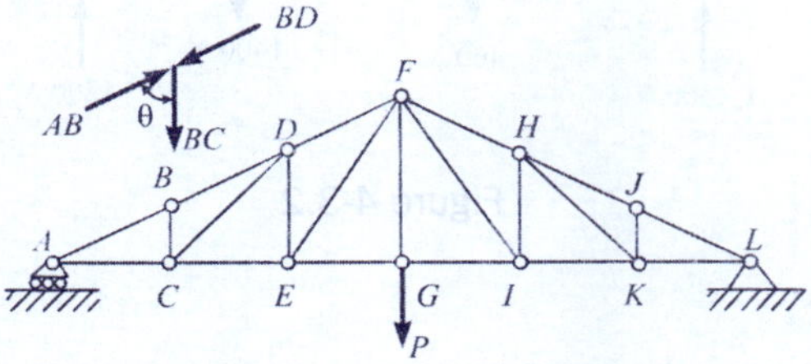

Figure 4-3.1

Occasionally, we encounter loading conditions illustrated by the truss shown in Fig. 4-3.1. Joint B of this truss is subjected to the forces exerted by the three members AB, BC, and BD. Members AB and BD are in the same straight line, so that a force summation taken perpendicular to them will eliminate them and result in $BC \sin \theta = 0$, where θ is the angle at which BC is inclined to AB. Obviously BC must carry zero load under these conditions which we generalize in the following rule: At a joint subjected to the action of three members but no other load, if two of the members are collinear, the force in the third member must be zero.

With $BC = 0$, joint C is subject to this rule, thereby giving $CD = 0$, whence at joint D we find $DE = 0$ and then at joint E, $EF = 0$. Similarly, starting at J, $JK = 0$, and thence proceeding from K to H to I, we also find HK, HI, and FI to carry zero load.

The obvious question is why have such members if they carry no load? The answer is that they are not useless; they would carry load under a different loading such as vertical loads also applied at B and J. They also are necessary to support the weight of the truss and the secondary loads induced by the tendency of long compression members to buckle.

ILLUSTRATIVE PROBLEM

4.3.1 For the pin-connected truss, loaded and supported as shown in Fig. 4-3.2, determine the force in each member.

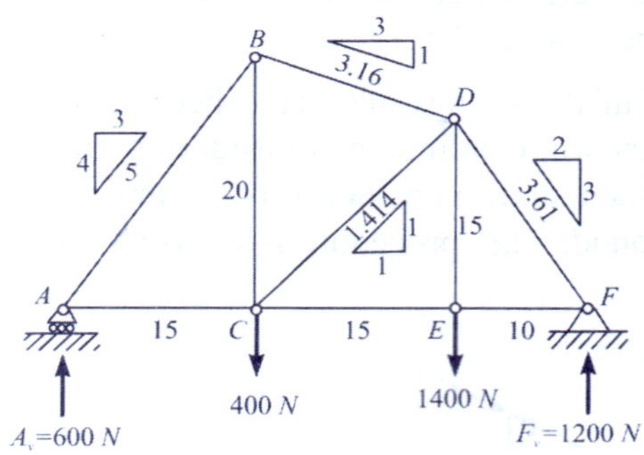

Figure 4-3.2

Solution

We start by determining the reactions, using moment summations about F and A to obtain respectively $A_v = 600$ N and $F_v = 1200$ N. A horizontal force summation shows F_h to be zero. Using the truss dimensions, compute and draw the slope triangles adjacent to each sloping member. These slope triangles will establish the relations among the force and its components in the corresponding member.

Start with joint A which, as shown in Fig. 4-3.2, has only two unmarked members (AB and AC) acting upon it. The FBD of joint A is shown in Fig. 4-3.3 which also shows for later discussion the FBD's of all other joints in their original spatial position.

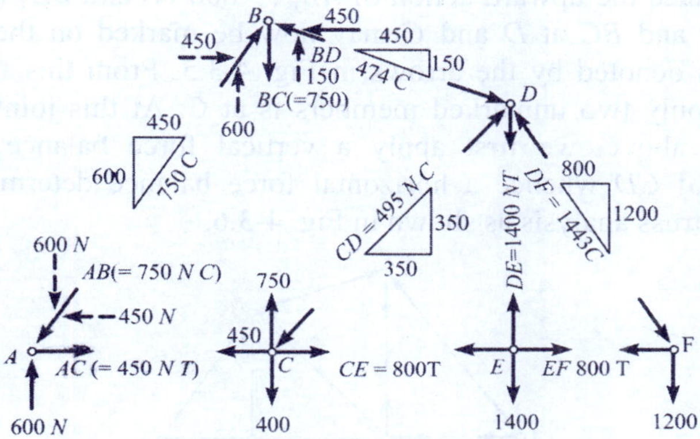

Figure 4-3.3 Free-body diagrams of all joints.

For joint A to be in equilibrium, it is evident that AB must be directed toward pin A (thereby denoting compression) in order to have a downward vertical component of 600 N that balances the upward reaction. From the proportional relations between AB and its slope, we find AB_h = 450 N and AB = 750 N (C). Horizontal force balance requires that AC = 450 N pulling away from pin A and therefore denoting tension. The action of members AB and AC, indicating respectively compression and tension, on their end pins may now be shown by the arrows in Fig. 4-3.4. (These arrows should be inserted on the original truss diagram, but to indicate the marked and unmarked members more clearly, the truss is here redrawn, but to a reduced size to conserve space.)

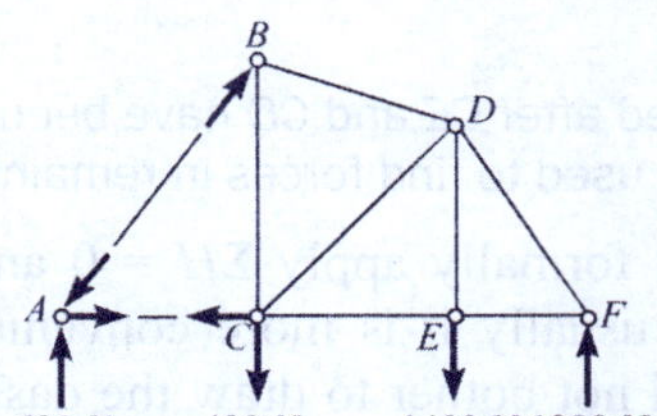

Figure 4-3.4 Truss marked to show effect of members AB and AC on their end pins.

From Fig. 4-3.4, the next joint at which no more than two unmarked members act is seen to be B. On its FBD which is part of Fig. 4-3.3, the components of the now known value of AB act as shown, opposite to the action of *AB* on *A*. Clearly, horizontal balance requires that the horizontal component of *BD* be 450 N directed leftward on *B*. This requires *BD* to push toward joint *B* (i.e., be in compression) and the slope relations give $BD_v = 150$ N and $BD = 474$ N *C*. Vertical force balance at *B* is achieved by *BC* pulling down with a value of 750 N to balance the upward action of AB_v (= 600 N) and BD_v (= 150 N). The action of *BD* and *BC* at *D* and *C* may now be marked on the original truss diagram as is denoted by the arrows in Fig. 4-3.5. From this figure, the next joint having only two unmarked members is at *C*. At this joint in Fig. 4-3.3, reasoning as above, we first apply a vertical force balance to obtain the components of *CD* whence a horizontal force balance determines *CE*. Now complete the truss analysis as shown in Fig. 4-3.6.

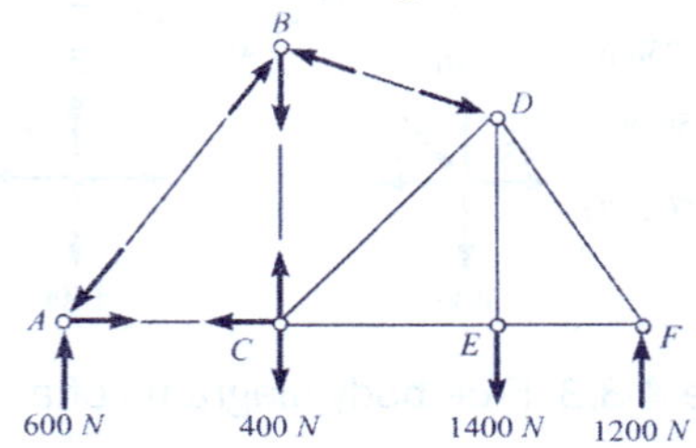

Figure 4-3.5 After marking effect of members *BD* and *BC* on their end pins, the next joint having only two unmarked members is joint *C*.

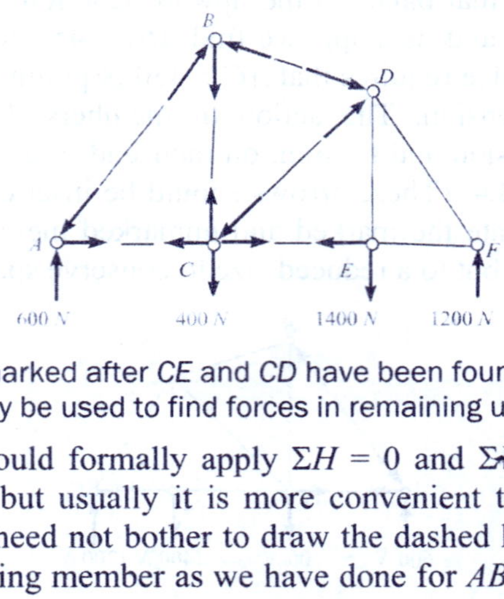

Figure 4-3.6 Truss marked after *CE* and *CD* have been found. Any combination of joints *D, F*, or *F* may be used to find forces in remaining unmarked members.

If we wish, we could formally apply $\Sigma H = 0$ and $\Sigma V = 0$ at each joint to obtain these results, but usually it is more convenient to reason as discussed above. Actually, we need not bother to draw the dashed horizontal and vertical components in a sloping member as we have done for *AB* and *BD* in Fig. 4-3.3. Instead, indicate them on a force triangle drawn between the end joints of the member on the original truss diagram, corresponding to those shown in Fig. 4-3.3 for members *AB*, *BD*, *CD*, and *DF*. Obviously each such force triangle is similar to its corresponding slope triangle. Omit directional arrows of the components on these force triangles, since their direction at each joint is evident from the arrow denoting the action of the member upon its end joints.

As a final observation, it will usually be faster to carry out the above analysis directly upon an enlarged drawing of the truss rather than draw a separate FBD of each joint in its proper spatial position as we did for explanatory purposes in Fig. 4-3.3.

4-3.2 Determine the force in each member of the pin-connected truss loaded and supported as shown in Fig. 4-3.7.

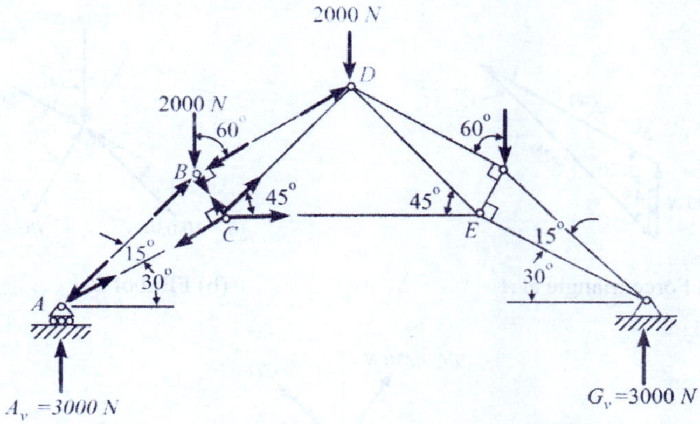

Figure 4-3.7

Solution

The given truss is symmetrical and also symmetrically loaded so that the forces need be found in only one-half of it. However, since the direction of each member is defined here by angles instead of its slope, the equilibrium of each joint is best determined here by using force triangles or the method of rotated axes rather than horizontal and vertical force summations as we found convenient in the preceding problem.

From symmetry, the reaction equals 3000 N at A which is the first joint having only two unmarked members (AB and AC) acting upon it. For the three forces in equilibrium at A, we draw the force triangle shown at (a) of Fig. 4-3.8. Applying the sine law, we find AC = 8190 N (T) and AB = 10,030 N (C). Tensile or compressive action is determined by marking the action of these members upon joint A to correspond with the directions of these forces in the force triangle. Clearly, AC pulls away from A and hence is in tension while AB pushes toward A and is in compression. The effect of AB and AC on B and C is of course opposite to their effect on A.

It is suggested that you make a sketch of the truss and mark on it the action of each member on its end joints as the force in it is found. Then you will see that the next joint having only two unmarked members is B. Here four forces act so that a force triangle cannot be used; instead we use the concept of rotated axes as shown in the FBD of B at (b) of Fig. 4-3.8. Assuming BD and BC to be in compression and therefore acting toward B, we rotate the X axis to coincide with the unknown force BD, whence we obtain

$$[\Sigma Y = 0] \qquad BC + 10{,}030 \sin 15^\circ - 2000 \sin 60^\circ = 0 \qquad BC = -870 \text{ N}$$

$$[\Sigma X = 0] \qquad 10{,}030 \cos 15^\circ - 2000 \cos 60^\circ - BD = 0 \qquad BD = 8690 \text{ N}$$

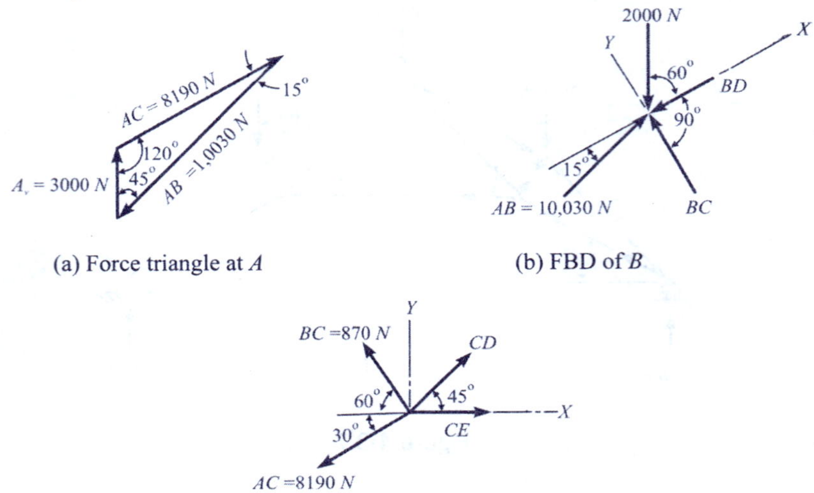

(a) Force triangle at A

(b) FBD of B

(b) FBD of C

Figure 4-3.8

The positive result for BD confirms that it is in compression as assumed, whereas the negative result for BC means that we assumed the wrong direction; BC therefore pulls away from B and is in tension. Now we mark the correct action of BD and BC upon their end pins and proceed to consider the equilibrium of the next joint C at which there are only two unmarked members.

Assume both CD and CE to be in tension. The FBD of joint C can now be drawn as in Fig. 4-3.8c. Note that we show BC in its correct tensile action. The X axis coincides with CE, and we have

$$[\Sigma Y = 0] \qquad CD \sin 45^\circ + 870 \sin 60^\circ - 8190 \sin 30^\circ = 0$$

$$CD = 4720 \text{ N (T)}$$

$$[\Sigma X = 0] \qquad CE + (CD = 4720) \cos 45^\circ - 870 \cos 60^\circ - 8190 \cos 30^\circ = 0$$

$$CE = 4185 \text{ N (T)}$$

The positive values obtained for *CD* and *CE* confirm the assumption that these members are in tension.

As mentioned previously, the loading and the truss are symmetrical so that the forces in all members are now determined. If the truss or loading were not symmetrical, however, the solution would be continued by proceeding to the next unmarked pin. This pin is *D*, but another pin having only two unmarked members acting upon it is pin *G*. It is preferable to avoid pin *D*, start anew from pin *C* and determine the forces in *FG* and *EG*. After the action of *FG* and *EG* upon their end pins is marked on the truss, the next pin to be selected for analysis will be pin *F*. From the FBD of *F*, the forces *DF* and *EF* can be found and their action marked on the truss. Next we consider the FBD of pin *E* to determine the forces in *DE* and *CE*. The force in *CE* will then have been determined independently from the FBD of pins *C* and *E*. A check on the accuracy of the computations is assured if the force in *CE* as found from pin *C* agrees with that found from pin *E*.

PROBLEMS

4-3.3 The members of the truss shown in Fig. P-4-3.3 are pin connected only at the lettered joints. When the truss carries the vertical loads *P* and *Q* shown, determine by inspection which members transmit no force.

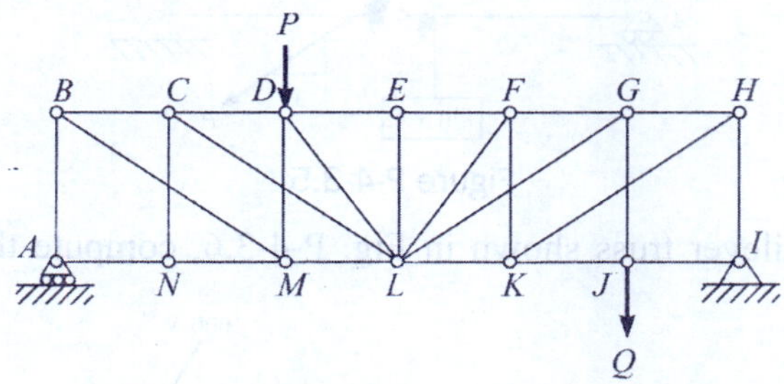

Figure P-4-3.3

AN, NM, KJ, JI, CN, CL, EL ***Ans.***

4-3.4 Determine the force in each bar of the truss shown in Fig. P-4-3.4.

Hint: First determine which bars carry no force.

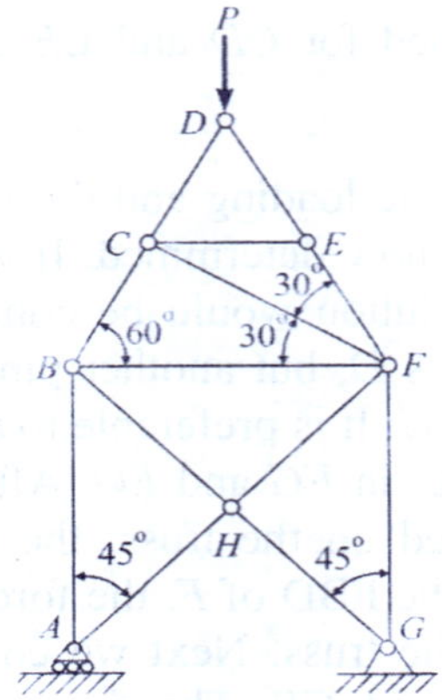

Figure P-4-3.4

$CD = DE = 0.577P$ (C); $BF = 0.289P$ (T) ***Ans.***

4-3.5 Determine the force in each bar of the truss shown in Fig. P-4-3.5 caused by lifting the 120 N load at a constant velocity of 8 m per sec. What change in these forces, if any, results from placing the roller support at D and the hinge support at A?

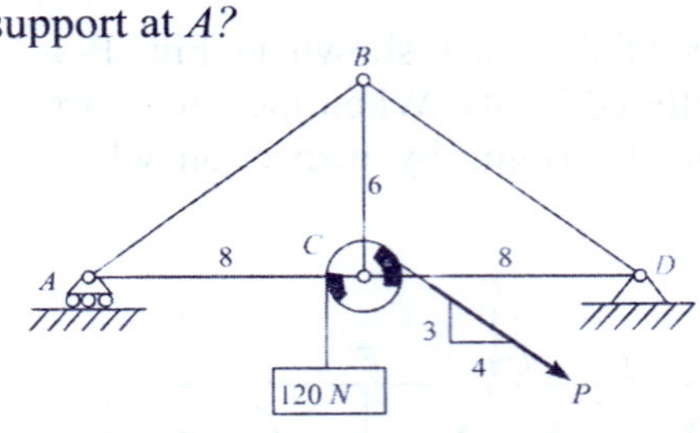

Figure P-4-3.5

4-3.6 In the cantilever truss shown in Fig. P-4-3.6, compute the force in each member.

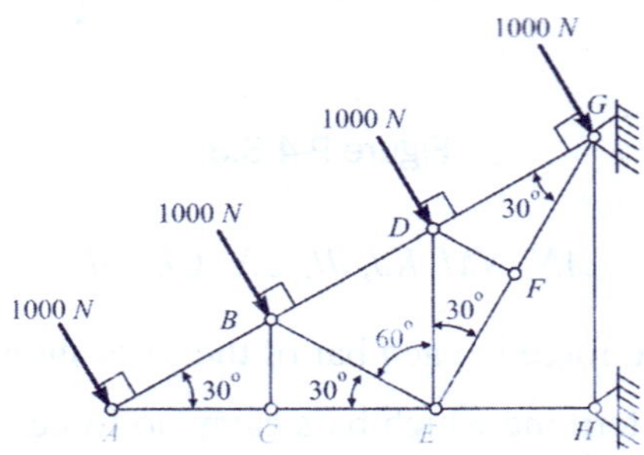

Figure P-4-3.6

$AB = 1732$ N (T); $BE = DE = 1154$ N (C) ***Ans.***

4-3.7 Determine the force in members *AB, BD, BE,* and *DE* of the Howe roof truss shown in Fig. P-4-3.7.

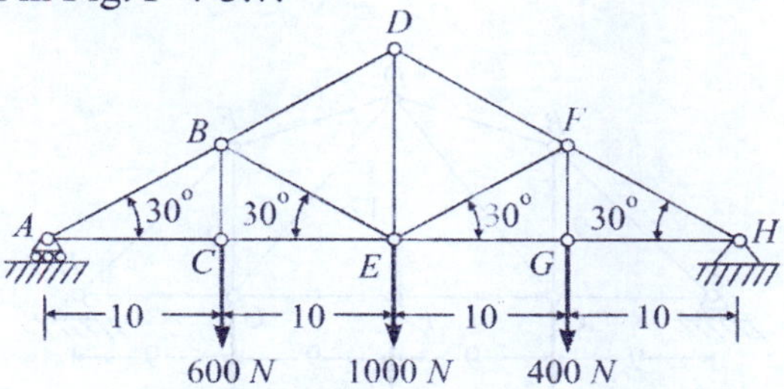

Figure P-4-3.7

4-3.8 Compute the force in each member of the truss shown in Fig. P4-3.8. If the loads at *B* and *D* are shifted vertically downward to add to the loads at *C* and *E*, which members, if any, would undergo change in internal force?

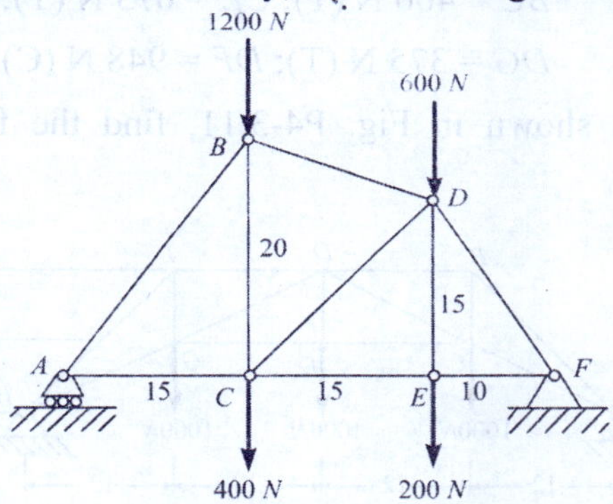

Figure P-4-3.8

AB = 1500 N (C); *BC* = 300 N (T); *BD* = 948 N (C);
CD = 141.4 N (T); *CE* = 800 N (T); *DF* = 1443 N (C) ***Ans.***

4-3.9 Determine the force in each member of the crane truss shown in Fig. P.4-3.9.

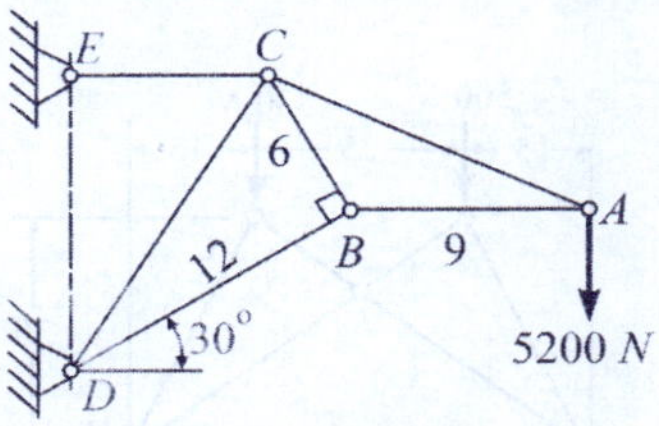

Figure P-4-3.9

AB = 12,000 N (C); *AC*=13,100 N (T);
BC = 6000 N (C); *BD* = 10,400 N (C); *CD* = 0 ***Ans.***

4-3.10 Determine the force in each member of the truss shown in Fig. P-4-3.10.

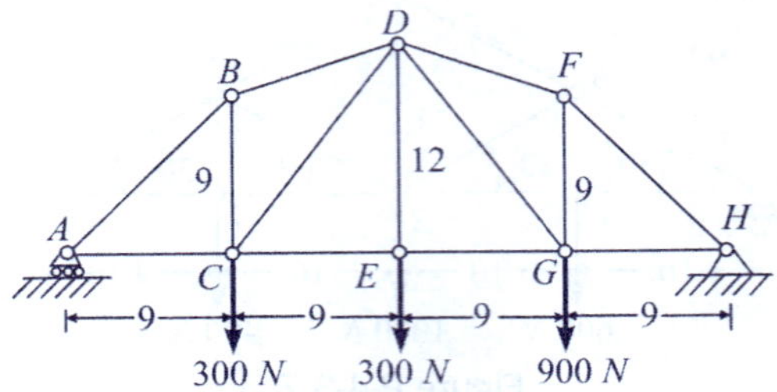

Figure P-4-3.10

$BC = 400$ N (T); $CE = 675$ N (T);

$DG = 375$ N (T); $DF = 948$ N (C) ***Ans.***

4-3.11 For the truss shown in Fig. P4-3.11, find the force acting in each member.

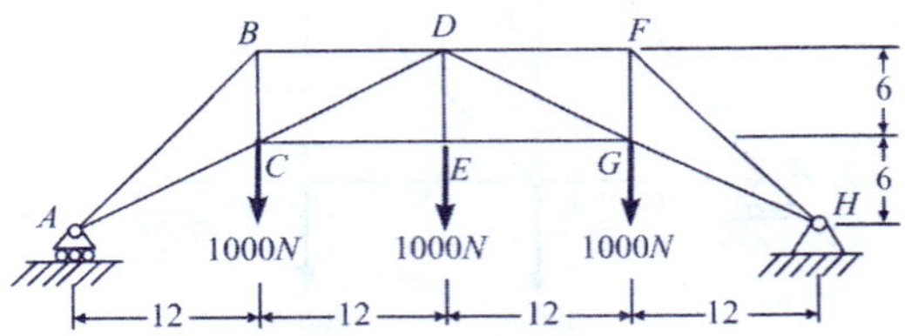

Figure P-4-3.11

$AB = 4240$ N (C); $BC = 3000$ N (T); $CD = 1120$ N (C) ***Ans.***

4-3.12 Determine the force in member AD of the truss shown in Fig. P-4-3.12.

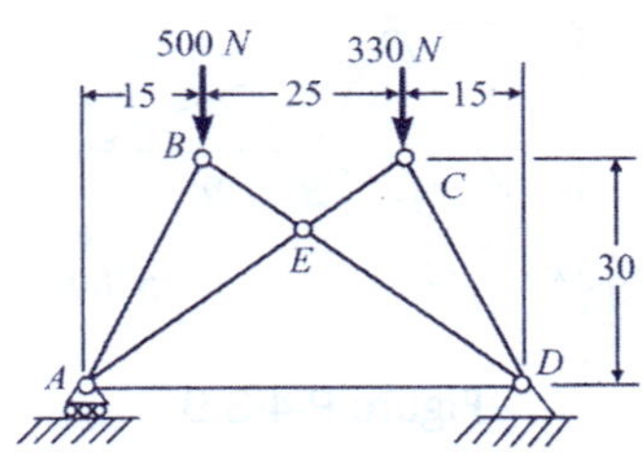

Figure P-4-3.12

4-3.13 Find the force acting in bar CE of the compound truss shown in Fig. P-4-3.13.

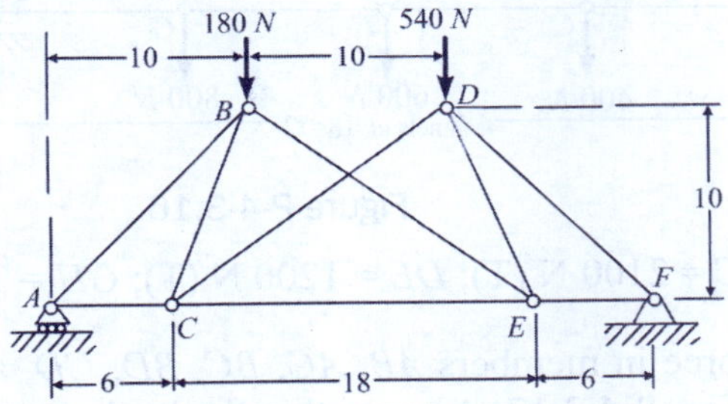

Figure P-4-3.13

$CE = 560$ N (T)

Ans.

4-3.14 For the truss shown in Fig. P-4-3.14, find the force in members BD and BE.

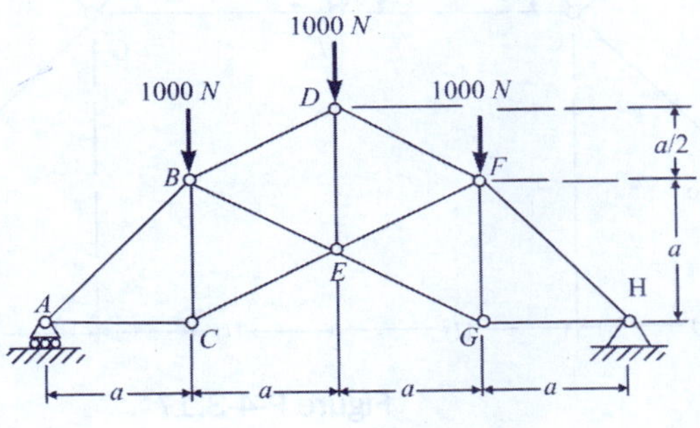

Figure P-4-3.14

$BD = 2240$ N (C); $BE = 783$ N (T) ***Ans.***

4-3.15 Solve for the force in members EF and EG for the truss shown in Fig. P-4-3.14.

4-3.16 The truss shown in Fig. P4-3.16 is pin-connected only at the lettered joints. Find the force acting in each member.

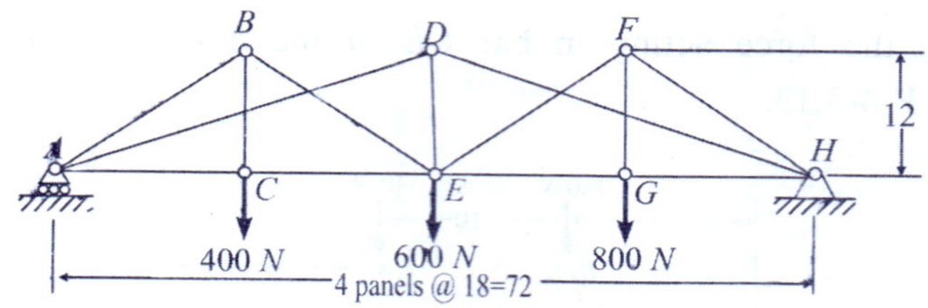

Figure P-4-3.16

$AC = 2100$ N (T); $DE = 1200$ N (T); $GH = 2400$ N (T) ***Ans.***

4-3.17 Find the force in members AB, AC, BC, BD, CD, and CE for the truss shown in Fig. P-4-3.17. Assume that the loads P are zero and that each load $Q = 4000$ N.

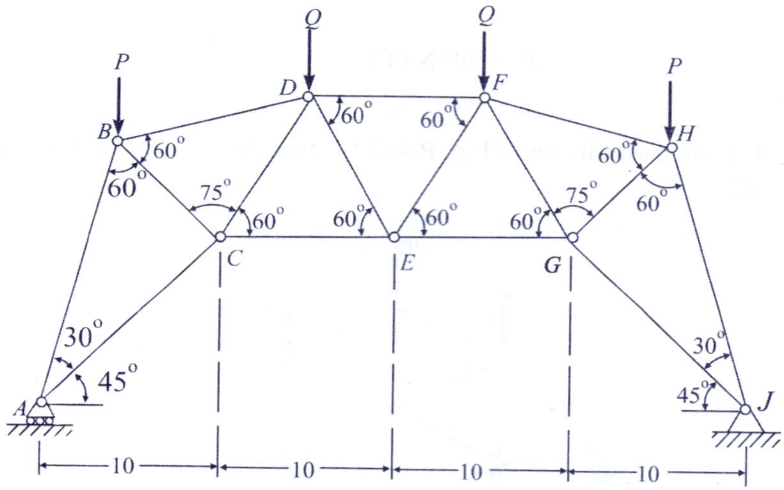

Figure P-4-3.17

$BD = 5660$ N (C); $CE = 6940$ N (T)

4-3.18 Repeat Prob. 4-3.17 if the loads Q are zero and each load $P = 2000$ N.

4-3.19 Repeat Prob. 4-3.17 if each load $P = 1500$ N and each load $Q = 2500$ N.

$BD = 4440$ N (C); $CE = 5065$ N (T)

4-4 METHODS OF SECTIONS

We have seen that the method of joints consists of analyzing trusses by applying the principles of equilibrium to the concurrent force systems acting at each joint. The principles of equilibrium of nonconcurrent force systems may also be applied to truss analysis; the procedure is known as the method of sections.[2] Its use permits us to determine *directly* the force in almost any member instead of proceeding to that member by a joint-to-joint analysis.

In the method of sections, a portion of the truss is isolated as a free body by passing an imaginary cutting plane through the entire truss, separating it into two parts. If possible, the separation should be done without cutting more than three members of the truss. We shall then have two isolated parts of the truss, each constituting a nonconcurrent system of forces in equilibrium under the action of the known loads that act on each part and the unknown forces that the cut members of one part exert on the other. For example, consider the truss shown in Fig. 4-4.1. If a cutting plane passes through the members *DF*, *EF*, and *EG*, the truss may be separated into two parts, each of which is acted upon by unknown forces equivalent to the loads being transmitted by these members.

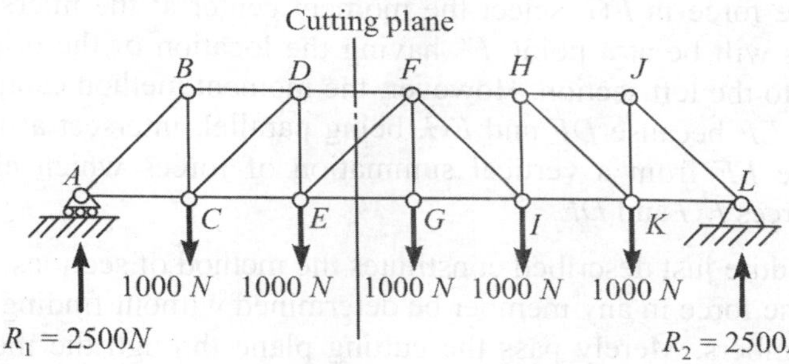

Figure 4-4.1 Selection through a truss.

These two parts are shown in Fig. 4-4.2. Each part in this figure constitutes a system of nonconcurrent forces in equilibrium. Since the unknowns in either system are the same, it is generally best to determine the unknowns from the equations of equilibrium applied to the simpler system. Here part (a) is obviously the simpler system since it involves fewer forces. Do not show the forces in the uncut members since they are internal to the free bodies shown. Observe that the action of all uncut members on their end joints occurs in equal opposite pairs at their end joints and hence cancels out of any calculation involving the entire free body.

[2]Strictly speaking, the method of joints is really a variation of the method of sections; the section is taken around the joint, thus isolating the joint as a free body. It is traditional, however, to use the expression *method of joints* for the equilibrium of concurrent forces at a joint, and the expression *method of sections* when considering the nonconcurrent force system set up by a section through the truss.

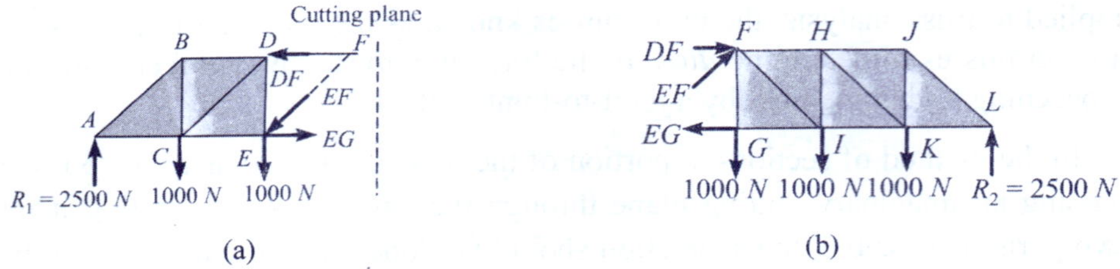

Figure 4-4.2 Each segment of the truss is held in equilibrium by forces equal to those in the cut members.

To simplify the calculations, use a condition of equilibrium that determines each unknown force independently of the other unknowns. Usually a moment summation is best, the center of moments being chosen at the intersection of the unknown forces to be eliminated from the moment summation. Thus, in Fig. 4-4.2a, if the force in *DF* is desired, select *E* as the center of moments. To determine the force in *EG*, select the moment center at the intersection of *DF* and *EF*. This will be at a point F' having the location of the original point *F* with respect to the left section. However, the moment method cannot be applied to determine *EF* because *DF* and *EG*, being parallel, intersect at infinity. Here we determine *EF* from a vertical summation of forces which eliminates the horizontal forces *EG* and *DF*

The procedure just described constitutes the method of sections. This method usually lets the force in any member be determined without finding the forces in the other members. Merely pass the cutting plane through the member whose force is desired. It is essential, however, that the plane does not cut more than three members whose internal forces are unknown, because the method of sections is based on the equilibrium of coplanar force systems for which only three independent equations of equilibrium are available to solve for three unknowns. An exception is the case where the cutting plane may pass through the desired member as well as several others, provided all these other members intersect at a common joint which is used as the center of moments. An example is the situation occurring in Prob. 4-4.20 where the force in *BC* can be found by using the curved section *a-a* and taking a moment summation about *G*.

Observe that the cutting section need not be straight; its purpose is to isolate the truss segment selected as the free body without involving more than three unknowns. This is illustrated by the curved section in Fig. 4-4.3 which isolates the truss segment *ADE* and exposes the forces in *AB*, *CD*, and *EF*. Although this curved section cuts through *CF* and *BF* twice, the equal opposite action of the exposed forces in these members cancels out of any force or moment summation. Then, after determining the forces in *AB*, *CD*, and *EF* from the isolated segment *ADE*, the method of joints may be used to find the forces in the remaining members.

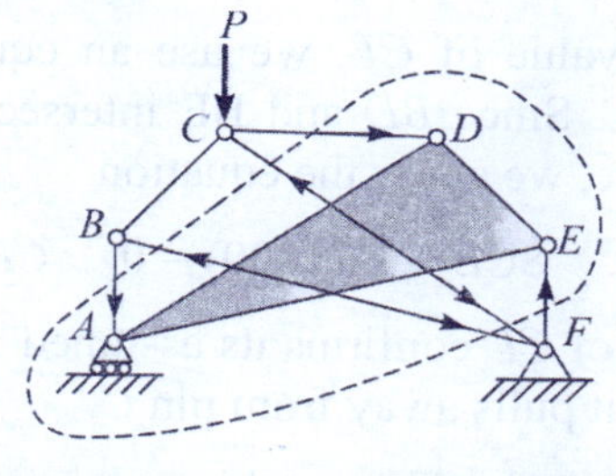

Figure 4-4.3 Isolation of segment *ADE* by curved section.

ILLUSTRATIVE PROBLEM

4-4.1.1 Use the method of sections to find the force acting in members *BD*, *BE*, and *CE* of the bowstring truss shown in Fig. 4-4.4. Determine each force by means of an equation which does not involve the other unknown forces.

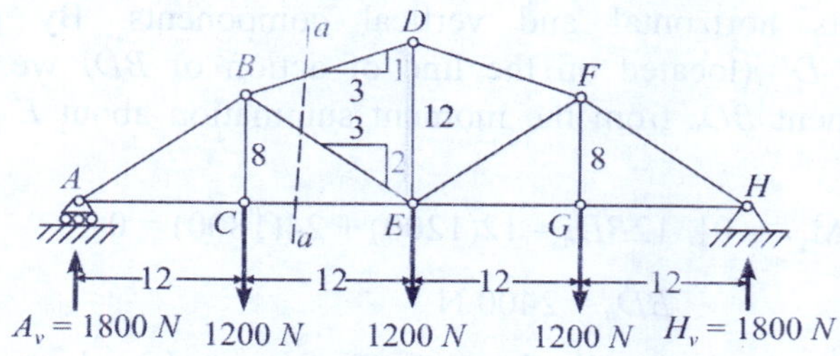

Figure 4-4.4

Solution

The required forces in *BD*, *BE*, and *CE* can be found by passing the section *a-a* through these members. Since the original truss is in equilibrium, the two segments into which it is divided by a cutting plane will also be in equilibrium if the loads transmitted by the cut members are replaced by external forces equal to these loads, These external forces are shown in Fig. 4-4.5 acting on the isolated left segment of the truss which involves a smaller number of forces than the right segment.

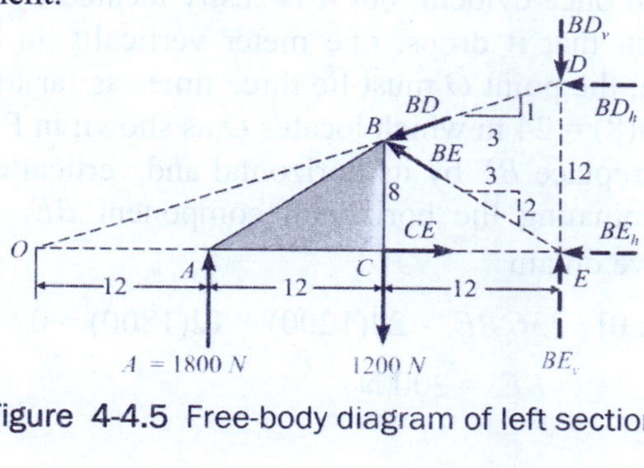

Figure 4-4.5 Free-body diagram of left section of truss.

To determine the value of *CE*, we use an equation of equilibrium which eliminates *BD* and *BE*. Since *BD* and BE intersect at *B*, these forces have no moment about *B*; hence, we apply the equation

$$[+\circlearrowleft \Sigma M_B = 0] \quad 8CE - 12(1800) = 0 \quad CE = 2700 \text{ N (T)} \qquad \textbf{\textit{Ans.}}$$

The positive value of *CE* confirms its assumed direction; hence, the member *CE* is in tension since it pulls away from pin *C*.

The value of *BD* is found by taking moments about E' which thereby eliminates *BE* and *CE* since their lines of action, if prolonged, intersect at E'. As a rule, the center of moments is chosen at the intersection of the two unknown forces to be eliminated from a moment summation. Note that here the spatial position of the moment center E' is determined from the dimensions of the original truss.

The moment of *BD* about E' is most easily found as the sum of the moments of its horizontal and vertical components. By placing these components at D' (located on the line of action of *BD)* we eliminate the vertical component BD_v from the moment summation about *E*. We therefore obtain

$$[+\circlearrowleft \Sigma M_{E'} = 0] \quad 12BD_h - 12(1200) + 24(1800) = 0$$

$$BD_h = 2400 \text{ N}$$

whence, using the proportionality between the force and its slope, we have

$$\frac{BD}{\sqrt{10}} = \frac{BD_v}{1} = \frac{BD_h}{3} = \frac{2400}{3}$$

or

$$BD_v = 800 \text{ N} \quad \text{and} \quad BD = 2530 \text{ N} \qquad \textbf{\textit{Ans.}}$$

To solve for *BE* independently of *BD* and *CE*, we take moments about the intersection of the forces to be eliminated. In this instance, the intersection of *BD* and *CE* is not at once evident, but it is easily located by observing that the slope of *BD* is such that it drops, one meter vertically in each three meters horizontally. Hence, the point *O* must lie three times as far to the left of *C* as *B* is above *C*; i.e., at 3(8) = 24 m which locates *O* as shown in Fig. 4-4.5. In taking moments about *O*, replace *BE* by its horizontal and vertical components acting at E', thereby eliminating the horizontal component BE_h from the moment summation. Thus, we obtain

$$[+\circlearrowleft \Sigma M_O = 0] \quad 36\,BE_v - 24(1200) + 12(1800) = 0$$

$$BE_v = 200 \text{ N}$$

whence, from the proportionality between the force *BE* and its slope, we get

$$\frac{BE}{\sqrt{13}} = \frac{BE_h}{3} = \frac{BE_v}{2} = \frac{200}{2}$$

or $$BE_h = 300 \text{ N and } BE = 361 \text{ N (C)} \qquad \textit{Ans.}$$

We check the accuracy of our computations by taking a horizontal summation of the forces acting on the FBD of Fig. 4-4.5. Thus,

$$[\Sigma H = 0] \qquad CE - BE_h - BD_h = 0$$

or

$$2700 - 300 - 2400 = 0 \qquad \textit{Check}$$

PROBLEMS

4-4.2 The Warren truss loaded as shown in Fig. P-4-4.2 is supported by a hinge at *G* and by a roller at *C*. Use the method of sections to compute the force in bars *BC*, *DF*, and *CE*.

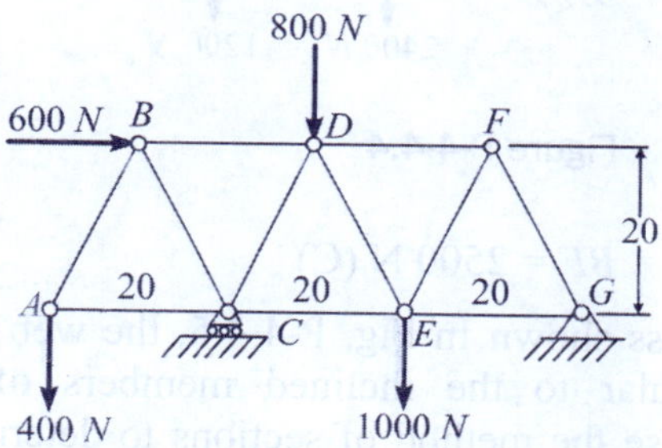

Figure P-4-4.2

$$BC = 448 \text{ N (C)};\ DF = 800 \text{ N C};\ CE = 100 \text{ N (T)} \qquad \textit{Ans.}$$

4-4.3 Use the method of sections to determine the force in bars *DF*, *DG*, and *EG* of the Howe truss shown in Fig. P4-4.3.

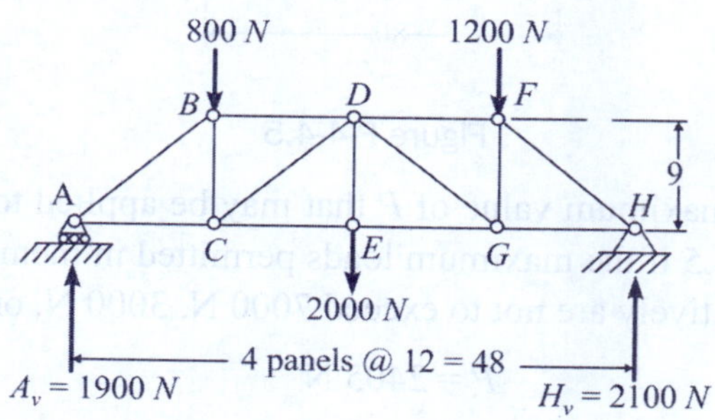

Figure P-4-4.3

4-4.4 For the truss shown in Fig. P-4-4.4, determine the force in BF by the method of joints and then check this result using the method of sections.

Hint: To apply the method of sections, first obtain the value of BE by inspection.

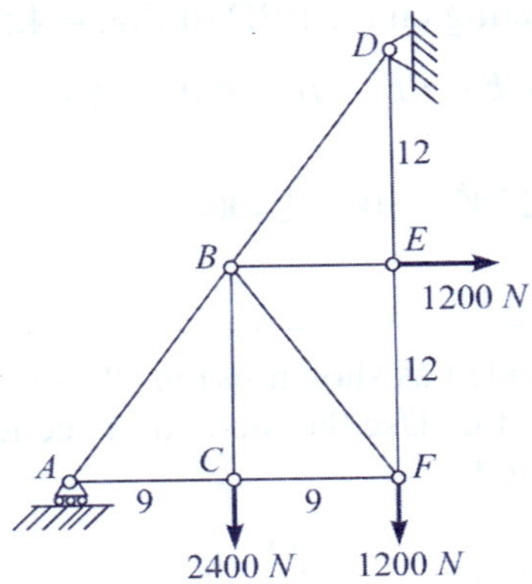

Figure P-4-4.4

$BF = 2500$ N (C) ***Ans.***

4-4.5 In the Fink truss shown In Fig. P-4-4.5, the web members BC and EF are perpendicular to the inclined members of their midpoints. If $P = 2000$ N, use the method of sections to determine the force in DF, DE, and CE.

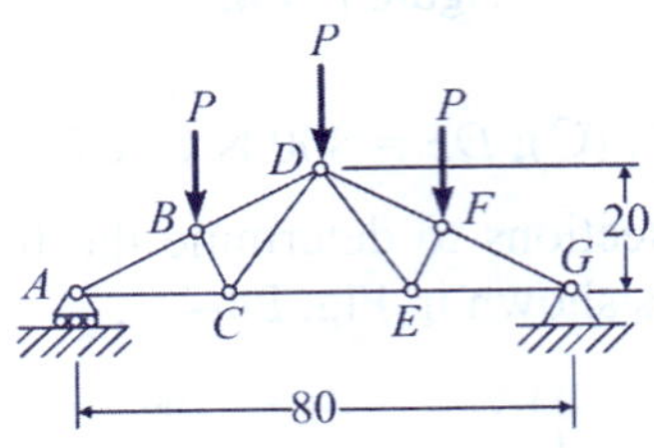

Figure P-4-4.5

4-4.6 Find the maximum value of P that may be applied to the Fink truss of Fig. P-4-4.5 if the maximum loads permitted in members BD, CD, and CE respectively are not to exceed 7000 N, 3000 N, or 5000 N.

$P = 2405$ N ***Ans.***

With this value of P, what will be the actual loads in these members?

4-4.7 Determine the force in bars *BD*, *CD*, *DE*, *BE*, and *CH* of the truss shown in Fig. P-4-4.7.

Hint: After finding *DE*, consider a FED of joint *D*.

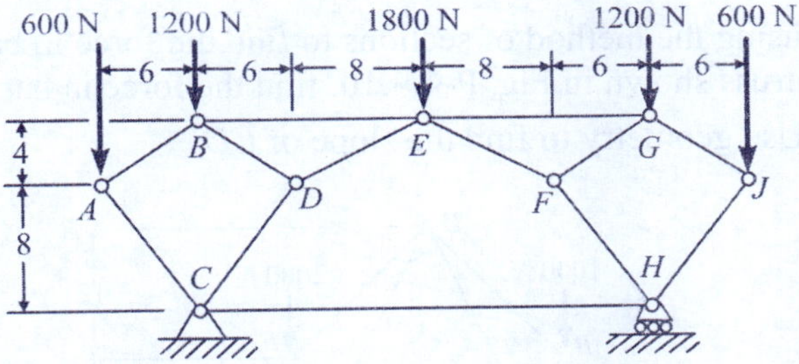

Figure P-4-4.7

$BD = 878$ N (C); $CD = 1750$ N (C); $DE = 2015$ N (C);

$BE = 1050$ N (T); $CH = 750$ N (T) ***Ans.***

4-4.8 Find the force in member *FG* of the triangular Howe truss shown in Fig. P-4-4.8. Wind pressure creates the loads which are perpendicular to the inclined members.

Hint: First use the method of sections to determine *DF* and then consider an FBD of joint *F*.

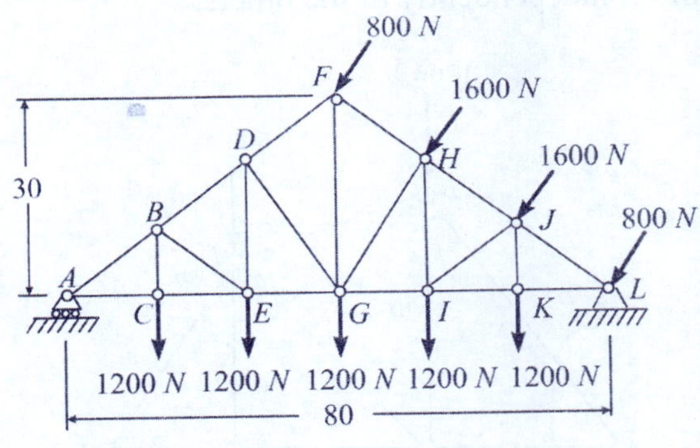

Figure P-4-4.8

$FG = 5600$ N (T) ***Ans.***

4-4.9 Assume that the group of vertical 1200 N loads are removed from the triangular Howe truss of Fig. P.4-4.8, leaving only the wind loads which are perpendicular to the inclined members. Use the method of sections to determine the force in members *FH*, *FG*, and *EG*.

Hint: Determine by inspection the forces in the web members of the left side of the truss.

4-4.10 After using the method of sections to find the force in bars *BD* and *CD* of the truss shown in Fig. P-4-4.10, find the force in bar *AB*.

Hint: Use geometry to find the slope of *CD*.

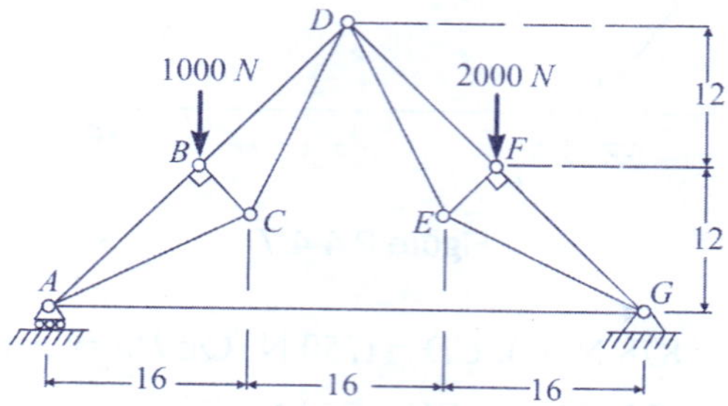

Figure P-4-4.10

AB = 2477 N (C); *BD* = 1770 N (C); *CD* = 1118 N (T) ***Ans.***

4-4.11 For the truss shown in Fig. P4-4.11, determine each of the forces in bars *BD*, *CD*, and *CE* by means of an equilibrium equation that determines it independently of the others.

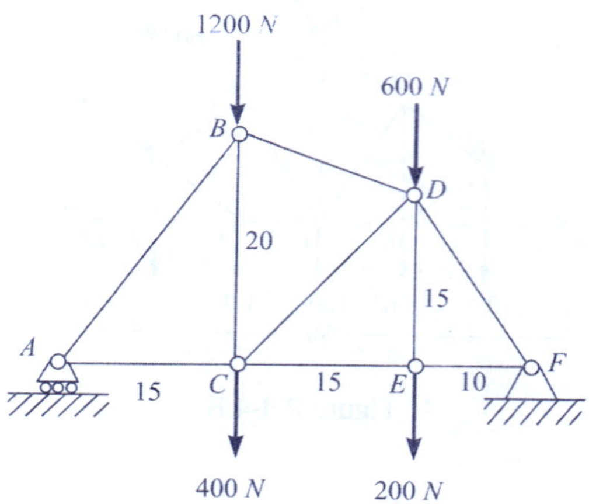

Figure P-4-4.11

4-4.12 Use the method of sections to determine independently the force in bars *AD*, *BE*, and *CE* of the truss shown in Fig. P-4-4.12.

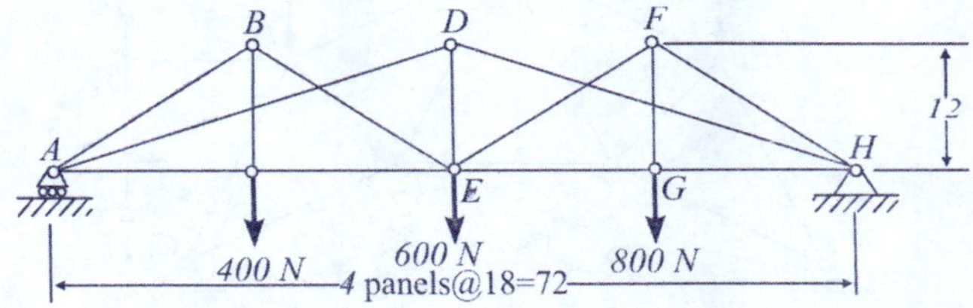

Figure P-4-4.12

AD = 1896 N (C); *BE* = 361 N (C); *CE* = 2100 N (T) ***Ans.***

4-4.13 The loads on the Parker truss shown in Fig. P-4-4.13 (p. 132) are in kilo newtons. Determine the forces in members *BD*, *BE*, *CE*, and *DE*.

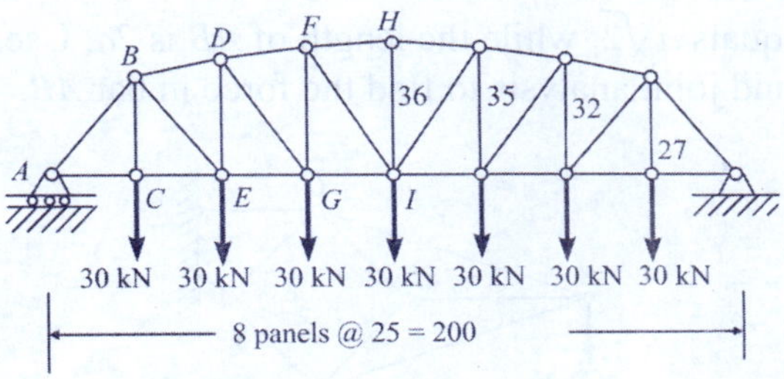

Figure P-4-4.13

4.4.14 Determine the force in members *DF*, *DG*, and *EG* for the Parker truss shown in Fig. P-4-4.13.

DF = 162 KN (C); *DC* = 32.7 KN (T); *EG* = 140.6 KN (T) ***Ans.***

4-4.15 For the truss shown in Fig. P-4-4.15, determine independently the forces in bars *BD*, *CD*, *CE*, and *ER*.

Hint: To find *EF* independently, dot a force summation with a unit vector perpendicular to *EG*.

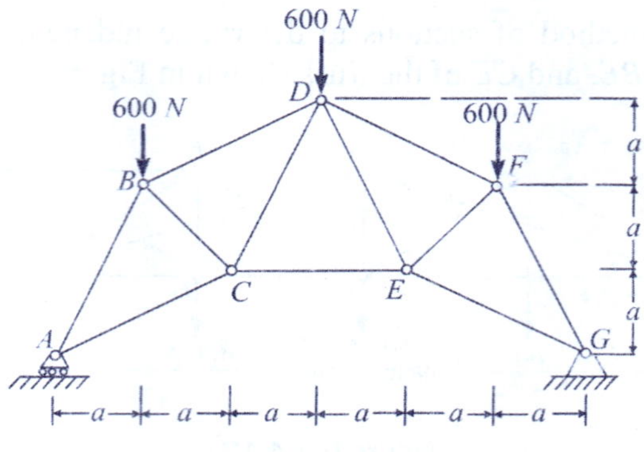

Figure P-4-4.15

$BD = 896$ N (C); $CD = 112$ N (T);

$CE = 750$ N (T); $EF = 283$ N (T) ***Ans.***

4-4.16 In the cantilever truss shown in Fig. P-4-4.16, each side of the square *BCFD* equals $a\sqrt{2}$ while the length of *AB* is *2a*. Use a combination of section and joint analysis to find the force in bar *AB*.

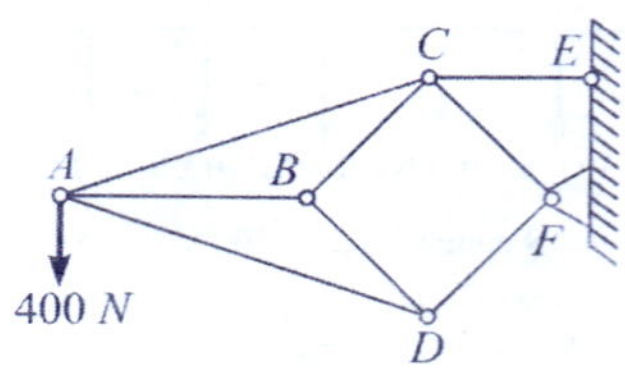

Figure P-4-4.16

$AB = 2400$ N (T) ***Ans.***

4-7.17 The structure shown in Fig. P-4.4.17 consists of two simple trusses joined by bars *BE*, *CF*, and *DG*. Find the force in bar *FG*.

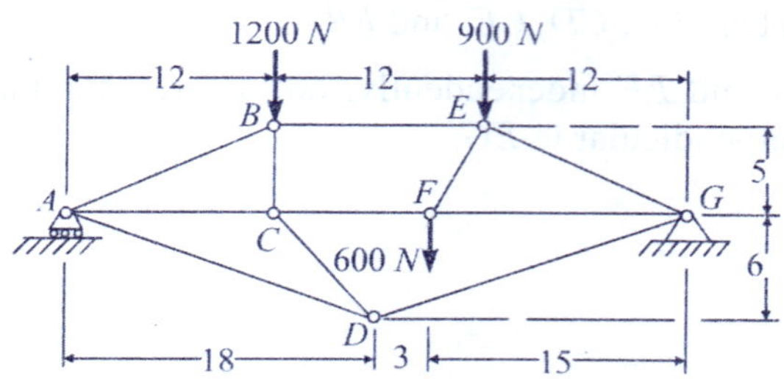

Figure P-4-4.17

4-4.18 Compute the force in bars *AB*, *AC*, *DF*, and *DE* of the scissors truss shown in Fig. P-4-4.18.

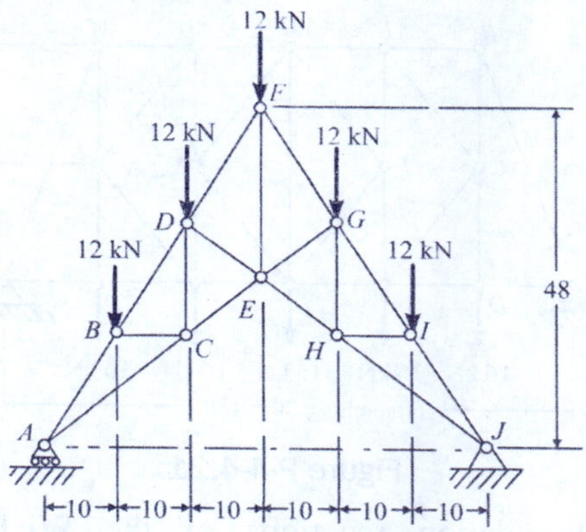

Figure P-4-4.18

4-4.19 Compute the force in bars *GI*, *GH*, *EH*, and *HI* for the scissors truss shown in Fig. P-4-4.18.

GI = 56.6 KN (C); GH = 6 KN (T);

EH = 38.4 KN (T); HI = 7.5 KN (C) ***Ans.***

4-4.20 For the transmission tower shown in Fig. P-4-4.20, determine the force in member *CJ*.

Hint: First use section *a-a* to find the force in *BC*.

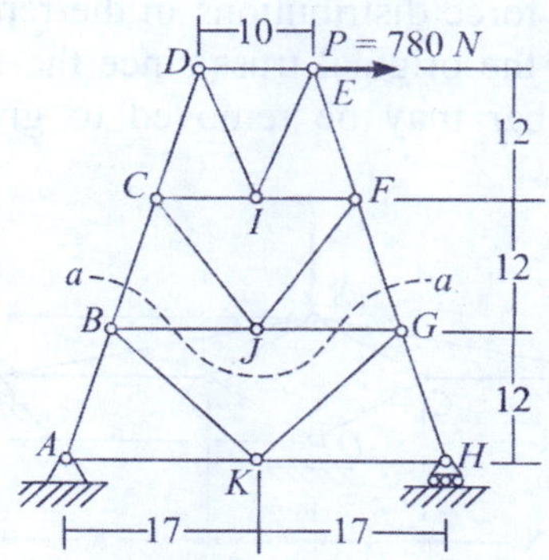

Figure P-4-4.20

CJ = 250 N (C) ***Ans.***

4-4.21 Find the force acting in members *EH*, *FH*, *FI*, and *GI* of the K-type bridge truss shown in Fig. P-4-4.21. Refer to the hint of the previous problem.

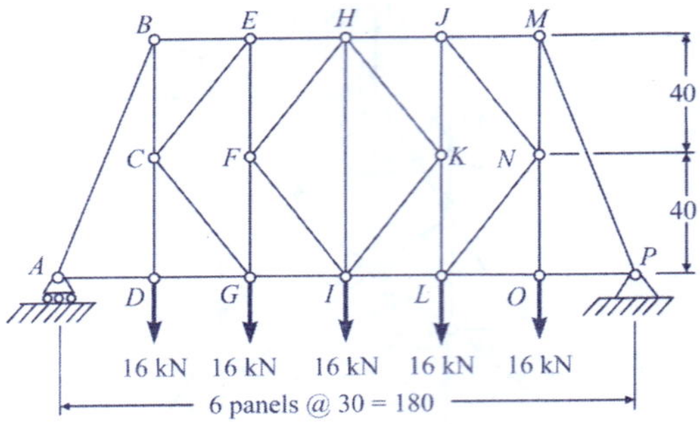

Figure P-4-4.21

4.4.22 Find the foundation reactions of the Wichert truss shown in Fig. P-4-4.22.

Suggestion: When the methods of joints or sections are not directly applicable although the structure as a whole is statically determinate, a method devised by Henneberg[3] may be used. Briefly, this method consists of removing a bar or reaction from the original truss and substituting for it a bar that produces a solvable replacement truss. The specified loading is then applied to the replacement truss and the force in the substitute bar computed. Next, the replacement truss is loaded by a force *X* in place of the removed bar or reaction, and again the force in the substitute bar is computed due to this force *A*. The value of *X* is determined by equating to zero the sum of the forces in the substitute bar under these two loadings of the replacement truss. A superposition of the two force distributions in the replacement truss then gives the forces in the bars of the original truss since the force in the substitute bar becomes zero and this bar may be removed to give the original truss and loading.

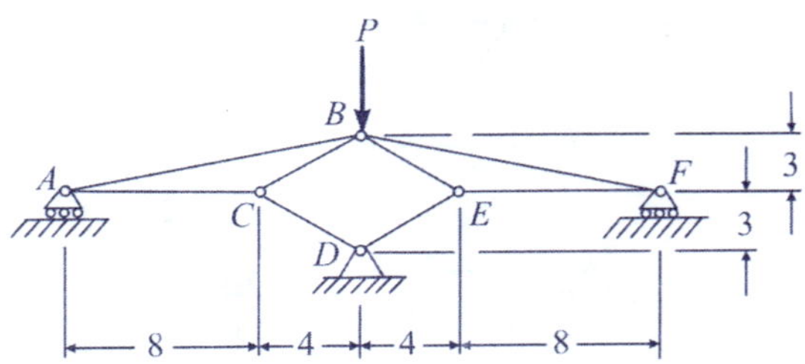

Figure P-4-4.22

[3]For a complete discussion of Henneberg's method4, see Timoshenko and Young, *Theory of Structures*, McGraw-Hill, p. 82.

In the present problem, apply Henneberg's method by removing the support at D and replacing it by the substitute bar BD. Solve for the force in this bar under the two loadings shown in Fig. P-4-4.22A and equate their sum to zero to obtain the value of R which you should find to be $3P$.

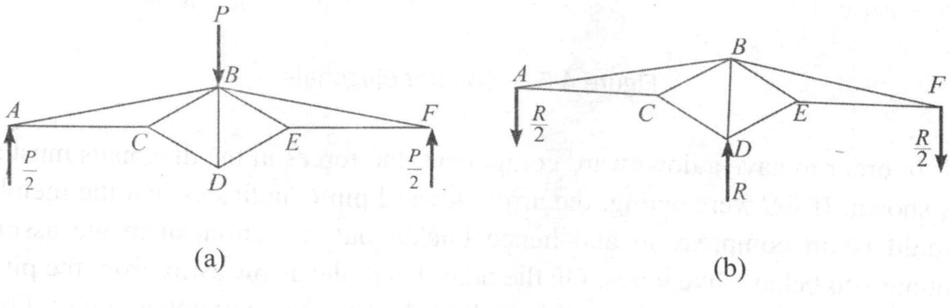

Figure P-4-4.22A

4-5 REDUNDANT MEMBERS; COUNTER DIAGONALS

In trusses subjected to moving loads, such as railroad bridges, the diagonal members may undergo a reversal of stress; that is, a diagonal normally in tension may become subject to a compressive load. If the diagonal is composed of eyebars or is otherwise so slender compared with its length that it will buckle under compressive loads, an additional diagonal sloping in the opposite direction must be provided in the truss panel to prevent collapse of the structure.

Additional diagonal of this type might be called redundant, i.e., unnecessary; but as only one of these diagonals acts at a time, the term redundant does not apply in this case, although it is frequently used. These additional diagonals are usually known as counter diagonals, or more briefly, as counters. They are often represented by dashed lines on a truss. In considering the action of counters, it is convenient to regard them as wires which can support tension but will buckle instantly if subjected to compression.

As an example, consider the truss shown in Fig. 4-5.1. It is required to determine which of the tension diagonals AD or BC in the third panel is acting under the given loading. This is accomplished by passing the section a-a through the truss. Although four members are cut, the horizontal forces in AB and CD are eliminated by applying a vertical force summation. Considering the part of the truss to the left of section a-a, we see that the upward effect of R_1 must be balanced by a downward component exerted by the diagonals.

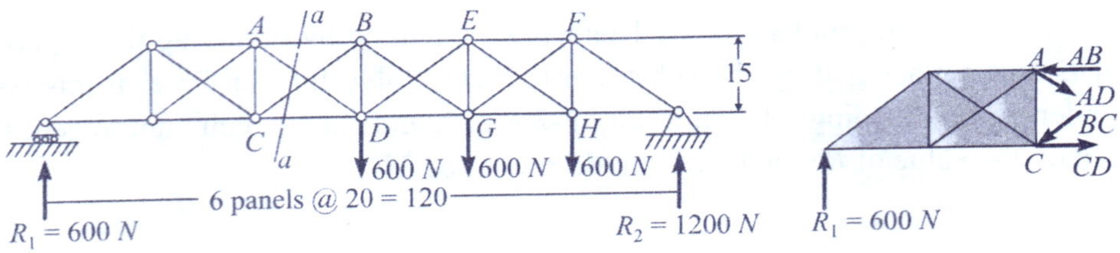

Figure 4-5.1 Counter diagonals.

In order to have a downward component, the forces in the diagonals must act as shown. If BC were acting, the arrow toward pin C indicates that the member would be in compression and hence buckle out of action since we assume counters to behave like wires. On the other hand, the arrow away from the pin at A indicates AD to be under tension and AD is the tension member acting. From the dimensions of the truss and by equating a vertical summation of forces to zero, we find the force in member AD to be

$$[\Sigma Y = 0] \qquad 600 - AD(15/25) = 0 \qquad \text{AD} = 1000 \text{ N (T)} \qquad \textbf{Ans.}$$

It is sometimes asked why both of the counters cannot act as tension diagonals. If this were so, the force in AD would be increased and this would violate the axiom of conservation of energy; i.e., forces are always distributed in the most efficient manner possible.

As an exercise, determine by inspection which of the other diagonals are acting in the other panels of the truss.

ROBLEMS

1 In Fig. P-4-4.3, assume that counter diagonals act from B to E and from E to F in addition to the counter diagonals CD and DG as shown in the figure. Assuming these counter diagonals can support only tension, determine which diagonals are acting and the force in each.

The center panel of the truss in Fig. P-4-5.2 contains two flexible ables. What load P will cause a compressive force of 2000 N in BD? hen determine which tension diagonal BE or CD is acting and the ce in it.

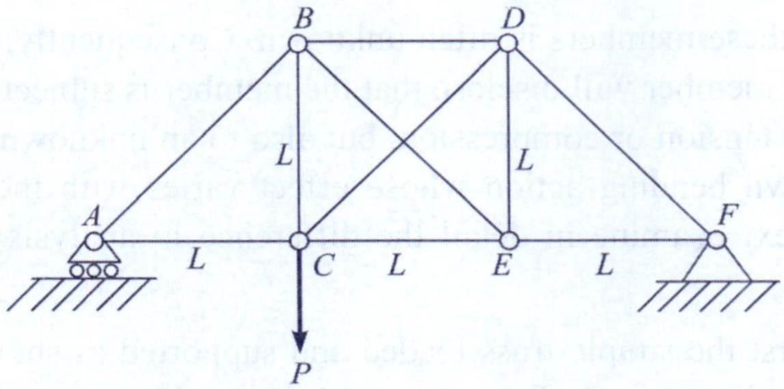

Figure 4-5.2

$CD = 1414$ N **Ans.**

4-5.3 The center diagonals of the truss in Fig. P-4-5.3 can support only tension. Compute the force in each center diagonal and the force in bars *BC*, *DE*, and *FG*.

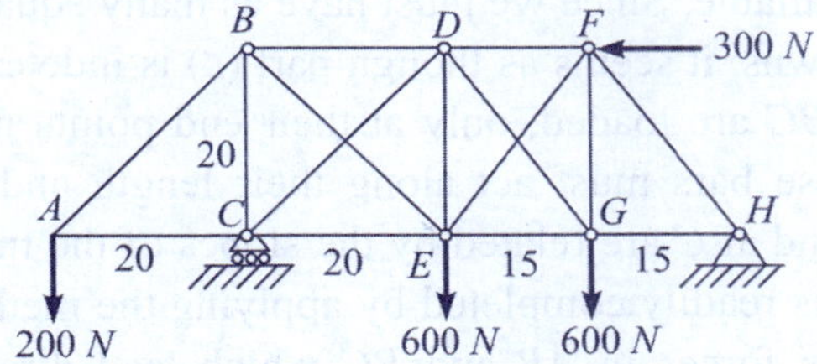

Figure 4-5.3

$BE = 1050$ N (T); $DG = 175$ N (T) **Ans.**

4-6 METHOD OF MEMBERS. FRAMES

If some or all of the members of a pin-connected structure are subjected to bending action because of loads applied along their length in addition to loads at their ends, the structure is called a frame. As we shall see, it is not feasible to analyze a frame by passing a section through the members of such structures as we did for trusses. Instead, we consider the separate free-body diagram of each member subjected to bending; the procedure is called the *method of members*.

It is important to note that in trusses, the various members are assumed to be loaded only at their ends so that the end forces in such two-force members are directed along their axes. However, in multiloaded members, the end forces will be shown *not* to act along the axis of the member; in fact, the direction of the end forces on these members is often unknown. Consequently, a section passed through such a member will disclose that the member is subjected not only to an unknown axial tension or compression, but also to an unknown transverse force and an unknown bending action whose effect varies with the location of the section. We next examine in detail the difference in analysis between trusses and frames.

Consider first the simple truss loaded and supported as shown in part (a) of Fig. 4-6.1. The dimensions of the truss and the load are assumed to be known. The three reaction components are readily found by applying the three equations of equilibrium. If the bar AC is removed to give the structure shown in part (b), it is clear that the structure will collapse since the roller at A is free to move horizontally. However, rigidity will be restored as shown in part (c) if the roller at A is replaced by hinge. Obviously each hinge reaction in (c) may be replaced by two components as shown, thus giving a structure with apparently four unknown reaction components for which only three equations of equilibrium are available. Since we must have as many equations of equilibrium as there are unknowns, it seems as though part (c) is indeterminate, but the fact that bars AB and BC are loaded, only at their end points means that the force transmitted by these bars must act along their length and hence the reaction components at A and at C are related by the slopes of the members AB and BC. In fact, a solution is readily completed by applying the method of joints around B, thus finding the forces in AB and BC which lead directly to the reaction components. Observe that in both parts (a) and (c), the equation of stability $2j = m + R$ is satisfied. In (a), we have 3 members and 3 reactions for 3 joints, whereas in (c) there are 2 members and 4 reactions for 3 joints.

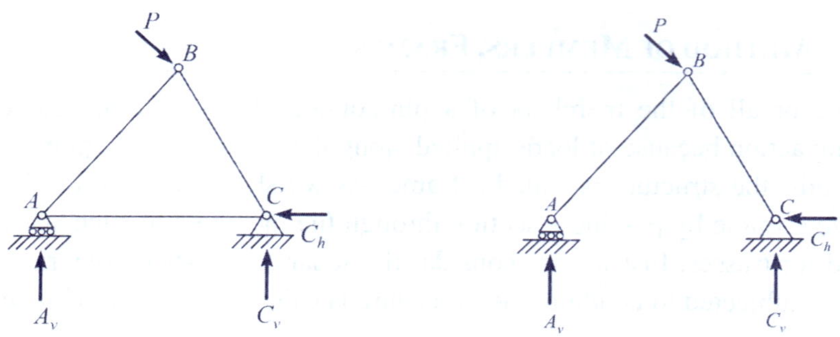

(a) (b)

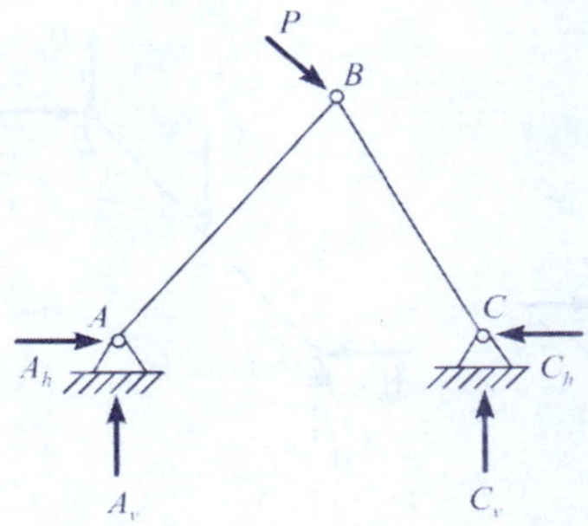

(c)

Figure 4-6.1 Removal of bar *AC* from (a) requires at *A* in (c) to maintain stability.

Consider next the structure shown in Fig. 4-6.2 where known loads *P* and *F* are applied to the members at specified positions along their lengths. The result is that members *AB* and *BC* are subjected to bending action instead of simple axial loading as was the case in Fig. 4-6.1. Obviously each hinge supplies two reaction components, but this time the direction of the total hinge reaction is not directed along the member as the free-body diagrams of *AB* and *BC* in parts (b) and (c) clearly show. Moment balance about *B* in each part requires the total hinge reactions to act as shown at unknown angles θ_A and θ_C. There are therefore four unknown elements of the hinge reactions consisting of the unknown hinge reactions *A* and *C* plus their unknown angles θ_A and θ_C or the four unknown magnitudes of their components, i.e., A_h, A_v, C_h, and C_v.

The three equations of equilibrium are not sufficient to solve for all four of these unknowns. However, member *AB* as a free body is held in equilibrium by two of these unknowns plus two additional unknown hinge components at *B* exerted through this hinge by member *BC* upon *AB*. Since three equations of equilibrium are available for the FBD of member *AB*, we have a total of 6 equations of equilibrium (3 for the original FBD of the given structure and 3 for member *AB*) to solve for the 6 unknown components acting upon these two free bodies. As a general guide to applying the equations of equilibrium most efficiently, it is best to start with moment summations which will solve for the unknown quantities common to both free-body diagrams, in this case for A_v and A_h. Thus in part (a) of Fig. 4-6.2, a moment summation about *C* determines A_v. When we use this result in part (b), a moment summation about *B* then determines A_h after which two force summations ($\Sigma H = 0$ and $\Sigma V = 0$) are available for each free body to determine the remaining unknown components at *C* and *B*.

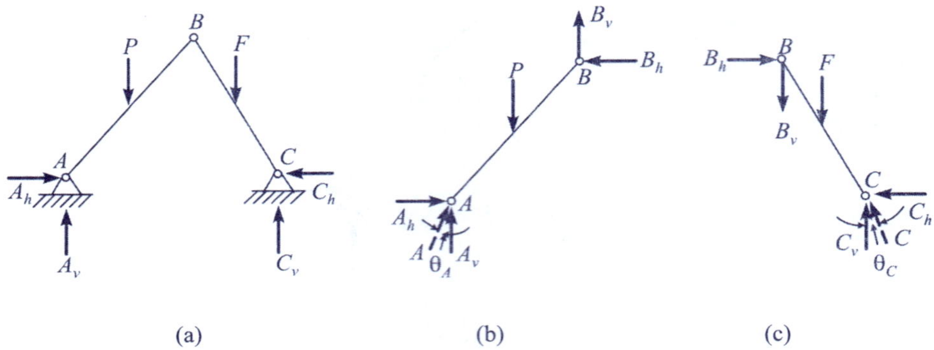

Figure 4-6.2 In members subjected to bending, the end forces are *not* directed along the member.

If desired, we could just as well have solved for the unknown hinge components by considering a combination of the free-body diagrams of the original structure and of member *BC*. For this combination, the common unknowns are C_v and C_h which are readily determined by first applying $\Sigma M_A = 0$ in part (a) to find C_v and then using this result in $\Sigma M_B = 0$ in part (c) to find C_h, after which two force summations applied to each free body, to determine the rest of the unknowns.

Finally, it is possible to solve for the unknown hinge components by considering the combination of the free-body diagrams of members *AB* and *BC*. Here the two common unknowns would be B_v and B_h whose values would now be obtained by applying $\Sigma M_A = 0$ to the FBD of *AB* and $\Sigma M_C = 0$ to the FBD of *BC*. Then, as before, two force summations may be applied to each FBD to determine the remaining unknowns. This time, however, we would have the disadvantage of solving a pair of simultaneous equations for B_V and B_h rather than finding each unknown directly. On the other hand, the moment arms involved in setting up these equations will be smaller than before. It is a matter of opinion which combination of free-body diagrams to use; they all yield the same result.

The essential principle involved in applying the method of members is to isolate and draw a separate FBD for each member subjected to bending action. When the members for analysis are isolated, Newton's law of action and reaction requires that the forces (usually represented by their components) exerted by the connecting hinges or pins must be consistently represent as acting in *opposite* directions on the separated members. Even if either component of a force is incorrectly assumed in direction, the solution will still give its correct magnitude but with a negative sign. This would mean that the component of the force on *both* ends of the separated member acts in the opposite direction to that originally assumed. Further discussion of the method of members is given in the following illustrative problem.

ILLUSTRATIVE PROBLEM

4-6.1 A billboard BC weighing 1000 N is subjected to a wind load of 300 N per linear meter as shown in Fig. 4-6.3. Neglecting the weights of the supporting members, determine the hinge forces at A and F.

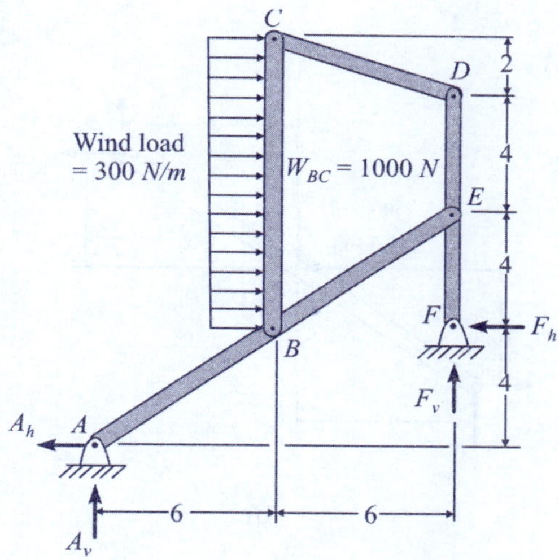

Figure 4-6.3

Solution

Since the directions of the hinge forces at A and F are unspecified, we replace the hinge forces by components assumed to be acting as shown. Hence, Fig. 4-6.3 shows four unknowns acting on a system for which only three equations of equilibrium are available. Considering this system, none of these unknowns can be found and the system seems to be indeterminate. Additional equations are needed and can be obtained by considering a free-body diagram of each of the members as is shown in Fig. 4-6.4. Then an analysis consisting of a progressive comparison of the number of unknowns on each free body with the available number of equations of equilibrium will disclose whether a solution is possible.

To make this analysis, we list the unknowns acting on each free body considered and for each successive free body we insert the additional unknowns not previously listed. If the sum of the listed unknowns just equals the total of available equations, a solution will be possible.

Observe that in Fig. 4-6.4, we have drawn a FBD of each member subjected to a bending action. At each hinge we represent the unknown hinge force by two components. It is unimportant whether we correctly assume the direction of a hinge force component. If we are correct, we shall obtain a positive answer; an incorrect direction is denoted by a negative result. The components of the hinge forces at C and D exerted by member CD are related by the slope of CD in the original diagram since CD is loaded only at its ends and hence transmits an axial force between C and D.

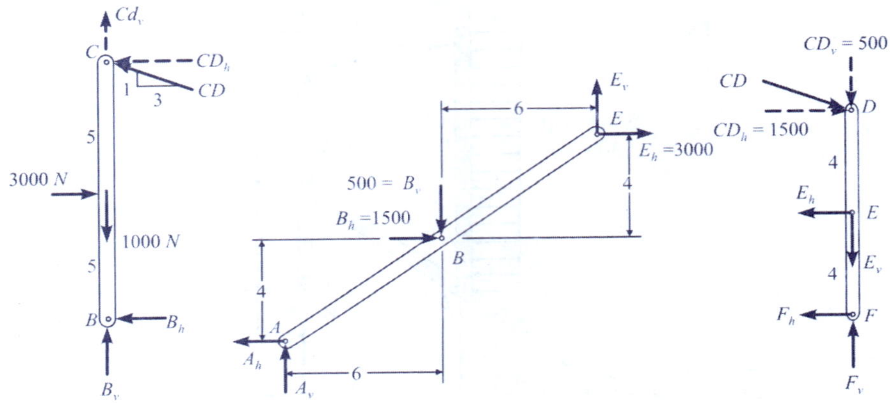

Figure 4-6.4

Note carefully that the action of the components of the hinge force at B upon BC in part (a) is equal but opposite to the action of these components upon AE in part (b); also that the components of the hinge force at E upon AE in part (b) are equal but opposite to these components acting upon DF in part (c). This is due to Newton's third law that action and reaction forces are equal but oppositely directed. *Apply this principle carefully when setting up the free-body diagrams.*

We are now ready to set up the following table:

Free body	Unknowns	Equations available
BC	B_h, B_v, CD	3
DF	E_h, E_v, F_h, F_v	3
AE	A_h, A_v	3
Totals	9	9

Solve for B_h, B_v, and CD

} Solve first for common unknowns E_h and E_v (DF and AE)

For the 9 unknowns listed, we now have 9 equations available and a solution is possible. As a general rule, if a body has only 3 unknowns (as does *BC*), we solve first for these unknowns. Next we solve for the unknowns common to the other free bodies, taking moment summations about centers which eliminate the other unknowns.

Applying this analysis, we start first with body *BC*. The equations of equilibrium and the slope relations for *CD* give

$[+\uparrow \Sigma_B = 0]$ $\quad 10(CD_h) - 3000(5) = 0$

$CD_h = 1500$ N whence $\quad CD_v = 500$ N

$[\Sigma H = 0]$ $\quad 3000 - 1500 - B_h = 0$ $\quad B_h = 1500$ N

$[\Sigma V = 0]$ $\quad B_v + 500 - 1000 = 0$ $\quad B_v = 500$ N

These results are now inserted in parentheses, in parts (b) and (c) as shown to clarify subsequent computations.

We next solve for the unknowns E_h and E_v which are common to the free bodies of *DF* and *AE*. We consider first *DF* (Fig. 4-6.4c) since a moment summation about *F* eliminates all unknowns except E_h. We obtain

$[+\uparrow \Sigma M_F = 0]$ $\quad 4E_h - 8(1500) = 0$ $\quad E_h = 3000$ N

Inserting this value of E_h on the FBD of *AE* (Fig. 4-6.4b), a moment summation about *A* eliminates all unknowns except E_v and yields

$[+\uparrow \Sigma M_A = 0]$ $\quad 12E_v - 8(3000) - 6(500) - 4(1500) = 0$

$$E_v = 2750 \text{ N}$$

For each of the members of *AE* and *DF*, there remain two equations of equilibrium which may now be applied to determine the remaining unknowns. Thus, for *AE*, we have

$[\Sigma H = 0]$ $\quad 3000 + 1500 - A_h = 0$ $\quad A_h = 4500$ N

$[\Sigma V = 0]$ $\quad A_v - 500 + 2750 = 0$ $\quad A_v = -2250$ N

To the FBD of *DF* we apply

$[\Sigma H = 0]$ $\quad 1500 - 3000 - F_h = 0$ $\quad F_h = -1500$ N

$[\Sigma V = 0]$ $\quad A_v - 2750 - 500 = 0$ $\quad F_v = 3250$ N

The negative results for A_v and F_h indicate that their assumed directions must be reversed. Therefore, combining the components of the hinge forces at A and F, we obtain

$$[A = \sqrt{A_h^2 + A_v^2}\,] \quad A = \sqrt{(4500)^2 + (2250)^2}$$

$$A = 5030 \text{ N} ; \theta_h = 26.6^\circ$$ **Ans.**

$$[F = \sqrt{F_h^2 + F_v^2}\,] \quad F = \sqrt{(1500)^2 + (3250)^2}$$

$$F = 3580 \text{ N} ; \theta_h = 65.3^\circ$$ **Ans.**

PROBLEMS

4.6.2 The structure shown in Fig. P-4-6.2 is hinged at A and C. Find the horizontal and vertical components of the hinge force at A, B, and C.

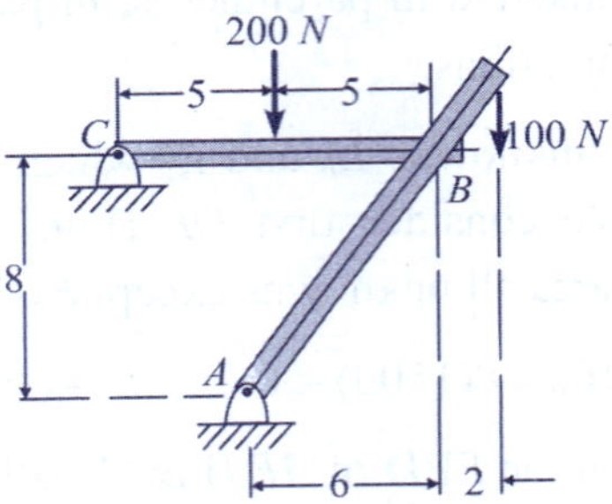

Figure P-4-6.2

$B_v = 100$ N; $B_h = 175$ N **Ans.**

4.6.3 The pin-connected frame in Fig. P-4-6.3 supports a load $W = 120$ N. Find the horizontal and vertical components of the pin force at C as it acts upon member CE. Neglect the weights of the members.

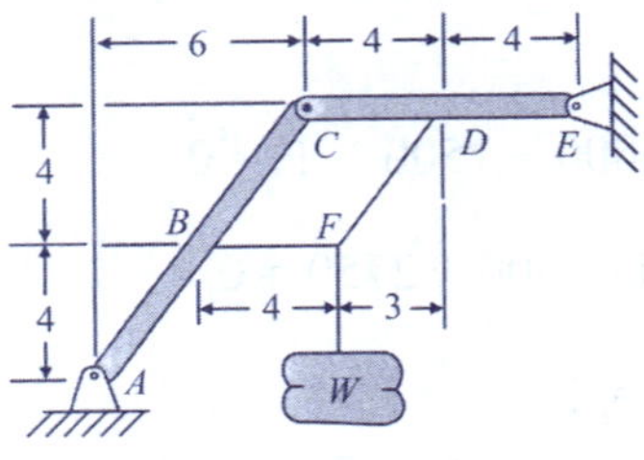

Figure P-4-6.3

$C_h = 90$ N right; $C_v = 60$ N up **Ans.**

4-6.4 Resolve Prob. 4-6.3 if, in addition to the load $W = 120$ N, members AC and CE weigh 10 N per linear meter.

4-6.5 In the system shown in Fig. P-4-6.5, find the reaction at D caused by the 1200 N load. Neglect the weight of the members.

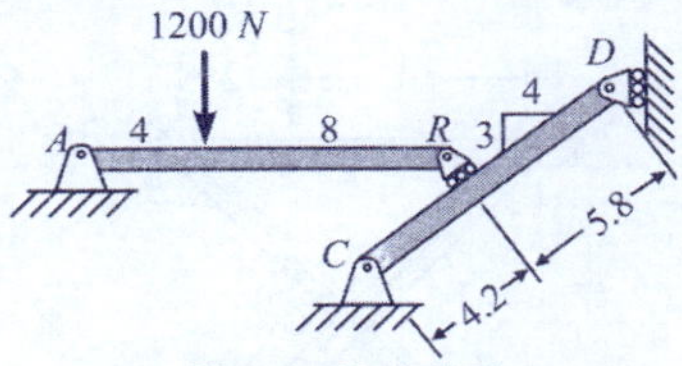

Figure P-4-6.5

$D = 350$ N **Ans.**

4-6.6 For the structure loaded as shown in Fig. P-4-6.6, find the horizontal and vertical components of the pin force at B as they act upon member CD.

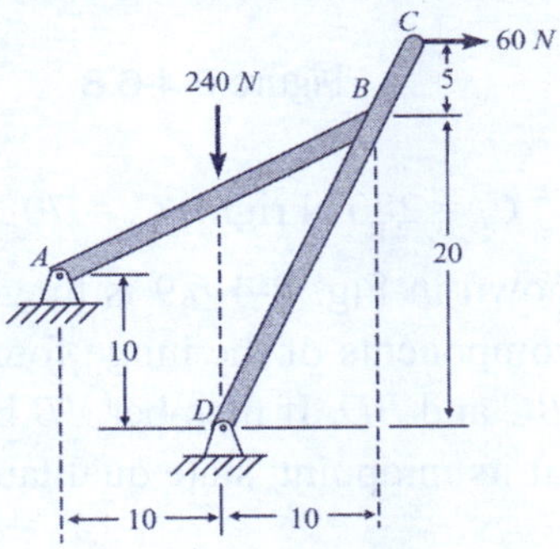

Figure P-4-6.6

$B_h = 180$ N left; $B_v = 210$ N down **Ans.**

4.6.7 The A frame shown in Fig. P.4-6.7 is supported by a hinge at A and a roller at E. Compute the horizontal and vertical components of the hinge forces at B and C as they act upon member AC

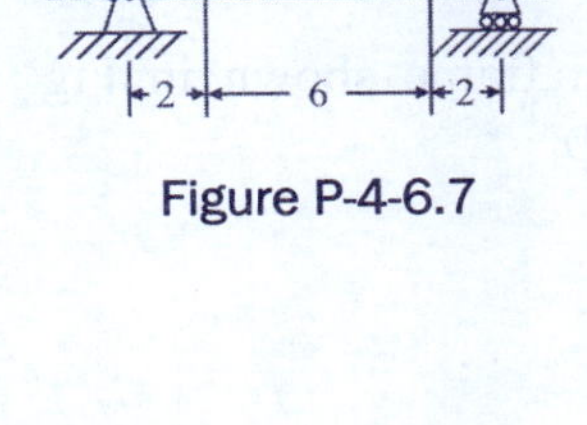

Figure P-4-6.7

4-6.8 The frame shown in Fig. P-4-6.8 is supported by a hinge at *E* and a roller at *D*. Determine the horizontal and vertical components of the hinge force at *C* as it acts upon member *BD*.

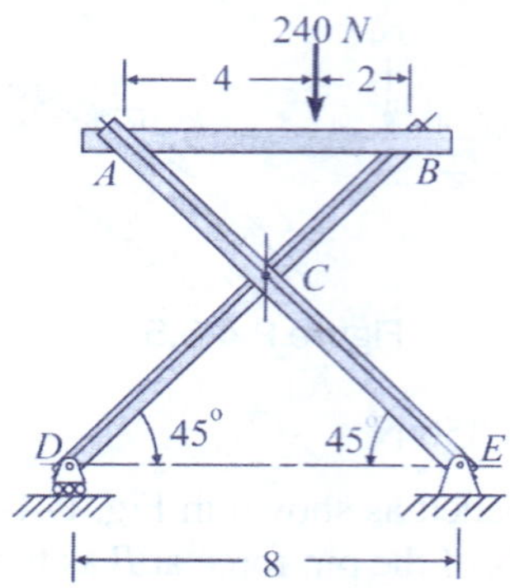

Figure P-4-6.8

$C_h = 280$ N right; $C_v = 70$ N up **Ans.**

4-6.9 The frame shown in Fig. P-4-6.9 is hinged to rigid supports at *A* and *E*. Find the components of the hinge forces at *A* and *E* and the forces in members *BC* and *BD*. If member *AB* had a different length but still supported *P* at its midpoint, state qualitatively how the answers would change.

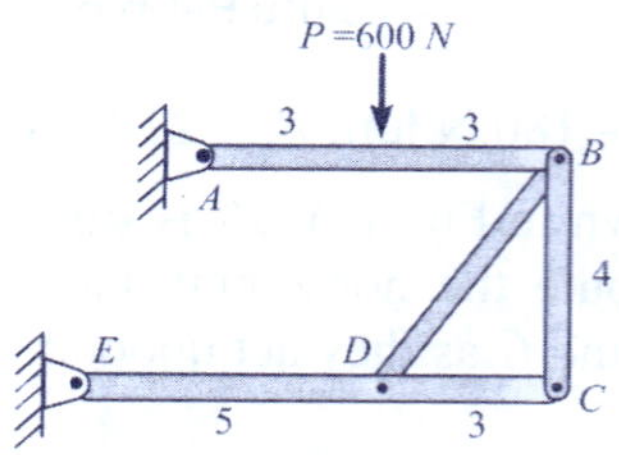

Figure P-4-6.9

$A_v = E_v = 300$ N; $A_h = E_h = 600$ N;

$BC = 500$ N (T); $BD = 1000$ N (C) **Ans.**

4.6.10 For the rhomboidal frame shown in Fig. P-4-6.10, find the forces acting in *AC* and *AD*.

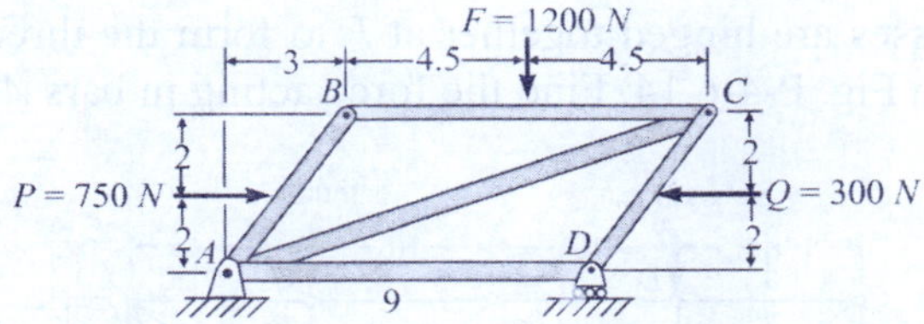

Figure P-4-6.10

$AC = 1580$ N (T); $AD = 975$ N (C) ***Ans.***

4.6.11 Resolve Prob. 4-6.10 if the magnitudes of the forces P and Q are interchanged.

4-6.12 A three-hinged arch is composed of two trusses hinged together at B as shown in Fig. P-4-6.12. Determine the type and magnitude of force in bars BD and BE.

Hint: First isolate each truss as a free body.

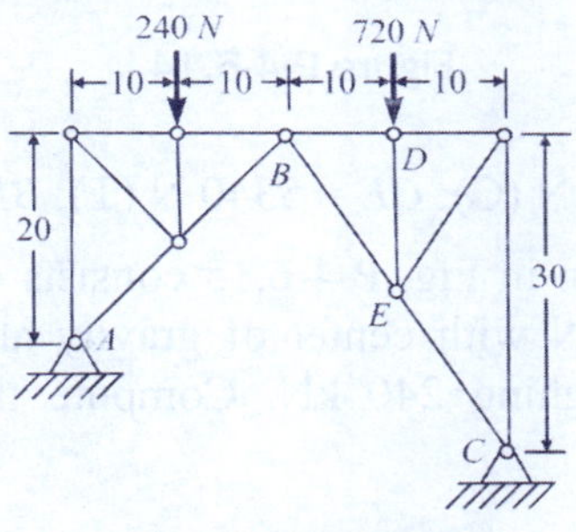

Figure P-4-6.12

4-6.13 As shown in Fig. P-4-6.13, two trusses ABC and DEF are supported respectively by hinges at A and F and joined by bars BE and CD. Determine the force in bar BE.

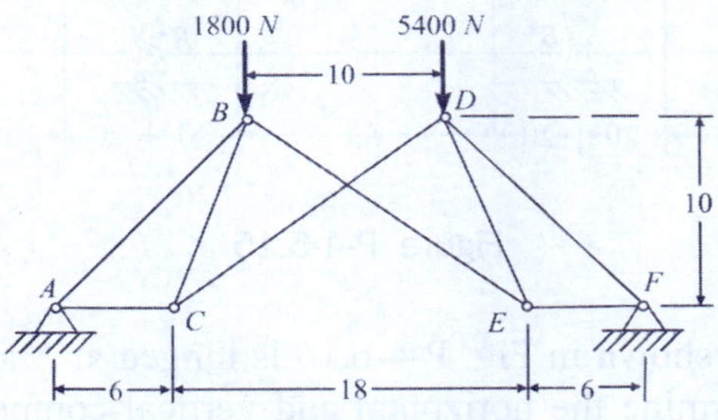

Figure P-4-6.13

$BE = 2410$ N (C) ***Ans.***

4-6.14 Two trusses are hinged together at F to form the three-hinged arch as shown in Fig. P-4-6.14. Find the force acting in bars AC, BE, and CE.

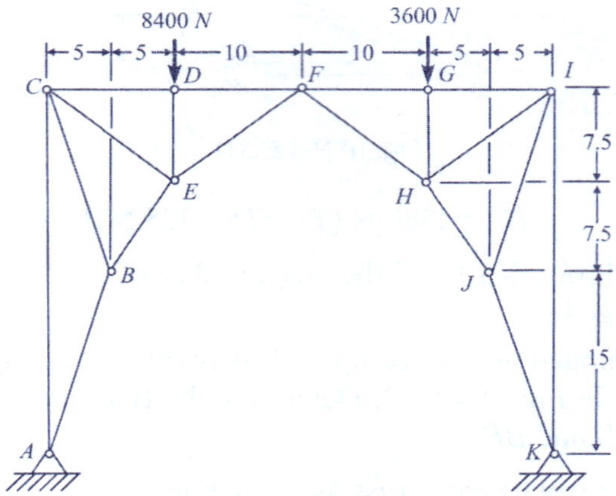

Figure P-4-6.14

$AC = 1200$ N (C); $CE = 5340$ N (T); $BE = 4810$ N (C) ***Ans.***

4-6.15 The bridge shown in Fig. P-4-6.15 consists of two end sections, each weighing 400 KN with center of gravity at G, hinged to a uniform center span weighing 240 kN. Compute the reactions at A, B, E, and F.

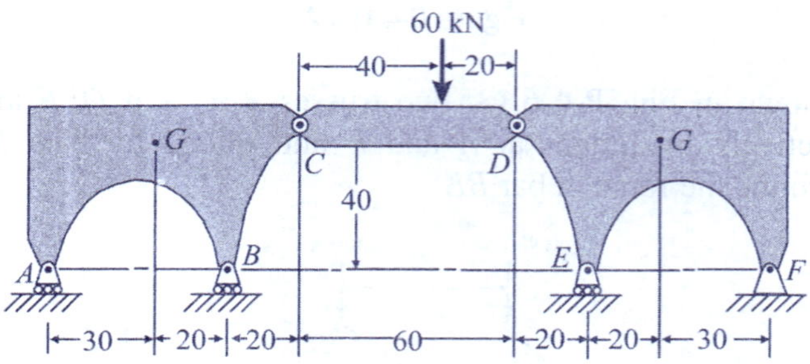

Figure P-4-6.15

4-6.16 The frame shown in Fig. P-4-6.16 is hinged at E and roller supported at A. Determine the horizontal and vertical components of the hinge forces at B, C, and D. Neglect the weights of the members.

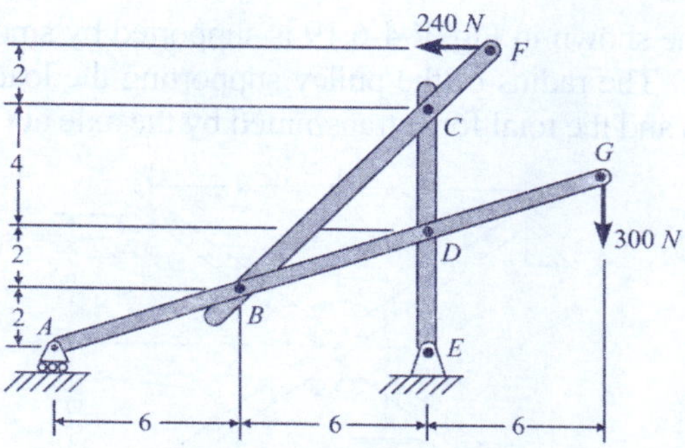

Figure P-4-6.16

4-6.17 The frame shown in Fig. P-4-6.17 is hinged to rigid supports at D and E and roller supported at A. Find the components of the hinge forces at D and E caused by the given loads.

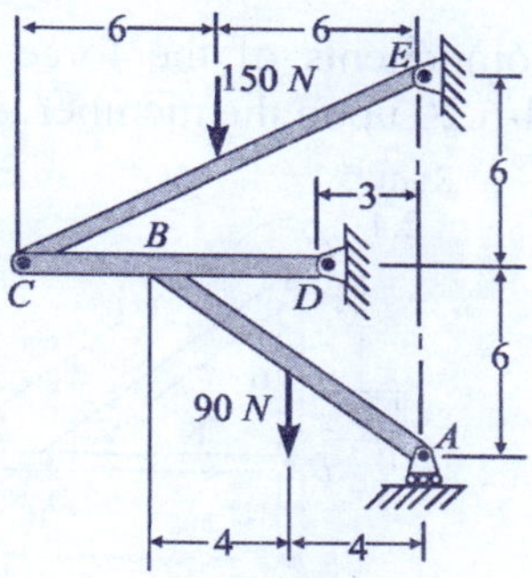

Figure P-4-6.17

$D_h = 200$ N; $D_v = 20$ N; $E_h = 200$ N; $E_v = 175$ N **Ans.**

4-6.18 For the frame shown in Fig. P-4-.6.18, determine the horizontal and vertical components of the hinge force at B as it acts upon member AC.

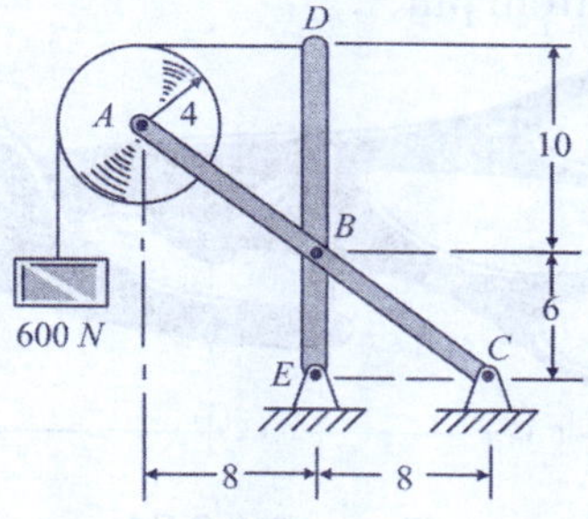

Figure P-4-6.18

4.6.19 The frame shown in Fig. P-4-6.19 is supported by smooth surfaces at A, B, D, and E. The radius of the pulley supporting the load W is 2 m. Find the reactions and the total force transmitted by the axle at C to member BE.

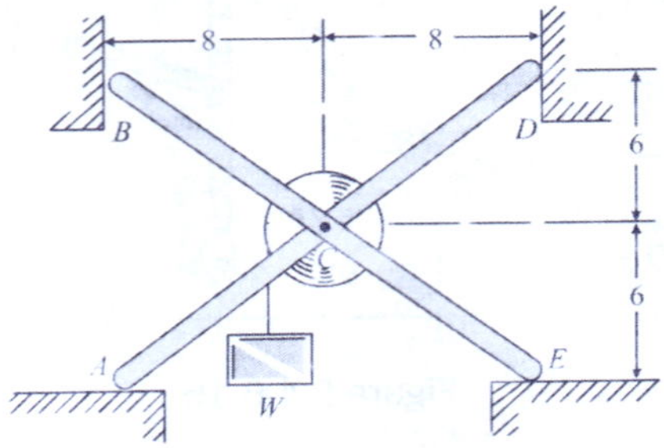

Figure P-4-6.19

$$C = \frac{5}{8} W$$ ***Ans.***

4.6.20 Determine the components of the force exerted by the axle of the pulley in Fig. P-4-6.20 upon the member AC and upon member CF.

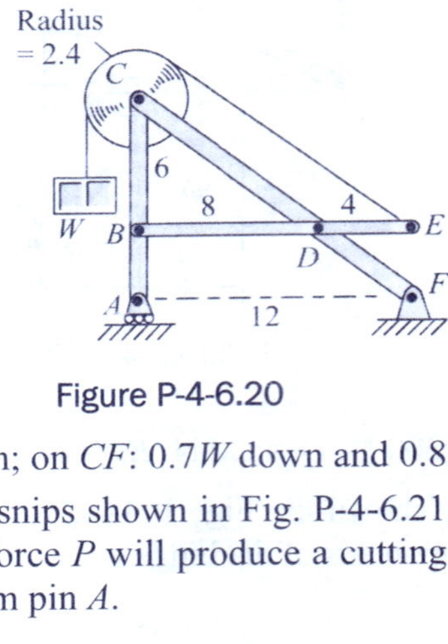

Figure P-4-6.20

On AC: $0.9W$ down; on CF: $0.7W$ down and $0.8W$ right ***Ans.***

4-6.21 For the compound snips shown in Fig. P-4-6.21, find the dimension b so that a gripping force P will produce a cutting force of $9P$ at 0.15 m along the blade from pin A.

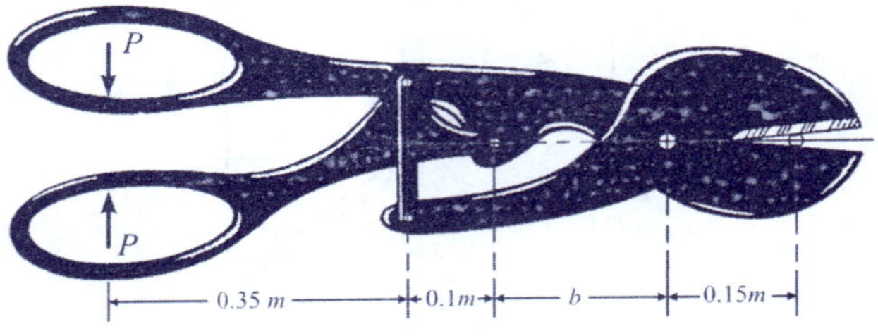

Figure P-4-6.21

$b = 0.2$ m ***Ans.***

4-6.22 Find the gripping force developed in the vise grip pliers by two 75 N forces applied as shown in Fig. P-4-6.22.

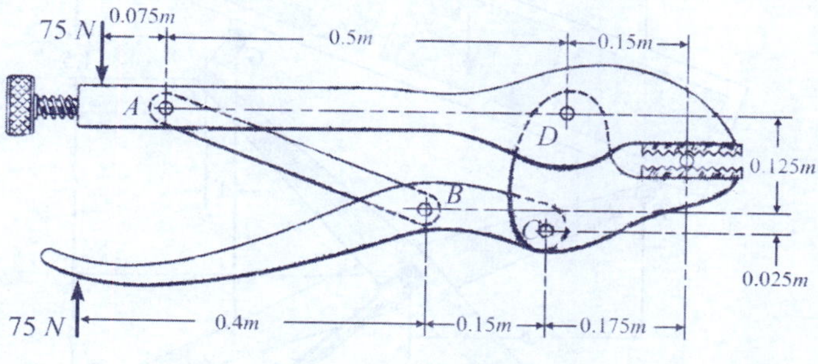

Figure P-4-6.22

1329 N *Ans.*

4-6.23 The scoop of a bulldozer is loaded with 4 KN of dirt with center of gravity at G. Duplicate linkages exist on each side of the machine. For the particular position in Fig. P-4-6.23, find the force P acting in the hydraulic cylinder AB and the force on the bolts at C and D. Neglect the weights of the linkage.

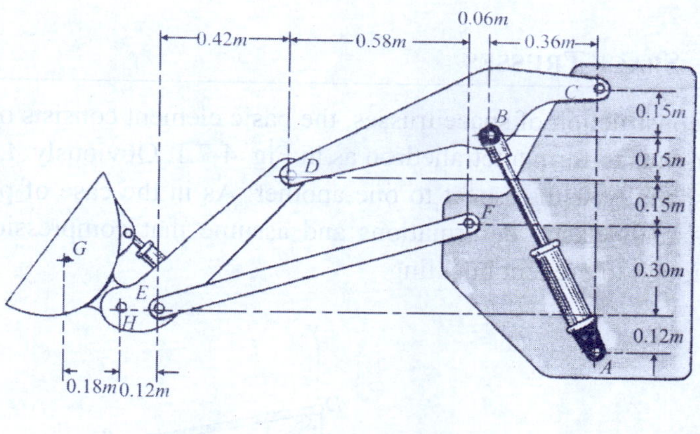

Figure P-4-6.23

$P = 5140$ N; $C = 6860$ N; $D = 5550$ N *Ans.*

4-8.24 An elevating dump truck body is activated by duplicate linkages on each side of the truck. At the position in Fig. P4-6.24, find the force P in the hydraulic cylinder BC and in the links CD and FG if the total load is $W = 10{,}520$ N. Neglect the weights of the linkage and note that each set of linkages supports half the total load.

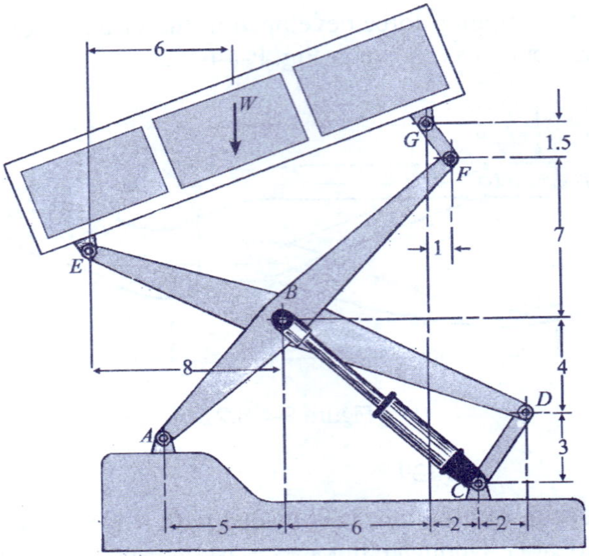

Figure P-4-6.24

$P = 7300$ N; $CD = 3000$ N; $PG = 2160$ N ***Ans.***

4-7 SPACE TRUSSES

In the construction of space trusses, the basic element consists of six bars joined at their ends to form a tetrahedron as in Fig. 4-7.1. Obviously *A*, *B*, *C*, and *D* are fixed spatially with respect to one another. As in the case of plane trusses, we ignore small elastic deformations and assume that compression members are stiff enough to prevent buckling.

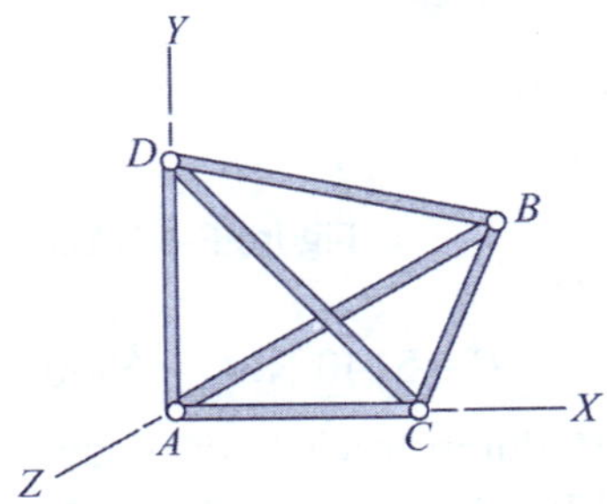

Figure 4-7.1 Basic tetrahedron.

Each additional joint is located by joining three more noncoplanar bars emanating from previously fixed joints. Thus, to the base tetrahedron *ABCD* in Fig. 4-7.2, joint *E* is determined by adding the three noncoplanar bars emanating from *B*, *C*, and *D*. In a similar manner, joints *F* and *G* have been formed. Usually the joints are denoted in the alphabetical order in which they are located since this makes it easier to visualize the construction of the truss and aids in analyzing the truss.

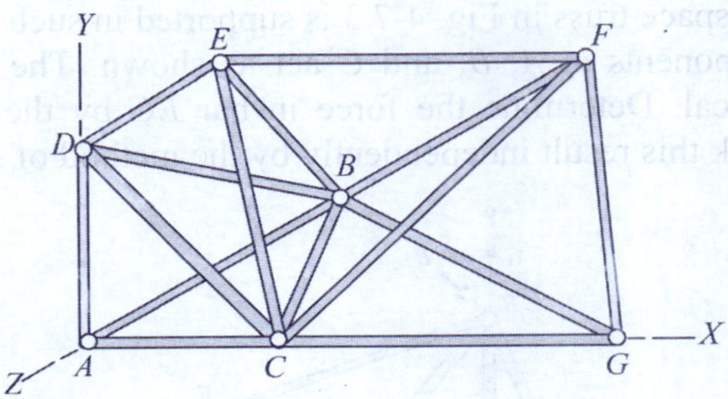

Figure 4-7.2 Space truss formed on base tetrahedron *ABCD*.

Space trusses constructed in this manner are called *simple trusses*. Two simple trusses connected by six noncoplanar bars form a *compound truss*. Also, as in plane trusses, the loads are assumed to be applied at the joints, and the various joints are assumed to be ball-and-socket connections so that the force transmitted by each bar is axial. To justify these assumptions in the case of riveted or welded connections, due care must be taken in assembling the truss to ensure that the centerlines of the joined bars intersect at a common point.

To anchor a space truss to the ground, six support reactions are needed to satisfy the six scalar equations of equilibrium. The supports may consist of a combination of six noncoplanar bars attaching the truss to at least three points not in the same straight line, or may be a ball-and-socket support and three other noncoplanar bars.

The criterion for stability of the space truss is obtained by noting that for the concurrent force system at each joint, there are three scalar equations of equilibrium. Hence, at j joints, there are a total of $3j$ equations that can be solved simultaneously for the forces in m members plus six external reactions; i.e.

$$3j = m + 6 \tag{4-7.1}$$

As in planar trusses, two methods are available for determining the forces in the members of a space truss: the method of joints and the method of sections. We shall apply each in the illustrative problem at the end of this discussion.

Analogous to the discussion concerning zero load members for planar trusses is the rule for determining zero load members in a space truss: At any joint where the load and all members except one lie in the same plane, that one is a zero load member. Also at a joint subjected to the action of three noncoplanar members but no other load, the force in each member must be zero.

ILLUSTRATIVE PROBLEM

4-7.1 The space truss in Fig. 4-7.3 is supported in such a way that the reaction components at A, B, and C act as shown. The loads at F and G are vertical. Determine the force in bar EC by the method of joints and check this result independently by the method of sections.

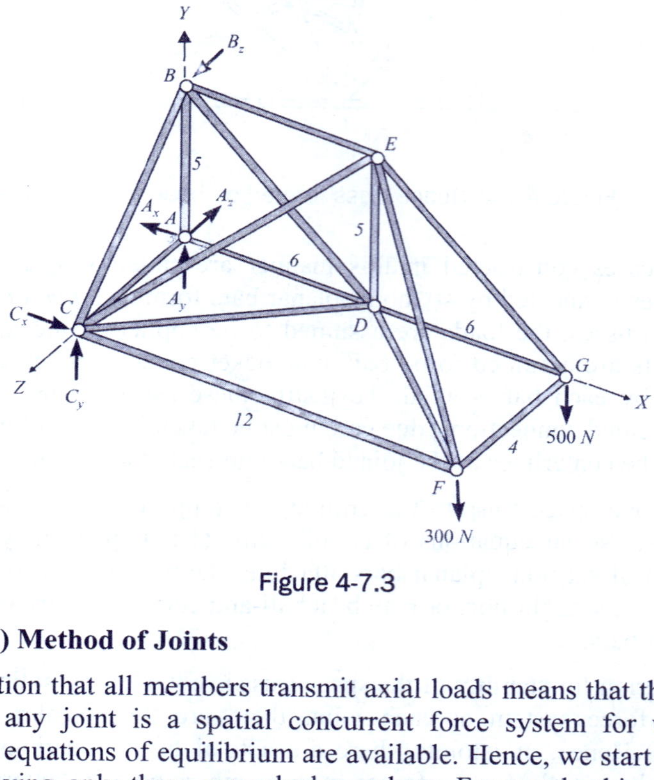

Figure 4-7.3

Solution: (a) Method of Joints

The assumption that all members transmit axial loads means that the free-body diagram of any joint is a spatial concurrent force system for which three independent equations of equilibrium are available. Hence, we start the analysis at a joint having only three unmarked members. Frequently this can be done without first finding the reactions since such a joint is usually the last one formed on the base tetrahedron. In this case, it is G. Once the forces at a starting joint have been found, their effects on adjacent joints are known. Continue by considering the next joint at which only three unknown forces act, and so on successively until the unknown force in each member has been found. Usually, each joint is analyzed in the inverse order in which it is located.

We start with joint G which has only three unknown bar forces acting on it. We need not draw a FBD of the joint if we assume each bar acting on the joint to be in tension and hence to pull away from it as shown in the line drawing of the truss in Fig. 4-7.4. Thus, at G, each bar will act from G toward its other end and will be designated as GE, GD, and GF. The components of these forces will be proportional to the corresponding coordinate dimensions of each bar and its length. Taking force summations as positive along the positive directions of the coordinate axes, we have

$$[\Sigma X = 0] \qquad -\frac{6}{\sqrt{61}} GE - GD = 0$$

$$[\Sigma Y = 0] \qquad \frac{5}{\sqrt{61}} GE - 500 = 0$$

$$[\Sigma Z = 0] \qquad GF = 0$$

Solving, we obtain $GE = 100\sqrt{61}$ N; $GD = -600$ N; and $GF = 0$.

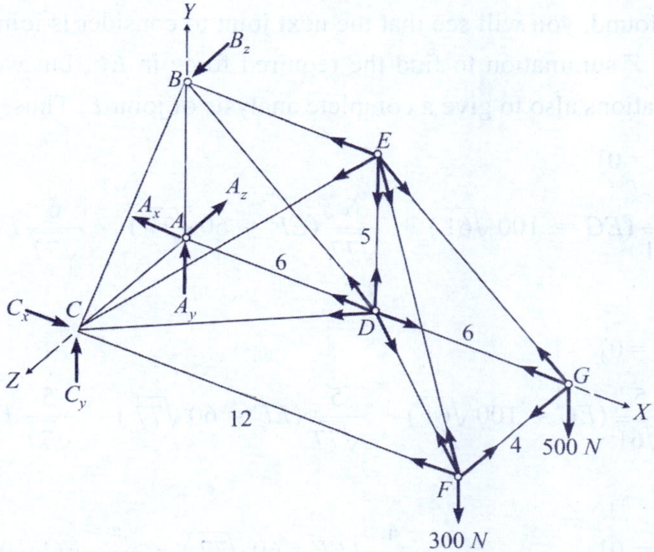

Figure 4-7.4 Joints analyzed in the order G, F and E.

The positive result for GE confirms our assumption of tension whereas the negative result for GD means it is in compression. Do not neglect the negative sign of a bar force in subsequent computations since we are assuming that all bar forces pull away from their end joints. Hereafter, the plus and minus sign are sufficient to indicate tension or compression respectively.

The next joint having three unknown bar forces acting upon it is joint F. Since $FG = GF = 0$, we observe that all loads except FC lie in the same plane DFE; hence, we conclude that $FC = 0$. This conclusion will be confirmed by formally applying the force summations of equilibrium to joint F. We obtain

$$[\Sigma X = 0] \qquad -FC - \frac{6}{\sqrt{52}} FD - \frac{6}{\sqrt{77}} FE = 0$$

$$[\Sigma Y = 0] \qquad \frac{5}{\sqrt{77}} FE = 300$$

$$[\Sigma Z = 0] \qquad -\frac{4}{\sqrt{52}} FD - \frac{4}{\sqrt{77}} FE = 0$$

Solving, we obtain $FE = 60\sqrt{77}$ N; $FD = -60\sqrt{52}$ N; $FC = 0$.

If you have marked each end of a bar on the truss diagram after the force in it has been found, you will see that the next joint to consider is joint E. Here we need only a Z summation to find the required force in EC, but we include the other summations also to give a complete analysis of joint E. Thus, we have

$$[\Sigma X = 0]$$

$$\frac{6}{\sqrt{61}}(EG = 100\sqrt{61}) + \frac{6}{\sqrt{77}}(EF = 60\sqrt{77}) - \frac{6}{\sqrt{77}} EC - EB = 0$$

$$[\Sigma Y = 0]$$

$$-\frac{5}{\sqrt{61}}(EG = 100\sqrt{61}) + \frac{5}{\sqrt{77}}(EF = 60\sqrt{77}) - \frac{5}{\sqrt{77}} EC - ED = 0$$

$$[\Sigma Z = 0] \qquad -\frac{4}{\sqrt{77}}(EF = 60\sqrt{77}) + \frac{4}{\sqrt{77}} EC = 0$$

From the last equation, we obtain $EC = -EF = -60\sqrt{77} = -527$ N as our required answer. In each of the above equations, we have inserted in parentheses the equivalent value of previously computed bar forces to make it simpler for you to follow the procedure. Although not required here, the values of EB and ED are 1320 N and – 500 N respectively.

(b) Method of Sections

To apply the method of sections, we isolate a rigid segment of the truss by passing a cutting surface through not more than six bars. The isolated segment is a free body in equilibrium under the action of the applied loads acting upon the isolated segment and supported by the forces acting in each of the cut bars. If you assume these forces to be tensile and hence to pull away from their end joints in the isolated segment, a separate free-body diagram usually need not be drawn, although we do so here to ensure a clearer explanation.

To find the force in *CE* by the method of sections, isolate the truss segment *DEFG* as shown in Fig. 4-7.5 by imagining a cutting surface passing through *EC*, *EB*, *DA*, *DC*, *DB*, and *FC*. We apply the concept of taking a moment summation about a center that will eliminate as many unknowns as possible. Since three of the cut bars (i.e., *DB*, *DA*, and *DC*) intersect at *D*, we apply $\Sigma \mathbf{M}_D = 0$. Of the remaining bars, *EB* and *FC* are parallel to the *X* axis and hence will be eliminated from a moment summation about the *X* axis through *D* which therefore solves directly for *EC*. Assuming the components of *EC* to act at *E*, we obtain

$$[\Sigma M_{XD} = 0] \qquad 5(EC_z) + (4)(300) = 0 \qquad EC_z = -240 \text{ N}$$

whence $EC_Z = (4/\sqrt{77})\, EC$ yields $EC = -527$ N as before with the method of joints.

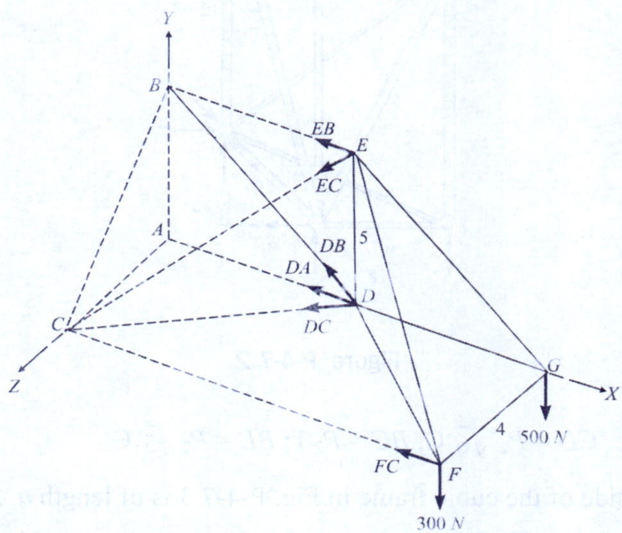

Figure 4-7.5 Free-body diagram of truss segment *DEFG*.

It may be observed that $\Sigma M_{YD} = 0$ eliminates all forces on the isolated body except *FC* and confirms that the force in *FC* is zero as we found from the method of joints.

A final observation is in order. Here we were able to eliminate three unknowns which had a common moment center. In general, a moment summation can always be taken about the intersection of two unknowns, but then the components of such a moment summation would involve the unknown forces in the other four bars. To complete the solutions, an additional equation must then be obtained from one of the force summations, or a moment summation about another center.

PROBLEMS

4-7.2 The space truss in Fig. P-4-7.2 is supported in such a way that the reaction components at *A*, *B*, and *C* act as shown. It carries the loads $-P_1\hat{\mathbf{j}}$ at *F* and $P_2\hat{\mathbf{i}}$ at *E*. Find force in bars *CD*, *BC*, and *BE*.

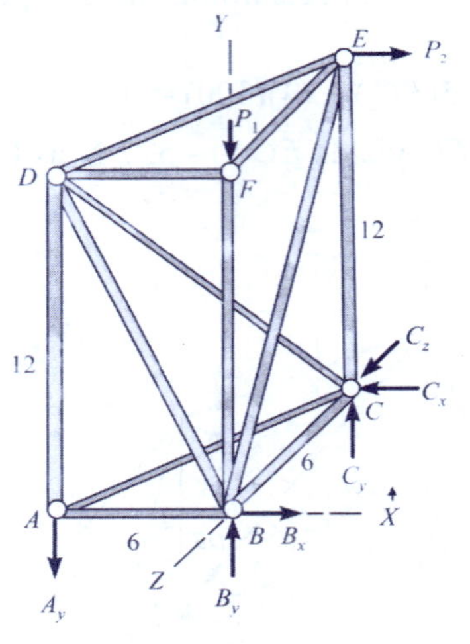

Figure P-4-7.2

$$CD = P_2\sqrt{6}\text{ C};\ BC = P_2\text{ T};\ BE = P_2\sqrt{5}\text{ C} \qquad \textbf{\textit{Ans.}}$$

4-7.3 Each side of the cubic frame in Fig. P-4-7.3 is of length *a*. A force $P\sqrt{3}$ is applied at *C* directed toward *B* and an equal force is applied at *B* directed toward *G*. Find the force in bars *DF*, *CD*, *CF*, and *BF*.

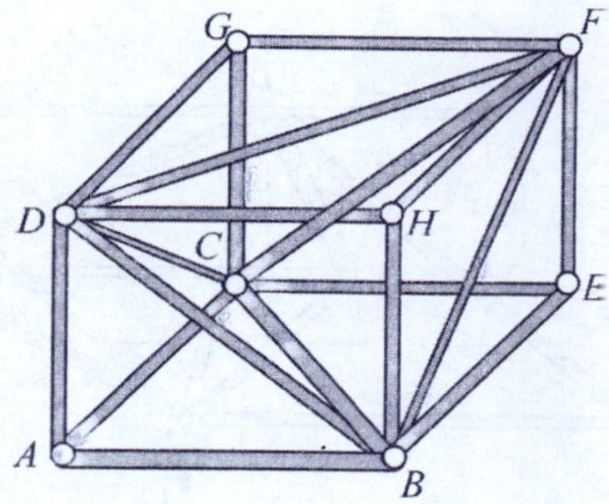

Figure P-4-7.3

Each equals $P/\sqrt{2}$ ***Ans.***

4-7.4 The reaction components of the space truss in Fig. P-4-7.4 act as shown. The applied loads in KN are $\mathbf{P} = 2\,\hat{\mathbf{i}}$ and $\mathbf{F} = 3\,\hat{\mathbf{j}}$. Determine the force in each bar.

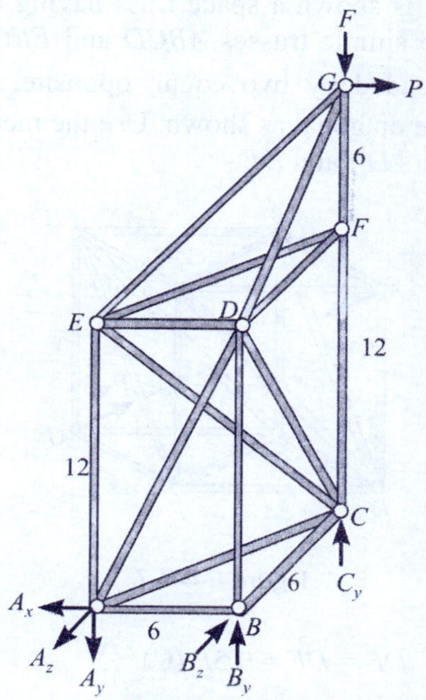

Figure P-4-7.4

In kN: $AC = 2.83$ T; $CD = 4.48$ T; $EC = 3.46$ T ***Ans.***

4-7.5 Find the force in bars *BE*, *CE*, and *CF* of the space truss shown in Fig. P-4-7.5.

Hint: Isolate the truss portion *DEFG* as a free body and apply $\Sigma M_D = 0$.

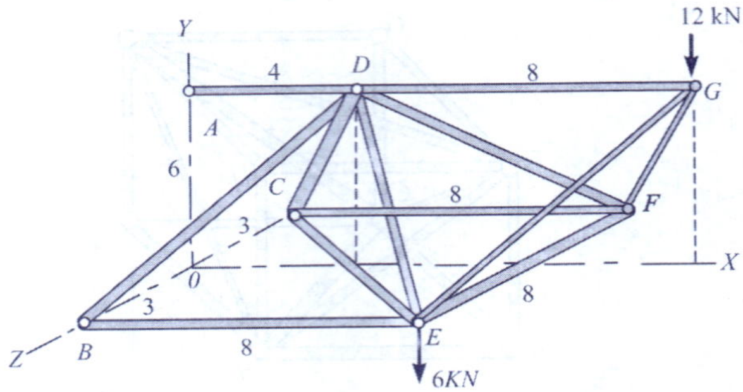

Figure P-4-7.5

$CF = 8$ KN (C); $CE = 5$ KN (C) ***Ans.***

4-7.6 Find the force in bars *BD* and *BE* of Fig. P-4-7.5 by isolating the truss portion *CDEFG* as a free body.

4-7.7 In Fig. P-4-7.7 is shown a space truss having the shape of a cube. It consists of two simple trusses *ABCD* and *EFGH* joined by six bars. The truss is loaded by two equal, opposite, collinear forces $P\sqrt{3}$ acting along the diagonal as shown. Use the method of sections to find the forces in bars *DF* and *DE*.

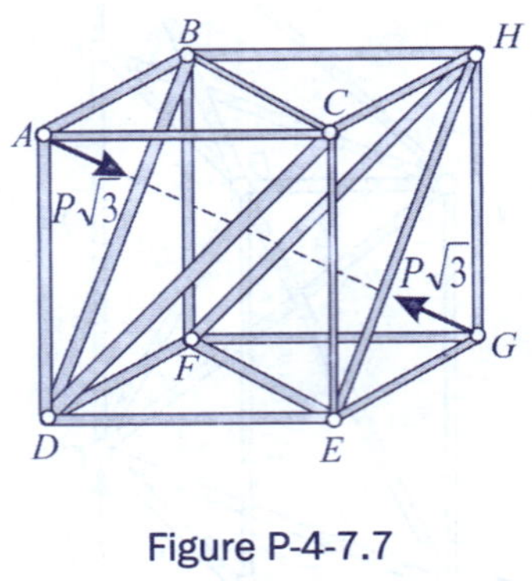

Figure P-4-7.7

$DE = DF = 0.5P$ (C) ***Ans.***

4-7.8 In Fig. P-4-7.8 is shown a compound space truss consisting of two tetrahedrons *ABCD* and *EFGH* joined by six bars. The truss is supported by reactions at *A*, *B*, and *C* acting as shown. Use the method of sections to find the force in bars *DF* and *CF*.

Hint: Isolate the tetrahedron carrying the simpler loads.

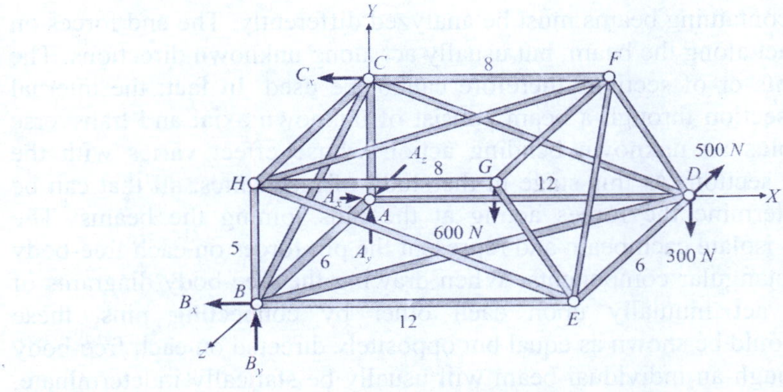

Figure P-4-7.8

$DF = -307$ N; $CF = +384$ N ***Ans.***

SUMMARY

Structures are composed of members which are commonly divided into two groups: those subjected to axial tensile or compressive loads only and those subjected to bending. *A* member of the first type is called a two-force member; a member of the second type is called a beam.

In two-force members, the forces are usually applied at the ends of the member and tend either to lengthen or to shorten it, thereby producing respectively tensile or compressive forces in the member. The action of a tensile member on its end pins is to pull away from them; the effect of a member subjected to compression is to push back toward its end pins.

The forces distributed throughout pin-connected trusses are determined on the assumption that the truss members are two-force members. Hence, the forces acting on any joint form a concurrent system. The conditions of equilibrium for concurrent force systems form the basis of the method of joints (Section 4-3). Once the forces are found at a starting joint where only two members act, their effects on adjacent joints are known. Continue by considering the next joint at which only two unknown forces act and so on successively until the force in each member has been found.

The method of sections (Section 4-4) is used to determine the force directly in any member. The method consists of isolating a portion of the truss by passing an imaginary cutting plane through that member and usually not more than two other members. Assuming that external reactions have been found, this establishes a nonconcurrent system in which appear the three unknown forces in the cut members. Each of these forces can then be found from the conditions of equilibrium for nonconcurrent force systems.

Structures containing beams must be analyzed differently. The end forces on beams do *not* act along the beam, but usually act along unknown directions. The method of joints or of sections therefore cannot be used. In fact, the internal forces over a section through a beam consist of unknown axial and transverse shear forces plus an unknown bending action whose effect varies with the location of the section. At this stage in the study of mechanics, all that can be done is to determine the forces acting at the pins joining the beams. The procedure is to isolate each beam and represent the pin forces on each free-body diagram by rectangular components. When drawing the free-body diagrams of beams which act mutually upon each other by connecting pins, these components should be shown as equal but oppositely directed on each free-body diagram. Although an individual beam will usually be statically indeterminate, the simultaneous application of the conditions of equilibrium to the nonconcurrent forces on each free-body diagram will determine the unknown components of the forces at the connecting pins (refer to Section 4-6).

For space trusses consisting of two-force members, two methods are available for determining the forces in the members: the method of joints and method of sections as in planar trusses. To apply the method of joints, start with a free-body diagram of a joint where only three members act and apply the conditions of equilibrium for concurrent spatial force systems. Then proceed to the next joint at which three unknown forces act, and so on. Usually the order in which each joint is analyzed is the inverse order by which it was located. To apply the method of sections, isolate a portion of the truss by using an imaginary cutting surface passing through not more than six members and apply the equilibrium conditions for nonconcurrent spatial force systems. Generally, the method of joints is preferable for the analysis of space trusses.

CHAPTER 5

FRICTION

5-1 INTRODUCTION

Friction may be defined as the contact resistance exerted by one body upon a second body when the second body moves or tends to move past the first body. From this definition, it should be observed that friction is a retarding force always acting opposite to the motion or the tendency to move. As we shall see, friction exists primarily because of the roughness of the contact surfaces. If the contact surfaces are perfectly smooth, as assumed in some earlier problems, there is no friction.

In machines, friction is both a liability and an asset. Where it causes loss of power and/or wear it is undesirable. On the other hand, friction is essential for various holding and fastening devices as well as for friction drives and brakes and, indeed, even for supplying the traction for walking or driving an automobile. In this chapter we shall consider the application of the principles of friction to engineering problems.

5-2 THEORY OF FRICTION

The following experiment is useful in discussing the principles of friction as applied to dry unlubricated surfaces. Let a block of weight W rest on a rough horizontal surface and assume a horizontal force P to be applied to the block as shown in Fig. 5-2.1. When P is zero, the block is in equilibrium and the frictional resistance F is also zero. When P is given increasing values that are insufficient to cause motion, the frictional resistance F increases correspondingly to maintain equilibrium. Eventually, the block is on the verge of moving and at this instant, F attains its maximum available value. Any further increase in P causes motion, but surprisingly, the value of F does not stay at its maximum value but decreases rapidly to a kinetic value which remains fairly constant, as depicted on the graph in Fig. 5-2.2. This graph is not rigorously accurate; it is intended only as a guide for visualizing the results of this experiment.

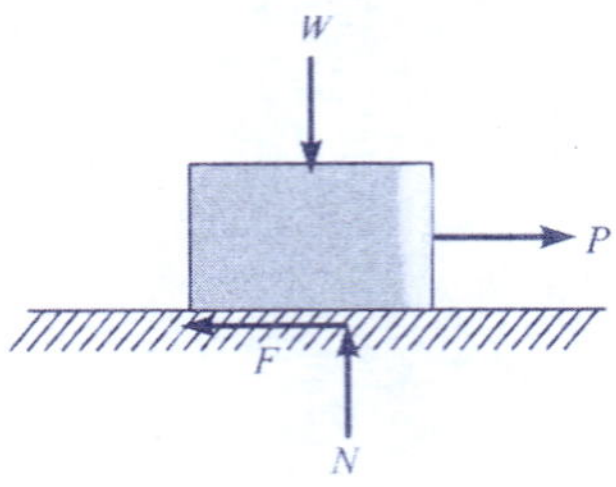

Figure 5-2.1 Frictional resistance.

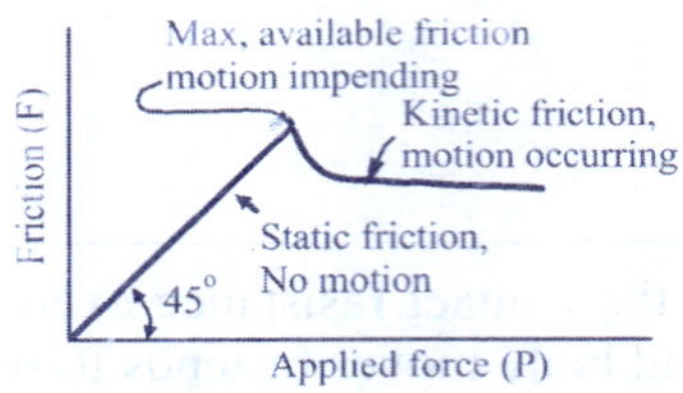

Figure 5-2.2 Variation of frictional resistance.

One way of understanding these results is to examine a magnified view of the contact surfaces. These are shown in Fig. 5-2.3 together with a FBD of the block. The surfaces are assumed to be composed of irregularities (which can be likened to hills and valleys) which mesh together. The frictional resistance F is developed by the effort of P to pull the hills of the block out of the meshing valleys of the supporting surface. In an ideal smooth surface there are no hills and vales to mesh together, and no frictional resistance is possible. This is why we have assumed the reaction to be normal to the surface when discussing ideal smooth surfaces in preceding sections. Actually an ideal smooth surface does not exist; it may be approximated by superfinishing ground surfaces.[1]

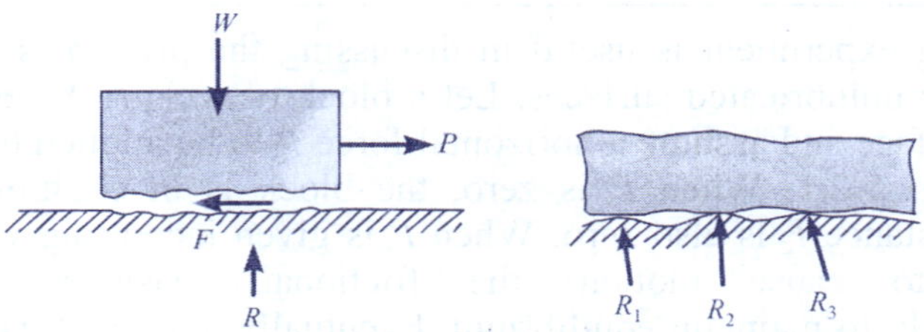

Figure 5-2.3 Magnified view of contact surfaces.

---

[1]In some cases, such as carefully finished flat surfaces of gage blocks, the surfaces adhere because of molecular adhesion. Where adhesion exists, it will be assumed to be part of the frictional resistance and its effect to be included in the coefficient of friction.

It is apparent that frictional resistance depends on the amount of wedging action between the hills and vales of the contact surfaces. The measure of this wedging action depends on the normal pressure R between the surfaces. As a result, the maximum frictional resistance that may exist is proportional to the normal pressure and is expressed symbolically as

$$F \propto R \tag{a}$$

where the sign $\propto$ is read as "is proportional to." This relation may be reduced to an equation by putting in a constant of proportionality, say f, which depends on the roughness of the contact surfaces. This constant is called the coefficient of friction, and Eq. (a) may be rewritten as

$$F_{max} = fR \tag{5-2.1}$$

When we use this equation, F is the maximum available frictional resistance developed when motion is impending.

In an actual situation, the equations of equilibrium will determine the value of F to maintain equilibrium. Of course, if F as determined from equilibrium conditions is less than or just equal to the maximum available friction, equilibrium will exist,; i.e., Eq. (5-2.1) determines only how much friction is available to maintain equilibrium. See Illus. Prob. 5-2.1 for additional explanation.

Obviously, if the resultant applied force in the direction of motion exceeds the maximum available friction, motion of the block ensues. In motion, there is less chance for the wedging action described above to take place; i.e., the hills and vales are not as free to mesh as when the block was at rest. This is the reason for a decrease in frictional resistance when motion occurs.

In the case of moving bodies, the resistance is defined as the kinetic-friction force. It always acts at its maximum value. Note the contrast with static friction which adjusts itself to the value required to resist motion (Refer to the graph discussed before). The magnitude of the kinetic friction is also expressed by Eq. (5-2.1) by inserting the coefficient for kinetic friction f_k in place of the coefficient for static friction f_s. In many instances, the subscripts k and s for the coefficient of friction are omitted when the statement of the data makes the meaning clear.

When motion first begins, the hills and vales of the contact surfaces mesh with less frequency, causing the static friction to decrease rather abruptly to the value of kinetic friction as motion starts. For high speeds, the kinetic friction decreases still more. When a body in motion is being brought to rest, the converse is true. It is usually assumed, however, that the kinetic friction is constant in value.

Some additional explanation of the phenomenon of dry friction is pertinent. The total normal pressure R and friction F are really the respective resultants of the forces existing at each of the high spots of the irregularities of the mating surfaces. When the surfaces are in relative motion, the individual forces are nearer the tops of the contact points and the components of these forces along the direction of motion are smaller than when the contact surfaces are at rest. Experiments also show that the friction force is essentially independent of the apparent contact area. The true contact area is much smaller, occurring as it does at the peaks of the contact irregularities. The high intensity of normal pressure at these peaks results in localized yielding. Consequently, if the normal pressure is doubled, the area of real contact must also double. The apparent increase of a greater friction force when using wide tires on automobiles is due more to a shearing of more rubber in contact with the ground than due to a larger area of contact. The physical differences of the surfaces in contact also are significant. Although a definitive explanation of the friction of dry surfaces has yet to be achieved, the principles described above and summarized later give good engineering results, especially when the correct coefficient of friction is used.

The foregoing discussion applies only to friction between dry unlubricated surfaces. If a lubricant such as oil is used, the hills and vales of the contact surfaces are separated by the lubricant, thereby materially reducing the frictional resistance. Indeed, if the thickness of the lubricating film is sufficient to separate the contact surfaces completely, the only frictional resistance remaining will be the internal friction in the lubricant itself, measured by its viscosity. The viscosity varies with the temperature, being less at high than at low temperatures. For an extended discussion of the theory of lubricated surfaces, which is beyond the scope of this book, the reader is referred to a standard work upon this subject.

5-3 ANGLE OF FRICTION

Fig 5-2.1 is redrawn in Fig. 5-3.1 to show that F and R are really components of the total reaction R_F exerted by the plane surface upon the block. The size of the angle between R_F and R depends on the value of the frictional resistance F. If F is zero, this angle will be zero. As F increases, so does the angle. The particular value of this angle when maximum frictional resistance is acting is defined as the *angle of friction. It acts at its maximum value of* ϕ *only when motion is impending.*

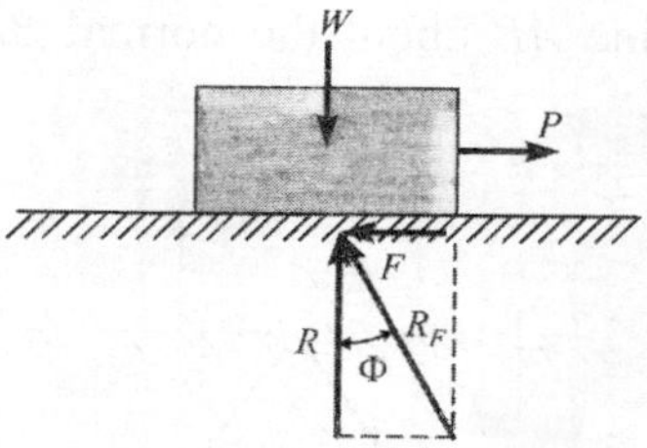

Figure 5-3.1 Angle of friction.

From Fig. 5-3.1, it is apparent that the angle of friction may be defined by the relation

$$\tan \phi = \frac{F}{R} \tag{b}$$

Comparing Eq. (*b*) with Eq. (5-2.1), we see that

$$\tan \phi = f \tag{5-3.1}$$

In other words, the tangent of the angle of friction is equal to the coefficient of friction.

A simple experiment for determining the angle of friction is to place a block of weight W upon an inclined plane for which the angle of inclination θ can be gradually increased from zero to a maximum value at which the block is on the verge of sliding down the incline. Fig 5-3.2 shows this condition. In order for equilibrium to exist, it is necessary that the lines of action of the weight W and the reaction R_F be collinear. From the geometry of the figure, it is evident that when motion impends, the angle of inclination θ determines the angle of friction ϕ.

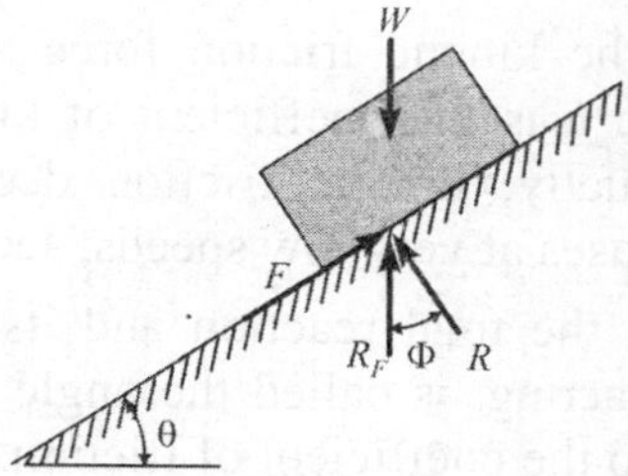

Figure 5-3.2 At impending motion, angle of friction equals angle of incline.

When we consider coplanar forces, in order for motion not to occur, the static reaction R_F must lie within the angle *ABC* in Fig. 5-3.3, whereas for problems involving noncoplanar forces, it must be contained within the cone generated by revolving line AB about the normal BN. The cone so formed is called the *cone of friction*.

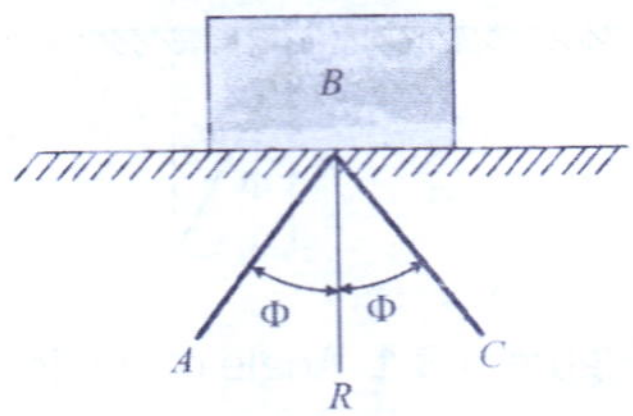

Figure 5-3.3 Cone of friction.

5-4 LAWS OF FRICTION

The principles discussed in the preceding sections summarize the results of experiments by Coulomb in 1781 on the friction of plane dry surfaces and those conducted by Morin in 1831. They constitute the laws of dry friction (sometimes called Coulomb friction) which may be stated as follows:

1. If friction is neglected, the reactions are always normal to the surfaces in contact.
2. Friction always acts to oppose the relative motion of the free body (or its tendency to move). It is tangential to the surfaces in contact.
3. If static friction is acting, the value of the friction force may vary from zero to its maximum available value, adjusting itself to the resultant force tending to cause motion.
4. The maximum available value of static friction (i.e., the limiting friction, when motion impends) is equal to $f_s R$ where f_s is the coefficient of static friction and R is the normal force.
5. If motion occurs, the kinetic friction force always acts at its constant value of $f_k R$ where f_k is the coefficient of kinetic friction and R is the normal force (Actually, kinetic friction decreases somewhat at high velocities and increases at very low speeds; see Section 5-2).
6. The angle between the total reaction and its normal component, when limiting friction is acting, is called the angle of friction. The tangent of this angle is equal to the coefficient of friction.

ILLUSTRATIVE PROBLEMS

5-4.1 A 200 N block is in contact with a plane inclined at 30° to the horizontal. A force P, parallel to and acting up the plane, is applied to the body. If the coefficient of static friction is 0.20, (a) find the value of P to just cause motion to impend up the plane, and (b) find P to just prevent motion down the plane. (c) If P = 80 N, determine the magnitude and direction of the friction force.

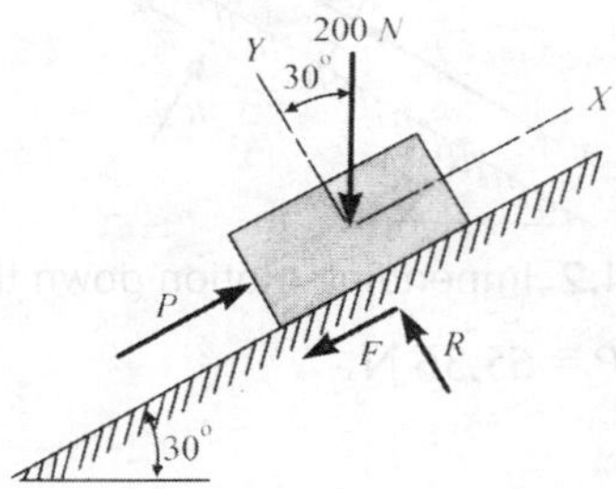

Figure 5-4.1 Impending motion up the plane.

Solution: Part (a)

If the block has impending motion up the plane, the friction force, acting at its maximum available value, will act down the plane to resist motion. The FBD of the block is shown in Fig. 5-4.1. While reference axes may be oriented at will, it is usually best to select them so that the X axis is parallel to and positive in the direction of impending motion. This choice of reference axes has the advantage of separating R and F. In this example, moreover, it also eliminates P from a Y summation of equilibrium and determines R directly;

thus,

$[\Sigma Y = 0]$ $\quad R - 200 \cos 30° = 0 \quad R = 173.2$ N

Since limiting friction acts, Eq. (5-2.1) determines F to be

$[F = fR]$ $\quad F = 0.2(173.2) \quad F = 34.64$ N

To determine P, we apply the equilibrium equation

$[\Sigma X = 0]$ $\quad P - (F = 34.64) - 200 \sin 30° = 0$

$P = 134.64$ N *Ans.*

Part (b)

When the block has impending motion down the plane, the friction force, acting at its maximum available value, will act up the plane to resist the motion. The FBD is shown in Fig. 5-4.2. Note that the X axis is chosen parallel to and positive down the plane in the direction of impending motion. $\Sigma Y = 0$ and the friction relation give the same values of R and F as in Part (a). Hence to find P, we need only apply the equation

$[\Sigma X = 0]$ $\quad 200 \sin 30° - P - (F = 34.64) = 0$

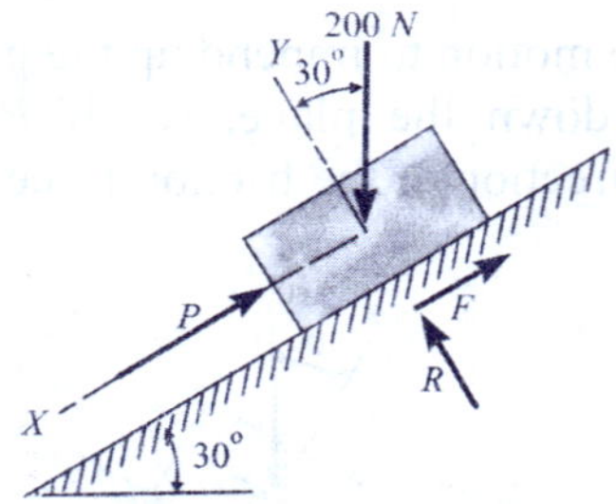

Figure 5-4.2 Impending motion down the plane.

$P = 65.36$ N **Ans.**

Part (c)

The results obtained for Parts (a) and (b) indicate that the 200 N block will remain at rest for all values of P between 65.36 N and 134.64 N, i.e., for P between 200 sin 30° $\pm F$ = 100 $\pm$ 34.64. Hence, when P = 80 N, the block will tend to slide down the plane, although limiting friction will not be developed. Since equilibrium is present, from Fig. 5-4.2, we have

$[\Sigma X = 0]$ $\quad$ $200 \sin 30^\circ - (P = 80) - F = 0$

$F = 20$ N **Ans.**

This result can be reached by another method of reasoning. When we temporarily neglect friction, a summation of forces parallel to the plane indicates that the down plane component of the block's weight (i.e., 200 sin 30° = 100 N) could not be balanced by an up plane force P of 80 N. For balance to exist, a frictional resistance of F = 20 N would be required acting up plane as shown in Fig. 5-4.2. Since the available static friction is 34.64 N, we conclude that limiting friction will not be developed. Similarly, if P were 130 N a down plane friction force of 30 N would be necessary for equilibrium.

5.4.2 A 200 N block is at rest on a 30° incline. The coefficient of friction between the block and the incline is 0.20. Compute the value of a horizontal force P that will cause motion to impend up the incline.

Solution

The FBD of the block is shown in Fig. 5-4.3a. Since motion is impending up the incline, the maximum static friction F is directed down the incline. A point diagram of the forces is formed by first selecting X and Y axes with the X axis parallel to and positive in the direction of impending motion, and then imagining the block squeezed to a point coincident with the origin of the axes. The forces on the body are then applied to this point to form the concurrent force system shown in Fig. 5-43b. The point diagram is sometimes more convenient than the FBD for computing components, especially here where it is evident that R is dependent on the Y components of both P and the 200 N weight.

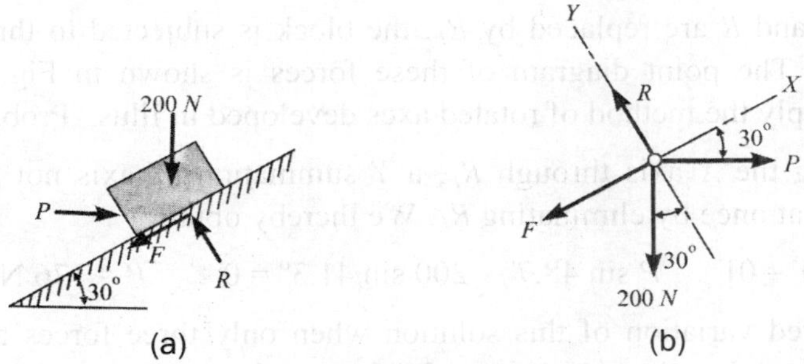

(a) (b)

Figure 5-4.3

The three unknowns R, F, and P are now found from the friction relation Eq. (5-2.1) and the two equations of equilibrium. We obtain

$[\Sigma Y = 0]$ $\qquad R - 200 \cos 30^\circ - P \sin 30^\circ = 0$

$\qquad R = 173.2 + 0.5P \qquad (a)$

$[F = fR]$ $\qquad F = 0.2(173.2 + 0.5P) = 34.64 + 0.1P \qquad (b)$

$[\Sigma X = 0]$ $\qquad P \cos 30^\circ - 200 \sin 30^\circ - F = 0 \qquad (c)$

On substituting the value of F from (b) into (c), we obtain

$$P = 176 \text{ N}$$ ***Ans.***

Alternate Solution

Whenever the normal pressure R must be expressed in terms of another unknown force, such as P in the preceding solution, it will generally be simpler to use the total reaction R_F instead of components R and F.

Since motion is impending, R_F will be inclined at the angle of friction ϕ with the normal pressure R as shown in Fig. 5-4.4a. The value of ϕ is found from Eq. (5-3.1) to be

$[\tan \phi = f]$ $\qquad \tan \phi = 0.20 \qquad \phi = 11.3^\circ$

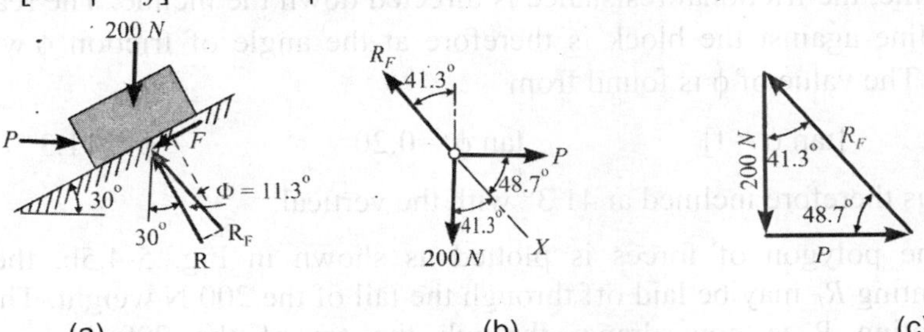

(a) (b) (c)

Figure 5-4.4

When F and R are replaced by R_F, the block is subjected to three forces in equilibrium. The point diagram of these forces is shown in Fig. 5-4.4(b) to which we apply the method of rotated axes developed in Illus. Prob. 3-4.1.

By taking the X axis through R_F, a Y summation (Y axis not shown) will determine P at once by eliminating R_F. We thereby obtain

$$[\Sigma Y = 0] \qquad P \sin 48.7^\circ - 200 \sin 41.3^\circ = 0 \qquad P = 176 \text{ N} \qquad \textit{Ans.}$$

A preferred variation of this solution when only three forces are involved consists of applying the sine law to the force polygon shown in Fig. 5-4.4(c). Since equilibrium exists, the force polygon will close. The 200 N weight is represented by the vertical vector shown. Through the tip of this vector, a horizontal line of indeterminate length is drawn to represent the known direction of P. From Fig. 5-4.4a, the known direction of R_F is $30^\circ + \phi = 41.3^\circ$ with the vertical. A line representing R_F may be drawn through the tail of the 200 N vector to intersect P as shown.

Values may now be obtained graphically by scaling from the polygon or analytically by applying the sine law. Using the latter, we have

$$\frac{P}{\sin 41.3^\circ} = \frac{200}{\sin 48.7^\circ}$$

whence as before

$$P = 176 \text{ N} \qquad \textit{Check}$$

5-4.3 Determine the minimum value and the direction of a force P required to cause motion of a 200 N block to impend up 30° incline. The coefficient of friction is 0.20.

Solution

The FBD of the block is shown in Fig. 5-4.5(a). Since motion is impending up the incline, the frictional resistance is directed down the incline. The reaction of the incline against the block is therefore at the angle of friction ϕ with R as shown. The value of ϕ is found from

$$[\tan \phi = f] \qquad \tan \phi = 0.20 \qquad \phi = 11.3^\circ$$

and R_F is therefore inclined at 41.3° with the vertical.

If the polygon of forces is plotted as shown in Fig. 5-4.5b, the vector representing R_F may be laid off through the tail of the 200 N weight. The vector representing P is now drawn through the tip of the 200 N vector and perpendicular to R_F. This obviously will give the minimum length of P (and hence the minimum value) to complete the polygon of forces. From Fig. 5-4.5b, we now have

$$P = 200 \sin 41.3^\circ = 132 \text{ N} \qquad \textbf{\textit{Ans.}}$$

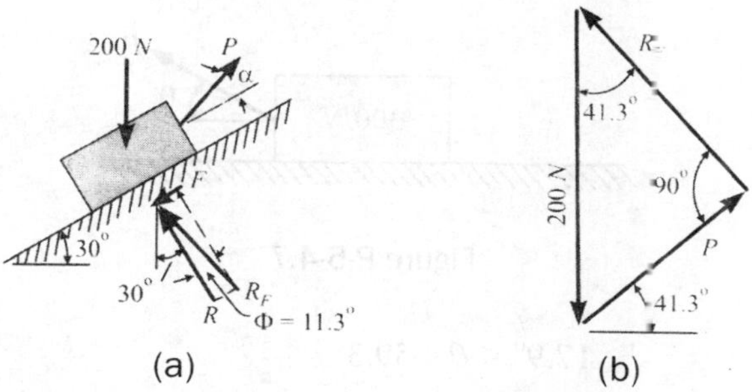

Figure 5-4.5

The inclination of P with the horizontal is also 41.3° whence the angle α is found to be

$$41.3^\circ = \alpha + 30^\circ \qquad \alpha = 11.3^\circ \qquad \textbf{\textit{Ans.}}$$

Observe that this value of α is equal to the angle of friction ϕ.

PROBLEMS

5-4.4 A block weighing W N is placed upon a plane inclined at an angle θ with the horizontal. Discuss what will happen if the angle of friction ϕ is (a) greater than θ, (b) equal to θ, (c) less than θ.

5-4.5 A 400 N block is resting on a rough horizontal surface for which the coefficient of friction is 0.40. Determine the force P required to cause motion to impend if applied to the block (a) horizontally or (b) downward at 30° with the horizontal. (c) What minimum force is required to start motion?

(a) 160 N; (b) 240 N; (c) 148.7 N ***Ans.***

5-4.6 The 500 N block shown in Fig. P-5-4.6 is in contact with a 45° incline. The coefficient of static friction is 0.25. Compute the value of the horizontal force P necessary to (a) just start the block up the incline or (b) just prevent motion down the incline. (c) If P = 400 N, what is the amount and direction of the friction force?

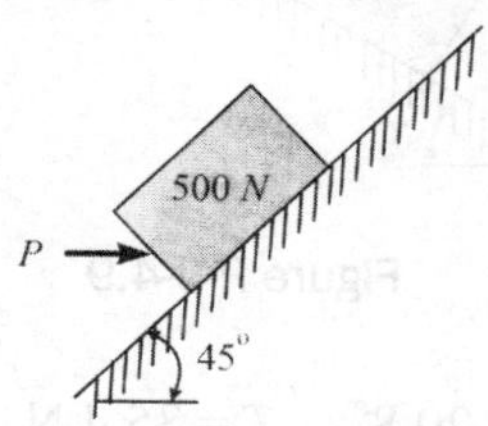

Figure 5-5-4.6

5-4.7 Referring to Fig. P-5-4.7, determine the range of values of θ for which the force $P = 200$ N will move the 300 N block.

Figure P-5-4.7

$17.9° < \theta < 59.3°$ ***Ans.***

5-4.8 Find the least horizontal force P to start motion of any part of the system of three blocks resting upon one another as shown in Fig. P-5-4.8. The weights of the blocks are $W_A = 300$ N, $W_B = 100$ N, and $W_c = 200$ N. Between A and B, $f = 0.30$; between B and C, $f = 0.20$; and between C and the ground, f = 0.10.

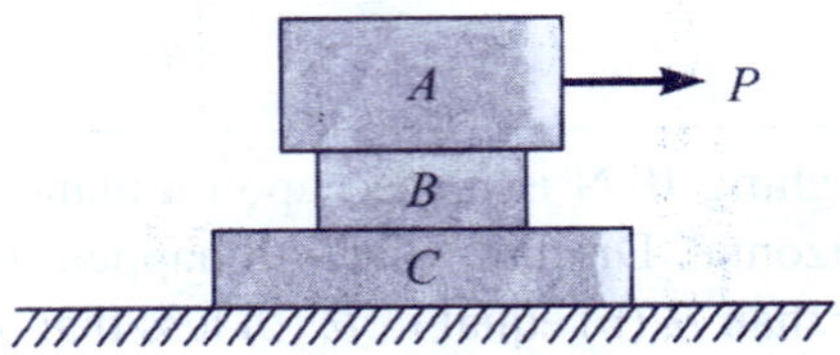

Figure P-5-4.8

5-4.9 Bodies A and B are joined by a cord parallel to the inclined plane as shown in Fig. P-5-4.9. Under body A which weighs 200 N, $f = 0.20$ while $f = 0.50$ under body B which weighs 300 N. Determine the angle θ at which motion impends. What is then the tension in the cord?

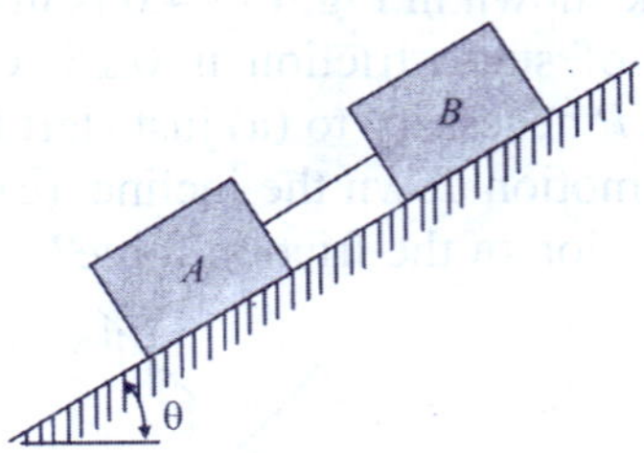

Figure P-5-4.9

$\theta = 20.8°$; $T = 35.4$ N ***Ans.***

5.4.10 Repeat the preceding problem if body A weighs 300 N, body B weighs 200 N. All other data remain the same.

5-4.11 The three blocks with weights as shown in Fig. P-5-4.11 are placed on a 20° incline so that they are in contact with each other and at rest. Determine which, if any, of the blocks will move and the friction force acting under each. Assume that under blocks A and C, the coefficients of friction are $f_s = 0.50$ and $f_k = 0.40$ while under B they are $f_s = 0.30$ and $f_k = 0.20$.

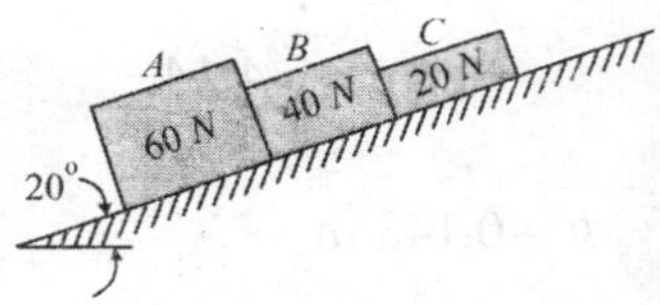

Figure P-5-4.11

$F_A = 22.9$ N; $F_B = 11.28$ N; $F_C = 6.84$ N **Ans.**

5-4.12 Solve the preceding problem if the coefficients of friction under A and C are $f_s = 0.30$ and $f_k = 0.20$ while under B they are $f_s = 0.50$ and $f_k = 0.40$.

5-4.13 Find the *least* value of P that will just start the system of blocks shown in Fig. p-5-4.13 moving to the right. The coefficient of friction under each block is 0.30.

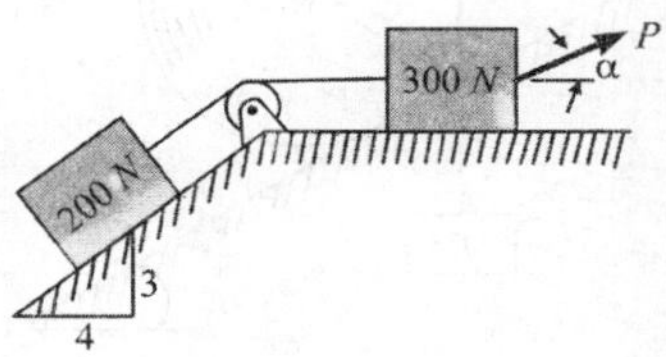

Figure P-5-4.13

$P = 247$ N; $\alpha = 16.7°$ **Ans.**

5-4.14 A homogeneous block of weight W rests upon the incline as shown in Fig. p-5-4.14. If the coefficient of friction is 0.30, determine the greatest height h at which a force P parallel to the incline may be applied so that the block will slide up the incline without tipping over.

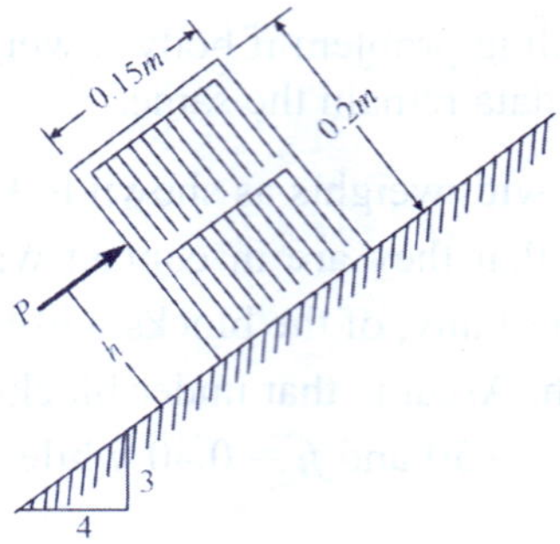

Figure P-5-4.14

$$h = 0.143 \text{ m}$$ ***Ans.***

5.4.15 In Fig. P-5-4.14, the homogeneous block weighs 400 N and the coefficient of friction is 0.40. If $h = 0.15$ m, determine the force P to cause motion to impend.

5-4.16 The 100 N cylinder shown in Fig. P-5-4.16 is held at rest by a weight P suspended from a cord wrapped around the cylinder. If slipping impends between the cylinder and the incline, determine P and the coefficient of friction.

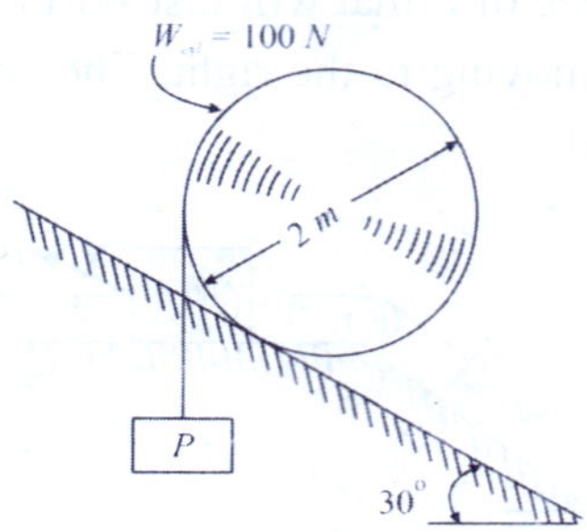

Figure P-5-4.16

5.4.17 Block A in Fig. P-5-4.17 weighs 120 N, block B weighs 200 N, and the cord is parallel to the incline. If the coefficient of friction for all surfaces in contact is 0.25, determine the angle θ of the incline for which motion of B impends.

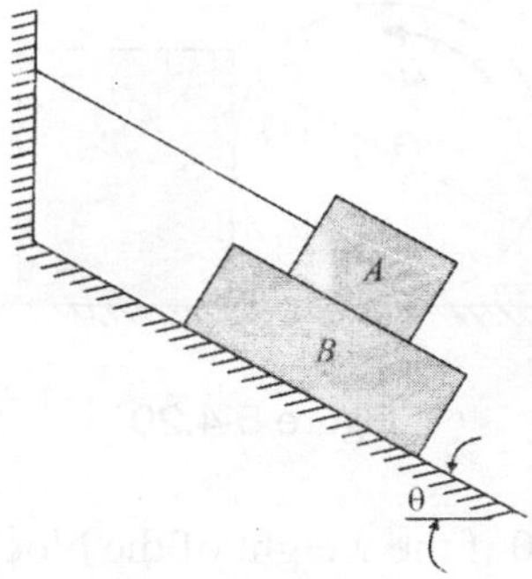

Figure P-5-4.17

5.4.18 Referring to Prob. 5-4.17, if the coefficient of friction is 0.60 and $\theta = 30°$, what force P applied to B acting down and parallel to the incline will start motion? What will be the tension in the cord attached to A?

$P = 128.6$ N; $T = 122.4$ N **Ans.**

5-4.19 A homogeneous cylinder of weight W rests on a horizontal floor in contact with a vertical wall as shown in Fig. P-5-4.19. If the coefficient of friction at all contact surfaces is f, determine the couple M acting on the cylinder which will start counterclockwise rotations.

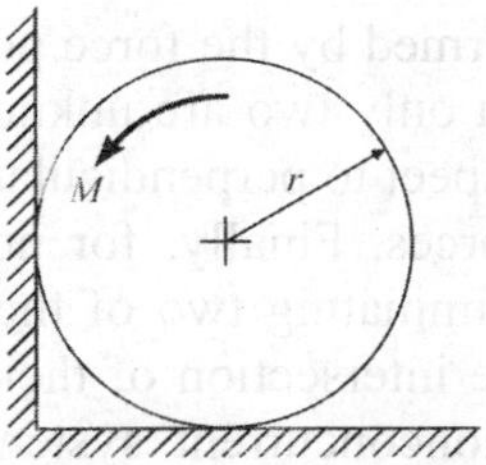

Figure P-5-4.19

$$M = \frac{rfW}{1+f^2}(1+f) \text{ N-m}$$ **Ans.**

5-4.20 As shown in Fig. P-5-4.20, a homogeneous cylinder weighing 232 N rests on a horizontal surface in contact with a block which weighs 180 N. If $f = 0.40$ at all contact surfaces, find the couple M acting on the cylinder which will start it rotating clockwise.

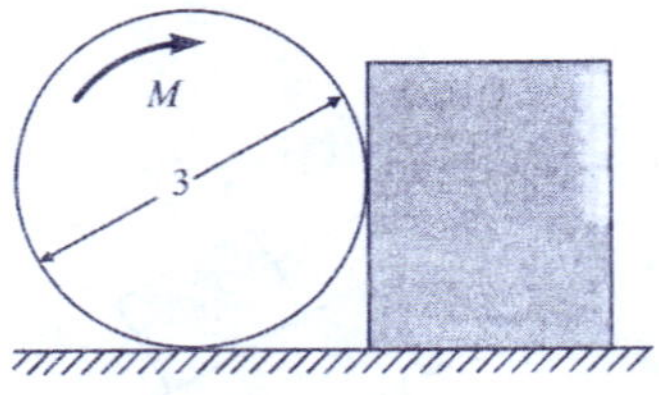

Figure 5-4.20

5-4.21 Solve Prob. 5-4.20 if the weight of the block is changed to 150 N.

M = 150 N-m *Ans.*

5-5 FURTHER APPLICATIONS OF FRICTION

The problems in the preceding section were concerned, for the most part, with only one body and were introductory in character. Here we consider more complex applications and concentrate on the development of the proper technique to solve them with the minimum of effort. In some of the problems, it will be simpler to use the frictional and normal components of a surface reaction; in others we should use the total surface reaction. A general rule is this: When the normal pressure can be computed directly in terms of known forces, use the frictional and normal components of the surface reaction, but for other cases use the total surface reaction.

Further, when *only* three forces act on a free body, it is usually best to apply the sine law to the triangle formed by the force polygon. When *more* than three forces are involved, of which only two are unknown, it is suggested that force summations be taken with respect to perpendicular axes, one of which coincides with one of the unknown forces. Finally, for nonconcurrent systems, always consider the possibility of eliminating two of the unknown forces by taking a moment summation about the intersection of their lines of action. However, in many situations involving nonconcurrent systems, a moment summation of equilibrium cannot be used because the location of the normal force is not known.

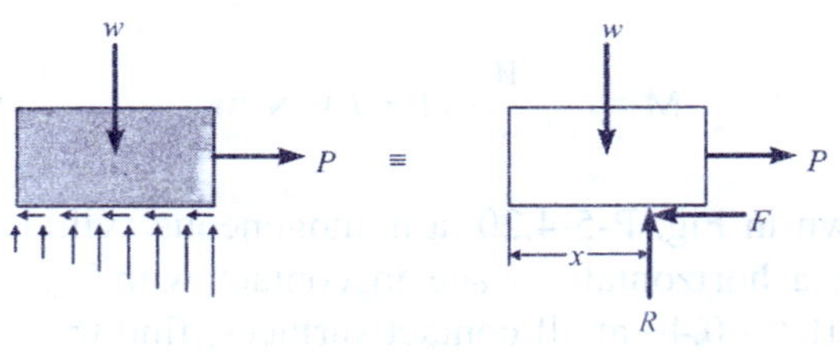

Figure 5-5.1 Actual frictional and normal distributions replaced by their resultants.

For example, consider the equivalent systems shown in Fig. 5-5.1. The application of a force P to the block of weight W causes the variable ground reactions and frictional components shown in part (a). These may be replaced by the resultant normal force N and friction force F shown in part (b), but the location x of the normal force can only be determined when the dimensions of the block and the position of P are specified. This concept and a method of avoiding the need to locate the normal force are discussed further in Illus. Prob. 5-5.1 that follows.

ILLUSTRATIVE PROBLEMS

5.5.1 The bodies shown in Fig. 5-5.2 are separated by a uniform strut weighing 100 N which is attached to the bodies with frictionless pins. The coefficient of friction under each body is 0.30. Determine the value of the horizontal force P that will start the system rightward.

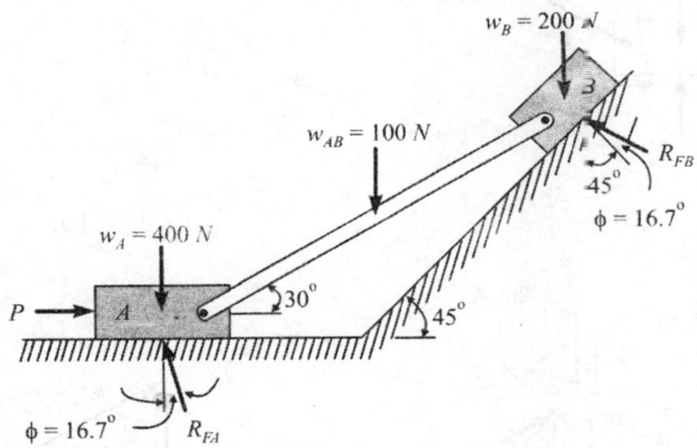

Figure 5-5.2

Solution

Since motion is impending rightward, the reactions under each body are inclined at the angle of friction $\phi = \tan^{-1} 0.30 = 16.7°$ with the normals as shown so that their frictional components resist the tendency to move. Apparently the FBD of the system involves only the unknowns P, R_{FA}, and R_{FB} so that we might attempt a solution for P from a moment summation about the intersection of R_{FA} and R_{FB}. However, this is not feasible because the location of R_{FA} and R_{FB} on their respective bodies is also unknown. Hence, even if the dimensions of the bodies were specified, there would be five unknowns in the given system. Thus, of the three equations normally available for this system, only two force summations can be used to relate P, R_{FA}, and R_{FB}. However, we can obtain additional equations by dismembering the system and considering the FBD of each part of the system as shown in Fig. 5-5.3. This dismemberment introduces four more unknowns consisting of the two components of the pin force at each end of the connecting strut. All told then, the unknowns consist of these four components plus the values of P, R_{FA}, and R_{FB}. To relate these seven unknowns, we can apply three equations of equilibrium to the FBD of the strut and two force summations to each of the bodies. Thus, we obtain sufficient equations to determine all the unknowns. Incidentally, this procedure does not involve determining the lines of action of R_{FA} and R_{FB} and hence the dimensions of the bodies are not necessary.

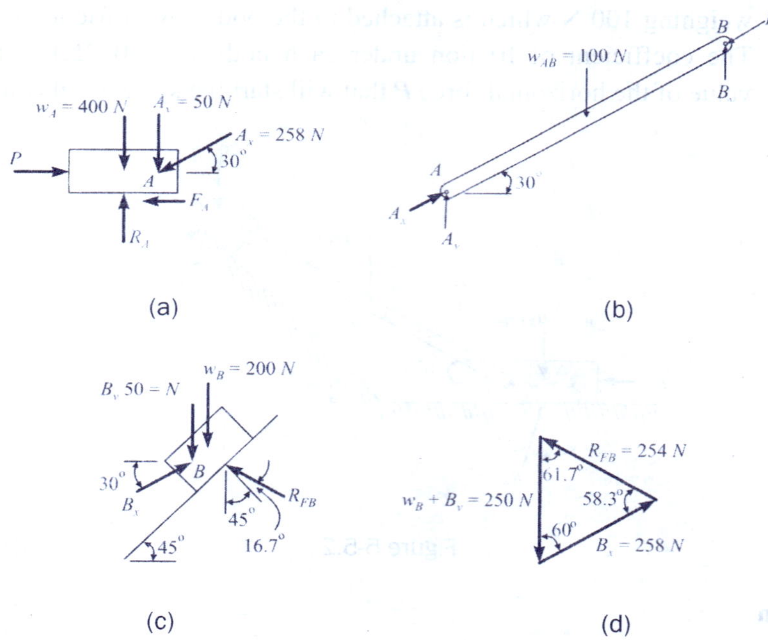

Figure 5-5.3

Consider now the FBD of the strut in part (b) of Fig. 5-5.3. Since the strut is a multiforce member, the directions of the pin forces at A and B are unknown. Instead of selecting the components of these pin forces as horizontal, and vertical as we usually do in analyzing multiforce members, here we take them as vertical and along the strut. This has the advantage of permitting moment summations about A and B to eliminate three of the four unknowns and determine B_v and A_v directly and independently. Since the weight of the strut is at its midpoint, its length is immaterial and we obtain $B_v = A_v = 50$ N at once $\Sigma X = 0$ taken along the strut gives $B_x = A_x$.

Inserting the values of B_v and A_v on the FBDs of the bodies, we see that the only unknowns on the 200 N body are B_x and R_{FB}. The corresponding force polygon of the FBD in part (c) is the force triangle shown in part (d). Applying the sine law, we obtain

$$\frac{250}{\sin 58.3^\circ} = \frac{B_x}{\sin 61.7^\circ} = \frac{R_{FB}}{\sin 60^\circ}$$

or

$$B_x = 258 \text{ N} \quad \text{and} \quad R_{FB} = 254 \text{ N} \qquad \textbf{\textit{Ans.}}$$

Consider now the FBD of the 400 N body in part (a) on which, since A_v and $A_x = B_x$ are now known, there are only the unknown forces P and R_{FA}. In this instance, however, it is easier to replace R_{FA} (not shown here) by its components F_A and R_A. The reason for replacing R_{FA} by its components is that R_A can be expressed directly in terms of known forces. Thus, we obtain

$$[\Sigma V = 0] \qquad R_A - 400 - 50 - 258 \sin 30^\circ = 0 \qquad R_A = 579 \text{ N}$$

Since motion impends, the friction force is found from the relation

$$[F = fR] \qquad F_A = 0.30(579) \qquad F_A = 174 \text{ N}$$

Finally, P is determined from the horizontal summation

$$[\Sigma H = 0] \qquad P - (F_A = 174) - 258 \cos 30^\circ = 0 \qquad P = 398 \text{ N} \quad \textbf{\textit{Ans.}}$$

An alternate approach (which may be used after $R_{FB} = 254$ N has been found from the FBD of body B) is to consider the point diagram of all the forces acting on the original system FBD of Fig. 5-5.2. The point diagram of these forces is shown in Fig. 5-5.4 from which by applying the friction relation and vertical and horizontal force summations, we obtain

$$[\Sigma V = 0] \qquad R_A + 254 \cos 61.7^\circ - 700 = 0 \qquad R_A = 579 \text{ N}$$

$$[F = fR] \qquad F_A = 0.30(579) \qquad F_A = 174 \text{ N}$$

$$[\Sigma H = 0] \qquad P - (F_A = 174) - 254 \sin 61.7^\circ = 0 \qquad P = 398 \text{ N} \quad \textbf{\textit{Check}}$$

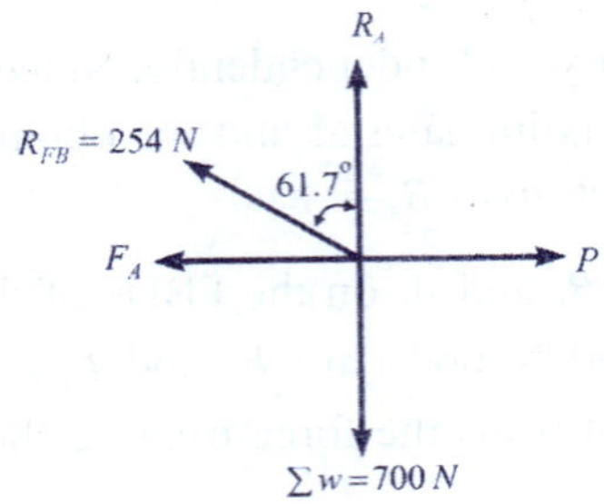

Figure 5-5.4 Point diagram of forces in Figure 5-5.2.

5-5.2 The block A in Fig. 5-5.5 supports a load $W = 1000$ N and is to be raised by forcing the wedge B under it. The angle of friction for all surfaces in contact is $\phi = 15°$. Determine the force P which is necessary to start the wedge under the block. The block and wedge have negligible weight.

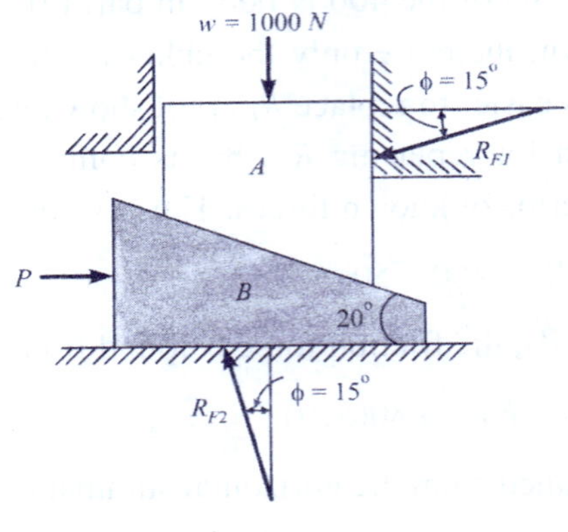

Figure 5-5.5 Block raised by wedge.

Solution

This wedge problem is very similar to the previous problem. The length of the strut in that problem is here reduced to zero; also the shape of the bodies is changed so that they make contact along a common surface. The contact reactions between the blocks at this common surface are not only equal and oppositely directed on the free-body diagram of each block, they also act so that their tangential or frictional components along the common contact surface oppose the impending motion of each block relative to the other.

The original system of block and wedge in Fig. 5-5.5 is acted upon by the three unknown forces P, R_{F1}, and R_{F2}. The reactions R_{F1} and R_{F2} make the angle ϕ with their normals and are directed as shown so their frictional components oppose motion. Since the position of the forces is unknown and dimensions are unspecified, the equation of equilibrium $\Sigma M = 0$ cannot be applied. It will be necessary to consider the FBD of each body as shown in Fig. 5-5.6.

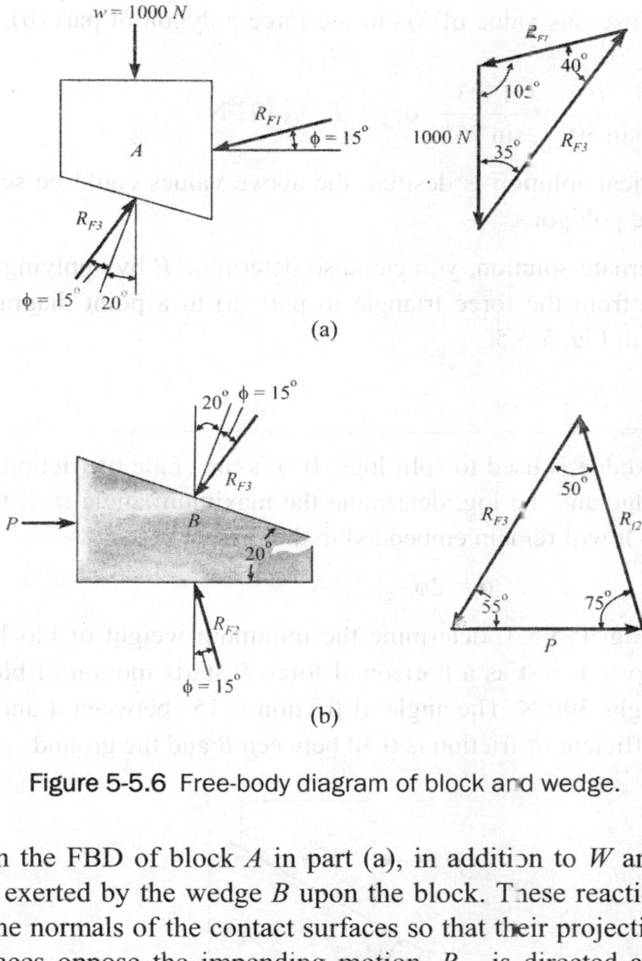

Figure 5-5.6 Free-body diagram of block and wedge.

Acting on the FBD of block A in part (a), in addition to W and R_{F1}, is the reaction R_{F3} exerted by the wedge B upon the block. These reaction forces are inclined to the normals of the contact surfaces so that their projections upon the contact surfaces oppose the impending motion. R_{F3} is directed at the friction angle $\phi = 15^\circ$ with the normal or, as shown, at an angle of 35° with the vertical. Adding these forces tip-to-tail gives the triangular force polygon as shown.

The FBD of wedge B in part (b), in addition to P and R_{F2}, also has the force R_{F3} exerted upon it by the block A. Note that R_{F3} acting on the wedge is numerically equal but oppositely directed to R_{F3} acting on the block in accordance with Newton's third law relating action and reaction forces. The triangular force polygon for the forces acting on the wedge is also shown.

By applying the sine law to the force polygon shown in part (a), we obtain

$$\frac{R_{F3}}{\sin 105^\circ} = \frac{1000}{\sin 40^\circ} \quad \text{or} \quad R_{F3} = 1503 \text{ N}$$

When we use this value of R_{F3} in the force polygon of part (b), the sine law now gives

$$\frac{P}{\sin 50^\circ} = \frac{1503}{\sin 75^\circ} \quad \text{or} \quad P = 1192 \text{ N}$$ ***Ans.***

If a graphical solution is desired, the above values could be scaled directly from the force polygons.

As an alternate solution, you can also determine P by applying the value of R_{F1} as found from the force triangle in part (a) to a point diagram of all the forces acting in Fig. 5-5.5.

PROBLEMS

5-5.3 A wedge is used to split logs. If ϕ is the angle of friction between the wedge and the log, determine the maximum angle α of the wedge so that it will remain embedded in the log.

$$\alpha = 2\phi$$ ***Ans.***

5-5.4 In Fig. P-5.5.4, determine the minimum weight of block B that will keep it at rest as a horizontal force P starts motion of block A which weighs 300 N. The angle of friction is 15° between A and B while the coefficient of friction is 0.30 between B and the ground.

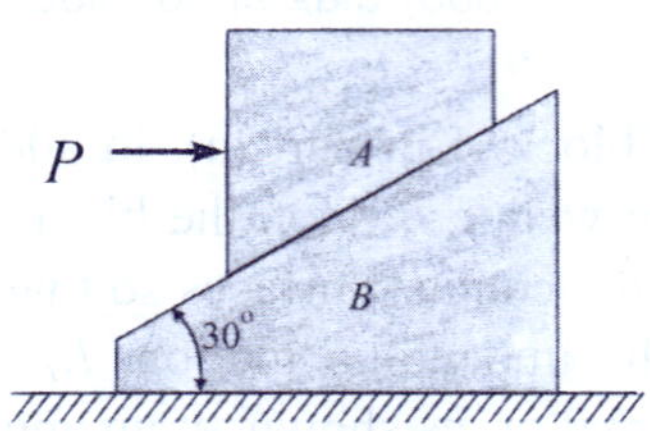

Figure P-5-5.4

5-5.5 A 400 N weight is suspended from the compound pulley shown in Fig. P-5-5.5: Find the smallest force P exerted by the hydraulic cylinder on the brake lever to keep the pulley at rest. The coefficient of friction at the brake surface is 0.40.

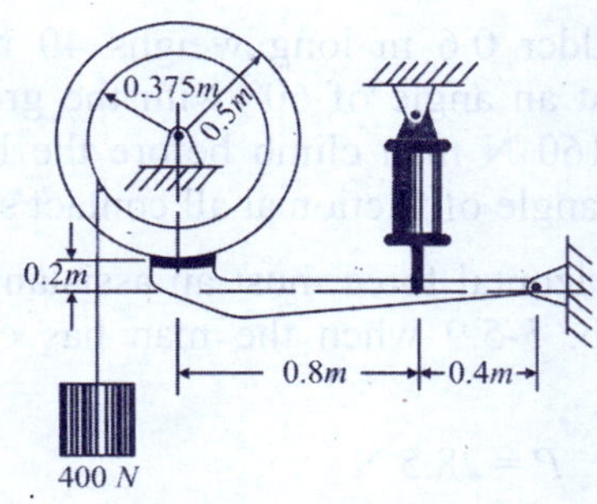

FIGURE P-5-5.5

$P = 2100$ N **Ans.**

5.5.6 Solve Prob. 5-5.5 if the 400 N weight is removed and a clockwise couple of 60 N-m is applied to the pulley. Would the value of P be changed if the 60 N-m couple acted counterclockwise?

5.5.7 A horizontal arm having a bushing 0.05 m long is slipped over a 0.05 m diameter vertical rod as shown in Fig. P-5-5.7. The coefficient of friction between the bushing and the rod is 0.20. Compute the minimum length L at which a weight W can be placed to prevent the arm from slipping down the rod. Neglect the weight of the arm.

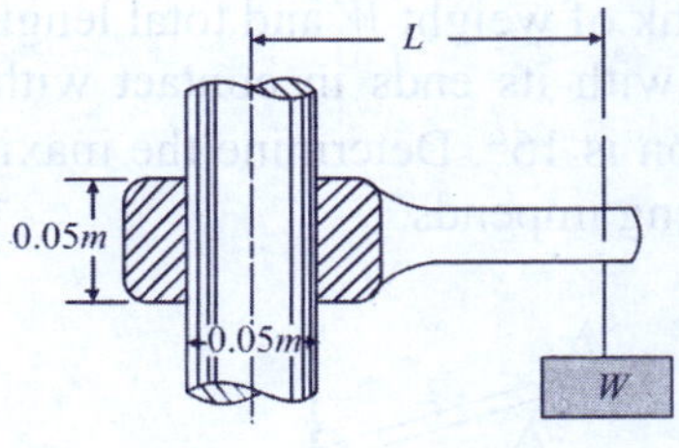

Figure P-5-5.7

L = 0.125 m **Ans.**

5-5.8 A uniform ladder of length L and weight W is placed with one end on the ground and the other against a vertical wall. The angle of friction at all contact surfaces is 20°. Find the minimum angle θ at which the ladder can be inclined with the horizontal before slipping occurs.

$\theta = 50°$ **Ans.**

5-5.9 A uniform ladder 0.6 m long weighs 40 N. It is placed against a vertical wall at an angle of 60° with the ground. How far along the ladder can a 160 N man climb before the ladder is on the verge of slipping? The angle of friction at all contact surfaces is 15°.

5-5.10 With what horizontal force must an assistant brace the bottom of the ladder in Prob. 5-5.9 when the man has climbed 0.5 m along the ladder?

$P = 28.5$ N ***Ans.***

5-5.11 A horizontal bar, 0.25 m long and of negligible weight, rests on rough inclined planes as shown in Fig. P-5-5.11. If the angle of friction is 15°, how close to *B* may the 200 N force be applied before motion impends?

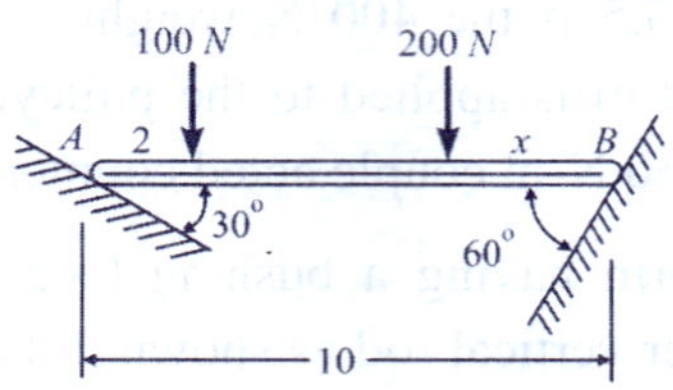

Figure P-5-5.11

$x = 0.0875$ m ***Ans.***

5-5.12 A uniform plank of weight *W* and total length 2*L* is placed as shown in Fig. P-5-5.12 with its ends in contact with the inclined planes. The angle of friction is 15°. Determine the maximum value of the angle α at which slipping impends.

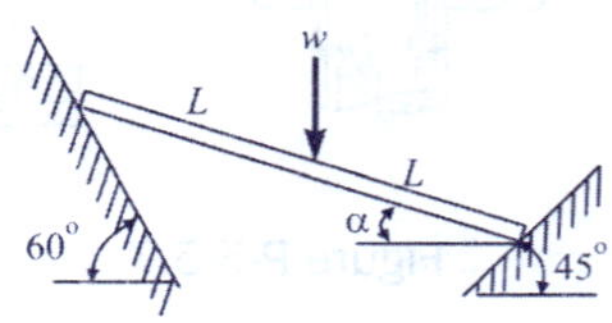

Figure P-5-5.12

$\alpha = 36.2°$ ***Ans.***

5-5.13 A plank 12 m long and of negligible weight, is supported and carries the horizontal and vertical loads as shown in Fig. P-5-5.13. The coefficient of friction is 0.30. If $T = 300$ N, determine the value of *P* to start motion downward.

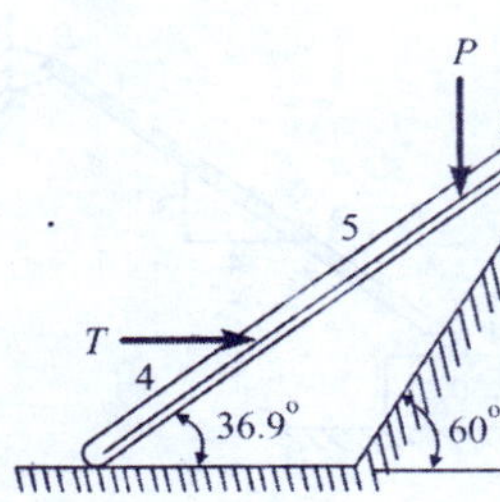

Figure P-5-5.13

$P = 990$ N **Ans.**

5-5.14 A homogeneous cylinder 3 m in diameter and weighing 400 N is resting on two rough inclined surfaces as shown in Fig. P-5-5.14. If the angle of friction is 15°, find the couple C applied to the cylinder that will start it rotating clockwise.

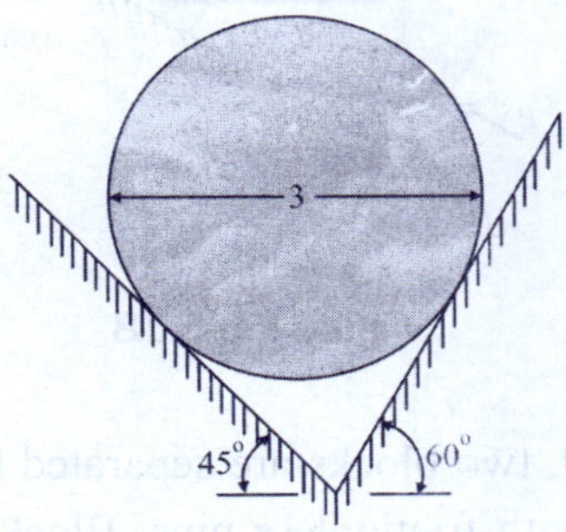

Figure P-5-5.14

$C = 251$ N-m **Ans.**

5-5.15 Find the couple applied to the cylinder of Prob. 5-5.14 that will start counterclockwise rotation.

5-5.16 For the cylinder described in Prob. 5-5.14, determine the minimum horizontal force P applied tangentially to the left at the top that will start counterclockwise rotation.

5-5.17 A uniform bar AB, 10 m long and weighing 280 N, is hinged at B and rests upon a 400 N block at A as shown in Fig. P-5-5.17. If the coefficient of friction is 0.40 at all contact surfaces, find the horizontal force P required to start moving the 400 N block

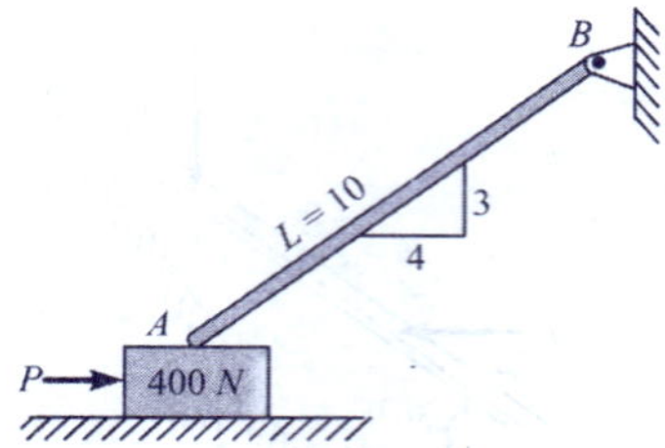

Figure P-5-5.17

$P = 320$ N **Ans.**

5-5.18 The two crates shown in Fig. P-5-5.18 are pin-connected by the horizontal strut *AB*. Find the minimum force *P* parallel to the incline which will maintain equilibrium.

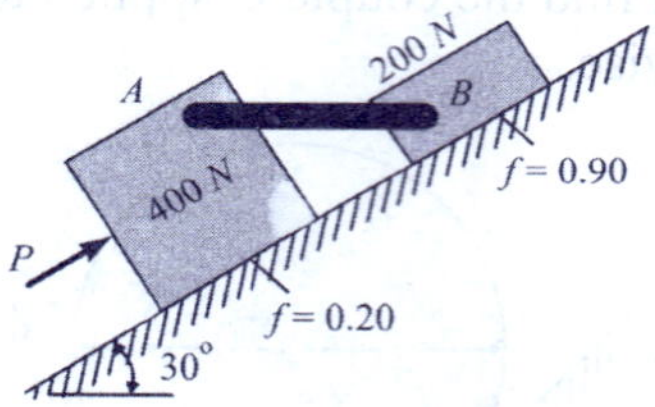

Figure P-5-5.18

5-5.19 In Fig. P.5-5.19, two blocks are separated by a uniform strut attached to each block with frictionless pins. Block *A* weighs 400 N, block *B* weighs 300 N and the strut *AB* weighs 200 *N*. If $f = 0.25$ under *B*, determine the minimum coefficient of friction under *A* to prevent motion.

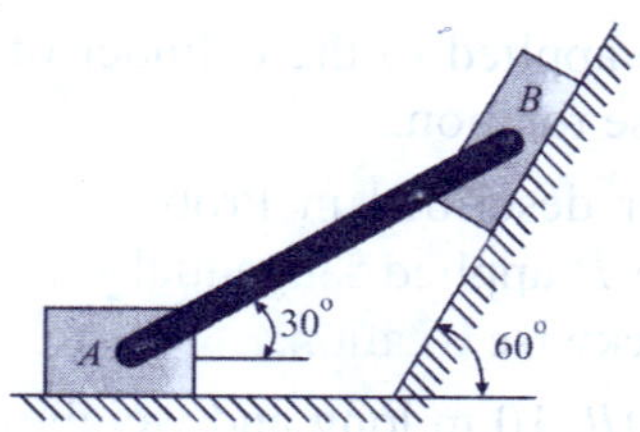

Figure P-5-5.19

$f = 0.40$ **Ans.**

5-5.20 In Fig. P-5-5.19, the uniform strut AB weighs 100 N and block A weighs 350 N. If $f = 0.30$ under both blocks, find the maximum weight of B that can be started up the incline by applying to A a rightward horizontal force P of 500 N.

5.5.21 What force P must be applied to the weightless wedges shown in Fig. P-5-5.21 to start them under the 1000 N block? The angle of friction for all contact surfaces is 10°.

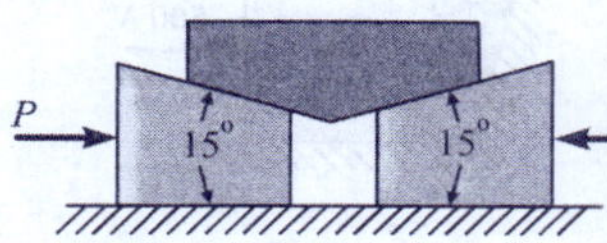

Figure P-5-5.21

$P = 321$ N *Ans.*

5-5.22 To adjust the vertical position of a column supporting a 2000 N load, two 5° wedges are used as shown in Fig. P-5-5.22. Determine the force P necessary to start the wedges if the angle of friction at all surfaces is 25°. Neglect friction at the rollers.

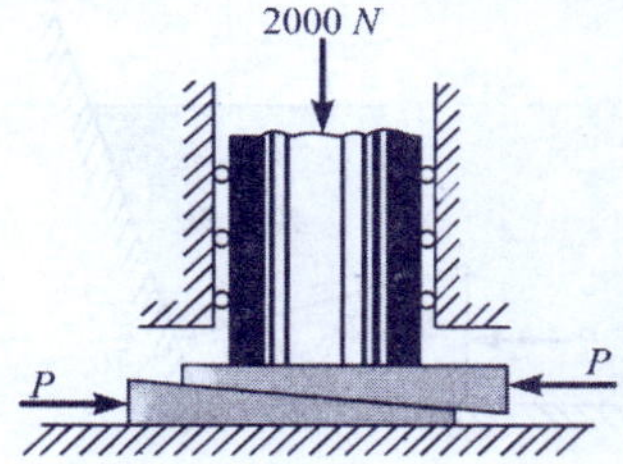

Figure P-5-5.22

5-5.23 As shown in Fig. P-5-5.23, two blocks, each weighing 200 N, are to be pushed apart by a 30° wedge. The angle of friction is 15° for all contact surfaces. What value of P is required to start movement of the blocks? How would this answer be changed if the weight of one of the blocks were increased to 300 N?

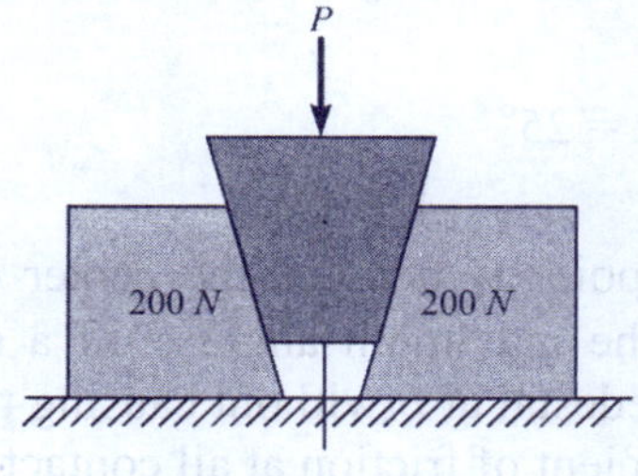

Figure P-5-5.23

5-5.24 Determine the force P required to start the wedge shown in Fig. P-5-5.24. The angle of friction for all surfaces in contact is 15°.

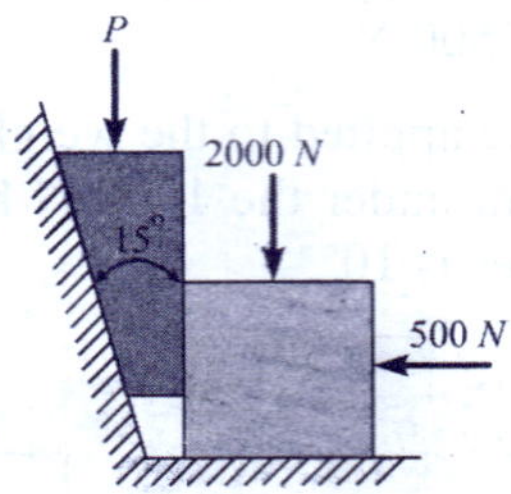

Figure P-5-5.24

$P = 944$ N **Ans.**

5-5.25 Determine the horizontal force P in Fig. P-5-5.25 to start the 400-N wedge (a) moving to the right and (b) moving to the left. The angle of friction is 20° at all contact surfaces.

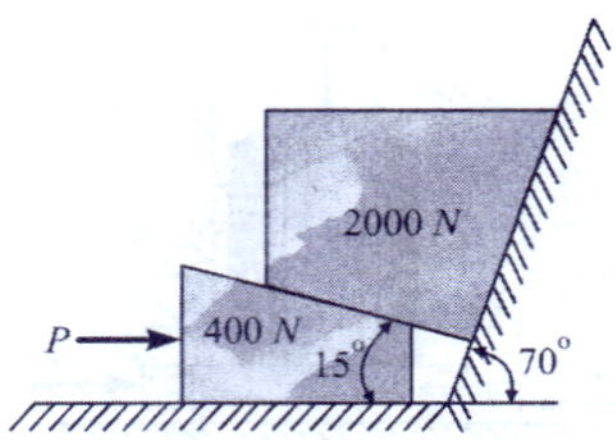

Figure P-5-5.25

(a) $P = 2495$ N; (b) $P = 872$ N **Ans.**

5-5.26 In Illus. Prob. 5-5.2, if the angle of friction is 10° at all surfaces in contact, determine the maximum wedge angle α that will give the wedge a mechanical advantage; i.e., make P less than the weight W of the block.

$\alpha = 25°$ **Ans.**

5-5.27 In Fig. P-5-5.27, point G denotes the center of gravity of a 2400 N lathe. Determine the maximum angle α of a steel wedge that can be driven under the pedestal at B without causing the lathe to slip along the ground. The coefficient of friction at all contact surfaces is 0.20.

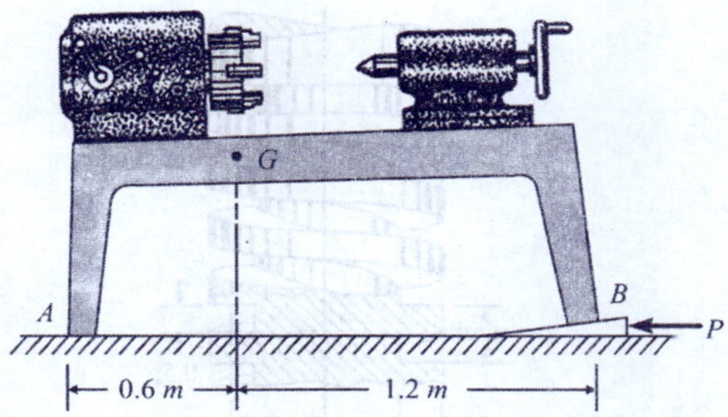

Figure P-5-5.27

$\alpha = 10.5^{\circ}$ ***Ans.***

5-5.28 Solve Prob. 55.27 if the wedge is to be driven under the pedestal at A instead of at B.

5-6 SQUARE-THREADED SCREWS

A square-threaded screw is essentially an inclined plane wrapped around a cylinder. Thus in Fig. 5-6.1 the inclined plane of height L and base b, if wrapped around the cylinder, would create the helix shown. An actual square-threaded screw results when the inclined plane has a certain width and constant thickness as it is wrapped around the cylinder. The result is pictured in Fig. 5-6.2. As before, the height of the equivalent inclined plane is the distance L, which is called the lead of the screw. This is the distance that a nut will advance along the screw in one revolution. In the case of a single threaded screw, the lead is synonymous with the pitch, which is the distance between similar points on adjacent threads. In a multiple threaded screw, the lead is mp, where m denotes the multiplicity of threading. The base length of the equivalent inclined plane is taken as the circumference of the mean radius of the thread and is expressed as $b = 2\pi r$. The mean radius is equal to one-half the sum of the outer radius and the root radius of the thread. The pitch angle θ of the equivalent inclined plane is determined from the relation $\tan \theta = L/2\pi r$.

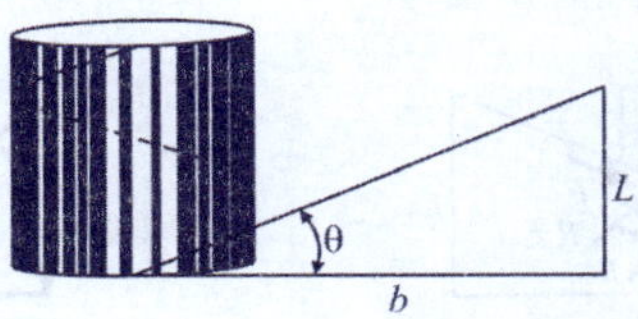

Figure 5-6.1 Helix formed by inclined plane.

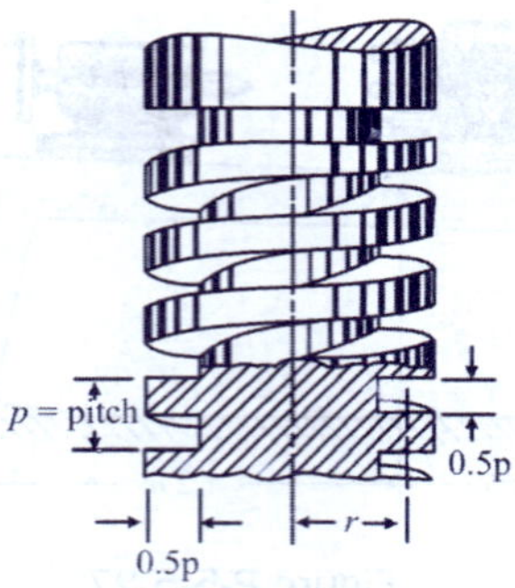

Figure 5-6.2 Square-threaded screw.

When the screw is used to lift a weight, as in a jackscrew, the weight may be assumed to be concentrated on one small element of the thread, as shown in Fig. 5-6.3. The force Q, acting in a plane perpendicular to the axis of the thread and at the mean radius of the thread, is determined from the free-body diagram of the weight shown on the equivalent inclined plane in Fig. 5-6.4(a).

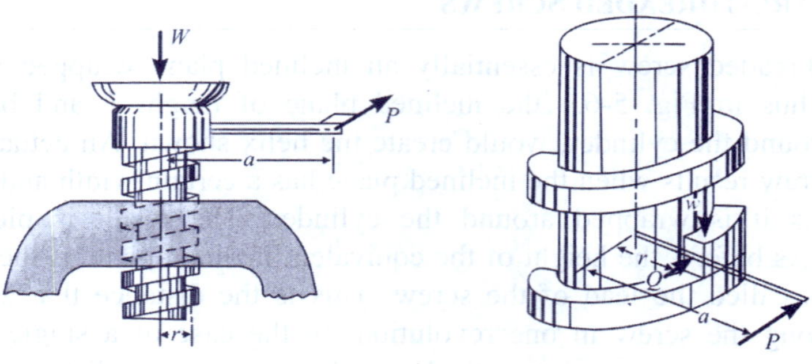

Figure 5-6.3 Forces on square-threaded screw.

With motion impending up the incline, the value of Q is obtained from the force triangle (Fig. 5-6.4b) to be

$$Q = W \tan(\phi + \theta) \qquad (a)$$

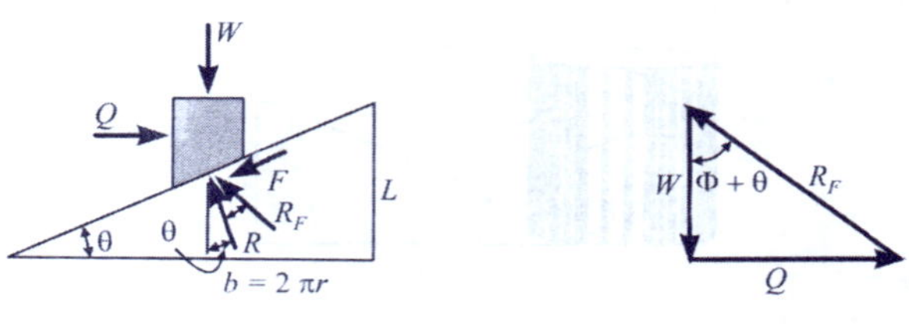

Figure 5-6.4 Motion impending up equivalent incline.

With motion impending down the free-body diagram and the force diagram are as shown in Fig. 5-6.5, from which we obtain

$$Q = W \tan(\phi - \theta) \qquad (b)$$

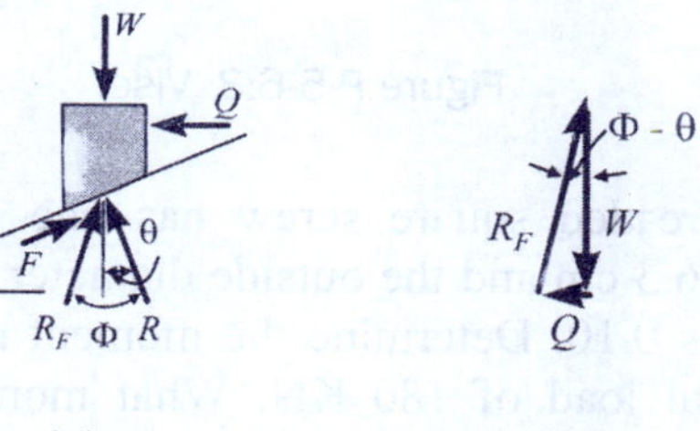

Figure 5-6.5 Motion impending down equivalent incline.

It is evident that if the screw is to be self-locking, the angle of friction ϕ must be larger than the pitch angle θ.

The force P exerted at the end of a lever arm of length a (see Fig. 5-6.3) is determined from the principle that the moment of P with respect to the axis of the screw must equal the moment effect of Q. We obtain $Pa = Qr$ or

$$P = \frac{Qr}{a} = \frac{Wr}{a} tan(\phi \pm \theta) \qquad (5\text{-}6.1)$$

PROBLEMS

5-6.1 A single-threaded jackscrew has a pitch of 0.5 cm and a mean radius of 1.75 cm. The coefficient of static friction is 0.15, and of kinetic friction, 0.10. (a) Determine the force P applied at the end of a lever 0.5 m long which will start lifting a weight of 2 KN. (b) What value of P will keep the jackscrew turning?

5.6.2 The distance between adjacent threads on a triple-threaded jackscrew is 5/3 cm. The mean radius is 5 cm. The coefficient of friction is 0.10. What load can be raised by exerting a moment of 9000 N-cm?

5-6.3 As shown in Fig. P-5-6.3, s square-threaded screw is used in a vise to exert a pressure of 2 KN. If the screw is double-threaded and has a pitch of 0.625 cm and a mean radius of 3.75 cm, determine the torque that must be applied at B to create this pressure. Assume the coefficient of friction to be 0.15.

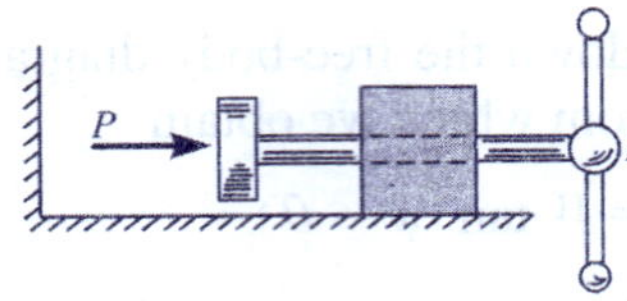

Figure P-5-6.3 Vise.

5.6.4 A single-threaded square screw has one thread per cm. The root diameter is 6.5 cm and the outside diameter is 7.5 cm. The coefficient of friction is 0.10. Determine the moment necessary to start lifting a vertical axial load of 180 KN. What moment is necessary to start lowering the load?

5-7 BELT FRICTION

The transmission of power by means of belt or rope drives, or the braking of large loads by means of band brakes, depends on the frictional resistance developed between the belt and the driving or resisting surface with which it is in contact. If a driving pulley is perfectly smooth, no driving torque is developed because no frictional resistance exists, and consequently the tension throughout the belt will be constant and will have the same value on both sides of the pulley. If the surface of the pulley is rough, however, the tension in the belt will vary throughout the length of contact, the difference in the belt tensions being caused by the incremental sum of the frictional resistances. We shall show that the frictional resistance is not constant nor does it even increase linearly; rather, it increases in exponential fashion.

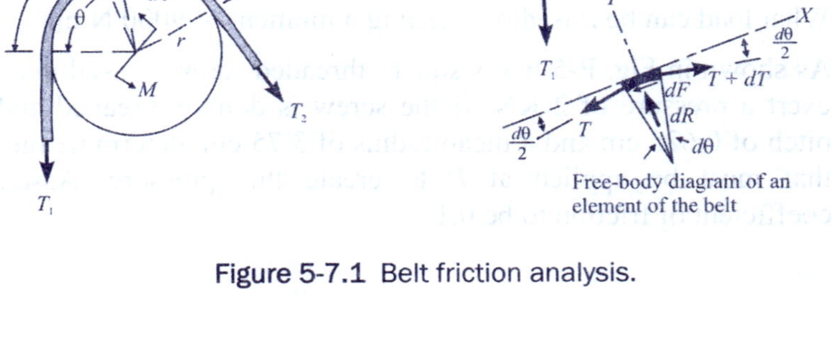

Figure 5-7.1 Belt friction analysis.

Consider the pulley of radius r in Fig. 5-7.1 where T_2 is greater than T_1 to overcome the resistance M of the pulley to clockwise rotation. The flexible belt is in contact with the pulley throughout an angle β determined by radii drawn to points of tangency of the belt. The adjacent diagram indicates how the friction forces cumulatively make up the difference between T_2 and T_1. To determine the relation between T_2 and T_1, consider a free-body diagram of a small element of the belt at the location θ from the tangent point of the smaller tension T_1. On this small element, the belt tension increases from T at the angle θ to $(T + dT)$ at the angle $(\theta + d\theta)$. These tensions are balanced by the differential normal force dR and the differential friction force $dF = f\,dN$.

Projecting these forces upon axes tangent and normal to the element gives

$$[\Sigma X = 0] \qquad (T + dT)\cos\frac{d\theta}{2} - T\cos\frac{d\theta}{2} - dF = 0$$

whence

$$dF = dT \tag{a}$$

since the cosine of differential angle is unity. We also obtain

$$[\Sigma Y = 0] \qquad dN - (T + dT)\sin\frac{d\theta}{2} - T\sin\frac{d\theta}{2} = 0$$

or

$$dR = 2T\sin\frac{d\theta}{2} + dT\sin\frac{d\theta}{2}$$

whence, neglecting the product of differential in the last term as negligibly small compared to first-order differentials and noting that for small angles the sine is practically equal to the angle expressed in radians, we have

$$dR = 2T\left(\frac{d\theta}{2}\right) + dT\,d\theta \tag{b}$$

When slipping impends, the friction relation gives

$$[F = fR] \quad dF = f\,dN = fT\,d\theta \tag{c}$$

Eliminating dF between Eqs. (a) and (c), we obtain

$$dT = fT\,d\theta$$

whence, separating the variables and integrating over the angle of contact of the belt, we have

$$\int_{T_1}^{T_2}\frac{dT}{T} = f\int_0^{\beta} d\theta$$

Whence

$$\log_e \frac{T_2}{T_1} = f\beta$$

$$\frac{T_2}{T_1} = e^{f\beta} \qquad (5\text{-}7.1)$$

Note that the angle of contact β must be expressed in radians. The value of $e^{f\beta}$ may be conveniently computed using the log–log scales on most slide rules. All that is needed is to set the hairline over the value of $f\beta$ on the D scale; then read the value of $e^{f\beta}$ on scale LL3 for values of $f\beta$ between 1 and 10, or on scale LL2 for values between 0.1 and 1.0, and so on. However, if log–log scales are not available, an alternate method using common logs is the following relation:

$$\log_{10} \frac{T_2}{T_1} = \log_{10} T_2 - \log_{10} T_2 = 0.434 f\beta \qquad (5\text{-}7.2)$$

ILLUSTRATIVE PROBLEM

5-7.1 A differential band brake is used to measure the torque output of an engine. The dimensions are shown in Fig. 5-7.2. Determine the torque M on the brake when P = 10 N. Assume the coefficient of kinetic friction to be 0.20.

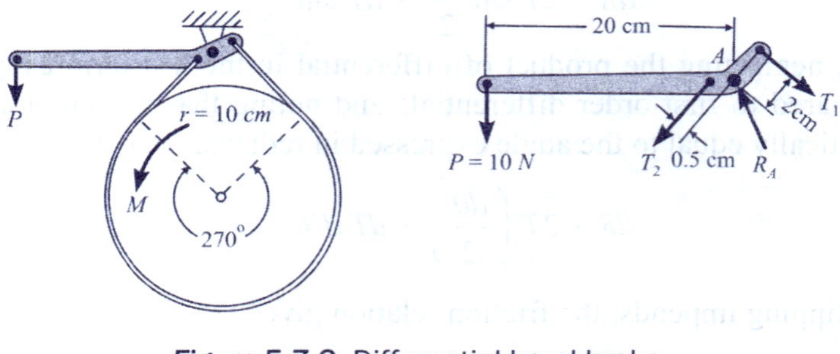

Figure 5-7.2 Differential band brake.

Solution

The tension T_2 acting on the free-body diagram of the brake arm must be greater than T_1 in order for the frictional moment $(T_2 - T_1)r$ to resist M. A summation of moments about the hinge A gives

$$[\Sigma M_A = 0] \quad 2T_1 - \frac{1}{2}T_2 - 10(20) = 0 \qquad (a)$$

Another relation between T_2 and T_1 is found by applying Eq. (5-7.1). Noting that $\beta = 270^\circ = 270(\pi/180) = 1.5\pi$ radians, we have

$$(T_2 = T_1\, e^{f\beta}) \Rightarrow \quad T_2 = T_1\, e^{0.2(1.5\pi)}$$

from which

$$T_2 = 2.565\ T_1 \qquad (b)$$

Solving Eqs. (*a*) and (*b*) simultaneously, we obtain

$$T_2 = 713 \text{ N} \quad \text{and} \quad T_1 = 278 \text{ N}$$

Then the torque exerted upon the brake is

$$[M = (T_2 - T_1)r] \qquad M = (713 - 278)(10)$$

$$M = 43500 \text{ N-cm}$$

$$= 435 \text{ Nm} \qquad \textit{Ans.}$$

PROBLEMS

5.7.2 A rope making 1¼ turns around a stationary horizontal drum is used to support a heavy weight. If the coefficient of friction is 0.4, what weight can be supported by exerting a 50 N force at the other end of the rope?

5-7.3 A boat exerts a pull of 4000 N on its hawser which is wrapped around a capstan on the pier. If the coefficient of friction is 0.3, how many turns must the hawser make around the capstan so that the pull at the other end does not exceed 50 N?

$$n = 2.32 \text{ turns} \qquad \textit{Ans.}$$

5-7.4 An acrobat weighing 700 N supports himself by wrapping a rope around one leg as shown in Fig. P.5-7.4. From his leg there hangs 1.5 m of rope which weighs 6 N per m. What is the minimum coefficient of friction between his leg and the rope? Neglect his pull on the rope.

Figure P-5-7.4

5-7.5 A torque of 320 N-m acts on the brake drum shown in Fig. P-5-7.5. If the brake band is in contact with the brake drum through 250° and the coefficient of friction is 0.3, determine the force P at the end of the brake lever.

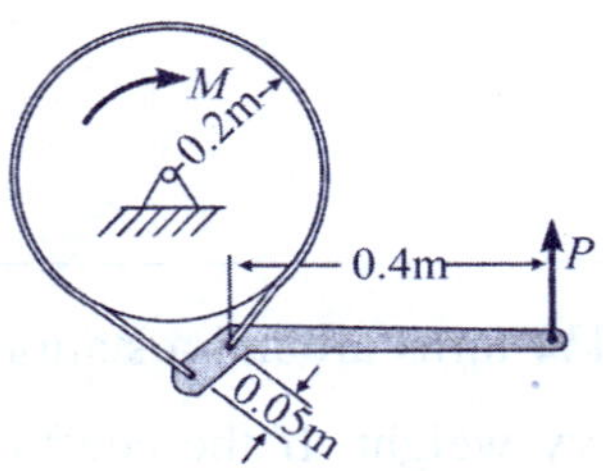

Figure P-5-7.5

$P = 61.7$ N **Ans.**

5-7.6 In Fig. P-5-7.6, the coefficient of friction is 0.20 between the rope and the fixed drum and between all surfaces in contact. Determine the minimum weight W to prevent downplane motion of the 1000 N body.

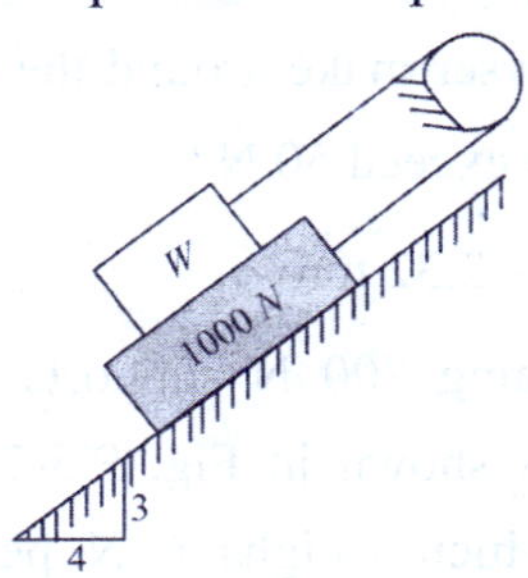

Figure P-5-7.6

$W = 253$ N **An**

5.7.7 Solve Prob. 5-7.6 for the minimum weight W that will cause the 1000 N body to move up the incline.

5-7.8 For the V-belt drive and matching grooved pulley shown in Fig. P-5-7.8, show that Eq. (5-7.1) becomes $T_2 = T_1\, e^{f\beta'}$ where $\beta' = \beta/\sin\frac{\alpha}{2}$.

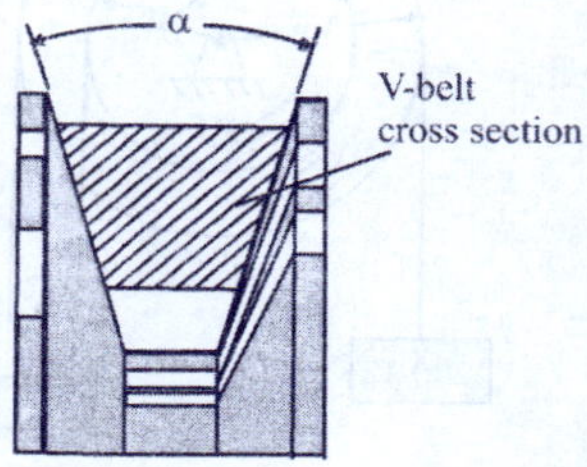

Figure P-5-7.8

5-7.9 A rope is looped over two fixed posts each of 1.2 m, diameter as shown in Fig. P-5-7.9. If $f = 1/\pi$, determine the maximum and minimum values of P that will prevent motion of the load $W = 1000$ N.

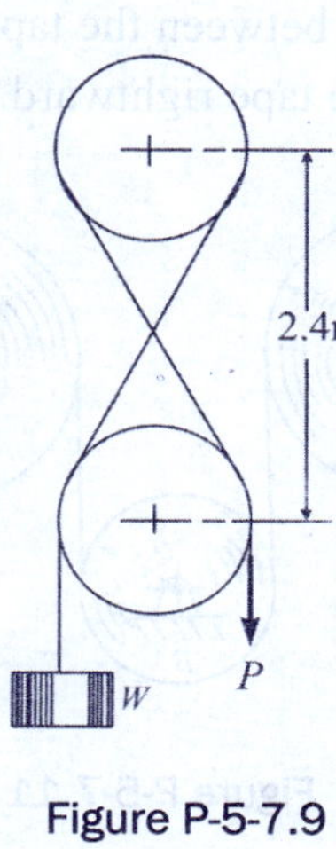

Figure P-5-7.9

Max. $P = 5300$ N; min. $P = 189$ N ***Ans.***

5.7.10 In Fig. P-5-7.10, a flexible belt runs from A over the compound pulley P, around the floating pulley B and back over P to a 200 N weight. The coefficient of friction is $1/\pi$ between the belt and the compound pulley P. Find the maximum weight W that can be supported without rotating the pulley P or slipping the belt on the pulley P.

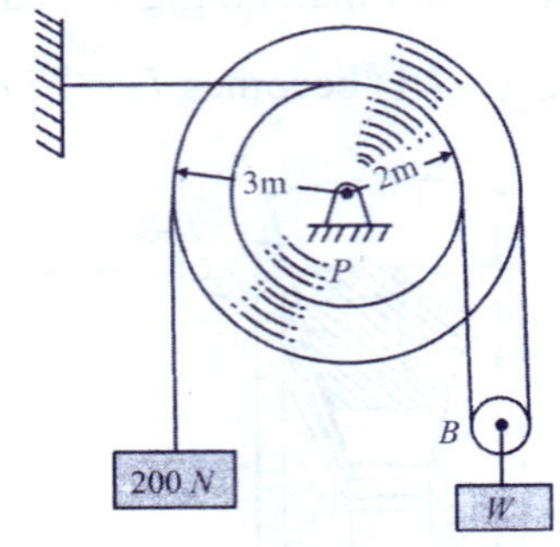

Figure P-5-7.10

$W = 704$ N **Ans.**

5-7.11 A tape is threaded over three pulleys as shown in Fig. P-5-7.11. Each of the pulleys requires a torque of 30 N-m, to make it rotate. If $P = 10$ N and $f = 0.20$ between the tape and each pulley, find the force T required to move the tape rightward.

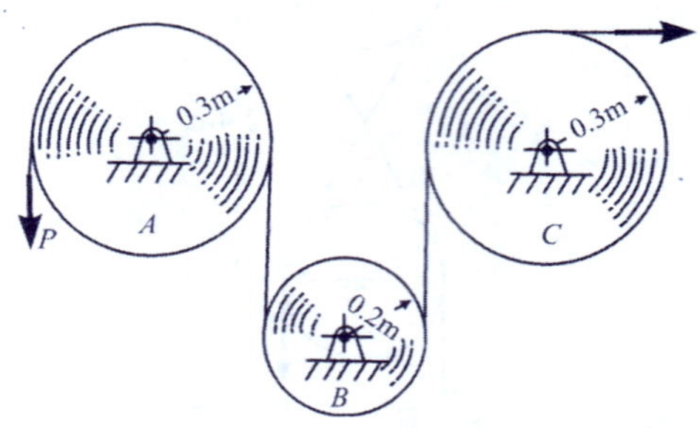

Figure P-5-7.11

5-7.12 In the preceding problem, if T = 10 N and $f = 0.20$, find the force P to start the tape moving leftward.

5-8 DISK FRICTION

This is the type of friction created when mating surfaces under normal pressure rotate or tend to rotate past one another. Typical examples are thrust bearings, pivot bearings, or disk clutches. To obtain a general relation between the axial force P and the moment M to start or maintain rotation, consider a hollow shaft supported by the step bearing shown in Fig. 5-8.1. The torque M must equal the frictional moments developed over the contact area.

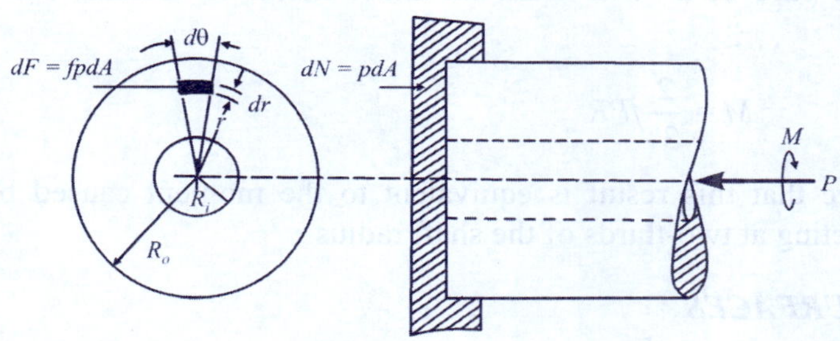

Figure 5-8.1 Hollow shaft in step bearing.

If p is the intensity of contact pressure at a typical elemental area dA, the normal force on this element is $dN = p\,dA$ and the corresponding friction force when slipping impends or occurs is $dF = fp\,dA$. On equating the moment sum of the elemental friction forces about the shaft axis to the applied torque, we obtain

$$M = \int r\,df = \int_{Area} fpr\,dA \qquad (a)$$

To evaluate this integral, we must know the variation of f and p with dA. Usually f is assumed to be constant since the mating surfaces are generally of the same material under the same condition. However, p will depend on the relative deformation of the mating surfaces. In addition, the pressure distribution may be assumed to be axially symmetrical and hence constant over an elemental ring area $dA = 2\pi r dr$. Thus, we obtain

$$M = 2\rho f \int_{Ri}^{R_o} pr^2\,dr \qquad (b)$$

NEW SURFACES

With new, well-supported and relatively rigid flat surfaces, it is reasonable to assume that p is constant and uniformly distributed so that $p = P/A = P/\pi\left(R_o^2 - R_i^2\right)$. Substituting this value of p in Eq. (b) gives

$$M = \frac{2\pi fP}{\pi\left(R_o^2 - R_i^2\right)} \int_{R_i}^{R_o} r^2\,dr = \frac{2fP}{3}\frac{\left(R_o^3 - R_i^3\right)}{\left(R_o^2 - R_i^2\right)} \tag{5-8.1}$$

For the case of a solid shaft, the inner radius R_i is zero, and Eq. (5-8.1) reduces to

$$M = \frac{2}{3} fPR \tag{5-8.2}$$

Observe that this result is equivalent to the moment caused by a friction force fP acting at two-thirds of the shaft radius.

WORN SURFACES

With continued rotation of the shaft in its bearing, the mating surfaces gradually wear into a new relative shape which stabilizes as the wear becomes constant over the surface. Since wear varies with the pressure and the rubbing velocity, we conclude that the product of pressure and velocity is constant. Denoting this wear constant by K, we have $K = pr$ since the velocity at any point is proportional to the radial distance r. Then $p = K/r$ and inserting this value of p in Eq. (b), we obtain

$$M = 2\pi fK \int_{R_i}^{R_o} r\,dr = \pi fk\left(R_o^2 - R_i^2\right) \tag{c}$$

To relate K to P, we equate an axial summation of forces to zero, thereby obtaining

$$P = \int p\,dA = \int_{R_i}^{R_o} \frac{K}{r}\,(2\pi r\,dr) = 2\pi K(R_o - R_i)$$

or

$$K = \frac{P}{2\pi\,(R_o - R_i)}$$

which, when substituted into Eq. (c), reduces it to

$$M = \frac{1}{2} fP\,(R_o - R_i) \tag{5-8.3}$$

For the case of a solid shaft, R_i is zero and Eq. (5-8.3) reduces to

$$M = \frac{1}{2} fPR \tag{5-8.4}$$

We conclude that, after wear has stabilized, for both solid and hollow shafts, the torque M is due to a frictional force fP acting at the average radius of the shaft.

PROBLEMS

5-8.1 An automobile clutch consists of a single annular facing of 1 m outside diameter and 0.6 m inside diameter. If $f = 0.60$, what axial force on the clutch is required to transmit a torque of 250 N-m?

5.8.2 Assuming a uniform pressure and constant f, determine the torque M to rotate the pivot bearing shown in Fig. P-5-8.2.

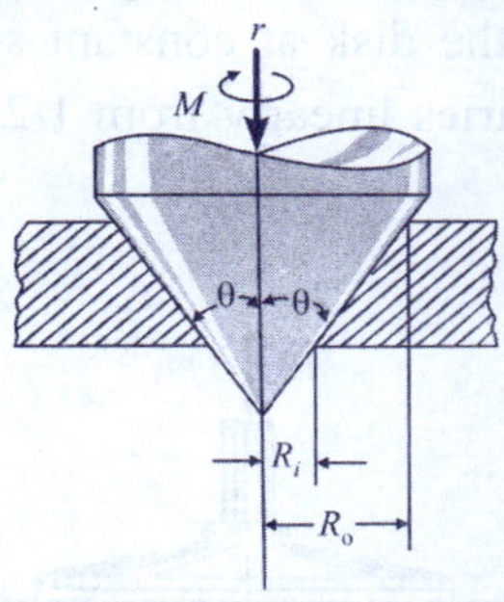

Figure P-5-8.2

$$M = \frac{2fP}{3 \sin \theta} \frac{\left(R_o^3 - R_i^3\right)}{\left(R_o^2 - R_i^2\right)} \qquad \textit{Ans.}$$

5-8.3 Resolve the previous problem assuming that the pivot bearing has been worn in.

5.8.4 A shaft of 0.25 m diameter has a collar of 0.75 m diameter. It passes through a circular opening of 0.3 m diameter and is pressed against a spring-supported plate as shown in Fig. P-5-8.4. The bolts prevent rotation of the bearing plate. Each of the two springs which support the bearing plate has a constant of 50 N/m. In the position shown, a torque M = 10 N-m is required to start rotation of the shaft. What torque will start rotation when $P = 400$ N? Assume $f = 0.25$ and a uniform pressure distribution.

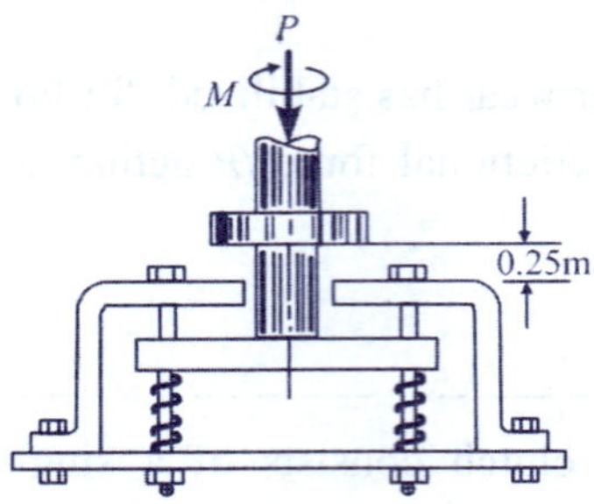

Figure P-5-8.4

5-8.5 In the flat sanding disk shown in Fig. P-5-8.5, find the torque M required to rotate the disk at constant speed if f is constant and the contact pressure varies linearly from 1/2 p_0 at the outside to p_0 at the center.

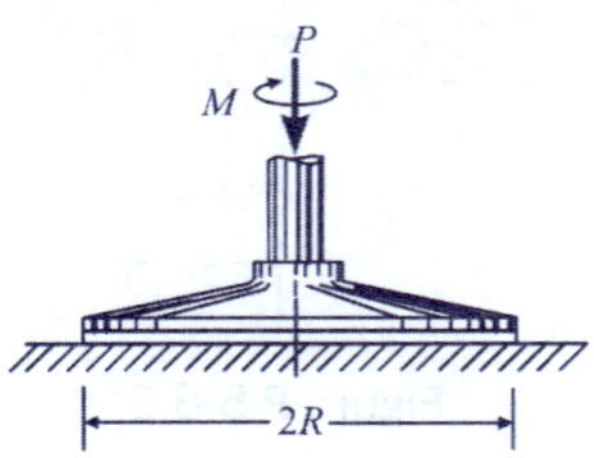

Figure P-5-8.5

$$M = \frac{5}{8} fPR$$ ***Ans.***

5-8.6 An axial load P presses a solid shaft of radius R into a flexible elastic surface so that the contact pressure decreases parabolically from p_0 at the center to zero at the outside. Show that the pressure distribution is given by

$$P = \frac{2P}{\pi R^2}\left(1 - \frac{r^2}{R^2}\right)$$

and then determine the torque M to start rotation. Assume f is constant.

$$M = \frac{8}{15} fPR \qquad \textbf{Ans.}$$

5.8.7 Determine the torque M to rotate the shaft having the hemispherical bearing shown in Fig. P-5-8.7. Assume f is constant but that the contact pressure varies according to $p = p_0 \cos\theta$ from a maximum p_0 at the base of the bearing to zero at the upper edge.

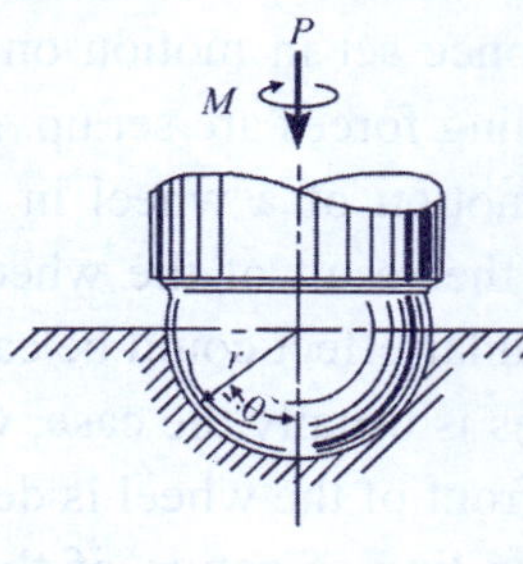

Figure P-5-8.7

$$M = fPr \qquad \textbf{Ans.}$$

5-8.8 A spherical steel bearing is pressed with a force P into a wooden surface and deforms it as shown in Fig. P-5-8.8. Assuming a uniform contact pressure and constant f, find the torque M to start rotation about the vertical axis.

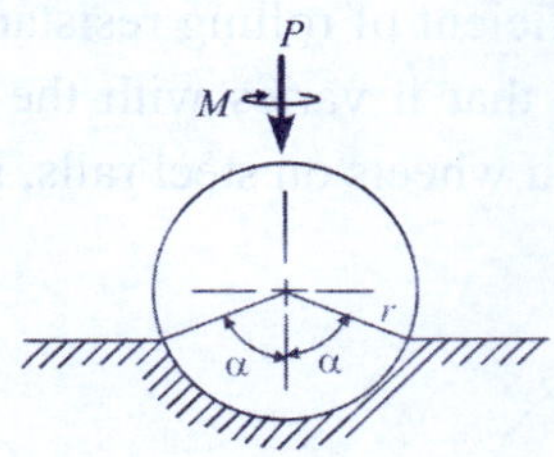

Figure P-5-8.8

$$M = \frac{fPr}{\sin^2\alpha}\left(\alpha - \frac{1}{2}\sin 2\alpha\right) \qquad \textbf{Ans.}$$

5-9 ROLLING RESISTANCE

When wheels, cylinders, or similar objects roll without slipping along flat surfaces, the frictional resistances are generally assumed to be static. As is shown in Section 13-4, the unbalanced moment causing motion is determined in these cases by taking a moment summation about the line of contact with the ground.

For example, in Fig. 5-9.1, if the ground is rigid and unyielding, a moment summation about C becomes zero because the weight W passes through C. Thus, theoretically, a wheel once set in motion on a horizontal surface would roll forever, because no retarding forces are set up. Actually we know this to be false. The resistance to the motion of a wheel in such cases is called rolling resistance. This resistance is the result of the wheel rolling against a yielding surface, as in Fig. 5-9.2. A similar effect could be caused when a yielding wheel rolls upon a rigid surface or, as is usually the case, when both the wheel and the surface yield. The ground in front of the wheel is depressed, causing the normal pressure R to act ahead of the line of action of the weight W. When we take moments about the line of contact C, we find that a resisting moment of magnitude Wb is created which resists the forward rolling of the wheel. The small distance b is known as the coefficient of rolling resistance. Since the dimension of this coefficient is expressed in meters, this coefficient is not synonymous with the coefficient of friction, which was defined in Section 5-2 as a ratio between the frictional resistance and the normal pressure. Moreover, it differs from the usual definition of the coefficient of friction in that it is not constant. Although the coefficient of rolling resistance is usually assumed to be constant, experiment shows that it varies with the radius of the wheel. In the case of ordinary steel railroad wheels on steel rails, it is about 0.508 mm.

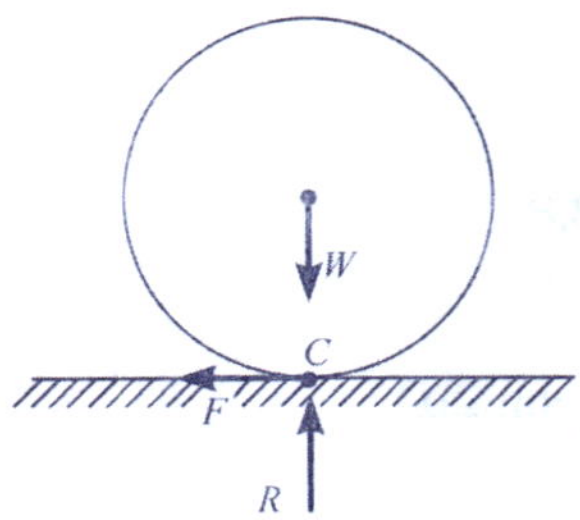

Figure 5-9.1

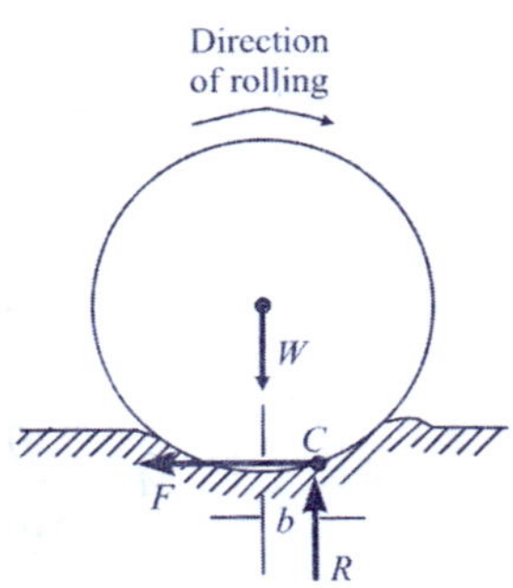

Figure 5-9.2

SUMMARY

Friction is defined as the contact resistance exerted by one body upon a second body when the second body moves or tends to move past the first body. In Section 5-2, the relation between the frictional resistance F, the coefficient of friction f, and the normal pressure R at the surface of contact was expressed by

$$F = fR \tag{5-2.1}$$

Note carefully that Eq. (5-2.1) determines only the maximum resistance which can exist. When motion is not impending, the friction may have any value from zero up to F, depending on the resultant force tending to cause motion.

It is frequently desirable to consider the contact reaction R_F rather than its components F and R (see Section 5-3). When limiting friction is acting, this contact reaction will be inclined at the angle of friction ϕ with the normal pressure. The angle of friction is related to the coefficient of friction by the equation

$$\tan\phi = f \tag{5-3.1}$$

When motion impends or occurs and the normal component R of a contact reaction must be expressed in terms of another unknown force, it will usually be simpler to use the total reaction R and the angle of friction instead of the frictional and normal components F and R.

The torque required to turn a square-threaded screw (Section 5-6) against a load W is expressed in terms of the pitch angle θ, the angle of friction ϕ, and the mean radius r by the equation

$$M = Wr\tan(\phi \pm \theta) \tag{5-6.1}$$

The plus sign is used when W resists the torque, the minus sign, when W aids the torque.

The difference in belt tensions T_2 and T_1 caused by wrapping a belt through a contact angle of β radians results in the following relation:

$$\frac{T_2}{T_1} = e^{f\beta} \tag{5-7.1}$$

or

$$\log_{10}\frac{T_2}{T_1} = 0.434\, f\beta$$

where f is the coefficient of friction (see Section 5-7).

The torque M required to start or maintain rotation between mating surfaces in a new hollow step bearing subjected to an axial load P is given by

$$M = \frac{2}{3} fP \frac{\left(R_o^3 - R_i^3\right)}{\left(R_o^2 - R_i^2\right)} \tag{5-8.1}$$

Where R_o and R_i are respectively the outer and inner radii of the step bearing. With continued rotation and wear, the torque reduces to

$$M = \frac{1}{2} fP \left(R_o - R_i\right) \tag{5-8.3}$$

CHAPTER 6

CENTROIDS AND CENTERS OF GRAVITY

6-1 INTRODUCTION

A body of weight W is supported by a string attached at A, as shown in Fig. 6-1.1. The only external forces acting on the body are its weight and the reaction exerted by the string. Equilibrium of the body can exist only if these two forces are equal, opposite, and collinear. The line of action of the weight W can be determined, therefore, by the line of action of the support reaction or the line of the string.

Let the body be supported in a new position by the string now attached to B. The body will shift its position so that the line of action of the weight is again collinear with the string. Thus two positions of the line of action of the weight are determined experimentally. The intersection of these positions of the line of action determines a point which is defined as the center of gravity of the body; this is the point through which the action line of the weight always passes.

From the above discussion it is apparent that the problem of locating the center of gravity of a body reduces to determining the point through which the resultant force of gravity of the body acts.

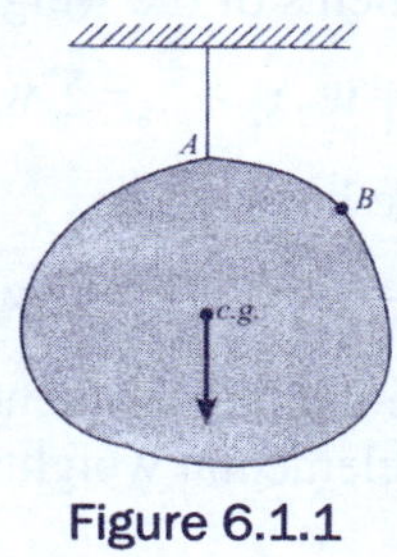

Figure 6.1.1

6-2 CENTRE OF GRAVITY OF FLAT PLATE

The analytical location of the center of gravity is simply an application of the principle of moments; that is, the moment of the resultant is equal to the moment sum of its parts. As an example, consider the flat plate of irregular section shown in Fig. 6-2.1. *A* pictorial as well as front and side views is shown. The network shown divides the plate into small elements having weights w_1, w_2, etc., which act at the center of each element. These gravity forces form a parallel force system, the resultant of which is the total weight *W* of the plate.

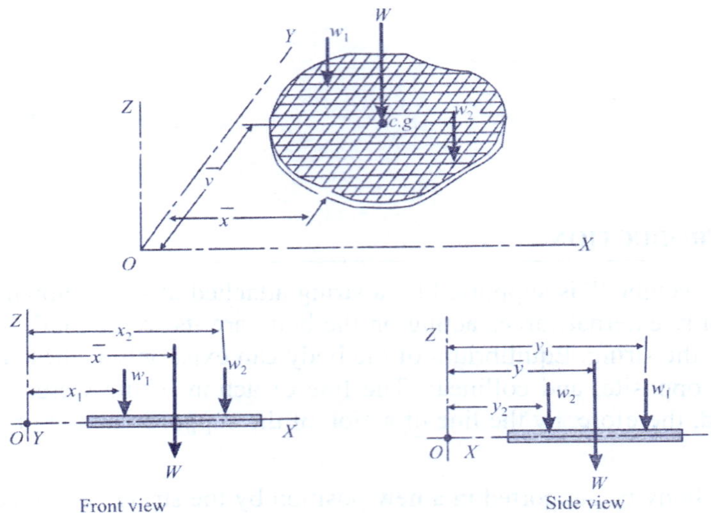

Figure 6-2.1 Coordinates of the center of gravity.

Let the coordinates of each elemental weight be (x_1, y_1), (x_2, y_2), etc., and the coordinates of the resultant weight be $(\bar{x}, \bar{y})$, as shown in Fig. 6-2.1. Note the use of the bar sign. In this book the coordinates of a resultant—be it force, weight, or area—are always distinguished by a bar sign. The coordinates are read as "bar *x*," etc. Taking moments of the weights about the Y axis, we get

$$W\,\bar{x} = w_1 x_1 + w_2 x_2 + \ldots = \sum w_i x_i \tag{a}$$

With respect to the *X* axis, we have

$$W\,\bar{y} = w_1 y_1 + w_2 y_2 + \ldots = \sum w_i y_i \tag{b}$$

These equations merely state that the moment of a weight *W* about an axis is equal to the moment sum of its elemental weights.

6-3 CENTROIDS OF AREAS AND LINES

If the material of the plate in Fig. 6-2.1 is homogeneous, the weight W may be expressed as the product of its density γ (i.e., weight per unit volume) multiplied by tA, where t is the thickness of the plate and A is its area. Similarly the weight w of an element is given by γta, where a is the cross-sectional area of the element. Substituting these values in Eq. (a) in Section 6-2 results in

$$\gamma tA\bar{x} = \gamma ta_1x_1 + \gamma ta_2x_2 + \ldots = \gamma t\Sigma ax$$

whence, canceling the constant terms γ and t, we get

$$\left.\begin{aligned} A\bar{x} &= \Sigma a_i x_i \\ A\bar{y} &= \Sigma a_i y_i \end{aligned}\right\} \tag{6-3.1}$$

By analogy with Eqs. (a) and (b) in Section 6-2, the expression $A\bar{x}$, as well as $A\bar{y}$, is called the moment of area. It is equivalent to the sum of the moments of the elemental areas composing the total area. Note that the moment of area is therefore defined as the product of the area multiplied by the perpendicular distance from the center of area to the axis of moments.

If Eq. (6-3.1) is rewritten in the following form:

$$\left.\begin{aligned} \bar{x} &= \frac{\Sigma a_i x_i}{A} \\ \bar{y} &= \frac{\Sigma a_i y_i}{A} \end{aligned}\right\} \tag{6-3.1a}$$

this gives a method of locating a point called the centroid of area. The *centroid* of area is defined as the point corresponding to the center of gravity of a plate of infinitesimal thickness. The term "centroid" rather than "center of gravity" is used when referring to areas (as well as to lines and volumes) because such figures do not have weight. The term "center of gravity" is widely used, although it is a misnomer. Strictly speaking, it should refer to the center of weight of actual bodies.

When referring to lines, we may determine the centroid by similar means. A line may be assumed to be the axis of a homogeneous slender wire. Thus Fig 6-3.1 represents the center line of a homogeneous wire of length L and constant cross-sectional area a lying in the XY plane. The weight W is given by the equation $W = \gamma aL$ and the weight w of an elemental length l by $w = \gamma al$.

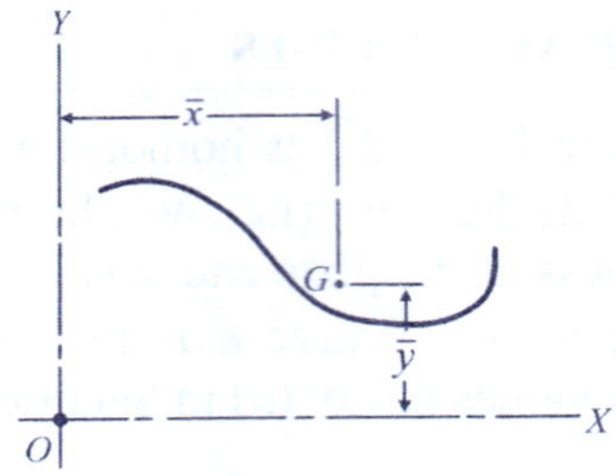

Figure 6.3.1 Homogeneous, slender and uniform wire.

Substituting these values in Eqs. (*a*) and (*b*) in Section 6-2, we have

$$\gamma a L\bar{x} = \gamma a l_1 x_1 + \gamma a l_2 x_2 + \ldots = \gamma a \sum lx$$

$$\gamma a L\bar{y} = \gamma a l_1 y_1 + \gamma a l_2 y_2 + \ldots = \gamma a \sum ly$$

whence, canceling the constant terms γ and a, we get

$$\left.\begin{aligned} L\bar{x} &= \sum l_i x_i \\ L\bar{y} &= \sum l_i y_i \end{aligned}\right\} \qquad (6\text{-}3.2)$$

6-4 IMPORTANCE OF CENTROIDS AND MOMENTS OF AREA

In subsequent work on strength of materials, the student will find the location of the centroid of an area is of great importance. For example, he will learn that in order to produce uniform stress distribution, the loads must be placed so that the line of action of their resultant coincides with the centroid of the cross section of the member. The position of the centroid of an area is also important for determining the location of the neutral axis in the bending of beams, for in strength of materials it is shown that the neutral axis (line of zero stress) passes through the centroid of the cross section of the beam.

An axis passing through the centroid of an area is known as a centroidal axis. The next chapter, which deals with moments of inertia, will make clear the great importance of the position of centroidal axes of areas. Many other instances of their importance will come to the student's attention in his engineering studies.

Of equal importance to the position of a centroid is the moment of an area. We recall that the moment of an area with respect to an axis was defined as the product of the area multiplied by the perpendicular distance from its centroid to the axis. In dynamics, the moment of area is used to determine the displacement of a body subjected to variable forces (see Prob. 11-3.1). In strength of materials, it is used to determine shearing stresses in beams. In addition, the moment of an area is extensively used for determining the deflection of beams by the area-moment method. These instances, as well as many others that the student will encounter, should indicate the importance of a permanent, not a temporary, knowledge of centroids and moments of area.

6-5 CENTROIDS DETERMINED BY INTEGRATION

We recall that integration is the process of summing up infinitesimal quantities. Except for a change in symbols and procedure, integration is equivalent to a finite summation. In the preceding sections, for example, if the area of an element had been expressed as the differential dA (i.e., a small part of the total area A), the equations for determining the centroid of an area would have become

$$\left.\begin{aligned} A\bar{x} &= \int x\,dA \\ A\bar{y} &= \int y\,dA \end{aligned}\right\} \qquad (6\text{-}5.1)$$

and for determining the centroid of a line, we could have used

$$\left.\begin{aligned} L\bar{x} &= \int x\,dL \\ L\bar{y} &= \int y\,dL \end{aligned}\right\} \qquad (6\text{-}5.2)$$

When we determine the centroid by integration, the figure is divided into differential elements so that:

1. All points of the element are located the same distance from the axis of moments, or
2. The position of the centroid of the element is known so that the moment of the element about the axis of moments is the product of the element and the distance of its centroid from the axis.

If a plane figure has a line of symmetry, its centroid is located on that line. This statement may be demonstrated by balancing a plate on its line of symmetry, whence the moments of the weights (also areas if the plate has constant thickness) on either side of the line of symmetry must be numerically equal and of opposite sign. If a plane figure has two lines of symmetry, the centroid is located at the point of intersection of the lines.

ILLUSTRATIVE PROBLEMS

6-5.1 *Centroid of the Arc of a Circle.*

Determine the centroid of the line which is an arc of a circle, as shown in Fig. 6-5.1.

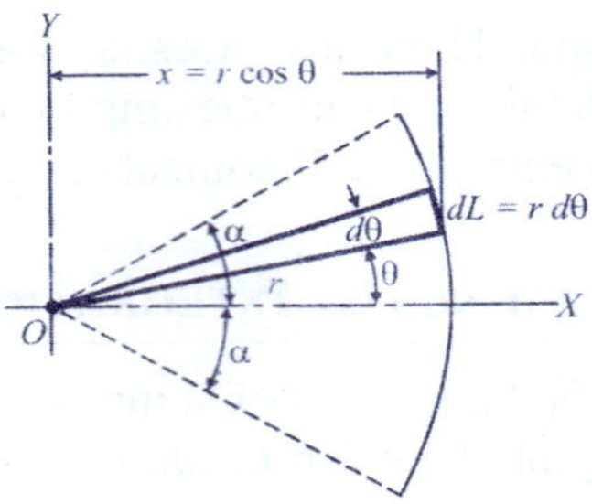

Figure 6.5.1

Solution

Let the axis of symmetry be chosen as the X axis. Then $\bar{y} = 0$. If the radius of the arc is denoted by r and the subtended angle by 2α, the element of arc dL and its distance from the Y axis are $dL = r\,d\theta$ and $x = r\cos\theta$. Applying Eq. (6-5.2), we have

$$[L\bar{x} = \int x\,dL] \qquad (2\alpha r)\bar{x} = \int_{-\alpha}^{+\alpha} (r\cos\theta)\,r\,d\theta = r^2 \int_{-\alpha}^{+\alpha} \cos\theta\,d\theta$$

$$= 2r^2 \sin\alpha$$

Finally,

$$\bar{x} = \frac{2r^2 \sin\alpha}{2\alpha r} = \frac{r\sin\alpha}{\alpha} \qquad \textbf{Ans.}$$

If the arc is a semicircle, as in Fig. 6-5.2, $\alpha = 90^{\circ} = \pi/2$ radians and $\sin\alpha = 1$. Substituting these values in the above result gives

$$\bar{x} = \frac{2r}{\pi}$$

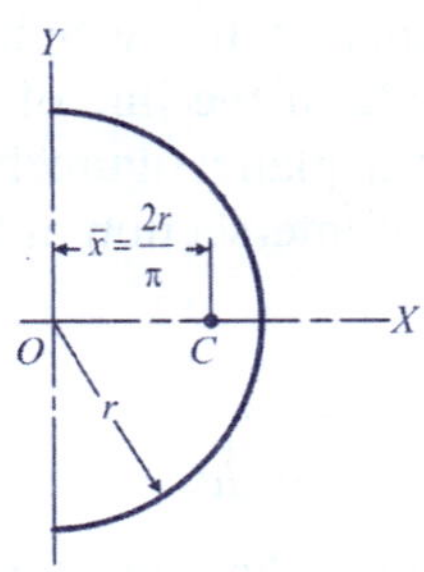

Figure 6.5.2

6-5.2 *Centroid of the Area of a Triangle.*

The triangle shown in Fig. 6-5.3 has a base b and an altitude h. Locate the centroid of the triangular area with respect to the base.

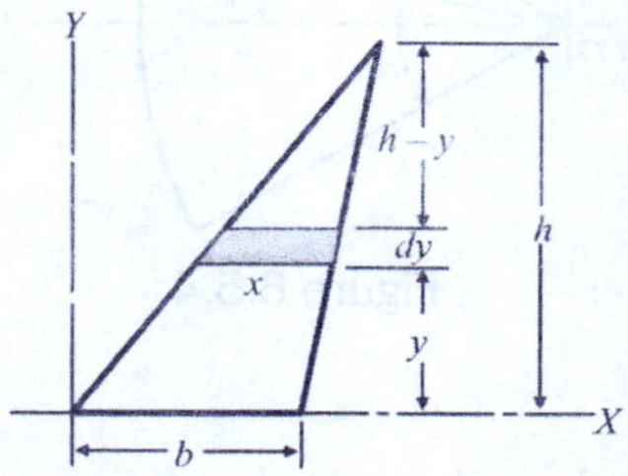

Figure 6.5.3

Solution

In accordance with Rule 1 given earlier, select strips parallel to the base as the differential elements of area. The area of any differential element is then $dA = x\,dy$. Applying Eq. (6-5.1), we obtain

$$[A\bar{y} = \int y\,dA] \qquad \left(\frac{1}{2}bh\right)\bar{y} = \int_0^h xy\,dy \qquad (a)$$

From similar triangles, $x = \dfrac{b}{h}(h - y)$, so that Eq. (a) becomes

$$\left(\frac{1}{2}bh\right)\bar{y} = \frac{b}{h}\int_0^h (h-y)y\,dy = \frac{1}{6}bh^2$$

$$\bar{y} = \frac{1}{3}h \qquad \textit{Ans.}$$

Observe that for any triangle the distance from the centroid to any side is equal to one-third of the altitude measured from that side. Furthermore, the centroid of a triangle is located on a median because the median to any side contains the centroids of all strips drawn parallel to that side. Therefore the centroid is at the intersection of the median

6-5.3 *Centroid of the Area of a Circular Sector.*

Determine the location of the centroid of the area of the sector of the circle shown in Fig. 6-5.4. Let the radius of the circle be r and the subtended angle be 2α.

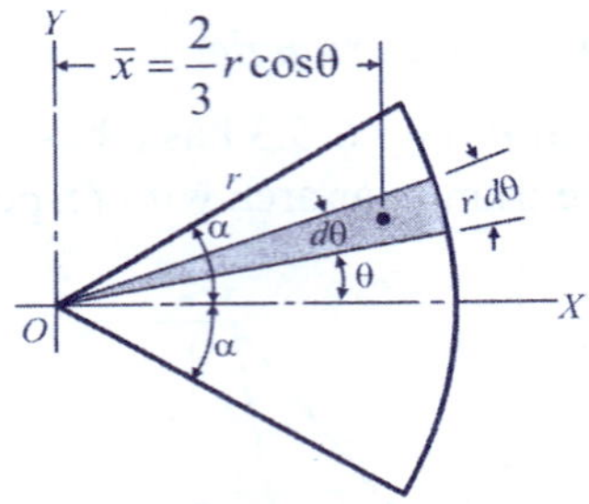

Figure 6.5.4

Solution

Let the axis of symmetry be taken as the X axis; then $\bar{y} = 0$. Select as the element of area the shaded triangle the position of whose centroid is known from the answer to Prob. 6-5.2 to be $x = \frac{2}{3} r \cos\theta$. The area of the element is $dA = \left(\frac{1}{2} r\right) r\, d\theta = \frac{1}{2} r^2 d\theta$. Applying Eq. (6-5.1), we obtain

$$[A\bar{x} = \int x\, dA] \qquad \bar{x} \times \int_{-\alpha}^{+\alpha} \frac{1}{2} r^2\, d\theta = \int_{-\alpha}^{+\alpha} \frac{2}{3} r \cos\theta \left(\frac{1}{2} r^2\, d\theta\right)$$

$$\bar{x}(r^2\alpha) = \frac{1}{3} r^3 \int_{-\alpha}^{+\alpha} \cos\theta\, d\theta = \frac{2}{3} r^3 = \sin\alpha$$

$$\bar{x} = \frac{2}{3} \frac{r\sin\alpha}{\alpha} \qquad \textit{Ans.}$$

If the sector is a *semicircular area* as in Fig. 6-5.5, $\alpha = 90^\circ = \pi/2$ radians, whence by substituting in the last equation above we find the distance of the centroid from the diameter to be

$$\bar{x} = \frac{4r}{3\pi} = 0.424r$$

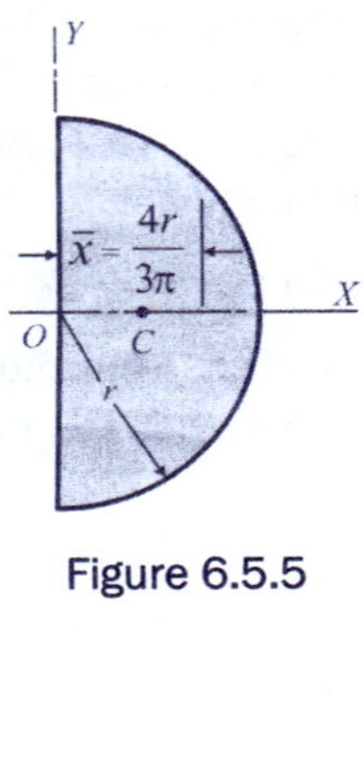

Figure 6.5.5

6-5.4 *Centroid of the Area of a Parabolic Segment.*

In Fig 6-5.6 is shown a parabolic segment bounded by the X axis, the line $x = a$, and the parabola $y = kx^2$. Determine the coordinates of the centroid.

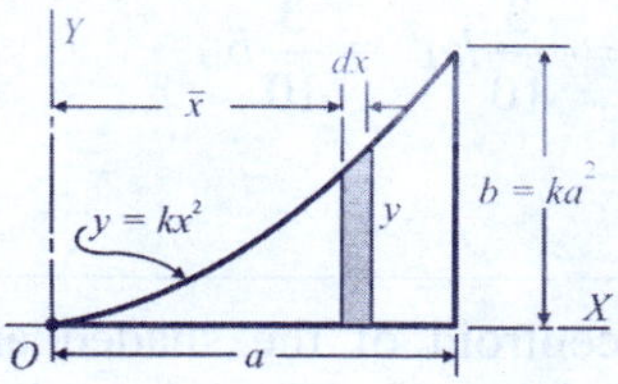

Figure 6.5.6

Solution

Select the element of area as the shaded strip parallel to the Y axis. All points in this element are the same distance from the Y axis. The area of the element is $dA = y\,dx$. The area A of the entire parabolic segment is found from

$$[A = \int dA] \qquad A = \int_0^a y\,dx = \int_0^a kx^2\,dx = \frac{1}{3}ka^3$$

To determine $\bar{x}$ we apply Eq. (6-5.1) as follows:

$$[A\bar{x} = \int x\,dA] \qquad \left(\frac{1}{3}ka^3\right)\bar{x} = \int_0^a xy\,dx = \int_0^a kx^3\,dx$$

$$\left(\frac{1}{3}ka^3\right)\bar{x} = \frac{1}{4}ka^4$$

$$\bar{x} = \frac{3}{4}a$$

To determine $\bar{y}$, we use the same elementary strip; but since each point of the element is not the same distance from the X axis, we must use Rule 2 above; i.e., the moment of the differential element about the X axis is the product of its centroidal coordinate, $\frac{1}{2}y$ multiplied by the area $y\,dx$. Applying Eq. (6-5.1) again, we obtain

$$[A\bar{y} = \int y\, dA] \qquad \left(\frac{1}{3}ka^3\right)\bar{y} = \int_0^a \frac{1}{2}y\,(y\,dx) = \frac{1}{2}k^2\int_0^a x^4\,dx$$

$$\left(\frac{1}{3}ka^3\right)\bar{y} = \frac{k^2a^5}{10}$$

$$\bar{y} = \frac{3}{10}ka^2 = \frac{3}{10}b \qquad \textit{Ans.}$$

PROBLEMS

6-5.5 Determine the centroid of the shaded area shown in Fig. P-6-5.5, which is bounded by the X axis, the line $x = a$ and the parabola $y^2 = kx$.

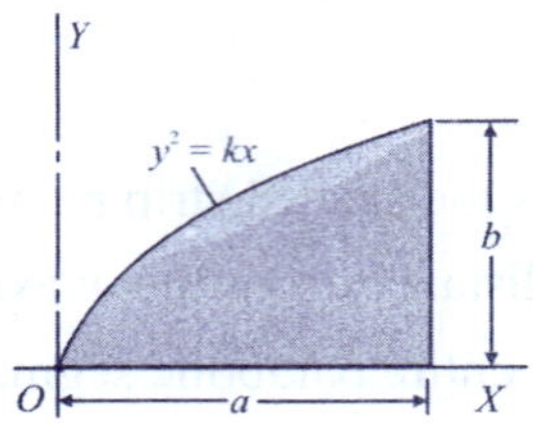

Figure P-6.5.5

$$\bar{x} = \frac{3}{5}a;\ \bar{y} = \frac{3}{8}b \qquad \textit{Ans.}$$

6-5.6 Determine the centroid of the quarter circle shown in Fig. P-6-5.6 whose radius is r.

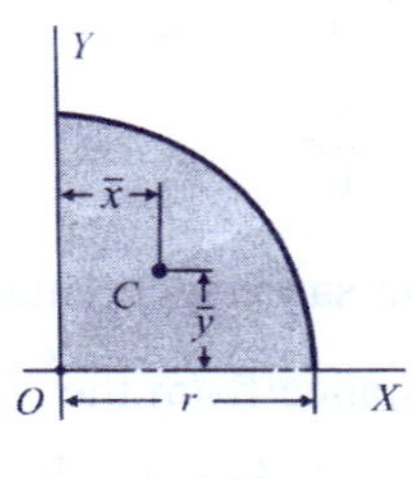

Figure P-6.5.6

$$\bar{x} = \bar{y} = \frac{4r}{3\pi} \qquad \textit{Ans.}$$

6-5.7 Determine the centroid of the quadrant of the ellipse shown in Fig. P- 6-5.7. The equation of the ellipse is $\frac{x^2}{a^2}+\frac{y^2}{b^2}=1$.

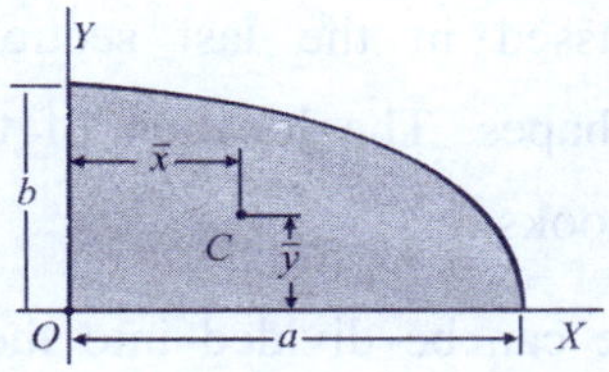

Figure P-6.5.7

$$\bar{x} = \frac{4a}{3\pi}\ ;\ \bar{y} = \frac{4b}{3\pi}$$ ***Ans.***

6-5.8 Compute the area of the spandrel in Fig. P-6-5.8 bounded by the X axis, the line x = b, and the curve $y = kx^n$ where $n \geq 0$. What is the location of its centroid from the line $x = b$? Prepare a table of areas and location of centroid for values of n = 0, 1, 2, and 3.

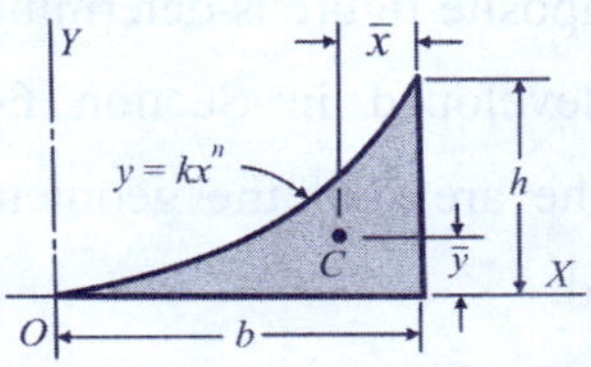

Figure P-6.5.8

$$A=\frac{1}{n+1}(bh);\ \bar{x} = \frac{b}{n+2};$$ refer to Table 9-4.1. ***Ans.***

6-5.9 Determine the y coordinate of the spandrel described in Prob. 6-5.8.

$$\bar{y} = \frac{n+1}{4n+2}\ (h)$$ ***Ans.***

6-5.10 Locate the centroid of the area bounded by the X axis and the sine curve $y = a \sin \pi x/L$ from $x = 0$ to $x = L$.

$$\bar{x} = \frac{L}{2}\ ;\ \bar{y} = \frac{\pi a}{8}$$ ***Ans.***

6-6 CENTROIDS OF COMPOSITE FIGURES

Many of the figures used in engineering are composed of combinations of the geometrical shapes discussed in the last section. Still other figures are composed of structural shapes. The location of the centroids for structural elements is given in handbooks.

When the given figure can be divided into the finite elements discussed above, these elements can be treated in the same manner as were the infinitesimal elements. When this is done, the process is called finite summation, as contrasted to integration, or the summation of infinitesimal elements.

If given area can be divided into parts, each centroid of which is known, the moment of the total area will be the sum of the moments of area of its parts. This statement is really an extension of the principle of moments (Section 2-9).

The centroid of the composite figure is determined by applying the following equations, which were developed in Section 6-3. In these equations the elemental areas become the areas of the geometrical shapes into which the entire area has been divided.

$$\left.\begin{aligned} A\bar{x} &= \sum a_i x_i \\ A\bar{y} &= \sum a_i y_i \end{aligned}\right\} \qquad (6\text{-}3.1)$$

A similar process may be applied to lines. The given line may be divided into finite segments whose centroids are known, and the following equations may be used:

$$\left.\begin{aligned} L\bar{x} &= \sum l_i x_i \\ L\bar{y} &= \sum l_i y_i \end{aligned}\right\} \qquad (6\text{-}3.2)$$

Before these equations are applied to illustrative problems, it will be convenient to summarize the location of centroids for common geometrical shapes (determined in preceding problems) given in Table 6-6.1.

Table 6-6.1 Centroids for Common Geometric Shapes.

Shape	Area or length	$\bar{x}$	$\bar{y}$
Rectangle 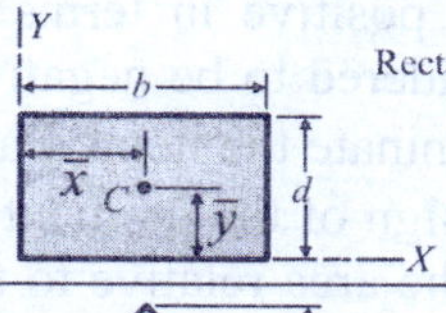	bd	$\frac{1}{2}b$	$\frac{1}{2}d$
Any triangle 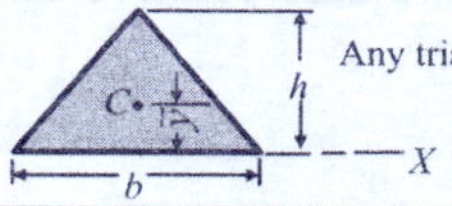	$\frac{1}{2}bh$		$\frac{1}{3}h$
Semicircle 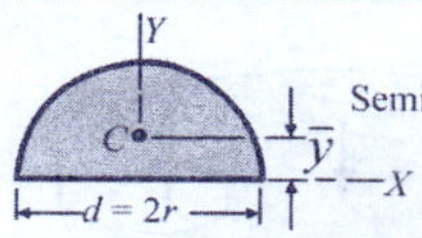	$\frac{\pi r^2}{2}$	0	$\frac{4r}{3\pi}$ or $0.424r$
Quarter circle 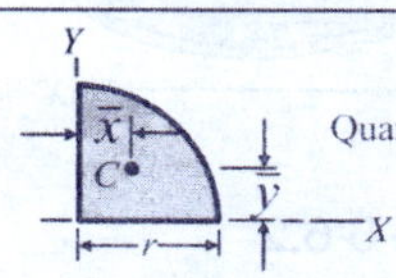	$\frac{1}{4}\pi r^2$	$\frac{4r}{3\pi}$ or $0.424r$	$\frac{4r}{3\pi}$ or $0.424r$
Circular sector 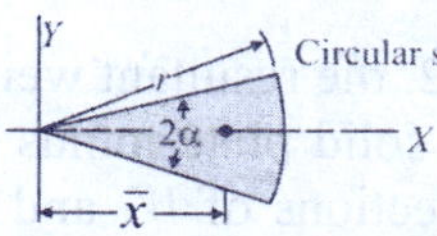	$r^2\alpha$	$\frac{2}{3}\frac{r\sin\alpha}{\alpha}$	0
Segment of arc 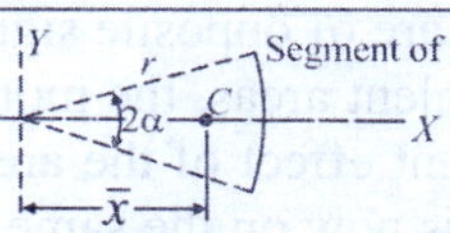	$2r\alpha$	$\frac{r\sin\alpha}{\alpha}$	0
Semicircular arc 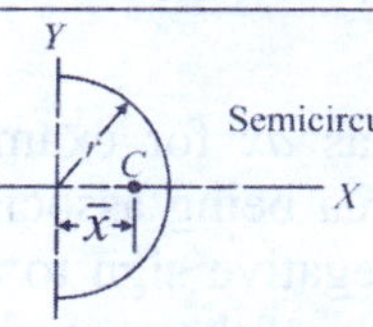	πr	$\frac{2r}{\pi}$	0
Area under spandrel 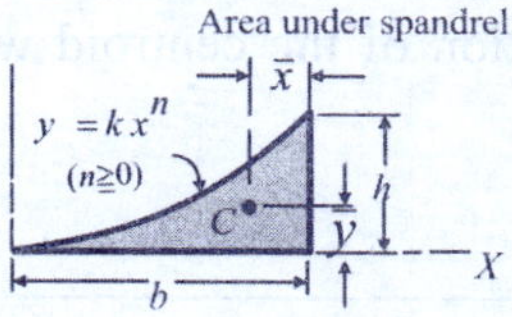	$\frac{1}{n+1}bh$	$\frac{1}{n+2}b$	$\frac{n+1}{n+2}h$

SIGNS

The following discussion indicates the method of determining which sign to give to an area or coordinate. Consider a plate balanced on a knife edge denoted by Y-Y in Fig. 6-6.1. It is evident that the moment of W_1 about Y-Y is opposite in effect to the moment of W_2. Since the directions of W_1 and W_2 are both down, the difference in moment effect is caused by W_1 and W_2 being located on opposite sides of the axis. Or if x_1 is considered to be positive in terms of coordinates with respect to the X-Y axes, x_2 must be considered to be negative. Proceeding to replace weight in terms of area, we may eliminate the items γ and t from the product $W = \gamma\, tA$, whence we observe that the sign of the moment of area similarly depends on the position of the centroid of the area relative to the coordinate axes.

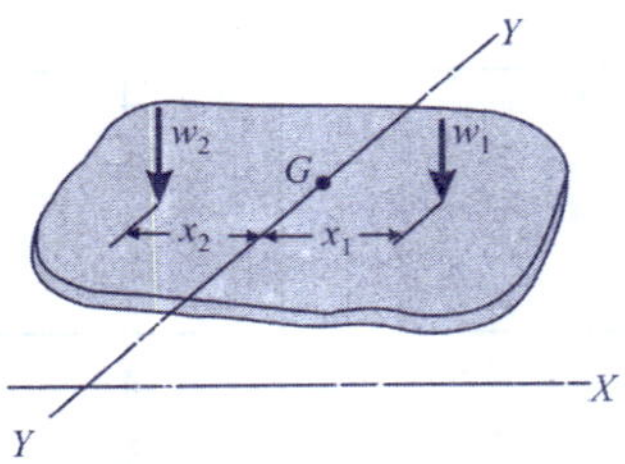

Figure 6-6.1

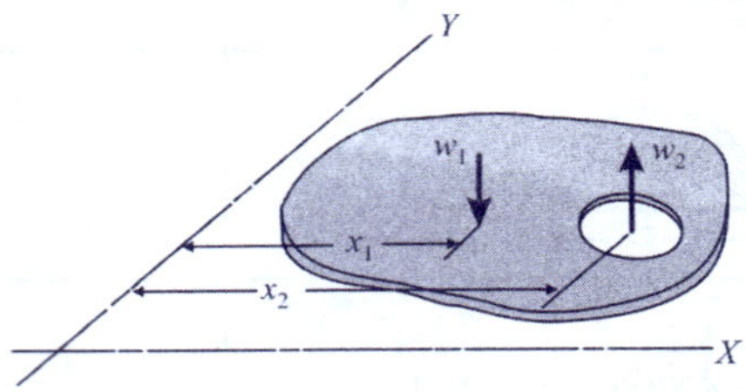

Figure 6-6.2

In the case of a plate with a hole cut in it as in Fig. 6-6.2, the resultant weight W may be considered equivalent to the weight W_1 of the solid plate minus the weight W_2 of the cut-out portion. In this instance, the directions of W_1 and W_2 are opposite, so their moments with respect to the Y axis are of opposite sign. If, as before, these weights are expressed in terms of equivalent areas, the moment effect of the area of the cut-out is opposite to the moment effect of the area of the original plate, even though the centroid of each area is now on the same side of the axis.

Hence the sign of the moment of an area, expressed as ax for example, depends on the signs of a and of x, the positive sign for area being associated with area that adds to the net area of the figure, and the negative sign to area that reduces the net area. Moreover, the sign of the coordinate of the centroid of an area may be plus or minus, depending on the location of the centroid with respect to the axis of moments.

ILLUSTRATIVE PROBLEMS

6-6.1 A slender homogeneous wire of uniform cross section is bent into the form shown in Fig. 6-6.3. Determine the position of the centroid of the wire with respect to the given axes.

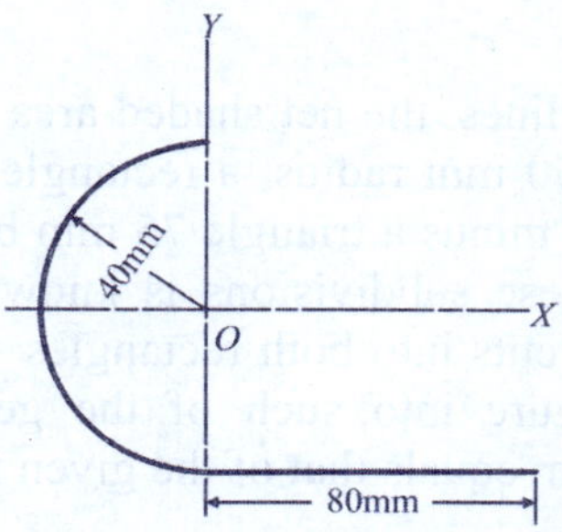

Figure 6-6.3

Solution

The wire is considered equivalent to two line segments, one a semicircle of 40 mm radius and the other a straight line 80 mm long. In Prob. 6-5.1 the centroid of the semicircular line was shown to be located a distance $\bar{x} = 2r/\pi$ from the Y axis. By applying Eq (6-3.2) and paying careful attention to signs, we obtain

$$[L\bar{x} = \Sigma lx] \qquad (40\pi + 80)\bar{x} = (40\pi)\left(-\frac{2\times 40}{\pi}\right) + 80\times 40$$

$$205.6\bar{x} = -3200 + 3200 = 0$$

$$\therefore \quad \bar{x} = 0$$

$$[L\bar{y} = \Sigma ly] \qquad (40\pi + 80)\bar{y} = (40\pi)\times 0 + (80)(-40)$$

$$205.6\bar{y} = -3200$$

$$\bar{y} = -15.58 \text{ mm}$$

The centroid of the wire lies on the Y axis

15.58 mm below the X axis. ***Ans.***

6.6.2 Determine the position of the centroid of the shaded area shown in Fig. 6-6.4.

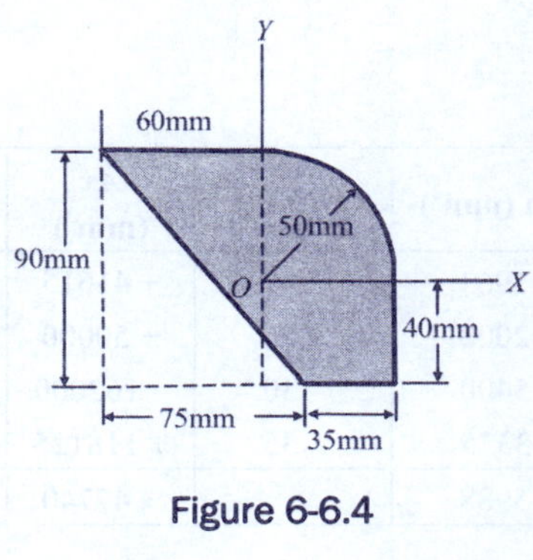

Figure 6-6.4

Solution

As indicated by the dashed lines, the net shaded area may be considered as the sum of a quarter circle of 50 mm radius, a rectangle 50 mm by 40 mm and a rectangle 60 mm by 90 mm minus a triangle 75 mm by 90 mm. The location of the centroid for each of these subdivisions is known from Table 6-6.1. It is immaterial that the triangle cuts into both rectangles – the important concept is to subdivide the given figure into such of the geometric shapes listed in Table 6-6.1 that their net sum equals that of the given figure.

Select reference axes as shown. When there are a number of elements to be considered, it is convenient to apply Eq. (6-3.1) in the form of the tabulation shown below. This simplifies the summations of the various terms. Before the quantities are entered in this tabulation, the proper sign should be put before each term. An area is considered positive or negative according to whether it adds to or subtracts from the net area of the figure. The signs of the coordinates of the centroids follow the usual mathematical convention; i.e., x coordinates are positive to the right of the Y axis and y coordinates are positive above the X axis.

For a specific example of how the tabulation is made, consider the triangle. Its area is $\frac{1}{2} \times 90 \times 75 = 3375$ mm^2 which is entered as a negative value because it reduces the net area of the figure. Its centroid is $\frac{1}{3} \times 75 = 25$ mm from its left edge and $\frac{1}{3} \times 90 = 30$ mm above its base. Hence, with respect to the selected reference axes, its x coordinate is $60 - 25 = 35$ mm to the left of the Y axis and therefore negative, while its y coordinate is $40 - 30 = 10$ mm below the X axis and hence also negative. Its moments of area, ax and ay, are both positive because they are the product of two negative quantities. The student is urged to pay particular attention to the signs because *incorrect signs are the greatest source of error.*

Computation of Results

	Area (mm^2)	X (mm)	Ax (mm^3)	y (mm)	ay (mm^3)
Quarter circle	+ 1963	+ 21.2	+ 41615	+ 21.2	+ 41615
Rectangle 50 × 40	+ 2000	+ 25	+ 50000	– 20	– 40000
Rectangle 60 × 90	+ 5400	– 30	– 162000	+ 5	+ 27000
Triangle	– 3375	– 35	+ 118125	– 10	+ 33750
Totals	+ 5988		+ 47740		62365

Taking the sums from the tabulation, Eq. (6-3. la) gives the following results:

$$\left[\bar{x} = \frac{\Sigma ax}{a}\right] \quad \bar{x} = \frac{47740}{5988} = 7.97 \text{ mm}$$

$$\left[\bar{y} = \frac{\Sigma ay}{a}\right] \quad \bar{y} = \frac{62365}{5988} = 10.41 \text{ mm}$$

Ans.

6-6.3 Determine the moment of area about the Y axis of the shaded area in Fig. 6-6.5 contained between the line OA and the curve $y^2 = kx^3$.

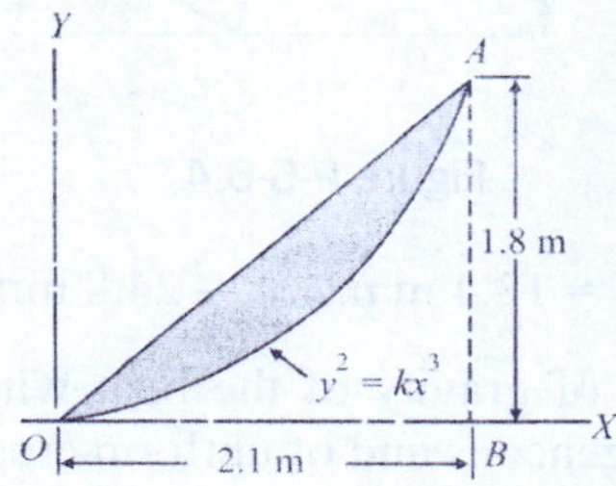

Figure 6-6.5

Solution

This problem may be solved by direct integration, but a simpler solution is to use the method of composite parts. The shaded area may be resolved into the triangle OAB from which is subtracted the area under the curve $y^2 = kx^3$. The equation of the curve may be rewritten as $y = (\sqrt{k})x^{3/8}$ from which it is evident that its area is that of the spandrel listed in Table 6-6.1 with $n = \frac{3}{2}$ and equal to $\frac{1}{n+1}bh = \frac{1}{\frac{3}{2}+1}bh = \frac{2}{5}bh$. Its centroid is $\frac{1}{n+2}b = \frac{1}{\frac{3}{2}+2}b = \frac{2}{7}b$ from AB and hence at $\frac{5}{7}b$ from the Y axis.

Applying Eq. (6-3.1) which states that the moment of area is equal to the sum of the moments of area of its parts, we obtain

$$[A\bar{x} = \Sigma ax] \; A\bar{x} = \left(\frac{1}{2} \times 1.8 \times 2.1\right)\left(\frac{2}{3} \times 2.1\right) - \left(\frac{2}{5} \times 2.1 \times 1.8\right)\left(\frac{5}{7} \times 2.1\right)$$

$$A\bar{x} = 2.646 - 2.268 = 0.378 \text{ m}^3$$ ***Ans.***

PROBLEMS

6-6.4 A slender homogeneous wire of uniform cross section is bent into the shape shown in Fig. P.6-6.4. Determine the coordinates of its center of gravity.

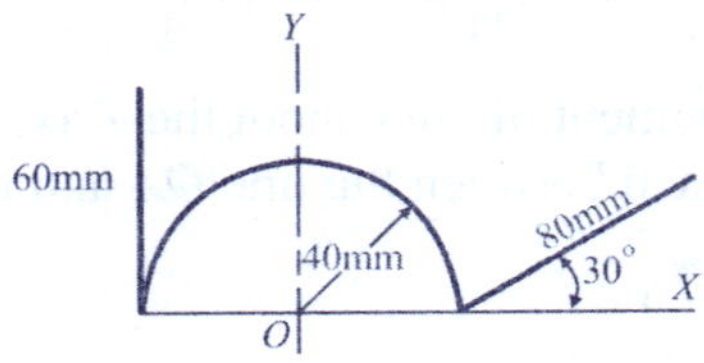

Figure P-6-6.4

$\bar{x}$ = 13.4 mm.; $\bar{y}$ = 24.8 mm

6-6.5 Locate the center of gravity of the bent wire shown in Fig. P.6-6.5. The wire is homogeneous and of uniform cross section.

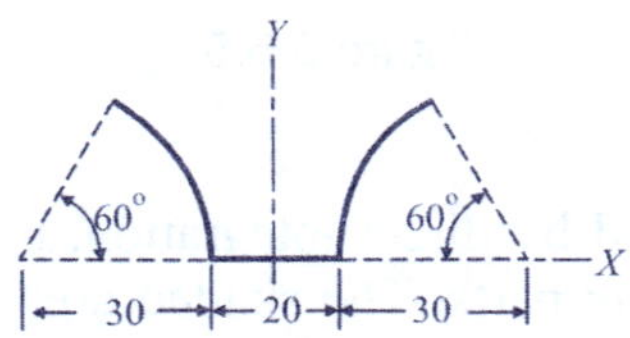

Figure P-6-6.5

6.6.6 Determine the centroid of the lines that form the boundary of the shaded area in Fig. P-6.6.6.

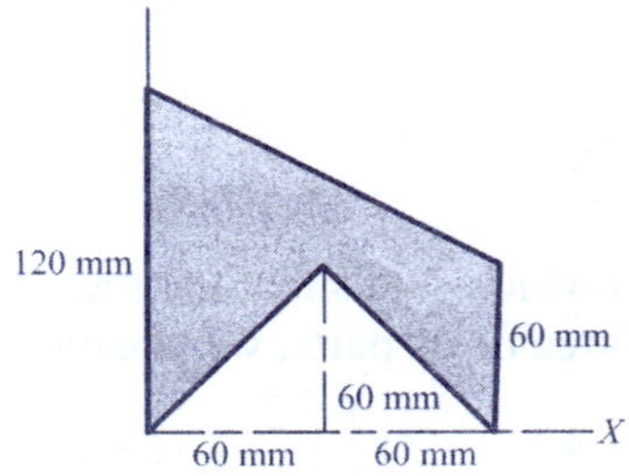

Figure P-6-6.6

$\bar{x}$ = 52.6 mm; $\bar{y}$ = 54.1 mm *Ans.*

6-6.7 Locate the centroid of the shaded area shown in Fig. P.6-6.6.

6-6.8 Locate the centroid of the shaded area in Fig. P-6-6.8 created by cutting a semicircle of diameter r from a quarter circle of radius r.

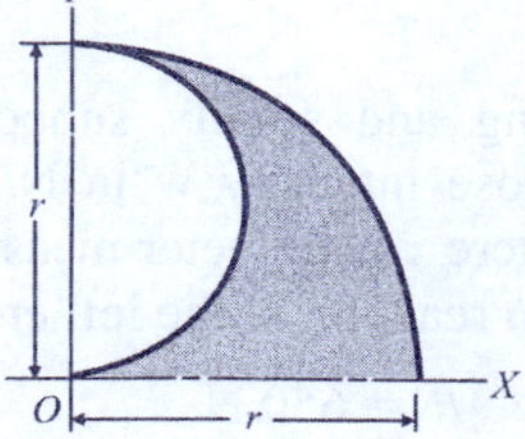

Figure P-6-6.8

$\bar{x} = 0.636\ r;\ \bar{y} = 0.348\ r$ ***Ans.***

6-6.9 In order to fully utilize the different values of the compressive stress S_c and the tensile stress S_t in cast iron beams, it is desirable to locate the centroidal axis so that the ratio of its distance to the top of a section to its distance from the bottom equals S_c/S_t. Using the section shown in Fig. P.6-6.9, find the dimension b to satisfy this criterion if $S_c/S_t = 3$.

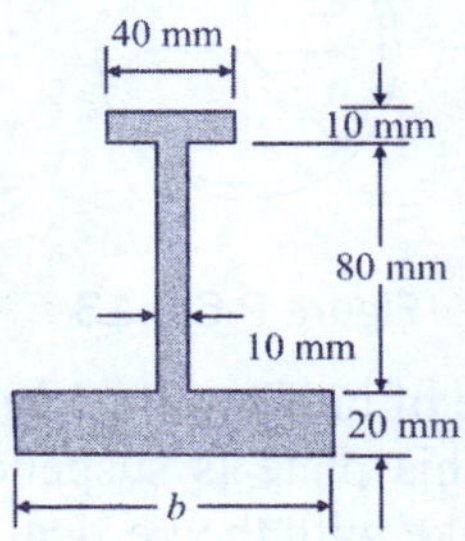

Figure P-6-6.9

6-6.10 Determine the dimension b that will locate the centroidal axis at 40 mm above the base of the section shown in Fig. P-6.6.10.

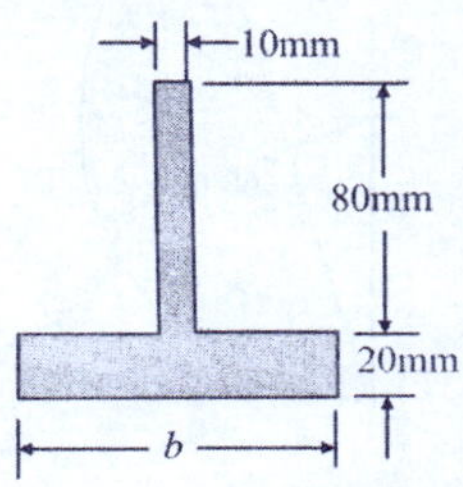

Figure P-6-6.10

$b = 70$ mm ***Ans.***

6-6.11 The 16 m wing of a small airplane is subjected to a lift which varies from zero at the tip to 360 N/m at the fuselage according to $w = 90\,x^{1/2}$ where x is measured in meter from the tip. Compute the resultant lift and its location from the wing tip. Avoid integration by using Table 6-6.1.

6-6.12 A beam, 10 m long and simply supported at its ends, carries a distributed load whose intensity w in N/m is given by the equation $w = 100 + 0.3\,x^3$ where x is in meter measured from the left end of the beam. Find the beam reaction at the left end.

$R_1 = 650$ N ***Ans.***

6-6.13 A slender homogeneous wire is bent into the shape shown in Fig. P-6-6.13 and suspended from a hinge at A. Find the angle the neck of the hook will make with the vertical.

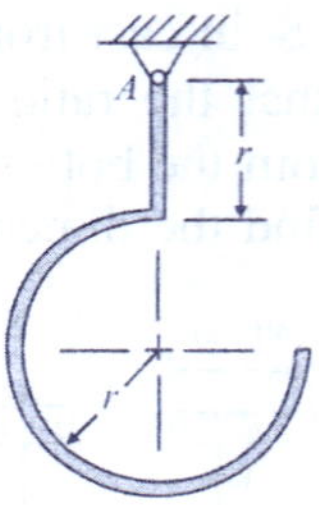

Figure P-6-6.13

6-6.14 The surface of a plate of uniform thickness is given by the shaded area in Fig. P.6-6.14. If this plate is suspended from a hinge at A, what angle will line AB make with the vertical?

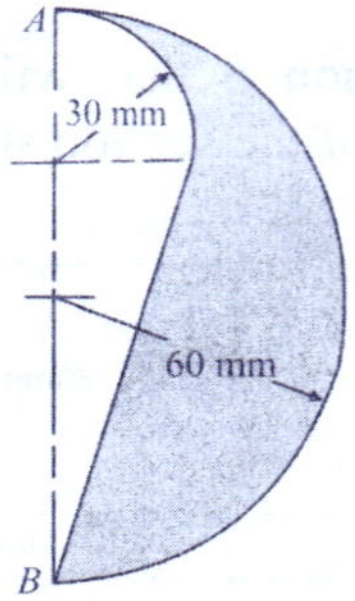

Figure P-6-6.14

$\theta = 23.7°$ ***Ans.***

6.6.15 With respect to the given axes, locate the centroid of the shaded area shown in Fig. P-6-6.15.

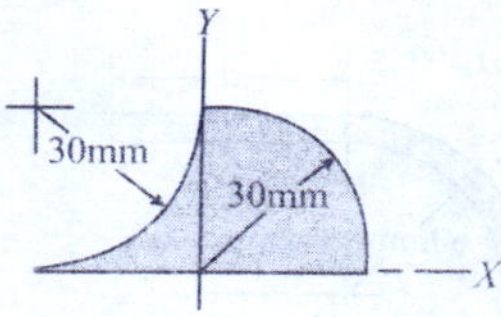

Figure P-6-6.15

$\bar{x} = 8.58$ *mm.*; $\bar{y} = 11.42$ *mm* **Ans.**

6-6.16 Locate the centroid of the shaded area shown in Fig. P.6-8.16 resulting from removing the circular segment of 60 mm radius from the circular plate of 80 mm radius.

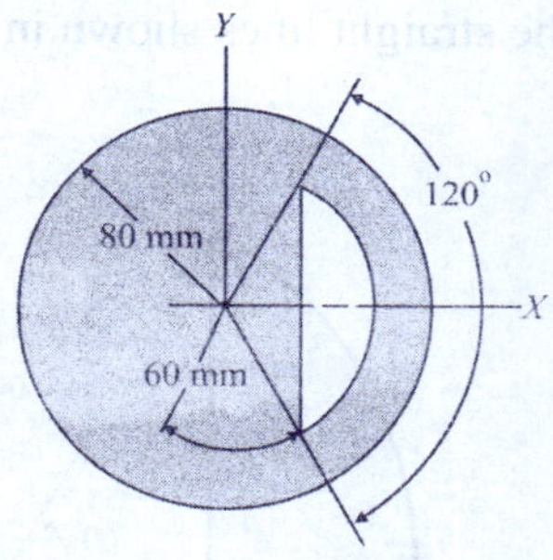

Figure P-6-6.16

$\bar{x} = -\ 3.35$ *mm*

6-6.17 Locate the centroid of the shaded area shown in Fig. P-6-6.17.

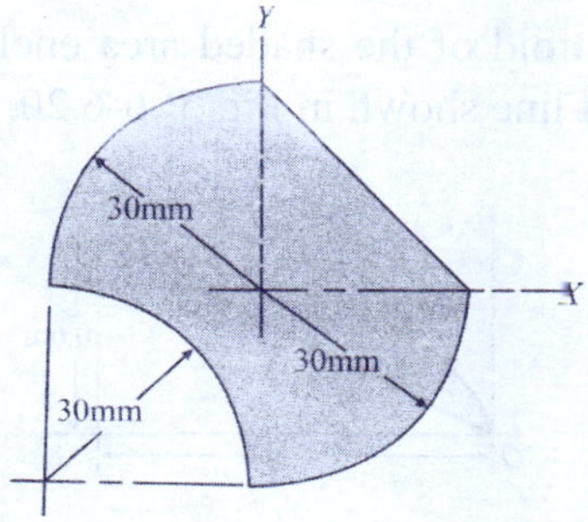

Figure P-6-6.17

6-6.18 For the shaded area shown in Fig. P-6-618, find the length a so that the centroid is at O.

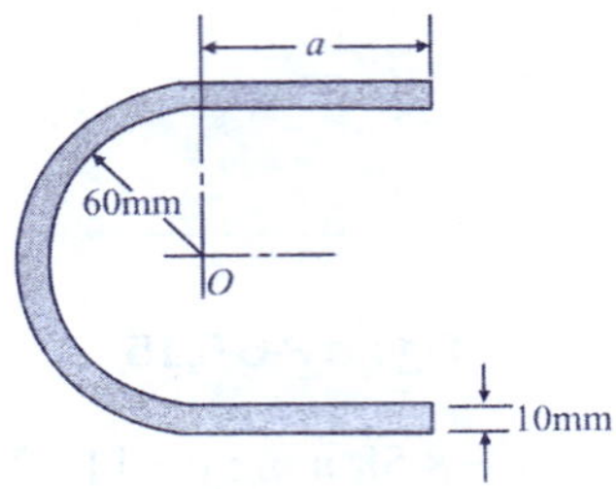

Figure P-6-6.18

6-6.19 Determine the centroid of the shaded area which is bounded by the circular arc and the straight lines shown in Fig. P-6-6.19.

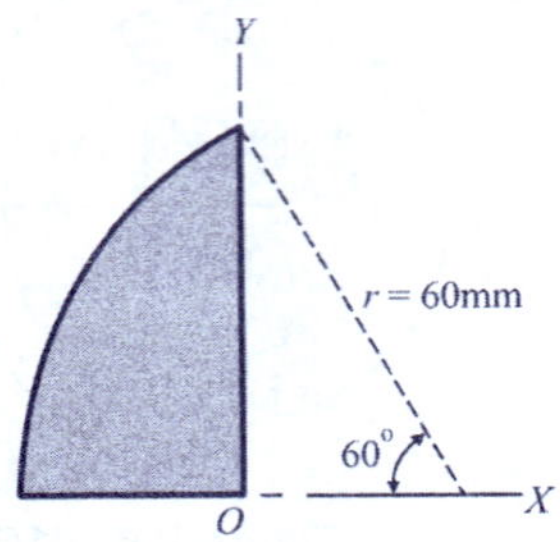

Figure P-6-6.19

$\overline{x} = -12.3\ mm.; \overline{y} = 20.4\ mm$ **Ans.**

6-6.20 Locate the centroid of the shaded area enclosed by the curve $y^2 = ax$ and the straight line shown in Fig. P-6-6.20.

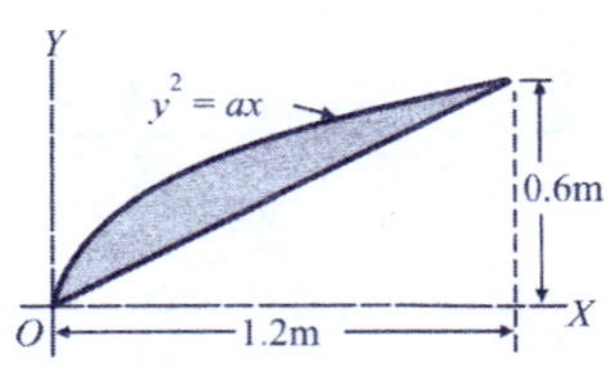

Figure P-6-6.20

$\overline{x} = 0.48\ m.; \overline{y} = 0.3\ m$ **Ans.**

6-6.21 Locate the centroid of the cross section of the culvert shown in Fig. P-6-6.21.

Hint: Integration is unnecessary if the area is subdivided into elements to be found in Table 6-6.1.

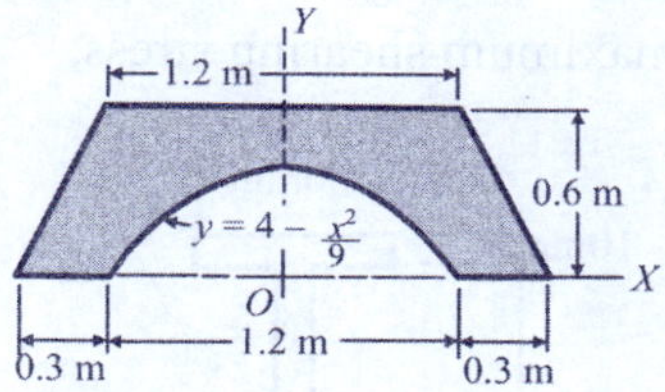

Figure P-6-6.21

$\bar{y} = 0.346$ m **Ans.**

6-6.22 A rectangle is divided into two parts by the curve $y = kx^n$ as shown in Fig. P-6-6.22. Using the known location of the centroid of the lower part A as given in Table 6-6.1, show that the centroid of the upper part B is located by $\bar{x}_B = \frac{1}{2}\bar{x}_A$ and $\bar{y}_B = 2\bar{y}_A$.

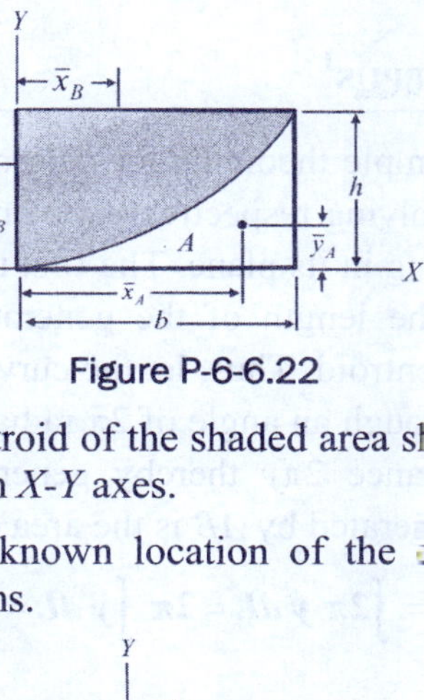

Figure P-6-6.22

66.23 Determine the centroid of the shaded area shown in Fig. P.6-6.23 with respect to the given *X-Y* axes.

Hint: Project the known location of the centroid of the semicircle upon X-Y directions.

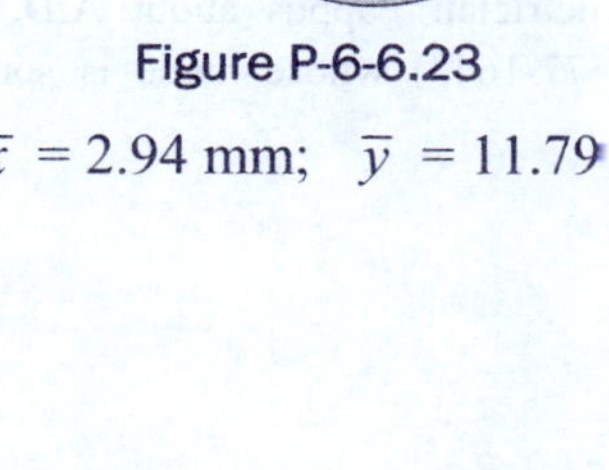

Figure P-6-6.23

$\bar{x} = 2.94$ mm; $\bar{y} = 11.79$ mm **Ans.**

6-6.24 A beam has the cross section shown in Fig. P-6-6.24. Compute the moment of area of the shaded portion about the horizontal centroidal axis X_o of the entire section.

Note: It is shown in strength of materials that this result is used in computing the maximum shearing stress.

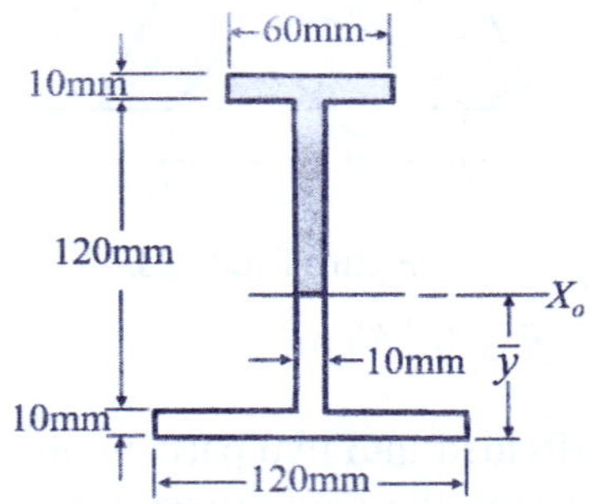

Figure P-6-6.24

$\bar{y}$ = 57 mm; Q = 73.4 mm^3. ***Ans.***

6-7 THEOREM OF PAPPUS[1]

Pappus developed two simple theorems for determining the surface area or the volume generated by revolving respectively a plane curve or a plane area about a *nonintersecting* axis lying in its plane. The first theorem states that the surface area is the product of the length of the generating curve multiplied by the distance traveled by its centroid. Thus, let the curve *AB* of length *L* in Fig. 6-7.1 be revolved about *OX* through an angle of 2π radians. The differential length dL sweeps through the distance $2\pi y$ thereby generating a hoop whose area is $2\pi y\, dL$. The total area generated by *AB* is the area of all such hoops or

$$A = \int 2\pi\, y\, dL = 2\pi \int y\, dL = 2\pi \bar{y} L \tag{6-7.1}$$

[1]According to W. W. Rouse Ball's A *Short History of Mathematics*, the theorems were discovered by the Greek geometrician Pappus about A.D. 300 and restated by the Swiss mathematician Paul Guldin (1577-1643) whose name is sometimes also associated with the theorems.

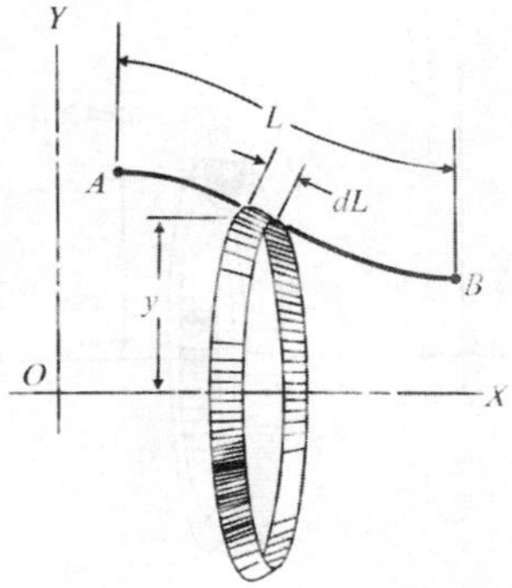

Figure 6-7.1 First theorem of Pappus surface generated by revolving a line about a nonintersecting axis

If the generating line L is composed of several segments, the centroid $\bar{y}$ of that line need not be found since the product $L\bar{y}$ is equivalent to and may be replaced by the sum of the moments of length (i.e., $\sum ly$) cf those segments.

When the centroid of an area is known, the volume generated by revolving the area about a *nonintersecting* axis can be found by applying the second theorem. This theorem states that the volume is the product of the area of the figure multiplied by the length of the path described by the centroid of the area. To demonstrate this, let the area A in Fig. 6-7.2 be rotated about the axis OX through an angle of 2π radians. The differential area dA sweeps through the distance $2\pi y$ and generates a ring whose volume is $2\pi y\, dA$. The total volume generated is the sum of the volumes of all such rings or

$$V = \int 2\pi\, y\, dA = 2\pi \int y\, dA = 2\pi A\bar{y} \qquad (6\text{-}7.2)$$

If the generating area A is composed of several parts, the centroid $\bar{y}$ of that area need not be found since the product $A\bar{y}$ is equivalent to the sum of the moments of area (i.e., Σay) of the several parts.

If the generating line or area is revolved through an angle θ less than 2π radians, the generated surface or volume may be found by substituting θ for 2π in Eqs. (6-7.1) and (6-7.2).

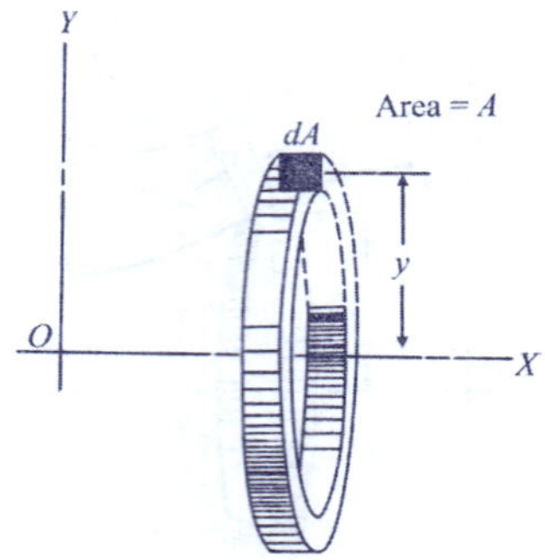

Figure 6-7.2 Second theorem of pappus. Volume generated by revolving an area about a nonintersecting axis.

ILLUSTRATIVE PROBLEMS

6.7.1 Compute the surface area of the cone generated by revolving the line in Fig. 6-7.3 about the Y axis.

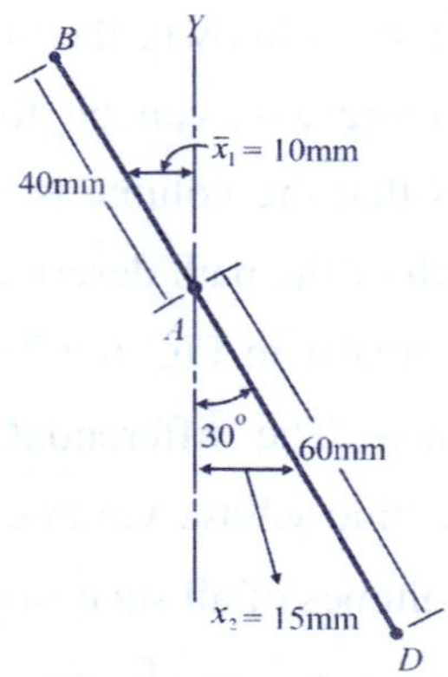

Figure 6-7.3

Solution

Two cones are generated by the line; one by the 40 mm, length above A and the other by the 60 mm length below A. The x coordinate of the centroid for each segment of the line is given by $\bar{x} = \frac{1}{2} L \sin 30°$; hence $\bar{x}_1 = 10mm$ and $\bar{x}_2 = 15mm$. Applying Eq. (6-7.1), we obtain

$[A = 2\pi\bar{x}L]$ 40 mm segment: $A_1 = 2\pi \times 10 \times 40 = 2513\ 2\ \text{mm}^2$

60 mm segment: $A_2 = 2\pi \times 15 \times 60 = 5654.8\ \text{mm}^2$

Total surface area $= A_1 + A_2 = 2513.2 + 5654.8$

$= 8168.0\ \text{mm}^2$ ***Ans.***

6-7.2 The shaded area in Fig. 6-7.4 is composed of a second-degree parabola and a semicircle. Determine the volume generated by rotating the area through one revolution about the Y_1-Y_1 axis.

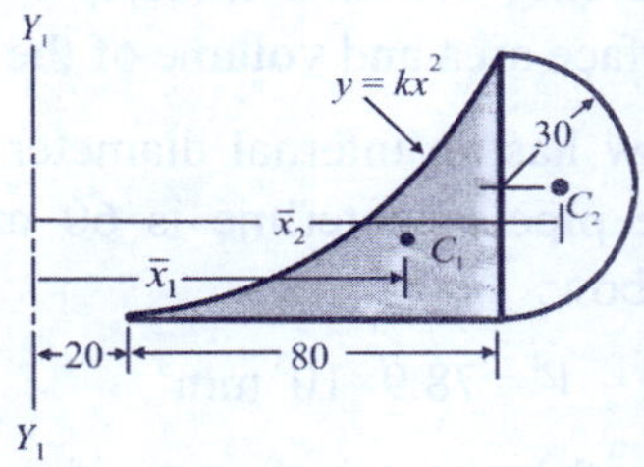

Figure 6-7.4

Solution

From Table 6-6.1, the x coordinate of the centroid for each part of the total area is

$$\bar{x}_1 = 20 + \frac{3}{4} \times 80 = 80\ \text{mm}$$

$$\bar{x}_2 = 100 + 0.424 \times 30 = 112.7\ \text{mm}$$

Applying Eq. (6-7.2) we now find the volume to be

$$[V = 2\pi A\bar{x} = 2\pi \sum ax] \qquad V = 2\pi\left[\left(\frac{80 \times 60}{3}\right)(80) + \left(\frac{\pi \times 30^2}{2}\right)(112.7)\right]$$

$$V = 2\pi\,[128 \times 10^3 + 159 \times 10^3] = 1805 \times 10^3\ \text{mm}$$ ***Ans.***

PROBLEMS

6.7.3 Show that the volume of a paraboloid of revo.ution is equal to one-half the volume of the circumscribing cylinder.

6-7.4 Determine the volume of an ellipsoid of revolution generated by rotating an ellipse about (a) its major axis (prolate ellipsoid) and (b) its minor axis (oblate ellipsoid). Take the larger semiaxis as a and the smaller semiaxis as b.

(a) $V = \frac{4}{3}\pi ab^2$; (b) $V = \frac{4}{3}\pi a^2 b$ ***Ans.***

6-7.5 A circle of radius r lies in the XY plane with its center at distance a above the X axis. Revolving it about the X axis will generate a doughnut-shaped ring called a torus, provided a is greater than r. Compute the surface area and volume of the torus.

6-7.6 A 60° pipe elbow has an internal diameter of 40 mm. The radius of curvature of the pipe's centerline is 60 mm. Compute the internal volume of the elbow.

$V = 78.9 \times 10^3\ \text{mm}^3$ ***Ans.***

6.7.7 Find the volume of the spherical wedge formed by rotating through an angle of 45° a semicircle of radius r about its base diameter. What is the total surface area?

$V = \frac{\pi r^3}{6}$; $A = \frac{3}{2}\pi r^2$ ***Ans.***

6-7.8 Determine the volume and surface area of the solid shown in Fig. P- 6-7.8.

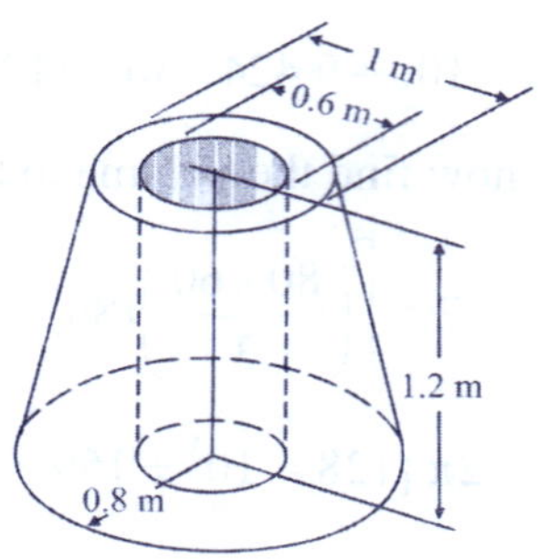

Figure P-6-7.8

6-7.9 The rim of a V-belt drive pulley has the cross section shown in Fig. P-6-7.9. If the rim is made of steel weighing 7860 kg/m^3 determine the weight of the rim.

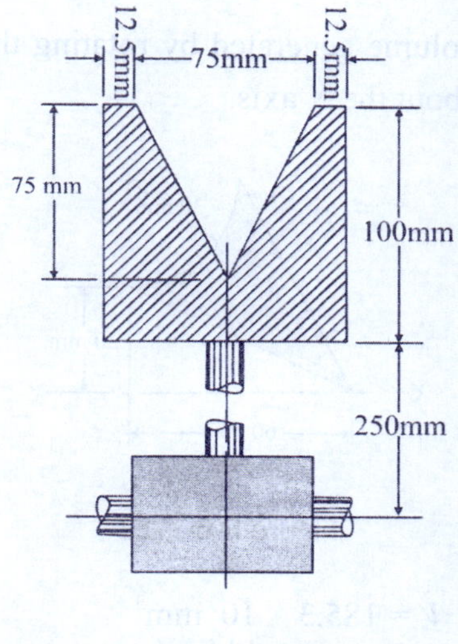

Figure P-6-7.9

6-7.10 Determine the difference in volume (if any) generated by rotating the shaded area shown in Fig. P-6-7.10 about the *X-X* axis and then about the *Y-Y* axis.

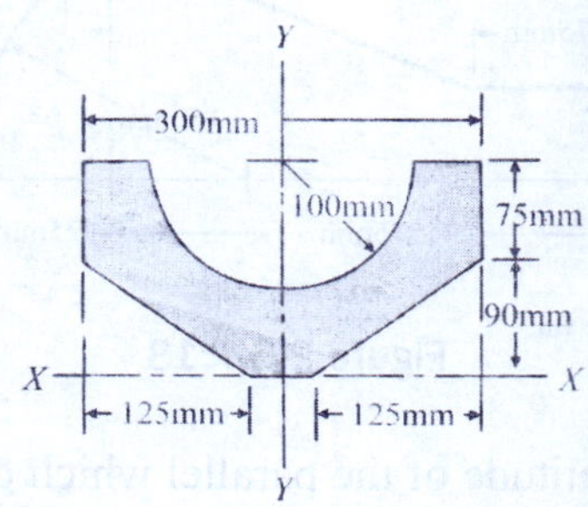

Figure P-6-7.10

6-7.11 The area contained between two concentric semicircles of radii 15 mm and 30 mm is rotated about an axis 40 mm away and parallel to the base diameters of the semicircles. Compute the surface area and volume generated by a complete revolution.

$A = 573 \times 10^2$ mm^2 ; $V = 365 \times 10^3$ mm^3 ***Ans.***

6-7.12 Compute the volume generated by rotating the shaded area shown in Fig. P-6-7.12 about the X axis.

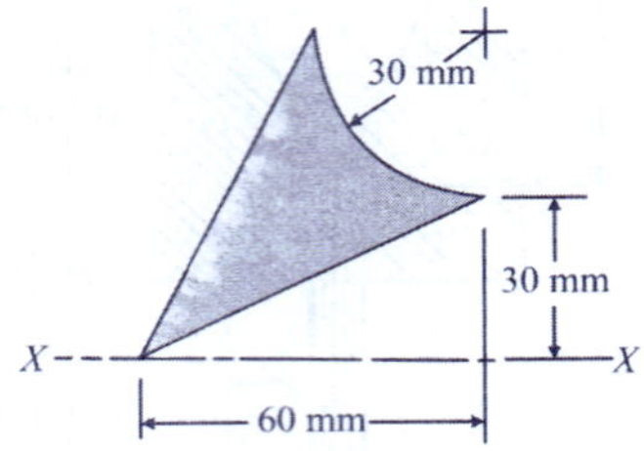

Figure P-6-7.12

$$V = 185.3 \times 10^3 \text{ mm}^3 \qquad \textbf{Ans.}$$

6-7.13 The shape of a space capsule is generated by rotating the line *ABCDE* about the *X* axis as shown in Fig. P-6-7.13. Determine the volume inside the capsule and the area of the best shield generated by line *DE*.

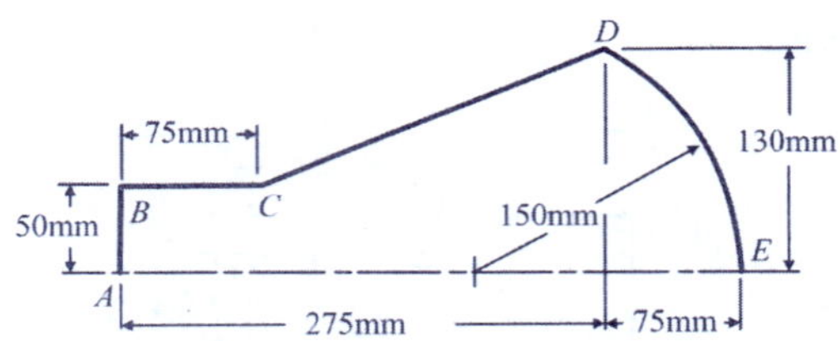

Figure P-6-7.13

6-7.14 Determine the latitude of the parallel which divides the surface area of the northern hemisphere of the northern hemisphere of the earth into two equal parts.

$$\phi = 30^\circ \qquad \textbf{Ans.}$$

6-8 CENTER OF GRAVITY OF BODIES. CENTROIDS OF VOLUMES

The principle of moments, by which the centroids of areas and lines was determined in the preceding sections, is readily extended to locate the center of gravity of bodies or the centroids of volumes.

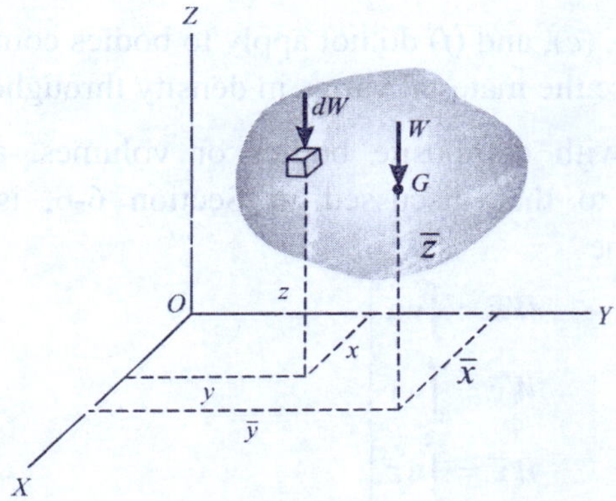

Figure 6.8.1

Thus, consider the body shown in Fig. 6-8.1. The total weight W of the body is the sum of its differential elements dW or $W = \int dW$. Also, the moment of the resultant weight W is equal to the moment sum of its elements. Hence, with respect to the Y and X axes respectively, we obtain.

$$W\bar{x} = \int x\, dW \qquad (a)$$

$$W\bar{y} = \int y\, dW \qquad (b)$$

When we rotate the body and its reference frame through 90° (either about the X or Y axis) so that the Z axis becomes horizontal, a moment summation about the Z axis gives

$$W\bar{z} = \int z\, dW \qquad (c)$$

If the body in Fig. 6-8.1 is composed of the same material throughout, its weight W is the product of its density γ (i.e., weight per unit volume) multiplied by its volume V. Similarly the weight dW of an elemental prism is the product of the density γ and its elemental volume dv. Substituting these values in Eq. (a.) results in

$$\left[W\bar{x} = \int x dW\right] \qquad (\gamma V)(\bar{x}) = \gamma \int x dV$$

whence, by canceling γ, we obtain

$$V\bar{x} = \int x\, dV \qquad (d)$$

and similarly,

$$V\bar{y} = \int y\, dV \qquad (e)$$

$$V\bar{z} = \int z\, dV \qquad (f)$$

Note that Eqs. (*d*), (*e*), and (*f*) do not apply to bodies composed of more than one material, or where the material varies in density throughout the body.

When we deal with composite bodies or volumes, a process of finite summation, similar to that discussed in Section 6-6, is used. The above equations then become

$$\left.\begin{aligned} W\bar{x} &= \int wx \\ W\bar{y} &= \int wy \\ W\bar{z} &= \int wz \end{aligned}\right\} \qquad (68.1)$$

And

$$\left.\begin{aligned} V\,\bar{x} &= \sum vx \\ V\,\bar{y} &= \sum vy \\ V\,\bar{z} &= \sum vz \end{aligned}\right\} \qquad (6\text{-}8.2)$$

If a body has a plane of symmetry, the center of gravity lies in that plane. If it has two planes of symmetry, the center of gravity is on the line of intersection of the planes. If it has three planes of symmetry, the center of gravity is at the point of intersection of the three planes.

ILLUSTRATIVE PROBLEMS

6-8.1 *Centroid of the Volume of a Right Circular Cone.*

The radius of the base of the right circular cone shown in Fig. 6-8.2 is denoted by *r* and its altitude by *h*. The axis of the cone is taken as the *X* axis. Determine the location of the centroid of the volume of the cone.

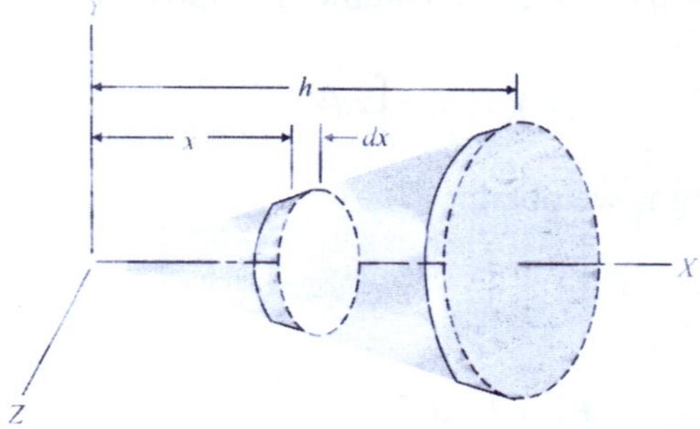

Figure 6-8.2

Solution

From symmetry, $\bar{y} = 0$ and $\bar{z} = 0$. To determine $\bar{x}$, select a thin plate parallel to the base of the cone as the differential volume dv. The centroid of the thin plate is at its center and at a distance x from the YZ plane, as shown in the figure. Applying Eq. (d), we obtain

$$[V\bar{x} = \int x\, dV] \qquad \left(\frac{1}{3}\pi r^2 h\right)\bar{x} = \int_0^h x\pi y^2\, dx \qquad (a)$$

From similar triangles, $\frac{y}{x} = \frac{r}{h}$, or $y = \left(\frac{r}{h}\right)x$; hence, Eq. ($a$) becomes

$$\left(\frac{1}{3}\pi r^2 h\right)\bar{x} = \frac{\pi r^2}{h^2}\int_0^h x^3\, dx = \frac{1}{4}\pi r^2 h^2$$

$$\bar{x} = \frac{3}{4}h \qquad \textit{Ans.}$$

6.8.2 The wooden block shown in Fig: 6-8.3 is 0.3×0.3× 0.225 m. A hole 0.1m in diameter and 0.15 m deep is drilled in it. The hole is filled by a metal pin weighing 225 N. If wood weighs 674 kg/m^3, locate the center of gravity of the body.

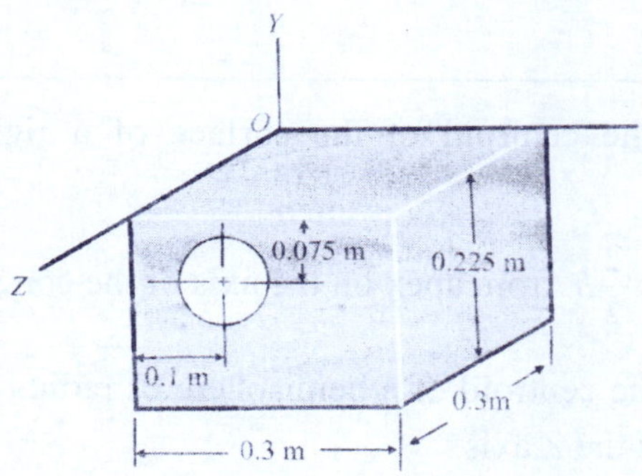

Figure 6.8.3

Solution

Assume the body to be composed of the fallowing parts: (1) the full-size block, less (2) a wooden pin 0.1 m diameter and 0.15 m long, plus (3) the metal pin which replaces the wooden pin. The weights of the wooden block and pin are

$$[W = \gamma AL] \qquad \text{Block}: W_1 = 674 \times 9.80 \times 0.3 \times 0.3 \times 0.225 = 133.75\ N$$

$$\text{Pin}: W_2 = 674 \times 9.8 \times \frac{\pi \times 0.1^2}{4} \times 0.15 = 7.78\ N$$

The following tabulation is now made:

Item	*W*	*x*	*Wx*	*y*	*Wy*	*z*	*Wz*
Wood block	+ 133.75	0.15	+ 20.06	0.1125	+ 15.05	0.15	+ 20.06
Wood pin	− 7.78	0.1	− 0.778	0. 075	− 0.583	0.225	− 1.75
Steel pin	+ 225	0.1	+ 22.5	0.075	+ 16.875	0.225	+ 50.62
Totals	350.97		41.782		31.342		68.93

From the sums in the table, we obtain

$$\left[\bar{x} = \frac{\Sigma Wx}{\Sigma x}\right] \qquad \bar{x} = \frac{41.782}{350.97} = 0.119 \text{ m}$$

$$\left[\bar{y} = \frac{\Sigma Wy}{\Sigma W}\right] \qquad \bar{y} = \frac{31.342}{350.97} = 0.089 \text{ m} \qquad \textbf{\textit{Ans.}}$$

$$\left[\bar{z} = \frac{\Sigma Wz}{\Sigma W}\right] \qquad \bar{z} = \frac{68.93}{350.97} = 0.196 \text{ m}$$

PROBLEMS

6-8.3 Determine the centroid of the surface of a right circular cone of altitude *h*.

$\frac{2}{3}h$ from apex on the axis of the cone ***Ans.***

6-8.4 Determine the centroid of a hemisphere of radius *r*, taking the axis of symmetry as the *Z* axis.

$\bar{z} = \frac{3}{8}r$ from the base ***Ans.***

6-8.5 A uniform wire is bent into the shape shown in Fig. P-6-8.5. The straight segments lie in the *X-Z* plane, and the 80 mm length makes an angle of 30° with the *X* axis. The semicircular segment is in the *X-Y* plane. Locate the center of gravity of the wire.

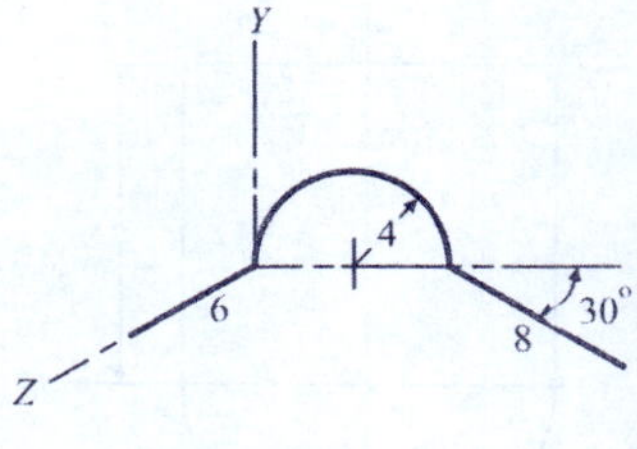

Figure P.6-8.5

$\bar{x}$ = 53.3 mm.; $\bar{y}$ = 12.04 mm.; $\bar{z}$ = 12.8 mm *Ans.*

6-8.6 A thin plate of sheet metal is composed of two rectangles and a triangle bent as shown in Fig: P-6-8.6. A hole whose area is 1200 sq. mm has been cut in the plate. Determine the coordinates of the center of gravity.

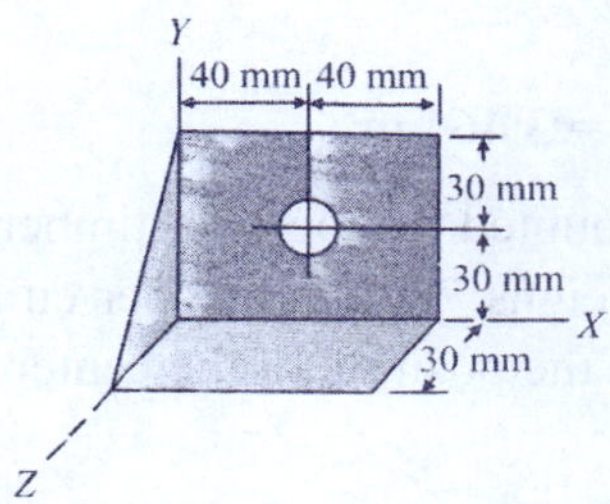

Figure P.6-8.6

6-8.7 Locate the center of gravity of a steel rivet having a cylindrical body 100 mm in diamèter and 200 mm long with a hemispherical head of 100 mm radius. Use the result of Problem 6-8.4.

6.8.8 A body consists of a right circular cone whose base is 120 mm and whose altitude is 160 mm. A hole 80 mm in diameter and 40 mm deep has been drilled from the base. The axis of the hole coincides with the axis of the cone. Locate the centroid of the net volume. Use the result of Prob. 6-8.1.

50 mm from base *Ans.*

6-8.9 Determine the height *h* of the cylinder mounted on the hemispherical base shown in Fig. P-6-8.9 so that the composite body will be in stable equilibrium on its base.

Hint: As long as the center of gravity does not lie above the *X-X* plane, there will exist a restoring couple when the body is tipped.

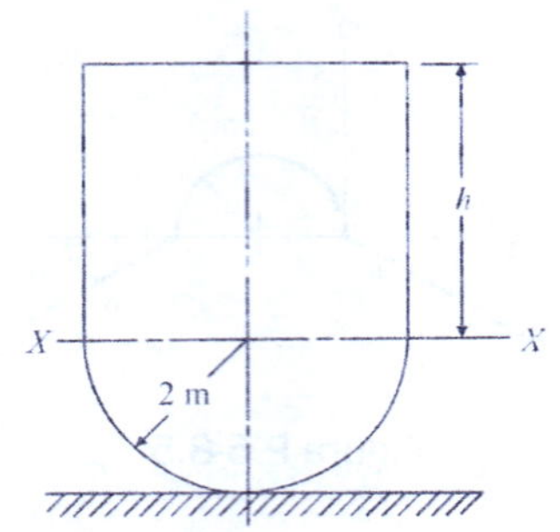

Figure P.6-8.9

$h = 1.414$ m **Ans.**

6-8.10 Repeat Prob. 6-8.9 if the cylindrical portion of the body in Fig. P-6-8.9 is replaced by a right conical portion with a 2 m radius base and altitude h.

$h = 3.464$ m **Ans.**

6-8.11 A steel ball is mounted on top of a timber cylinder as shown in Fig. P-6-8.11. Steel weighs 7860 kg/m^3 per cu m and timber weighs 1580 kg/m^3. Determine the position of the center of gravity.

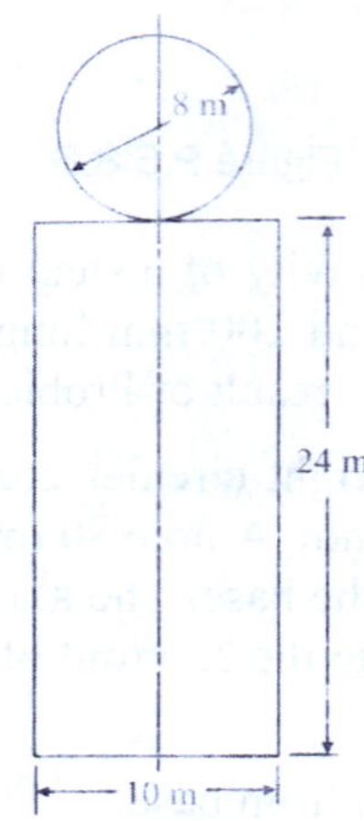

Figure P.6-8.11

SUMMARY

The center of gravity of a body is the point through which the line of action of the weight always passes. The term centroid replaces center of gravity when referring to a weightless figure such as a line, an area, or a volume. Centroids and centers of gravity are determined from the following equations:

$$\left.\begin{aligned} L\bar{x} &= \int x\, dL = \sum lx \\ L\bar{y} &= \int y\, dL = \sum ly \end{aligned}\right\} \qquad (6\text{-}3.2)$$

$$\left.\begin{aligned} A\bar{x} &= \int x\, dA = \sum ax \\ A\bar{y} &= \int y\, dA = \sum ay \end{aligned}\right\} \qquad (6\text{-}3.1)$$

$$\left.\begin{aligned} V\bar{x} &= \int x\, dV = \sum vx \\ V\bar{y} &= \int y\, dV = \sum vy \\ V\bar{z} &= \int z\, dV = \sum vz \end{aligned}\right\} \qquad (6\text{-}8.1)$$

$$\left.\begin{aligned} W\bar{x} &= \int x\, dW = \sum wx \\ W\bar{y} &= \int y\, dW = \sum wy \\ W\bar{z} &= \int z\, dW = \sum wz \end{aligned}\right\} \qquad (6\text{-}8.2)$$

When we apply these equations to composite figures (Sections 6-6 and 6-8), the figure is resolved into its component geometric parts to locate the centroids of these parts.

Particular attention is called to the form $A\bar{x}$ which is called moment of area. When the moment of area is desired, as in many problems in dynamics and strength of materials, it is found as the sum of the moments of areas of the parts composing the figure.

The theorems of Pappus (Section 6-7) provide a means of determining the surface area generated by revolving a line about a nonintersecting axis, and also the volume generated by rotating an area about a nonintersecting axis of revolution. They also help to locate centroids of lengths and areas. The theorems are expressed by the following equations:

$$a = \theta\bar{x}L = \theta\sum lx \qquad (6\text{-}7.1)$$

$$V = \theta\bar{x}A = \theta\sum ax \qquad (6\text{-}7.2)$$

in which the angle θ is equal to or less than 2π radians.

CHAPTER 7

MOMENTS OF INERTIA

In this chapter we shall discuss both area moment of inertia and mass moment of inertia. Area moment of inertia is applied in structural design while mass moment of inertia appears in the dynamics of rigid bodies. Their development is similar in many respects, but their application is so different that we consider each separately.

AREA MOMENTS OF INERTIA

7-1 DEFINITION OF MOMENT OF INERTIA

Many engineering formulas, such as those relating to strength of beams, columns, deflection of beams, involve the use of a mathematical expression of the form $\int \rho^2 dA$, where ρ is the perpendicular distance from dA to the axis of inertia. This integral appears so frequently that it has been named *moment of inertia*[1]. Moment of inertia applied to areas has no physical meaning when examined by itself; it is merely a mathematical expression usually denoted by

[1]The term *moment* of *inertia* is derived as follows: Force is related to the mass (i.e., inertia) of a body and its acceleration by the equation $F = ma$. The equation relating applied forces to the angular acceleration α of rotating bodies is $Fd = \left[\int \rho^2 dm\right]\alpha$. If the first equation is stated as force equals inertia times acceleration, then by analogy the second equation may be stated as moment of force equals moment of inertia times acceleration. By comparison of the above statements, the expression $\int \rho^2 dm$ is termed moment of inertia. Similarly, for areas, the expression $\int \rho^2 dA$ is known as the moment of inertia.

the symbol I. However, when used in combination with other terms, as in the flexure formula for beam stresses, $S = Mc/I$, it begins to have significance.

The mathematical definition of moment of inertia, $I = \int p^2 dA$, indicates that an area is divided into small parts such as dA, and each area is multiplied by the square of its moment arm about the reference axis. Thus, as shown in Fig. 7-1.1, if the coordinates of the differential area dA are (x, y), the moment of inertia about the X axis is the summation of the product of each area dA by the square of its moment arm y. This gives

$$I_x = \int y^2 dA \qquad (7\text{-}1.1)$$

Similarly, the moment of inertia about the Y axis is given by

$$I_y = \int x^2 dA \qquad (7\text{-}1.2)$$

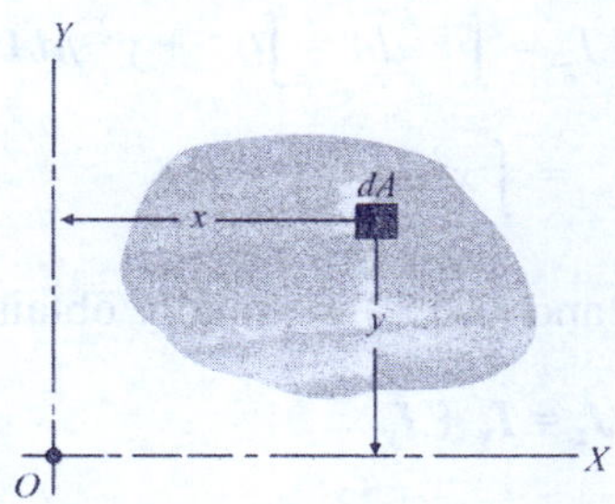

Figure 7-1.1

The moment of inertia (of area) is sometimes called the *second moment of area* because each differential area multiplied by its moment arm gives the moment of area; when multiplied a second time by its moment arm it gives the moment of inertia. The term *second moment of area* is preferable to the expression *moment of inertia*; the latter is confusing when applied to an area having no inertia. The term moment of inertia, however, is long established and is not likely to be superseded by the other.

Units and Signs

Examination of the integral $\int \rho^2 dA$ shows it to be a fourth-dimensional term because it is composed of a distance squared multiplied by an area. Thus if L is the unit of distance, the unit of I is $(L)^4$. A convenient unit of L is mm or m; this gives quartic milli meters (mm^4) or quartic meters (m^4) as the dimensional unit of I.

The sign of I is obviously independent of the sign of the moment arm ρ (since if ρ is minus, squaring it makes it plus); it depends entirely on the sign of the area. We shall define a positive area as one which adds to the area of a figure, and a negative area as one which reduces the area of the figure. For a net area, the moment of inertia must always be positive.

7-2 POLAR MOMENT OF INERTIA

The moment of inertia for an area relative to a line or axis perpendicular to the plane of the area is called the polar moment of inertia and is denoted by the symbol J. In Fig. 7-2.1 the moment of inertia of an area in the XY plane with respect to the Z axis is

$$\left[I = \int \rho^2 \, dA\right] \qquad J_z = \int r^2 dA = \int (x^2 + y^2) dA$$

$$= \int x^2 dA + \int y^2 dA$$

whence from Eqs. (7-1.1) and (7-1.2) we finally obtain

$$\boldsymbol{J_z = I_x + I_y} \tag{7-2.1}$$

Expressed in words, this equation states that the polar moment of inertia for an area with respect to an axis perpendicular to its plane is equal to the sum of the moments of inertia about any two mutually perpendicular axes in its plane which intersect on the polar axis.

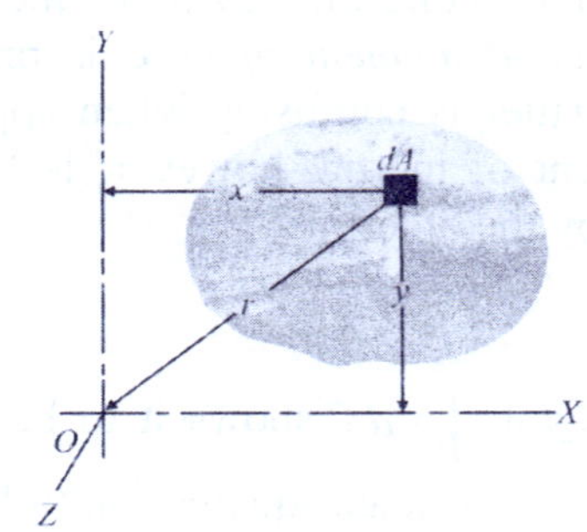

Figure 7-2.1

7-3 RADIUS OF GYRATION

The term radius of gyration is used to describe another mathematical expression and appears most frequently in column formulas. Radius of gyration is usually denoted by the symbol k (sometimes by r) and is defined by the relation

$$k = \sqrt{\frac{I}{A}} \quad \text{or } I = Ak^2 \tag{7-3.1}$$

where I is the moment of inertia and A the cross-sectional area.

The following is a geometrical interpretation of this relation. Assume the area of Fig. 7-1.1 to be squeezed into a long narrow strip as shown in Fig. 7-3.1. Each differential element of area dA will then be the same distance k from the axis of inertia. The moment of inertia is given by

$$I = \int \rho^2 \, dA = k^2 \int dA = Ak^2$$

because each differential element has the same moment arm. The strip may be placed on either side of the reference axis, since if k is minus, squaring it will automatically make it plus, or part of the strip may be at a distance k from one side of the reference axis and the rest of it at an equal distance k from the other side of the axis.

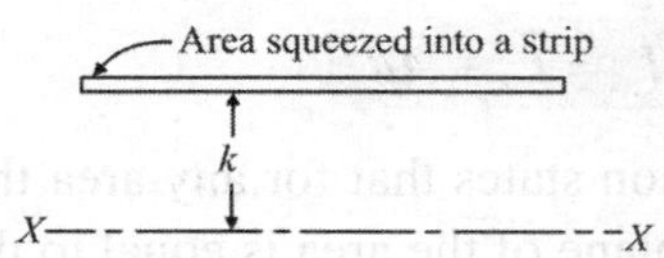

Figure 7-3.1 Concept of radius of gyrat on.

In view of this discussion, the radius of gyration is frequently considered to be the uniform distance from the reference axis at which the entire area may be *assumed* to be distributed. For an area whose dimensions perpendicular to a reference axis are negligibly small compared with its distance from that axis, radius of gyration is practically equivalent to the centroidal location of the area.

7-4 TRANSFER FORMULA FOR MOMENT OF INERTIA

It is often necessary to transfer the moment of inertia from one axis to another parallel axis. The transfer formula affords a method of doing this without further integration. For example, in Fig. 7-4.1, the moment of inertia with respect to a centroidal X axis (X_o) is given by the expression $\bar{I}_x = \int y^2 dA$. The moment of inertia for the same area with respect to a *parallel* axis (X) located a distance d from the centroidal axis is given by the equation

$$\left[I = \int \rho^2\, dA\right] \qquad I_x = \int (y+d)^2\, dA$$

$$= \int y^2\, dA + 2\, d \int y\, dA + d^2 \int dA \qquad (a)$$

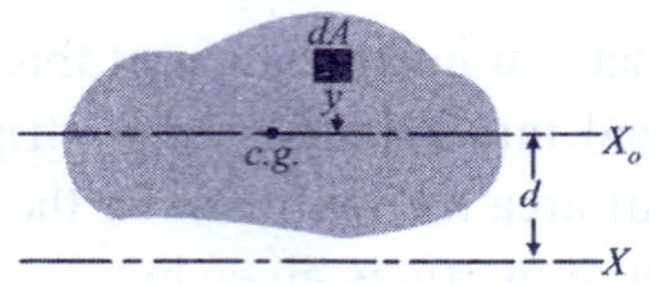

Figure 7-4.1 Moments of inertia between parallel axes.

The d is written outside the integral sign because it is a constant that represents the distance separating the axes. The second of the right-hand terms in Eq. (a) becomes zero because $\int y\, dA = A\bar{y}$, where $\bar{y}$ represents the distance from the reference axis X_0 to the centroid. In this instance $\bar{y}$ has the value of zero because X_0 passes through the centroid. We obtain finally,

$$\boldsymbol{I_x = \bar{I}_x + Ad^2} \qquad (7\text{-}4.1)$$

Put into words, this equation states that for any area the moment of inertia with respect to any axis in the plane of the area is equal to the moment of inertia with respect to a *parallel centroidal axis* plus a transfer term composed of the product of the area multiplied by the square of the distance between the axes. Evidently the least moment of inertia for any given direction of an axis is the centroidal moment of inertia. *Note carefully that the centroidal axis involved in the transfer formula is always the centroidal axis of the area used in the transfer term Ad^2.*

A similar relation exists between the radii of gyration with respect to parallel axes, one of which is a centroidal axis. Replacing I_x by Ak_x^2 and $\bar{I}_x$ by $A\bar{k}_x^2$ in Eq. (7-4.1), we obtain

$$Ak_x^2 = A\bar{k}_x^2 + Ad^2$$

whence

$$k_x^2 = \bar{k}_x^2 + d^2 \qquad (7\text{-}4.2)$$

In like fashion, for polar moments of inertia and polar radii of gyration we obtain the following analogous relations between any axis and a parallel centroidal axis:

$$\left.\begin{aligned} I &= \bar{I} + Ad^2 \\ k^2 &= \bar{k}^2 + d^2 \end{aligned}\right\} \qquad (7\text{-}4.3)$$

7-5 MOMENTS OF INERTIA BY INTEGRATION

In determining the moment of inertia by integration, it is desirable to choose the differential area so that either

1. All parts of the differential area are at the same distance from the reference axis[2], or
2. The moment of inertia of the differential area with respect to the reference axis is known. The moment of inertia of the area is then the summation of the moments of inertia of its elements.

As in the case of centroids, the moment of inertia of a composite figure may be found by combining the moments of inertia of its parts. When the evaluation of $\bar{I}$ for a particular part is known, the transfer formula (Section 7-4) is used to refer the moments of inertia of the various parts of the figure to a common reference axis.

ILLUSTRATIVE PROBLEMS

7.5.1 Determine the moment of inertia for a rectangle of base b and depth h with respect to (a) a centroidal axis parallel to the base and (b) an axis coinciding with the base.

Solution: Centroidal Axis

Select the differential element as shown in Fig. 7-5.1. All parts of the element are at the same distance from the centroidal X_0 axis. Applying Eq. (7-1.1), we find the centroidal moment of inertia to be

$$\left[I_x = \int y^2\, dA\right] \qquad \bar{I}_x = \int_{-h/2}^{h/2} y^2 b\, dy = b\left[\frac{y^3}{3}\right]_{-h/2}^{h/2} = \frac{bh^3}{12} \qquad \textit{Ans.}$$

[2] When all parts of an element are at the same distance, from an axis, this distance is really the radius of gyration for the element. See Fig. 7-3.1.

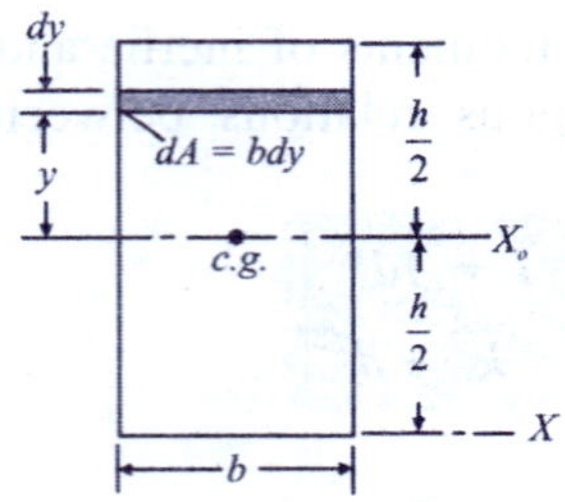

Figure 7-5.1

Axis Coinciding with the Base

The preceding result can be transferred through the distance *h*/2 to the parallel base axis by applying the transfer formula as follows:

$$\left[I = \bar{I} + Ad^2\right] \qquad I_x = \frac{bh^3}{12} + (bh)\left(\frac{h}{2}\right)^2 = \frac{bh^3}{3} \qquad \textbf{\textit{Ans.}}$$

The moment of inertia of the parallelogram in Fig. 7-5.2 has the same values as for a rectangle because the elemental strips composing the parallelogram have merely shifted their position laterally from the dashed rectangle of corresponding dimensions, but have not altered their distances from the corresponding axes of inertia.

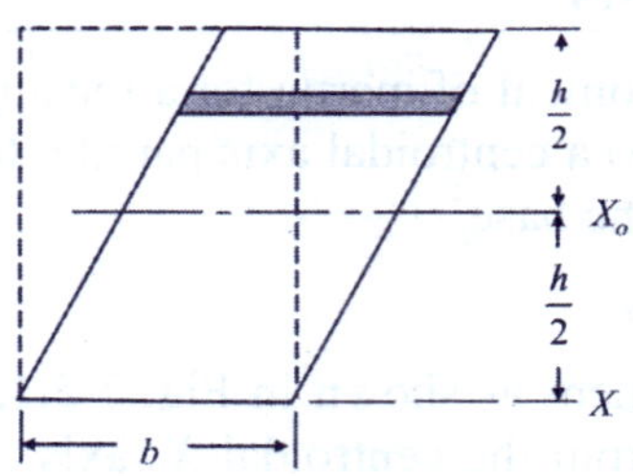

$$I_{x_o} = \frac{bh^3}{12}; \; I_x = \frac{bh^3}{3}$$

Figure 7-5.2

7-5.2 Determine the moment of inertia for a triangle of base *b* and altitude *h* with respect to (a) an axis coinciding with its base and (b) a centroidal axis parallel to its base.

Solution: Axis Coinciding with the Base

Select the differential element, as shown in Fig. 7-5.3 From similar triangles, the length $x = \frac{b}{h}(h - y)$. The moment of inertia with respect to the X axis is obtained from

$$\left[I_x = \int y^2\, dA\right] \qquad I_x = \int_0^h (y^2)x\, dy = \int_0^h (y^2)\frac{b}{h}(h-y)\, dy$$

$$= \frac{b}{h}\left[\int_0^h hy^2 dy - \int_0^h y^3 dy\right]$$

$$= \frac{b}{h}\left[\frac{hy^3}{3} - \frac{y^4}{4}\right]_0^h$$

$$I_x = \frac{bh^3}{12} \qquad \textit{Ans.}$$

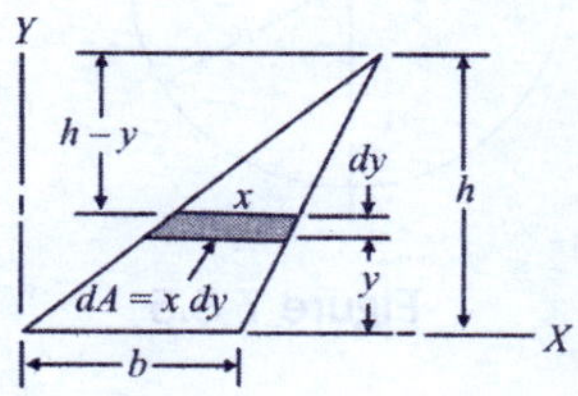

Figure 7-5.3

Centroidal Axis

To obtain the centroidal moment of inertia $\bar{I}_x$, we transfer the known value of I_x from the base axis X to the parallel centroidal axis X_0. Since the transfer distance is $h/3$ as shown in Fig. 7-5.4, we obtain

$$\left[I_x = \bar{I}_x + Ad^2\right] \qquad \frac{bh^3}{12} + \bar{I}_x + \left(\frac{bh}{2}\right) = \left(\frac{h}{3}\right)^2$$

$$\bar{I}_x = \frac{bh^3}{36} \qquad \textit{Ans.}$$

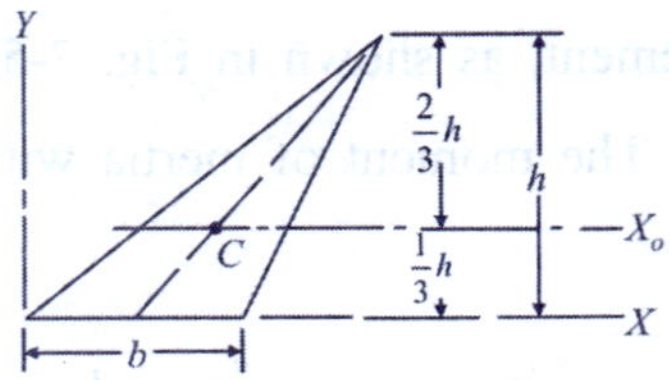

Figure 7-5.4

7-5.3 Determine the moment of inertia of a circular area of radius r with respect to a diametral axis.

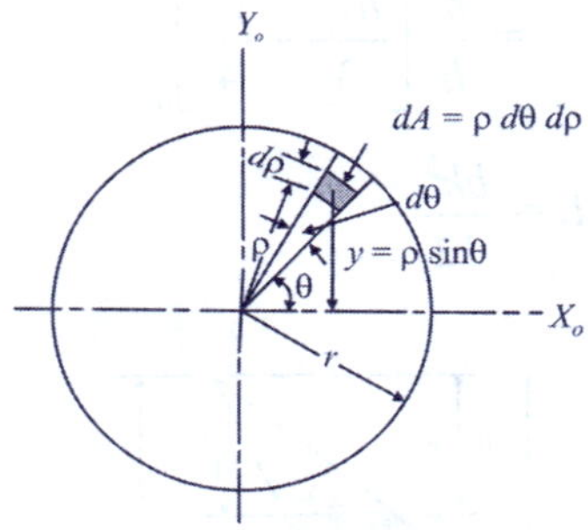

Figure 7-5.5

Solution

Using polar coordinates select the differential element, as shown in Fig. 7-5.5. From the figure, $y = \rho \sin \theta$. The moment of inertia with respect to the diameter is

$$\left[I_x = \int y^2 \, dA\right] \qquad \bar{I}_x = \int_0^r \int_0^{2\pi} \rho^2 sin^2\theta\rho \, d\theta \, d\rho$$

$$= \int_0^r \int_0^{2\pi} (\rho^3 \, \mathrm{d}\rho) \, \sin^2\theta \, \mathrm{d}\theta$$

$$= \frac{r^4}{4} \int_0^{2\pi} \sin^2\theta \, d\theta = \frac{r^4}{4}(\pi)$$

$$\bar{I}_x = \frac{\pi r^4}{4} \qquad \textbf{\textit{Ans.}}$$

An alternate and simpler solution is to use Fig. 7-5.6 in which the differential element is taken as the shaded ring of area $dA = (2\pi\rho)(d\rho)$. The polar moment of inertia is

$$\left[J = \int \rho^2\, dA\right] \qquad \bar{J} = \int_0^r (\rho^2)2\pi\rho\, d\rho = \frac{\pi r^4}{2}$$

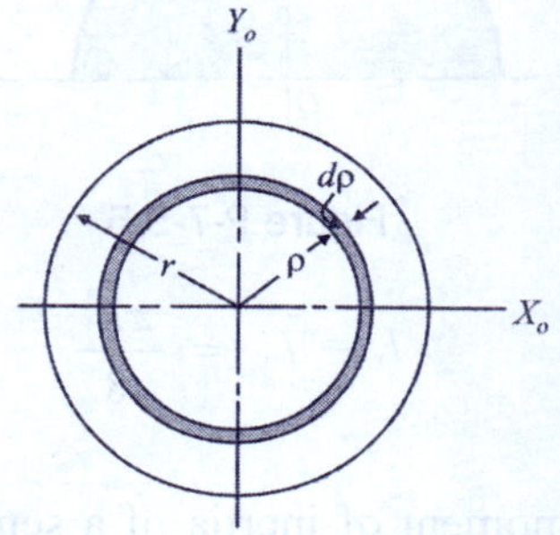

Figure 7-5.6

The rectangular moments of inertia $\bar{I}_x$ and $\bar{I}_y$ are obviously equal because of symmetry so that applying Eq. (7-2.1) yields

$$\left[\bar{J} = \bar{I}_x + \bar{I}_y\right] \qquad \frac{\pi r^4}{2} = \bar{I}_x + \bar{I}_x \quad \text{or} \quad \bar{I}_x = \frac{\pi r^4}{4} \qquad \textit{...Check}$$

PROBLEMS

7.5.4 Determine the moment of inertia of the quarter circle shown in Fig. P-7-5.4 with respect to the given axes.

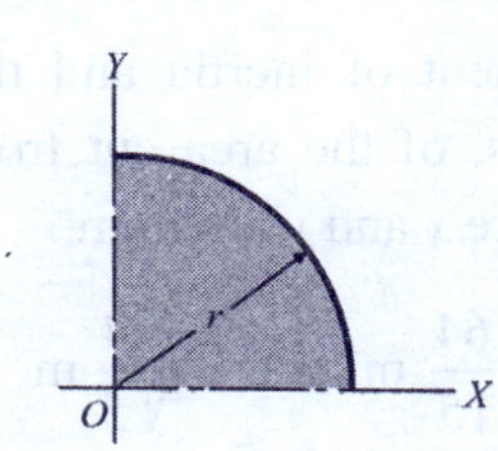

Figure P-7-5.4

$$I_x = I_y = \frac{\pi r^4}{16} \qquad \textit{Ans.}$$

7-5.5 Determine the moment of inertia of the semicircle shown in Fig. P-7-5.5 with respect to the given axes.

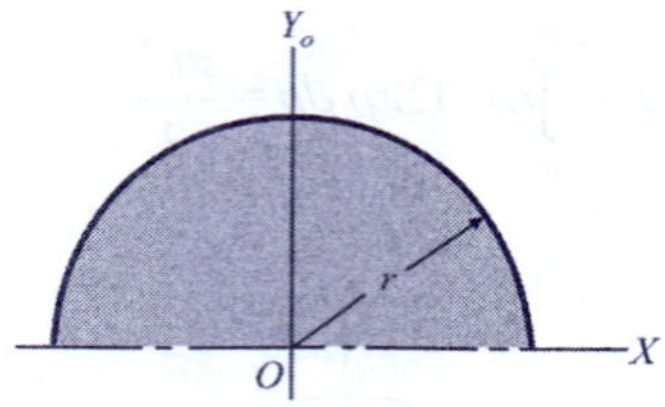

Figure P-7-5.5

$$I_x = \bar{I}_y = \frac{\pi r^4}{8}$$ ***Ans.***

7-5.6 Show that the moment of inertia of a semicircle of radius r is $0.11r^4$ with respect to a centroidal axis parallel to the diameter.

7-5.7 Determine the moment of inertia for the quarter circle shown in Fig. P-7-5.4 with respect to a centroidal X axis.

$$\bar{I}_x = 0.55r^4$$ ***Ans.***

7-5.8 Determine the moment of inertia with respect to the X axis for the area enclosed by the ellipse whose equation is $\frac{x^2}{a^2}+\frac{y^2}{b^2}=1$. Also determine the radius of gyration.

$$\bar{I}_x = \frac{\pi ab^3}{4};\ \bar{k}_x = \frac{b}{2}$$ ***Ans.***

7-5.9 Determine the moment of inertia and the radius of gyration, with respect to the Y axis, of the area cut from the first quadrant by the curve $y = 4 - x^2$ where x and y are in m.

$$I_y = \frac{64}{15}\ \text{m}^4;\ k_y = \sqrt{\frac{4}{5}}\ \text{m}$$ ***Ans.***

7-5.10 Determine the moment of inertia with respect to the X axis of the shaded parabolic area shown in Fig. P-7-5.10.

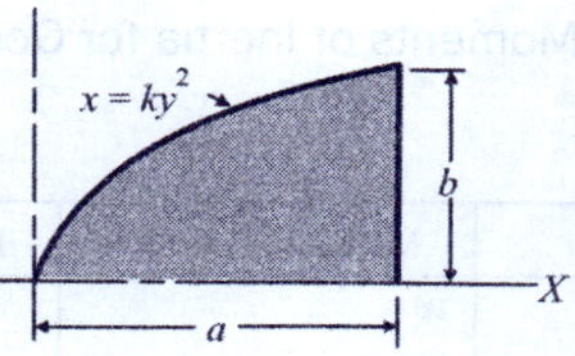

Figure P-7-5.10

$$I_x = \frac{2}{15} ab^3$$ **Ans.**

7-5.11 Determine I_y for the shaded parabolic area of Fig. P-7-5.10.

$$I_y = \frac{2}{7} a^3 b$$ **Ans.**

7-5.12 Determine the moments of inertia with respect to the X and Y axes for the shaded area under the spandrel $y = kx^n$ shown in Fig. P-7-5.12. Use these results to verify the answers to Probs. 7-5.10 and 7-5.11.

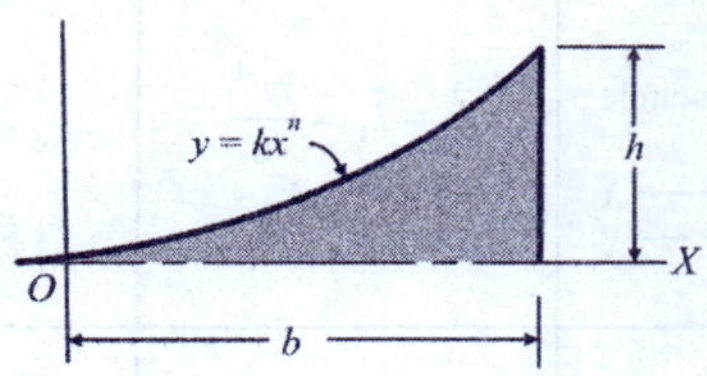

Figure P-7-5.12

7-6 MOMENTS OF INERTIA FOR COMPOSITE AREAS

When a composite area can be divided into geometric elements (rectangles triangles, etc.) for which the moments of inertia are known, the moment of inertia for the composite area is the sum of the moments of inertia for the separate elements. Before the moments of inertia of the elements can be added, however, they must all be found with respect to the same axis, using the transfer formula when necessary.

In the problems which follow, the values of the moments of inertia for geometric elements can be taken from the results of the problems in section 7-5; these are summarized in Table 7-6.1.

Table 7-6.1 Moments of Inertia for Geometric Shapes

Shape	Moment of inertia	Radius of gyration
Rectangle 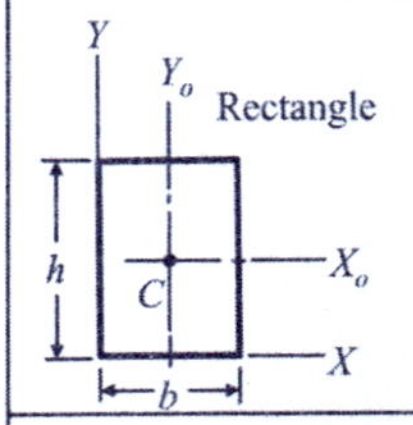	$\bar{I}_x = \frac{bh^3}{12}$ $I_x = \frac{bh^3}{3}$	$\bar{k}_x = \frac{h}{\sqrt{12}}$ $k_x = \frac{h}{\sqrt{3}}$
Any triangle 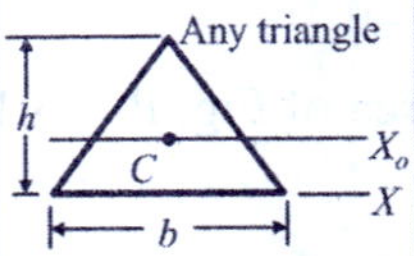	$\bar{I}_x = \frac{bh^3}{36}$ $I_x = \frac{bh^3}{12}$	$\bar{k}_x = \frac{h}{\sqrt{18}}$ $k_x = \frac{h}{\sqrt{6}}$
Circle 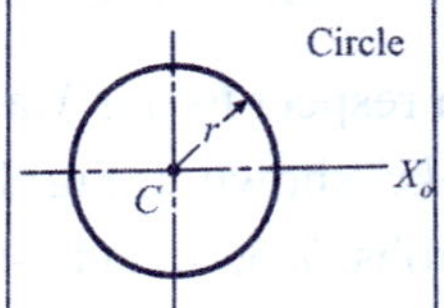	$\bar{I}_x = \frac{\pi r^4}{4}$ $\bar{J} = \frac{\pi r^4}{2}$	$\bar{k}_x = \frac{r}{\sqrt{2}}$ $\bar{k}_z = \frac{r}{\sqrt{2}}$
Semicircle 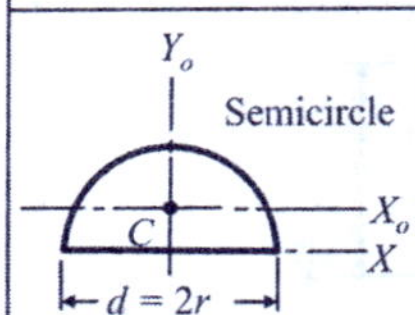	$\bar{I}_x = \bar{I}_y = \frac{\pi r^4}{8}$ $\bar{I}_x = 0.11 r^4$	$k_x = \bar{k}_y = \frac{r}{2}$ $\bar{k}_x = 0.264 r$
Quarter circle 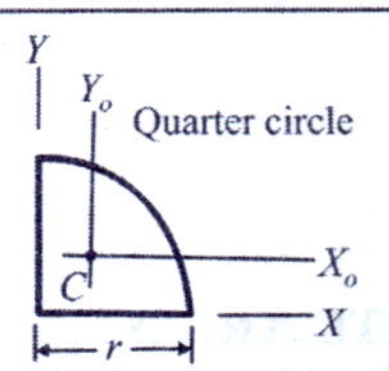	$I_x = I_y = \frac{\pi r^4}{16}$ $\bar{I}_x = \bar{I}_y = 0.55 r^4$	$k_x = k_y = \frac{r}{2}$ $\bar{k}_x = \bar{k}_y = 0.264 r$
Ellipse 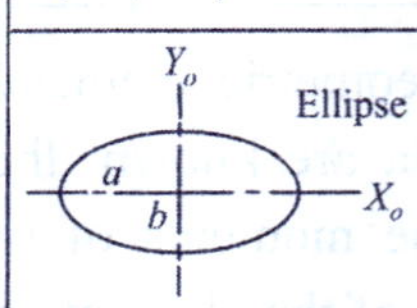	$\bar{I}_x = \frac{\pi ab^3}{4}$ $\bar{I}_y = \frac{\pi ba^3}{4}$	$\bar{k}_x = \frac{b}{2}$ $\bar{k}_y = \frac{a}{2}$
Spandrel 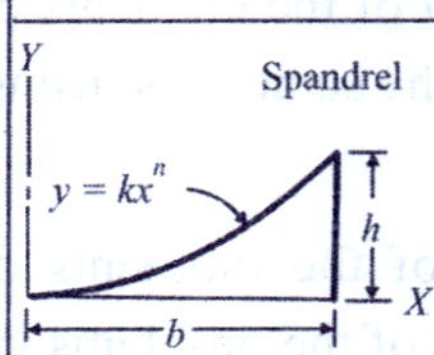	$I_x = \frac{bh^3}{3(3n+1)}$ $I_y = \frac{hb^3}{n+3}$	

ILLUSTRATIVE PROBLEMS

7-6.1 Determine the moments of inertia with respect to the centroidal X and Y axes of the wide flange beam section shown in Fig. 7-6.1.

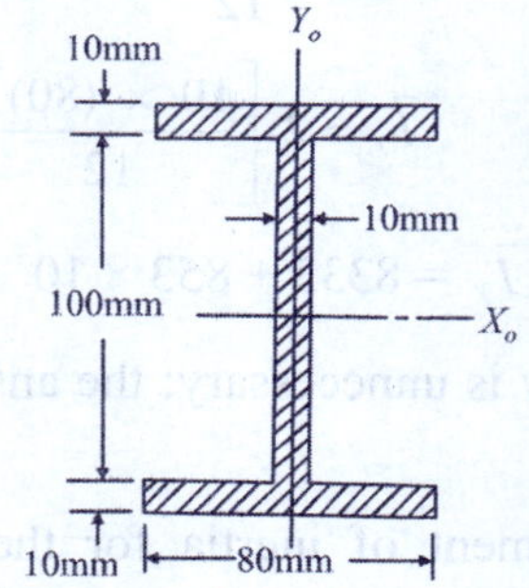

Figure 7-6.1

Solution

The moment of inertia of a composite area is the sum of the moments of inertia of the various parts of the area, all the moments of inertia being referred to the same axis of inertia before the addition is made.

With respect to the X_o axis, the simplest subdivision of the given area is to resolve it into a large rectangle 80 × 120 mm from which two smaller rectangles, each 35 × 100 are subtracted. The centroidal axis for each of these parts coincides with the X_o axis of the figure; hence the transfer formula need not be used. Using the result listed in Table 7-6.1, we therefore obtain

$$\left[\bar{I}_x = \frac{bh^3}{12}\right] \quad 80 \times 120 \text{ mm rectangle: } \bar{I}_x = \frac{80 \times (120)^3}{12} = 1152 \times 10^4 \text{ mm}^4$$

$$\text{Two} \quad 35 \times 100 \text{ mm rectangle: } \bar{I}_x = 2\left[\frac{35 \times (100)^3}{12}\right] = 583 \times 10^4 \text{ mm}^4$$

Hence for shaded area: $\bar{I}_x = 1152 - 583 = 569 \times 10^4 \text{ mm}^4$ ***Ans.***

With respect to the Y_o axis, assume the figure to be composed of a 10 × 100 mm rectangle and two 10 × 80 mm. rectangles. The Y_o axis is also the centroidal axis for each of these rectangles so that this subdivision of the area again eliminates the need for using the transfer formula. Referring again to Table 7-6.1, we have

$$\left[\bar{I}_y = \frac{hb^3}{12}\right]$$

10 × 100 mm rectangle: $\bar{I}_y \dfrac{100 \times (10)^3}{12} = 8333 \text{ mm}^4.$

Two 10 × 80 rectangles: $\bar{I}_y = 2\left[\dfrac{10 \times (80)^3}{12}\right] = 853 \times 10^3 \text{ mm}^4.$

Net area: $\bar{I}_y = 8333 + 853 \times 10^3 = 861.33 \times 10^3 \text{ mm}^4$ ***Ans.***

Such numerical accuracy is unnecessary; the answer would usually be given as $862 \times 10^3 \text{ mm}^4$

7-6.2 Compute the moment of inertia for the composite area shown in Fig. 7-6.2 with respect to the indicated X axis.

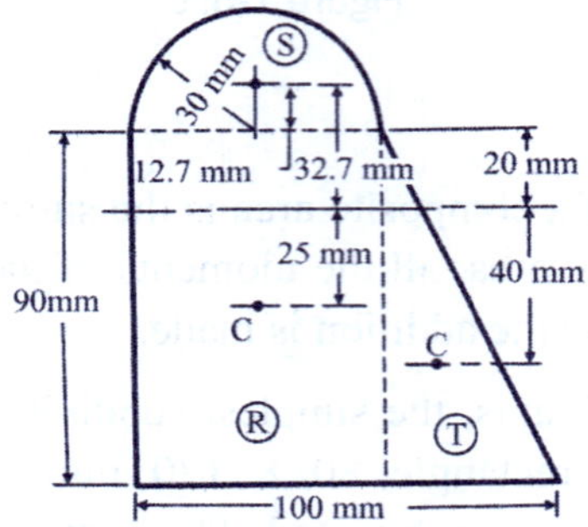

Figure 7-6.2

Solution

The area is composed of a semicircle (S) whose radius is 30 mm, 60 × 90 mm rectangle (R), and a 40 × 90 mm triangle (T). With respect to the X axis, the moment of inertia for the area is the sum of the moments of inertia of these elements, each moment of inertia being referred to the X axis before addition:

$$I_x = I_R + I_S + I_T \qquad (a)$$

Expressing the moment of inertia of each element in terms of its centroidal moment of inertia plus a transfer term, we obtain

$$I_R = \bar{I}_R + (Ad^2)_R$$
$$I_S = \bar{I}_S + (Ad^2)_S$$
$$I_T = \bar{I}_T + (Ad^2)_T$$

Adding the left- and right-hand members of these equations results in

$$I_x = \Sigma \bar{I} + \Sigma Ad^2 \qquad (b)$$

Equation (*b*) indicates that the moment of inertia of a composite figure is the summation of the centroidal moments of inertia of the elements plus the summation of the transfer terms for these elements. This equation is readily adapted to tabular computation, as shown below. From Table 7-6.1, the values of $\bar{I}$ for each of these elements are $\bar{I}_R = bh^3/12$, $\bar{I}_S = 0.11r^4$, and $\bar{I}_T = bh^3/36$. The transfer distances are indicated on Fig. 7-6.2.

Item	$\bar{I}$ mm^4	Area mm^2	d mm	d^2 mm^2	Ad^2 mm^3
Rectangle	365×10^4	54×10^2	25	625	337.5×10^4
Semicircle	8.91×10^4	14.14×10^2	32.7	1070	151.3×10^4
Triangle	81×10^4	18×10^2	40	1600	288×10^4
Total	454.91×10^4				776.8×10^4

Taking the summations from the table and substituting in Eq. (*b*), we obtain

$$\left[\bar{I}_x = \Sigma\bar{I} + \Sigma Ad^2\right]$$

$$I_x = 454.91 \times 10^4 + 776.8 \times 10^4$$

$$I_x \approx 1232 \times 10^4 \text{ mm}^4$$ ***Ans.***

PROBLEMS

7-6.4 The shaded section shown in Fig. P-4-6.4 has an area of 500 mm^2 and a moment of inertia of 30 × 10^4 mm. about axis X_1. Find the moment of inertia about axis X_2.

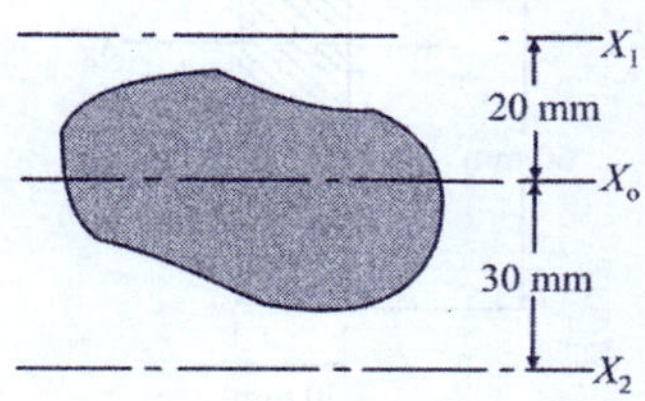

Figure P-7-6.4

7.6.5 Blocks of wood are glued together to form the shaded section shown in Fig. P-7-6.5. Find the moment of inertia of the shaded section about its horizontal and vertical centroidal axes.

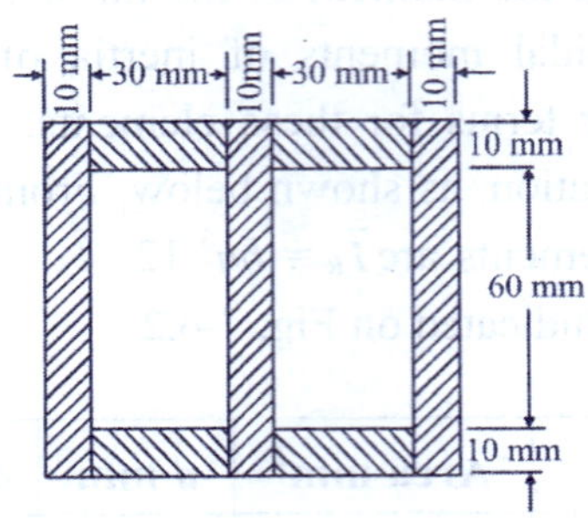

Figure P-7-6.5

$\bar{I}_x = 276 \times 10^4 \text{ mm}^4$; $\bar{I}_y = 315 \times 10^4 \text{ mm}^4$ ***Ans.***

7-6.6 A hollow square cross section consists of an 80 mm by 80 mm square from which is subtracted a concentrically located square 40 mm by 40 mm. Find the polar moment of inertia and the polar radius of gyration with respect to a Z axis Passing through one of the outside corners.

$J = 2176 \times 10^4 \text{ mm}^4$; $k = 67.3$ mm ***Ans.***

7-6.7 Determine the moment of inertia of the T-section shown in Fig. P-7-6.7 about its centroidal X_o axis.

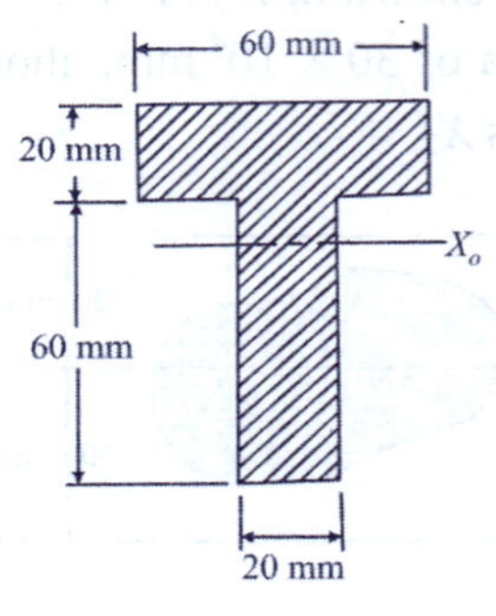

Figure P-7-6.7

7-6.8 Determine the moment of inertia of the area shown in Fig. P-7-6.8 with respect to its centroidal axes.

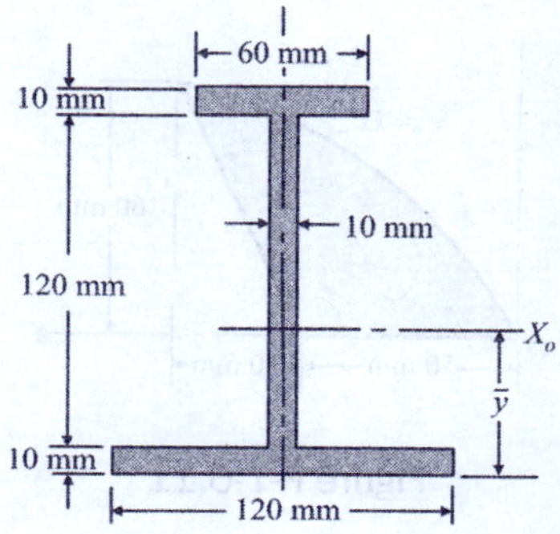

Figure P-7-6.8

$\bar{y} = 57$ mm; $\bar{I}_x = 855 \times 10^4$ mm^4; $\bar{I}_y = 163 \times 10^4$ mm^4 ***Ans.***

7.6.9 Find the moment of inertia about the indicated Y axis for the shaded area shown in Fig. P.7-6.9.

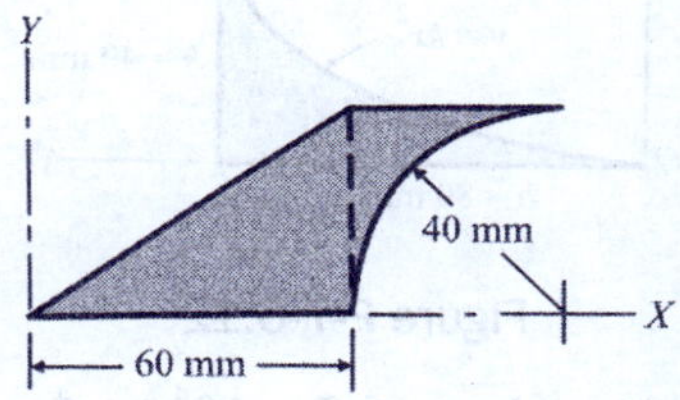

Figure P-7-6.9

7.6.10 The cross section of a culvert is shown in Fig. P.7-6.10. Compute the moment of inertia about the horizontal X axis.

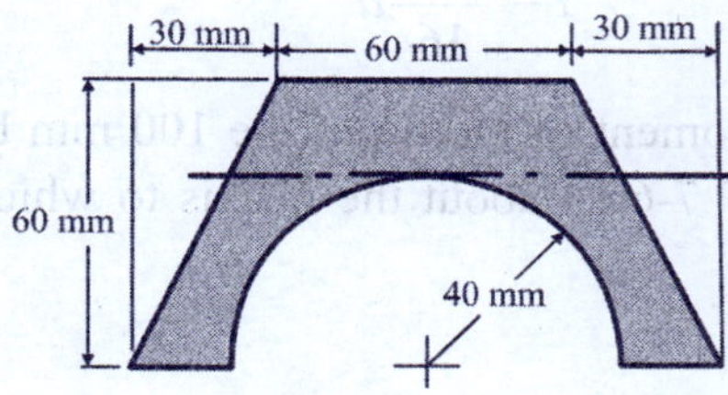

Figure P-7-6.10

$I_x = 90.8 \times 10^4$ mm^4 ***Ans.***

7-6.11 For the shaded area shown in Fig. P-7-6.11, compute the moment of inertia about the given X and Y axes.

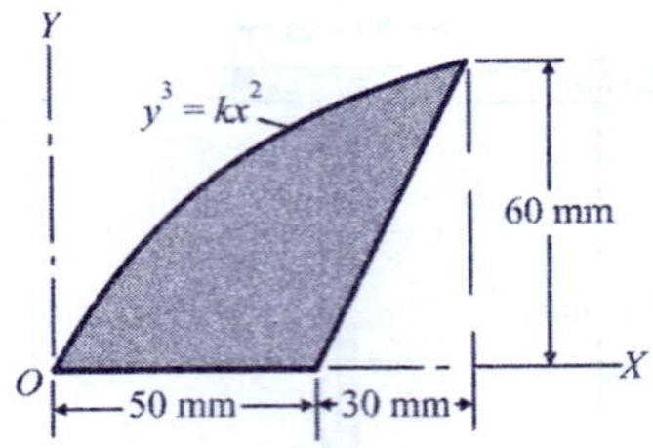

Figure P-7-6.11

7-6.12 Find the moment of inertia about the axis Y_1 of the shaded area under the second-degree curve shown in Fig. P-7-6.12.

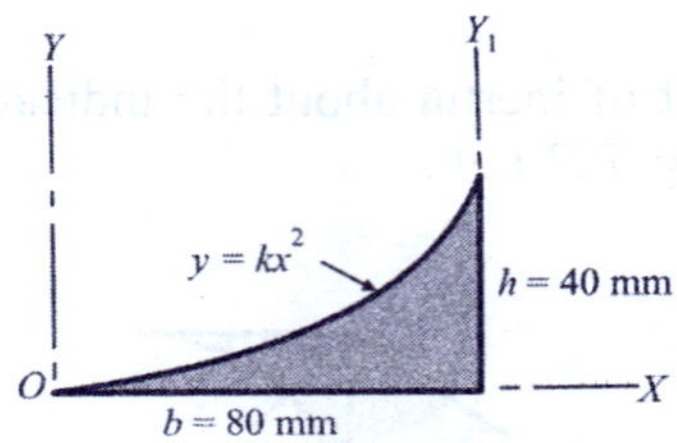

Figure P-7-6.12

$$I_{y1} = 68.7 \times 10^4 \text{ mm}^4 \qquad \textbf{\textit{Ans.}}$$

7-6.13 Compute the moment of inertia with respect to an axis passing through two opposite apexes of a regular hexagon of side a.

$$I = \frac{5\sqrt{3}}{16}a^4 \qquad \textbf{\textit{Ans.}}$$

7-6.14 Compute the moment of inertia of the 100 mm by 150 mm rectangle shown in Fig. P.7-6.14 about the X axis to which it is inclined at an angle $\theta = \sin^{-1}\frac{4}{5}$.

Hint: Resolve the figure into parts A, B, and C.

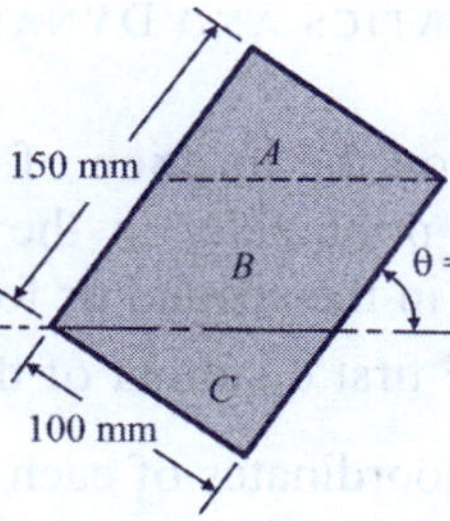

Figure P-7-6.14

7-6-15 Determine the distance d at which the two 20 mm by 60 mm rectangles shown in Fig. P-7-6.15 should be spaced so that $I_x = I_y$.

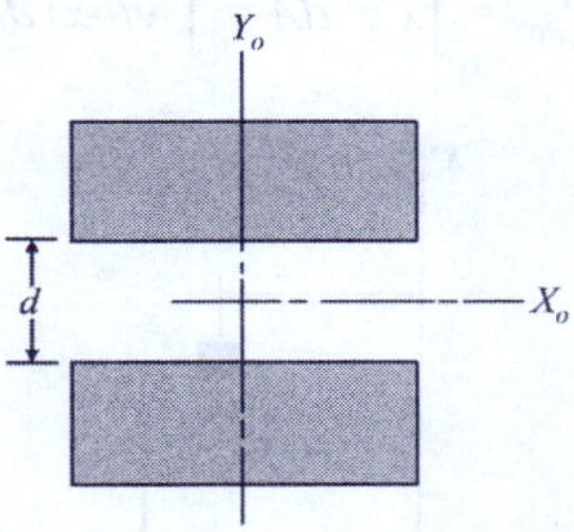

Figure P-7-6.15

7-7 PRODUCT OF INERTIA

The product of inertia for an area is the summation of the product of each differential area by its moment arms about two perpendicular reference axes. Thus, in Fig. 7-7.1, with respect to the X and Y axes, the product of inertia is the mathematical expression $\int xy\, dA$ and is denoted by the symbol P_{xy} wherein the subscripts denote the reference axes.[3] The product of inertia is not used as often as the moment of inertia but is needed in such problems as determining maximum and minimum moments of inertia, unsymmetrical bending of beams, and structural analysis of indeterminate frames.

Units and Signs

The unit of the product of inertia is of the same form as that for the moment of inertia, namely, $(\text{length})^4$. Unlike the moment of inertia, however, the sign for

[3] For uniformity of notation, frequently the symbol I_{xy} is used in place of P_{xy}, especially in advanced work. At this stage, however, the symbol P(being the first letter of "product of inertia") is less apt to cause confusion between product of inertial and moment of inertia.

the product of inertia depends on the location of the area relative to the axes, being positive if the area lies principally in the first or third quadrants and negative if area lies principally in the second or fourth quadrants. For example, the area in Fig. 7-7.1 lies in the first quadrant of the XY axes and $P_{xy} = \int xy\, dA$ is positive because all x and y coordinates of each differential area are positive. However, with respect to a new set of axes, marked as X' and Y' and rotated 90° counterclockwise from the original set of axes, the area is in the fourth quadrant. The new coordinates of dA are $x' = y$ and $y' = -x$ so that with respect to the new axes, the product of inertia is

$$P_{x'y'} = \int x'y'\, dA = \int y(-x)\, dA = -\int xy\, dA = -P_{xy}$$

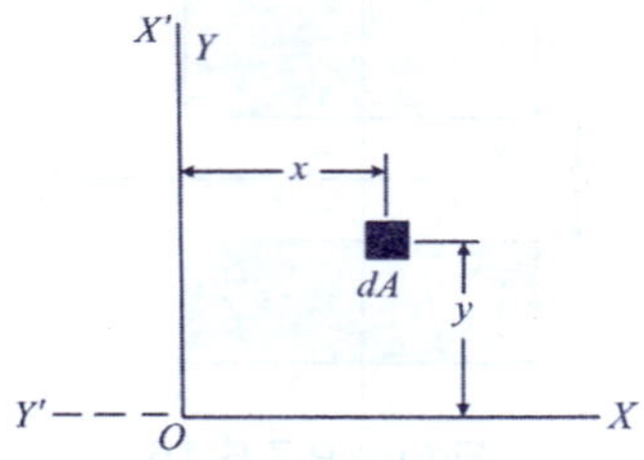

Figure 7-7.1

Not only does this result confirm the rule of sign stated previously, but it indicates that during the rotation of the axes, there will occur one critical position at which the product of inertia changes sign and will have a zero value. When in this position, the axes are known as the *principal axes* of inertia for the area. Their application is discussed in Section 7-12.

7-8 PRODUCT OF INERTIA IS ZERO WITH RESPECT TO AXES OF SYMMETRY

If an area has an axis of symmetry, this axis together with any axis perpendicular to it will form a set of axes for which the product of inertia is zero. Consideration of the symmetrical T-section shown in Fig. 7-8.1 I will disclose that for any differential area like that at A, there is a symmetrically placed equal differential area at B. With respect to the Y axis of symmetry, the x coordinates of A and B are equal but of opposite sign, whereas their y coordinates are equal and of the same sign regardless of the position of the X axis. Hence the sum of the products $xy\, dA$ for each such pair of symmetrically placed elements as A and B will be zero. It follows, therefore, that the value of $\int xy\, dA$ for the entire area will be zero if either or both reference axes are axes of symmetry.

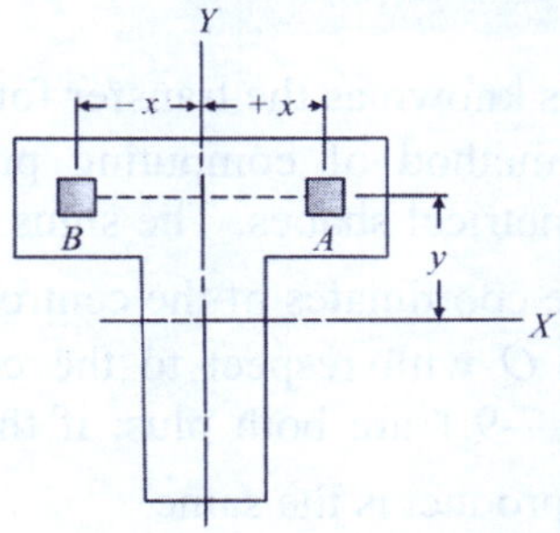

Figure 7-8.1

7-9 TRANSFER FORMULA FOR PRODUCT OF INERTIA

Consider any irregular area, such as that in Fig. 7-9.1, whose cross-sectional area is A and whose product of inertia relative to the centroidal axes is denoted by $\bar{P}_{xy}$. Let a parallel set of axes X and Y be located so that the coordinates of the centroid of the given irregular area are $\bar{x}$ and $\bar{y}$ relative to these axes.

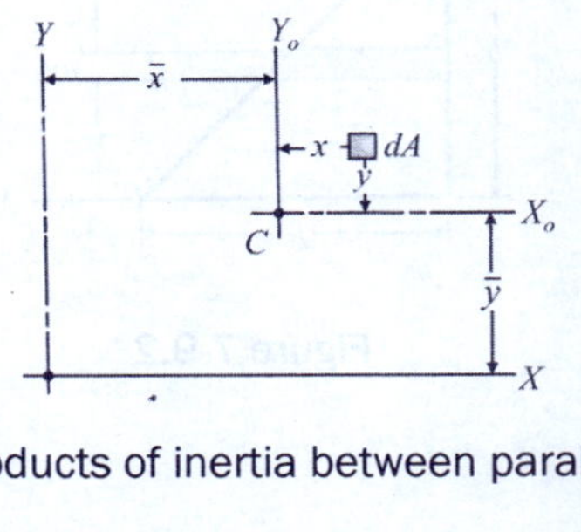

Figure 7-9.1 Products of inertia between parallel sets of axes.

From the fundamental definition of product of inertia we have, with respect to centroidal axes

$$\bar{P}_{xy} = \int (x' + \bar{x})(y + \bar{y})\, dA \qquad (a)$$

and with respect to any parallel set of X and Y axes

$$p_{xy} = \int (x' + \bar{x})\,(y' + \bar{y})dA \qquad (b)$$

Expanding Eq. (b) gives

$$p_{xy} = \int x'y'dA + \bar{x}\int y'dA + y\int x'dA + \bar{x}\,\bar{y}\int dA \qquad (c)$$

Note that the two middle terms represent the moment of area relative to the centroidal axes multiplied respectively by the constants $\bar{x}$ and $\bar{y}$. Since the moment of area relative to centroidal axes is zero, Eq. (*c*) finally reduces to

$$P_{xy} = \bar{P}_{xy} + A\bar{x}\,\bar{y} \tag{7-9.1}$$

This equation, which is known as the transfer formula for products of inertia, forms the basis of the method of computing products of inertia for areas composed of simple geometrical shapes. The signs of $\bar{x}$ and $\bar{y}$ in this equation may be taken either as the coordinates of the centroid C relative to the X-Y axes or as the coordinates of O with respect to the centroidal X_o-Y_o axes. If the former, $\bar{x}$ and $\bar{y}$ in Fig. 7-9.1 are both plus; if the latter, $\bar{x}$ and $\bar{y}$ are both minus; in each case their product is the same.

ILLUSTRATIVE PROBLEMS

7.9.1 Determine the product of inertia of the right triangle shown in Fig. 7-9.2 with respect to the X and Y axes.

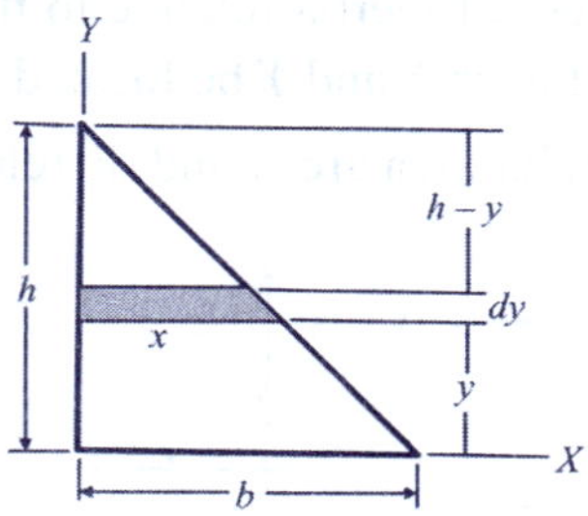

Figure 7-9.2

Solution

In applying the definition of product of inertia, $P = \int xy\, dA$, observe that x and y represent the coordinates of the centroid of the differential area dA. For the right triangle illustrated, select the differential area as the shaded strip parallel to the base. The area of this strip is $dA = x\, dy$, and the coordinates of its centroid are $\frac{1}{2}x$ and y.

From consideration of similar triangles, it is evident that $x = \frac{b}{h}(h-y)$.

Hence $dA = x\,dy = \frac{b}{h}(h-y)\,dy$. Applying the definition of product of inertia, we obtain

$$\left[P = \int xy\,dA\right] \qquad P_{xy} = \int_0^h \left[\frac{1}{2}\frac{b}{h}(h-y)\right](y)\left[\frac{b}{h}(h-y)\,dy\right]$$

$$= \frac{b^2}{2h^2}\int_0^h \left[h^2 y - 2hy^2 + y^3\right] dy$$

$$= \frac{b^2}{2h^2}\left[\frac{h^2y^2}{2} - \frac{2hy^3}{3} + \frac{y^4}{4}\right]_0^h$$

$$P_{xy} = +\frac{b^2h^2}{24} \qquad \textit{Ans.}$$

7-9.2 Determine the product of inertia of the 80 × 60 × 10 mm angle section shown in Fig. 7-9.3 with respect to the indicated X and Y axes.

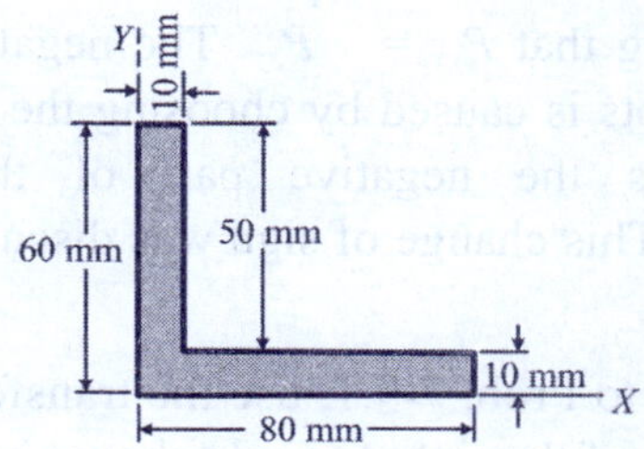

Figure 7-9.3

Solution

The angle section can be considered composed of a 50 × 10 mm rectangle plus an 80 × 10 mm rectangle. For the fist rectangle, the centroidal axes parallel to the X and Y axes are axes of symmetry; hence, from sect on 7-8, $\overline{P}_{xy}$ for this rectangle equals zero. This situation is similar for the other rectangle. Hence for the composite area, we obtain

$\left[P_{xy} = \bar{P}_{xy} + A\bar{x}\,\bar{y}\right]$

50 × 10 mm rectangle:	$P_{xy} = (50 \times 10) \times 5 \times 35 = 87.5 \times 10^3\ \text{mm}^4$
80 × 10 mm rectangle:	$P_{xy} = (80 \times 10) \times 40 \times 5 = 160 \times 10^3\ \text{mm}^4$
For the composite area:	$P_{xy} = 87.5 \times 10^3 + 160 \times 10^3 = 247.5 \times 10^3\ \text{mm}^4$

Ans.

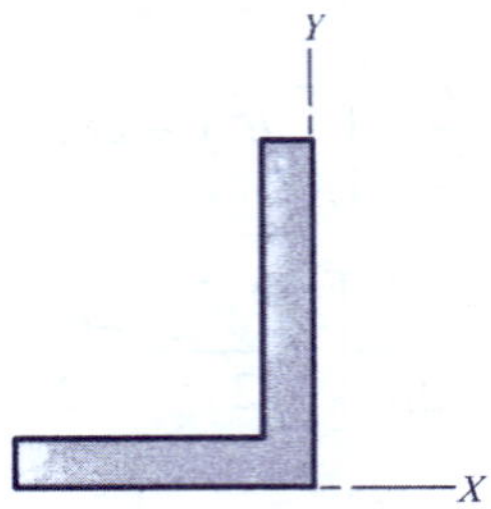

Figure 7-9.4 Rotation of Fig 7-9.3 through 90°.

If the angle is rotated 90° counterclockwise to the position shown in Fig. 7-9.4, the same value of P_{xy} but with opposite sign will be obtained; that is, $P_{xy} = -247.5 \times 10^3\ \text{mm}^4$. This result is equivalent to a 90° ration of the axes in Fig. 7-9.3, thereby showing that $P_{yx} = -P_{xy}$. The negative sign resulting from the interchange of subscripts is caused by choosing the *Y* axis as the first axis; then the second axis is the negative part of the *X* axis lying 90° counterclockwise from *Y*. This change of sign was discussed in connection with Fig. 7-7.1.

7-9.3 From the answer to Prob. 7-9.1, use the transfer formula to obtain the product of inertia of the right triangle shown in Fig. 7-9.5 with respect to the indicated centroidal axes.

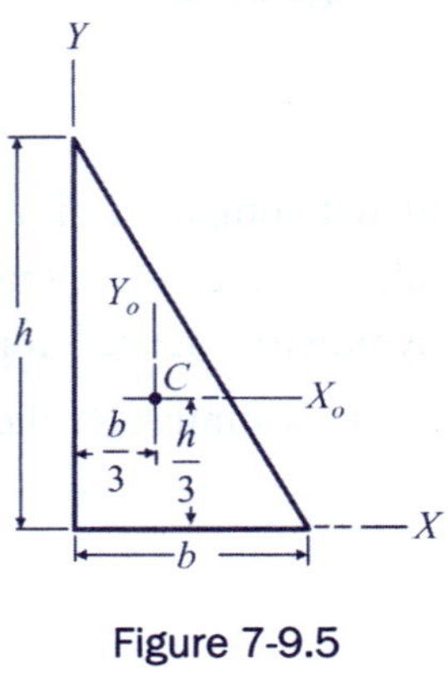

Figure 7-9.5

Solution

Form the answer to Prob. 7-9.1 we have $P_{xy} = b^2h^2/24$. Applying the transfer formula, we obtain

$$[P_{xy} = \bar{P}_{xy} + A\bar{x}\,\bar{y}] \qquad \frac{b^2h^2}{24} = \bar{P}_{xy} + \frac{bh}{2}\times\frac{b}{3}\times\frac{h}{3}$$

$$= \bar{P}_{xy} + \frac{b^2h^2}{18}$$

$$\bar{P}_{xy} = \frac{b^2h^2}{24} - \frac{b^2h^2}{18} = -\frac{b^2h^2}{72} \qquad \textit{Ans.}$$

Note the minus sign carefully. It confirms the rule of sign stated because here most of the area lies in the second and fourth quadrants of the centroidal axes. If the triangle is rotated through 90° from the position shown the sign of $\bar{P}_{xy}$ will change to plus, but its magnitude will still be $b^2h^2/72$.

PROBLEMS

7-9.4 Determine the product of inertia of an 80 by 60 by 10 mm, angle section with respect to centroidal axes parallel to the legs. Explain the significance of the plus or minus sign of the answer.

$$\bar{P}_{xy} = \pm\ 323.1 \times 10^3 \text{ mm}^4 \qquad \textit{Ans.}$$

7-9.5 Repeat Prob. 7-9.4 using an 80 by 40 by 10 mm angle section.

7.9.6 Find the product of inertia with respect to the specified *XY* axes for the shaded area of Fig. P.7-9.6.

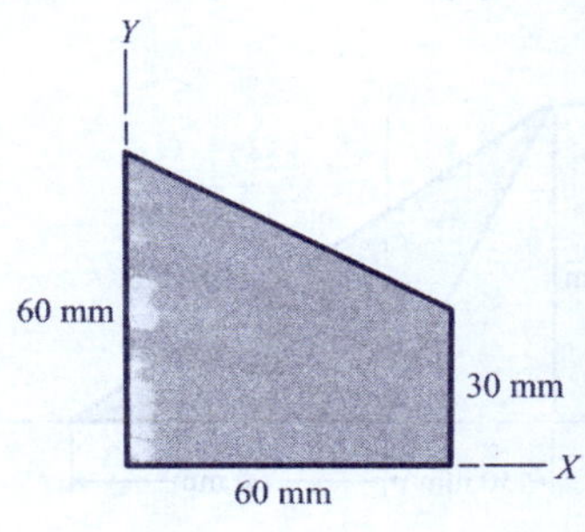

Figure P-7-9.6

$$P_{xy} = 1485 \times 10^3 \text{ mm}^4 \qquad \textit{Ans.}$$

7-9.7 Compute the product of inertia of the triangular area shown in Fig. P.7-9.7 with respect to the given X and Y axes.

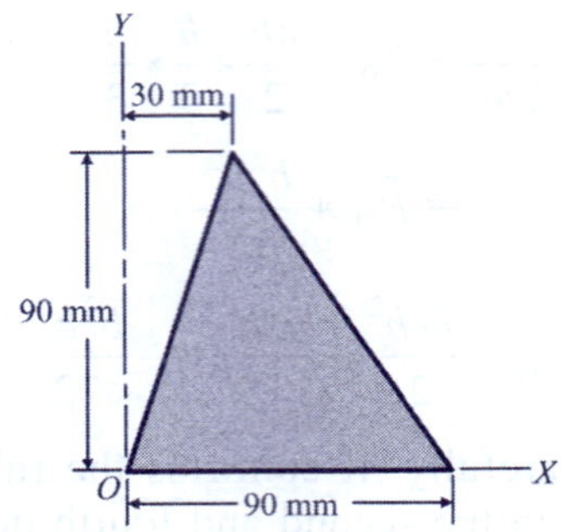

Figure P-7-9.7

7-9.8 Compute the product of inertia of the triangular area shown in Fig. P.7-9.7 with respect to centroidal axes parallel to the given X and Y axes. Can this result be used to obtain the answer to the preceding problem?

$$P_{xy} = -304 \times 10^3 \text{ mm}^4 \qquad \textbf{\textit{Ans.}}$$

7-9.9 Determine the product of inertia of the shaded triangular area shown in Fig. P.7-9.9 with respect to the given X and Y axes.

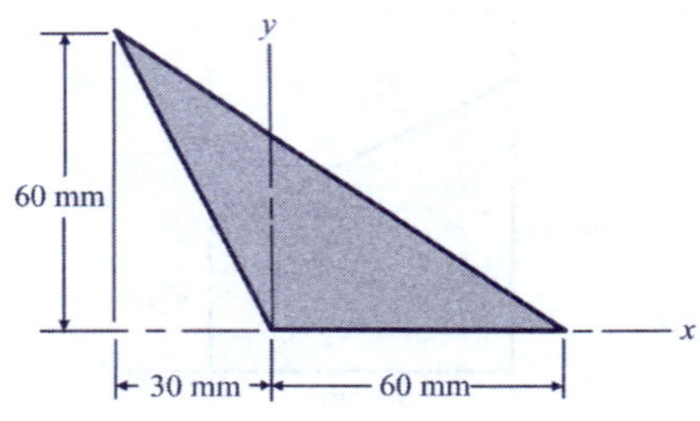

Figure P-7-9.9

7-9.10 Find the product of inertia of the shaded area in Fig. P-7-9.10 about the specified X and Y axes.

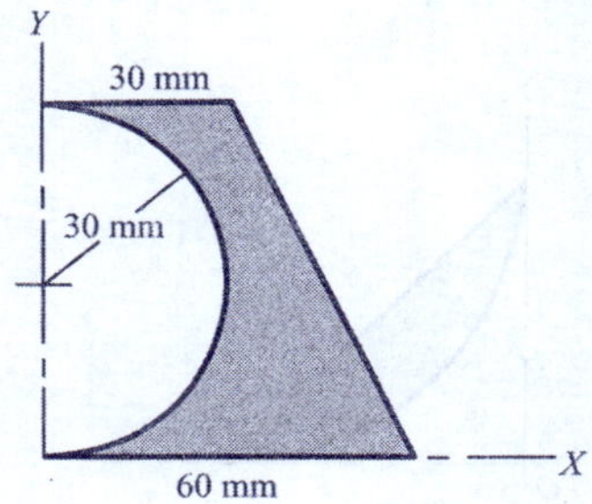

Figure P-7-9.10

$P_{xy} = 945 \times 10^3\ mm^4$ ***Ans.***

7-9.11 Determine the product of inertia of the quarter circular area shown in Fig. P-7-9.11 with respect to the given X and Y axes.

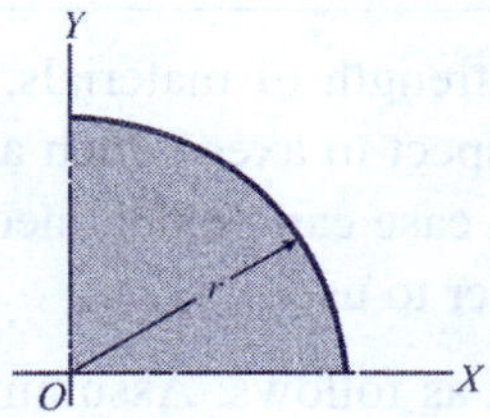

Figure P-7-9.11

$P_{xy} = r^4/8$ ***Ans.***

7-9.12 Use the result of the previous problem to determine the product of inertia of the shaded area shown in Fig. P-7-9.12 with respect to the given X and Y axes.

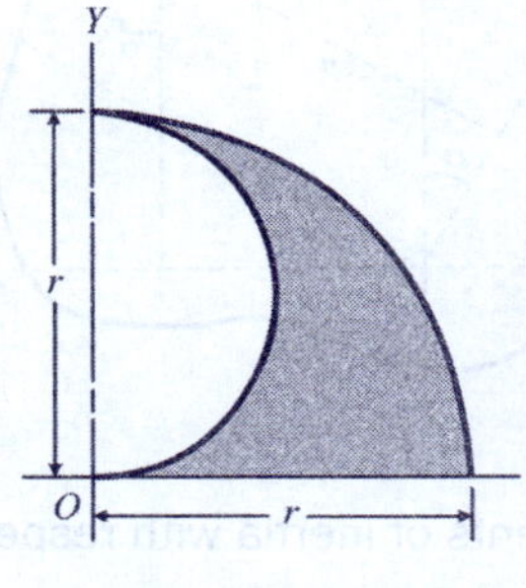

Figure P-7-9.12

$P_{xy} = r^4/12$ ***Ans.***

7-9.13 Compute the product of inertia of the shaded area shown in Fig. P-7-9.13 with respect to the specified X and Y axes.

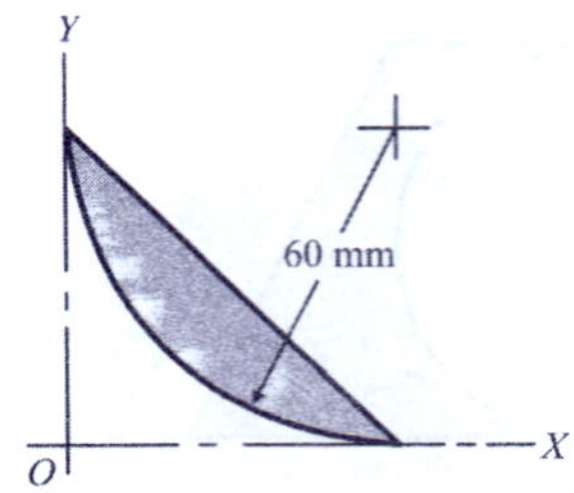

Figure P-7-9.13

$P_{xy} = 47 \times 10^4 \text{ mm}^4$ ***Ans.***

7-10 MOMENT OF INERTIA WITH RESPECT TO INCLINED AXES

In some cases, especially in strength of materials, it is necessary to determine the moment of inertia with respect to axes which are inclined to the usual axes. The moment of inertia in such case can be obtained by formal integration, but a general formula is usually easier to use.

The problem may be stated as follows: Assuming the values of I_x, I_y, and P_{xy} with respect to the X and Y axes to be known, determine the values of I_u, I_v, P_{uv} with respect to the U and V axes inclined at an angle α with the X and Y axes, as shown in Fig. 7-10.1.

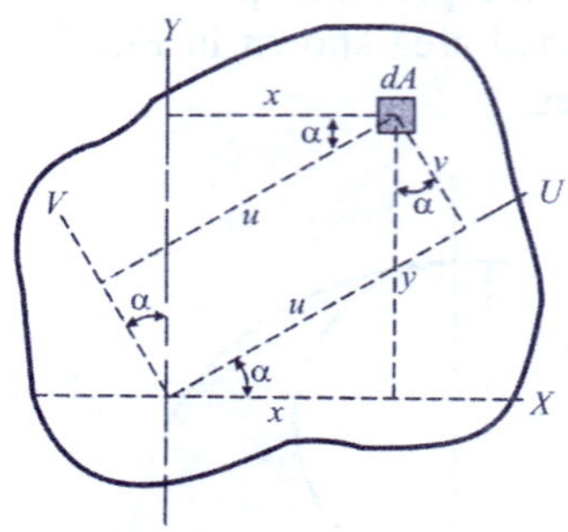

Figure 7-10.1 Moments of inertia with respect to inclined axes.

The coordinates for a typical differential area dA are given by x and y with respect to the X and Y axes, and by u and v relative to the U and V axes. The relations between these coordinates can be obtained by projecting x and y upon the U and V axes. This gives

$$v = y \cos \alpha - x \sin \alpha$$

$$u = y \sin \alpha + x \cos \alpha \qquad (a)$$

By definition ($I = \int \rho^2 \, dA$), the values of I_u and I_v are

$$I_u = \int v^2 \, dA \qquad (b)$$

$$I_v = \int u^2 \, dA \qquad (c)$$

Replacing v in Eq. (b) by its value from Eq. (a), we obtain

$$I_u = \int (y^2 \cos^2 \alpha - 2xy \sin \alpha \cos \alpha + x^2 \sin^2 \alpha) \, dA$$

Since

$$I_x = \int (y^2 \, dA, \; I_y = \int x^2 \, dA, \text{ and } P_{xy} = \int xy \, dA$$

this reduces to

$$I_u = I_x \cos^2 \alpha + I_y \sin^2 \alpha - P_{xy} \sin 2\alpha \qquad (d)$$

If the relations

$$\cos^2 \alpha = \frac{1 + \cos 2\alpha}{2} \quad \text{and} \quad \sin^2 \alpha = \frac{1 - \cos 2\alpha}{2} \qquad (e)$$

are substituted in Eq. (d), the result is

$$\boldsymbol{I_u = \frac{I_x + I_y}{2} + \frac{I_x - I_y}{2} \cos 2\alpha - P_{xy} \sin 2\alpha} \qquad (7\text{-}10.1)$$

Similarly, replacing u in Eq. (c) by its value from Eq. (a) gives

$$I_v = \int (y^2 \sin^2 \alpha + 2xy \sin \alpha \cos \alpha + x^2 \cos^2 \alpha) \, dA$$

This reduces to

$$I_v = I_x \sin^2 \alpha + I_y \cos^2 \alpha + P_{xy} \sin 2\alpha \qquad (f)$$

The relations in Eq. (e) transform Eq. (f) into

$$\boldsymbol{I_v = \frac{I_x + I_y}{2} - \frac{I_x - I_y}{2} \cos 2\alpha + P_{xy} \sin 2\alpha} \qquad (7\text{-}10.2)$$

When values of I_x, I_y and P_{xy} are known, Eqs. (7-10.1) and (7-10.2) permit the values of I_u and I_v, with respect to the U and V axes inclined at an angle α to the X and Y axes, to be determined without further integration. In a sense, these equations do for rotated axes what the transfer formula does for parallel axes.

Adding Eqs. (7-10.1) and (7-10.2) gives the relation

$$I_u + I_v = I_z + I_y$$

which shows that the sum of the moments of inertia with respect to any set of rectangular axes through the same point is a constant quantity. This conclusion could also have been obtained from Section 7-2, which shows that the polar moment of inertia J_z is the sum of the moments of inertia with respect to rectangular axes passing through the polar axis. Since J_z is a constant we obtain as before

$$J_z = I_x + I_y = I_u + I_v$$

To determine the product of inertia relative to the U and V axes, we note that P_{uv} is defined as

$$P_{uv} = \int uv\, dA \qquad (g)$$

Substituting the values of u and υ given in Eq. (a), We have

$$P_{uv} = \int (y^2 \sin\alpha \cos\alpha + xy\cos^2\alpha - xy\sin^2\alpha - x^2\sin\alpha\cos\alpha)\, dA$$

$$= \frac{I_x}{2}\sin 2\alpha + P_{xy}\cos^2\alpha - P_{xy}\sin^2\alpha - \frac{I_y}{2}\sin 2\alpha \qquad (h)$$

whence by using the relation $\cos^2\alpha - \sin^2\alpha = \cos 2\alpha$ we obtain

$$P_{uv} = \frac{I_x - I_y}{2}\sin 2\alpha + P_{xy}\cos 2\alpha \qquad (7\text{-}10.3)$$

The angles defining maximum and minimum moments of inertia may be found by differentiating Eq. (7-10.1) with respect to α and setting the derivative equal to zero. For these values of α, it will be found that the product of inertia is zero while the maximum and minimum moments of inertia are

$$I_{\substack{max \\ min}} = \frac{I_x + I_y}{2} \pm \sqrt{\left(\frac{I_x - I_y}{2}\right)^2 + (P_{xy})^2} \qquad (7\text{-}10.4)$$

7-11 MOHR'S CIRCLE FOR MOMENTS OF INERTIA

The formulas developed in the preceding section need not be memorized since a visual interpretation of them devised by the German engineer Otto Mohr in 1882 is much more useful. In this interpretation, a circle is used; accordingly the construction is called *Mohr's circle.*[4] This circle represents all the possible values of I and P with respect to all axes passing through a specified point in an area. If the circle is plotted to scale, the results can be obtained graphically; usually, however, only a rough sketch of the circle need be drawn, analytical results being obtained from it by the following given below.

We can easily show that Eqs. (7-10.1) and (7-10.3) define a circle by first rewriting them as follows:

$$I_u - \frac{I_x + I_y}{2} = \frac{I_x - I_y}{2} \cos 2\alpha - P_{xy} \sin 2\alpha \qquad (a)$$

$$P_{uv} = \frac{I_x - I_y}{2} \sin 2\alpha + P_{xy} \cos 2\alpha \qquad (b)$$

By squaring (*a*) and (*b*), adding the results, and simplifying, we obtain

$$\left(I_u - \frac{I_x + I_y}{2}\right)^2 + (P_{uv})^2 = \left(\frac{I_x - I_y}{2}\right)^2 + \left(P_{xy}\right)^2 \qquad (c)$$

Recall that I_x, I_y, and P_{xy} are known values with respect to a specified set of X and Y axes while I_u and P_{uv} vary with a rotation of the U and V axes. Consequently, $(I_x + I_y)/2$ is a constant, say C, and the right-hand member of Eq. (*c*) is another constant, say R. Using these substitutions, we transform Eq. (*c*) into

$$(I_u - C)^2 + P_{uv}{}^2 = R^2 \qquad (d)$$

This, being of the form $(x - C)^2 + y^2 = R^2$, is readily recognized as a circle of radius

$$R = \sqrt{\left(\frac{I_x - I_y}{2}\right)^2 + \left(P_{xy}\right)^2}$$

whose center is offset rightward a distance $C = (I_x + I_y/2)$ from the origin of axes.

[4]Mohr's circle for moments of inertia is similar to Mohr's circle for the variation in stresses and strains used in strength of materials. With a suitable change of notation, one can be understood from the other.

The following rules summarize the construction of Mohr's circle. At first it may seem more difficult for you to remember the rules for Mohr's circle than the equations it replaces. After you have used them, however, you will find the rules much easier to apply (and remember) than the equations. Mohr's circle has the further advantage of presenting a graphic visualization of the equations that the student interprets more readily than the equations themselves.

Rules for Applying Mohr's Circle

1. On a set of rectangular coordinate axes, choose one axis on which to plot numerical values of moments of inertia, and the other on which to plot products of inertia; call these the I and P axes, Plot points having the coordinates (I_x, P_{xy}) and (I_y, $-P_{xy}$). The values for I_x, I_y and P_{xy} are assumed to be already known. Note carefully that the value for P_{xy} with its real sign is associated with the value of I_x, and that the value of P_{xy} with the opposite sign is associated with I_y. Actually P_{yx} should be paired with I_y, but $P_{yx} = -P_{xya}$ as we discussed in Illus. Prob. 7-9.2.
2. Join the points just plotted by a straight line. This line is the diameter of Mohr's circle having its center on the I coordinate axis. Draw the circle.
3. As different axes are passed through the selected point in the given area, the values of I and P relative to these axes are represented by the coordinates of points whose positions shift around the circumference of Mohr's circle.
4. The radius of the circle to any point on its circumference represents the axis of inertia corresponding to the I coordinate of that point.
5. The angle between any two radii on Mohr's circle is double the actual angle between the two axes of inertia represented by these two radii. The rotational sense of this angle corresponds to the rotational sense of the actual angle between the axes; i.e., if the U axis of inertia is located at a counterclockwise angle α relative to the X of inertia, then on Mohr's circle, the U radius is laid off as a counterclockwise angle 2α from the X radius.

ILLUSTRATIVE PROBLEMS

7-11.1 For the rectangle shown in Fig. 7-11.1 compute the values of I_u, I_v. and P_{uv} with respect to the U and V axes. These axes are inclined 30° counterclockwise to the X and Y axes.

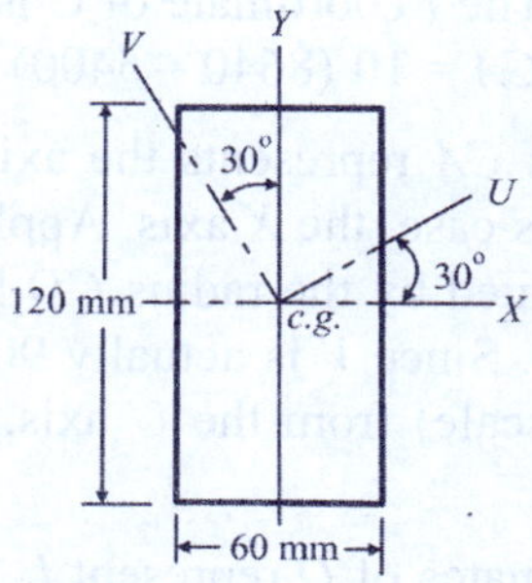

Figure 7-11.1

solution

The moments of inertia and the product of inertia are first found with respect to the X and Y axes as follows:

$$\left[I_x = \frac{bh^3}{12}\right] \qquad I_x = \frac{60 \times (120)^3}{12} = 8640 \times 10^3 \text{ mm}^4$$

$$\left[I_y = \frac{hb^3}{12}\right] \qquad I_y = \frac{120 \times (60)^3}{12} = 2160 \times 10^3 \text{ mm}^4$$

$P_{xy} = 0$ because X and Y are axes of symmetry

Following the rules given above, draw a set of rectangular coordinate axes and label them I and P, as shown in Fig. 7-11.2. Using the values of I_x, I_y and P_{xy}, plot the Points A and B whose coordinates are $(8640 \times 10^3, 0)$ and $(2160 \times 10^3, 0)$.

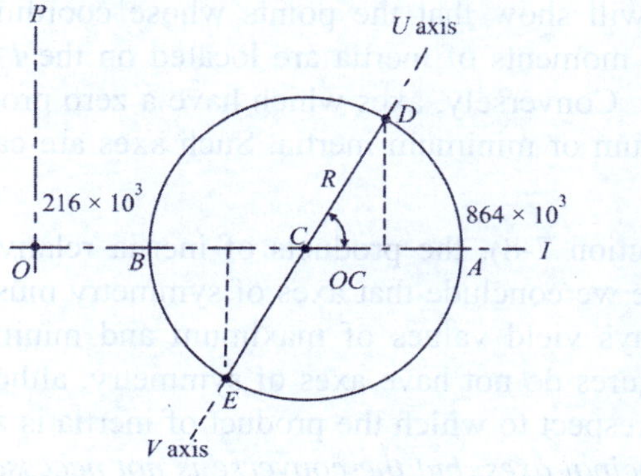

$R = 3240 \times 10^3$

$$OC = \frac{8640 \times 10^3 + 2160 \times 10^3}{2} = 5400 \times 10^3$$

Figure 7-11.2 Application of Mohr's circle.

According to Rule 2, the diameter of Mohr's circle is AB. Its center C is midway between A and B. The I coordinate of C is 5400×10^3 mm^4. The radius of the circle is the distance $CA = 10^3(8640 - 5400) = 3240 \times 10^3$ mm^4.

From Rule 4, the radius CA represents the axis of inertia corresponding to the I coordinate of A; in this case, the X axis. Applying Rule 5, we find that the U axis of inertia is represented by the radius CD laid off 60° counterclockwise from the X axis (CA). Also, Since V is actually 90° from U, the V axis (CE) is laid off 180° (i.e., double scale) from the U axis. D, C, and E form a straight line.

From Rule 3, the coordinates of D represent I_u, and P_{uv}; the coordinates of E represent I_v and P_{uv} with the opposite sign. Accordingly, from the diagram we obtain

$[I_u = OC + CD \cos 60°]$

$$I_u = 5400 \times 10^3 + 3240 \times 10^3 \cos 60° = 5400 \times 10^3 + 1620 \times 10^3$$

$$I_u = 7020 \times 10^3 \text{ mm}^4 \qquad \textbf{\textit{Ans.}}$$

$[I_v = OC - CE \cos 60°]$

$$I_v = 5400 \times 10^3 - 3240 \times 10^3 \cos 60°$$

$$I_v = 3780 \times 10^3 \text{ mm}^4 \qquad \textbf{\textit{Ans.}}$$

$[P_{uv} = CD \sin 60° \text{ N}]$

$$P_{uv} = 3240 \times 10^3 \sin 60°$$

$$P_{uv} = 2810 \times 10^3 \text{ mm}^4 \qquad \textbf{\textit{Ans.}}$$

7-12 MAXIMUM AND MINIMUM MOMENTS OF INERTIA. PRINCIPAL AXES

An inspection of Mohr's circle will show that the points whose coordinates indicate maximum and minimum moments of inertia are located on the I axis and have a zero product of inertia. Conversely, axes which have a zero product of inertia must be axes of maximum or minimum inertia. Such axes are called *principal axes of inertia.*

As we have already seen (Section 7-8), the products of inertia relative to axes of symmetry are zero. Hence we conclude that axes of symmetry must be principal axes because they always yield values of maximum and minimum moments of inertia. But many figures do not have axes of symmetry, although they do have principal axes with respect to which the product of inertia is zero. *Axes of symmetry are always principal axes, but the converse is not necessarily true.*

PROBLEMS

7-12.2 For a certain area it is known that $I_x = I_y = 1200 \times 10^3$ mm^4 and $P_{xy} = -600 \times 10^3$ mm^4. Determine the moments of inertia and product of inertia with respect to U and V axes which are rotated 60° clockwise to the X and Y axes respectively.

7-12.3 A certain area has the following properties: $I_x = 400 \times 10^3$, $I_y = 1000 \times 10^3$ mm^4, $P_{xy} = 400 \times 10^3$ mm^4. Determine the maximum and minimum moments of inertia and also the angle that the axis of maximum inertia makes with the X axis. Illustrate by a diagram:

max. $I = 1200 \times 10^3$ mm^4 ; min. $I = 200 \times 10^3$ mm^4; $\theta = 63.4°$

Ans.

7-12.4 In the preceding problem, determine the moments of inertia and product of inertia with respect to U and V axes inclined 45° clockwise to X and Y, respectively.

7-12.5 For the shaded area shown in Fig. P-7-12.5, determine the maximum and minimum moments of inertia for axes with origin at 0.

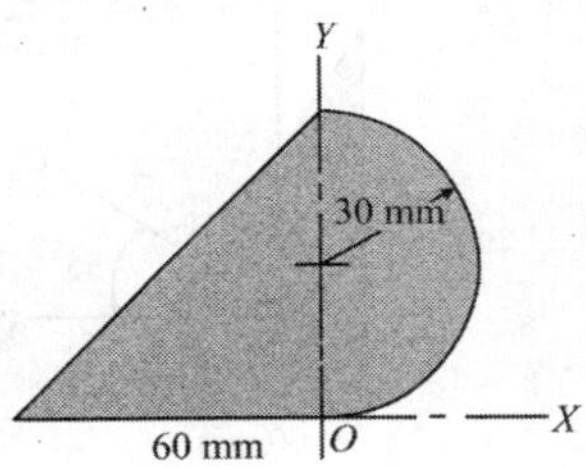

Figure P-7-12.5

7-12.6 Find the moment of inertia of the shaded area in Fig. P-7-12.6 with respect to a U axis through 0 which is inclined at 30° clockwise from the X axis.

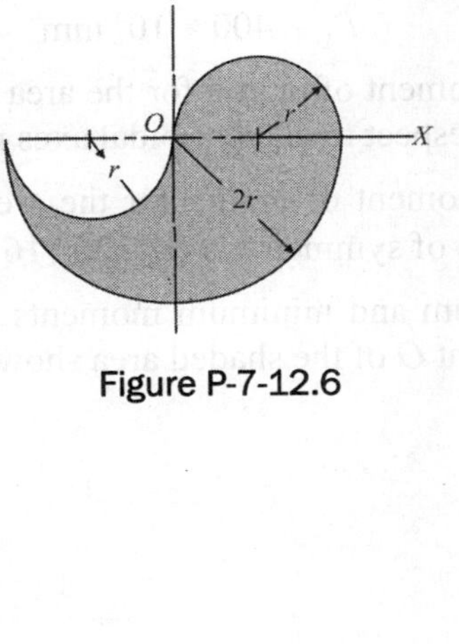

Figure P-7-12.6

7-12.7 For the 60 mm by 120 mm, rectangular area shown in Fig. P-7-12.7, find the moment of inertia about the indicated U axis.

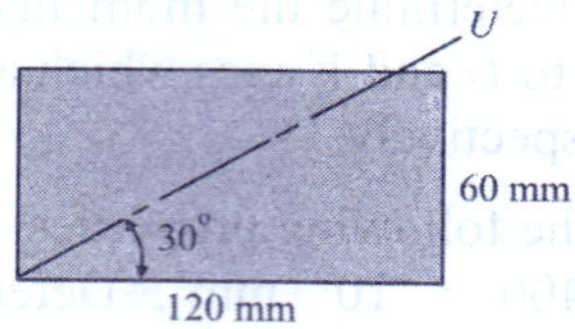

Figure P-7-12.7

$$I_u = 3900 \times 10^3 \text{ mm}^4$$ ***Ans.***

7-12.8 Find the maximum and minimum moment of inertia with respect to axes through the centroid of the 90 mm by 60 mm, triangular area shown in Fig. P-7-12.8. Also compute the moment of inertia with respect to the given centroidal U axis.

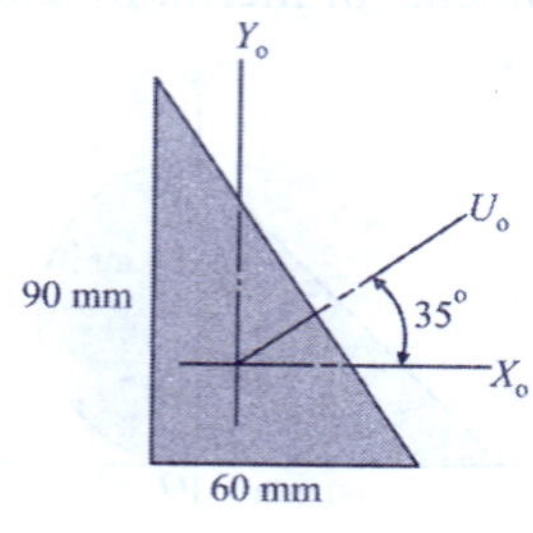

Figure P-7-12.8

$$\bar{I}_u = 1375 \times 10^3 \text{ mm}^4$$ ***Ans.***

7-12.9 For a certain area, it is known that $I_x = 1200 \times 10^3$ mm^4; $I_y = 600 \times 10^3$ mm^4, and min. $I = 400 \times 10^3$ mm^4. Find P_{xy}.

$$P_{xy} = 400 \times 10^3 \text{ mm}^4$$ ***Ans.***

7-12.10 Show that the moment of inertia for the area of any regular polygon is a constant with respect to all centroidal axes in the plane the area.

7-12.11 Show that the moment of inertia for the area of a quater circle with respect to its axis of symmetry is $(\pi - 2)r^4/16$ where r is radius.

7-12.12 Find the maximum and minimum moments of inertia with respect to axes through point O of the shaded area shown in Fig. P-7-12.12.

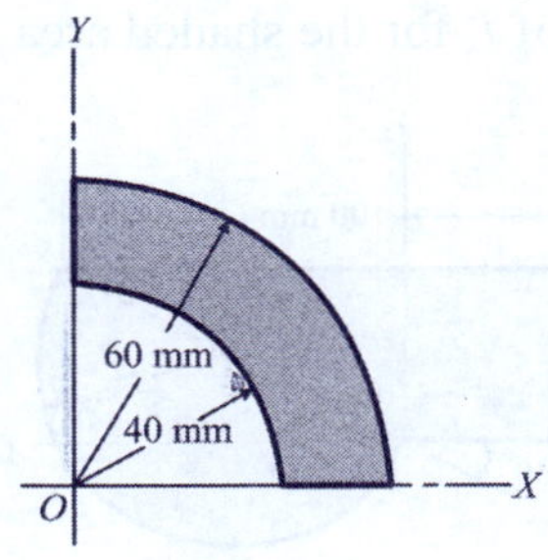

Figure P-7-12.12

7-12.13 The figure for Prob. 7-6.14 is redrawn here as shown in Fig. P-7-12.13. Check the answer to Prob. 7-6.14 (i.e., find I_x) by first computing $\bar{I}_u$ and $\bar{I}_v$, then using Mohr's circle to find $\bar{I}_x$, and finally transferring this value to the X axis.

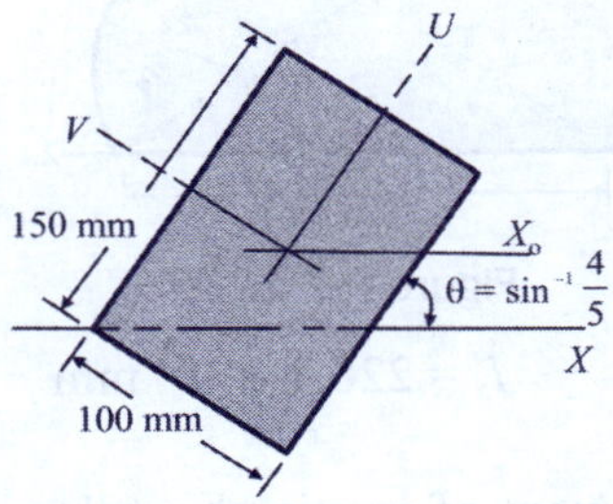

Figure P-7-12.13

7-2.15 Use the idea discussed in Prob. 7-12.13 to compute the value of I_x for the shaded area shown in Fig. P-7-12.15.

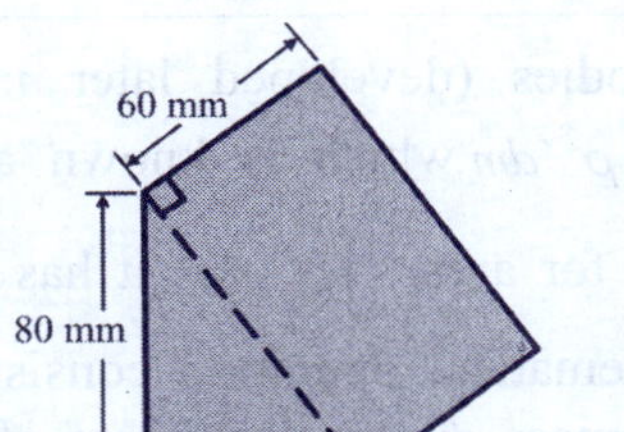

Figure P-7-12.15

7-12.16 Compute the value of I_x for the shaded area shown in Fig. P-7-12.16.

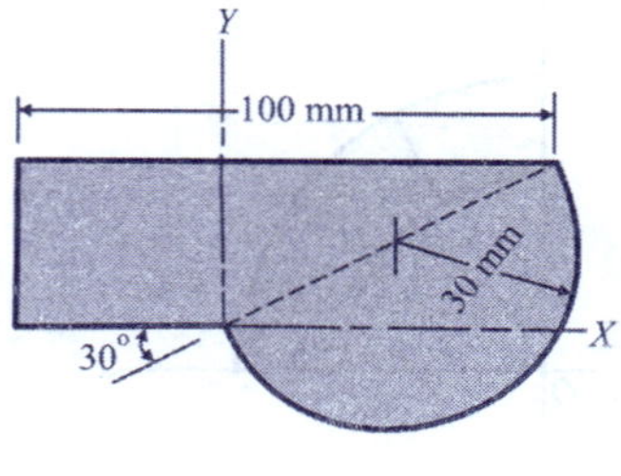

Figure P-7-12.16

$I_x = 952.4 \times 10^3\ \text{mm}^4$ ***Ans.***

7-12.17 Use the idea discussed in Prob. 7-12.13 to determine the moment of inertia for the shaded area shown in Fig. P-7-12.17 with respect to the X axis.

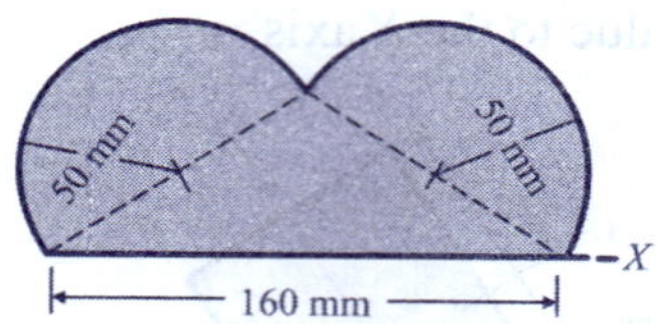

Figure P-7-12.17

$I_x = 228.4 \times 10^5\ \text{mm}^4$ ***Ans.***

7-12.18 Determine the moment of inertia about the vertical centroidal axis of the shaded area shown in Fig. P-7-12.17.

MASS MOMENTS OF INERTIA

7-13 MOMENTS OF INERTIA OF MASSES

The dynamics of rigid bodies (developed later in Part II) involves the mathematical expression $\int \rho^2\, dm$ which is known as the mass moment of inertia. Like its counterpart for area, $\int \rho^2\, dA$, it has no physical meaning by itself but is merely a mathematical operation consisting of summing up the product of each differential mass dm by the square of its moment arm ρ from the axis of inertia. In a sense, however, when used in connection with rotating bodies, the expression may be regarded as a measure of mass distribution or of the resistance of the body to angular acceleration.

The symbol I which has been used to denote area moment of inertia will also be used for mass moment of inertia. In cases where confusion may arise, the subscripts m and A will be used to distinguish between moment of inertia of mass and area respectively, namely, I_m and I_A.[5] The similarity between their definitions suggests that the computations for mass moments of inertia will parallel those for area moments of inertia. In fact, as we shall learn, the method of computing mass moments of inertia duplicates that of area moments of inertia as follows;

$$I_A = \int \rho^2 \, dA \qquad I_A = \Sigma(\bar{I} + Ad^2)$$

$$I_m = \int \rho^2 dm \qquad I_m = \Sigma(\bar{I} + md^2)$$

wherein m has replaced A in the fundamental definition and in the transfer formula. Of course, the centroidal mass moment of inertia for various shapes must be known before the transfer formula can be applied.

7-14 RADIUS OF GYRATION

Mass moment of inertia is sometimes expressed in terms of the product of its mass and the square of a distance k, according to the equation

$$I = mk^2 \text{ or } k = \sqrt{\frac{I}{m}} \qquad (7\text{-}14.1)$$

In this equation k is called the radius of gyration. It may be regarded as the distance from the axis of inertia to the point at which the mass of the body may be assumed to be concentrated and still have the same moment of inertia as does the actual distributed mass of the body.

7-15 TRANSFER FORMULA FOR MASS MOMENTS OF INERTIA

Frequently the mass moment of inertia about axes must be obtained when it is difficult to apply directly the integration of $\int \rho^2 \, dm$. In such cases the transfer formula is a useful method of transferring a known value of moment of inertia from one axis to a parallel axis without further integration.

[5]Occasionally it will be convenient to distinguish between them by using a double subscript for a reference axis for mass moment of inertia and a single subscript for area moment of inertia; thus, $I_m = I_{xx}$ and $I_A = I_x$.

To derive the transfer formula, consider Fig. 7-15.1 which represents the cross section of a body whose mass moment of inertia with respect to a centroidal axis perpendicular to the section is denoted by $\bar{I}$. The value of $\bar{I}$ in terms of the coordinates of a differential element dm is given by

$$\bar{I} = \int r^2\, dm = \int (x^2 + y^2)\, dm \qquad (a)$$

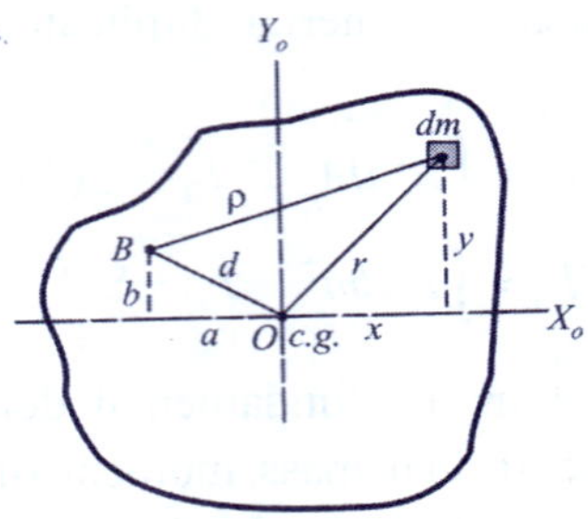

Figure 7-15.1

Let B be the point at which any parallel axis intersects the section. The mass moment of inertia with respect to the parallel axis through B is denoted by I and its value is given by

$$I = \int \rho^2\, dm \qquad (b)$$

where ρ is the moment arm of dm about the axis through B. However, from the figure

$$\rho^2 = (x + a)^2 + (y - b)^2 = x^2 + 2ax + a^2 + y^2 - 2by + b^2$$

which, since $d^2 = a^2 + b^2$ and $r^2 = x^2 + y^2$, becomes

$$\rho^2 = r^2 + d^2 + 2ax - 2by \qquad (c)$$

Substituting this value of ρ^2 in Eq. (b), we obtain

$$I = \int r^2 dm + d^2 \int dm + 2a \int x\, dm - 2b \int y\, dm \qquad (d)$$

Note that $\int x\, dm = m\bar{x}$ and $\int y\, dm = m\bar{y}$; also that $\bar{x}$ and $\bar{y}$ represent the distance from the center of mass to the reference axes (zero in this case because the reference axes contain the center of mass). Hence, the third and fourth terms in Eq. (d) become zero and we obtain as the transfer formula

$$\boldsymbol{I = \bar{I} + md^2} \qquad (7\text{-}15.1)$$

The formula states that for any body the moment of inertia with respect to any axis is equal to the sum of the moment of inertia with respect to a *parallel centroidal axis* and a transfer term composed of the product of the mass multiplied by the square of the distance between the axes. *Note carefully that the centroidal axis involved in the transfer formula is always the axis through the mass center of the mass used in the transfer term* md^2.

A similar relation exists between the radii of gyration with respect to parallel axes, one of which is a centroidal axis. Replacing I by mk^2 and $\bar{I}$ by $m\bar{k}^2$ in Eq. (7-15.1), we have

$$mk^2 = m\bar{k}^2 + md^2$$

whence

$$k^2 = \bar{k}^2 + d^2 \tag{7-15.2}$$

7-16 MASS MOMENTS OF INERTIA BY INTEGRATION

As a preliminary step to applying the transfer formula, especially to determine the mass moment of inertia of composite bodies, it is necessary to develop expressions for the mass moments of inertia of various fundamental geometric shapes by formal integration. When the definition $I = \int \rho^2\, dm$ is applied, the differential mass used in the integration should be selected so that either:

1. All points in the element are at the same distance from the reference axis,[6] or
2. The mass moment of inertia of the selected element with respect to the reference axis is already known. The mass moment of inertia of the body is then found by summing up the moments of inertia of all the elements.

ILLUSTRATIVE PROBLEMS

7-16.1 Find the mass moments of inertia with respect to axes perpendicular to or in the surface of the homogeneous thin plate shown in Fig. 7-16.1. Assume the thickness t to be constant and negligibly small and let γ be the mass per unit volume.

[6] This distance in effect is the radius of gyration of the element.

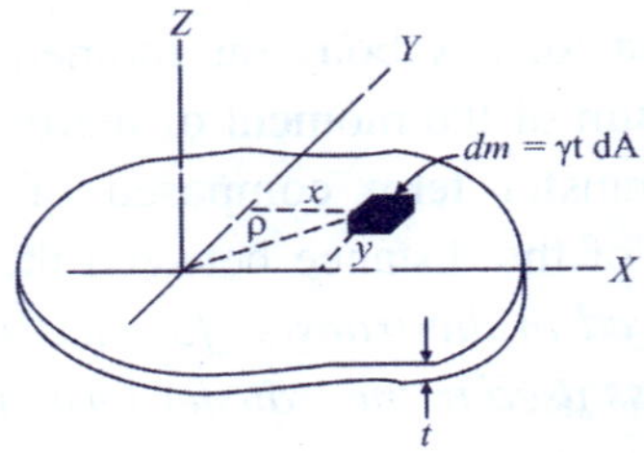

Figure 7-16.1

Solution

We will find it convenient here to use the notation of double subscripts for mass moments of inertia and single subscripts for area moments of inertia. Then with respect to the X axis in the plane of the plate, we have

$$[I = \int \rho^2 \, dm\,] \qquad I_{xx} = \int y^2 \, dm = \gamma t \int y^2 \, dA = \gamma t I_x \tag{a}$$

Similarly,

$$I_{yy} = \int x^2 \, dm = \gamma t \int x^2 \, dA = \gamma t I_y \tag{b}$$

and

$$I_{zz} = \int \rho^2 \, dm = \gamma t \int (x^2 + y^2)\, dA$$

$$= \gamma t\, (I_x + I_y) = \gamma t J_z \tag{c}$$

where J_z is the polar moment of inertia of the area of the plate. Note that this result is also valid for any homogeneous right prism with respect to an axis perpendicular to the base of the prism. Obviously, this is equivalent to adding up identical plates whose cumulative thickness is the height h of the right prism.

7.16.2 Use the results of the previous problem to determine the mass moments of inertia with respect to centroidal axes for the rectangular plate and circular plate shown in Fig. 7-16.2. Consider both plates to be quite thin.

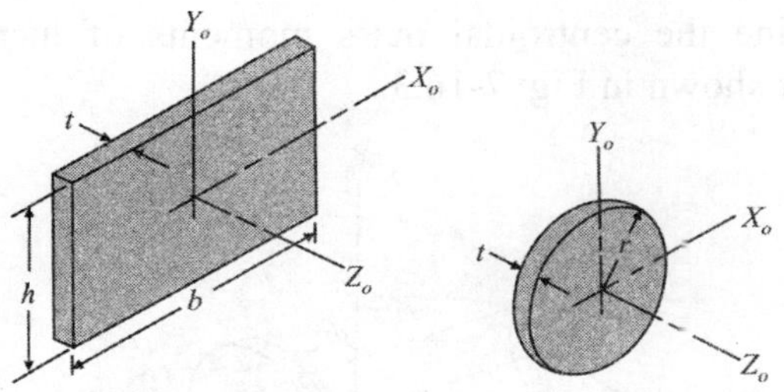

Figure 7-16.2

Solution

For the rectangular plate, we recall that $\bar{I}_x = bh^3/12$ so that we obtain

$$[I_{xx} = \gamma t I_x] \qquad \bar{I}_{xx} = \gamma t\left(\frac{bh^3}{12}\right) = \gamma t\left(\frac{bh^3}{12}\right)\left(\frac{m}{\gamma tbh}\right) = \frac{1}{12}mh^2 \qquad \textbf{Ans.}$$

Where the reduction has been achieved by multiplying by the mass m and dividing by its equivalent $m = \gamma V = \gamma tbh$. Similarly, we get

$$[I_{yy} = \gamma t I_y] \qquad \bar{I}_{yy} = \gamma t\left(\frac{b^3h}{12}\right)\left(\frac{m}{\gamma tbh}\right) = \frac{1}{12}mb^2 \qquad \textbf{Ans.}$$

and

$$[I_{zz} = \gamma t(I_x + I_y)] \quad \bar{I}_{zz} = \gamma t\left(\frac{bh^3}{12} + \frac{b^3h}{12}\right)\left(\frac{m}{\gamma tbh}\right)$$

$$= \frac{1}{12}m(b^2 + h^2) \qquad \textbf{Ans.}$$

For the circular plate, where $\bar{I}_x = \bar{I}_y = \pi r^4/4$ and $\bar{J}_z = \pi r^4/2$, we obtain

$$\bar{I}_{xx} = \bar{I}_{yy} = \gamma t\left(\frac{\pi r^4}{4}\right)\left(\frac{m}{\gamma t\pi r^2}\right) = \frac{1}{4}mr^2 \qquad \textbf{Ans.}$$

and

$$\bar{I}_{zz} = \gamma t\bar{J}_z = \gamma t\left(\frac{\pi r^4}{2}\right)\left(\frac{m}{\gamma t\pi r^2}\right) = \frac{1}{2}mr^2 \qquad \textbf{Ans.}$$

7-16.3 Determine the centroidal mass moments of inertia for the hollow cylinder shown in Fig. 7-16.3.

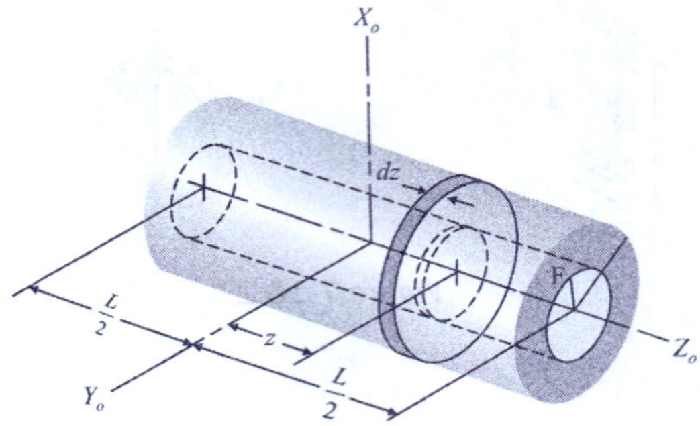

Figure 7-16.3

Solution

Select the differential mass as the slice shown having an infinitesimal thickness *dz*. Using the results of Prob. 7-16.1, the mass moments of inertia of this element of mass $dm = \gamma\pi(R^2 - r^2)\, dz$ are

$$[I_{xx} = \gamma t I_x]$$

$$dI_{xx} = dI_{yy} = \gamma\, dz \frac{\pi}{4}(R^4 - r^4)\left(\frac{dm}{\gamma\pi(R^2 - r^2)dz}\right) = \frac{1}{4}(R^2 + r^2)\, dm$$

Where $dI_{xx} = dI_{yy}$ because of symmetry. With respect to the Z axis, we have

$$[I_{zz} = \gamma t J_z]$$

$$dI_{zz} = \gamma\, dz\, \frac{\pi}{2}(R^4 - r^4)\left(\frac{dm}{\gamma\pi(R^2 - r^2)dz}\right) = \frac{1}{2}(R^2 + r^2)\, dm$$

We now apply the transfer formula to transfer these values to the specified centroidal axes of the entire cylinder and then sum up these transferred values as follows:

$$[I = \bar{I} + md^2]$$

$$\bar{I}_{xx} = \int \frac{1}{4}(R^2 + r^2)\, dm + \int z^2\, dm$$

$$= \frac{1}{4}(R^2 + r^2)m + \gamma\pi\,(R^2 - r^2)\int_{-L/2}^{L/2} z^2\, dz$$

$$= \frac{1}{4}(R^2 + r^2)m + \gamma\pi (R^2 - r^2)\left(\frac{L^3}{12}\right)\left(\frac{m}{\gamma\pi (R^2 - r^2)L}\right)$$

$$= \frac{1}{4}(R^2 + r^2)m + \frac{1}{12} mL^2$$

and finally

$$\bar{I}_{xx} = \bar{I}_{yy} = \frac{1}{12}m\left[L^2 + 3(R^2 + r^2)\right] \qquad \textit{Ans.}$$

With respect to the Z axis, no transfer term is involved and we obtain

$$\bar{I}_{zz} = \int dI_{zz} = \frac{1}{2}(R^2 + r^2)\int dm = \frac{1}{2}(R^2 + r^2)m \qquad \textit{Ans.}$$

For a solid cylinder where $r = 0$, these results become

$$\bar{I}_{xx} = \bar{I}_{yy} = \frac{1}{12}m\left[L^2 + 3R^2\right] \text{ and } \bar{I}_{zz} = \frac{1}{2}mR^2 \qquad \textit{Ans.}$$

In addition, if R is negligibly small compared with L, the cylinder becomes a slender rod for which, on neglecting R, we have

$$\bar{I}_{xx} = \bar{I}_{yy} = \frac{1}{12}mL^2 \quad \text{and} \quad \bar{I}_{zz} = 0$$

7-16.4 Determine the mass moment of inertia of a homogeneous sphere of radius r with respect to a diameter.

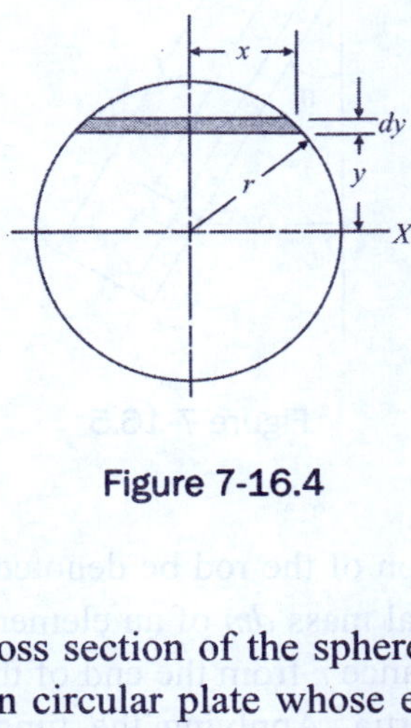

Figure 7-16.4

Solution

Figure 7-16.4 shows the cross section of the sphere in the XY plane. Select the differential mass as the thin circular plate whose edge is shaded in the figure. This plate has a radius x and a thickness dy. The mass dm of the elemental cylinder is $\gamma\pi x^2 dy$ and from Prob. 7-16.2, its moment of inertia relative to the Y

axis is $\frac{1}{2}r^2\,dm = \frac{1}{2}\gamma\pi x^4\,dy$. From Rule 2, the moment of inertia of the sphere is the sum of the moments of inertia of all such elemental circular plates. Thus, we obtain

$$\bar{I} = \frac{1}{2}\gamma\pi \int_{-r}^{+r} x^4\,dy = \frac{1}{2}\gamma\pi \int_{-r}^{+r} (r^2 - y^2)^2\,dy = \frac{8}{15}\gamma\pi r^5 \qquad (a)$$

The volume of a sphere is $\frac{4}{3}\pi r^2$ and its mass is therefore $m = \frac{4}{3}\gamma\pi r^3$. Multiplying the result in Eq. (*a*) by m and dividing by its equivalent, we have

$$\bar{I} = \frac{8}{15}\gamma\pi r^5 \left(\frac{m}{\frac{4}{3}\gamma\pi r^3} \right) = \frac{2}{5}mr^2 \qquad \textbf{\textit{Ans.}}$$

No axial subscript is given for $\bar{I}$ since this result is obviously valid for any diameter.

7-16.5 Determine the moment of inertia of a slender rod of length L and mass m with respect to an axis through one end which makes an angle θ with the length of the rod as shown in Fig. 7-16.5.

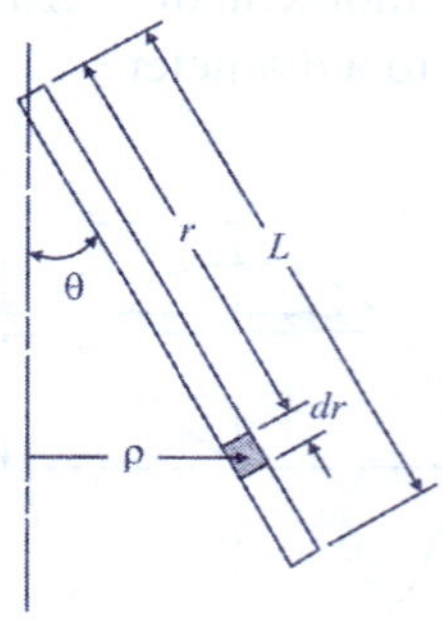

Figure 7-16.5

Solution

Let the constant cross section of the rod be denoted by A and the mass per unit volume by γ. The differential mass dm of an element of length dr is equal to $\gamma A\,dr$. This element is at a distance r from the end of the rod but at a distance $\rho = r \sin\theta$ from the axis of inertia. Applying the fundamental definition of mass moment of inertia, we have

$[I = \int \rho^2\, dm]$

$$I = \int_0^L r^2 \sin^2\theta(\gamma A\, dr) = \gamma A \sin^2\theta \int_0^L r^2\, dr = \frac{1}{3}\gamma A L^3 \sin^2\theta$$

Multiplying and dividing this result by the identity $m = \gamma A L$, we obtain finally

$$I = \frac{1}{3} mL^2 \sin^2\theta = \frac{1}{3} mL_e^2 \qquad \textit{Ans.}$$

where L_e is the equivalent length projected upon a direction perpendicular to the axis of inertia.

Using this result and the transfer formula, it is easy to show that the moment of inertia of the rod with respect to a parallel centroidal axis becomes

$$\bar{I} = \frac{1}{12} mL_e^2 \qquad \textit{Ans.}$$

Can you see the correlation between this result and that at the end of Prob. 7-16.3?

PROBLEMS

7-16.6 Find the mass moment of inertia of the homogeneous parallelepiped (Fig. P-7-16.6) with respect to centroidal axes parallel to the edges.

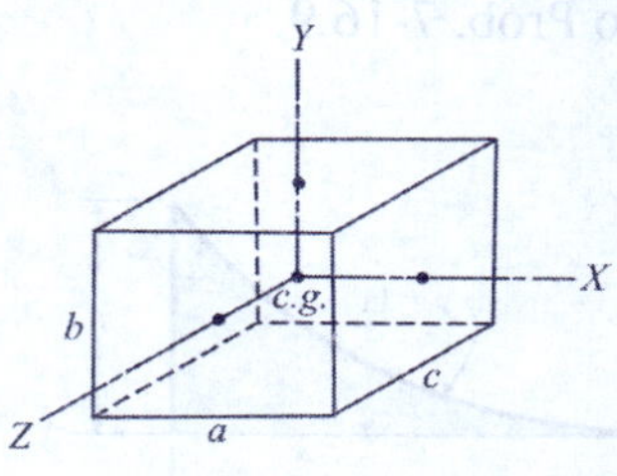

Figure p-7-16.6

$$\bar{I}_x = \frac{1}{12} m(a^2 + b^2) \text{ and similarly for } \bar{I}_x \text{ and } \bar{I}_y \qquad \textit{Ans.}$$

7-16.7 By using the transfer formula and the result of the previous problem, determine the moment of inertia of the rectangular parallelepiped shown in Fig. P-7-16.6 with respect to a median line in the Z face. Take the median line parallel to the X axis.

$$\bar{I}_x = \frac{1}{12} m(b^2 + 4c^2) \qquad \textit{Ans.}$$

7-16.8 Determine the moment of inertia of the rectangular parallelepiped shown in Fig. P-7-16.6 with respect to an axis through one edge parallel to the Y axis.

7.16.9 Derive the expression for the moment of inertia of a homogeneous right circular cone of mass m, base radius r, and altitude h, with respect to its geometric axis.

$$\bar{I} = \frac{3}{10} mr^2 \qquad \textbf{\textit{Ans.}}$$

7-16.10 Determine the moment of inertia of the right circular cone described in the previous problem with respect to any diametral axis in the base of the cone.

$$I = \frac{m}{20}(3r^2 + 2h^2) \qquad \textbf{\textit{Ans.}}$$

7-16.11 A right cylinder of length L has an elliptical cross section with semiaxes a and b. Determine its moment of inertia with respect to the longitudinal centroidal axis.

7.16.12 The shaded area under the spandrel $y = kx^n$ shown in Fig. P-7-16.12 is revolved about the X axis to form a solid of revolution. Determine the mass moment of inertia of this solid about the X axis. Use this result to check the answer to Prob. 7-16.9.

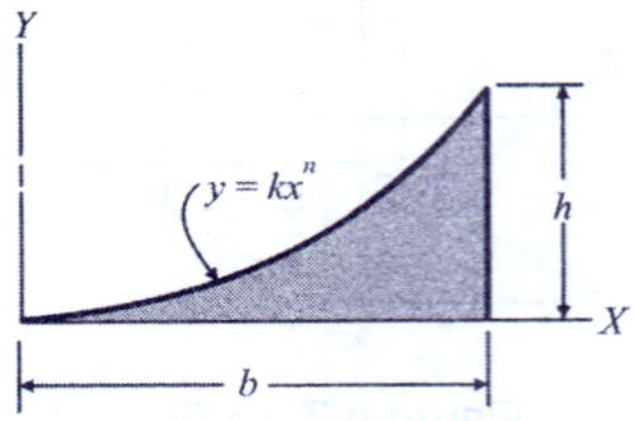

Figure p-7-16.12

$$\bar{I}_x = \frac{(2n+1)}{2(4n+1)} mh^2 \qquad \textbf{\textit{Ans.}}$$

7.16.13 Determine the mass moment of inertia about the X axis of the paraboloid of revolution shown in Fig. P-7-16.13.

Hint: Use the result of the previous problem.

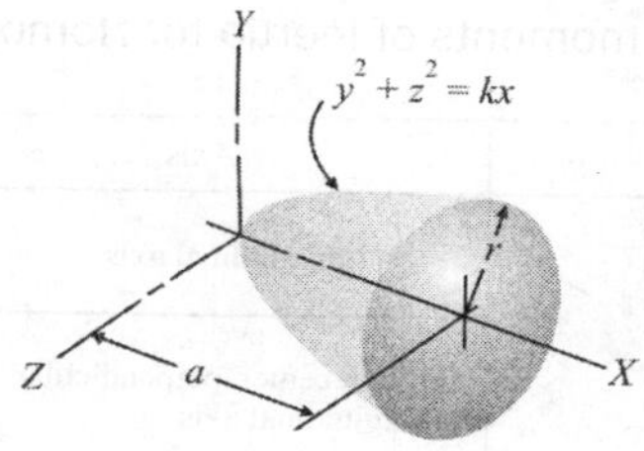

Figure p-7-16.13

7.16.14 Derive the mass moment of inertia of a thin-walled spherical shell of mean radius r and thickness t with respect to any diametral axis.

$$\bar{I} = \frac{2}{3}mr^2 \qquad \textbf{\textit{Ans.}}$$

7-17 MOMENTS OF INERTIA OF COMPOSITE BODIES

Many of the bodies used in engineering are combinations of the geometric elements discussed in the preceding section and accompanying problems. The moment of inertia of a composite body composed of these elements is the sum of the moments of inertia of these elements but, before they can be combined, each element's moment of inertia must be transferred to the same desired axis of inertia. As in area moments of inertia, elements which add to the net mass of the body are positive whereas elements which reduce the net mass are negative. For convenience, the centroidal mass moments of inertia of various shapes are summarized in Table 7-17.1.

Frequently it is preferable to compute *weight* moment of inertia, using weight in place of mass and any convenient unit of length (either millimeter or meters), the resulting unit being either kg-mm^2 or kg-m^2. However, this unit must be converted into kg-mm-sec^2 before using the equations of dynamics (developed later in Part II) which require the use of *mass* moments of inertia. This conversion is easily accomplished by dividing by $g = 9.81$ m / sec^2.

Table 7-17.1 Mass moments of inertia for Homogeneous bodies.

Description of Body	Axis	Moment of Inertia
Solid right circular cylinder or radius r and length L 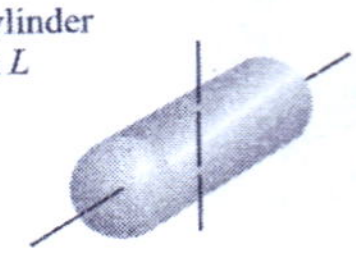	Longitudinal axis	$\frac{1}{2}mr^2$
	Through center, perpendicular to longitudinal axis	$\frac{1}{12}m[L^2+3r^2]$
Hollow right circular cylinder of outer radius R and inner radius r, and length L 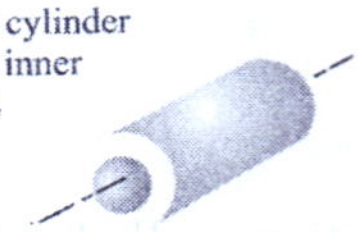	Longitudinal axis	$\frac{1}{2}m[R^2+r^2]$
	Through center, 0perpendicular to longitudinal axis	$\frac{1}{12}m\left[L^2+3\left(R^2+r^2\right)\right]$
Sphere of radius r 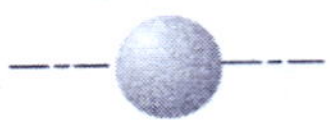	Any diameter	$\frac{2}{5}mr^2$
Spherical shell of mean radius r 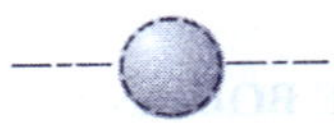	Any diameter	$\frac{2}{3}mr^2$
Uniform slender rod of length L 	At end, perpendicular to rod	$\frac{1}{3}mL^2$
	Through center, perpendicular to rod	$\frac{1}{12}mL^2$
Rectangular prism 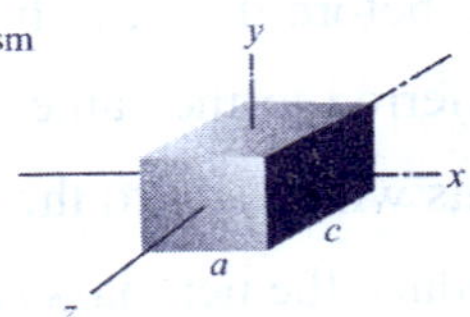	Centroidal axes	$\bar{I}_x=\frac{1}{12}m\left(b^2+c^2\right)$ $\bar{I}_y=\frac{1}{12}m\left(c^2+a^2\right)$ $\bar{I}_z=\frac{1}{12}m\left(a^2+b^2\right)$
Right circular cone of base radius r and altitude h 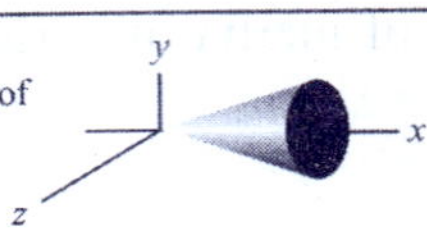	Axis of revolution	$\frac{3}{10}mr^2$
	Through vertex perpendicular to axis symmetry	$I_y=I_z=\frac{3}{5}m\left(\frac{r^2}{4}+h^2\right)$

ILLUSTRATIVE PROBLEMS

7-17.1 The right circular cylinder shown in Fig. 7-17.1 is made of steel weighing 7860 kg/m^3; it is 0.15 m long and has a diameter of 0.6 m. Four holes, each 0.15 m in diameter and equally spaced around a circle of 0.25 m diameter, are drilled from the cylinder. Compute the mass moment of inertia about the geometric axis of the body.

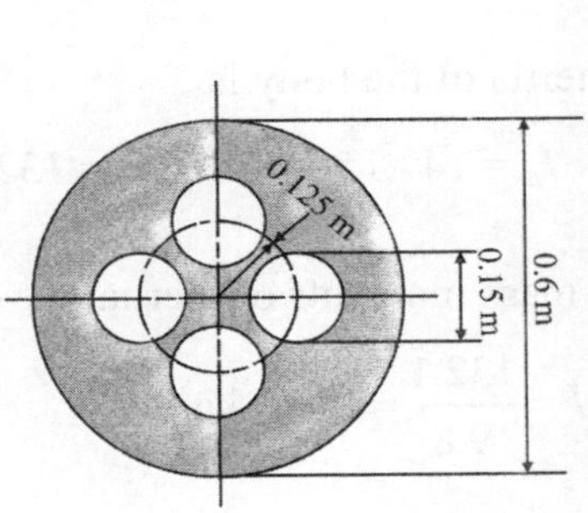

Figure 7-17.1

Solution

The moment of inertia of the body is equal to the moment of inertia of the cylinder minus the moments of inertia of the masses drilled from the cylinder. We start by computing the weight of each element as follows:

$[W = \omega AL]$ Cylinder: $W = 7860 \times 9.8 \times \dfrac{\pi}{4} \times 0.6^2 \times 0.15 = 3266.87$ N

Each weight drilled out: $W = 7860 \times 9.8 \times \dfrac{\pi}{4} \times (0.15)^2 \times (0.15) = 204.18$ N

Dimensional conversions are simplified by first computing weight moment of inertia and finally converting to mass moment of inertia. Hence replacing *m* by *W*, we have for the cylinder

$$[I = \frac{1}{2}Wr^2] \qquad I_c = \frac{1}{2} \times \frac{3266.87}{9.8} \times (0.3)^2 = 15 \times 9.8 \text{ N. m}^2$$

The weight moment of inertia for each drilled out mass with respect to its geometric axis is

$$[I = \frac{1}{2}Wr^2] \; \bar{I}_h = \frac{1}{2} \times \frac{204.18}{9.8} \times (0.075)^2 = 0.0586 \times 9.8 \text{ N. m}^2 = 8.535 \text{ N. m}^2$$

Transferring this value to the geometric axis of the body, we obtain for one drilled out mass,

$$[I = \bar{I} + Wd^2] \; I_h = 0.575 + \frac{204.18}{9.8} = (0.125)^2 = 0.384 \times 9.8 \text{ N.m}^2$$

The weight moment of inertia of the body is

$$[I = I_c - 4I_h] \qquad I_\omega = 147.15 - 4(3.765) = 132.1 \text{ N.m}^2$$

whence converting to mass moments of inertia, we obtain

$$\left[I_m = \frac{I_w}{9.8}\right] \qquad I = \frac{132.1}{9.8} = 13.48 \text{ kg m}^2$$ ***Ans.***

7-17.2 The system shown in Fig. 7-17.2 is part of a governor assembly and consists of a hub to which is hinged two rods 400 mm long with 100 mm diameter spheres welded to their ends. The hub weighs 225 N, each rod weighs 135 N, and each sphere weighs 180 N. For the given position of the system, find the moment of inertia about the *Y* axis and the corresponding radius of gyration.

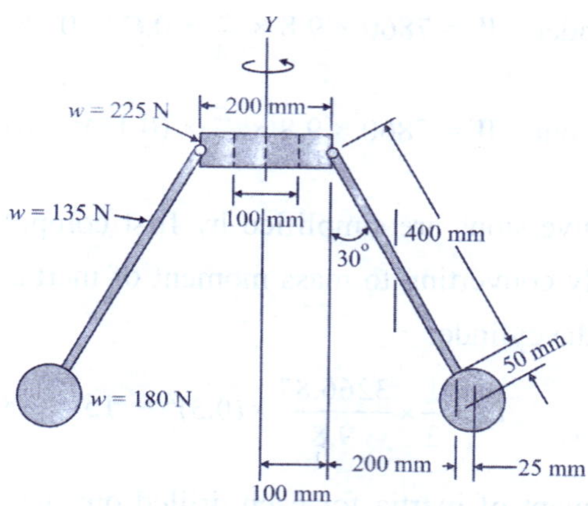

Figure 7-17.2

Solution

The moment of inertia of the system is equal to the sum of the moments of inertia of the hub, two rods, and two spheres, or

$$I = I_{\text{hub}} + 2I_{\text{rod}} + 2I_{\text{sphere}}$$

Since the weights of the various elements are specified, we proceed directly to computing the various weight moments of inertia.

$$\left[I_{hub} = \frac{1}{2}W(R^2 + r^2)\right] \quad I_{hub} = \frac{1}{2} \times \frac{225}{9.8}[0.1^2 + 0.05^2] = 0.143 \times 9.8 \text{ N m}^2$$

$$= 3.402 \text{ Nm}^2$$

For each rod, its centroidal axis parallel to the common Y axis is at 30° with its length. Using the equivalent length projected on a perpendicular to the Y axis and the result from Prob. 7-16.5, we obtain

$$\left[\begin{aligned} I_r &= \frac{1}{12}WL_e^2 + Wd^2 \\ &= W\left(\frac{1}{12}L_e^2 + d^2\right)\end{aligned}\right] \quad I_{rod} = \frac{135}{9.8}\left[\frac{0.2^2}{12} + 0.2^2\right] = 0.597 \times 9.8 \text{ N m}^2$$

$$= 5.857 \text{ N m}^2$$

For each sphere, we have

$$\left[\begin{aligned} I_s &= \frac{2}{5}Wr^2 + Wd^2 \\ &= W\left(\frac{2}{5}r^2 + d^2\right)\end{aligned}\right] \quad I_{sphere} = \frac{180}{9.8}\left[\frac{2}{5}(0.05)^2 + (0.325)^2\right] = 1.956 \times 9.8 \text{ N m}^2$$

$$= 19.588 \text{ N m}^2$$

The total weight moment of inertia then is

$$[I_w = I_h + 2I_r + 2I_s] \quad I = 0.143 + 2\,(0.597) + 2\,(1.956) = 5.249 \times 9.8 \text{ N m}^2$$

$$= 51.493 \text{ N m}^2$$

Finally, converting to mass moment of inertia, we obtain

$$\left[I_m = \frac{I_\omega}{g}\right] \qquad I = \frac{5.249}{9.8} \text{kg.m}^2 \qquad \textbf{\textit{Ans.}}$$

To compute the radius of gyration, we note that

$$k = \sqrt{\frac{I_m}{m}} = \sqrt{\frac{I_\omega}{W}}$$

Since the weight of the entire system is $W = 855$ N, we get

$$\left[k = \sqrt{\frac{I_\omega}{W}}\right] \qquad k = \sqrt{\frac{5.249}{855} \times 9.8} = 0.245 \text{ m} \qquad \textbf{\textit{Ans.}}$$

7-17.3 Compute the moment of inertia about the X axis of the steel link shown in Fig.7-17.3.

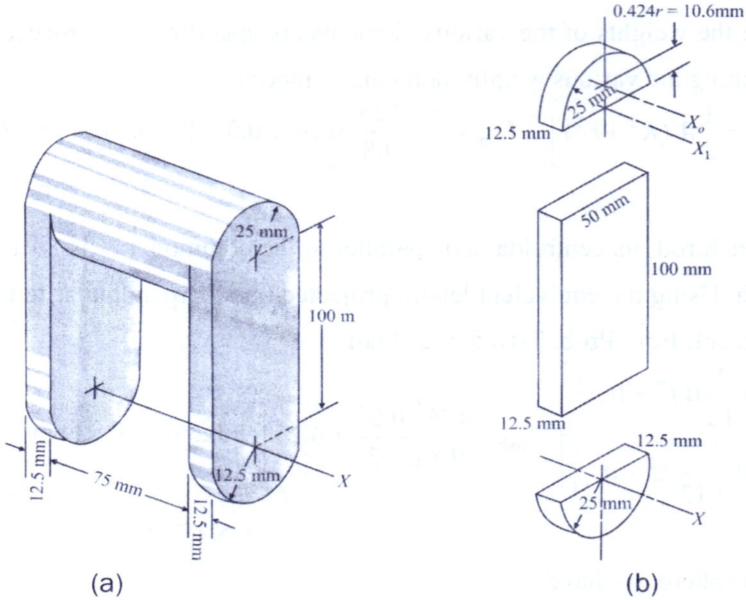

Figure 7-17.3

Preliminary

To subdivide the system into elements appearing in Table 7-17.1, note that each vertical side arm is equal to the sum of two half-cylinders and the rectangular prism shown in part (b). Since two half-cylinders combine into a solid cylinder, the moment of inertia of each half-cylinder with respect to an axis in its base coinciding with the geometric axis of a solid cylinder will be $\frac{1}{2}\left[\frac{1}{2}(2m)r^2\right] = \frac{1}{2}mr^2$ where m is the mass of the half-cylinder. For the lower half-cylinder, this axis coincides with the X axis, but for the upper half-cylinder we must first transfer this value to its centroidal axis before making a second transfer from its centroidal axis to the common X axis.

Solution

The moment of inertia of the link is equal to the sum of the moment of inertia of the cylinder plus that of each side arm. Considering first one side arm, we start by computing the weights of its elements.

$[W = \omega AL]$

Each half-cylinder: $\quad m = 7860 \times \frac{\pi}{2} \times (0.025)^2 \times (0.0125) = 0.0964$ kg

Rectangular prism: $\quad m = 7860 \times (0.0125)(0.05) \times (0.1) = 0.4912$ kg

Now computing the mass moment of inertia of each of these elements, start with the upper half-cylinder. For future use, a general value of its centroidal moment of inertia is

$$\left[I = \bar{I} + md^2\right] \qquad \bar{I}_x = \frac{1}{2}mr^2 - m(0.424r)^2 \approx 0.32\, mr^2$$

Now transferring this centroidal value to the specified X axis, we have

$$\left[I_x = 0.32mr^2 + md^2\right]$$

$$I_x = 0.32\,(0.0964)\,(0.025)^2 + 0.0964\,(0.1106)^2 = 1.198 \times 10^{-3}\ \text{kg-m}^2$$

For the prism in which $\bar{I}_x = \frac{1}{12}m(b^2 + c^2)$ we have about the X axis

$$I_x = \bar{I}_x + md^2 \qquad I_x = \frac{0.4912}{12}[0.05^2 + 0.1^2] + 0.4912 \times (0.05)^2$$

$$= 1.645 \times 10^{-3}\ \text{kg-m}^2$$

For the lower half-cylinder, we obtain directly

$$\left[I_x = \frac{1}{2}mr^2\right] \qquad I_x = \frac{1}{2}(0.0964)(0.025)^2 = 3.013 \times 10^{-5}\ \text{kg-m}^2$$

For both side arms, we double the sum of the above values to obtain

Two side arms: $I_x = 5.746 \times 10^{-3}\ \text{kg-m}^2$

The mass of the cylinder connecting the two side arms is $7860 \times \pi \times (0.025)^2 \times (0.075) = 1.157$ kg and its moment of inertia about the specified X axis is $\left[I_x = \frac{1}{2}mr^2 + md^2\right]$ $I_x = \frac{1}{2} \times (1.157)(0.025)^2 + 1.157(0.1)^2 \approx 0.371 \times 10^{-3}\,\text{kg.m}^2$

Finally, the total mass moment of inertia of the assembly is

$$I_x = 5.746 \times 10^{-3} + 0.371 \times 10^{-3} = 6.117 \times 10^{-3}\ \text{kg-m}^2$$

PROBLEMS

7.17.4 For a hollow cast iron sphere of 500 mm outside diameter and 400 mm inside diameter, compute the moment of inertia and the radius of gyration with respect to any diameter. Cast iron weighs 7860 kg / m.^{3}

7.17.5 A sphere of 150 mm, diameter weighing 9 kg is attached to the end of a slender rod 500 mm long weighing 4.5 kg. The assembly rotates about an axis perpendicular to the rod and passing through a point on the rod which is 25 mm from the center of the sphere. Compute the radius of gyration in mm about the specified axis.

7.17.6 Solve the previous problem if the specified axis is inclined at 60° with the rod.

7-17.7 A uniform steel rod weighing 2 kg is bent into the shape as shown in Fig. P-7-17.7. Compute the radius of gyration about an axis through *O* perpendicular to the plane of the figure.

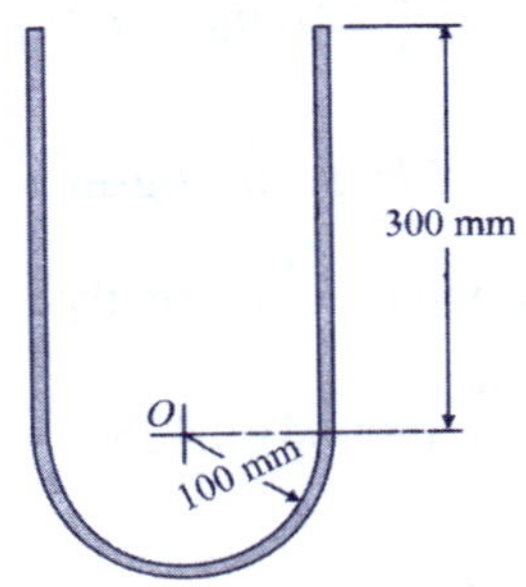

Figure P-7-17.7

7-17.8 Three slender rods are pinned together as shown in Fig. P-7-17.8. If the assembly weighs 3 kg, find the moments of inertia about the given *X*, *Y*, and *Z* axes. The *XZ* axes are in the plane of the figure.

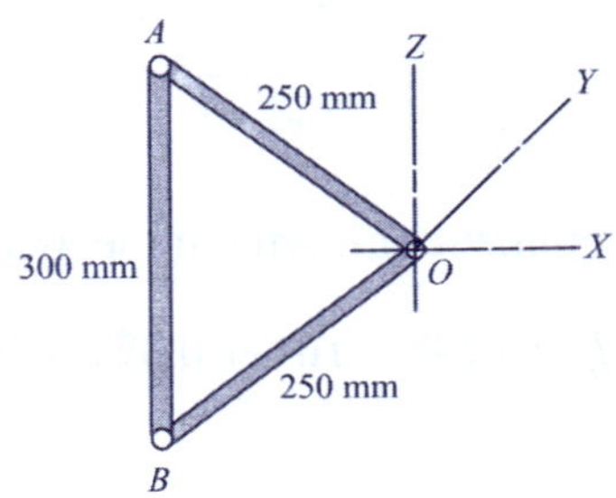

Figure P-7-17.8

7-17.9 A slender rod weighing 1.8 g/mm is bent into the shape shown in Fig. P-7-17.9. The leg *AB* is in the *XZ* plane and the leg *CD* is in the *XY* plane. Compute the moment of inertia about the *X*, *Y*, and *Z* axes.

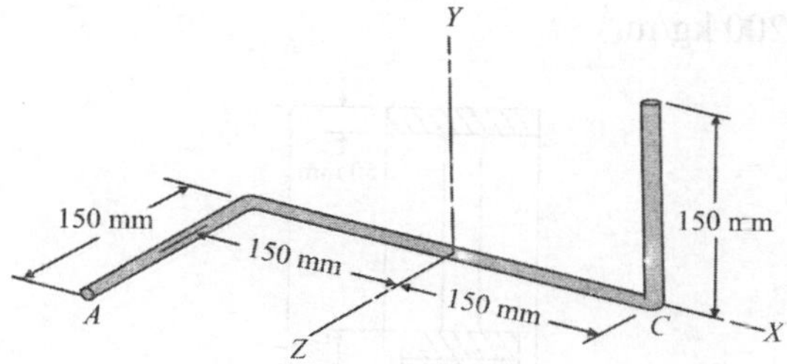

Figure P-7-17.9

7-17.10 A model of a communications satellite consists of a thin-walled sphere 0.9 m in diameter weighing 180 kg with four slender bars, each 0.3 m long and weighing 1 kg, protruding from it in a diametral plane as shown in Fig. P-7-17.10. Compute the mass moment of inertia with respect to the *X*, *Y*, and Z axes.

Hint: See the result of Prob. 7-16.14.

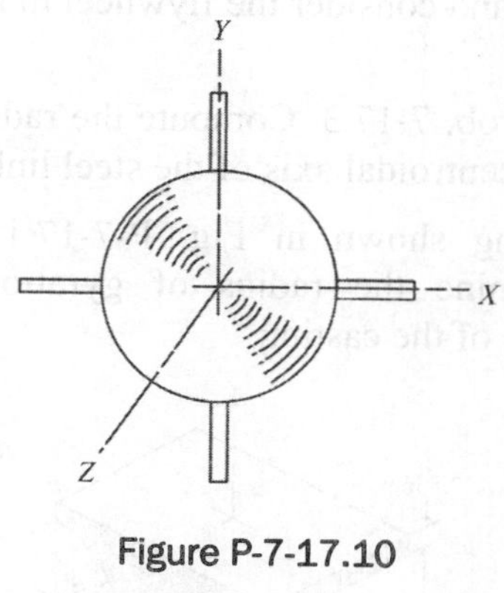

Figure P-7-17.10

7-17.11 A right circular cone made of steel has an altitude of 0.5 m and a base diameter of 0.6 m. A hole 0.15 m deep and of 0.2 m diameter is drilled from the center of the base of the cone and filled with lead. Lead weighs 11400 kg/m^3 and steel weighs 7860 kg/m^3. Determine the mass moment of inertia of the resulting solid with respect to its geometric axis.

7-17.12 Determine the mass moment of inertia of the cast iron flywheel whose cross section is shown in Fig. P-7-17.12 with respect to its axis of rotation. The flywheel has six elliptical spokes 75 mm × 100 mm in cross section, which may be considered as slender rods. Cast iron weighs 7200 kg/m.[3]

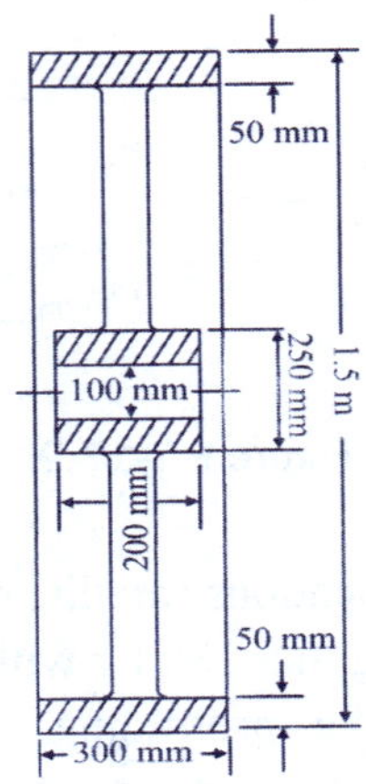

Figure P-7-17.12

7-17.13 Compute the mass moment of inertia of the flywheel described in the previous problem by using the following approximate method: Neglect the hub and consider the flywheel to have three double spokes each 1.4 m long.

7-17.14 Refer to Illus. Prob. 7-17.3. Compute the radius of gyration in meters about a vertical centroidal axis of the steel link.

7-17.15 The steel casting shown in Fig. P-7-17.15 has two semicircular grooves. Determine the radius of gyration in meters about the centroidal X axis of the casting.

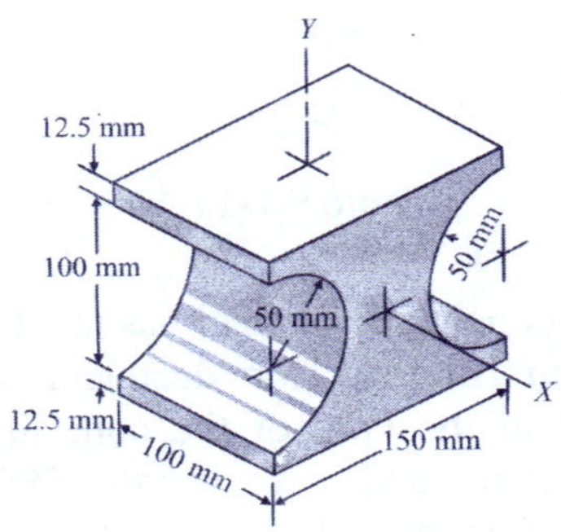

Figure P-7-17.15

7-17.16 For the steel casting of the previous problem, determine the radius of gyration in meters about the centroidal Y axis.

SUMMARY

The moment of inertia with respect to an axis in the plane of an area is defined by the equation

$$I = \int \rho^2 \, dA$$

where ρ is the perpendicular distance from dA to the axis of inertia. The radius of gyration k is defined by the relation

$$I = Ak^2 \qquad (7\text{-}3.1)$$

where A is the total area of the figure.

The polar moment of inertia J with respect to an axis perpendicular to the plane of a figure is the sum of the plane moments of inertia with respect to two mutually perpendicular axes which intersect on the polar axis; i.e.,

$$J = I_x + I_y \qquad (7\text{-}2.1)$$

The transfer formula is useful to transfer the moment of inertia from a centroidal axis to any parallel axis, or vice versa. The formula is

$$I = \bar{I} + Ad^2 \qquad (7\text{-}4.1)$$

The moment of inertia for a composite area is the sum of the moments of inertia of its component parts. Note that the moments of inertia of each part must be found with respect to a common axis before they can be added (see Section 7-6).

The product of inertia for an area is defined by the mathematical expression $\int xy \, dA$ and is denoted by the symbol P. With respect to an axis of symmetry, the product of inertia is zero. The transfer formula below is useful for transferring the product of inertia between a centroidal and a parallel set of axes.

$$P_{xy} = \bar{P}_{xy} + A\bar{x}\,\bar{y} \qquad (7\text{-}9.1)$$

When it is necessary to compute moments of inertia with respect to axes inclined at an angle α with a set of reference axes for which I and P are known, the following formulas may be used:

$$I_{x'} = \frac{I_x + I_y}{2} + \frac{I_x - I_y}{2} \cos 2\alpha - P_{xy} \sin 2\alpha \qquad (7\text{-}10.1)$$

$$I_{y'} = \frac{I_x + I_y}{2} - \frac{I_x - I_y}{2} \cos 2\alpha + P_{xy} \sin 2\alpha \qquad (7\text{-}10.2)$$

Mohr's circle (Section 7-11) may be used as an alternative to these equations. This circle is especially convenient when determining maximum and minimum moments of inertia and principal axes of inertia.

Moments of inertia for bodies, also known as mass moments of inertia, are similar to moments of inertia for area. The essential difference lies in the use of *dm* in place of *dA*, as may be seen from the definition $I_m = \int \rho^2 \, dm$. The formulas for I_m for several bodies are given in Table 7-17.1 and are applied to composite bodies in section 7-17.

CHAPTER 8

SPECIAL TOPICS

In this chapter we present several topics which are supplementary to or are an extension to the basic topics of Statics. These optional topics are shear and bending moment in beams, virtual work, and graphic statics. Each of these topics is presented as a complete entity and is independent of the others.

SHEAR AND BENDING MOMENT IN BEAMS

8-1 INTRODUCTION

It is shown in strength of materials[1] that the stresses in beams vary directly with the resultant internal forces transmitted from section to section of the beam. These internal forces reduce to a force-couple system which, for the case of coplanar vertical loads considered here, consist of a transverse shearing force and a bending couple usually called bending moment. For brevity, the terms shear and moment are often used. As a preliminary to the study of stresses in beams, therefore, we must understand how shear and moment vary in beams subjected to various combinations of loadings under different conditions of support and, especially, be able to determine the maximum values of shear and moment.

Methods of supporting some types of beams are shown in Fig. 8-1.1. A simple beam is supported by a hinged support at one end and a roller support at the other, but is not otherwise restrained. A cantilever beam is supported at one

[1]See, for example, the author's *Strength of Materials,* published by Harper & Row.

end only, with suitable restraint to prevent rotation of that end. An overhanging beam is supported by a hinge and a roller reaction, with either or both ends extending beyond the supports. These beams are all statically determinate; their reactions can be determined directly from the equations of static equilibrium.

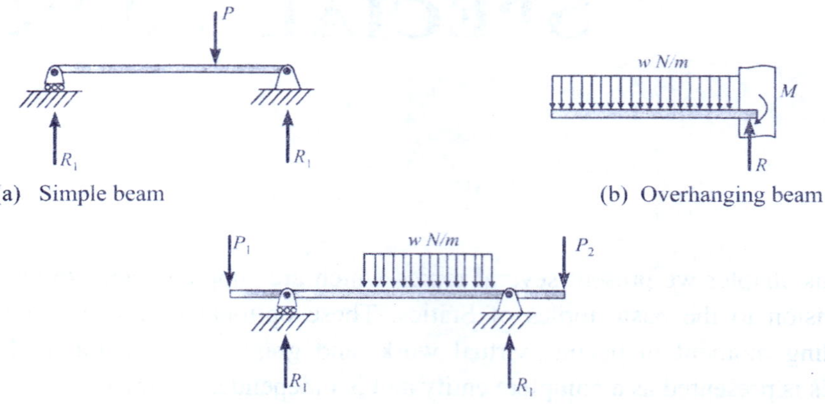

(a) Simple beam (b) Overhanging beam

(c) Cantilever beam

Figure 8-1.1 Statically determine beams.

Other methods of supporting beams are shown in Fig. 8-1.2. The propped beam, the fixed-ended or restrained beam, and the continuous beam each have at least one more reactive element than is absolutely necessary to support it. Such beams are statically indeterminate; they will not be considered here since the presence of excess supports requires the use of additional equations obtained from considering the elastic deformations of the beam as discussed in strength of materials.

A *concentrated load* is one that acts over so small a distance that it can be assumed to act at a point, as in Fig. 8-1.1a. In contrast, a *distributed load* acts over a considerable length of the beam. It may be distributed uniformly over the entire length, as in Fig. 8-1.lb, or over part of the length as in Fig. 8-1.1c. Distributed loads may also be uniformly varying or nonuniform. In a uniformly varying or triangular load, the intensity of loading increases or decreases at a constant rate, as is Fig. 8-1.2a and 8-1.2b; this condition might result from water pressure acting on the face of a dam or from the dumping of a pile of sand. The trapezoidal loading in the right segment of Fig. 8-1.2c is a combination of a uniform and a uniformly varying load. The loading may also be nonuniform, as in the left segment of Fig. 8-1.2c; this may result from the haphazard piling of sand bags.

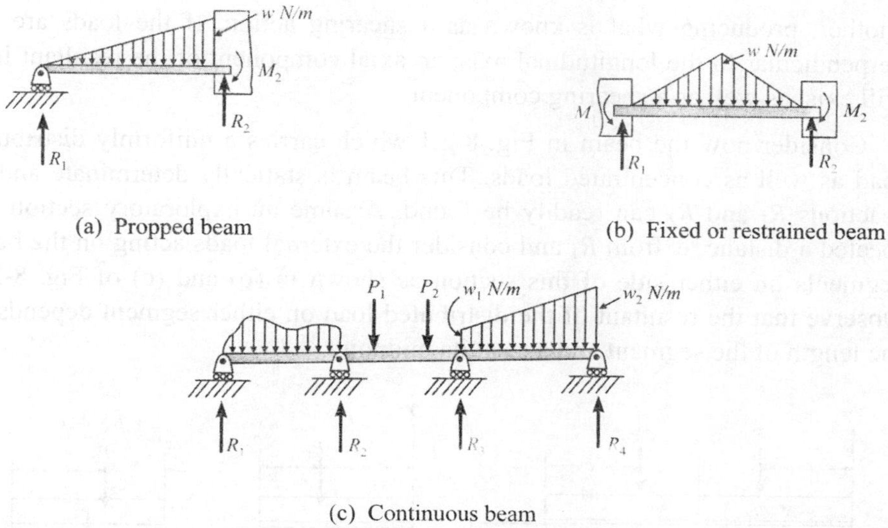

Figure 8-1.2 Statically indeterminate beams.

8-2 SHEAR AND MOMENT EQUATIONS AND DIAGRAMS

We define shear and moment in a beam as equivalent respectively to the resultant force-couple effect of the loading acting on a portion of the beam that lies to either side of an exploratory section through the beam. However, for reasons to be clarified later, usually the portion to the left side of an exploratory section is chosen for analysis. The resultant force-couple of the loads on this segment is computed with respect to a reference point which should always be the centroid of the beam cross section.[2] Since the resultant force-couple, i.e., the equivalent shear and moment, changes with the location of the exploratory section, our objective is to express shear and moment as a function of position of the exploratory section so that we can determine maximum shear and maximum bending moment. In other words, we shall learn how to write shear and moment equations. The graphs of these equations are called shear and moment diagrams.

Usually the loads on a beam are perpendicular to its longitudinal axis so that the resultant of these loads tends to move adjacent sections in a beam past one

[2]It may be useful to review Section 2-13 dealing with the reduction of any force system to a resultant force-couple.

another, producing what is known as a shearing action. If the loads are not perpendicular to the longitudinal axis, an axial component of the resultant load will exist as well as a shearing component.

Consider now the beam in Fig. 8-2.1 which carries a uniformly distributed load as well as concentrated loads. This beam is statically determinate and its reactions R_1 and R_2 can readily be found. Assume an exploratory section *b-b* located a distance x from R_1 and consider the external loads acting on the beam segments on either side of this section as shown in (b) and (c) of Fig. 8-2.1. Observe that the resultant of the distributed load on either segment depends on the length of the segment and acts at its midpoint.

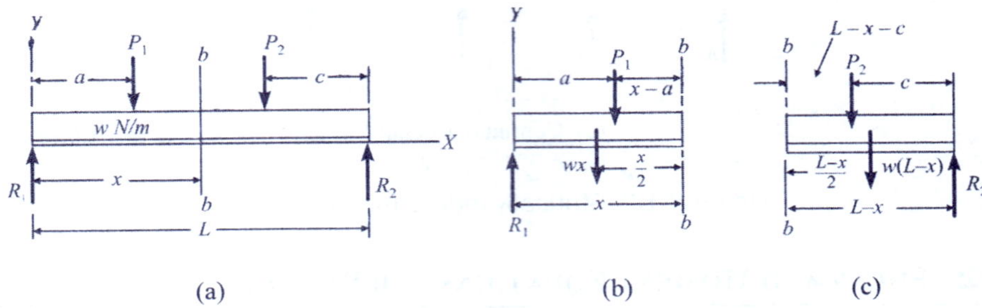

Figure 8-2.1 Beam loading and corresponding to external force diagram.

The shear V is equal to the summation of the vertical components[3] of the external loads acting on either side of the exploratory section. When we compute V, positive forces are those which tend to move the left segment upward with respect to the right segment, and vice versa, as shown in Fig. 8-2.2. A better rule of sign, which produces the same result but is easier to remember, is to give a positive sign to those forces acting upward on the *left* segment. Restricting ourselves to this convention, we define shear mathematically as

$$V = (\Sigma Y)_L \tag{8-2.1}$$

where the subscript L emphasizes that the vertical summation includes only the external loads acting on the beam segment to the left of the exploratory section. Applying this definition to part (b) of Fig. 8-2.1, we find that the vertical shear at section *b-b* is

$$V = R_1 - P_1 - wx \tag{a}$$

---

[3]The beam is assumed to be horizontal. With the beam in any other position, the shearing force is computed from the summation of components perpendicular to the axis of the beam.

Figure 8-2.2 Relative movements corresponding to signs of shearing force.

Bending Moment

When the external forces acting to either side of an exploratory section are reduced to a force-couple system referred to the centroid of the section, the resulting couple is defined as the bending moment at that section. The bending moment is equivalent to the summation of moments about the *centroidal axis* of any selected section of all the loads acting either to the left or right of the section. Mathematically, bending moment M is expressed by

$$M = (\Sigma M)_L = (\Sigma M)_R \tag{8-2.2}$$

where the subscript L indicates that the bending moment is computed in terms of the loads acting to the left of the section, and the subscript R similarly refers to loads acting to the right of the section.

Sign of Bending Moment

To many engineers, bending moment is positive if it produces bending of the beam concave upward as in Fig. 8-2.3. We prefer an equivalent convention which states that *upward-acting forces cause positive bending moments with respect to any section; downward forces cause negative bending moments.* Insofar as the left segment of a beam is concerned (Fig. 8-2.1*b*), this is equivalent to taking clockwise moments about the bending axis as positive, as indicated by the moment sense of R_1. With respect to the right segment of a beam (Fig. 8-2.1*c*), this convention means that the moment sense of the upward reaction R_2 is positive in a counterclockwise direction. This convention has the advantage of permitting bending moment to be computed, without any confusion in sign, in terms of the forces to either the left or the right of a section, depending on which requires the least arithmetical work. We never need think about whether a moment is clockwise or counterclockwise; upward-acting forces always cause positive bending moments regardless of whether they act to the left or the right of the exploratory section.

Figure 8-2.3 Curvatures corresponding to signs of bending moment.

As an illustration, the bending moment at section *b-b* of Fig. 8-2.1 may be written as

$$[M=(\Sigma M)_L] \qquad M=R_l x-P_l(x-a)-wx\left(\frac{x}{2}\right)$$

or as

$$[M=(\Sigma M)_R]$$

$$M=R_2(L-x)-P_2(L-x-c)-w(L-x)\left(\frac{L-x}{2}\right)$$

To summarize, transverse shear V and bending moment M represent the force-couple effect of the external loading acting either to the left or the right of an exploratory section. They are obtained by drawing external force diagrams as we did in (b) and (c) of Fig. 8-2.1; not from free-body diagrams in equilibrium. By adding resisting shear V_r and resisting moment M_r to them, they will become free-body diagrams. But basically, we are interested in the resultant force-couple and its equivalent transverse shear and bending moment which must be resisted by the internal stress distributions which compose V_r and M_r.

The definitions of shearing force and bending moment expressed mathematically are

$$V=(\Sigma Y)_L \tag{8-2.1}$$

$$M=(\Sigma M)_L=(\Sigma M)_R \tag{8-2.2}$$

in which positive effects are produced by upward forces and negative effects by downward forces. This rule of sign[4] for beam analysis will be used exclusively hereafter, and it will be further extended to give a positive sign to any quantity or expression in which such adjectives as "up," "above," etc. are used, and vice versa for negative signs. Remember that the subscripts L and R refer to the beam segment lying respectively to the left and right of the exploratory section. With experience, the external force diagrams from which V and M are found can be visualized directly from the given beam loading.

---

[4]To avoid conflict with this rule, it is necessary to compute vertical shear in terms of the forces lying to the *left* of the exploratory section. If the forces to the *right* of the section were used, it would be necessary to take downward forces as positive so as to agree with the sign convention shown in Fig. 8-2.2.

ILLUSTRATIVE PROBLEMS

8-2.1 Write shear and moment equations for tie beam loaded as shown in Fig. 8-2.4, and sketch the shear and moment diagrams.

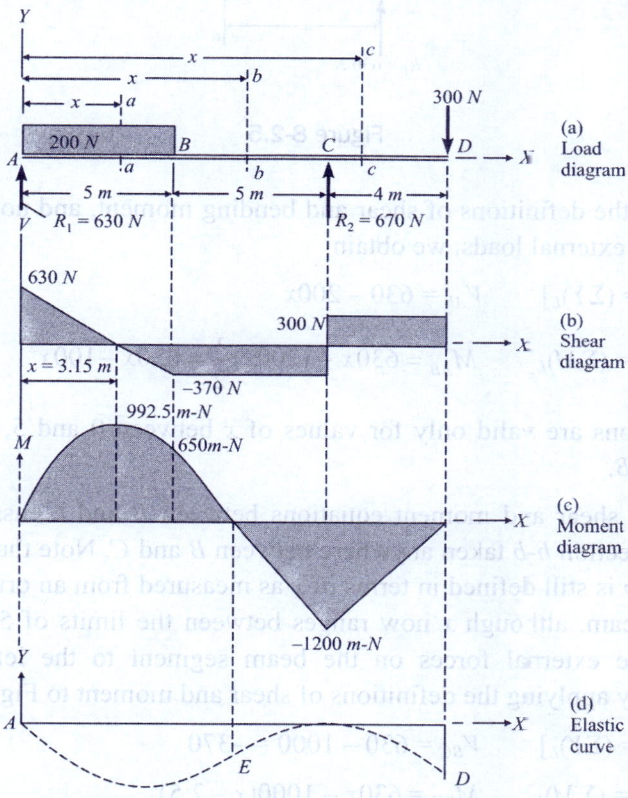

Figure 8-2.4

Solution

Begin by computing the reactions. Applying $\Sigma M_{R_2} = 0$ gives $R_1 = 630$ N and $\Sigma M_{R_1} = 0$ yields $R_2 = 670$ N. A check of these values is given by $\Sigma Y = 0$. The sections in the beam at which the loading conditions change are called *change of load points* and are designated by the letters *A*, *B*, *C*, and *D*.

For the beam segment to the left of a section *a-a* taken through the beam anywhere between *A* and *B*, the external loads on it appear as in Fig. 8-2.5.

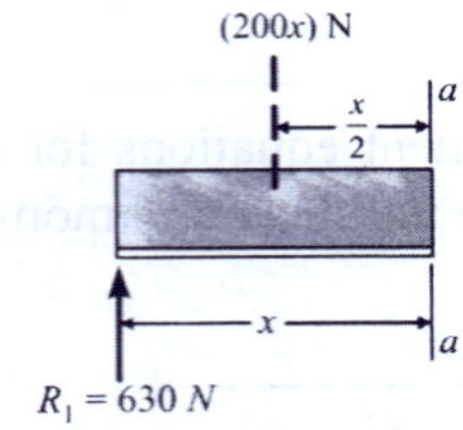

Figure 8-2.5

Applying the definitions of shear and bending moment, and noting that they apply only to external loads, we obtain

$$[V = (\Sigma Y)_L] \qquad V_{Ab} = 630 - 200x \qquad (a)$$

$$[M = (\Sigma M)_L] \qquad M_{AB} = 630x - (200x)\frac{x}{2} = 630x - 100x^2 \qquad (b)$$

These equations are valid only for values of x between 0 and 5, i.e., between points A and B.

To obtain shear and moment equations between B and C, assume another exploratory section b-b taken anywhere between B and C. Note that the location of section b-b is still defined in terms of x as measured from an origin at the left end of the beam, although x now ranges between the limits of 5 and 10. The effects of the external forces on the beam segment to the left of b-b are determined by applying the definitions of shear and moment to Fig. 8-2.6.

$$[V = (\Sigma Y)_L] \qquad V_{BC} = 630 - 1000 = -370 \qquad (c)$$

$$[M = (\Sigma M)_L] \qquad M_{BC} = 630x - 1000(x - 2.5)$$

$$= -370x + 2500 \qquad (d)$$

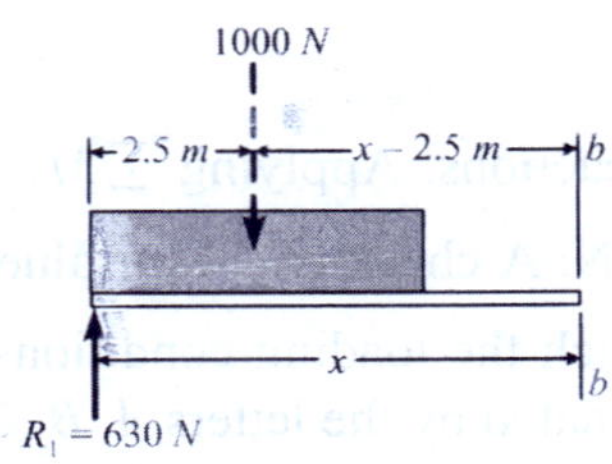

Figure 8-2.6

The shear and moment equations for segment *CD* are obtained similarly by passing a section *c-c* anywhere between *C* and *D*. The external loads acting on the beam to the left of this section are shown in Fig. 8-2.7 whence we obtain

$$[V = (\Sigma Y)_L] \qquad V_{CD} = 630 - 1000 + 670 = +300 \qquad \text{(e)}$$

$$[M = (\Sigma M)_L] \qquad M_{CD} = 630x - 1000(x - 2.5) + 670(x - 10)$$

$$= 300x - 4200 \qquad (f)$$

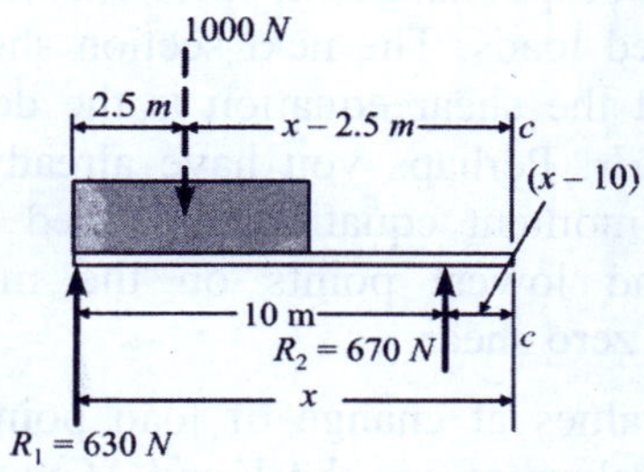

Figure 8-2.7

A simpler method of obtaining M_{CD} is to consider the external forces lying to the right of section *c-c* as is shown in Fig. 8-2.8, from which, noting that downward forces produce negative bending moment, we also obtain

$$[M = (\Sigma M)_R] \qquad M_{CD} = -300(14 - x) = 300x - 4200 \qquad (f')$$

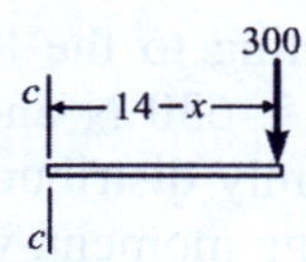

Figure 8-2.8

Summarizing, we have computed the shear equations by considering only the external forces lying to the left of any exploratory section, whereas the moment equations may be computed by taking moments about the exploratory section caused by the external loads which lie *either* to the left or to the right of the section. We have been careful to assign plus signs to *V* and *M* caused by upward-acting loads, and minus signs to *V* and *M* caused by downward-acting loads. Note further that Figs. 8-2.5, 8-2.6, 8-2.7, and 8-2.8 have been drawn only for explanatory purposes; you usually need not draw them since you will soon learn to visualize such diagrams directly from the original beam loading.

SHEAR AND MOMENT DIAGRAMS

Shear and moment diagrams are merely the graphic visualization of the shear and moment equations plotted on V-x and M-x axes, usually located below the loading diagram as in parts (b) and (c) of Fig. 8-2.4. We shall learn in the next section how to sketch them rapidly without using the corresponding equations. The diagrams are included here as a prelude to the following discussion.

The discontinuities in the shear diagram are joined by vertical lines drawn up or down to represent the abrupt changes in shear caused respectively by upward or downward concentrated loads. The next section shows why this is correct. We shall also prove that the shear equation is the derivative of the moment equations, i.e., $V = dM/dx$. Perhaps you have already observed this relation between the shear and moment equations obtained above. Because of this relation, the highest and lowest points on the moment diagram always correspond to sections of zero shear.

Shear and moment values at change of load points can be computed by substituting appropriate values of x in the V and M equations (a) to (f), but it is simpler and more direct to compute these numerical values by applying the fundamental definitions of V and M to specific sections. For example, the section of zero shear between A and B occurs because the weight of x meters of load applied at 200 N/m must balance the vertical shear of 630 N at A. Hence, we have

$$630 = 200x \quad \text{or} \quad x = 3.15 \text{ m}$$

The moment at this section of zero shear is computed by taking moments about the section of the forces acting to the left of the section. These forces consist of the upward reaction $R_1 = 630$ N and the downward load of 630 N caused by the length of the uniformly distributed load necessary to cause zero shear. From the definition of bending moment, we obtain

$$[M = (\Sigma M)_L]$$

$$\text{at} \qquad x = 3.15,\ M = 630(3.15) - 630\left(\frac{3.15}{2}\right) = 992.5 \text{ N-m}$$

A final point of interest is brought out in part (d) of Fig. 8-2.4 which shows the shape taken by the beam under the given loading, assuming the beam to be quite flexible. The beam between A and E is concave up, and between E and D it is concave down. Since for beam analysis, anything associated with *up* has a positive sign, it is not surprising that the moment diagram has positive values corresponding to the region AE, while for the portion ED, where the beam flexes concave down, the moment diagram has negative values. Sketching the shape of the flexed beam therefore provides a check of the sign of bending moment.

At point E, where the flexed beam changes its shape from concave up to concave down, we have what is called a *point of inflection*; it corresponds to the section of zero bending moment. Its position may be calculated by setting Eq. (d) equal to zero, which yields

$$[M_{BC} = 0] \qquad -370x + 2500 = 0 \qquad x = 6.76 \text{ m}$$

8-2.2 Write the shear and moment equations for the beam shown in Fig. 8-2.9 which carries a uniformly varying load increasing in intensity from zero at A to 240 N/m at B, a uniformly distributed load of 60 N/m over the region BC, and a concentrated load of 100 N at D. Also sketch shear and moment diagrams, specifying values at all change of loading positions at all points of zero shear. Neglect the weight of the beam itself.

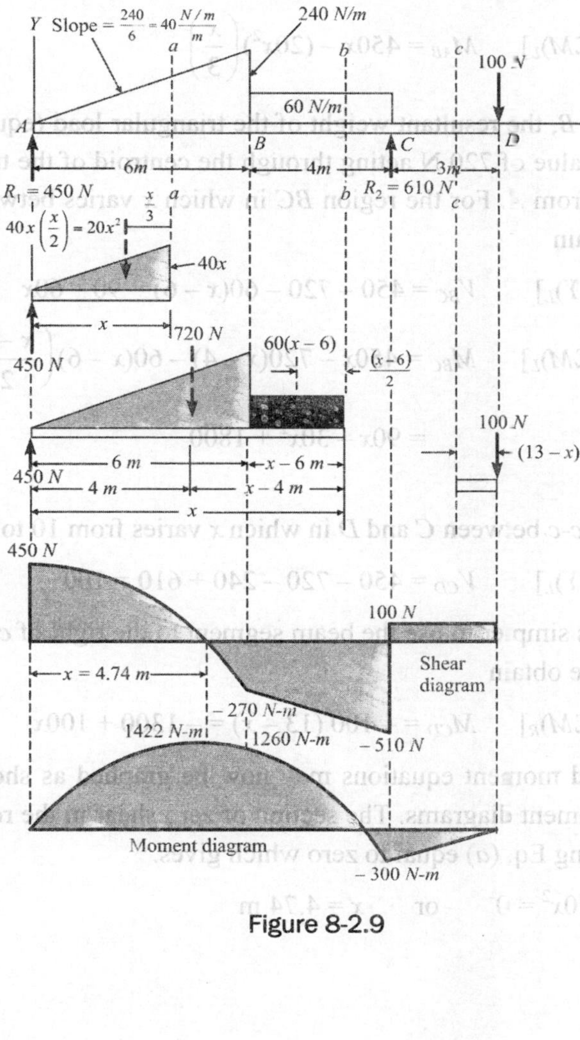

Figure 8-2.9

Solution

The reactions R_1 = 450 N and R_2 = 610 N are determined from a free-body diagram of the beam on which the distributed loads are replaced by their resultants and taking moment summations about R_2 and R_1.

As an aid in writing the shear and moment equations between each change of loading position, we show the external force diagram of loads lying to the left of sections *a-a* and *b-b* and also the load acting to the right of section *c-c*. Study them until you are sure you can visualize their counterparts on the original loading diagram and thereby eliminate drawing them in other problems.

For the region *AB* in which *x* varies from 0 to 6 m, we have

$$[V = (\Sigma Y)_L] \qquad V_{AB} = 450 - 20x^2 \tag{a}$$

$$[M = (\Sigma M)_L] \qquad M_{AB} = 450x - (20x^2)\left(\frac{x}{3}\right) \tag{b}$$

After passing *B*, the resultant weight of the triangular load (equal to its area) is constant at a value of 720 N acting through the centroid of the triangular load diagram at 4 m from *A*. For the region *BC* in which *x* varies between 6 and 10, we therefore obtain

$$[V = (\Sigma Y)_L] \qquad V_{BC} = 450 - 720 - 60(x - 6) = 90 - 60x \tag{c}$$

$$[M = (\Sigma M)_L] \qquad M_{BC} = 450x - 720(x - 4) - 60(x - 6)\left(\frac{x - 6}{2}\right)$$

$$= 90x - 30x^2 + 1800 \tag{d}$$

For a section *c-c* between *C* and *D* in whicn *x* varies from 10 to 13, we have

$$[V = (\Sigma Y)_L] \qquad V_{CD} = 450 - 720 - 240 + 610 = 100 \tag{e}$$

However, it is simpler to use the beam segment to the right of *c-c* to compute *M*. Doing this, we obtain

$$[M = (\Sigma M)_R] \qquad M_{CD} = -\,100\,(13 - x) = -\,1300 + 100x \tag{f}$$

The shear and moment equations may now be graphed as shown to obtain the shear and moment diagrams. The section of zero shear in the region *AB* may be found by setting Eq. (*a*) equal to zero which gives.

$$450 - 20x^2 = 0 \qquad \text{or} \qquad x = 4.74 \text{ m}$$

Since this value lies within the region *AB*, it is valid and may be substituted in Eq. (*b*) to find the bending moment of 1422 N-m at this position. Note that the bending moment at $x = 6$ m may also be found from Eq. (*b*) or more directly from the fundamental definition of bending moment as follows:

$$[M = (\Sigma M)_L] \qquad M_B = 450(6) - 720(2) = 1260 \text{ N-m}$$

Similarly, instead of using Eq. (*f*) to find the moment at *C*, we obtain directly

$$[M = (\Sigma M)_R] \qquad M_C = -100(3) = -300 \text{ N-m}$$

PROBLEMS

Determine the shear and moment equations for the beams in the following problems. Also draw shear and moment diagrams, specifying values at all change of loading positions and at all sections of zero shear. Neglect the weight of the beam itself.

8-2.3 Beam loaded as shown in Fig. P-8-2.3.

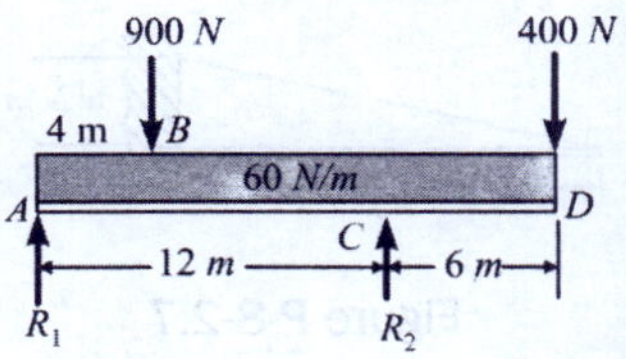

Figure P-8-2.3

$M_{BC} = 3600 - 230x - 30x^2$; max. $M = -3480$ N-m **Ans.**

8-2.4 Beam loaded as shown in Fig. P-8-2.4.

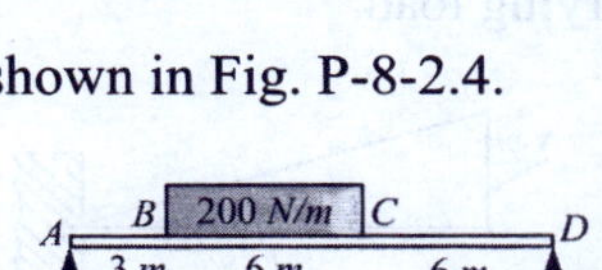

Figure P-8-2.4

max. $M = 3456$ N-m **Ans.**

8-2.5 Beam loaded as shown in Fig. P-8-2.5.

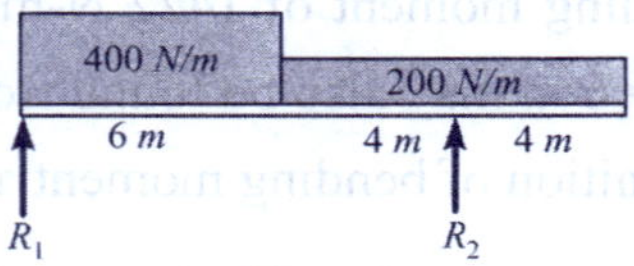

Figure P-8-2.5

max. $M = 3530$ N-m **Ans.**

8-2.6 Cantilever beam loaded as shown in Fig. P-8-2.6.

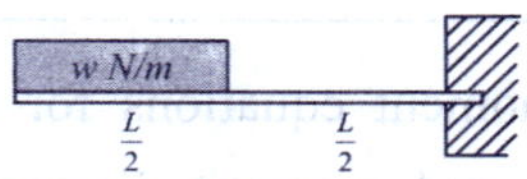

Figure P-8-2.6

8-2.7 Cantilever beam carrying the uniformly varying load as shown in Fig. P-8-2.7.

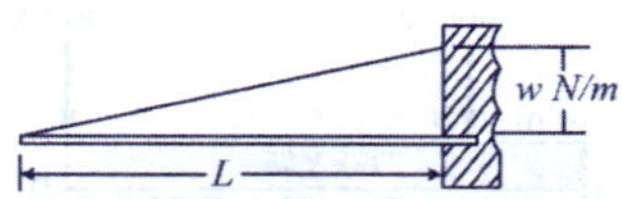

Figure P-8-2.7

8-2.8 Cantilever beam carrying a distributed load varying from w N/m at the free end to zero at the wall, as shown in Fig. P-8-2-8.

Hint: Given load is equivalent to a uniformly distributed load reduced by a uniformly varying load.

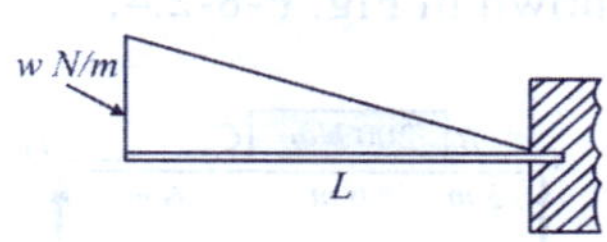

Figure P-8-2.8

$$M = \frac{wx^3}{6L} - \frac{wx^2}{2}$$ **Ans.**

8-2.9 Beam carrying the uniformly varying load as shown in Fig. P-8-2.9.

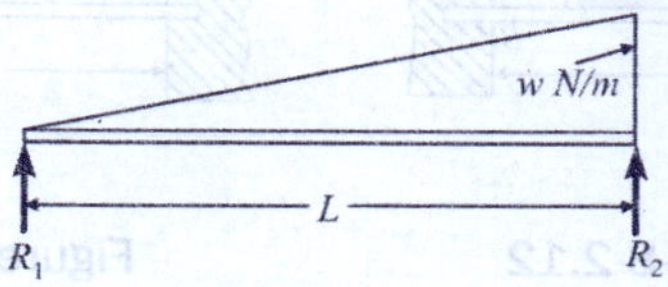

Figure P-8-2.9

$$M = \frac{wL^2}{9\sqrt{3}} = \frac{2WL}{9\sqrt{3}} \qquad \textit{Ans.}$$

8-2.10 Beam loaded as shown in Fig. P-8-2.10.

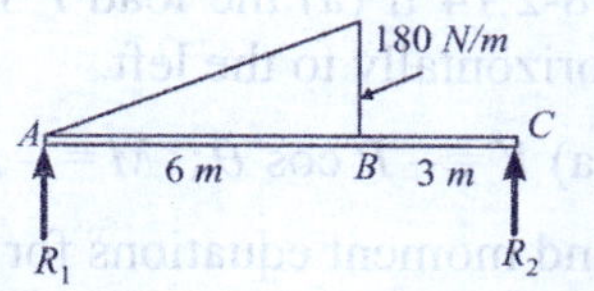

Figure P-8-2.10

$$M_{AB} = 300x - 5x^3 \qquad \textit{Ans.}$$

8-2.11 A total distributed load of 30,000 N is supported by a beam with a uniformly distributed reaction as shown in Fig. P-8-2.11.

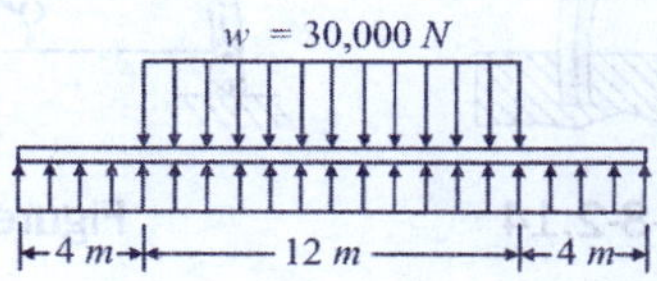

Figure P-8-2.11

$$M = 30{,}000 \text{ N-m} \qquad \textit{Ans.}$$

8-2.12 The bending moments at the ends of the perfectly restrained beam shown in Fig. P-8-2.12 are $M_A = -4150$ M-m and $M_B = -5450$ N-m. Determine the maximum positive bending moment developed in the beam.

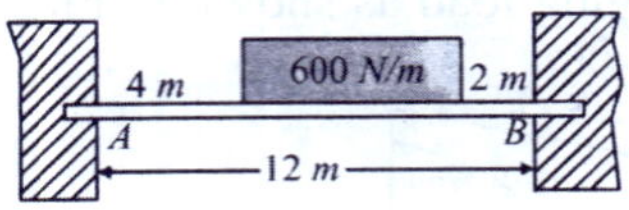

Figure P-8-2.12

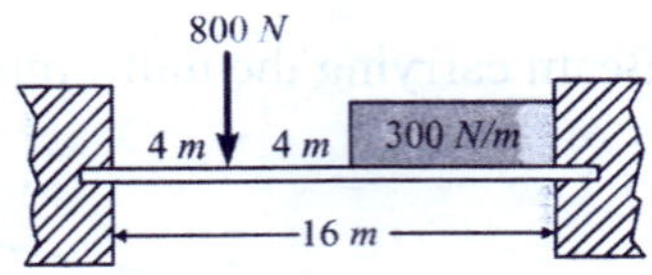

Figure P-8-2.13

8-2.13 Determine the maximum positive bending moment in the perfectly restrained beam shown in Fig. P-8-2.13 if the bending moments at the ends are $M_A = -3130$ m-N and $M_B = -3530$ m-N.

$M = 1546$ N-m **Ans.**

8-2.14 Write the shear and moment equations for the built-in circular bar shown in Fig. P-8-2.14 if (a) the load P is vertical as shown and (*b*). the load P acts horizontally to the left.

(a) $V = -P\cos\theta$; $M = -PR\sin\theta$ **Ans.**

8-2.15 Write the shear and moment equations for the semicircular arch shown in Fig. P-8-2.15 if (a) the load P is vertical as shown and (*b*) the load P is applied horizontally to the left at the top of the arch.

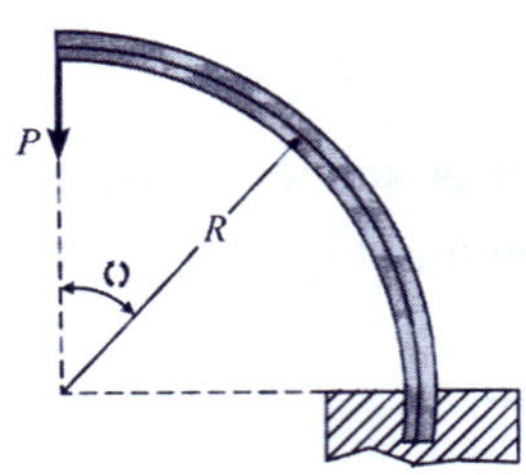

Figure P-8-2.14

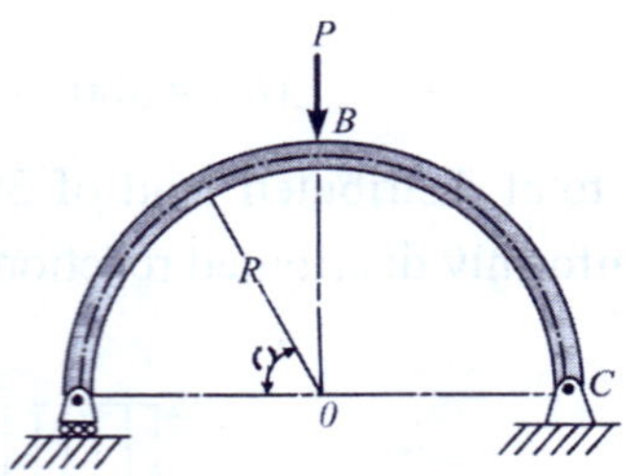

Figure P-8-2.15

(a) $M_{AB} = \frac{1}{2}PR(1 - \cos\theta)$; $M_{BC} = \frac{1}{2}PR(1 + \cos\theta)$ **Ans.**

8-3 RELATIONS BETWEEN LOAD, SHEAR AND BENDING MOMENT

The relations existing between the loads, shears, and bending moments in any beam provide a method of constructing shear and moment diagrams without writing shear and moment equations. These relations are not independent of the basic definitions of shear and moment; instead they supplement and are used in conjunction with them.

We begin by considering the beam in Fig. 8-3.1*a* which is assumed to carry any general loading. The free-body diagram of a segment of this beam of length *dx* is shown magnified in part (b). As we showed in the preceding section, the effect of the loads to the left of this segment reduces to the shear V and the moment M, and the loads to the right of this segment produce the slightly different values of shear and moment $V + dV$ and $M + dM$. Although the loading is variable, it may be assumed constant at the intensity of w N/m over the small length dx, thereby producing the upward load $w\,dx$ which completes the free-body diagram.

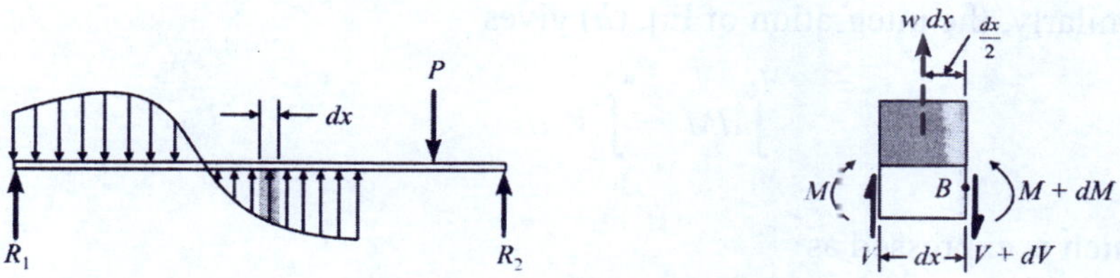

(a) (b)

Figure 8-3.1

Applying the conditions of static equilibrium to Fig. 8-3.1*b*, a summation of vertical forces yields

$$[\Sigma Y = 0] \qquad V + w\,dx - (V + dV) = 0$$

which reduces to

$$dV = w\,dx \tag{a}$$

From a moment summation about point B, we have

$$[\Sigma M_B = 0] \qquad M + V\,dx + (w\,dx)\frac{dx}{2} - (M + dM) = 0$$

The third term in this equation is the square of a differential which is negligible in comparison with the other terms; hence, the equation reduces to

$$dM = V\,dx \tag{b}$$

Integrating Eq. (*a*), we obtain

$$\int_{V_1}^{V_2} dV = \int_{x_1}^{x_2} w\,dx$$

in which the limits are the shear V_1 at position x_1 and the shear V_2 at position x_2. The left-hand term is easily integrable; it reduces to $V_2 - V_1$ and represents the change in shearing force between sections x_2 and x_1. We denote this change in shear as ΔV. In the right-hand term the product $w\ dx$ represents the area of an element of the load diagram like that shown shaded in Fig. 8-3.1*a*.

Hence, the definite integral $\int_{x_1}^{x_2} w\ dx$, which means the summation of such terms, represents the area under the load diagram between positions x_1 and x_2. Therefore, the integration of Eq. (a) is interpreted as

$$\mathbf{V_2 - V_1 = \Delta V = (Area)_{load}} \qquad (8\text{-}3.1)$$

Similarly, the integration of Eq. (*b*) gives

$$\int_{M_1}^{M_2} dM = \int_{x_1}^{x_2} V\ dx$$

which is expressed as

$$\mathbf{M_2 - M_1 = \Delta M = (Area)_{shear}} \qquad (8\text{-}3.2)$$

inasmuch as the product $V\ dx$ in the right-hand integral represents the area of an element under the shear diagram. Therefore, the integral itself is equivalent to the area under the shear diagram between positions x_1 and x_2. Expressed in words, Eq. (8-3.2) expresses that the change in bending moment ΔM between any two sections is equal to the area of the shear diagram for this interval.

Positive shearing forces are plotted upward from the X axis; hence, positive shear areas are those which lie above the X axis and represent increases in the bending moment. The load diagram, however, is usually drawn with the loads on top of the beam because this is their natural position; as a consequence, the area of such downward-acting loads is considered negative and represents *decreases* in the shearing force,

Equations (8-3.1) and (8-3.2) provide a convenient method of computing the changes in shear and moment and also the numerical values of shear and moment at any section, as will be demonstrated in the illustrative problems. Of almost equal importance are the following variations of Eqs. (*a*) and (*b*) which enable us to sketch the shapes of the shear and moment diagrams:

$$w = \frac{dV}{dx} = \textbf{slope of shear diagram} \qquad (8\text{-}3.3)$$

$$V = \frac{dM}{dx} = \text{slope of moment diagram} \tag{8-3.4}$$

As an application of these equations, consider the simply supported beam carrying the varying loading shown in Fig. 8-3.2a. Since positive slopes are directed up to the right and negative slopes are directed down to the right, i.e.,

$$+\text{ slope} = \nearrow$$
$$-\text{ slope} = \searrow$$

we observe from Eq. (8-3.3) that the shear diagram in Fig. 8-3.2b must slope continuously down to the right to correspond with the downward or negative ordinates of the load diagram. The slope varies directly with the corresponding ordinate of the load diagram, being steepest where the load ordinate is maximum and horizontal (or of zero slope) at the ends where the intensity of loading is zero.

Similarly, by means of Eq. (8-3.4), the slope and shape of the moment diagram in Fig. 8-3.2c are determined by the corresponding ordinates of the shear diagram which, being positive but decreasing in magnitude for the left portion, determine that the moment diagram slopes continuously up to the right with decreasing inclination. The slope becomes zero where the shear ordinate is zero. Note also that after the shear ordinates change sign and become increasingly larger negatively, the moment diagram slopes correspondingly more steeply down to the right. These conditions establish a maximum moment at the ordinate of zero shear.

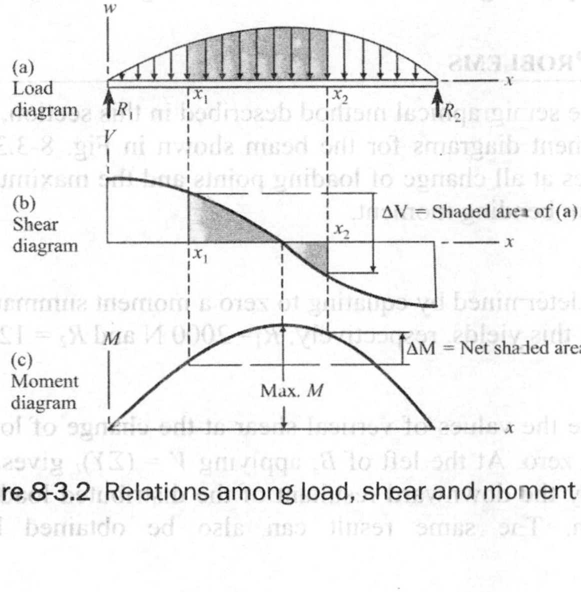

Figure 8-3.2 Relations among load, shear and moment diagrams.

The changes in shear (ΔV) and moment (ΔM) defined by Eqs. (8-3.1) and (8-3.2) are indicated in (b) ad (c) of Fig. 8-3.2. The shaded negative area of the load diagram determines ΔV to be negative or directed downward; in the shear diagram, the excess of positive over negative area in the region between x_1 and x_2 determines the positive change in moment ΔM, which is therefore directed upward.

A summary of the principles presented here and in Section 8-2 suggests the following procedure for the construction of shear and moment diagrams:

1. Compute the reactions.
2. Compute values of shear at the change of load positions, using either $V = (\Sigma Y)_L$ or $\Delta V = (\text{Area})_{\text{load}}$.
3. Sketch the shear diagram, determining its shape from Eq. (8-3.3); i.e., the intensity of the load ordinate equals the slope at the corresponding ordinate of the shear diagram.
4. Locate the sections of zero shear.
5. Compute values of bending moment at the change of load points and at the sections of zero shear, using either $M = (\Sigma M)_L = (\Sigma M)_R$ or $\Delta M = (\text{Area})_{\text{shear}}$, whichever is more convenient.
6. Sketch the moment diagram through the ordinates of the bending moments computed in step 5. The shape of the diagram is determined from Eq. (8-3.4); i.e., the intensity of the shear ordinate equals the slope at the corresponding ordinate of the moment diagram.

ILLUSTRATIVE PROBLEMS

8.3.1 Using the semigraphical method described in this section, sketch shear and moment diagrams for the beam shown in Fig. 8-3.3, computing the values at all change of loading points and the maximum shear and maximum bending moment.

Solution

The reactions are determined by equating to zero a moment summation about R_2 and then about R_1; this yields, respectively, $R_1 = 2000$ N and $R_2 = 1200$ N.

Shear Values

We next determine the values of vertical shear at the change of load positions. At A, the shear is zero. At the left of B, applying $V = (\Sigma Y)_L$ gives the shear as – 600 N caused by the downward resultant of the distributed load of 200 N/m applied for 3 m. The same result can also be obtained by applying

$\Delta V =$ (Area)$_{load}$ in the interval AB. The area of the downward load is – 200(3) = – 600 N and hence the shear ordinate to the left of B has decreased 600 N from the zero shear at A to yield a net value of – 600 N. The concentrated load reaction at B causes the shear at B to increase abruptly by 2000 N to a net positive shear ordinate of 1400 N at the right of B.

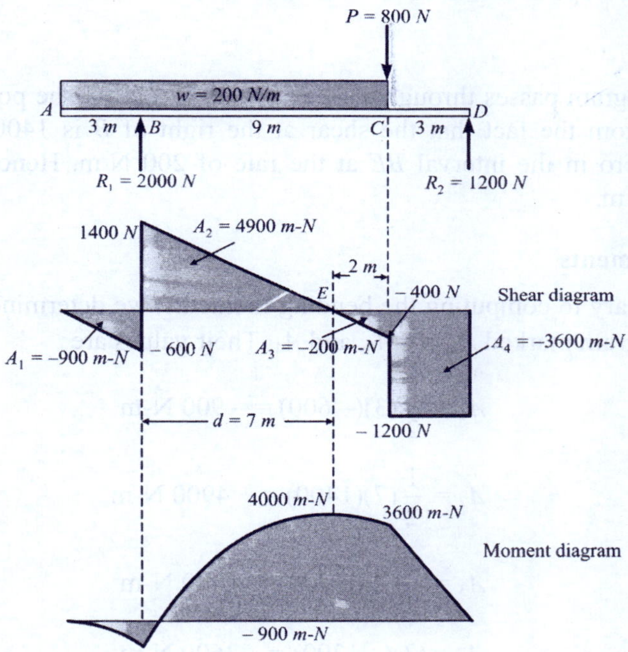

Figure 8-3.3 Load, shear and moment diagrams.

Between B and C, the area of the load diagram is – 200(9) = – 1800 N. which by Eq. (8-3.4) represents the change in shear between B and C. The net shear ordinate at the left of C is therefore $V_c = V_B + \Delta V$ = 1400 – 1800 = – 400 N. At C, the concentrated load of 800 N changes the shear ordinate to – 1200 N at the right of C. The shear ordinate stays constant at this value between C and D since there is no load in this interval. At D, the upward reaction of 1200 N reduces the shear ordinate to zero.

Shape of Shear Diagram

The shape of the diagram connecting these shear ordinates is determined from Eq. (8-33) which specifies that the slope of the shear diagram is equal to the corresponding ordinates of the load diagram. Thus, between A and B, the load intensity is constant and downward (or negative); hence, the slope in this interval is constant and down to the right. Similarly, between B and C, the load intensity is constant and negative; therefore, the slope of the shear diagram here also is constant and down to the right. Note that the slopes in the intervals AB and BC are parallel because they are each determined by the same load intensity. Finally, between C and D, the intensity of loading is zero and the corresponding slope is also zero (a horizontal line).

Zero Shear

The shear diagram passes through zero at B, and also at E. The position of E is determined from the fact that the shear at the right of B is 1400 N which is reduced to zero in the interval BE at the rate of 200 N/m. Hence $BE = d$ = 1400/200 = 7 m.

Bending Moments

As a preliminary to computing the bending moments, we determine the areas of the shear diagram marked A_1, A_2, A_3, and A_4. Their values are

$$A_1 = \frac{1}{2}(3)(-600) = -900 \text{ N-m}$$

$$A_2 = \frac{1}{2}(7)(1400) = +4900 \text{ N-m}$$

$$A_3 = \frac{1}{2}(2)(-400) = -400 \text{ N-m}$$

$$A_4 = (3)(-1200) = -3600 \text{ N-m}$$

According to Eq. (8-3.2), the change in bending moment between any two sections equals the corresponding area of the shear diagram; hence, since the bending moment is zero at A (there are no loads to the left of A to cause a bending moment), the bending moment at B is given by A_1, or $M_B = -900$ m-N.

Similarly, the bending moment at E is

$$M_E = M_B + \Delta M = A_1 + A_2 = -900 + 4900 = +4000 \text{ N-m}$$

The bending moment at C can also be computed as the sum of the areas A_1, A_2, and A_3 giving M_C = 3600 m-N; however, small errors arising from neglecting sufficient significant figures in computing similar areas may cause a cumulative error. Hence, at sections near the right end of the beam, it is usually preferable to use the shear area to the right of such sections, or to apply the basic definition $M = (\Sigma M)_R$ whence in terms of the loads acting to the right of C we obtain $M_C = 1200(3) = 3600$ m-N. The correlation between this result and the area A_4 is evident if we observe that A_4 represents the amount by which the moment at C changes to become zero at D; i.e.,

$$M_D = M_C + \Delta M = M_C + A_4$$

or

$$0 = M_C + (-3600) \quad \text{and} \quad M_C = 3600 \text{ N-m}$$

Whenever the change in bending moment between the ends of a beam is zero, as in this problem, the net area under the shear diagram is also zero; in other words, there must be as much positive as there is negative shear area. This provides a useful check on the accuracy of all intermediate values of bending moments computed from the area of a shear diagram.

Shape of Moment Diagram

After these bending moments are plotted as ordinates, the shape of the moment diagram connecting them is determined from Eq. (8-3.4); i.e., the intensity of the shear ordinate equals the slope at the corresponding ordinate of the moment diagram. Thus, as the shear ordinates between A and B change linearly from zero to – 600 N, the slopes at corresponding ordinates of the moment diagram change from horizontal or zero slope at A to increasingly steeper negative slopes as we pass from A to B, thereby producing a second-degree curve concave downward as shown. In other words, the tangents the moment curve have increasingly steeper slopes directed down ward to the right.

The sudden change in shear at B from – 600 N to 1400 N causes the moment curve there to slope abruptly upward to the right; it becomes less steep, and eventually horizontal, as we move from B to F because the corresponding positive shear ordinates decrease uniformly to zero. From E to C, however, the slope is again increasingly steeper and directed downward to the right as the corresponding shear ordinates change to increasingly negative values.

At C, the shear ordinate changes abruptly from – 400 N to – 1200 N, then remains constant until D. The slope of the moment diagram is correspondingly abruptly steeper and constant, resulting in the straight line joining the moment ordinates at C and D.

The above discussion has been presented in detail in order to emphasize the correlation between ordinates of the shear diagram and corresponding slopes of the moment diagram. Several conclusions for future use should be evident. An abrupt change in shear creates an abrupt change in the slope of the moment curve as at C, and where the shear changes sign abruptly, there is a cusp in the moment diagram as at B. Also where the shear diagram is constant because only concentrated loads are involved, as in the region CD, the moment diagram consists of straight lines, whereas for an interval in which the shear diagram varies uniformly because of a uniformly distributed load, the moment diagram is a parabolic arc with a vertical axis of symmetry at the section of zero shear, as at E.

8.3.2 Without writing shear or moment equations, sketch shear and moment diagrams for the beam in Fig. 8-3.4, and compute values at all change of loading positions and the values of maximum shear and maximum moment.

Solution

We begin by computing the reactions. Replacing the uniformly distributed reaction between C and D by its resultant R_2 and equating a moment summation about R_1 to zero, we obtain $R_2 = 420$ N. Dividing R_2 by the length of 4 m over which it is assumed to be uniformly distributed gives the upward intensity of this reaction as 105 N/m. We next set a moment summation about R_2 equal to zero to obtain $R_1 = 240$ N.

Shear Diagram

The shear diagram stats with an ordinate of 240 N caused by the reaction R_1. Applying $\Delta V = (\text{Area})_{\text{load}}$ between A and B, the change in shear is equal to the area of the triangular load diagram $\frac{1}{2}(9)(-120) = -540$ N; this reduces the vertical shear at B to $-$ 300 N. The shape of the shear diagram from A to B varies from zero slope at A to increasing steeper negative slopes corresponding to the increasingly greater downward intensity of the loading.

Between B and C no load is applied hence the shear stays constant and the corresponding portion of the shear diagram is horizontal.

From C to D, the loading is at a constant upward rate of 105 N/m, creating a positive change of shear of 420 N equal to the area of the load diagram in this interval, and a constant upward slope of the shear diagram as shown. This shear change of + 420 N is added to the shear of $-$ 300 N at C to produce the shear of + 120 N at D. From D to E, the loading is zero which means that the shear stays constant and the shear diagram is horizontal. The concentrated load of 120 N at E reduces the shear abruptly to zero.

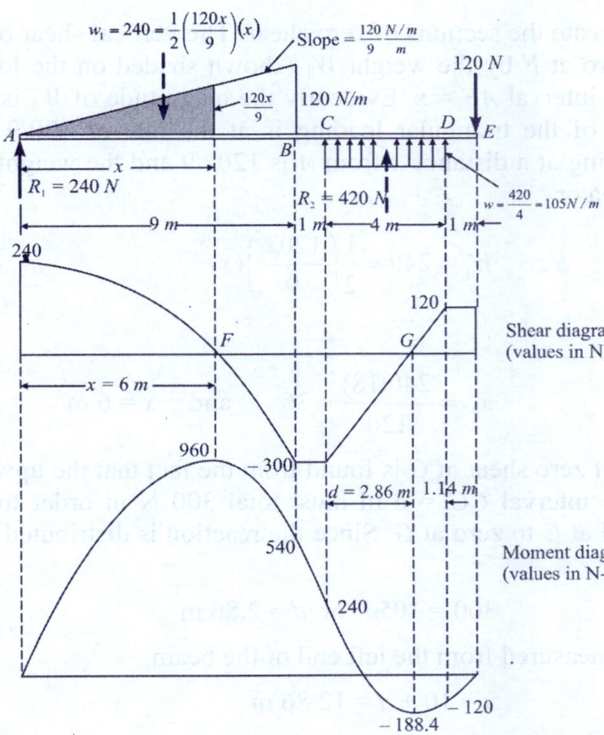

Figure 8-3.4 Load, shear and moment diagrams.

Zero Shear

Before we locate the positions of zero shear at F and G on the shear diagram, consider the effect of narrowing the distance over which the reaction R_2 is distributed. If points C and-D are moved an equal amount toward each other, this narrowing will not change the magnitude or position of R_2. However, a reduction in the distance CD will increase the intensity at which R_2 is distributed and cause a corresponding increase in the slope of the shear diagram in the narrowed interval CD. For the extreme case in which the reaction is distributed over an infinitesimal width – i.e., becomes a concentrated force – the intensity of loading is infinitely upward, and the corresponding slope of the shear diagram is vertically upward, as is the case at the reaction R_1. Similarly, for a concentrated downward load at E, the intensity of loading is infinitely downward and the corresponding slope of the shear diagram is vertically downward. *This explains why a concentrated load causes an abrupt vertical change in the shear diagram.*

Let us now locate the sections of zero shear. The vertical shear of 240 N at A is reduced to zero at F by the weight W_1 (shown shaded on the load diagram) applied over the interval $AF = x$. Evidently the magnitude of W_1 is also 240 N. Since the slope of the triangular loading is at the rate of 120/9 N/m/m, the intensity of loading at a distance x from A is $120x/9$ and the weight W_1 is equal to the shaded area or

$$W_1 = 240 = \frac{1}{2}\left(\frac{120x}{9}\right)(x)$$

from which

$$x^2 = \frac{240(18)}{120} = 36 \qquad \text{and} \qquad x = 6 \text{ m}$$

The section of zero shear at G is found from the fact that the upward reaction applied over the interval CG = d m must total 300 N in order to change the shear of – 300 N at C to zero at G. Since the reaction is distributed at 105 N/m, we have

$$300 = 105d \text{ or } d = 2.86 \text{ m}$$

or in terms of x measured from the left end of the beam,

$$x = 10 + d = 12.86 \text{ m}$$

Bending Moment

The moment at F where $x = 6$ m is found by applying the definition of bending moment.[5] Note that the shaded part of the triangular load applied between A and F totals 240 N and acts at the centroid of the triangular area, i.e., at $\frac{1}{3}$ of 6 m from F. Thus, we obtain

$$[M = (\Sigma M)_L] \qquad M_F = 240(6) - 240\left(\frac{6}{3}\right) = 960 \text{ N-m}$$

Similarly, the moment at B where $x = 9$ m is found to be

$$[M = (\Sigma M)_L] \qquad M_B = 240(9) - 540\left(\frac{9}{3}\right) = 540 \text{ N-m}$$

---

[5]It is better not to use ΔM (Area)$_{shear}$ when the shear diagram is curved since then the areas under such curved portions of the shear diagram are not too easily computed, especially in the interval FB.

The moment at C can also be computed from this basic definition, but whenever the shear diagram consists of straight lines (either horizontal or inclined), it is usually simpler to apply $\Delta M = (\text{Area})_{\text{shear}}$. For example, between B and C, the area of the shear diagram is a rectangle from which $\Delta M = -300(1) = -300$ N-m. Hence, if the bending moment changes by -300 N-m in the interval BC, the bending moment at C is

$$[M_C = M_B + \Delta M] \qquad M_C = 540 + (-300) = 240 \text{ N-m}$$

The bending moment at D may be found from $M_D = (\Sigma M)_R = -120(1) = -120$ N-m; the negative sign comes from the downward load to the right of D. However, to emphasize the significance of the shear diagram, note that the area of the shear diagram between D and E equals 120 N-m which represents the change in moment in this interval. Since this change in moment is positive, it follows that to cause zero moment at the free end E, the moment at D must be -120 N-m.

Similarly, the area of the shear diagram between G and D equals $\frac{1}{2}(120)(1.14) = 68.4$ N-m. Since this is a positive increase in bending moment, the moment at G must be smaller by this amount than the moment at D or

$$M_G = -120 - 68.4 = -188.4 \text{ N-m}$$

Sketching the flexed shape of the beam, as in Fig. 8-2.4d for Prob. 8-2.1, will show it to be concave downward at G and D, which is further verification of the negative bending moment signs at these sections.

Shape of the Moment Diagram

After we plot the moment values obtained above, the shape of the moment diagram connecting them is determined from E.q (8-3.4).

Since the shear ordinates are positive and decrease to zero as we move from A to F, the moment curve has correspondingly positive slopes (i.e., directed upward to the right) that decrease to zero slope at F.

Between F and B, the vertical shear is increasingly negative, and hence the corresponding slopes of the moment curve become increasingly steeper downward to the right until the steepest slope is reached at section B. Between B and C, the shear stays constant; therefore, the slope of the moment curve is constant, being represented by the straight line that joins sections B and C.

Between sections C and G, the shear is negative, becoming zero at G; hence, the slope of the moment curve is negative and gradually reduces to zero (i.e., horizontal) at G. Similarly, the increasing positive shear between G and D results in an increasingly positive slope (directed upward to the right), up to D. Between D and E the slope remains constant because the vertical shear is constant between D and E.

The moment curve between C and D is a symmetrical parabola with its vertex at G because, at equal distances to either side of G, the shear is numerically equal but of opposite sign, thereby producing equal slopes oppositely directed. The moment curve between A and B, however, is *not* symmetrical about F because the shear ordinates do not have equal values at equal distances on either side of F; here the moment curve is actually a third-degree curve.

PROBLEMS

Without writing shear or moment equations, draw shear and moment diagrams for the beams loaded as described in the following problems. Give numerical values at all change of loading positions and at all sections of zero shear. (Problems 8-2.3 to 8-2.15 may also be assigned for solution by the semigraphical method described in this section.)

8-3.3 A beam loaded as shown in Fig. P-8-3.3.

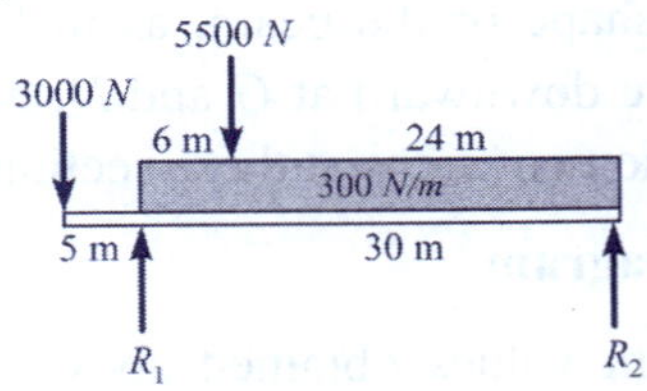

Figure P-8-3.3

8-3.4 Beam loaded as shown in Fig. P-8-3.4.

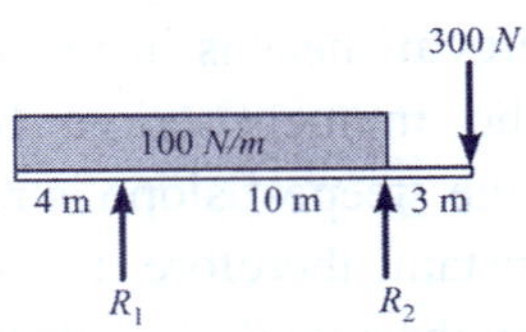

Figure P-8-3.4

8-3.5 Beam loaded as shown in Fig. P-8-3.5.

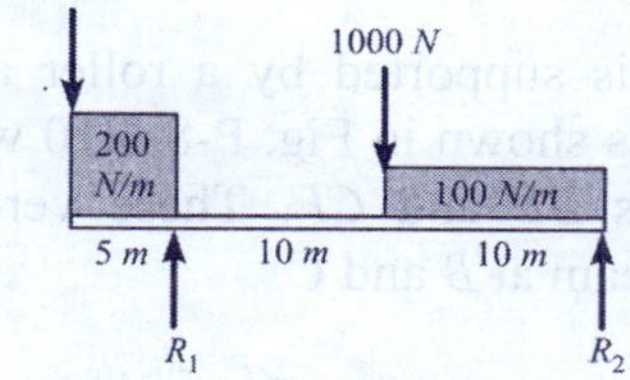

Figure P-8-3.5

max. $M = 5000$ N-m ***Ans.***

8-3.6 A beam loaded as shown in Fig. P-8-3.6 consists of two segments joined by a frictionless hinge at which the bending moment is zero.

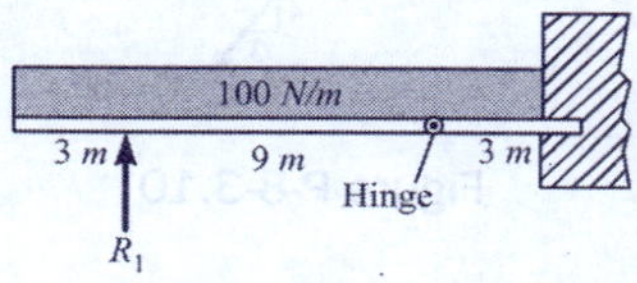

Figure P-8-3.6

max. $M = -1650$ N-m ***Ans.***

8-3.7 A beam supported on three reactions as shown in Fig. P-8-3.7 consists of two segments joined by a frictionless hinge at which the bending moment is zero.

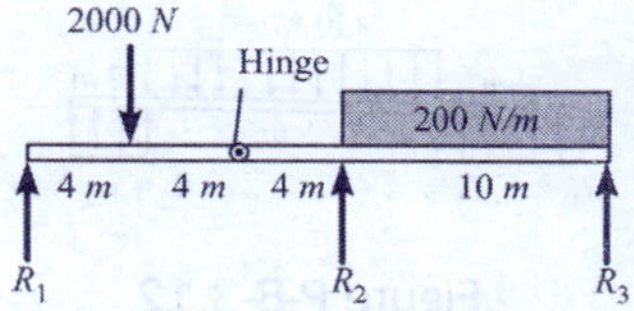

Figure P-8-3.7

8-3.8 The beam in Fig. P-8-3.8 is loaded as shown by clockwise couples M_A 300 N-m and $M_B = 600$ N-m. Find the maximum bending moment and sketch the shape of the flexed beam.

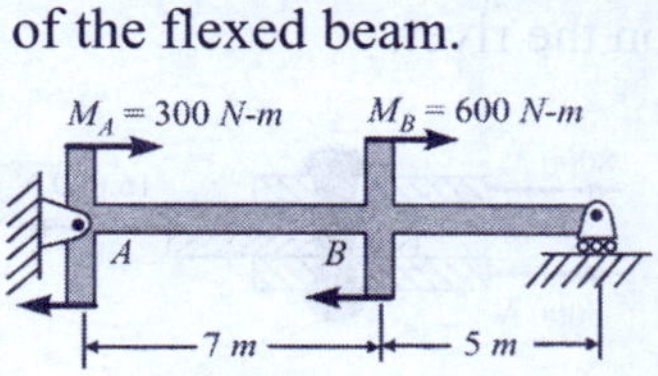

Figure P-8-3.8

max. $M = 375$ N-m ***Ans.***

8.3.9 Repeat the previous problem if the couple at A acts counterclockwise instead of clockwise.

8.3.10 A beam $ABCD$ is supported by a roller at A and a hinge at D. It supports the loads shown in Fig. P-8-3.10 which act at the ends of the vertical members BE and CF. These vertical members are rigidly attached to the beam at B and C.

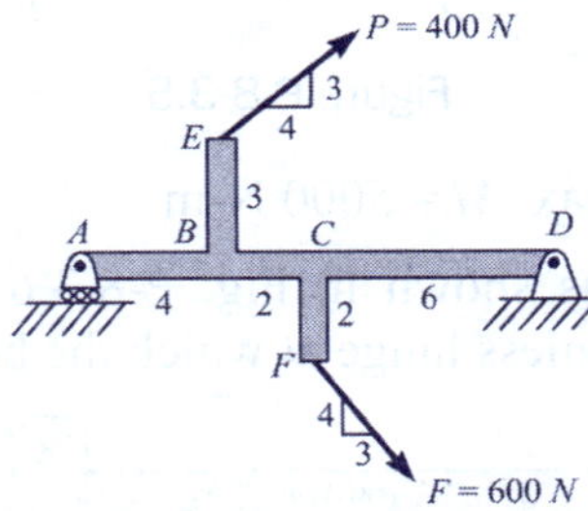

Figure P-8-3.10

max. $M = 1800$ N-m ***Ans.***

8-3.11 Solve the previous problem if P is changed to 600 N and F is changed to 400 N, all other data remaining unchanged.

8-3.12 A distributed load is supported by two distributed reactions as shown in Fig. P-8-3.12.

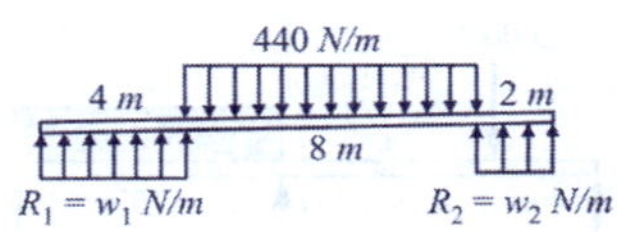

Figure P-8-3.12

8-3.13 In the portion of a butt joint shown in Fig. P-8-3.13, a rivet secures 15 mm cover plates to a 20 mm main plate. Assuming the loads are uniformly distributed along the rivet, determine the maximum bending moment acting on the rivet.

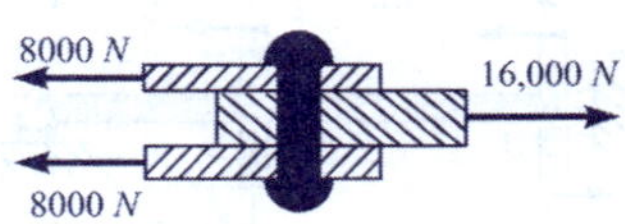

Figure P-8-3.13

8.3.14 A railroad tie is supported by a uniformly distributed reaction and carries two uniformly distributed loads each totaling 12 KN applied as shown in Fig. P-8-3.14. Find the maximum bending moment in the tie.

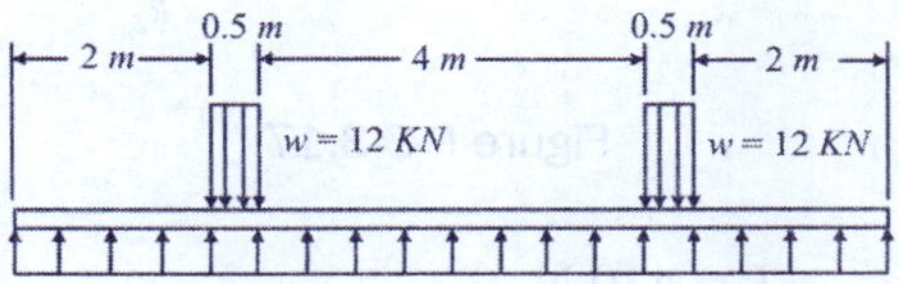

Figure P-8-3.14

8-3.15 In the beam shown in Fig. P-8-3.15, determine P so that the moments over the supports equal the moment at midspan.

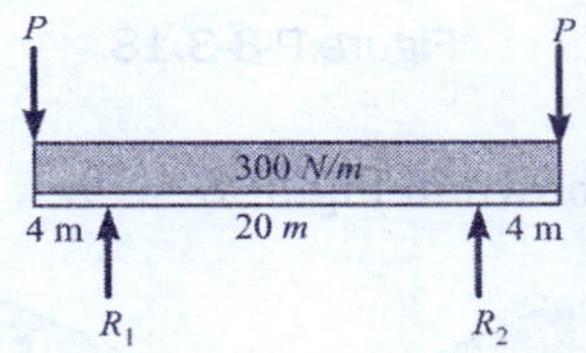

Figure P-8-3.15

$P = 1275$ N; max. $M = 7500$ N-m ***Ans.***

8-3.16 Determine the values of P and w in the beam shown in Fig. P.8-3.16 so that the maximum positive and negative bending moments are each numerically equal to 14,400 N-m.

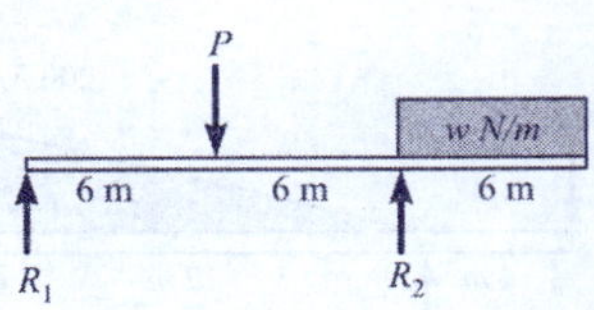

Figure P-8-3.16

8.3.17 For the beam loaded as shown in Fig. P-8-3.17, find the value of P to cause a maximum bending moment of 1800 N-m.

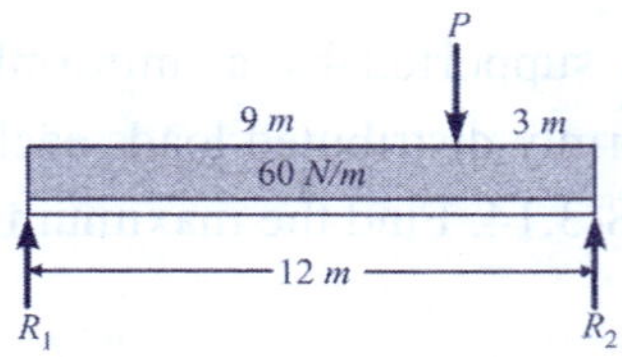

Figure P-8-3.17

$$P = 420 \text{ N} \qquad \textbf{\textit{Ans.}}$$

8.3.18 Beam carrying the triangular load shown in Fig. P-8-3.18.

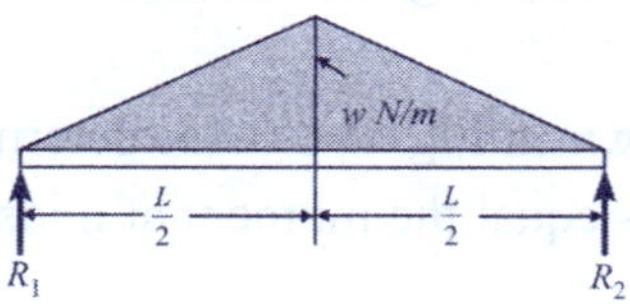

Figure P-8-3.18

8.3.19 Beam loaded as shown in Fig. P-8-3.19.

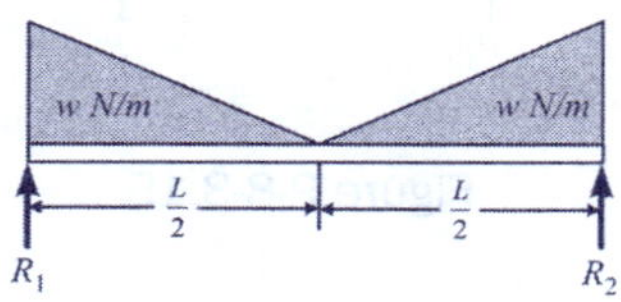

Figure P-8-3.19

$$\text{max. } M = wL^2/24 \qquad \textbf{\textit{Ans.}}$$

8.3.20 Find the maximum moment in the beam shown in Fig. P-8-3.20.

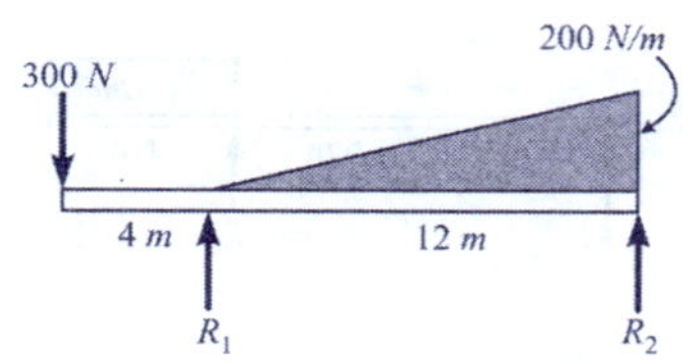

Figure P-8-3.20

8-3.21 The beam shown in Fig. P-8-3.21 carries a uniformly varying load and is simply supported at the left end and by a uniformly distributed reaction at the right end. Sketch shear and moment diagrams.

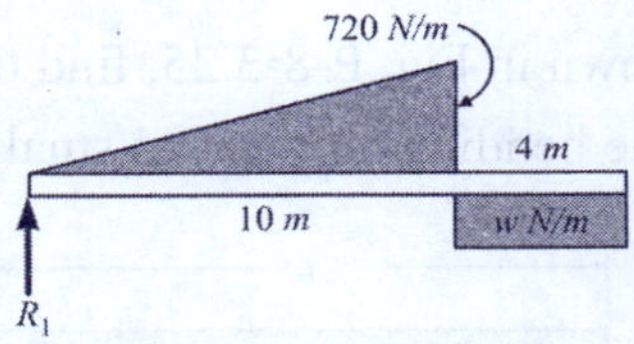

Figure P-8-3.21

max. $M = 7110$ N-m **Ans.**

8-3.22 A beam carrying the triangular load shown in Fig. P-8-3.22 is supported on a uniformly distributed reaction.

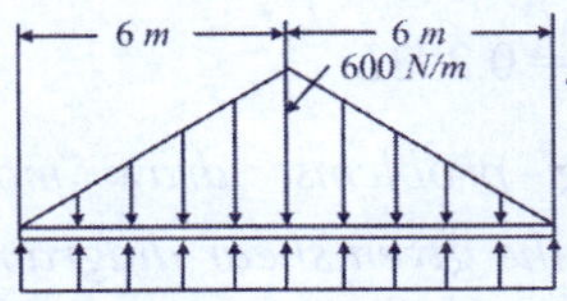

Figure P-8-3.22

8.3.23 Beam loaded and supported as shown in Fig. P-8-3.23.

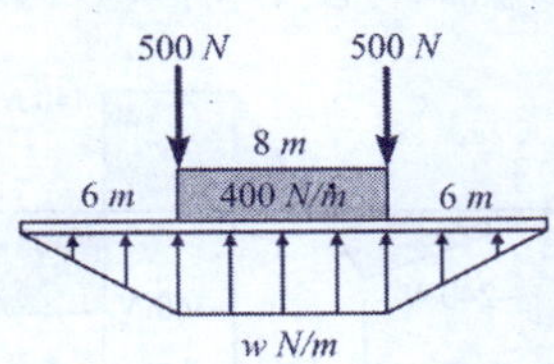

Figure P-8-3.23

8-3.24 In the overhanging beam shown in Fig. P-8-3.24, determine the length a of the overhang so that the bending moment in the beam will be as small as possible.

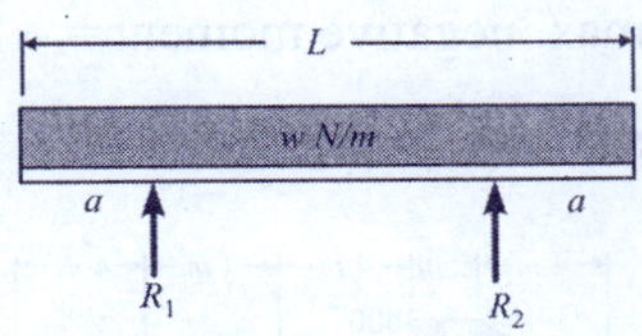

Figure P-8-3.24

$a = 0.207L$

8-3.25 For the beam shown in Fig. P-8-3.25, find the length a of the overhang that will make the bending moment as small as possible.

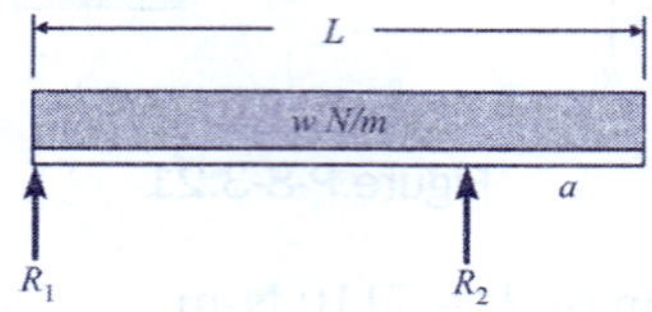

Figure P-8-3.25

$$a = 0.293L \qquad \textbf{Ans.}$$

In the following problems, draw moment and load diagrams corresponding to the given shear diagrams where values are given in newtons. Specify values at all change of load positions at all sections of zero shear.

8.3.26 Shear diagram shown in Fig. P-8-3.26.

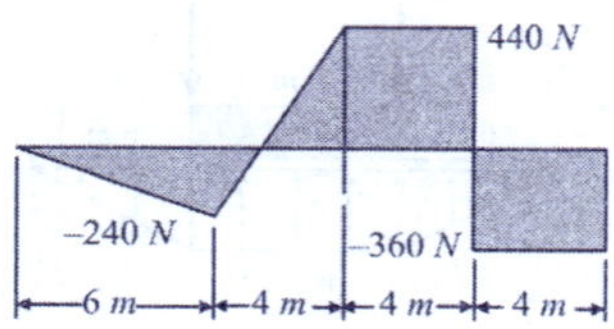

Figure P-8-3.26

max. positive moment = 1440 N-m;

max. negative moment = – 890 N-m **Ans.**

8-3.27 Shear diagram as shown in Fig. P-8-3.27.

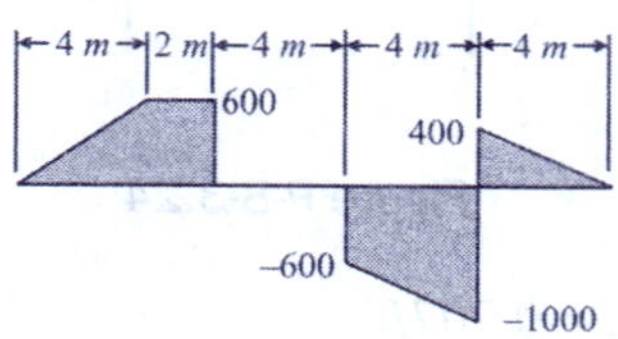

Figure P-8-3.27

8.3.28 Shear diagram shown in Fig. P-8-3.28.

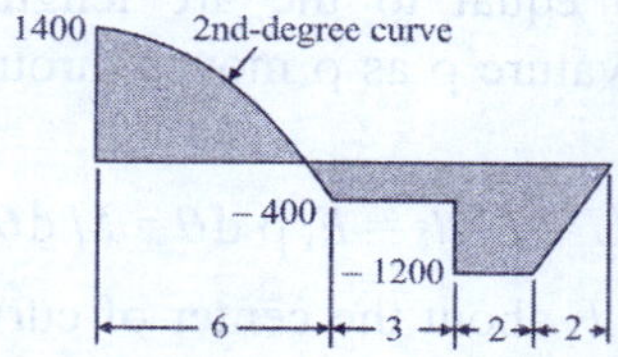

Figure P-8-3.28

max. $M = 4940$ N-m **Ans.**

VIRTUAL WORK

8-4 DEFINITION AND MEANING OF WORK

For the reasons that will be discussed in Section 14.2, the differential work done by a force **F** upon a particle that moves through a differential displacement dr is given by the dot product $dU = \mathbf{F}.\, dr$. Recalling that the dot product is a scalar quantity consisting of the product of the magnitudes of the two vectors by the cosine of their included angle, we then have

$$dU = F(ds \cos\theta) = (F\cos\theta)ds \qquad \text{(a)}$$

in which we have replaced for convenience the magnitude of dr by ds measured along the path of motion.

By referring to Fig. 8-4.1b, we see that the work dU may be interpreted as the product of the component $F \cos\theta$ of the force in the direction of the displacement multiplied by the displacement ds. Note especially that in accordance with this definition, forces which are perpendicular to the displacement do no work because they have no component in the direction of displacement. Also if dU is rewritten as the product $F(ds \cos\theta)$, we see that work may also be interpreted as the product of the force multiplied by the component of displacement in the direction of the force. Work may be positive or negative depending on whether the force component $F \cos\theta$ is directed respectively along or opposite to the direction of displacement.

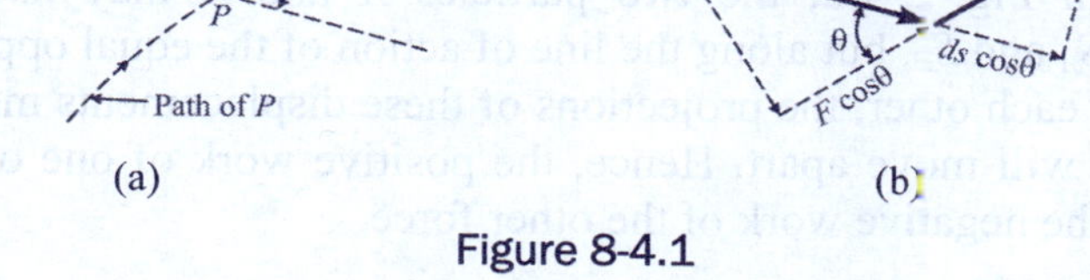

Figure 8-4.1

Further, if we interpret F cos θ in Eq. (*a*) as the tangential component F_t along the path and *ds* as equal to the arc length $\rho\, d\theta$ subtended by the instantaneous radius of curvature ρ as ρ moves through a differential angle $d\theta$, we have

$$dU = F_t\, ds = F_t\, \rho\, d\theta = M\, d\theta \qquad (b)$$

where *M* is the moment of *F* about the center of curvature. Can you show that this result also represents the work done by a couple *M* on a rigid body?

The unit of work is the dimensional product of force and distance; i.e., N-m, N-mm, dyne-cm, KN-m, etc. Since the units of work and moment are dimensionally the same, some engineers distinguish between them by using m-N for work and N-m for moment, but this is usually an unnecessary refinement.

When we consider the work done by external forces upon a rigid body, recall that in Section 2-13 we showed that these forces can always be reduced to a force-couple system acting at any convenient reference point *O*. The resultant force **R** is independent of the reference point whereas the resultant couple $\mathbf{M}_O$ depends on the location of the reference point. By taking advantage here of a principle of rigid-body motion that we shall discuss at length in Chapter 12, we state that any rigid-body motion is equivalent to combining geometrically the differential displacement dx_O of any reference point *O* with a rotation about *O* caused by an angular displacement $d\boldsymbol{\theta}$ of the body. The work of all external forces on a rigid body can now be stated as

$$dU = \mathbf{R} \cdot d\mathbf{r}_O + \mathbf{M}_O \cdot d\boldsymbol{\theta} \qquad (8\text{-}4.1)$$

Observe the beautiful duality of this result. **R,** which is independent of the reference point *O*, is dot multiplied by the infinitesimal displacement dr_O whereas M_O, which depends on the location of *O*, is dot multiplied by the angular displacement $d\boldsymbol{\theta}$ which is independent of the choice of a reference point.

Only the work done by external forces on a rigid body need be considered. No work is done by the internal forces distributed among the particles of a rigid body. This is so because any two particles exert equal, oppositely directed forces on each other. Although the displacements of these two particles may be different, the requirement of rigidity means that their displacement components are equal along the line of action of each pair of equal, oppositely directed forces. Thus, in Fig. 8-4.2, the two particles *A* and *B* may have different displacements S_A and S_B, but along the line of action of the equal opposite forces **F** they exert on each other, the projections of these displacements must be equal or the particles will move apart. Hence, the positive work of one of the paired forces cancels the negative work of the other force.

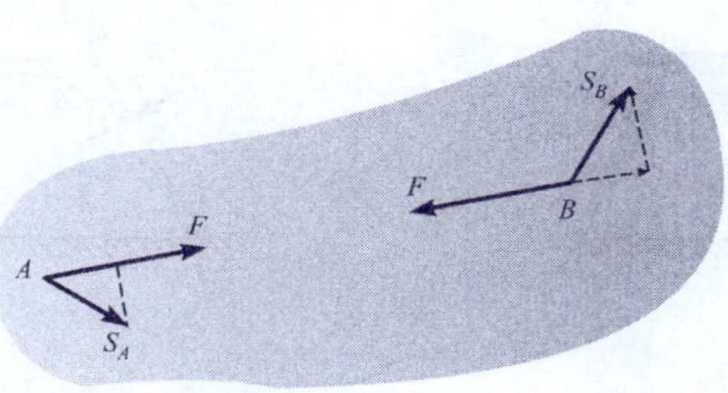

Figure P-8-4.2 Because of rigidity, displacement s_A and s_B have equal components along line of action of paired forces F.

Similarly, in a system of connected bodies, there is no work done by internal connecting forces since these occur in equal opposite pairs at points which have the same displacement. One important exception to this rule is the case of a nonrigid or an elastic connecting member such as a spring. Here work can be absorbed or given up by the spring, depending on its initial and final configurations.

8-5 INTRODUCTORY CONCEPTS OF VIRTUAL WORK METHOD

Instead of using force and moment summations to solve equilibrium problems, another approach is the method of virtual work. Although this method offers no advantage over previous methods when applied to a single body, it greatly simplifies problems involving connected systems of rigid bodies or of establishing criteria for stability of such systems. As we shall learn, the method is based on the relations between the applied forces and the movements of these forces that correspond to a *possible* small motion of the system. We emphasize "possible" to indicate that the motion does not in fact take place but instead is any arbitrary infinitesimal motion that is compatible with the constraints of the system. Such a motion is called a *virtual displacement*.[6] We define *virtual work* of a force as the dot product of a force with its corresponding virtual displacement, viz., $\delta U = \mathbf{F} \cdot \delta\mathbf{r}$ where the symbol dU replaces δU and $\delta\mathbf{r}$ is used in place of $d\mathbf{r}$ to distinguish a virtual or hypothetical differential from a real or actual one.

---

[6]If the virtual displacement is assumed to occur in a differential time, we obtain a virtual velocity having the direction of the virtual displacement. This concept is useful in understanding the kinematics of the system, i.e., its geometry of motion. In this connection, the instant center (see Section 12-5) is very helpful.

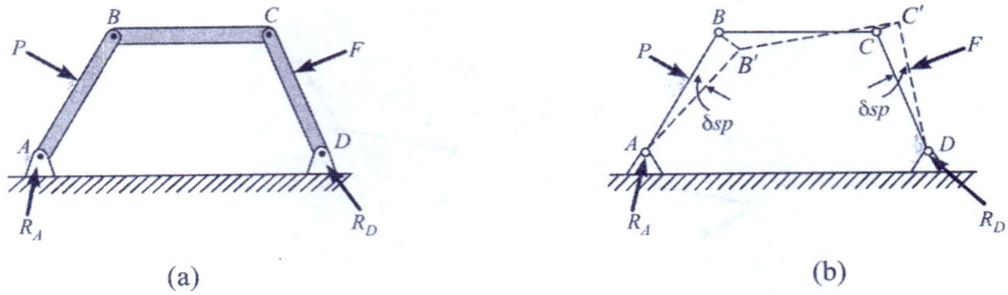

Figure 8.5.1 Virtual work of P balances that of F.

As, a preview of the method of virtual work, consider the pin-connected mechanism shown in Fig. 8-5.la. Suppose we must determine the force F on member CD to maintain equilibrium when a known force P is applied to member AB. Ordinarily we would have to dismember the system and apply force and moment summations to the free-body diagram of each member. But in the method of virtual work, we need only equate to zero the sum of the product of each applied force by its corresponding component of virtual displacement. Thus, if the mechanism is displaced slightly from the equilibrium position to the dashed position shown in part (b), we obtain $P(\delta s_P) = F(\delta s_F)$ from which F is found provided we know δs_P and δs_F. Thus, the equilibrium problem is transformed into a geometric problem. In many cases it will be relatively easy to analyze the geometry of the motion, especially if combined with the study of kinematics discussed later in Chapter 12.

We have stated the essence of the virtual work method without discussing any of the details so that you can see where we are heading. In subsequent sections, complete explanations, proofs, and associated details of the geometry of motion will be covered. One great advantage of the method is that it eliminates reactive forces such as R_A and R_D which act at points which do not move. Another advantage is that we deal with scalar quantities which thereby eliminates consideration of the directional aspects associated with vector quantities.

The method of virtual work is a subdivision of a general "energy" approach that culminates in the Lagrangian and Hamiltonian principles in dynamics and also is extensively applied to continuum mechanics and to structural analysis. Here we confine our attention to connected systems of rigid bodies in equilibrium.

8-6 PRINCIPLE AND APPLICATION OF VIRTUAL WORK

We now consider the application of the virtual work method to two types of equilibrium problems:

1. the determination of forces necessary to maintain equilibrium or

2. finding the equilibrium position of a system when all the forces are known.

In either case, it is evident that no actual motion occurs and that no real or actual work is done. However, by changing some of the specified constraints or conditions of a given situation, we can assume a movement to occur. Only *small* movements are permitted; otherwise we would change the geometry of the original configuration. Hence, to distinguish a virtual or hypothetical quantity from a real or actual one, a differential is indicated by the symbol δ instead of *d*. Thus, except for a different prefix, both δU and $\delta \mathbf{r}$ are first-order differentials and are determined mathematically the same as would be dU and $d\mathbf{r}$.

Consider now the restatement of Eq. (8-4.1) in terms of virtual differentials. We have then

$$\delta U = \mathbf{R} \,.\, \delta \mathbf{r}_O + \mathbf{M}_O \,.\, \delta \boldsymbol{\theta} \qquad (a)$$

Substituting into this equation the conditions of equilibrium ($\mathbf{R} = 0$ and $\mathbf{M}_O = 0$ where $\mathbf{R}$ is the resultant force and $\mathbf{M}_O$ the resultant couple about an arbitrary reference point O), we obtain

$$\delta U = 0 \qquad (8\text{-}6.1)$$

which states that the virtual work done on a rigid body or a system of rigid bodies in equilibrium is zero for any virtual displacements compatible with the constraints on the system. This statement is known as the *principle of virtual work.*

If Eq. (8-6.1) is expanded[7] in terms of the dot products of Eq. (*a*), there appear to be six possible coordinates of translational and rotational displacement. Some comments on how the constraints affect the number of independent virtual displacements are pertinent. In general, the reference point O may have three components of the virtual displacement $\delta \mathbf{r}_O$ and the body rotation $\delta \boldsymbol{\theta}$ may likewise be resolved into three components of rotation about three axes through the reference point. Actually, the constraints of a system limit the number of independent displacements to the degrees of freedom of the

---

[7]In Eq. (a), note that in terms of their components, we can write $R = (\Sigma X)\,\hat{\mathbf{i}} + (\Sigma Y)\,\hat{\mathbf{j}} + (\Sigma Z)\,\hat{\mathbf{k}}$ and $M_0 = (\Sigma M_x)\,\hat{\mathbf{i}} + (\Sigma M_y)\,\hat{\mathbf{j}} + (\Sigma M_z)\,\hat{\mathbf{k}}$; also $dr_0 = \delta x\hat{\mathbf{i}} + \delta y\hat{\mathbf{j}} + \delta z\hat{\mathbf{k}}$ and $d\theta = \delta\theta_x\hat{\mathbf{i}} + \delta\theta_y\hat{\mathbf{j}} + \delta\theta_z\hat{\mathbf{k}}$. Thus, by expanding the dot products $\mathbf{R} \,.\, \delta r_0$ and $M_0 \,.\, \delta\theta$, we obtain

$$\delta U = (\Sigma X)\,\delta x + (\Sigma Y)\,\delta_y + (\Sigma Z)\,\delta_z + (\Sigma M_x)\,\delta\theta_x + (\Sigma M_y)\,\delta\theta_y + (\Sigma M_z)\delta\theta_z$$

system. The degrees of freedom of a system is defined as the number of independent coordinates required to give a unique configuration of the system.

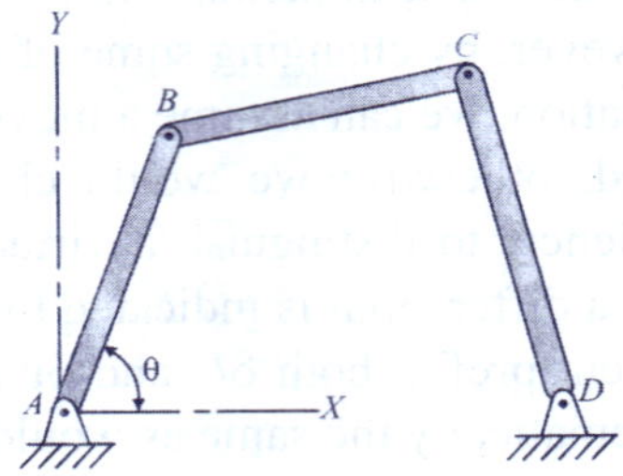

Figure 8-6.1 One degree of freedom.

For example, if the reference point is fixed in position, this will permit only three components of small body rotations which would be further reduced to only one if the motion is restricted to motion in a plane. Thus, the constraints of a fixed reference point and planar motion yield only one degree of freedom. This is illustrated by Fig. 8-6.1 where the position of the four-link mechanism (one link is the ground) is fixed by the value of θ. Alternatively, either the x or y coordinate of B may be used in place of θ to locate the position of B. Note that x and y are not independent but are related by the length of link AB. For the double pendulum in Fig. 8-6.2, consisting of two links pinned smoothly together, we have a system with two degrees of freedom since the value of θ_1 does not uniquely determine the position of BC nor does θ_2 uniquely determine the position of AB.

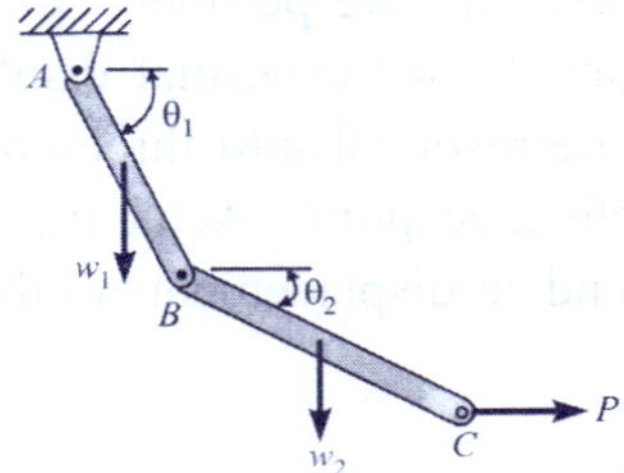

Figure 8-6.2 Two degrees of freedom.

The number of *unknowns*, consisting of active forces or equilibrium positions, that can be determined by the virtual work method is equal to the degrees of freedom of the system since for each independent coordinate of position we can write a corresponding equation of virtual work. For multiple degrees of freedom, we can express the total virtual work as

$$\delta U = \frac{\partial U}{\partial x}\delta x + \frac{\partial U}{\partial y}\delta y + \ldots \qquad (b)$$

and obtain various independent equations of virtual work by separately varying each independent displacement.

Reviewing this discussion of the virtual work method, we see that its application to connected systems of rigid bodies has these advantages:

1. Only forces that are external to the system do work since internal forces occur in equal opposite pairs whose net work is zero. Consequently, it is not necessary to dismember a system in order to determine relations between external forces as in the equilibrium method of force and moment summations.
2. No work is done by reaction forces at fixed hinges or forces that are perpendicular to a virtual displacement. Therefore, instead of a free-body diagram where all external forces are shown, we use an *active-force diagram* which shows only the forces doing work during a virtual displacement.

The method has little or no advantage when applied to a single body. Moreover, it cannot easily be applied to systems which involve frictional forces unless the normal forces upon which friction depends can be easily found by conventional methods.

The following illustrative problems demonstrate the application of the virtual work method. Note particularly Prob. 8-6.2 where a constraint condition is violated so that an external reaction may be found. This problem also explains in detail how to determine virtual displacements both mathematically and geometrically. It will be instructive also for you to solve these problems by the conventional methods of equilibrium and compare the effort required.

ILLUSTRATIVE PROBLEMS

8-6.1 The parallelogram linkage shown in Fig. 8-6.3 carries forces P, F, and Q applied at the midpoints of the links. For a given value of θ, determine the force Q in terms of P and F that will maintain equilibrium of the linkage.

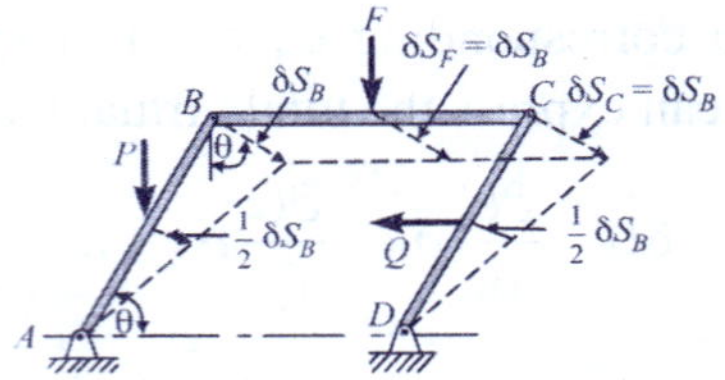

Figure 8-6.3

Solution

Only the active forces doing work on the system are shown. The dashed lines in the figure exaggerate the effect of a virtual rotation of AB about A. This rotation causes the virtual displacement δs_B of B to be directed perpendicular to AB. Clearly, the virtual displacement of C is equal to that of B while the virtual displacements of P and Q are one-half of δs_B. Applying the concept that work is the product of a force by its component of displacement in the direction of the force, the virtual work principle gives

$$[\delta U = 0] \qquad P\left(\frac{1}{2}\delta s_B\right)\cos\theta + F(\delta s_B \cos\theta) - Q\left(\frac{1}{2}\delta s_B \sin\theta\right) = 0$$

which, by cancelling the common displacement δs_B, reduces to

$$Q = (P + 2F)\cot\theta \qquad \textit{Ans.}$$

Observe the relative ease with which this result was obtained as compared to the conventional method of force and moment summations. Notice also that this result is independent of the position of F on BC since, all points on this link have the same virtual displacement and thereby produce the same work expression for F. Conversely, this result will determine the equilibrium position when all the forces are specified.

8-6.2 Use the method of virtual work to determine the horizontal component of the reaction at hinge A of the frame shown in Fig. 8-6.4.

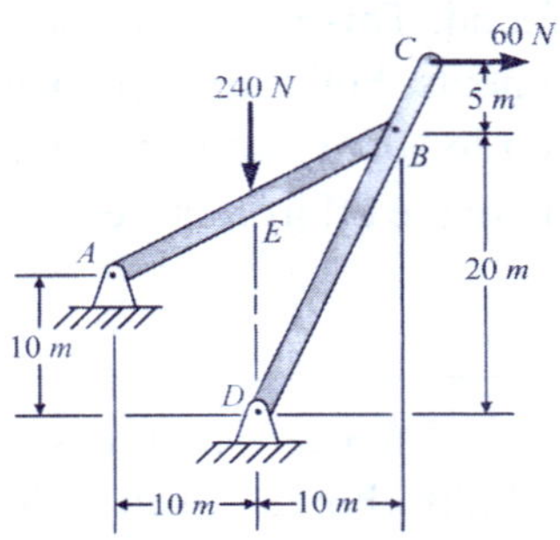

(a)

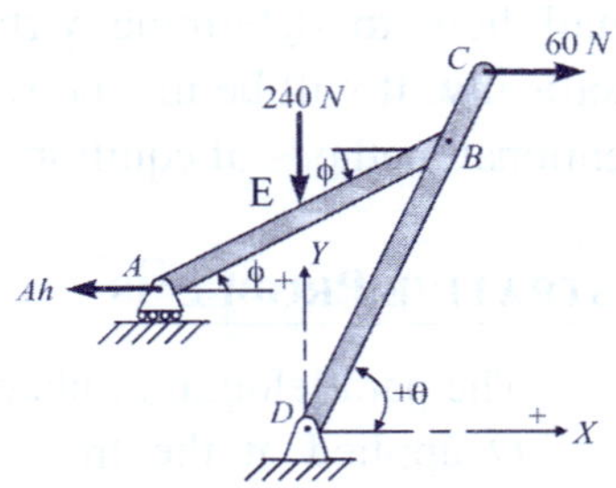

(b) Active force diagram for roller at A instead of hinge

Figure 8-6.4

Solution

Here no motion is physically possible so virtual displacements cannot occur. However, by changing a constraint so that a movement in the direction of a desired constraining force becomes possible, the constraining force becomes an active force which may be determined from a virtual work equation consistent with the assumed movement. Thus, to determine A_h, assume the hinge at A to be replaced by a roller support on a horizontal surface as shown in part (b).

We now show how to determine the virtual displacements of the points of application of all active forces consistent with a virtual rotation $\delta\theta$ of CD about D. We shall use first a mathematical method and then apply an independent geometrical approach. The mathematical method is straightforward, but the geometrical approach will require a somewhat lengthy explanation since it is based on as yet unfamiliar concepts. Nonetheless, the geometrical approach gives physical meaning to the mathematical method and may be even easier to apply when you understand it.

Mathematically, we choose a coordinate system with its origin at D as shown in Fig. 8-6.4b. All coordinates can be expressed in terms of the angle θ so that we have a one degree of freedom system. In this coordinate system, the positive senses of coordinates and forces are rightward and upward. The virtual work equation for all active forces is

$$[\delta U = 0] \qquad -A_h\,\delta x_A - 240\,\delta y_E + 60\,\delta x_C = 0 \qquad (a)$$

The virtual displacements are found by taking the derivatives of the corresponding coordinates expressed in terms of θ. Remember that the symbol δ is used instead of d to denote a virtual differential, but otherwise the process of differentiation is unchanged. Temporarily using ϕ to denote the inclination of AB, we obtain

$$x_A = \overline{DB}\cos\theta - \overline{BA}\cos\phi$$

$$y_E = \overline{DB}\sin\theta - \overline{BE}\sin\phi$$

$$x_C = \overline{DC}\cos\theta$$

Differentiating, we obtain

$$\delta x_A = -\,\overline{DB}\sin\theta\,\delta\theta + \overline{BA}\sin\phi\,\delta\phi$$

$$\delta y_E = \overline{DB}\cos\theta\,\delta\theta - \overline{BE}\cos\phi\,\delta\phi \qquad (b)$$

$$\delta x_C = -\,\overline{DC}\sin\theta\,\delta\theta$$

To express $\delta\phi$ in terms of $\delta\theta$, note that

$$y_B = \overline{DB}\sin\theta = 10 + \overline{AB}\sin\phi$$

whence differentiating we obtain

$$\overline{DB}\ \cos\theta\,\delta\theta = \overline{AB}\ \cos\phi\,\delta\phi$$

or
$$\delta\phi = \frac{\overline{DB}\cos\theta}{\overline{AB}\cos\phi}\delta\theta \qquad (c)$$

From the dimensions of Fig. 8-6.4a, we see that $\cos\theta = 10/BD$ and $\cos\phi = 20/AB$ which is substituted into Eq. (c) to give

$$\delta\phi = \frac{1}{2}\delta\theta \qquad (d)$$

Now we can evaluate the various virtual displacements of Eq. (b) by substituting dimensions from Fig. 8-6.4a. We thereby obtain

$$\delta x_A = -\overline{DB}\left(\frac{20}{\overline{DB}}\right)\delta\theta + \overline{BA}\left(\frac{10}{\overline{BA}}\right)\left(\frac{1}{2}\delta\theta\right) = -15\ \delta\theta$$

$$\delta y_E = \overline{DB}\left(\frac{10}{\overline{DB}}\right)\delta\theta - \overline{BE}\left(\frac{10}{\overline{BE}}\right)\left(\frac{1}{2}\delta\theta\right) = 5\ \delta\theta$$

$$\delta x_C = -\overline{DC}\left(\frac{25}{\overline{DC}}\right)\delta\theta = -25\ \delta\theta$$

These values of virtual displacements may now be substituted back into Eq. (a) to give

$$[\delta U = 0] \qquad -A_h\,(-15\ \delta\theta) - 240(5\ \delta\theta) + 60(-25\ \delta\theta) = 0$$

which reduces to

$$A_h = 180\text{ N}$$ ***Ans.***

The positive sign of this answer confirms the assumed direction of A_h.

This virtual displacements may also be found from a purely geometrical approach which will confirm and give physical significance to the mathematical values. We use the concept that the virtual motion of any point on a rigid body is the geometric sum of a rigid-body rotation about *any* point in the body plus the motion of that point. Thus to express the motion of a point A in terms of the motion of a reference point B, we use the vector equation

$$\delta s_A = \delta s_B \nrightarrow \delta s_{A/B}$$

where δs_B is the motion of the reference point and $\delta s_{A/B}$ is to be read as the displacement of A rotating about B; it is equal to $r\ \delta\theta$ where r is the distance between A and B, and is perpendicular to r in the sense of $\delta\theta$.

Consider now point B on CD. As is shown in Fig. 8-6.5a, a clockwise rotation $\delta\theta$ of CD about D gives B the displacement δs_B = (DB) $\delta\theta$ which is directed perpendicular to CD in the sense of $\delta\theta$. Its components are $\delta y_B = x\ \delta\theta = 10\ \delta\theta$ and $\delta x_B = y\ \delta\theta = 20\ \delta\theta$, respectively perpendicular to the x and y components of DB. Similarly, we obtain the shown components of δs_C.

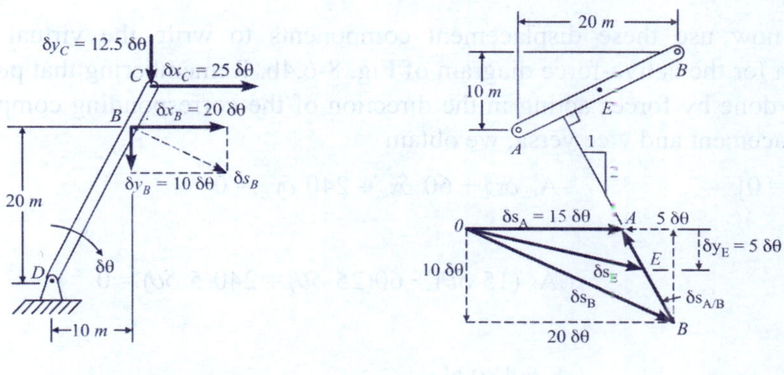

(a) Displacement of components for points on CD

(b) Displacement components for points on AB

Figure 8-6.5

This was relatively simple because the reference point D was stationary. The displacements of points on AB is more complex since no point on it is fixed. However, since the motion of B is known now, we may use B as a reference point to relate the displacements of A and E using the following relative motion equations:

$$\delta s_A = \delta s_B \nrightarrow \delta s_{E/B} \qquad (e)$$

and

$$\delta s_E = \delta s_B \nrightarrow \delta s_{A/B} \qquad (f)$$

To apply them, we use any convenient scale to plot δs_B as shown in Fig. 8-6.5b. Through the tip of δs_B, draw $\delta s_{A/B}$ perpendicular to BA and, through the tail of δs_B, draw δs_A horizontally as prescribed by the virtual movement of the imaginary roller support at A. The intersection of these lines determines the vectors δs_A and $\delta s_{A/B}$ whose directions are such that δs_A is the vector sum of δs_B and $\delta s_{A/B}$ in accordance with Eq. (*e*). Using Eq. (*f*), the displacement δ_{s_E} of E can now be drawn from the tail of δs_B to the midpoint of $\delta s_{A/B}$ since the rotational component $\delta s_{E/B}$ is one-half that of $\delta s_{A/B}$. Knowing that $\delta s_{A/B}$ has the inverse slope of AB, its horizontal component is $5\ \delta\theta$ whence $\delta s_A = 15\ \delta\theta$. Similarly, the vertical component of δs_E is one-half that of $\delta s_{A/B}$ and equals $5\ \delta\theta$ as shown.

We now use these displacement components to write the virtual work equation for the active-force diagram of Fig. 8-6.4b. Remembering that positive work is done by forces acting in the direction of the corresponding component of displacement and vice versa, we obtain

$$[\delta U = 0] \qquad -A_h\ \delta s_A + 60\ \delta x_c + 240\ \delta y_E = 0$$

or

$$-A_h\ (15\ \delta\theta) + 60(25\ \delta\theta) + 240(5\ \delta\theta) = 0$$

whence

$$A_h = 180 \text{ N}$$ ***Ans.***

Note that the numerical values of displacements by the geometric method are identical with those of the mathematical method. Their signs are different, however, because we assumed a clockwise $\delta\theta$ in the geometric method whereas a positive mathematical increment $\delta\theta$ is counterclockwise.

One possible drawback to the virtual work method should be mentioned. It is necessary to write a separate virtual work equation for each unknown active force. If the forces acting at all hinges were required, the effort of computing the corresponding virtual displacements would make it preferable to apply force and moment summations to separate free-body diagrams of each member. Note, however, that more sophisticated methods of finding virtual displacements are given in texts on kinematics which greatly simplify the force analysis of complex mechanisms.

8.6.3 The double pendulum in Fig. 8-6.6 consists of two links pinned smoothly together and held in position by a horizontal force P. If link AB is twice as long and heavy as link BC, find the equilibrium position as defined by the angles θ_1 and θ_2.

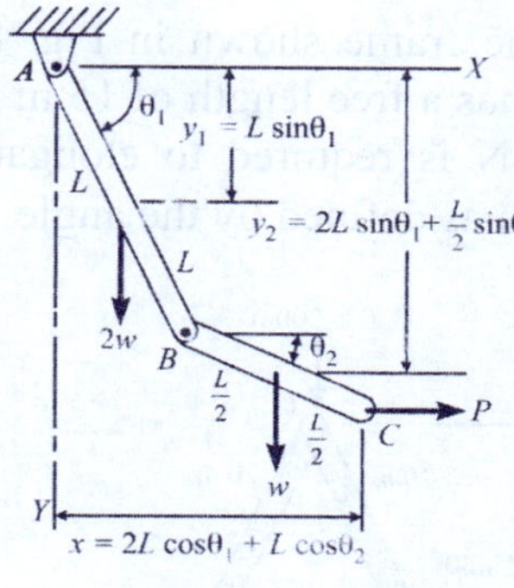

Figure 8-6.6

Solution

With respect to the coordinate system shown, the coordinates in the direction of each active force are as shown. Since we have a two degree of freedom system in which θ_1 and θ_2 can vary independently, we write the principle of virtual work in the following form:

$$\delta U = \frac{\partial U}{\partial \theta_1}\delta\theta_1 + \frac{\partial U}{\partial \theta_2}\delta\theta_2 = 0$$

Keeping θ_1 constant and permitting θ_2 to vary gives

$$[\delta U = 0] \qquad W\,\delta y_2 + P\,\delta x = 0$$

which, taking the partial derivatives with respect to θ_2, reduces to

$$\left[\frac{\partial U}{\partial \theta_2} = 0\right] \qquad W\frac{L}{2}\cos\theta_2\;\delta\theta_2 - PL\sin\theta_2\delta\theta_2 = 0$$

whence

$$\tan\theta_2 = \frac{W}{2P} \qquad \textit{Ans.}$$

Now keeping θ_2 constant while permitting θ_1 to vary, we obtain

$$[\delta U = 0] \qquad 2W\,\delta y_1 + W\,\delta y_2 + P\,\delta x = 0$$

for which partial differentiation with respect to θ_1 yields

$$\left[\frac{\partial U}{\partial \theta_1} = 0\right] \qquad 2W(L\cos\theta_1)\delta\theta_1 + W(2L\cos\theta_1)\;\delta\theta_1$$

$$- P\,(2L\sin\theta_1)\;\delta\theta_1 = 0$$

whence

$$\tan\theta_1 = \frac{2W}{P} \qquad \textit{Ans.}$$

8-6.4 The two legs of the frame shown in Fig. 8-6.7 are connected by a spring. The spring has a free length of 15 m and a constant of 30 N/m, i.e. a force of 30 N is required to elongate it 1 m. Determine the equilibrium position as defined by the angle θ.

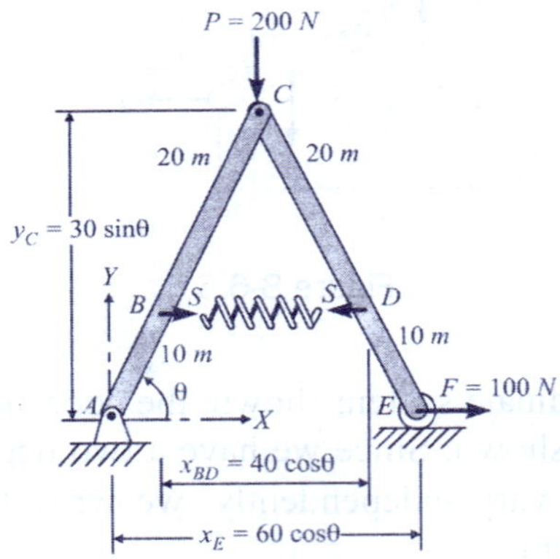

Figure 8-6.7

Solution

This example illustrates the work done by an elastic member (i.e., the spring) on an otherwise rigid-body system. In terms of the coordinate axes at A and the angle θ which determines the position of the system, the virtual work equation is

$$[\delta U = 0] \qquad F\,\delta x_E - P\,\delta y_C - S\,\delta x_{BD} = 0 \qquad \text{(a)}$$

Observe that the negative work done by the spring is the result of combining the work done by the equal and opposite spring force S acting on each leg; Viz., $S\,\delta x_B - S\,\delta x_D = S(\delta x_B - \delta x_D) = -S\,\delta x_{BD}$. Thus, the negative work done by the spring depends only on the virtual change in displacement of its overall length. The spring force **S** is given by $\mathbf{S} = k\,(\Delta L)$ where k is the spring constant and ΔL is the stretch of the spring. Since here the spring has a free length of 15 m, its net stretch is $40\cos\theta - 15$ and the spring force is $S = 30(40\cos\theta - 15) = (1200\cos\theta - 450)$ N. Observe that the spring force S is essentially constant during an infinitesimal change dx_{BD}.

The procedure now is mere routine. Performing the indicated differentiations of Eq. (*a*) and substituting numerical data. we obtain

$$100(-60\sin\theta)\,\delta\theta - 200(30\cos\theta)\,\delta\theta$$
$$-(1200\cos\theta - 450)(-40\sin\theta)\,\delta\theta = 0$$

which reduces to the transcendental equation.

$$8 \sin \theta - 4 \tan \theta = 1$$

Solving this equation by trial, we obtain

$$\theta = 53.6° \qquad \textbf{Ans.}$$

PROBLEMS

8.6.5 In Roberval's balance shown in Fig. P-8-6.5, determine the force P to balance the weight W.

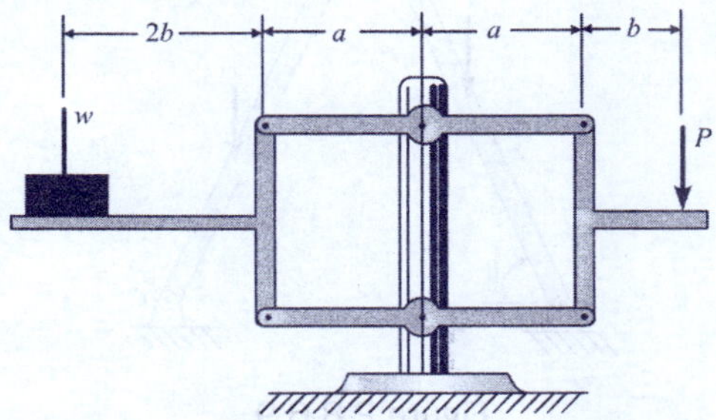

Figure P-8-6.5 Roberval's balance.

8.6.6 The rear end differential of an automobile has a ratio of 2.90: 1 which means that the drive shaft turns 2.90 revolutions for each revolution of the driving wheels. If the automobile weighs 1600 kg and has 0.7 m diameter wheels, how much torque acts on the drive shaft when the car is moving at constant speed up a 5% grade? Neglect air resistance and all internal mechanical friction.

8.6.7 For the lazy tongs mechanism shown in Fig. P.8-6.7, determine the relation between P and R for any position angle θ. How is this relation changed if the weights of the links is considered? Each link weighs 2w N.

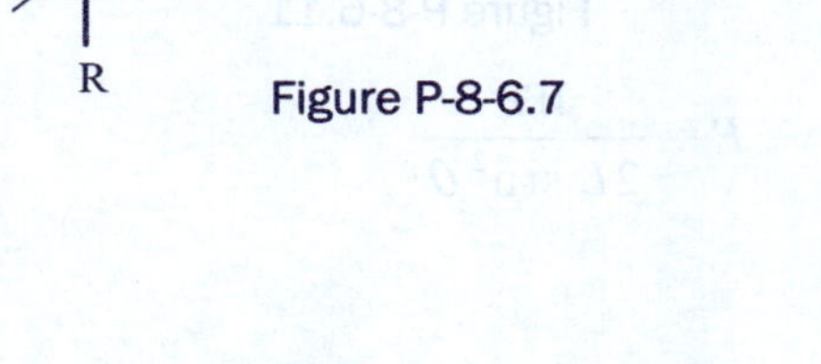

Figure P-8-6.7

8-6.8 Solve the previous problem if the mechanism is rotated 90° counterclockwise so that it is above the supports.

$$R = \frac{7P + 49w}{2\tan\theta}$$ ***Ans.***

8-6.9 Use the method of virtual work to determine the horizontal component of the reaction at C for the frame shown in Fig. P-8-6.9.

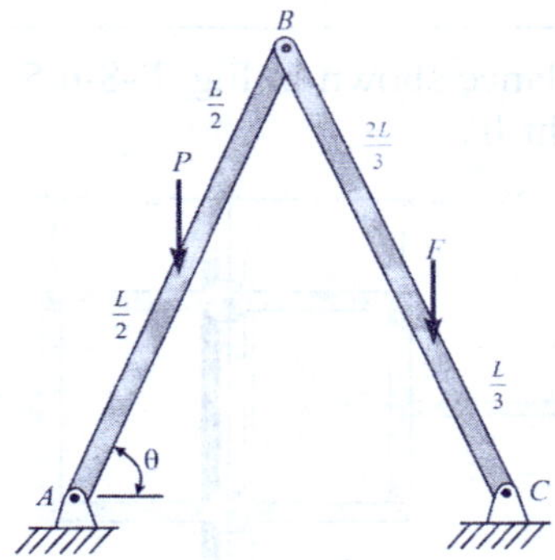

Figure P-8-6.9

$$C_h = \left(\frac{P}{4} + \frac{F}{6}\right)\cot\theta$$ ***Ans.***

8-6.10 Solve the previous problem if the force F acts horizontally to the right instead of vertically.

8-6.11 Find the force P acting on each of the two links shown in Fig. P-8-6.11 that will support the cylinder of weight W. Neglect friction and the weights of the links.

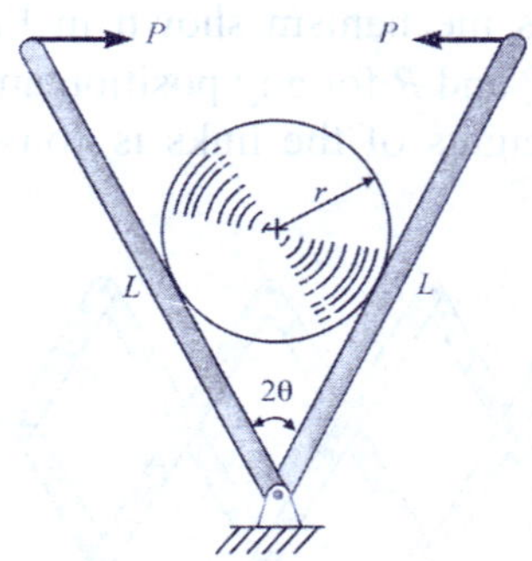

Figure P-8-6.11

$$P = \frac{Wr}{2L\,\sin^2\theta}$$ ***Ans.***

8-6.12 Find the couple M applied at A that will maintain equilibrium at the given position of the four-link mechanism shown in Fig. P-8-6.12. The loads act at the midpoints of the links.

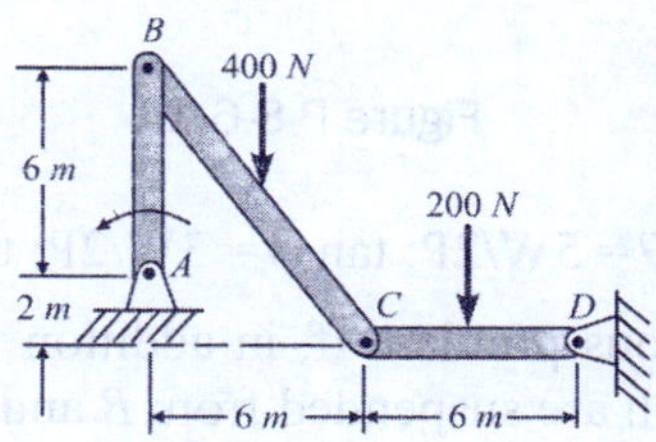

Figure P-8-6.12

$M = 1350$ N-m **Ans.**

8-8.13 In Fig. P-8-6.13, the link BD is pinned at C to a collar which is free to slide along the fixed vertical shaft. Find the force P supplied by a hydraulic cylinder on the collar to maintain W for any value of the angle θ.

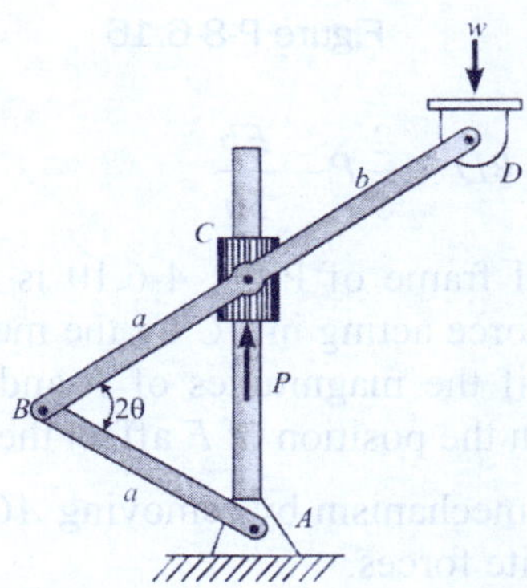

Figure P-8-6.13

8-6.14 As shown in Fig. P-8-6.14, a horizontal force P is applied to end D of three identical links smoothly pinned together. If each link is of weight W, determine the equilibrium position as defined by the angles θ, ϕ, and α.

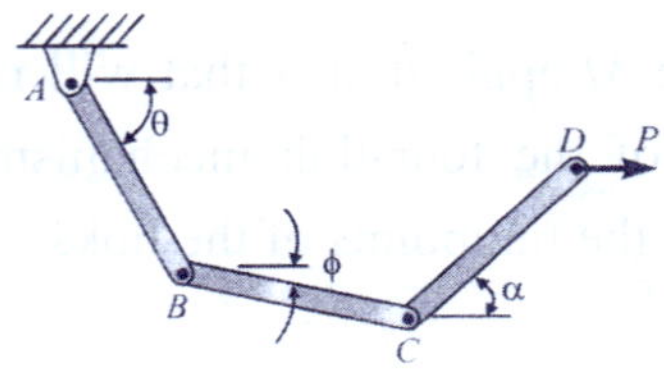

Figure P-8-6.14

$$\tan\theta = 5W/2P;\ \tan\phi = 3W/2P;\ \tan\alpha = -W/2P \qquad \textbf{\textit{Ans.}}$$

8-6.15 Solve the previous problem if, in addition to the weights of the links, loads of $2W$ each are suspended from B and C.

8-6.16 Without solving first for the reactions, use the virtual work method to find the force in bar BD of the truss shown in Fig. P-8-6.16.

Hint: Expose BD as an active force on a mechanism obtained by removing bar BD.

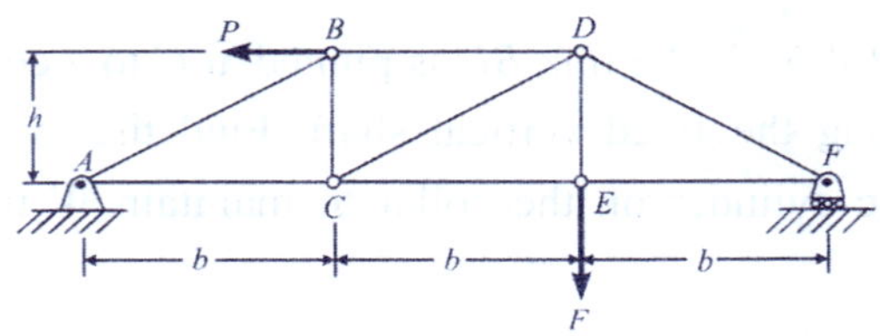

Figure P-8-6.16

$$BD = \frac{2}{3}P - \frac{Fb}{3h} \qquad \textbf{\textit{Ans.}}$$

8-6.17 The rhomboidal frame of Prob. 4-6.10 is repeated here as Fig. P-8-6.17. Find the force acting in AC by the method of virtual work. Then find AC again if the magnitudes of P and Q are interchanged. How does a change in the position of F affect the solution?

Hint: Obtain a mechanism by removing AC and replacing it by a pair of equal, opposite forces.

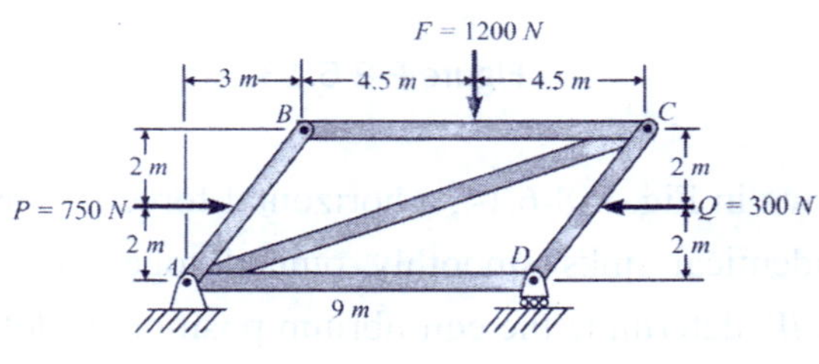

Figure P-8-6.17

8-6.18 In Fig. P-8-6.18, use the method of virtual work to determine the vertical component of the hinge force at A caused by a windload of 300 N/linear m acting on the billboard BC which weighs 1000 N. Neglect the weights of other supporting members.

Hint: Obtain a mechanism by replacing hinge A by a roller free to move vertically.

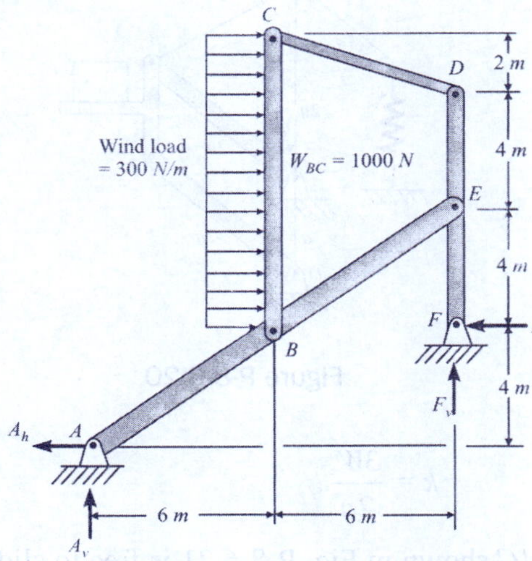

Figure P-8-6.18

$A_v = 2250$ N **Ans.**

8-6.19 The bulldozer of Prob. 4-6.2.3 is repeated here as Fig. P-8-6.19. Duplicate linkages exist on each side of the machine. Use the virtual work method to find the force P in the hydraulic cylinder AB when the scoop is loaded with 2 KN of dirt with center of gravity at G.

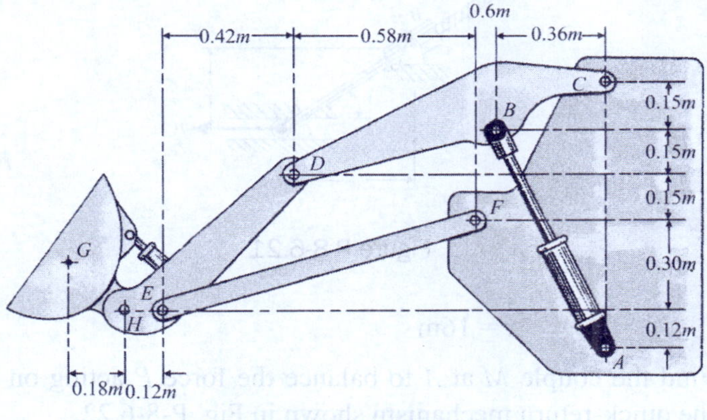

Figure P-8-6.19

8-6.20 Determine the constant k of the spring in Fig. P-8-6.20 that will maintain the hoisting frame in equilibrium. When $\theta = 0$, the spring is unstretched. The links AB and CD are uniform bars each weighing W N while BC with its load weighs $2W$ N.

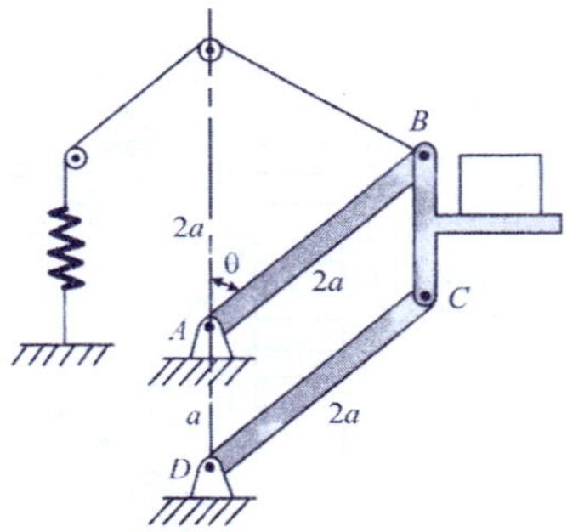

Figure P-8-6.20

$$k = \frac{3W}{2a}$$ ***Ans.***

8-6.21 The rod AC shown in Fig. P-8-6.21 is free to slide in the pivoted collar at B. The end C is constrained by the roller to move horizontally. The spring is undeformed when AC is vertical. Show that the equilibrium position is defined by $\frac{P}{kh}\tan\theta + \sin\theta = 1$. What is the value of x if $P = 192$ N; $k = 30$ N/m., and h = 12 m?

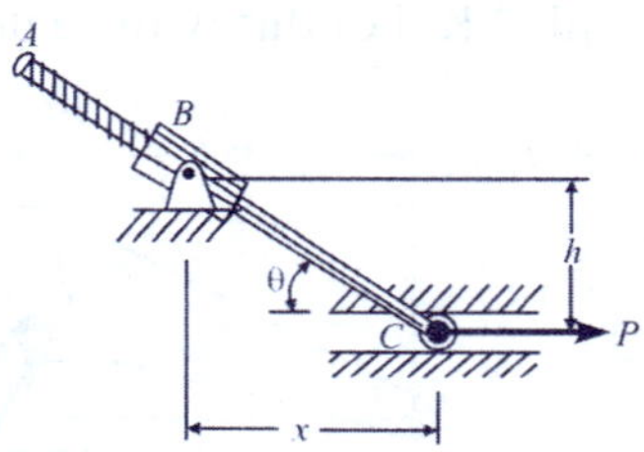

Figure P-8-6.21

$$x = 16\text{m}$$ ***Ans.***

8-6.22 Find the couple M at A to balance the force P acting on the slider of the quick-return mechanism shown in Fig. P-8-6.22.

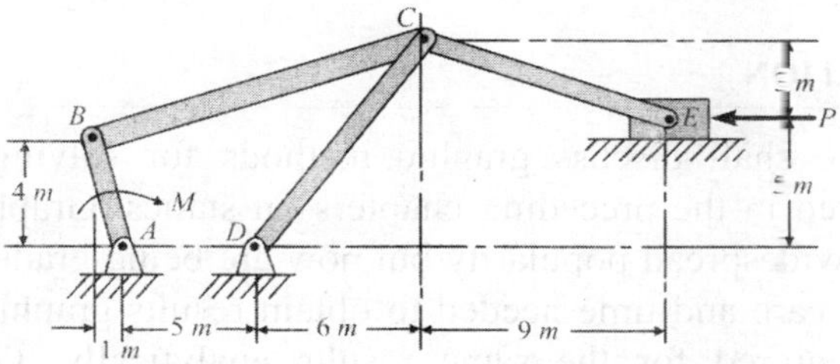

Figure P-8-6.22

$M = 3250$ N-m **Ans.**

8-6.23 The bars of a truss are not ideally rigid but actually undergo an axial deformation $\Delta = SL/AE$ where S is the bar force, L its length, A its cross sectional area, and E is the modulus of elasticity. As in the case of a spring, the work absorbed by each bar is the product $S(\Delta)$. Assuming $L/AE = 24 \times 10^{-6}$ m/N is constant for all bars of the truss shown in Fig. P-8-6.23, determine the vertical displacement of joint D.

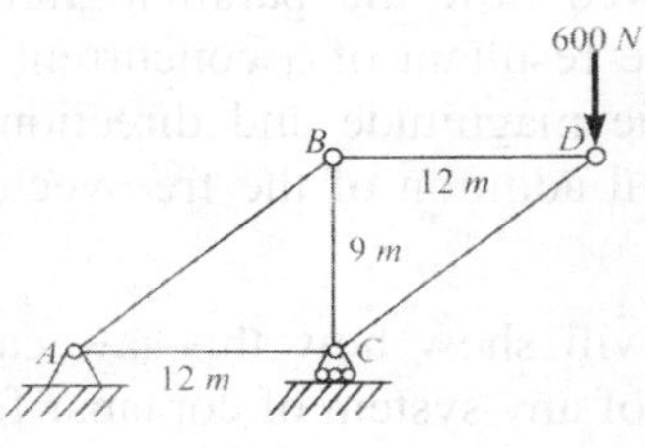

Figure P-8-6.23

$v_D = 146 \times 10^{-2}$ m **Ans.**

8-6.24 Find the horizontal displacement of joint D of the truss in the preceding problem. Use the following method: Assume an imaginary load Q acts horizontally at D in addition to the actual loads, and equate $Q\,(h_D)$ to the work absorbed by bar forces due to Q acting over the bar deformations SL/AE caused by the actual load.

$h_D = 6.0 \times 10^{-2}$ m **Ans.**

GRAPHIC METHODS

8-7 INTRODUCTION

In this portion we shall discuss graphic methods for solving many of the problems considered in the preceding chapters on statics. Graphic solutions at one time enjoyed widespread popularity but now are being gradually discarded. In most cases, the care and time needed to obtain results graphically are much greater than is required for the same results analytically. Graphic results, moreover, are frequently not as accurate as the analytical.

In spite of these objections, graphic methods are still used for the following reasons: They are valuable as an aid in checking analytical calculations. This check is rapid and is often sufficiently accurate even when done freehand to approximate scale. Also, in some cases of stress analysis analytical calculations are long and cumbersome. The care required in plotting a diagram to scale is justified in such instances. Moreover, although graphic solutions are not as accurate as analytical ones, the degree of accuracy obtained from them is often sufficiently accurate for many engineering purposes.

8-8 APPLICATION OF PARALLELOGRAM LAW

In Section 2-4 we showed how the parallelogram law could be applied to determine graphically the resultant of a concurrent system of coplanar forces. We showed also that the magnitude and direction of the resultant could be obtained from a tip-to-tail addition of the free vectors of the system (See Fig. 2-4.2.).

The present section will show how this law can be applied to determine graphically the resultant of any system of coplanar forces. It will also show that the magnitude and direction of the resultant can be determined from a *force polygon* formed by a tip-to-tail addition of the free vectors of the forces comprising the system.

In Fig. 8-8.1*a* is shown a body acted upon by the nonconcurrent forces F_1, F_2, F_3, and F_4. Extend the lines of action of F_1 and F_2 to intersect at A and construct the parallelogram shown. The resultant R_1 of F_1 and F_2 can be similarly extended to intersect the line of action of F_3 at B, and the resultant R_2 determined by the parallelogram law. R_2 is evidently the resultant of the forces F_1, F_2, and F_3. In the same way, the resultant of R_2 and F_4, and consequently of the given system of forces, is determined to be acting through point C.

Comparing the polygon of forces created by the tip-to-tail addition of their free vectors (Fig. 8-8.1*b*) with the successive applications of the parallelogram law in Fig. 8-8.1*a*, we conclude that a force polygon can be used to determine the magnitude and direction of the resultant of any system of forces. The position of the resultant force, however, cannot be found from the force polygon. One method of determining this position has just been discussed, but it is tedious and cannot be used when the action lines of the forces do not intersect within the limits of the paper. Accordingly, other and simpler methods are necessary. Some of them will be discussed.

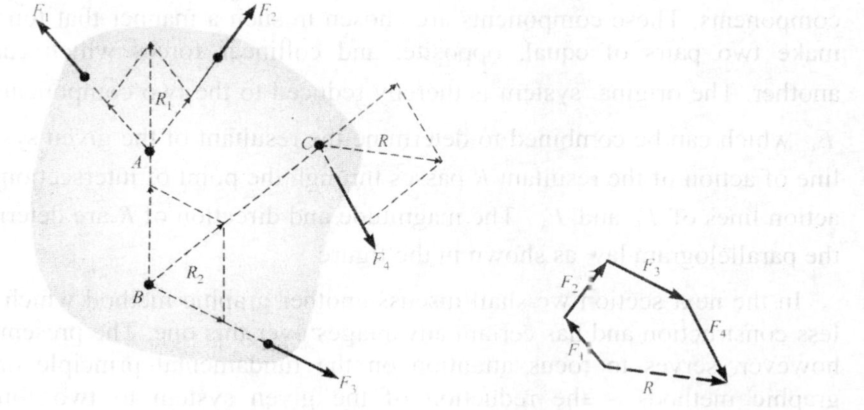

Figure 8-8.1 Resultant determined from paralle ogram law.

8-9 RESULTANT OF COPLANAR FORCE SYSTEMS

Let it be required to determine the resultant of the nonconcurrent system of coplanar forces shown in Fig. 8-9.1. The system consists of forces F_1, F_2, and F_3 having the lines of action *ab*, *bc*, and *cd*. (This is in accordance with Bow's notation, which is described in the next section.)

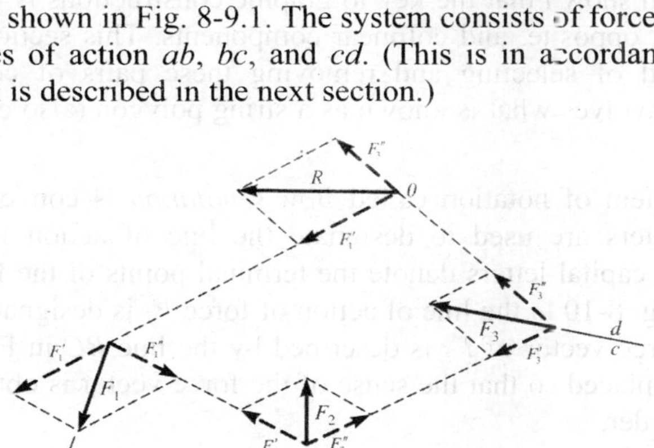

Figure 8-9.1 Another method of using the paral elogram law.

We begin by using the parallelogram law to resolve F_1 into any two components F_1' and F_1''. Now resolve F_2 into the components F_2' and F_2'' so that F_2' is equal, opposite, and collinear with F_1'' as shown. Similarly, resolve F_3 into components F_3' and F_3'' so that F_2'' and F_3' are also equal, opposite, and collinear.

Thus the original system of forces is replaced by an equivalent system of six components. These components are chosen in such a manner that four of them make two pairs of equal, opposite, and collinear forces which cancel one another. The original system is thereby reduced to the two components F_1' and F_3'' which can be combined to determine the resultant of the given system. The line of action of the resultant R passes through the point of intersection O of the action lines of F_1' and F_3''. The magnitude and direction of R are determined by the parallelogram law as shown in the figure.

In the next section we shall discuss another graphic method which involves less construction and has certain advantages over this one. The present method, however, serves to focus attention on the fundamental principle underlying graphic methods – the reduction of the given system to two forces. The intersection of these two forces determines a point on the line of action of the resultant. For the special case in which the system reduces to two equal, opposite, parallel forces, the resultant is a couple.

8-10 STRING POLYGON

It has just been shown that the key to graphic constructions is the cancelling of pairs of equal, opposite, and collinear components. This section will present a simple method of selecting and removing these pairs of components. The construction involves what is known as a string polygon (also called a funicular polygon).

Here a system of notation called *Bow's notation* is convenient. In it, two lower-case letters are used to designate the line of action for a force. The corresponding capital letters denote the terminal points of the force vector. For instance, in Fig. 8-10.1*a* the line of action of force F_2 is designated by the letters *bc*, and the force vector of F_2 is described by the line *BC* in Fig. 8-10.1*b*. The letters *BC* are placed so that the sense of the force vector is obtained from their alphabetical order.

Consider the system of nonconcurrent coplanar forces F_1, F_2, and F_3 in Fig. 8-10.1. We first construct a force polygon *ABCD* (Fig. 8-10.1*b*) by adding the free vectors of the forces in tip-to-tail fashion. The magnitude and direction of the resultant R are determined by the closing side AD of the force polygon. Only the line of action of R remains to be determined.

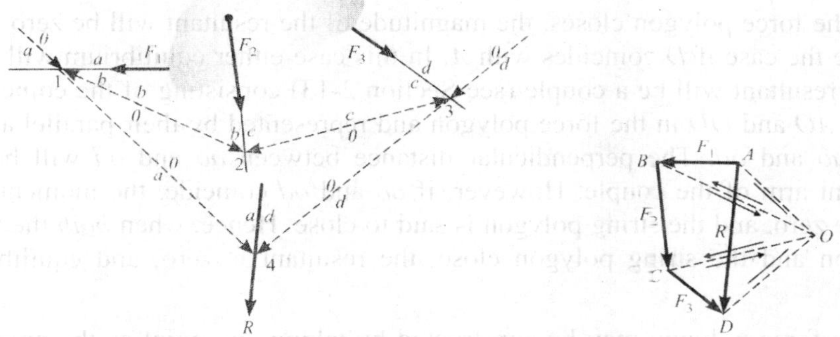

Figure 8-10.1 String and force polygons.

To accomplish this, we draw lines from any arbitrarily chosen point O (either inside or outside of the force polygon) to points A, B, C, and D on the polygon. The arbitrary point O is called a *pole*, and lines OA, OB, etc., are culled *rays*. The rays of the force polygon are the components into which the forces of the given system are to be resolved. For example, from the arrows inside triangle AOB, it is evident that F_1 is the vector sum of AO and OB. Similarly, F_2 can be considered as the resultant of BO and OC, and F_3 as the resultant of CO and OD. Likewise, R is the resultant of AO and OD.

Referring to Fig. 8-10.la, which respresents the position of the action lines of the forces, we replace each of the given forces by the components determined from the rays of the force polygon. Starting at any point *1* on the action line of F_1, we represent AO and OB by their lines of action ao and ob, respectively parallel to AO and OB. At point *2*, where ob intersects the action line of F_2, we draw bo and oc parallel respectively to BO and OC. Since the components BO and OB are laid off on the same line of action, they form a pair of equal, opposite, and collinear forces. (Compare with F_1'' and F_2'' in Fig. 8-9.1.) Similarly at point *3*, where oc intersects the line of action for F_3, we draw co and od parallel to CO and OD. The intersection of ao and od determines point *4* through which passes the line of action of the resultant.

Lines ao, bo, co, and do in Fig. 8-10.1*a* may be compared with a string held at the ends and stretched taut by the action of the forces F_1, F_2, and F_3. The figure described by these lines is therefore called a *string polygon* and the lines are spoken of as *strings*. The reason for the other name, *funicular polygon*, is evident, from the fact that funicular means "consisting of a rope."

We conclude that the resultant of a force system can be completely determined by constructing a force polygon and a string polygon. The force polygon determines the magnitude and direction of the resultant, and the string polygon determines its line of action.

If the force polygon closes, the magnitude of the resultant will be zero. This will be the case if *D* coincides with *A*. In this case either equilibrium will exist or the resultant will be a couple (see Section 2-13) consisting of the coincident forces *AO* and *OD* in the force polygon and represented by their parallel action lines *ao* and *od*. The perpendicular distance between *ao* and *od* will be the moment arm of the couple. However, if *ao* and *od* coincide, the moment arm will be zero, and the string polygon is said to close. Hence, when *both* the force polygon and the string polygon close, the resultant is zero, and equilibrium exists.

The force polygon may be constructed by taking any point as the pole; the string polygon may be started at any point on the line of action of any one of the forces. A change in the shape of either polygon only locates a different point on the line of action of the resultant, unless the resultant is a couple. In this case, such a change will produce an equivalent couple consisting of a different moment arm with a corresponding change in the magnitude of the forces composing the couple.

ILLUSTRATIVE PROBLEMS

8-10.1 Determine graphically the resultant of the parallel force system shown in Fig. 8-10.2.

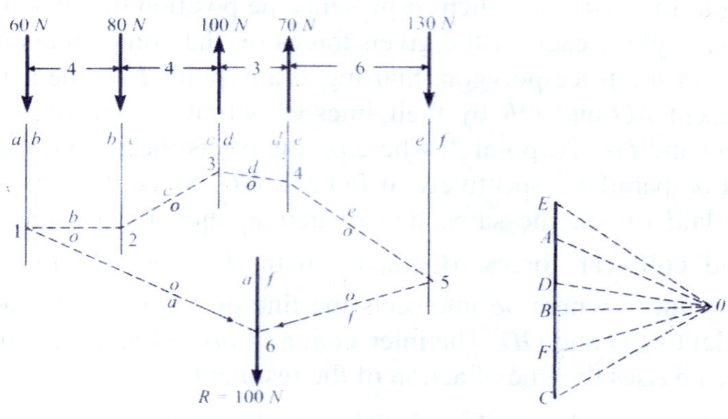

(a) (b)

Figure 8-10.2

Solution

The force polygon in Fig. 8-10.2b is obtained by drawing to scale a tip-to-tail addition of the forces vectors comprising the given system. It is a straight line because the fortes are parallel. The sense of a force is incicated by reading the letters in alphabetical order. For instance, the 60 N force, whose line of action is *ab* in Fig. 8-10.2a, is read from *A* to *B* in Fig. 8-10.2b, the downward alphabetical sequence of the letters indicating the direction of the force. The upward 70-N force whose action line is *de* is read alphabetically upward from *D* to *E* in the force polygon. Since the direction of a force is indicated by the alphabetical sequence of the letters describing it, it is usually unnecessary to use arrows on the force polygon. The resultant *R* is represented by the vector *AF* drawn from the tail *A* of the first vector to the tip *F* of the last one. From the force polygon, *AF* scales 100 N, and its direction is downward from *A* to *F*.

Choose a pole *O* and draw rays *AO*, *BO*, etc. The string polygon is started at any point say point *1* on the action line of *ab*. Through this point draw strings *ao* and *bo* parallel to rays *AO* and *BO*. At point *2*, the intersection of *bo* with *bc*, draw string co parallel to ray *CO*. Continuing in this fashion, construct the string polygon. All the components of the forces neutralize each other except *AO* and *OF*. These two intersect at point *6* which is a point on the action line of the resultant. Accordingly, show the resultant passing through point *6*. From Fig. 8-10.2a the distance from the line of action of *ab* to the line of action of *R* is found, by scaling, to be 9.6 m.

8-10.2 Determine graphically the resultant of the nonconcurrent system of coplanar forces shown in Fig. 8-10.3.

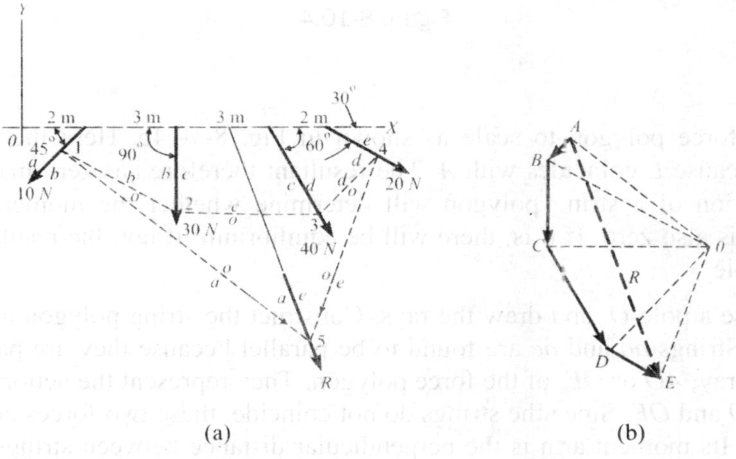

Figure 8-10.3

Solution

Draw the force polygon to scale as shown in Fig. 8-10.3b. The resultant R is drawn from A to E and scales 87 N. Choose a convenient pole O, and draw rays AO, BO, etc. Starting at any point 1 on the line of action of ab in Fig. 8-10.3a, draw strings ao, bo, etc., parallel to the respective rays. All the components of the forces neutralize each other except AO and OE. The strings of these components intersect at point 5, through which is drawn the line of action of R. If we scale from Fig. 8-10.3a, the action line of the resultant intersects the X axis at a point 6.62 m to the right of the origin of the axes.

8-10.3 Determine graphically the resultant of the nonconcurrent system of forces shown in Fig. 8-10.4.

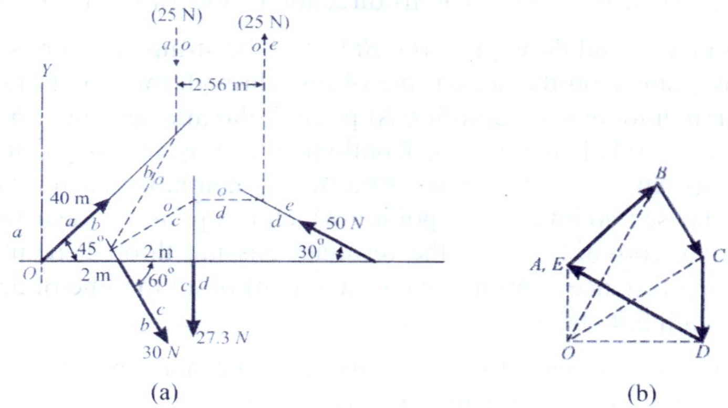

Figure 8-10.4

Solution

Plot the force polygon to scale as shown in Fig. 8-10.4b. Here the polygon closes because E coincides with A. The resultant therefore has zero magnitude. Construction of a string polygon will determine whether the moment of the resultant is also zero. If it is, there will be equilibrium; if not, the resultant will be a couple.

Choose a pole O, and draw the rays. Construct the string polygon as in Fig. 8-10.4a. Strings ao and oe are found to be parallel because they are parallel to the same ray, AO or OE, of the force polygon. They represent the action lines of forces AO and OE. Since the strings do not coincide, these two forces constitute a couple. Its moment arm is the perpendicular distance between strings ao and oe, which scales 2.56 m. Each force of this couple is found to be 25 N by scaling AO in Fig. 8-10.4b. Therefore the moment of the resultant couple is $25 \times 2.56 = 64$ N-m acting in a counterclockwise sense, as shown in Fig. 8-10.4a.

8-11 EQUILIBRIUM OF FORCE SYSTEMS BY GRAPHIC METHODS

The graphic condition for equilibrium is that both the force polygon and the string polygon must close. This is sufficient to reduce the resultant of any coplanar force system to zero because a closed force polygon means that the resultant has zero magnitude, and a closed string polygon means that the resultant has zero moment.

The force polygon may be constructed in any order and the pole O chosen at any point. Although the string polygon can be started at any point in the plane of the forces, it is usually best to begin it on the line of action of one of the unknown forces. In case one of the unknown forces is unknown as far as both magnitude and direction are concerned, the string polygon *must* be started at the point of application of the force. The reason is that the apexes of the string polygon lie on the action line of the forces; hence, if only the point of application of an unknown force is specified, the string polygon must be started at that point of application in order to have the apex of the string polygon lie on the action line of this undetermined force.

ILLUSTRATIVE PROBLEMS

8-11.1 Determine graphically the reactions on the beam loaded and supported as shown in Fig. 8-11.1.

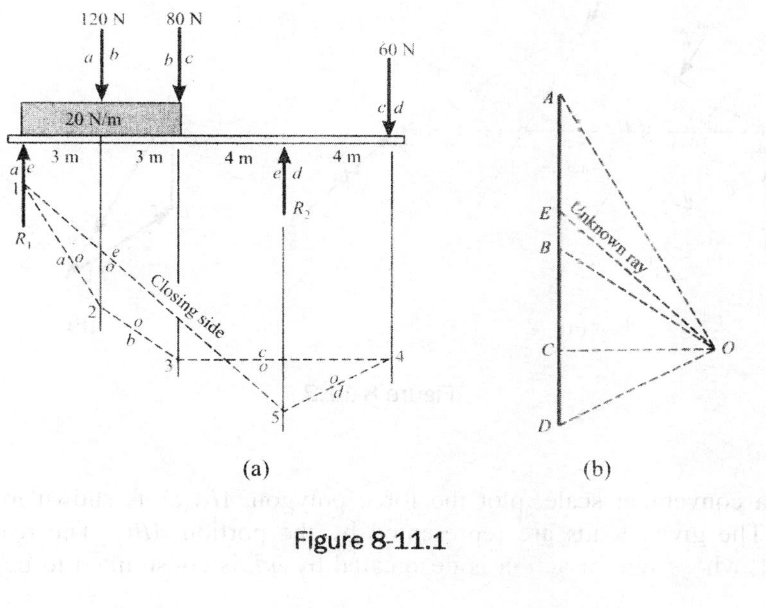

Figure 8-11.1

Solution

To determine the reactions, we replace the uniformly distributed load by its resultant of 120 N having the line of action *ab*. We begin by plotting to scale the force polygon in Fig. 8-11.1*b*. Reactions *DE* and *EA* close the force polygon, but the position of point *E* is still undetermined so that ray *OE* cannot be drawn yet. Its direction will be determined from the closing side of the string polygon.

A characteristic of the string polygon is that its apexes lie on the action lines of the forces. Further, the strings intersecting any action line (such as *ab*) are those defined by the letters (*a* and *b*) of the action line of the force combined with the letter *o* (thus, *ao* and *bo*). Therefore, starting at point *1* on the action line of the unknown reaction *ea*, draw strings *ao*, *bo*, etc., respectively parallel to rays *AO*, *BO*, etc. Close the polygon *1, 2, 3, 4, 5* by drawing the closing side *eo* from point *5* to point *1*.

The position of point *E* in the force polygon can now be found by drawing the unknown ray *OE* parallel to the closing string *oe*. Vector *DE* represents the magnitude and sense of reaction R_2, and vector EA similarly determines reaction R_1. By scaling from the force polygon, we obtain

$$R_1 = EA = 92 \text{ N};\ R_2 = DE = 168 \text{ N} \qquad \textbf{\textit{Ans.}}$$

8-11.2 The Fink truss shown in Fig. 8-11.2 carries the given loads and is supported by a roller at *A* and a hinge at *B*. Determine graphically the reactions at *A* and *B*.

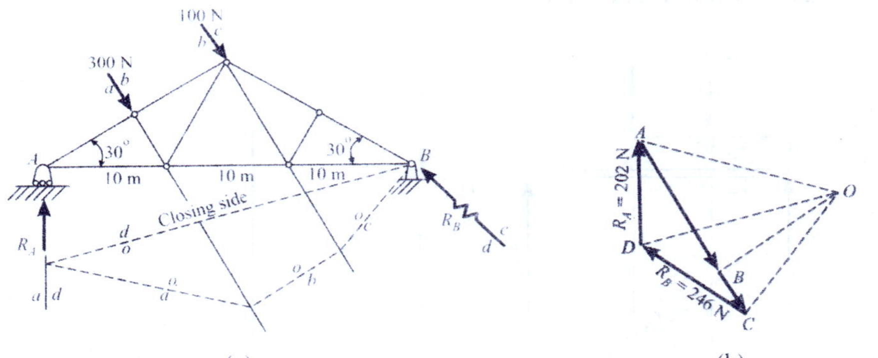

Figure 8-11.2

Solution

Using a convenient scale, plot the force polygon *ABCD* as shown in Fig. 8-11.2b. The given loads are represented by the portion *ABC*. The reaction at roller *A*, whose line of action is designated by *ad*, is constrained to be vertical because of the roller support. Represent this reaction in the force polygon by a vertical line through A for vector DA. However, the magnitude of this vector is unknown; consequently the position of point D remains to be determined. Once this is determined, the magnitude and direction of the unknown hinge reaction R_B will be given by vector CD which closes the force polygon. Choose a pole O, and draw the known rays AO, BO, and CO. The direction of the unknown ray OD, which locates D on the force polygon, must be determined from the closing side of the string polygon.

As discussed in Illus. Prob. 8-11.1, a characteristic of the string polygon is that its apexes must lie on the action lines of the forces. In the present problem, hinge B is the only known point on the action line of R_B, and therefore the hinge must be an apex of the string polygon. In other words, to make B an apex, the string polygon must be begun at B.

Beginning at point B in Fig. 8-11.2, draw strings co, bo, and ao, thereby establishing the closing side do. The unknown ray DO in the force polygon can now be drawn parallel to the closing side do of the string polygon. The intersection of ray DO with the vertical line previously drawn through point A determines point D in Fig. 8-11.2b. The magnitudes of the reactions are obtained by scaling from the force polygon. These values are

$$R_A = DA = 202 \text{ N}; R_B = CD = 246 \text{ N} \qquad \textit{Ans.}$$

The directions and senses of the reactions can also be obtained from the force polygon.

PROBLEMS

8-11.3 Determine graphically the reactions on the simply supported beam loaded as shown in Fig. P-8-11.3.

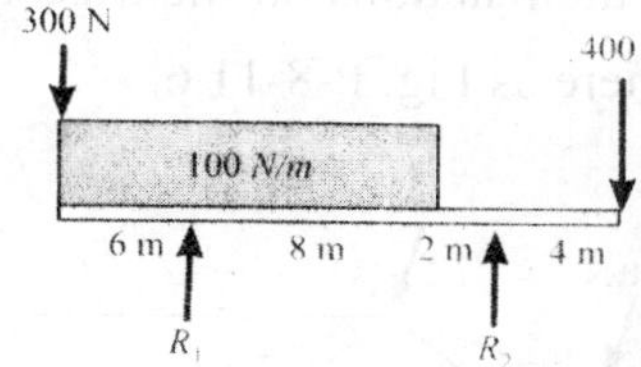

Figure P-8-11.3

$$R_1 = 1580 \text{ N}; R_2 = 520 \text{ N} \qquad \textit{Ans.}$$

8-11.4 Determine graphically the reactions of the truss shown in Fig. P-8-11.14.

Hint: The reactions are unchanged if the load at F is applied at E.

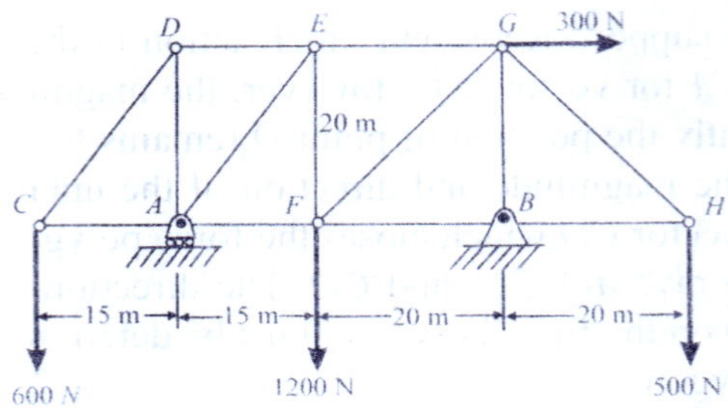

Figure P-8-11.4

$R_A = 1085$ N; $R_B = 1250$ N up to the left at $\theta_h = 76.1°$ ***Ans.***

8.11.5 Check graphically the reactions to the truss described in Prob. 3-5.18 and shown again here as Fig. P-8-11.5.

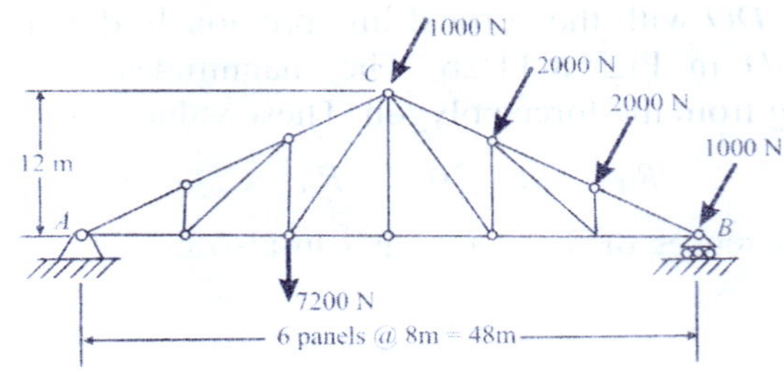

Figure P-8-11.5

8.11.6 Check graphically the reactions on the truss described in Prob. 3-5.17 which is repeated here as Fig. P-8-11.6.

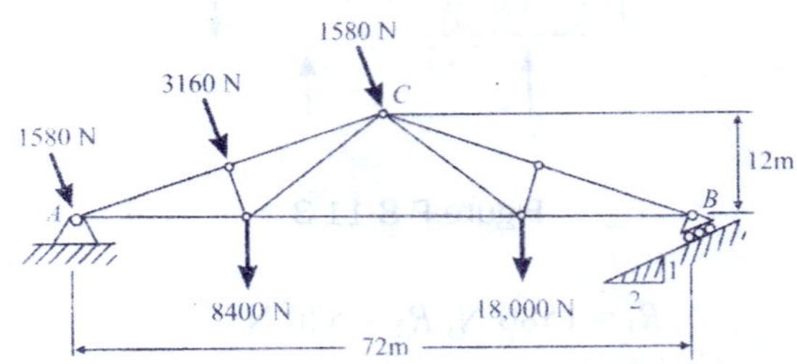

Figure P-8-11.6

8-12 GRAPHIC ANALYSIS OF TRUSSES MAXWELL DIAGRAMS

Consider the truss loaded as in Fig. 8-12.1 and supported by a roller at A and a hinge at D. The values of the reactions are assumed to have been determined. The stresses in the various members of the truss may be determined graphically by constructing the polygon of forces acting on each joint. From Section 4-3, the forces at a joint are concurrent, the unknown stresses being directed along the axes of the members. Since no more than two unknown forces can be determined to maintain equilibrium of a concurrent force system, the solution is begun at a joint at which there are only two unknown stresses. Selecting joint A, lay out R_1 to some convenient scale. Through the tip of R_1 draw a line parallel to AB and, through the tail of R_1, a line parallel to AC. The intersection of these two lines determines the magnitudes of the stresses in members AB and AC. Proceeding next to joint B and starting with the now known stress AB, determine the stresses in BC and BD. Continue to draw force polygons until the stresses at all joints have been found.

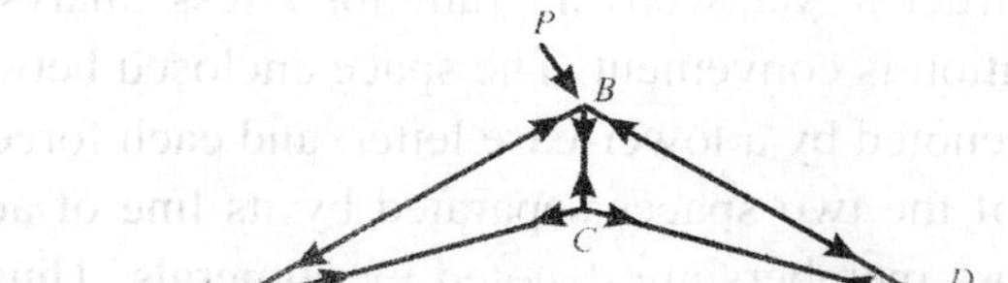

Figure 8-12.1 Maxwell diagram.

Notice that the force polygons for joints A and B have the common stress AB representing the reaction of member AB upon the pins at A and B. These reactions are equal but oppositely directed. Also observe that the force polygons for joints A and B can be superimposed because of the common vector for stress AB. Hence the stress in AB need be represented by only one line.

The force polygon for joint D can be similarly superimposed on the common diagram for joints A and B to form the single stress diagram shown in the figure. One procedure must be observed if the force polygons are to be superimposed on each other – when we construct the polygons, the vectors of loads and stresses must be drawn consecutively in the order in which the loads and stresses are met in a clockwise direction around each joint.[8] It is immaterial whether the order of drawing the loads and stresses is taken by a clockwise or counterclockwise direction around a joint. *But the same, order must be taken around each joint in order to combine the several force polygons into a common one.* This combined diagram of loads and stresses is known as a *Maxwell diagram*, after James Clerk Maxwell, who published the method in 1864.

When we construct a Maxwell diagram for stress analysis, the following type of Bow's notation is convenient. The space enclosed between two adjacent external forces is denoted by a lower-case letter, and each force is designated by the capital letters of the two spaces separated by its line of action. The spaces enclosed by the truss members are denoted by numerals. Thus any vector on a Maxwell diagram which includes a numeral in its designation is automatically recognized as a stress in a truss member as distinguished from an applied load whose designation includes only letters.

For example, consider the truss shown in Fig. 8-12.2a, Load P denoted by the letters *a-b*; load Q by *b-c*; reaction R_h by *c-d*, etc. The members of the truss are similarly named; thus the upper chord member is b-2. When constructing a Maxwell diagram, each load or member stress is designated by the corresponding upper-case letters or numerals at the end of the force vector.

To construct a Maxwell diagram, we begin by drawing the force polygon for the truss, considered as a free body; this is shown in Fig. 8-12.2b. Let us assume

---

[8]As shown in Section 2-4, the order in which the vectors of a force polygon are added is immaterial. Unless a special sequence is observed, however, the vectors for one joint will not be in the correct order so that those for succeeding joints can be attached without redrawing some of them.

that the *clockwise* order is to be used throughout in drawing the vectors acting on a free body. (The reactions on the truss are assumed to have been previously determined.) Considering joint *e-a-1* as a free body, we construct the polygon of forces *E-A-1* by drawing through point *A* vector *A-1* parallel to member *a-1*, and through point *E* vector *1-E* parallel to member *1-e*. Note that these three vectors follow each other in the clockwise order in which they are met in going around joint *e-a-1*, starting with the known force *E-A*.

Considering next joint *1-a-b-2*, we begin with the known vectors *1-A* and *A-B* (Fig. 8-12.2c) and, in clockwise order, draw through point *B* vector *B-2* parallel to member *b-2*, and through point *1* vector *2-1* parallel to member *2-1*. The intersection of these two vectors establishes point 2.

At joint *2-b- c-3*, the known vectors (in clockwise order) are *2-B* and *B-C*. These are already established. The force polygon *2-B-C-3* (Fig. 8-12.2d) is completed by drawing through points *C* and *2* vectors *C-3* and *3-2* parallel respectively to members *c-3* and *3-2*, thereby establishing point *3* on the Maxwell diagram.

Finally, joint *3-c-d-e* is considered as a free body. Here there is only one unknown, that in member *e-3*. Since points *E* and *3* are already established, the accuracy of the construction may be checked by drawing vector *E-3* (Fig. 8-12.2e) and noting whether it is parallel to member *e-3* as it should be. Although joint *e-1-2-3* has not been considered a free body, it is represented by the force polygon *E-1-2-3*.

The stress in any member can now be obtained by scaling the length of the corresponding line in the completed Maxwell diagram (Fig. 8-12.2e). The character of the stress in any member is obtained by using Bow's notation in the following way. Select a joint and name a member by reading the letters in a clockwise direction around the joint. Now, *maintaining the same order of letters*, read the letters which designate this member's stress in the Maxwell diagram. The reaction of the member upon this joint is in the direction in which the stress is read. For example, consider the upper chord member which is read as *b-2* at joint *1-a-b-2* and as *2-b* at joint *2-b-c-3*. In the Maxwell diagram (Fig. 8-12.2e), reading this member in direction *B* toward *2* shows that the stress acts toward joint *1-a-b-2*, whereas reading it in direction *2* toward *B* shows that it acts toward joint *2-b-c-3*. In either case, the stress in member *b-2* is compressive, according to the convention established in Section 4-2.

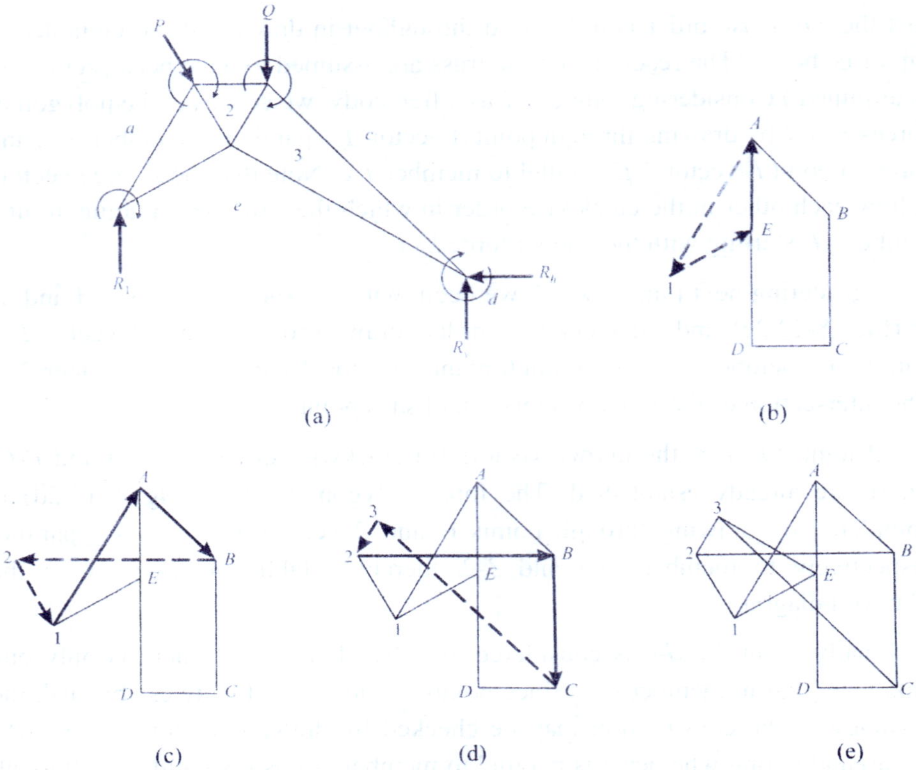

Figure 8-12.2 Construction of a Maxwell diagram.

A convenient aid in drawing a Maxwell diagram is to observe that there is a corresponding point in the diagram for each space in the truss diagram. Furthermore, through each point on the Maxwell diagram must pass lines parallel to all the lines that bound the corresponding space in the truss diagram. Thus the Maxwell diagram can be drawn by starting with the force polygon *A-B–C-D-E* and drawing through point *E* lines parallel to all those bounding space *e* in Fig. 8-12.2a. By doing the same thing at points *A* and *C*, points *1* and *3* will be established. Then lines through points *1* and *3* drawn parallel to all those that bound spaces *1* and *3* will determine point *2* on the Maxwell diagram. The accuracy of the construction can be checked by observing whether stress *B-2* is parallel to member *b-2*.

ILLUSTRATIVE PROBLEM

8-12.1 The triangular Fink truss, loaded as in Fig. 8-12.3, is supported by a roller at *A* and a hinge at *O*. Determine graphically the stress in each member of the truss.

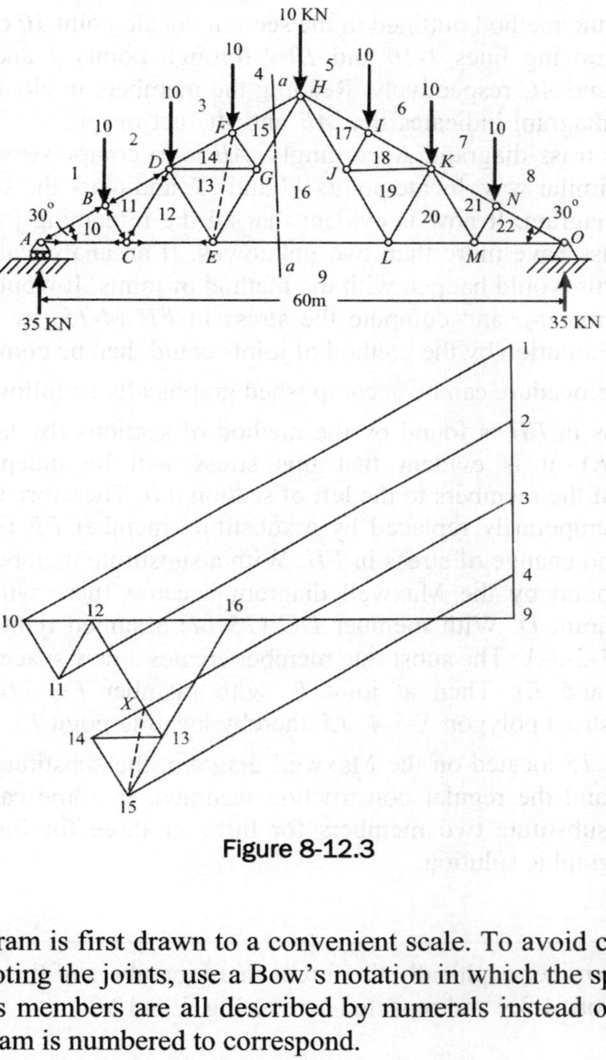

Figure 8-12.3

Solution

The truss diagram is first drawn to a convenient scale. To avoid confusion with the letters denoting the joints, use a Bow's notation in which the spaces between loads and truss members are all described by numerals instead of letters.[9] The Maxwell diagram is numbered to correspond.

Since the loading is symmetrical, the reaction at *A*, denoted by *9-1* is 35 KN. Also as a result of symmetry, only the stresses in one-half of the truss need be found. Accordingly, start the Maxwell diagram with the force polygon *9-1-2-3-4*, using a clockwise order.

---

[9]This departure from the accepted convention described above is adopted here because of the convenience of describing a particular joint in the discussion by a single letter, and because of the student's familiarity with defining a member in terms of the joints that it connects. This shows that there is nothing sacred about the type of notation adopted, although some admittedly have advantages over others.

Following the method outlined in the section, locate point *10* on the Maxwell diagram by drawing lines *1-10* and *10-9* through points *1* and *9* parallel to members *AB* and *AC* respectively. Reading the members in clockwise order on the Maxwell diagram indicates that *AB* and *AC* act on pin *A* as shown by the arrows on the truss diagram. Accordingly, *AB* is in compression and *AC* is in tension. In a similar way, locate points *11* and *12*, and mark the stress directions on the truss diagram. It now is evident that all the remaining joints of the left half of the truss have more than two unknowns. If an analytical solution were being made, this would happen with the method of joints. It would be necessary to take a section *a-a* and compute the stress in *FH* (*4-15*) by the method of moments. The solution by the method of joints could then be completed.

The same procedure can be accomplished graphically as follows:

If the stress in *FH* is found by the method of sections (by taking moments about point *E*), it is evident that this stress will be independent of the arrangement of the members to the left of section *a-a*. Therefore if members *FG* and *DC* are temporarily replaced by a substitute member *FE* (shown dotted), there will be no change of stress in *FH*. With a substitute member, the stress in *FH* can be found by the Maxwell diagram because there will be only two unknowns at joint *D*. With member *DG* (*13-14*) assumed removed, construct polygon *12-11-2-3-X* (The substitute member creates a new space *x*, bounded by joints *D*, *F*, and *E*). Then at joint *F*, with member *FG* (*14-15*) assumed removed, construct polygon *X-3-4 -15*, thereby locating point *15*.

With point *15* located on the Maxwell diagram, the substitute member may be discarded and the regular construction resumed. In some cases, it may be necessary to substitute two members for three or three for four, in order to complete the graphic solution.

PROBLEMS

8-12.2 Determine graphically the forces in all members of the truss described in Prob. 4-3.9 and repeated here as Fig. P-8-12.2.

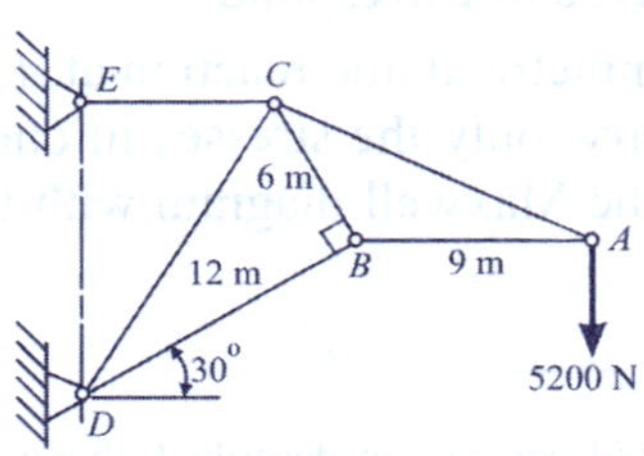

Figure P-8-12.2

8-12.3 Determine graphically the forces in all members of the truss described in Prob. 4-3.10 and shown here again as Fig. P-8-12.3.

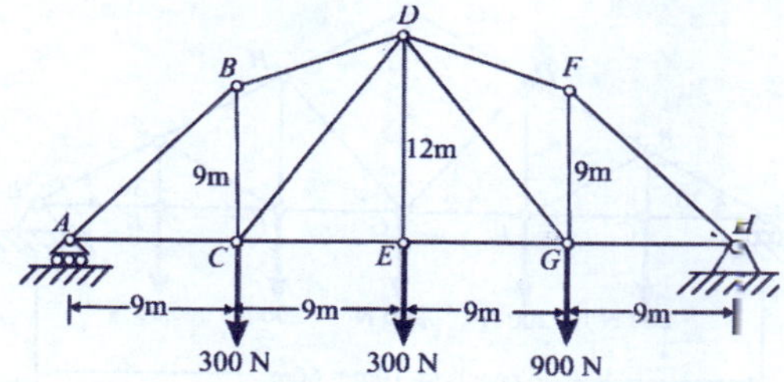

Figure P-8-12.3

8-12.4 Determine graphically the forces in all members of the truss described in Prob. 4-4.7 which is repeated here as Fig. P-8-12.4.

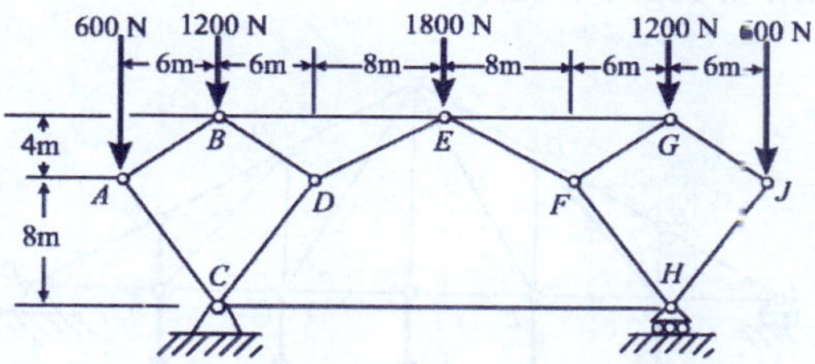

Figure P-8-12.4

8-12.5 Determine graphically the forces in all members of the fan Fink truss shown in Fig. P-8-12.5

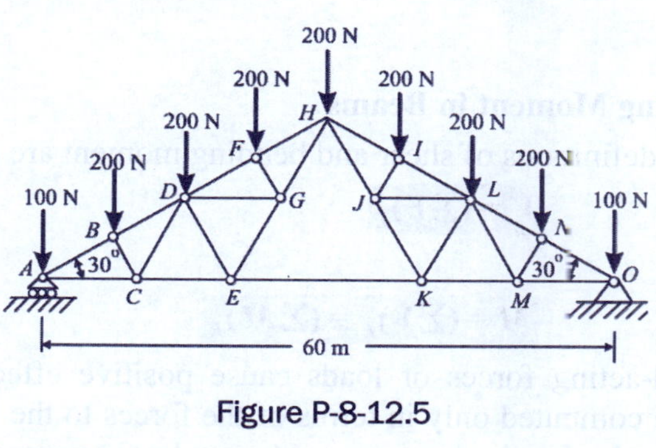

Figure P-8-12.5

$DF = 1200$ N C; $DG = 173$ N T; $EG = 346$ N T; $EK = 693$N T ***Ans.***

8-12.6 Determine graphically the force in each member of the triangular Howe truss shown in Fig P-8-12.6.

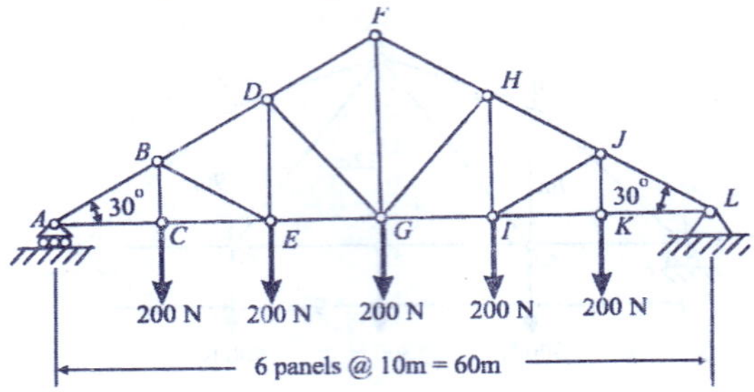

Figure P-8-12.6

$BD = 800$ N C; $DG = 264$ N C; $FG = 600$ N T **Ans.**

8-12.7 Determine graphically the force in each member of the triangular Pratt truss shown in Fig. P4-12.7;

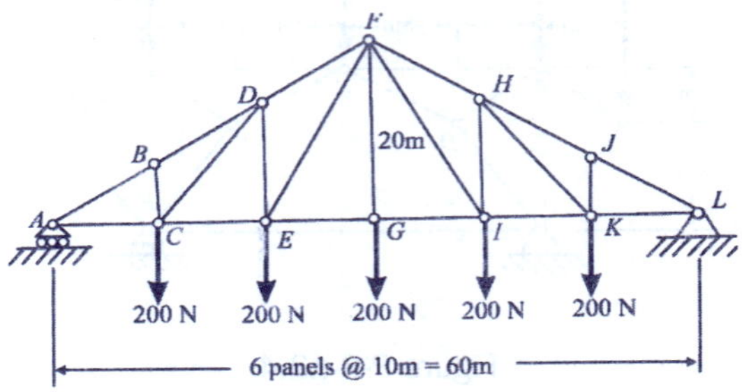

Figure P-8-12.7

$BD = 900$ N C; $EF = 336$ N T; $EG = 450$ N T **Ans.**

SUMMARY

Shear and Bending Moment in Beams

The fundamental definitions of shear and bending moment are expressed by

$$V = (\Sigma Y)_L \tag{8-2.1}$$

and

$$M = (\Sigma Y)_L = (\Sigma M)_R \tag{8-2.2}$$

in which upward-acting forces or loads cause positive effects. The shearing force V should be computed only in terms of the forces to the left of the section being considered; the bending moment M may be computed in terms of the forces either to the left or the right of the section, depending on which requires less arithmetical work.

Relations among load, shear, and bending moment are given by

$$w = \frac{dY}{dx} = \textbf{slope of shear diagram} \quad (8\text{-}3.3)$$

and

$$V = \frac{dM}{dx} = \textbf{slope of moment diagram} \quad (8\text{-}3.4)$$

which enable the proper shapes of the shear and moment diagrams to be sketched easily and rapidly. In their integrated form, we have

$$V_2 - V_1 = \Delta V = (\text{Area})_{\text{load}} \quad (8\text{-}3.1)$$

and

$$M_2 - M_1 = \Delta M = (\text{Area})_{\text{shear}} \quad (8\text{-}3.2)$$

which provide alternate methods of computing shear and bending moment, and supplement Eqs. (8-2.1) and (8-2.2).

Virtual Work

The differential work done by all external forces on a rigid body expressed in terms of their resultant force-couple is

$$dU = \mathbf{R} \cdot d\mathbf{r}_O + \mathbf{M}_O \cdot d\boldsymbol{\theta} \quad (8\text{-}4.1)$$

where $d\mathbf{r}_O$ is the differential displacement of any arbitrary reference point O and $d\boldsymbol{\theta}$ is the differential rotation of the body about O. Clearly, if $dU = 0$, either no motion occurs or both **R** and $\mathbf{M}_O$ are zero; either condition specifies that equilibrium exists.

Equation (8-4.1) can be used as an alternate approach to equilibrium problems if, by changing some of the specified constraints of a given situation, we permit *possible* small movements of the system to occur. Only *small* movements are permitted; otherwise, there would be a change in the geometry of the original configuration. If we use δ instead of d to distinguish a virtual or hypothetical quantity from a real or actual one, Eq. (8-4.1) becomes.

$$\delta U = \mathbf{R} \cdot \delta\mathbf{r}_O + \mathbf{M}_O \cdot \delta\boldsymbol{\theta}$$

Now if the differential virtual work is zero when these small movements are assumed to occur, it follows that both **R** and $\mathbf{M}_O$ must be zero. Hence an alternate statement of equilibrium is

$$\delta U = 0 \quad (8\text{-}6.1)$$

Refer to the illustrative problems of Section 8-6 for further details.

Graphic Statics

The resultant of any coplanar force system can be determined in magnitude and direction by the tip-to-tail addition of the free vectors of the forces. The force polygon thus created does not, however, determine the position of the resultant force. This position is determined by applying the parallelogram law in the following manner:

Select pairs of equal, opposite, and collinear components of forces composing the force polygon by drawing *rays* from a pole *O* to the ends of the force vectors. These rays form the pairs of components into which the given forces are resolved.

Start at any convenient point on the action line of a force and resolve it into components parallel to the rays drawn to its ends. These components are represented in direction, but not in magnitude, by lines called *strings*. At the intersection of a string with the action line of the second force for which it represents a common component, this second force is also resolved into components.

This procedure is continued until all forces have been replaced by pairs of components, all of which, except two, will cancel. These last two intersect in a point on the action line of the resultant. As previously noted, the magnitude and direction of the resultant are determined from the force polygon.

If the force polygon closes, the magnitude of the resultant is zero. In this case either equilibrium exists or the resultant is a couple consisting of the rays to the closing point of the force polygon and a moment arm which is the perpendicular distance between the parallel strings representing these rays. However, if the strings coincide, the moment arm will be zero and the string polygon is said to close. We conclude that for the case in which *both* the force polygon and the string polygon close, the resultant is zero and equilibrium exists.

A Maxwell diagram is used to determine the stress analysis of a truss. This is made by combining the force polygons acting at each joint into a common polygon. Note that the same order of loads and stresses must be taken around each joint in order to combine the several force polygons into a common one. A convenient help in drawing the Maxwell diagram is to observe that for each space in the truss diagram there is a corresponding point in the Maxwell diagram. Furthermore, through each point on this diagram there must pass lines parallel to all lines bounding the corresponding space in the truss diagram.

PART - II

DYNAMICS

CHAPTER 9

KINEMATICS OF A PARTICLE

9-1 INTRODUCTION

In this and in succeeding chapters, we shall develop rigorously the principles of dynamics. At this place, however, we present an overall preview of what these principles are and how they are interrelated. First recall that in Chapter 2 we showed that any force system could be reduced to a resultant force-couple system. In statics, both the resultant force and the resultant couple were zero, thus establishing the equations of static equilibrium. In dynamics, however, the resultant force-couple system is not zero and causes a change in the state of motion of the body on which they act.

Usually the body is rigid and the resultant force-couple system is applied at the center of gravity of the body. If the resultant of the applied force system consists only of a single force passing through the gravity center of a body starting from rest as in Fig. 9-1.1*a*, the body will move in the direction of the resultant **R**, but it will not rotate. If the direction of **R** is constant, the motion of the body is along a straight path and is called rectilinear translation. If the direction of **R** varies, although continuing to pass through the gravity center, so will the motion of the body, resulting in a curved path motion known as curvilinear translation.

If the resultant of the applied force system is a couple **M** as in Fig 9-1.1*b*, the body will spin about an axis passing through its center of gravity and directed perpendicular to the plane of the couple, but the center of gravity will remain stationary. All particles will describe circular paths about the axis of rotation. This type of motion is called centroidal rotation.

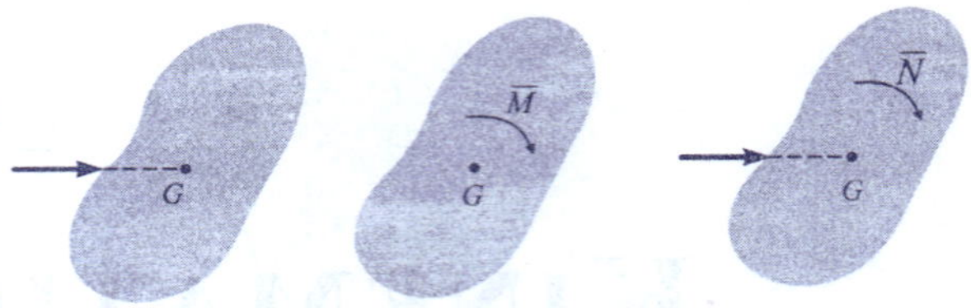

(a) Translation (b) Rotation (c) Combined translation and rotation

Figure 9-1.1 Types of rigid-body motion as determined by different resultants of applied forces.

When the resultant of the applied force system consists of both a centroidal force and a couple as in Fig. 9-11*c*, the body has motion consisting of a combined translation and a centroidal rotation. Further, if the applied forces always lie in the same plane, the resulting motion will be coplanar; otherwise, the motion will be spatial or three-dimensional.

The converse of the preceding discussion is also true; namely, the type of motion specifies the resultant of the applied force system. For example, a body constrained to have a motion of translation requires that the applied forces be so distributed that their resultant is a single force passing through the center of gravity. Or if a body is constrained so that it can only rotate about its centroidal axis, the resultant of the applied force system must be a couple.

The correlation between applied forces and the motion of body is governed by only two basic equations which we shall develop later but merely state now. They are

$$R = m\bar{a} = \frac{W}{g}\bar{a} \qquad (9\text{-}1.1)$$

which relates the resultant force **R** to the mass m of the body and the acceleration $\bar{\mathbf{a}}$ of its mass center. (Frequently, the alternate use of W/g for mass m is a useful stratagem.)

The other basic equation is

$$\Sigma\bar{M} = \dot{\bar{H}} \qquad (9\text{-}1.2)$$

which relates the moment sum $\Sigma\bar{\mathbf{M}}$ of the applied forces about the mass center, to the rate of change $\dot{\bar{\mathbf{H}}}$ of the angular momentum $\bar{\mathbf{H}}$ of the body about its mass center. This equation is valid for all angular motion, but it reduces to the simpler form

$$\Sigma\bar{\mathbf{M}} = \bar{I}\alpha \qquad (9\text{-}1.2a)$$

for coplanar motion of a rigid body which is symmetrical about a plane of motion containing its mass center. The term $\bar{I}$ is the mass moment of inertia of the body about a centroidal axis perpendicular to the plane of motion while α is the angular acceleration of the body.

These equations are valid only at a given instant of the body's motion. They are used to determine either the instantaneous forces acting when the motion of a body is known or conversely, when the forces are specified, they determine the accelerations from which the motion of the body may be calculated. Note that either $\mathbf{R}$ or $\Sigma\overline{\mathbf{M}}$ or both may be constant or variable. If variable, they may depend on position, velocity, or time or even a combination of these items. Such combinations may affect the complexity of the mathematical solution in a given situation, but should not obscure the fact that all problems in dynamics depend on only these two basic equations plus the correlation among acceleration and the motion. Such correlation is known as the time-geometry of motion and constitutes the area known as kinematics. It defines the motion of a particle or a body without consideration of the forces causing the motion. Since kinematics is the starting point for the study of dynamics, it is necessary that we begin with a clear understanding of what is meant by displacement, velocity, and acceleration and how they are related to each other.

9-2 MOTION OF A PARTICLE

The natural way of describing the motion of a particle is by lines of sight from an arbitrary origin to the successive positions of the particle. These lines of sight are called position vectors. Usually the position vectors emanate from an origin that is *fixed* in space, resulting in what is called absolute motion. With respect to a *moving* origin, the position vectors define what is known as relative motion. In most engineering situations, any origin on the earth's surface constitutes a fixed origin. For orbiting satellites, however, because of the rotation of the earth's surface, the fixed origin should be at the earth's center. For interplanetary motion, the origin may be taken at the mass center of the solar system, while for astronomical observations, reference axes are referred to the so-called fixed stars.

Although the following discussion is valid with respect to either a fixed or moving origin, for the present we take the origin as fixed. Later in Section 12-9, we shall discuss the correlation between motions with respect to fixed and moving origins. In Fig. 9-2.1, the position vectors $\mathbf{r}_A$ and $\mathbf{r}_B$ completely define positions A and B of a moving particle. Their difference $\Delta\mathbf{r}$ denotes the change in position (also known as displacement) occurring in the elapsed time Δt. Observe that the change in position $\Delta\mathbf{r}$ is independent of the choice of an origin as shown by the dashed position vectors to A and B from any, other fixed origin O_1. Another observer at O_1 simultaneously sees the same displacement $\Delta\mathbf{r}$ as an observer at O. Since the elapsed time interval is the same for both observers, both record the same displacement, velocity, and acceleration of the moving particle albeit in terms of different position vectors.

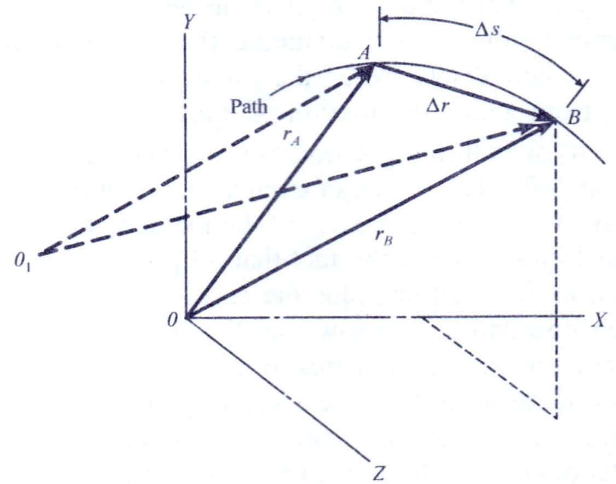

Figure 9-2.1 Relation between path and position vectors.

Observe carefully the distinction between a change in position and the distance traveled along the path. The change in position $\Delta\mathbf{r}$ is a vector involving both magnitude and direction whereas the arc distance Δs is a scalar measuring only the length of the path. The difference between the vector displacement (i.e., change in position) and the scalar distance is very evident if we assume the particle to move from A to B and back to A. The change in position, $\Delta\mathbf{r}$, would then be zero, but the distance traversed would be the accumulated length from A to B and back to A. This distinction is important because all the kinematic relations we shall develop involve vector change of position and *not* scalar distance.

VELOCITY

We define velocity as the time rate of change of position. If Δt is the elapsed time between A and B, the average velocity between these points will be

$$\mathbf{v}_{\text{ave}} = \frac{\Delta\mathbf{r}}{\Delta t}$$

and the instantaneous velocity is the limit of this ratio as Δt approaches zero, or

$$\mathbf{v} = \lim_{\Delta t \to 0} \frac{\Delta\mathbf{r}}{\Delta t} = \frac{d\mathbf{r}}{dt} \tag{9-2.1}$$

The geometric significance of this result is that in Fig. 9-2.1 as Δt approaches zero, B gets closer to A and chord $\Delta \mathbf{r}$ coincides more completely with the arc Δs so that, in the limit, $d\mathbf{r}$ coincides with ds and consequently the velocity $\mathbf{v}$ is tangent to the path.

We can carry this interpretation further by rewriting Eq. (9-2.1) as

$$\mathbf{v} = \lim_{\Delta t \to 0} \frac{\Delta \mathbf{r}}{\Delta s}\frac{\Delta s}{\Delta t} = \frac{d\mathbf{r}}{ds}\frac{ds}{dt}$$

The ratio $\dfrac{\Delta \mathbf{r}}{\Delta s}$ is the chord of an arc divided by the length of the arc. In the limit, the chord and the arc coincide and $\dfrac{d\mathbf{r}}{ds}$ becomes a vector of unit length directed tangent to the path. Denoting this unit tangent vector by $\hat{\mathbf{e}}_t$ and recognizing that $\dfrac{ds}{dt}$ denotes the *speed* (i.e., magnitude of velocity or rate of change of distance) of the particle along the path, we obtain

$$\mathbf{v} = \frac{ds}{dt}\hat{\mathbf{e}}_t = v\hat{\mathbf{e}}_t \tag{9.2.2}$$

ACCELERATION

We define acceleration as the time rate of change of velocity. If $\Delta \mathbf{v}$ is the change in velocity during the time Δt, the average acceleration will be

$$\mathbf{a}_{\text{ave}} = \frac{\Delta \mathbf{v}}{\Delta t}$$

and the instantaneous acceleration is the limit of this ratio as Δt approaches zero, or

$$\mathbf{a} = \lim_{\Delta t \to 0} \frac{\Delta \mathbf{v}}{\Delta t} = \frac{d\mathbf{v}}{dt} \tag{9-2.3}$$

Since $\mathbf{v} = \dfrac{d\mathbf{r}}{dt}$, the instantaneous acceleration may also be written as

$$\mathbf{a} = \frac{d\mathbf{v}}{dt} = \frac{d\left(\dfrac{d\mathbf{r}}{dt}\right)}{dt} = \frac{d^2 r}{dt^2} \tag{9-2.3a}$$

Since time is so frequently the independent variable in dynamics, it will often simplify notation if we use the dot convention devised by Newton to signify differentiation with respect to time. This consists of a single dot above a variable to denote its first derivative with respect to time, while two dots above a variable denote its second time derivative. When we use this convention, velocity and acceleration may be expressed as

$$\mathbf{v} = \frac{d\mathbf{r}}{dt} = \dot{\mathbf{r}} \tag{9-2.1}$$

$$\mathbf{a} = \frac{d\mathbf{v}}{dt} = \dot{\mathbf{v}} = \ddot{\mathbf{r}} \tag{9-2.3}$$

Let us now discuss more explicitly how the change in the velocity vector is correlated to acceleration. The change in a vector is easy to see if the vectors radiate from the same reference point. While this is natural for the position vectors as in Fig. 9-2.2a, it must be contrived for the velocity vectors. We do this by representing the velocities as free vectors, radiating from a common origin as in Fig. 9-2.2b where the path, for ease of representation, is assumed to be in the plane of the paper. A visual comparison between the simultaneous changes in the position and velocity vectors is achieved by drawing the position vector **r** to the particle at *equal* successive intervals of time. Then as shown in Fig. 9-2.2a, the velocity is tangent to the path and its magnitude is proportional to the distance from one position vector head to the next.

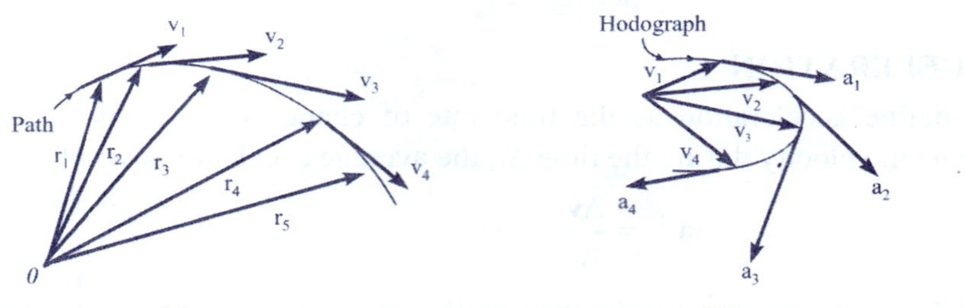

Figure 9-2.2 Relations among position, velocity, and acceleration for equal time intervals along a plane curve.

We now draw the successive velocity vectors from a common origin as in Fig. 9-2.2b, Then it is obvious that the change in velocity (drawn from one velocity vector head to the next) is caused by both the swinging and stretching of the velocity vector; that is, by changes in both the direction and the magnitude of the velocity vector. In addition, just as the path of the particle is generated by the head of the position vector, we can imagine a curve being generated simultaneously by the head of the velocity vector. This curve is called the *hodograph*. A chord of the hodograph, drawn from one velocity vector head to the next, is the change in velocity during the corresponding time interval. In the limit as the time interval approaches zero, this chord coincides with the hodograph so that the acceleration vector is tangent to the hodograph in the analogous manner that the velocity is tangent to the corresponding point of the actual path of the particle. As Fig. 9-2.2b demonstrates, the velocity and acceleration have different directions so that the acceleration is *not* tangent to the path of the particle except for the special case of straight-line motion.

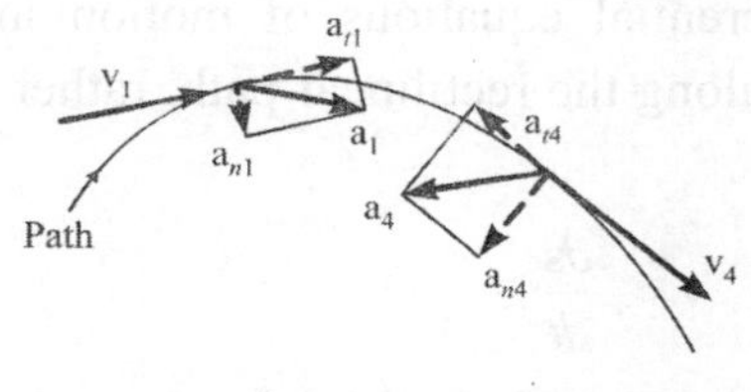

Figure 9-2.3 Velocity and acceleration related to path. Particle is speeding up at position 1, but slowing down at 4.

The correlation between velocity and acceleration is explained further in Fig. 9-2.3 where we have drawn the instantaneous velocity and acceleration vectors on the path itself for positions 1 and 4. At each position, the acceleration vector may be resolved into components that are tangent and normal to the path. In every instance, the normal component of acceleration will be directed inward toward the instantaneous center of curvature of the path. As we shall show later in Section 9-7, this normal component of acceleration occurs because of the changing *direction* of the velocity. The tangential component of acceleration coincides with the velocity and denotes the changing *magnitude* of the velocity. Indeed, if we consider the particle to be an automobile moving along a curving road, the car's speedometer measures the *speed* or magnitude of velocity while the tangential acceleration represents the variation in speed, being positive as we increase speed and vice versa.

UNITS

The units used for displacement, velocity, and acceleration depend on the units chosen to represent length and time, such as meter, mm, and kilometer for length; and second, minute, and hour for time. Accordingly, since displacement is synonymous with length, velocity with change of length per unit time, and acceleration with change of velocity per unit time, the common units for these terms are:

Displacement: meter, millimeter, kilometer
Velocity: meter per second (m/s), centimeters per sec (cm/sec), kilometer per hour (km/h) etc
Acceleration: meter per second per second (m/sec^2), mm per second per second (mm/sec^2), kilometer per hour per hour (km/h^2), etc.

9-3 RECTILINEAR MOTION

In rectilinear motion, the particle travels along a straight line. Since the motion is unidirectional, the sense of the position, velocity, and acceleration vectors may be denoted by a plus or minus sign as in Fig. 9-3.1. We may therefore rewrite the vector differential equations of motion in terms of the position coordinate s, measured along the rectilinear path, rather than the position vector $\mathbf{r}$ to obtain

$$\mathbf{v} = \frac{ds}{dt} \tag{9-3.1}$$

$$\mathbf{a} = \frac{d\mathbf{v}}{dt} \tag{9-3.2}$$

$$\mathbf{a}\,ds = \mathbf{v}\,d\mathbf{v} \tag{9-3.3}$$

where (9-3.3) has been obtained by eliminating dt from the others.

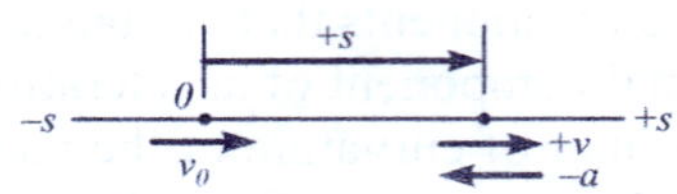

Figure 9-3.1

SIGNS

A convenient sign convention is to take the initial direction of motion as the positive sense of s, v, and a. Hence, a negative value of velocity obtained in applying the equations would mean that the velocity is directed oppositely to the initial direction of motion. A negative value for displacement would indicate that the position of the moving particle is to be measured back from the origin of displacement. Finally, if a moving particle returns to its starting position, the displacement s will be zero, *not* the distance actually traversed by the particle. Read the discussion of Fig. 9-2.1.

We now turn to the solution of the differential equations of motion. Our problem is to determine the relations among s, v, a, and t when a relation between any two of them is specified. Basically, we have three principal variables related by a common parameter t. Each of these principal variables may be known in terms of the time, or they may be specified in terms of each other or even a combination of the others. Let us consider here the simpler combinations indicated in the following box.

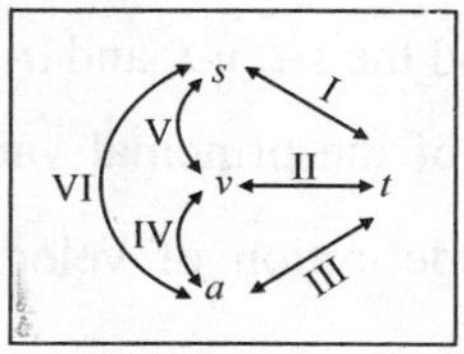

Case I: Given the displacement in terms of the time, i.e., $s = f(t)$, to find the v-t and a-t relations.

This is the simplest case and is easily solved by two successive differentiations of the s-t relation thus,

$$v = \frac{ds}{dt} \text{ and } a = \frac{dv}{dt} = \frac{d^2 s}{dt^2}.$$

Case II: Given the velocity in terms of the time, i.e., $v = f(t)$, to find the a-t and s-t relations.

Applying the definition $a = \frac{dv}{dt}$ gives the a-t relation directly, while separating the variables in $v = \frac{ds}{dt}$ gives $\int ds = \int f(t)dt$ which may be integrated to find the s-t relation.

Case III: Given the acceleration in terms of the time, i.e., $a = f(t)$, to find the v-t and s-t relations.

From $a = f(t) = \frac{dv}{dt}$, we separate the variables v and t to obtain $dv = f(t)\,dt$ which may be integrated to determine the v-t relation, whence we proceed as outlined in Case II to find the s-t relation

When the principal variables are not given as functions of the time, direct differentiation or integration cannot be performed without a preliminary treatment which we now discuss.

Case IV: Given $a = f(v)$, to find the a-t, v-t, and s-t relations.

Starting with $a = f(v) = \frac{dv}{dt}$, we separate the variables v and t to obtain $\frac{dv}{f(v)} = dt$ which is integrated to determine the v-t relation. Then we can proceed as outlined above in Case II.

Case V: Given $v = f(s)$, to find the s-t, v-t, and a-t relations.

Here, as in Case IV, one of the principal variables is given in terms of an adjacent variable. From the definition of velocity, we have $v = f(s) = \dfrac{ds}{dt}$ whence separating the variables s and t, we obtain $\dfrac{ds}{f(s)} = dt$ which upon integration determines the s-t relation.

Thus, we have Case I and two successive differentiations yield the v-t and a-t relations.

Case VI: Given $a = f(s)$, to find the s-t, v-t and a-t relations.

Here the principal variables are not adjacent as in the two previous cases. In this case, we substitute the given relation in $v\,dv = a\,ds$ to obtain $v\,dv = f(s)\,ds$ which is integrated and solved for $v = f(s)$. We now proceed as outlined in Case V.

The preceding outline is only a guide to mathematical analysis and should by no means be memorized or used as a set of rules. While the procedures are straightforward, some of the integrations sometimes may be complex enough to require reference to your calculus textbook.

UNIFORM ACCELERATION

One variation of Case III is so often used that we discuss it separately. This is the case of straight-line motion in which the acceleration is constant. As will be seen later (Section 11-2), this condition arises when a body is acted upon by forces which remain constant in magnitude and direction.

The usual assumptions are that at time $t = 0$, the initial velocity is v_o and $s = 0$. Applying $dv = a\,dt$ and integrating between the given limits, we have

$$\int_{v_o}^{v} dv = a\int_{0}^{t} dt \qquad \text{or} \qquad v = v_o + at \qquad \text{(9-3.4)}$$

Note that a is placed outside the integral sign because it is assumed constant.

We now replace the variable v, just found in terms of the time, in the differential equation $ds = v\,dt$ and again proceed to integrate between the given limits. This gives

$$\int_{0}^{s} ds = \int_{0}^{t} (v_o + at)\,dt \qquad \text{or} \qquad s = v_o t + \frac{1}{2}at^2 \qquad \text{(9-3.5)}$$

Finally, we use $v\,dv = a\,ds$ and again proceed to integrate between definite limits to obtain

$$\int_{v_o}^{v} v dv = a \int_{o}^{s} ds \qquad \text{or} \qquad v^2 = v_o^2 + 2as \qquad (9\text{-}3.6)$$

Remember that these equations can only be used when the acceleration is known to be constant. A common error is the attempt to apply them to all types of motion.

For bodies falling near the surface of the earth under the influence of gravity only, the acceleration may be assumed constant at the value $g = 9.81$ m/s^2 directed downward. In any motion involving freely falling bodies, therefore the equations of motion for constant acceleration may be applied directly by merely replacing a by g. The most convenient sign convention to use is that in which the initial direction of motion is the positive sense for displacement, velocity, and acceleration.

ILLUSTRATIVE PROBLEMS

9-3.1 As shown in Fig. 9-3.2, a stone is thrown vertically into the air from a tower 30 m high, at the same instant that a second stone is thrown upward from the ground. The initial velocity of the first stone is 15 m/s and that of the second stone is 23 m/s. When and where will the stones be at the same height from the ground?

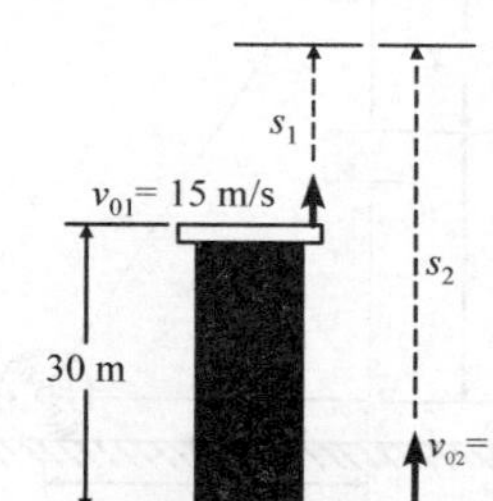

Figure 9-3.2

Solution

The initial direction of motion for each stone is upward which we therefore take as the positive sense for s, v, and a. Applying Eq. (9-3.5) and noting that the acceleration for freely falling bodies is $g = 9.81$ m/s^2 directed downward and in this case therefore is negative, we obtain

$[s = v_o t + \frac{1}{2} at^2]$ For stone 1: $s_1 = 15t - ½ \times 9.81t^2$ (*a*)

For stone 2: $s_2 = 23t - ½ \times 9.81t^2$ (*b*)

From Fig. 9-3.2, $s_2 - s_1 = 30$ m. Hence, subtracting Eq. (*a*) from Eq. (*b*) gives

$$s_2 - s_1 = 30 = 8t \quad \text{or} \quad t = 3.75 \text{ sec}$$

Substituting this value of t in Eqs. (*a*) and (*b*), we have

$s_1 = 15\,(3.75) - 4.905\,(3.75)^2 = 56.25 - 68.98$ $s_1 = -12.73$ m ***Ans.***

$s_2 = 23\,(3.75) - 4.905\,(3.75)^2 = 86.25 - 68.98$ $s_2 = 17.27$ m ***Ans.***

Hence, the stones pass each other 12.73 m below the top of the tower, or 17.27 m from the ground. Note that although we assumed that they would pass above the top of the tower, the negative sign of s_1 indicates otherwise. Since the terms involved in the equations are vector quantities, an incorrect assumption of direction merely results in a negative sign.

9-3.2 A rope of length L connects the wheel A and the weight B by passing over a pulley of negligible size at C as shown in Fig. 9-3.3. At the instant when $x = 3$ m, the center of wheel A has a velocity $v_A = 3.5$ m and an acceleration $a_A = 1.5$ m/s^2, both rightward. What is then the velocity and acceleration of B?

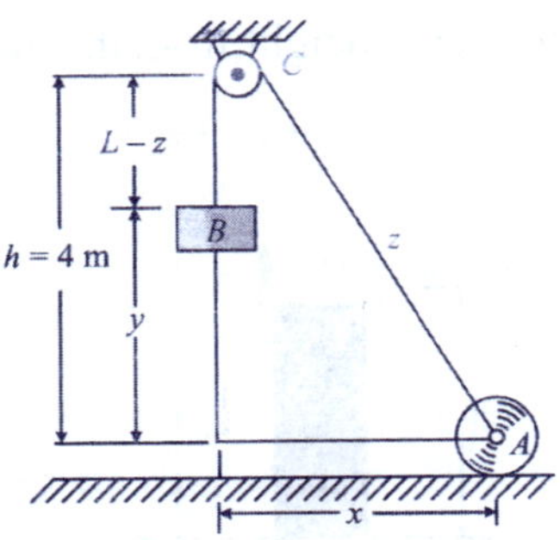

Figure 9-3.3

Solution
If we denote the variable distance AC by z, then the vertical length $CB = L - z$ and hence

$$h = L - z + y \tag{a}$$

From the figure, we also have

$$z^2 = x^2 + h^2 \tag{b}$$

By eliminating z from these relations, y may be expressed directly in terms of x so that, successive differentiation with respect to time will relate the velocity and acceleration of B to that of A. However, a preferable method is to retain z as a common parameter and proceed as follows:

Using the dot notation to indicate differentiation with respect to time, and noting that $\dot{y} = v_B$, we obtain from Eq. (*a*),

$$0 = -\dot{z} + \dot{y} \qquad \text{or} \qquad v_B = \dot{z} \tag{c}$$

This result is not surprising if we note that the change in length of z determines the rise (or fall) of B.

We next differentiate Eq. (*b*) with respect to the time which gives

$$2z\dot{z} = 2x\dot{x} \qquad \text{or} \qquad zv_B = xv_A \tag{d}$$

Another differentiation of Eq. (*d*) yields

$$\dot{z}\, v_b = z\dot{v}_B = \dot{x}\, v_A + x\dot{v}_A$$

or

$$v_B^2 + za_B = v_A^2 + xa_A \tag{e}$$

By substituting the given data in Eqs. (d) and (e) and noting that $z = 5$ m when $x = 3$ m, we obtain

[From Eq. (d)] $\quad 5v_B = 3(3.5) \qquad v_B = 2.1$ m/s up ***Ans.***

[From Eq. (e)] $\quad (2.1)^2 + 5\ a_B = (3.5)^2 + 3(1.5)$

$a_B = 2.47$ m/s^2 up ***Ans.***

An interesting aspect of this example is that to determine the accelerations here, the velocities must first be found.

9-3.3 The rectilinear motion of a particle is given by $s = v^2 - 3$ where s is in m and v in m/s. When $t = 0$, $s = 0$ and $v = 1.0$ m/s, determine the *s-t*, *v-t*, and *a-t* relations.

Solution

The given data suggest the procedure outlined in Case V; i.e., $v = f(s)$. However rather than follow this outline by expressing v in terms of s, we differentiate the given relation directly with respect to time, and obtain

$$\frac{ds}{dt} = 2v\frac{dv}{dt} \quad \text{or} \quad \text{v} = 2\text{va} \quad \text{or a} = \tfrac{1}{2}\ \text{m/s}^2$$ ***Ans.***

which is the *a-t* relation and shows *a* to be constant.

Since the acceleration is constant, we proceed directly to apply Eqs. (9-3.4) and (9-3.5) which are valid for constant acceleration. We therefore obtain

$$[v = v_o + at] \qquad v = 1 + \frac{1}{2}t \qquad \textbf{\textit{Ans.}}$$

$$[s = v_o t + \frac{1}{2}at^2] \qquad s = 1t + \tfrac{1}{4}t^2 \qquad \textbf{\textit{Ans.}}$$

9-3.4 The motion of a particle is defined by the relation $a = 2t$ where a is in m/s^2 and t in seconds. It is known that $s = 4$ m and $v = 2$ m/s when $t = 1$ sec. Find s and v at $t = 4$ sec.

Solution

Since a is given in terms of t, we start with $a = dv/dt$ in which we substitute $a = 2t$ to obtain $2t = dv/dt$. Separating the variables and integrating between the given limits we have,

$$[dv = a\,dt] \qquad \int_2^v dv = \int_1^t 2t\,dt$$

hence

$$v - 2 = t^2 - 1 \quad \text{or} \quad v = t^2 + 1 \qquad (a)$$

We now replace the variable v, just found in terms of the time, in the definition of velocity written in the form $ds = v\,dt$ and again proceed to integrate between the given limits, This gives

$$[ds = v\,dt] \qquad \int_4^s ds = \int_1^t (t^2 + 1)\,dt$$

hence

$$s = \frac{t^3}{3} + t + \frac{8}{3} \qquad (b)$$

Finally, substituting $t = 4$ sec in these *v-t* and *s-t* relations we obtain

$$[\text{From Eq. } (a)] \qquad v = 4^2 + 1 = 17 \text{ m/s} \qquad \textbf{\textit{Ans.}}$$

$$[\text{From Eq.}(b)] \qquad s = \frac{(4)^3}{3} + 4 + \frac{8}{3} = 28 \text{ m} \qquad \textbf{\textit{Ans.}}$$

A common error to be avoided is to substitute in $ds = v\,dt$ the particular value of v at $t = 4$ sec instead of using the *v-t* relation expressing v as a variable in terms of the time.

PROBLEMS

9.3.5 An automobile is driven at 15 m/s for 720 sec, then at 18 m/s for 1200 sec, and finally at 23 m/s for 480 sec. What is its average speed over this interval?

9-3.6 How fast must the automobile of the previous problem move in the last 480 sec to obtain an average speed of 16 ms/?

9.3.7 Car A at a gasoline station stays there for 600 sec after a car B passes at an average speed of 18 m/s. How long will it take car A moving at an average speed of 23 m/s to overtake car B?

9-3.8 On a certain stretch of track, trains run at 27 m/s. How far back of a stopped train should a warning torpedo be placed to signal an oncoming train? Assume that the brakes are applied at once and retard the train at the uniform rate of 1.2 m/s^2.

9-3.9 A stone is thrown vertically upward and returns to earth in 5 sec. How high does it go?

9-3.10 A ship being launched slides down the ways with a constant acceleration. She takes 4 sec to slide the first meter. How long will she take to slide down the ways if their length is 275 m?

9.3.11 A ball is thrown vertically into the air at 37 m/s. After 3 sec another ball is thrown vertically. What initial velocity must the second ball have to pass the first ball at 30 m from the ground?

9.3.12 A stone is dropped down a well and 5 sec later the sound of the splash is heard. If the velocity of sound is 242 m/s, what is the depth of the well?

9-3.13 A stone is dropped into a well with no initial velocity and 4.5 sec later the splash is heard. Then a second stone is thrown downward into the well with an initial velocity v_o and the splash is heard in 4.0 sec. If the velocity of sound is constant at 242 m/s, determine the initial velocity of the second stone.

9-3.14 A train moving with constant acceleration travels 7.5 m during the 10th sec of its motion and 5.5 m during the 12th sec of its motion. Find its initial velocity.

9.3.15 An automobile starting from rest speeds up to 12 m/s with a constant acceleration of 1.2 m/s^2, runs at this speed for a time, and finally comes to rest with a deceleration of 1.5 m/s^2. If the total distance traveled is 366 m, find the total time required.

9.3.16 An auto A is moving at 6 m/s and accelerating at 1.5 m/s^2 to overtake an auto B which is 18 m ahead. If auto B is moving at 18 m/s and decelerating at 0.9 m/s^2, how soon will A pass B?

9.3.17 A balloon rises from the ground with a constant acceleration of 0.9 m/s^2. Five seconds later, a stone is thrown vertically up from the launching site. What must be the minimum initial velocity of the stone for it to just touch the balloon? Note that the balloon and the stone have the same velocity at contact.

9-3.18 The rectilinear motion of a particle is governed by the equation

$s = r \sin \omega t$ where r and ω are constants. Show that the acceleration is $a = -\omega^2 s$.

9.3.19 The motion of a particle along a straight line is defined by

$$s = \frac{1}{3}t^3 - 36t.$$

(a) Find, the average acceleration during the fourth second. (b) When the particle reverses its direction, what is its acceleration?

9.3.20 A ladder of length L moves with its ends in contact with a vertical wall and a horizontal floor, if the ladder starts from a vertical position and its lower end A moves along the floor with a constant velocity v_A, show that the velocity of the upper end B is $v_B = -v_A \tan \theta$ where θ is the angle between the ladder and the wall. What does the minus sign mean? Is it physically possible for the upper end B to remain in contact with the wall throughout the entire motion? Explain.

9-3.21 In the previous problem, find the acceleration of the upper end B of the ladder as a function of θ.

9.3.22 The velocity of a particle moving along the X axis is defined by $v = kx^3 - 4x^2 + 6x$ where v is in m/s, x is in m, and k is a constant. If $k = 1$, compute the value of the acceleration when $x = 0.6$ m.

9-3.23 In the previous problem, find the smallest value of k that will make the acceleration equal to 5 m/s^2 at x = 0.9 m.

9-3.24 Because of the resistance exerted by a fluid, the rectilinear motion of a particle is given by $a = -kv$ where k is a constant. When $t = 0$, $s = 0$ and $v = v_{o,}$ determine the particle's velocity as a function of (a) the time t and (b) its position s. (c) what is the maximum distance the particle will move?

9-3.25 The motion of particle in rectilinear motion is defined by $a = 6\sqrt{v}$ where a is in m/s^2 and v in m/s. When $t = 2$ sec, $v = 10$ m/s and $s = 91$ m, determine the value of s at $t = 3$ sec.

9-3.26 The rectilinear motion of a particle is governed by $a = -8s^{-2}$ where a is in m/s^2 and s is in m. When $t = 1$ sec, $s = 1.2$ m, and $v = 0.6$ m/s, determine the acceleration of the particle at t = 2 sec.

9-3.27 The straight line motion of a particle is defined by $a = 3 + \frac{1}{2}t$. When $t = 0$, $s = 0.6$ m and $v = -1.2$ m/s find s at $t = 6$ sec.

9-3.28 The rectilinear motion of a particle is governed by $a = 12t - 6t^2$. It starts from rest when t = 0. Determine its velocity when it returns to its starting position.

9-4 MOTION CURVES

The use of motion curves which show the variation of s, v, and a with time frequently provides a preferred method of solving the problems considered before. The method is especially useful when the motion has distinct phases, each requiring its associated set of equations. The method also provides a means of using experimental data to determine the *s-t*, *v-t*, or *a-t* curves when any one of them is known. Especially useful is the *a-t* curve showing the variation of acceleration with time since it alone is sufficient to determine values of velocity and displacement at any instant of the entire motion.

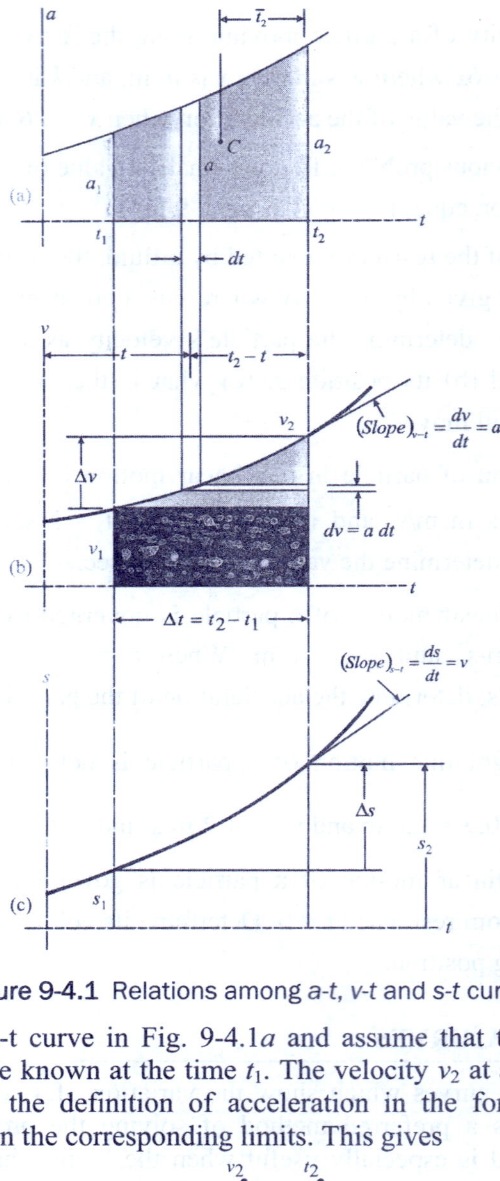

Figure 9-4.1 Relations among *a-t*, *v-t* and *s-t* curves.

Consider the a-t curve in Fig. 9-4.1*a* and assume that the velocity v_1 and displacement s_1 are known at the time t_1. The velocity v_2 at any other time t_2 is found by writing the definition of acceleration in the form $dv = a\ dt$ and integrating between the corresponding limits. This gives

$$\Delta v = \int_{v_1}^{v_2} dv = \int_{t_1}^{t_2} a\ dt$$

The geometric significance of the right-hand term is apparent from Fig. 9-4.1*a*. During the infinitesimal time interval *dt*, the acceleration *a* may be considered constant. Obviously *a dt* is the area of the unshaded vertical strip.

Since $\int_{t_1}^{t_2} a\,dt$ means the summation of such strips, we conclude that the shaded area under the a-t curve between the times t_1 and t_2 represents the velocity change Δv during this time interval, or

$$v_2 - v_1 = \Delta v = (\text{Area})_{v-t} \qquad (9\text{-}4.1)$$

Similarly, writing the definition of velocity in the form $ds = v\,dt$ and integrating between corresponding limits, we obtain

$$\int_{s_1}^{s_2} ds = \int_{t_1}^{t_2} v\,dt \quad \text{or} \quad s_2 - s_1 = \Delta s = (\text{Area})_{v-t} \qquad (9\text{-}4.2)$$

where in similar fashion in part (b) the shaded area under the v-t curve represents the corresponding change in displacement during the time interval from t_1 to t_2.

Instead of using a summation of vertical elements $v\,dt$ to determine the area under the v-t curve, the alternate subdivision shown in part (b) for this area leads to a very useful result. Thus, let the shaded area of part (b) be the sum of the rectangle of area $v_1(t_2 - t_1)$ and the sum of the unshaded horizontal strips of area $dv(t_2 - t)$. Note that v_1 is the velocity at the start of the time interval $(t_2 - t_1) = \Delta t$, and that dv is the increment of velocity at any time t during this interval. Then we have

$$s_2 - s_1 = \Delta s = (\text{Area})_{v-t} = v_1(\Delta t) + \int_{v_1}^{v_2} dv\,(t_2 - t)$$

Substituting $dv = a\,dt$ in the integral, we obtain

$$\Delta s = v_1(\Delta t) + \int_{t_1}^{t_2} (a\,dt)(t_2 - t)$$

To interpret the meaning of the second right-hand term, we note that $a\,dt$ is the area of an elemental strip under the a-t curve and that $(t_2 - t)$ is the moment arm of this elemental strip about an ordinate through t_2. The integral therefore is the summation of moments of area of all such strips and is equivalent to the moment of area about the ordinate at t_2 of the area under the a-t curve included in the time interval $\Delta t = t_2 - t_1$. Hence, we obtain

$$\Delta s = v_1(\Delta t) + (\text{Area})_{a-t}(\bar{t}_2) \qquad (9\text{-}4.3)$$

in which $\bar{t}_2$ is the moment arm to the centroid C of the shaded area in part (a). In applying this equation, note that v_1 is the velocity at the start of the time interval and that the moment of the area under the a-t curve during the time interval is taken about an ordinate at the end of the time interval. When the area under the a-t curve is resolved into parts like those listed in Table 9-4.1, the total moment of area is the sum of the moments of area of its composite parts.

The use of Eq. (9-4.3) in conjunction with Eq. (9-4.1) is especially convenient as only the *a-t* curve is necessary to find the changes in displacement and in velocity.

Next consider the shape of the *v-t* curve. Since $a = dv/dt$, the slope dv/dt of the *v-t* curve at any instant such as t_2 is determined by the corresponding ordinate of the *a-t* curve. Since the shown values of a are positive and increasing, the corresponding slopes of the *v-t* curve are positive (i.e., up to the right) and increasingly steeper.

Table 9-4.1 Properties of Areas.

Equation	Graph	Area	Location of centroid
(Zero degree) $y = h$	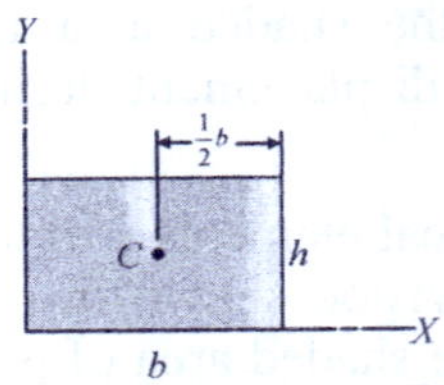	$\frac{1}{1}(bh)$	$\frac{1}{2}b$
(1st degree) $y = mx$	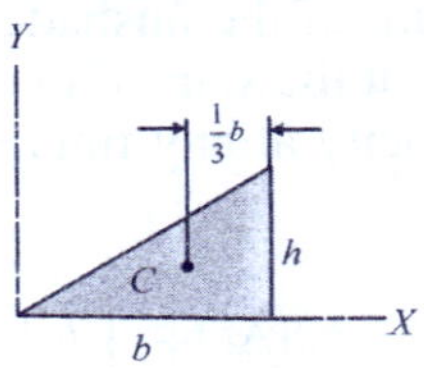	$\frac{1}{2}(bh)$	$\frac{1}{3}b$
(2nd degree) $y = kx^2$	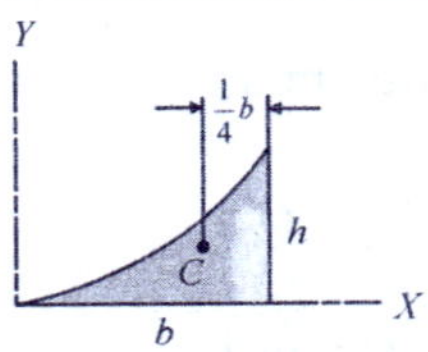	$\frac{1}{3}(bh)$	$\frac{1}{4}b$
(3rd degree) $y = kx^3$	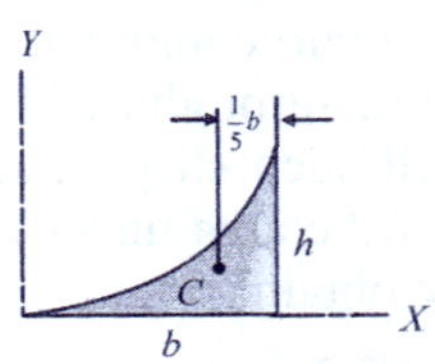	$\frac{1}{4}(bh)$	$\frac{1}{5}b$
(n^{th} degree) $y = kx^n$	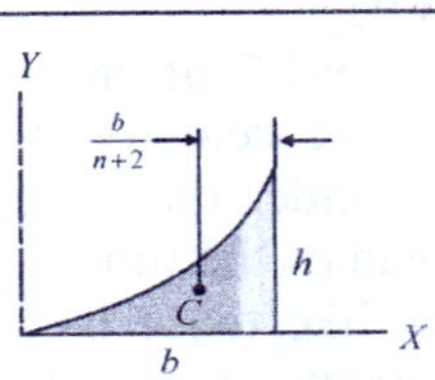	$\frac{1}{n+1}(bh)$	$\frac{1}{n+2}(b)$

In similar fashion, the shape of the *s-t* curve is determined by $v = ds/dt$. Since ds/dt is the slope of the *s-t* curve, it has increasingly positive steeper slopes to agree with the corresponding larger and positive ordinates of the *v-t* curve.

If you have studied or will study the relations between load, shear, and moment diagrams (see Chapter 8), you should observe that they are related in exactly the same manner as are the *a-t*, *v-t*, and *s-t* diagrams for motion curves. Except for a change in notation, one is exactly analogous to the other as shown in the following comparison:

Shear and moment diagrams	***Motion curves***
$w = \dfrac{dv}{dx} = (\text{slope})_{\text{shear}}$	$a = \dfrac{dv}{dt} = (\text{slope})_{(v-t)}$
$v = \dfrac{dM}{dx} = (\text{slope})_{\text{moment}}$	$v = \dfrac{ds}{dt} = (\text{slope})_{(a-t)}$
$\Delta V = (\text{Area})_{\text{load}}$	$\Delta v = (\text{Area})_{a-t}$
$\Delta M = (\text{Area})_{\text{shear}}$	$\Delta s = (\text{Area})_{v-t}$

Two other motion curves are of minor interest. For the *a-s* curve in which acceleration is plotted against displacement, application of $\int_{s_1}^{s_2} a\,ds = \int_{v_1}^{v_2} v\,dv$ results in

$$\frac{1}{2}\left(v_2^2 - v_1^2\right) = (\text{Area})_{a-s} \tag{9-4.4}$$

For the *v-s* curve showing the variation of velocity with displacement, writing $a = v(dv/ds)$ shows that the acceleration can be found by multiplying v at any instant by the corresponding slope of the *v-s* curve.

ILLUSTRATIVE PROBLEMS

9-4.1 A particle, starting with an initial velocity of 18 m/s, has a rectilinear motion with a constant deceleration of 3 m/s^2. Determine the velocity and displacement at the end of 9 sec by sketching the *a-t*, *v-t*, and *s-t* curves and using the relations between them.

Solution

This elementary problem is easily solved by using the equations of constant acceleration, but its very simplicity will develop confidence in applying the relations between motion curves to more complex cases. Since the acceleration is constant but negative, the slope of the *v-t* curve is likewise constant and negative or directed down to the right as shown in Fig. 9-.4.2. The velocity change in the 9-sec interval is

$$[\Delta v = (\text{Area})_{a-t}] \qquad \Delta v = -3(9) = -27 \text{ m/s}$$

which decreases the initial velocity of 18 m/s to –9 m/s.

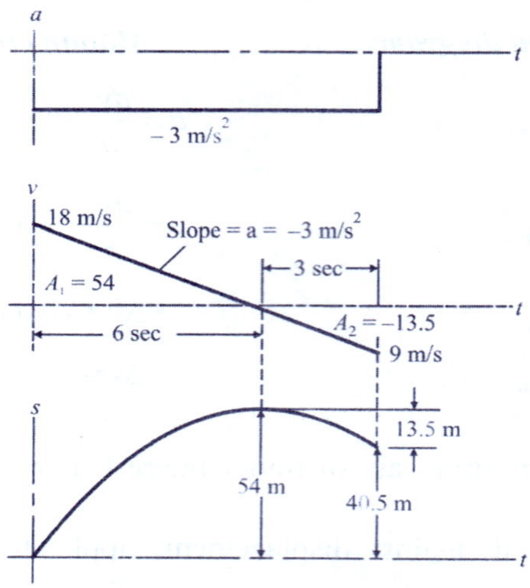

Figure 9-4.2

Evidently the velocity is zero at $t_1 = 6$ sec so that the change in displacement can be computed as the algebraic sum of the positive and negative areas A_1 and A_2 under the *v-t* curve. Hence,

$$[\Delta s = (\text{Area})_{v-t}]\ \Delta s = \frac{1}{2}(18)(6) - \frac{1}{2}(9)(3)$$

$$= 54 - 13.5 = 40.5 \text{ m} \qquad \textbf{\textit{Ans.}}$$

Alternatively, Eq. (9-4.3) can be used to obtain Δs in the 9-sec interval. This gives

$$[\Delta s = v_o(\Delta t) + (\text{Area})_{a-t}\ \bar{t}_2\ ]$$

$$\Delta s = 18(9) + (-3)(9)\left(\frac{9}{2}\right) = 40.5 \text{ m} \qquad \textbf{\textit{Check}}$$

which, in this case, is identical to applying $s = v_o t + \frac{1}{2}at^2$.

The shape of the *s-t* curve is determined by noting that the velocity ordinate equals the slope at the corresponding ordinate of the *s-t* curve. Thus, at $t = 0$, the tangent to the *s-t* curve is up to the right, becoming less steep and eventually horizontal at $t = 6$ sec as the corresponding velocity ordinates gradually reduce to zero. Thereafter the tangents to the *s-t* curve are increasingly steeper down to the right as the velocity ordinates become increasingly negative.

Observe that here the *s-t* curve is a symmetric parabola with its vertex at $t = 6$ sec because at equal intervals to either side of this instant, the velocities are numerically equal but of opposite sign, thereby producing equal slopes but of opposite inclinations.

Finally, we observe that the shape of the *s-t* curve indicates that the moving particle reaches a maximum rightward displacement of 54 m at $t = 6$ sec after which it returns leftward. The total *distance* traveled is the sum of these travels and is given by the sum of A_1 and A_2 or $54 + 13.5 = 67.5$ m.

9-4.2 Draw the *v-t* and *s-t* curves for the two stones having the motions described in Illus. Prob. 9-3.1. The first stone is thrown vertically upward from a tower 30 m high with a velocity of 15 m/s at the same instant the second stone is projected upward from the ground with a velocity of 23 m/s. When and where are the stones at the same level?

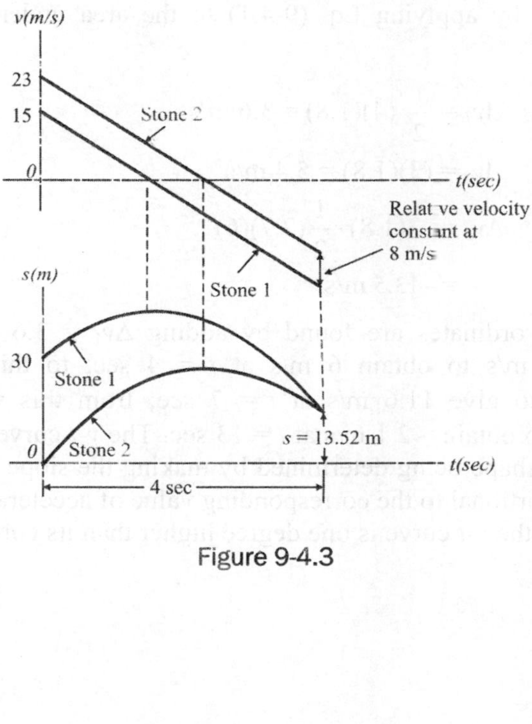

Figure 9-4.3

Solution

Both stones move with the same downward or negative acceleration of $g = -9.81$ m/s^2. Consequently from $a = (\text{Slope})_{v-t}$, both *v-t* curves have the same slope. But since the stones start with different velocities, their *v-t* curves are parallel as shown in Fig. 9-4.3. It is apparent that the stones have a constant relative velocity of 23 – 15 = 8 m/s.

Choosing the initial position of stone 2 as a common origin for displacement we draw the s-t curves as shown. Their shapes are determined from making their slopes proportional to their corresponding velocity ordinates. Since the initial *relative* displacement of 30 m between the stones is being reduced at the constant relative velocity of 8 m/s, the stones meet in 4 sec. This checks the value previously computed in Prob. 9-3.1 and explains the physical significance of that computation. The common height *s* at this instant is most easily computed from

$$s = v_0 t - \frac{1}{2} g t^2 = 23(4) - \frac{1}{2}(9.81)(4)^2 = 13.52 \text{ m} \qquad \textbf{\textit{Ans.}}$$

9.4.3 The *a-t* curve for a particle having rectilinear motion is shown in Fig. 9-4.4. At $t = 0$, the velocity is 2.5 m/s and the particle is 18 m to the left of the origin of displacement. Draw the *v-t* and *s-t* curves, specifying values of *v* and *s* at $t = 4$ sec, 7 sec. and 13 sec.

Solution

The changes in velocity during the specified time intervals of 4 sec, 3 see, and 6 sec are found by applying Eq. (9-4.1) to the area under the *a-t* curve as follows:

$$[\Delta v = (\text{Area})_{a-t}] \qquad \Delta v_1 = \frac{1}{2}(4)(1.8) = 3.6 \text{ m/s}$$

$$\Delta v_2 = (3)(1.8) = 5.4 \text{ m/s}$$

$$\Delta v_3 = -3(1.8) - \frac{1}{2}(2.7)(6)$$

$$= -13.5 \text{ m/s}$$

The velocity ordinates are found by adding $\Delta v_1 = 3.6$ m/s to the initial velocity of 2.5 m/s to obtain 6 m/s at $t = 4$ see; to this result is added $\Delta v_2 = 5.4$ m/s to give 11.6 m/s at $t = 7$ see; from this value we subtract $\Delta v_3 = 13.5$ m/s to obtain – 2.1 m/s at $t = 13$ sec. The *v-t* curve is drawn through these points, its shape being determined by making the slope of the *v-t* curve at any instant proportional to the corresponding value of acceleration. Observe that each segment of the *v-t* curve is one degree higher than its corresponding part of the *a-t* diagram.

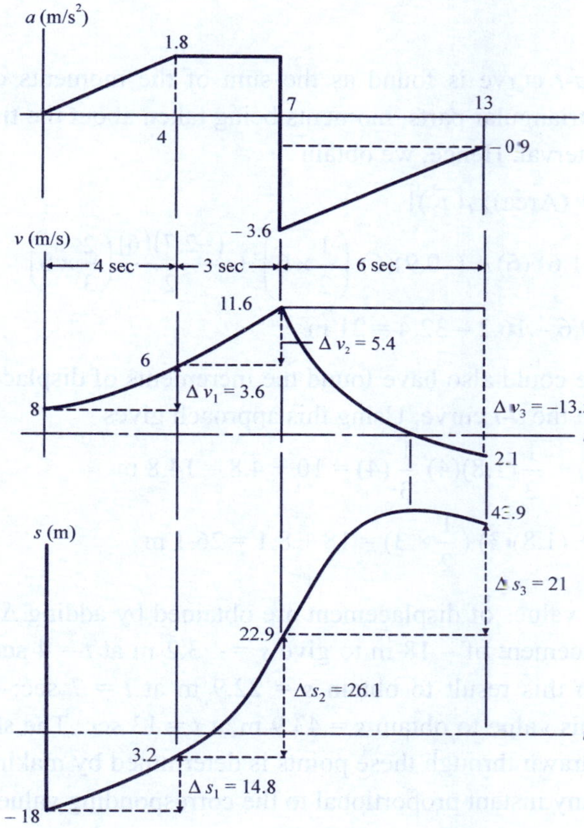

Figure 9-4.4

Now we may apply Eq. (9-4.2) to the area under this *v-t* curve to obtain the following changes in displacement. Observe that in the first 4-sec interval, the area is composed of a rectangle surmounted by a second-degree curve while in the second interval of 3 see, the area consists of a rectangle plus a triangle.

$[\Delta s = (\text{Area})_{v-s}]$ $\quad \Delta s_1 = 2.5(4) + \frac{1}{3}(3.6)(4) = 14.90 \text{ m}$

$$\Delta s_2 = 6(3) + \frac{1}{2}(5.4)(3) = 26.10 \text{ m}$$

The last increment of displacement Δs_3 cannot easily be found from the area under the *v-t* curve. Instead we find it by applying Eq. (9-4.3) to the last 6-sec interval. The velocity at the start of this interval is 11.6 m/s. The moment of area under the *a-t* curve is found as the sum of the moments of areas of its rectangular and triangular parts, moments being taken about the time ordinate at the end of the interval. Hence, we obtain

$$[\Delta s = v_1(\Delta t) + (\text{Area})_{a-t}(\bar{t}_2)]$$

$$\Delta s_3 = (11.6)(6) + (-0.9)(6)\left(\frac{1}{2}\times 6\right) + \frac{(-2.7)(6)}{2}\left(\frac{2}{3}\times 6\right)$$

$$= 69.6 - 16.2 - 32.4 = 21 \text{ m}$$

If desired, we could also have found the increments of displacement Δs_1 and Δs_2 directly from the *a-t* curve. Using this approach gives

$$\Delta s_1 = (2.5)(4) + \frac{1}{2}(1.8)(4)\ \frac{1}{3}(4) = 10 + 4.8 = 14.8 \text{ m} \qquad \textbf{\textit{Check}}$$

$$\Delta s_2 = (6)(3) + (1.8)(3)\left(\frac{1}{2}\times 3\right) = 18 + 8.1 = 26.1 \text{ m} \qquad \textbf{\textit{Check}}$$

The required values of displacement are obtained by adding Δs_1 = 14.8 m to the initial displacement of – 18 m to give s = – 3.2 m at t = 4 see; then adding Δs_2 = 26.1 m to this result to obtain s = 22.9 m at t = 7 sec; finally adding Δs_3 = 21 m to this value to obtain s = 43.9 m at t = 13 sec. The shape of the *s-t* curve which is drawn through these points is determined by making the slope of the *s-t* curve at any instant proportional to the corresponding value of velocity.

PROBLEMS

9-4.4 The elevators in a modern office building are designed to accelerate or decelerate at a constant rate to or from 800 m/s in 6 sec. With the aid of a *v-t* diagram, determine the minimum time for the elevator to rise 365 m.

9-4.5 An auto is accelerated from rest to a top speed of 27 m/s and then immediately, decelerated to a stop. If the total elapsed tinie is 20 see, determine the distance covered. The acceleration and deceleration are both constant but not necessarily of the same magnitude.

9-4.6 An auto starts from rest and comes to a stop 400 m away. If its acceleration and deceleration are limited to 3 m/s and 6 m/s^2, respectively, what maximum speed does it reach if the elapsed time is to be as small as possible?

9-4.7 A car and a truck are both moving at 20 m/s along a highway. To pass the truck safely, the car must move 60 m relative to the truck. Determine the minimum time to do this if the car can accelerate at 2 m/s^2 but must not exceed a speed of 27 m/s.

9.4.8 An auto is to travel a distance from A to B of 550 m in exactly 40 sec. The auto accelerates and decelerates at 2 m/s^2, starting from rest at A and coming to a stop at B. Find its maximum speed in m/sec.

9-4.9 The motion of a particle starting from rest is governed by the a-t curve shown in Fig. P-9-4.9. Sketch the v-t and s-t curves. Determine the displacement at $t = 9$ sec.

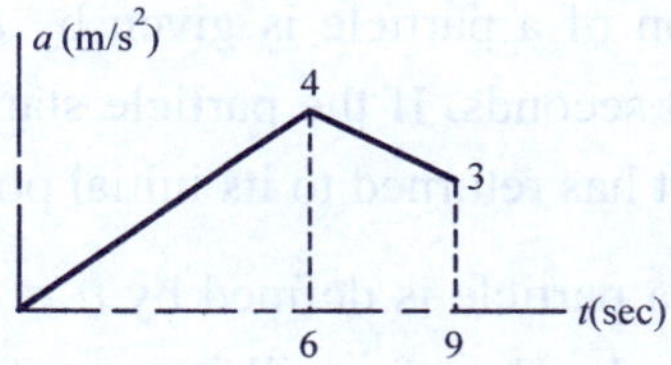

Figure P-9-4.9

9.4.10 The curved portions of the v-t curve shown in Fig. P-9-4.10 are second-degree parabolas with horizontal slope at $t = 0$ and $t = 12$ sec. Sketch the a-t and s-t curves if s_o is zero. Check values of s using both Eqs. (9-4.2) and (9-4.3).

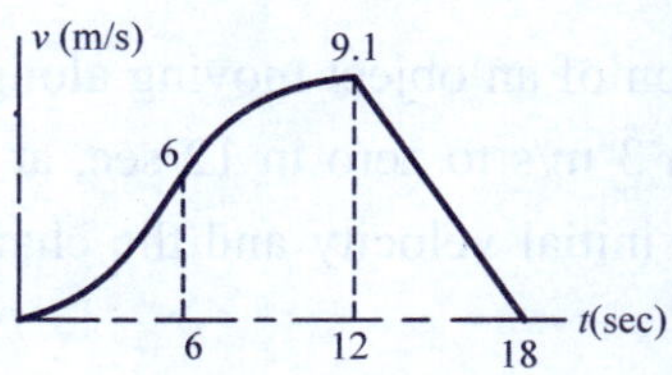

Figure P-9-4.10

9-4.11 One cycle of the s-t curve for a certain machine is shown in Fig. P-9-4.11. The curves are second-degree parabolas that are tangent at the indicated points of inflection and have zero slope at $t = 0$, 8, and 12 sec. Sketch the v-t and a-t curves and compute the maximum velocity and maximum acceleration.

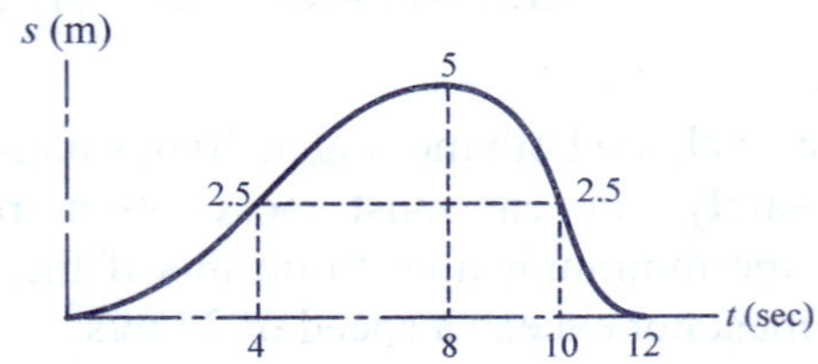

Figure P-9-4.11

9-4.12 The rate of change of acceleration, called "jerk" of an elevator is constant at ± 0.6 m/s^2. Find the shortest time for the elevator, starting from rest, to rise 10 m and stop.

9-4.13 The acceleration of a particle is given by $a = 18 - 3t$ where a is in m/s^2 and t is in seconds. If the particle starts from rest, determine its velocity when it has returned to its initial position.

9-4.14 The motion of a particle is defined by $a = 10t - t^2$ where a is in m/s^2 and t is in seconds. How far will it move from rest before starting to reverse its direction of motion?

9-4.15 An auto starts from rest and reaches a speed of 18 m/s in 15 sec. The acceleration increases uniformly from zero for the first 9 sec after which the acceleration reduces uniformly to zero in the next 6 sec. Compute the displacement in this 15 sec interval.

9-4.16 The acceleration of an object moving along a straight path decreases uniformly from 3 m/s to zero in 12 sec, at which time its velocity is 2 m/s. Find its initial velocity and the change in position during the 12-sec interval.

9-4.17 An object attains a velocity of 15 m/s with an acceleration which varies uniformly from 1.2 m/s^2 to 3.6 m/s^2 in 9 sec. Compute its initial velocity and the change in position during the 9-sec interval.

9-4.18 The velocity of a particle changes from – 3 m/s to 21 m/s during a 12 sec interval during which its acceleration increases uniformly with time from an initial value of 1 m/s^2. Determine the displacement covered in the 12 sec interval.

9-5 INTRODUCTION TO VECTOR CALCULUS

In all motions except rectilinear motion, the vector quantities involved change in direction as well as in magnitude. Any change in the direction of a vector results in its derivative having a direction different from that of the original vector. This directional consideration is automatically accounted for in vector calculus; indeed, it is the basic difference between vector and scalar calculus. We shall emphasize and explain this distinction in subsequent applications.

For the present, we need only the following rules for the differentiation of the sums and products of two vectors. They are almost exactly the same as those for scalar calculus:

$$\frac{d}{dt}(\mathrm{A}+\mathrm{B})=\frac{d\mathrm{A}}{dt}+\frac{d\mathrm{B}}{dt} \tag{9-5.1}$$

$$\frac{d}{dt}(\mathrm{A}\bullet\mathrm{B})=\frac{d\mathrm{A}}{dt}\bullet\mathrm{B}+\mathrm{A}\bullet\frac{d\mathrm{B}}{dt} \tag{9-5.2}$$

$$\frac{d}{dt}(\mathrm{A}\times\mathrm{B})=\frac{d\mathrm{A}}{dt}\times\mathrm{B}+\mathrm{A}\times\frac{d\mathrm{B}}{dt} \tag{9-5.3}$$

These rules will be very useful in theoretical derivations. Notice that it is essential to maintain the order of the terms in differentiating the cross product.

One of the most useful methods of distinguishing between changes in magnitude and direction of a vector is given by the following rule for differentiating the product of a scalar and a vector, both of which are functions of t:

$$\frac{d}{dt}(n\mathrm{A})=\frac{dn}{dt}\mathrm{A}+n\frac{d\mathrm{A}}{dt} \tag{9-5.4}$$

To integrate a vector function, we reverse the process of differentiation. Thus, if

$$\mathbf{A}=\frac{dB}{dt}$$

the integral of **B** is

$$\mathrm{B}=\int A\,dt+\mathrm{C} \tag{9-5.5}$$

where the constant of integration **C** is vector, constant in both magnitude and direction. One of the most useful methods of performing the integration is to express the integrand in terms of constant unit vectors directed along fixed coordinate directions. For example, if

$$\mathbf{A}=\mathrm{A}_x\hat{\mathbf{i}}+\mathrm{A}_y\hat{\mathbf{j}}+\mathrm{A}_z\hat{\mathbf{k}}$$

where A_x, A_y, A_z are scalar functions of t, the integral becomes

$$\mathbf{B} = \int \mathbf{A}dt = \int (A_x\hat{\mathbf{i}} + A_y\hat{\mathbf{j}} + A_z\hat{\mathbf{k}})dt + \mathbf{C} \qquad (9\text{-}5.6)$$

$$= \hat{\mathbf{i}}\int A_x dt + \hat{\mathbf{j}}\int A_y dt + \hat{\mathbf{k}}\int A_z dt + \mathbf{C}$$

in which the constant unit vectors have been extracted from the integral sign and C is a vector constant of integration. Additional concepts will be developed in subsequent sections as they are needed.

9-6 RECTANGULAR COMPONENTS OF CURVILINEAR MOTION

We now consider the motion of a particle along a space curve generated by the tip of a variable position vector. With respect to a fixed set of reference axes as shown in Fig. 9-6.1, the position vector **r** of the moving particle P may be expressed as

$$\mathbf{r} = x\hat{\mathbf{i}} + y\hat{\mathbf{j}} + z\hat{\mathbf{k}} \qquad (9\text{-}6.1)$$

where the rectangular coordinates x, y, z are functions of time. When we successively differentiate this expression to obtain velocity and acceleration, note that each of the right-hand terms are a product of a scalar and a vector. Hence; applying the rule expressed by Eq. (9-5.4), the time derivative of **r** is

$$\frac{d\mathbf{r}}{dt} = \frac{dx}{dt}\hat{\mathbf{i}} + x\frac{d\hat{\mathbf{i}}}{dt} + \frac{dy}{dt}\hat{\mathbf{j}} + y\frac{d\hat{\mathbf{j}}}{dt} + \frac{dz}{dt}\hat{\mathbf{k}} + z\frac{d\hat{\mathbf{k}}}{dt}$$

However, here the unit vectors $\hat{\mathbf{i}}$, $\hat{\mathbf{j}}$, $\hat{\mathbf{k}}$ are constant in magnitude as well as direction so that their derivatives are zero.

Hence, we find the velocity to be

$$\left.\begin{aligned} v &= \frac{d\mathbf{r}}{dt} = \frac{dx}{dt}\hat{\mathbf{i}} + \frac{dy}{dt}\hat{\mathbf{j}} + \frac{dz}{dt}\hat{\mathbf{k}} \\ &= v_x\hat{\mathbf{i}} + v_y\hat{\mathbf{j}} + v_z\hat{\mathbf{k}} \end{aligned}\right\} \qquad (9\text{-}6.2)$$

and the acceleration to be

$$\left.\begin{aligned} \mathbf{a} &= \frac{dv}{dt} = \frac{dv_x}{dt}\hat{\mathbf{i}} + \frac{dv_y}{dt}\hat{\mathbf{j}} + \frac{dv_z}{dt}\hat{\mathbf{k}} \\ &= \frac{d^2\mathbf{r}}{dt^2} = \frac{d^2x}{dt^2}\hat{\mathbf{i}} + \frac{d^2y}{dt^2}\hat{\mathbf{j}} + \frac{d^2z}{dt^2}\hat{\mathbf{k}} \\ &= a_x\hat{\mathbf{i}} + a_y\hat{\mathbf{j}} + a_z\hat{\mathbf{k}} \end{aligned}\right\} \qquad (9\text{-}6.3)$$

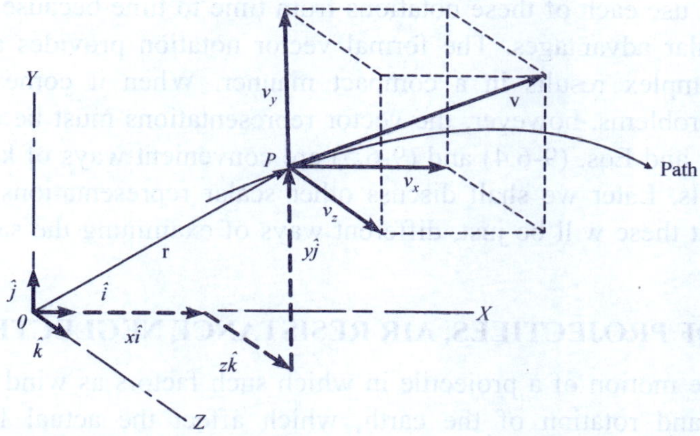

Figure 9-6.1 Rectangular components of velocity

The significance of these results is that the motion of a particle is equivalent to the simultaneous projection of the motion along fixed rectangular coordinate axes. Each such projection is then a rectilinear motion, the details of which have been presented in Section 9-3.

When the sense and direction of a vector component is completely defined by its subscript (as is the situation here), it is often useful to replace formal vector notation with the following alternative form:

$$r = x \nrightarrow y \nrightarrow z$$

$$v = v_x \nrightarrow v_y \nrightarrow v_z \qquad (9\text{-}6.4)$$

$$a = a_x \nrightarrow a_y \nrightarrow a_z$$

where the symbol $\nrightarrow$ is used to denote that we are dealing with the geometric sum of quantities whose magnitudes have known directions. The positive directions of these quantities are in the positive directions of the coordinate axes.

At other times, it may be convenient to ignore vector notation completely and consider only the scalar relations among components. Thus, for a moving particle whose position is defined by rectangular components x, y, and z which vary with time, the components of velocity and acceleration may be written as

$$v_x = \frac{dx}{dt},\ v_y = \frac{dy}{dt},\ v_z = \frac{dz}{dt}$$

and (9-6.5)

$$a_x = \frac{dv_x}{dt} = \frac{d^2x}{dt^2},\ a_y = \frac{dv_y}{dt} = \frac{d^2y}{dt^2},\ a_z = \frac{dv_z}{dt} = \frac{d^2z}{dt^2}$$

We shall use each of these notations from time to time because each has its own particular advantages. The formal vector notation provides a method of deriving complex results in a compact manner. When it comes to solving numerical problems, however, the vector representations must be expressed in scalar form, and Eqs. (9-6.4) and (9-6.5) are convenient ways of keeping track of the details. Later we shall discuss other scalar representations of a vector quantity, but these will be just, different ways of examining the same physical entity.

FLIGHT OF PROJECTILES, AIR RESISTANCE NEGLECTED

Consider the motion of a projectile in which such factors as wind velocity, air resistance, and rotation of the earth, which affect the actual flight of the projectile, will be neglected. Then the flight path of the projectile will be a plane curve such as *OBCD* shown in Fig 9-6.2. It is convenient to resolve the curvilinear motion along the path into rectilinear motions along *X* and *Y* axes having an origin at the initial point of flight. The initial velocity is denoted by v_O directed at an angle θ with the *X* axis.

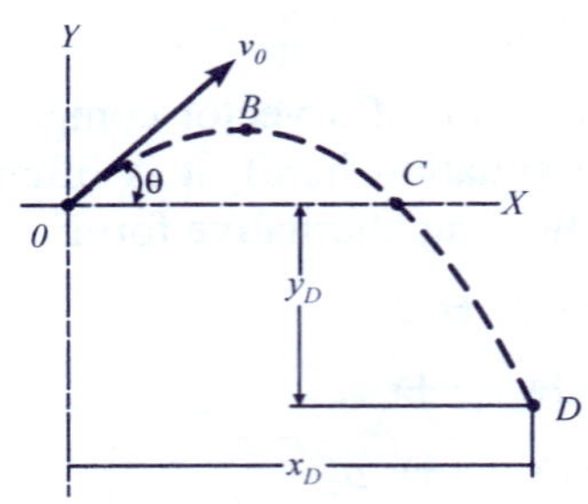

Figure 9-6.2 Flight of projectile

From the initial direction of motion rectilinear displacements will be positive both rightward and upward. Since the only force assumed to be acting on the projectile is its own weight, its total acceleration at all positions is due to gravity and is directed vertically downward with the value *g*. Hence, the rectangular components of this acceleration are constant at $a_x = 0$ and $a_y = -$ g.

Since these components of acceleration are constant, we may use the equations of rectilinear motion with constant acceleration to determine the corresponding rectangular components of the curvilinear motion. This procedure is explained in the accompanying table which should be clear enough to eliminate memorizing the results.

Flight of Projectiles[a]

Rectilinear motion with constant acceleration	X component of flight $(a_x = 0, v_{o_x} = v_o \cos\theta)$	Y component of flight $(a_y = -g, v_{o_y} = v_o \sin\theta)$
$v = v_o + at$	$v_x = v_{o_x} + a_x t$ or $v_x = v_o \cos\theta$	$v_y = v_{o_y} + a_y t$ or $v_y = v_o \sin\theta - gt$
$s = v_o t + \frac{1}{2}at^2$	$x = v_{o_x} t + \frac{1}{2}a_x t^2$ or $x = (v_o \cos\theta)t$	$y = v_{o_y} t + \frac{1}{2}a_y t^2$ or $y = (v_o \sin\theta)t - \frac{1}{2}gt^2$

[a] if air resistance, etc., were not neglected, a_x and a_y would become variable quantities, and the differential equations of motion — Eqs. (9-3.1), (9.3.2), and (9.3.3) – would have to be solved to obtain the proper equations for the flight of the projectile.

If, as in Fig. 9-6.2, the time of flight is less than that required to reach *C*, the projectile will be above its initial position and values of the *Y* displacement will be positive. If the time of flight is more than that required to reach *C*, the projectile will be on the path *CD* and values of the *Y* displacement will then be negative. At the topmost point of flight *B*, the value of v_y will be zero.

ILLUSTRATIVE PROBLEMS

9.6.1 In Fig. 9-6.3, the pivoted rigid link *AB* carries a pin whose position is controlled by the horizontal slotted bar. At the instant that $y = 0.2$ m, the slotted bar is moving upward at a constant velocity of 0.4 m/s. What is then the *x* components of velocity and acceleration of the pin P?

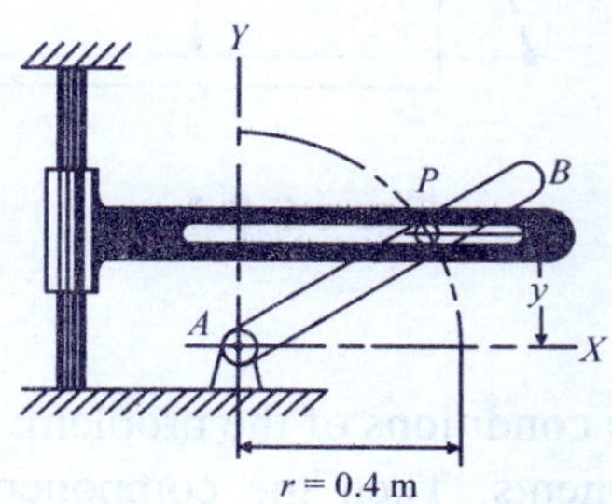

Figure 9-6.3

Solution

The motion of P is common to the upward motion of the bar and the circular path $x^2 + y^2 = 0.16$ where at $y = 0.2$ m $x = 0.35$ m. As determined by the bar, $\dot{y} = 0.4$ m/s and $\ddot{y} = 0$. On the circular path, the velocity and acceleration of P are determined by two successive time differentiations of the path to be

$$x\dot{x} + y\dot{y} = 0 \qquad (a)$$

$$x\ddot{x} + \dot{x}^2 + y\ddot{y} + \dot{y}^2 = 0 \qquad (b)$$

Into Eq. (a), we substitute the known values of x, y, and $\dot{y}$ to obtain

$$0.35\,\dot{x} + 0.2(0.4) = 0$$

whence $\dot{x} = -\,0.23$ m/s **Ans.**

With $\dot{x}$ known eq. (b) now gives

$$0.35\,\ddot{x} + (-\,0.23)^2 + 0.2(0) + (0.4)^2 = 0$$

whence

$\ddot{x} = -\,0.61$ m/s^2 **Ans.**

9-6.2 A projectile is fired from the top of a cliff 100 m high with a velocity of 430 m/s directed at 45° to the horizontal. Find the range on a horizontal plane through the base of the cliff.

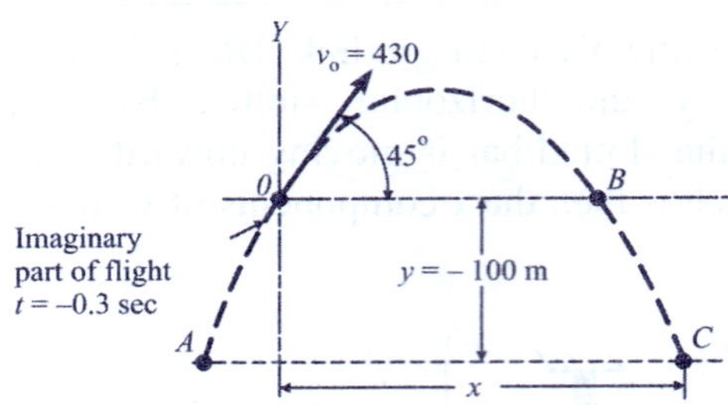

Figure 9-6.4

Solution

Figure 9-6.4 represents the conditions of the problem. The initial components of velocity are v_{ox} = components. Then the components of dis• v_{oy} = 304 m/s

directed rightward and upward. When we choose these directions as the positive senses of all vector components, Y displacement is positive upward and the downward acceleration g is negative; furthermore, the final position of the projectile is below the origin and hence negative. Therefore; we have

$$[y = v_{o_y} t - \frac{1}{2} gt^2] \qquad -100 = 304.t - \frac{1}{2} \times 9.81t^2 \quad \ldots\ldots (a)$$

$$t = 62.3 \text{ sec} \quad \text{or} \quad t = -0.3 \text{ sec}$$

Using the positive value of t, we obtain

$$[x = v_{o_x} t) \qquad x = 304 \times 62.4 = 18970 \text{ m} \qquad \textit{Ans.}$$

The negative value of t obtained from Eq. (a), i.e., $t = -0.3$ sec, may be interpreted as the time required for the projectile to leave the base of the cliff at A and rise to the origin O. This is also the time required for the projectile to travel from B to C. This observation may be checked by finding the time required to travel distance OB from

$$[y = v_{o_y} t - \frac{1}{2} gt^2] \qquad 0 = 304\, t - \frac{1}{2} \times 9.81t^2 \text{ or } t = 62 \text{ sec}$$

Adding 0.3 sec to this value gives the total time of flight to be 62.3 sec as before.

9.6.3 Determine the position at which a ball thrown up to the right will strike the inclined surface shown in Fig. 9-6.5. The initial velocity of the ball is 30 m/s directed at $\theta = \tan^{-1} \frac{4}{3}$ with the horizontal.

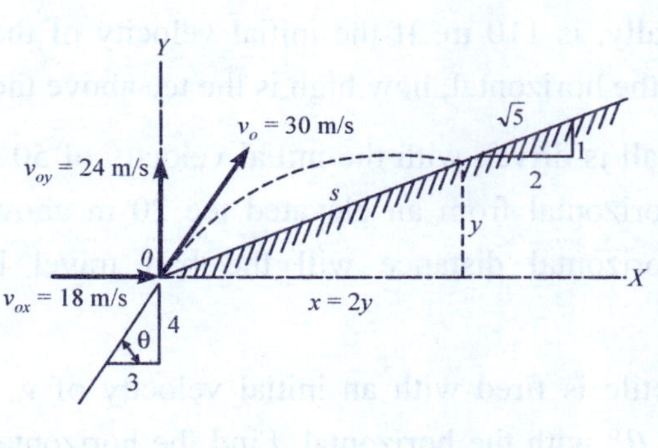

Figure 9-6.5

Solution

Because of the slope of the incline, the components of displacement are $x = 2y$. The initial components of velocity are $v_{o_x} = 18$ m/s and $v_{o_y} = 24$ m/s. Their directions establish the positive sense of all vector placement are

$[x = v_{o_x} t]$ $\qquad x = 2y = 18\,t$ $\qquad (a)$

$[y = v_{o_y} t - \frac{1}{2} gt^2]$ $\qquad y = 24t - \frac{1}{2} \times 9.81t^2$ $\qquad (b)$

Multiplying Eq. (*b*) by –2 and adding it to Eq. (*a*) eliminates *y* to give

$$0 = -\,30t + 9.81t^2 \qquad \text{or} \qquad t = 3.06 \text{ sec}$$

Substituting this value of *t* in Eq. (a) gives

$$x = 18.06(3.06) = 55.08 \text{ m}$$

from which the distance *s* along the incline is found to be

$\left[\frac{s}{\sqrt{5}} = \frac{x}{2}\right]$ $\qquad \frac{s}{\sqrt{5}} = \frac{55.08}{2}$ $\qquad s = 61.6\text{m}$ $\qquad$ ***Ans.***

If the incline had sloped down to the right instead of tip, the final coordinate of *y* displacement would be below the origin and hence negative. Therefore, by inserting a minus sign before y in Eq. (b), the time of flight down the incline will become $t = 6.73$ sec. The corresponding values of *x* and *s* become

$$x = 121.14 \text{ m and } s = 135.44 \text{ m.}$$

PROBLEMS

9.6.4 A golf ball is hit from an elevated tee to a green; the distance, horizontally, is 110 m. If the initial velocity of the ball is 30 m/s at 53.1° to the horizontal, how high is the tee above the green?

9-6.5 A golf ball is driven with the initial velocity of 50 m/s upward at 30° to the horizontal from an elevated tee 20 m above a level fairway. What horizontal distance will the ball travel before hitting the fairway?

9-6.6 A projectile is fired with an initial velocity of v_o m/s upward at an angle of θ° with the horizontal. Find the horizontal distance covered before the projectile returns to its original level. Also determine the maximum height attained by the projectile.

9-6.7 As an exmple of poetic license, consider the following passage from Longfellow's *The Song of Hiawatha:*

Swift of foot was Hiawatha:
He could shoot an arrow from him
And run forward with such swiftness
That the arrow fell behind him!
Strong of arm was Hiawatha:
He could shoot ten arrows upward,
Shoot them with such strength and swiftness
That the tenth had left the bowstring
Ere the first to earth had fallen.

Assuming that 1 sec elapses between the discharge of each arrow and that Hiawatha shot at his greatest range each time, how fast must he have been able to run?

9-6.8 A stunt man is to drive an auto across the water-filled gap shown in Fig. P-9-6.8. Determine the auto's minimum take-off velocity and the angle θ of the landing ramp.

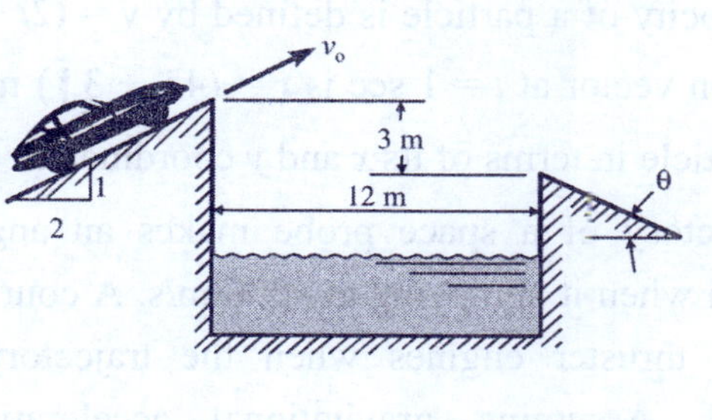

Figure P.9-6.8

9-6.9 A ball is thrown so that it just clears a 7.5 m wall 30 m away. If it left the hand 1.5 m above the ground and at an-angle of 60° to the horizontal, what was the initial velocity of the ball?

9-6.10 In Fig. P-9-6.10, a ball thrown down the incline strikes it at a distance s = 80 m. If the ball rises to a maximum height h = 20 m above the point of release, compute its initial velocity and inclination θ.

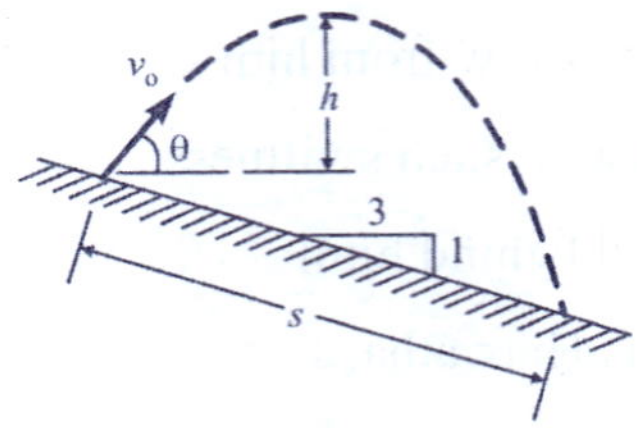

Figure P.9-6.10

9-6.11 A particle has an initial velocity of 30 m/s up to the right at 30° with the horizontal. The components of acceleration are constant at $a_x = -1.2$ m/s^2 and $a_y = -6$ m/s^2. Compute the horizontal distance covered until the particle reaches a point 18 m below its original elevation.

9-6.12 A particle moves along the path $y = x^2 - 4x + 100$, starting with an initial velocity of $v_o = (4\hat{i} - 16\hat{j})$ m/s. If v_x is constant, determine v_y and a_y at $x = 16$ m.

9-8.13 If the velocity of a particle is defined by $v = (2t + 1)\hat{i} + 3\hat{j}$ m/s, and its position vector at $t = 1$ sec is $\mathbf{r} = (4\hat{i} + 3\hat{j})$ m, determine the path of the particle in terms of its x and y coordinates.

9-6.14 The trajectory of a space probe makes an angle of 60° with the horizontal when it is moving at 4500 m/s. A course correction is to be made by thruster engines when the trajectory is at 30° to the horizontal. Assuming gravitational acceleration is constant at 9 m/s^2 during this unpowered flight, what time interval should elapse before the thruster engines are ignited? What is the increase in altitude in this interval?

9.8.15 A rocket is released from a jet fighter flying horizontally at 335 m/s at an altitude of 2400 m above its target. The rocket thrust gives it a constant horizontal acceleration of 0.6 g. Determine the angle between the horizontal and the line of sight to the target.

9-6.16 If the initial velocity of an object is 12 m/s, determine the horizontal distance it can cover without rising more than 3 m.

9-6.17 The telescoping rod shown in Fig. P-9-6.17 forces the pin P to move along the fixed path $9y = x^2$ where x and y are in meters. At any time t, the x coordinate of P is given by $x = t^2 - 5t$. Determine the y components of velocity and acceleration of P at $x = 0.15$ m.

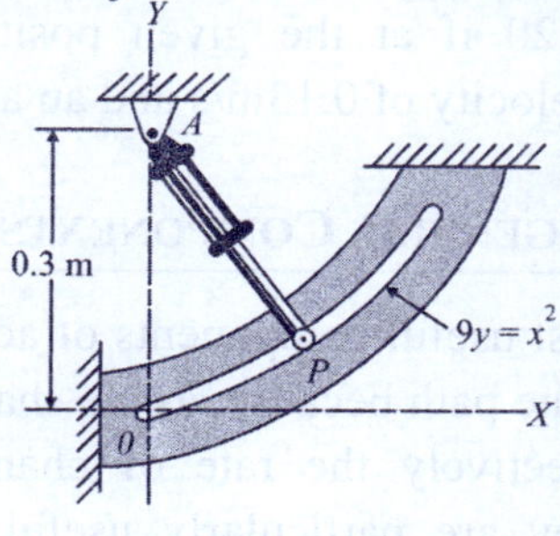

Figure P-9-6.17

9-6.18 A particle moves in the XY plane so that its x coordinate is defined by $x = 5t^3 - 105t$ where x is in meters and t is in seconds. When $t = 2$ sec, the total acceleration is 2 m/s². If the y component of acceleration is constant and the particle starts from rest at the origin when $t = 0$, determine its total velocity when $t = 4$ sec.

9.6.19 A particle is constrained to move upward to the right along the path $2y^2 = x^3 + 26$ where x and y are in meters. The x coordinate of the particle at any time is $x = t^2 - t + 4$. Determine the y components of velocity and acceleration at $x = 0.15$ m.

9.6.20 The position of pin P in the circular slot shown in Fig. P-9-6.20 is controlled by the inclined guide which is moving rightward at the constant rate of 0.1 m/s for an interval of its motion. Compute the velocity and acceleration of P at the given position.

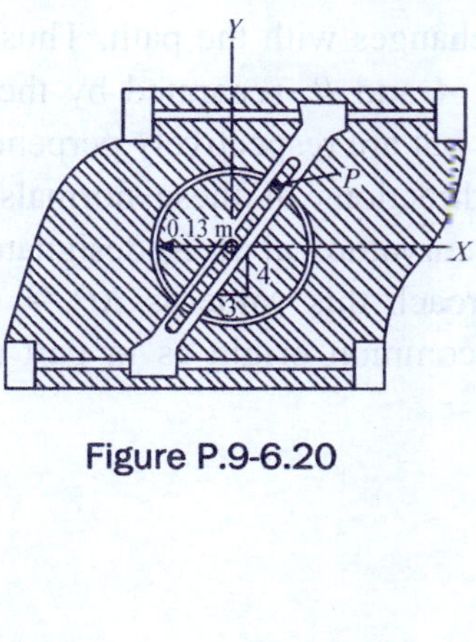

Figure P.9-6.20

Hint: By sketching the position of the guide a short time t after the given position, obtain the absolute coordinates of motion along the guide in terms of time. The absolute motion of P in the circular slot is equal to the geometric sum of the motion of the guide plus that of P along the guide.

9-6.21 Solve Prob. 9-6.20 if at the given position the guide is moving leftward with a velocity of 0.13m/s and an acceleration of 0.1 m/s^2.

9-7 NORMAL AND TANGENTIAL COMPONENTS OF ACCELERATION

In many situations, the most useful components of acceleration are those which are tangent and normal to the path because, as we shall show, these components separate and denote respectively the rate of change of *magnitude* and of *direction* of velocity. They are particularly useful when we need to relate velocity and acceleration directly with the path itself The following discussion refers to motion in a plane, but it is also applicable to three-dimensional spatial motion.

In Section 9-2, we have already denoted velocity in terms of its magnitude and direction by the expression $\mathbf{v} = v\hat{e}_t$. Since acceleration is the time derivative of velocity, we obtain

$$\mathbf{a} = \frac{dv}{dt} = \frac{d}{dt}(v\hat{\mathbf{e}}_t) = \frac{dv}{dt}\hat{\mathbf{e}}_t + v\frac{d\hat{\mathbf{e}}_t}{dt} \qquad (a)$$

Here $\frac{dv}{dt} = \frac{d^2s}{dt^2} = a_t$, represents the change in the magnitude of the velocity, directed tangent to the path as indicated by $\hat{e}_t$.

The directional change of v is expressed by $\frac{d\hat{\mathbf{e}}_t}{dt}$ which we now interpret. Unlike unit vectors associated with a fixed set of reference axes, $\hat{\mathbf{e}}_t$ is a unit vector whose direction changes with the path. Thus, in Fig. 9-7.1, unit tangent vectors at two positions A and B, separated by the time interval Δt, have the same magnitude of unity but are respectively perpendicular to the normals to the path at A and B. The directions of these normals differ by an angle of $\Delta\theta$ radians. The radius of curvature at A is designated as ρ and the radius of curvature at B will approach this value as $\Delta\theta \to 0$. When we draw the unit tangent vectors from a common origin as in part (b), they form an isosceles triangle whose vertex angle is $\Delta\theta$ and whose base $\Delta\hat{\mathbf{e}}_t$ denotes their directional change.

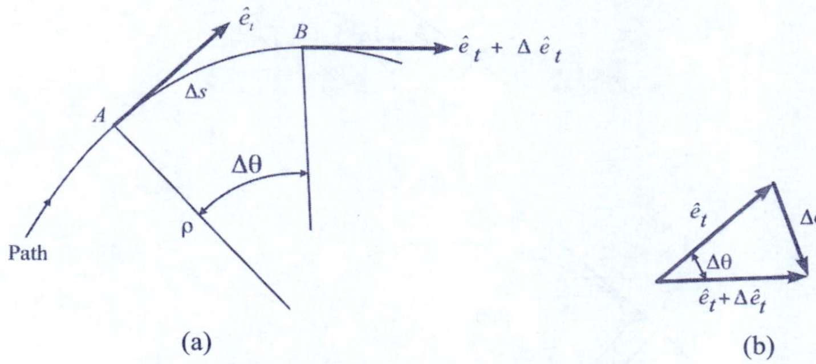

Figure 9-7.1 Change in $\hat{\mathbf{e}}_t$ during time Δt.

Consider now the geometric significance of permitting the time interval between positions A and B to approach zero as a limit. In Fig. 9-7.Ia, B approaches to within a differential distance ds of A; angle $\Delta\theta$ becomes $d\theta$; the radius of curvature at B also becomes ρ. In part (b) of Fig. 9-7.1, as $\Delta\theta \to 0$, the base angles approach 90° and hence in the limit, the direction of $\Delta\hat{\mathbf{e}}_t$ is normal to the path at A. As a result of $\Delta t \to 0$, therefore, Fig. 9-7.1 is transformed into Fig. 9-7.2. Note that in the time dt, the unit tangent vector $\hat{\mathbf{e}}_t$ swings through the angle $d\theta$ and its tip moves through the distance $|d\hat{\mathbf{e}}_t| = (1)\ d\theta = d\theta$ in a direction perpendicular to $\hat{\mathbf{e}}_t$ i.e., inward toward the center of curvature O in the direction defined by the unit normal vector $\hat{\mathbf{e}}_n$. Thus, we obtain

$$\frac{d\hat{\mathbf{e}}_t}{dt} = \frac{d\theta}{dt}\hat{\mathbf{e}}_n = \frac{1}{p}\frac{ds}{dt}\hat{\mathbf{e}}_n = \frac{v}{\rho}\hat{\mathbf{e}}_n \qquad (b)$$

Substituting this result in Eq. (a) then gives

$$\mathbf{a} = \frac{dv}{dt}\hat{\mathbf{e}}_t + \frac{v^2}{\rho}\hat{\mathbf{e}}_n = a_t + a_n \qquad (9\text{-}7.1)$$

where the magnitudes of the acceleration components are $a_t = \dfrac{dv}{dt}$ and $a_n = \dfrac{v^2}{\rho}$ and their directions are such that positive a_t is tangent to the path in the direction of increasing speed whereas a_n is always directed inward toward the center of curvature.

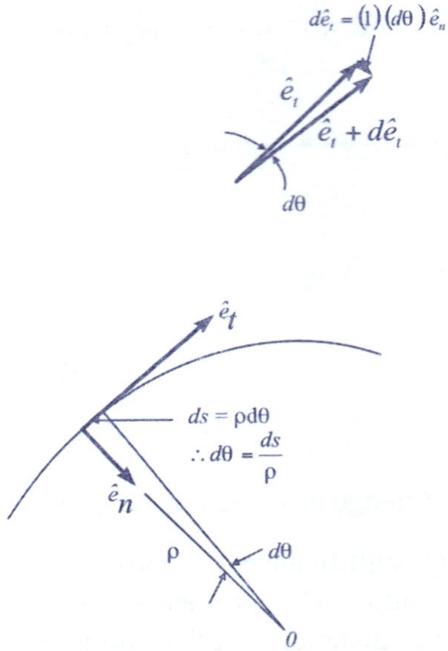

Figure 9-7.2 Change in $\hat{e}_t$ as $\Delta t \to 0$

Note that for measurements along the path, the same relations exist among s, v, and a_t as discussed previously for rectilinear motion in Section 9-3. Observe that the tangential component of acceleration at $a_t = \dfrac{dv}{dt}$ represents only the change in *magnitude* of velocity and that it will be zero if the speed is constant. On the other hand, the normal component of acceleration, $a_n = \dfrac{v^2}{\rho}$ is caused by the change in the *direction* of the velocity and will be zero only if $v = 0$ or if ρ is infinite as at a point of inflection of the path, or if the path is straight.

Alternate expressions for the magnitude of $\mathbf{a}_n$ are

$$a_n = v\omega = \rho\omega^2 = \frac{v^2}{\rho} \tag{9-7.2}$$

where $\omega = \dfrac{d\theta}{dt}$ represents the angular speed of the radius of curvature. One of the most useful applications of Eq. (9-7.2) occurs in motion along a circular path; for example, rotation of a rigid body about a fixed axis. Here ρ is the constant radius of a circular path described by a particle in the rotating body.

For other curved paths, ρ and its angular speed ω are generally difficult to compute as a preliminary to finding the normal component of acceleration.

The correlation between the total acceleration **a** of a particle and its rectangular components or its normal and tangential components is shown in Fig. 9-7.3. Either set of components may be derived from the other by projecting one set upon the other. Doing this gives

$$\left.\begin{aligned} a_n &= a_x \sin\theta + a_y \cos\theta \\ a_t &= a_x \cos\theta - a_y \sin\theta \end{aligned}\right\} \quad (9\text{-}7.3)$$

or

$$\left.\begin{aligned} a_x &= a_n \sin\theta + a_t \cos\theta \\ a_y &= a_n \cos\theta - a_t \sin\theta \end{aligned}\right\} \quad (9\text{-}7.4)$$

Once you understand these correlations among the components of acceleration, you need not bother to remember them since they are so easily obtained from a diagram appropriate to the situation.

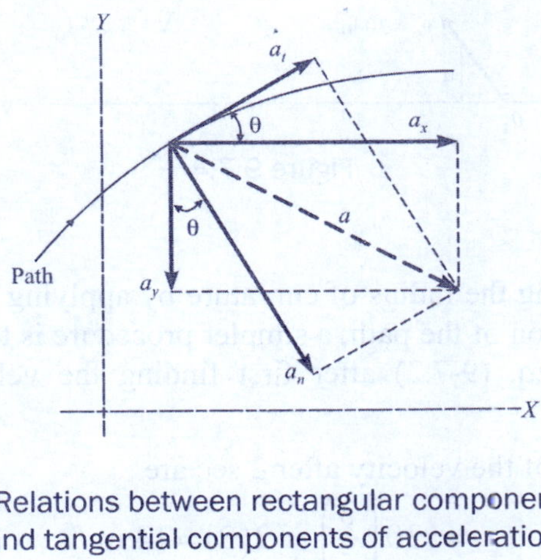

Figure 9.7.3 Relations between rectangular components and normal and tangential components of acceleration.

The preceding discussion of tangential and normal components of acceleration is also valid for three-dimensional spatial motion, but finds limited use because of the difficulty of specifying the directions of $\hat{e}_t$ and $\hat{e}_n$. The direction of the tangent to the spatial path is easily visualized, but there are an infinite number of perpendiculars to this tangent. The direction of the *principal normal* along which $\hat{e}_n$, is directed is that which passes through the center of curvature. Then $\hat{e}_t$ and $\hat{e}_n$ define a plane which is coincident with that described by the simultaneous stretching and swinging of the velocity vector. This plane is known as the *osculating plane*. The perpendicular to the osculating plane is the binormal direction. The unit binormal vector $\hat{\mathbf{e}}_b$, together with $\hat{\mathbf{e}}_t$ and $\hat{\mathbf{e}}_n$, forms a right-handed set of mutually perpendicular vectors, defined by $\hat{\mathbf{e}}_t \times \hat{\mathbf{e}}_n = \hat{\mathbf{e}}_b$, whose directions continually change in space with the motion of a particle along its path.

ILLUSTRATIVE PROBLEMS

9.7.1 A particle moves in the *XY* plane with $a_x = -2$ m/s^2 and $a_y = -10$ m/s^2. If its initial velocity is 30 m/s directed at a slope of 4 to 3 as shown in Fig. 9-7.4, compute the radius of curvature of the path 2 sec later.

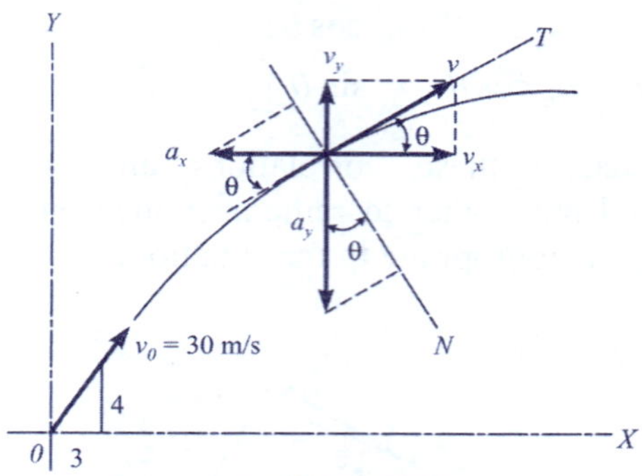

Figure 9.7.4

Solution

Instead of determining the radius of curvature by applying the familiar calculus formula to the equation of the path, a simpler procedure is to compute the radius of curvature from Eq. (9-7.2) after first finding the velocity v and normal acceleration a_n.

The components of the velocity after 2 sec are

$$[v = v_o + at] \qquad v_x = 30\left(\frac{3}{5}\right) - 2(2) = 14 \text{ m/s}$$

$$v_y = 30\left(\frac{4}{5}\right) - 10(2) = 4 \text{ m/s}$$

Combining these components yields the resultant velocity v and the inclination θ of the tangent to the path.

$$[V = \sqrt{V_x^2 + V_y^2}\,] \qquad v = \sqrt{(14)^2 + (4)^2} = 14.56 \text{ m/s}$$

$$\tan\theta = \frac{v_y}{v_x} = \frac{4}{14};\ \theta = 15.95^\circ$$

The normal component of acceleration is found by projecting the rectangular components of acceleration upon the normal to the path. As shown in the Fig. 9-7.4 these projections are oppositely directed so that a_n is equal to their difference.

$[a_n = a_y \cos\theta - a_x \sin\theta]$ $\quad a_n = 10\cos 15.95^\circ - 2\sin 15.95^\circ$

$$= 9.06 \text{ m/s}^2$$

Finally, by applying Eq. (9-7.2), we obtain

$\left[a_n = \dfrac{v^2}{\rho}\right]$ $\quad 9.06 = \dfrac{(14.56)^2}{\rho} \quad \rho = 23.4 \text{ m}$ ***Ans.***

9-7.2 The motion of a particle is defined by the position vector $\mathbf{r} = 5t\hat{\mathbf{i}} + 3t^2\hat{\mathbf{j}} + \frac{1}{3}t^3\hat{\mathbf{k}}$ where r is in metres and t is in seconds. At the instant when $t = 2$ sec, find the tangential and normal components of acceleration and the principal radius of curvature.

Solution

The velocity and acceleration at any instant are determined by successive differentiation of the position vector to be

$$\mathbf{v} = 5\hat{\mathbf{i}} + 6t\hat{\mathbf{j}} + t^2\hat{\mathbf{k}}$$

$$\mathbf{a} = \quad + 6\hat{\mathbf{j}} + 2t\hat{\mathbf{k}}$$

so that at $t = 2$ sec.

$$\mathbf{v} = 5\hat{\mathbf{i}} + 12\hat{\mathbf{j}} + 4\hat{\mathbf{k}} \qquad v = \sqrt{185} = 13.6 \text{ m/s}$$

$$\mathbf{a} = 6\hat{\mathbf{j}} + 4\hat{\mathbf{k}}, \qquad a = \sqrt{52} = 7.21 \text{ m/s}^2$$

Since the velocity is tangent to the path, the unit tangent vector at $t = 2$ sec is

$$\hat{\mathbf{e}}_t = \frac{\mathbf{v}}{v} = \frac{1}{13.6}(5\hat{\mathbf{i}} + 12\hat{\mathbf{j}} + 4\hat{\mathbf{k}})$$

whence the tangential component of acceleration at this instant is

$$a_t = \mathbf{a}\cdot\hat{\mathbf{e}}_t = (6\hat{\mathbf{j}} + 4\hat{\mathbf{k}}) \bullet \frac{1}{13.6}(5\hat{\mathbf{i}} + 12\hat{\mathbf{j}} + 4\hat{\mathbf{k}})$$

$= \dfrac{72 + 16}{13.6} = 6.47 \text{ m/s}^2$ ***Ans.***

Applying the concept that $\mathbf{a}_t$ and $\mathbf{a}_n$ are perpendicular components of $\mathbf{a}$, we find the magnitude of the normal component of acceleration from

$$a_n^2 = a^2 - a_t^2 = 52 - (6.47)^2 = 10.2; \; a_n = 3.19 \text{ m/s}^2 \qquad \textit{Ans.}$$

Finally, the principal radius of curvature is found from the relation

$$\rho = \frac{v^2}{a_n} = \frac{185}{3.19} = 58.0 \text{ m} \qquad \textit{Ans.}$$

A general method of finding ρ in any coordinate system is to take the cross product of the velocity v and the acceleration **a**. Thus,

$$\mathbf{v} \times \mathbf{a} = \mathbf{v} \times (\mathbf{a}_t + \mathbf{a}_n) = \mathbf{v} \times \mathbf{a}_n$$

where $v \times a_t = 0$ because **v** and a_t are collinear tangents to the path. Then the magnitude of $\mathbf{v} \times \mathbf{a}_n$ is equal to that of $\mathbf{v} \times \mathbf{a}$, or

$$|\mathbf{v} \times \mathbf{a}| = |\mathbf{v} \times \mathbf{a}_n| = va_n \sin 90^\circ = va_n = v\frac{v^2}{\rho} = \frac{v^3}{\rho}$$

whence

$$\rho = \frac{v^3}{|\mathbf{v} \times \mathbf{a}|} \qquad (9\text{-}7.5)$$

To illustrate this general procedure, let us check the value of ρ previously obtained. Noting that at $t = 2$ sec, $\mathbf{v} = 5\hat{\mathbf{i}} + 12\hat{\mathbf{j}} + 4\hat{\mathbf{k}}$ and $\mathbf{a} = 6\hat{\mathbf{j}} + 4\hat{\mathbf{k}}$, we obtain

$$\mathbf{v} \times \mathbf{a} = \begin{vmatrix} 5 & 12 & 4 \\ 0 & 6 & 4 \\ \hat{\mathbf{i}} & \hat{\mathbf{j}} & \hat{\mathbf{k}} \end{vmatrix} = \hat{\mathbf{i}}(48-24) + \hat{\mathbf{j}}(-20) + \hat{\mathbf{k}}(30)$$

whence

$$|\mathbf{v} \times \mathbf{a}| = \sqrt{(24)^2 + (20)^2 + (30)^2} = 43.2 \text{ m}^2/\text{sec}^3$$

Applying Eq. 9-7.5 we obtain

$$\left[\rho = \frac{v^3}{|\mathbf{v} \times \mathbf{a}|}\right] \qquad \rho = \frac{(13.6)^3}{43.2} = 58.0 \text{ m}$$

Also observe that this general method determines the magnitude of $\mathbf{a}_n$ as follows:

$[|\mathbf{v} \times \mathbf{a}| = va_n]$ $\qquad 43.2 = 13.6a_n \qquad a_n = 3.19 \text{ m/s}^2$ ***Check***

As a final comment, whenever the unit normal vector is needed to specify the direction of the normal acceleration, the following procedure may be used. Using the data previously found at $t = 2$ sec, we apply

$$\mathbf{a} = a_t\hat{\mathbf{e}}_t + \mathbf{a}_n\hat{\mathbf{e}}_n$$

$$6\hat{\mathbf{j}} + 4\hat{\mathbf{k}} = 6.47\left(\frac{1}{13.6}\right)(5\hat{\mathbf{i}} + 12\hat{\mathbf{j}} + 4\hat{\mathbf{k}}) + 3.19\hat{\mathbf{e}}_n$$

which reduces to

$$3.19\hat{\mathbf{e}}_n = -2.48\hat{\mathbf{i}} + 0.28\hat{\mathbf{j}} + 2.097\hat{\mathbf{k}}$$

whence

$$\hat{\mathbf{e}}_n = -0.778\hat{\mathbf{i}} + 0.0888\hat{\mathbf{j}} + 0.658\hat{\mathbf{k}}$$ ***Ans.***

The unit binormal vector is $\hat{\mathbf{e}}_b = \hat{\mathbf{e}}_t \times \hat{\mathbf{e}}_n$ as defined. It may readily be determined now that $\hat{\mathbf{e}}_t$ and $\hat{\mathbf{e}}_n$ are known. In this example, we obtain

$$\hat{\mathbf{e}}_b = \frac{1}{13.6}\begin{vmatrix} 5 & 12 & 4 \\ -0.778 & 0.0888 & 0.658 \\ \hat{\mathbf{i}} & \hat{\mathbf{j}} & \hat{\mathbf{k}} \end{vmatrix} = 0.554\hat{\mathbf{i}} - 0.471\hat{\mathbf{j}} + 0.719\hat{\mathbf{k}}$$

PROBLEMS

9-7.3 A particle moves on a circular path of 12 m radius so that its arc distance from a fixed point on the path is given by $s = (1.2t^3 - 3t)$ where s is in metres and t is in seconds. Compute the total acceleration at the end of 2 sec.

9-7.4 The pin P moves along a curved path which is determined by the motions of two slotted links A and B. At the instant shown in Fig. P-9-7.4, A has a velocity of 3 m/s and an acceleration of 2.5 m/s² both to the right, while B has a velocity of 4 m/s and an acceleration of 1.25 m/s² both vertically downward. Find the radius of curvature of the path of P at this instant and sketch how the path curves.

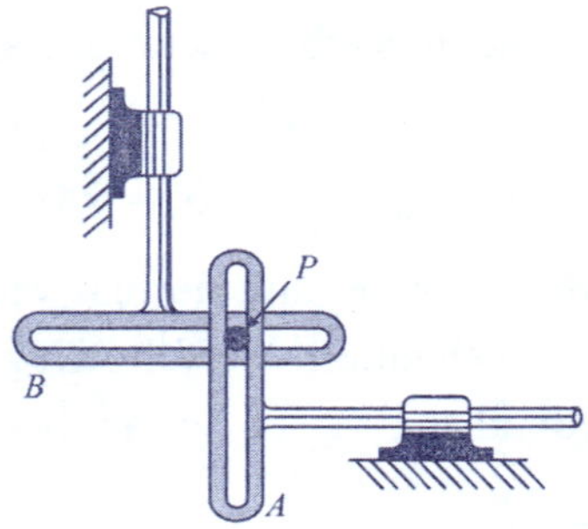

Figure P-9-7.4

9.7.5 A stone is thrown with an initial velocity of 30 m/s upward at 60° to the horizontal. Compute the radius of curvature of its path at the position where it is 15 m horizontally from its initial position.

9.7.6 A particle has an initial velocity of 60 m/s up to the right at a slope of 0.75. The component of acceleration are constant at $a_x = -3.5$ m/s^2 and $a_y = -6$ m/s^2. Compute the radius of curvature at the start and at the top of the path.

9.7.7 A particle moves counterclockwise on a circular path of 120 m radius. It starts from a position which is horizontally to the right of the center of the path and moves so that $s = (3t^2 + 6t)$ where s is the arc distance in metres and t is in seconds. Compute the horizontal and vertical components of acceleration at the end of 3 sec.

9-7.8. The velocity of a particle is defined by $v_x = 100 - t^{3/2}$ and $v_y = 100 + 10t - 2t^2$ where v is in m/s and t in seconds. Determine the radius of curvature at the top of its path.

9-7.9 Using the data of the preceding problem, determine the radius of curvature of the path of the particle at $t = 12$ sec.

9.7.10 The rectangular components of acceleration for a particle are $a_x = 3t$ and $a_y = 30 - 10t$ where a is in m/s^2. If the particle starts from rest at the origin, find the radius of curvature of the path at $t = 2$ sec.

9-7.11 The telescoping rod shown in Fig. P-9-7.11 forces the pin P to move along the fixed path $9y = x^2$ where x and y are in m. At any time t, the

x coordinate of P is given by $x = t^2 - 5t$. At what rate is the speed of P changing at $t = 6$ sec? Solve without using the value of a_y.

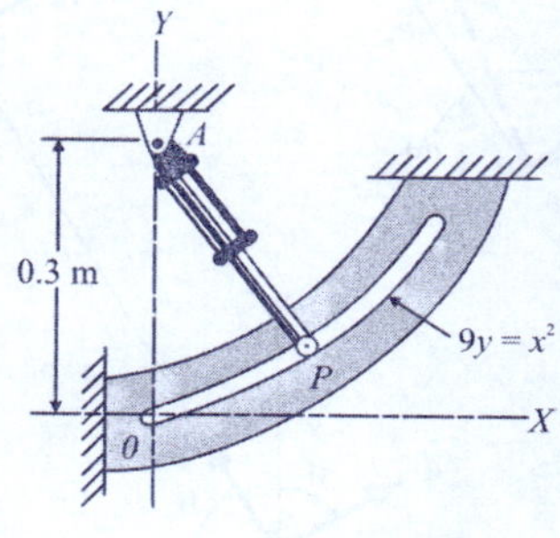

Figure P-9-7.11

9-7.12 The position vector of a particle is defined by $\mathbf{r} = (2t^3 - 12t)\ \hat{i} + (6t^2)\ \hat{j}$ where $\mathbf{r}$ is in metre and t is in seconds. Find the principal radius of curvature of the path of the particle at $t = 2$ sec.

9-7.13 The position vector of a particle is given by $\mathbf{r} = (2t^2\hat{\mathbf{i}} + 10t\,\hat{\mathbf{j}} + 1/3\ t^3\hat{\mathbf{k}}$ where $\mathbf{r}$ is in metre and t is in seconds. Determine the normal and tangential components of acceleration and the principal radius of curvature of the path of the particle at $t = 3$ sec.

9-7.14 Repeat Prob. 9-7.13 using $\mathbf{r} = 9\,t\,\hat{\mathbf{i}} - \dfrac{2}{3}\,t^3\,\hat{\mathbf{j}} + 3\,t^2\,\hat{\mathbf{k}}$ m and $t = 2$ sec.

9-7.15 A particle starting from rest moves with the acceleration $\mathbf{a} = 4\,t\,\hat{\mathbf{i}} - 3\,t^2\,\hat{\mathbf{j}} + 6\,\hat{k}$ m/s^2. Determine the principal radius of curvature of its path at $t = 2$ sec.

9-8 RADIAL AND TRANSVERSE COMPONENTS CYLINDRICAL COORDINATES

Usually the problems of dynamics are solved by means of rectangular or normal and tangential coordinate systems. Less frequently, the motion of a particle is more conveniently described by polar coordinates and its three-dimensional extension to cylindrical coordinates. We now consider how velocity and acceleration are expressed in these coordinate systems.

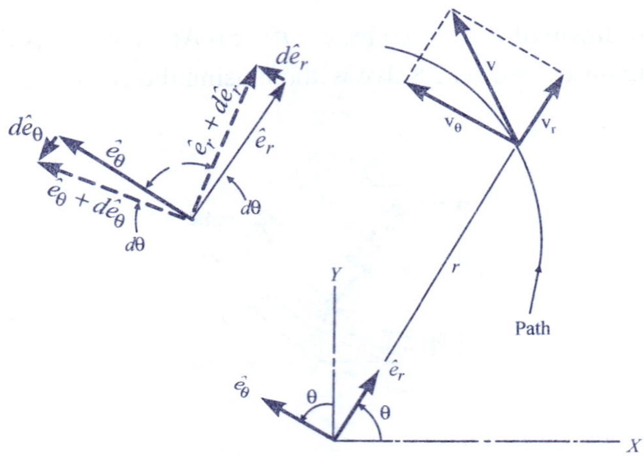

Figure 9-8.1 Radial and transverse components of velocity.

We start with plane curvilinear motion using polar coordinates to denote the position vector **r** in terms of its magnitude r and its angular coordinate θ measured from a fixed reference axis as shown in Fig. 9-8.1. Let the components of velocity be taken along and perpendicular to the position vector as shown, where $\mathbf{v}_r$ and $\mathbf{v}_\theta$ denote respectively the radial and transverse components of velocity. Note that the total velocity v is not perpendicular to **r** except for the special case when **r** is of constant length and the path therefore is circular.

Let us also establish unit vectors $\hat{\mathbf{e}}_r$ and $\hat{\mathbf{e}}_\theta$ acting in the positive directions of increasing r and θ, i.e., the unit vectors $\mathbf{e}_r$ and $\mathbf{e}_\theta$ are parallel to the positive senses of $\mathbf{v}_r$ and $\mathbf{v}_\theta$. Both these unit vectors will swing through the angle $d\theta$ as the position vector **r** swings through the angle $d\theta$ in the time *dt*. As shown in Fig. 9.8.1 (and explained in detail in Section 9-7), their tips move through the distances $|d\hat{\mathbf{e}}_r| = d\theta$ in the direction of $\hat{\mathbf{e}}_\theta$, and though $|d\hat{\mathbf{e}}_r| = d\theta$ in the direction of $-\hat{\mathbf{e}}_r$. Hence, their time rates of change (due to a swing but no stretch) are respectively

$$\left.\begin{aligned}\dot{\hat{\mathbf{e}}}_r &= \frac{\mathbf{d\hat{e}_r}}{\mathbf{dt}} = \frac{d\theta}{dt}\hat{\mathbf{e}}_\theta = \dot{\theta}\,\hat{\mathbf{e}}_\theta \\ \dot{\hat{\mathbf{e}}}_\theta &= \frac{d\hat{\mathbf{e}}_\theta}{dt} = \frac{d\theta}{dt}(-\,\hat{\mathbf{e}}_r) = -\,\dot{\theta}\,\hat{\mathbf{e}}_r\end{aligned}\right\} \tag{9-8.1}$$

Another more direct method of obtaining these results is discussed later in Section 12-6 where we derive the omega theorem.

We may now proceed to obtain expressions for velocity and acceleration by successive differentiation of the position vector $\mathbf{r} = r\hat{e}_r$. The velocity is found to be

$$\mathbf{v} = \frac{dr}{dt} = \frac{d}{dt}(r\hat{e}_r) = \dot{r}\,\hat{e}_r + r\dot{\hat{e}}_r$$

$$\mathbf{v} = \dot{r}\hat{e}_r + r\dot{\theta}\,\hat{e}_\theta \qquad (9\text{-}8.2)$$

or

$$v = v_r \nrightarrow v_\theta$$

where $v_r = \dot{r}$ and $v_\theta = r\dot{\theta}$ are respectively the magnitudes of the radial and transverse components of velocity acting in the positive senses of $\hat{e}_r$ and $\hat{e}_\theta$.

The acceleration is the time rate of change of velocity which results in

$$\mathbf{a} = \frac{dv}{dt} = \frac{d}{dt}(\dot{r}\,\hat{e}_r) + \frac{d}{dt}(r\,\dot{\theta}\,\hat{e}_\theta)$$

$$= \ddot{r}\,\hat{e}_r + \dot{r}\dot{\hat{e}}_r + \dot{r}\dot{\theta}\,\hat{e}_\theta + r\ddot{\theta}\,\hat{e}_\theta + r\dot{\theta}\,\dot{\hat{e}}_\theta$$

into which we substitute the values of $\dot{\hat{e}}_r$ and $\dot{\hat{e}}_\theta$ and from Eq. (9-8.1) to obtain

$$a = \ddot{r}\,\hat{e}_r + \dot{r}(\dot{\theta}\,\hat{e}_\theta) + \dot{r}\dot{\theta}\,\hat{e}_\theta + r\ddot{\theta}\,\hat{e}_\theta + r\dot{\theta}\,(-\dot{\theta}\,\hat{e}_r)$$

or

$$= (\ddot{r} - r\dot{\theta}^2)\,\hat{e}_r + (r\ddot{\theta} + 2\dot{r}\dot{\theta})\,\hat{e}_\theta \qquad (9\text{-}8.3)$$

$$a = a_r \nrightarrow a_\theta$$

where $a_r = \ddot{r} - r\dot{\theta}^2$ and $a_\theta = r\ddot{\theta} + 2\dot{r}\dot{\theta}$ are respectively the magnitudes of the radial and transverse components of acceleration acting in the positive senses of $\hat{e}_r$ and $\hat{e}_\theta$.

An alternate form of the transverse component is

$$a_\theta = \frac{1}{r}\frac{d}{dt}\left(r^2\dot{\theta}\right)$$

which is useful when it is necessary to integrate a_θ. This equivalence may be verified by direct differentiation of the alternate form.

On occasion, it may be required to transform the radial and transverse components of acceleration into normal and tangential components. Since either set of components defines the same acceleration vector, this transformation is readily accomplished by projecting the r-θ components upon the normal and tangential directions. Thus in Fig. 9-8.2, the tangential direction (*T*) is along the velocity vector **v** while the normal direction (*N*) is perpendicular to **v**. The algebraic sums of the projections of a_r and a_θ along the *T* and *N* directions then gives

$$a_t = a_r \cos\beta + a_\theta \sin\beta$$

$$a_n = a_\theta \cos\beta - a_r \sin\beta$$

which, since the relations among the velocity components is seen to give $\cos\beta = v_r/v$ and sin v_θ/v, readily reduces to

$$\left.\begin{aligned} a_t &= \frac{a_r v_r + a_\theta v_\theta}{v} \\ a_n &= \frac{a_\theta v_r - a_r v_\theta}{v} \end{aligned}\right\} \qquad (9\text{-}8.4)$$

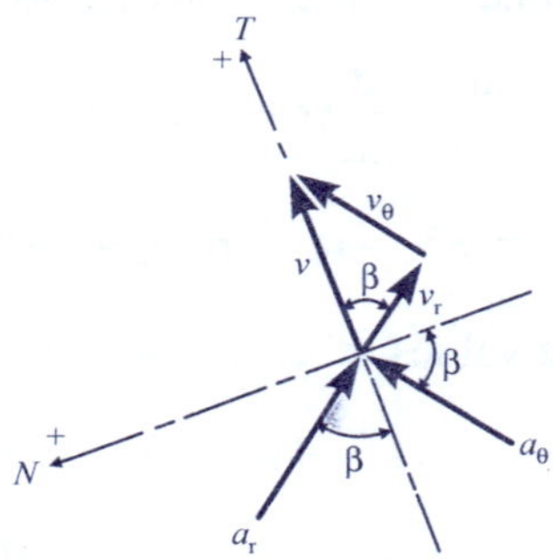

Figure 9-8.2 Algebraic summation of projections of a_r and a_θ along *N* and *T* determine a_n and a_t.

These results are not important enough to memorize; they are discussed only to illustrate the geometric relations among the several sets of components and to emphasize a method of reasoning whereby they can be easily developed when needed.

A more elaborate approach which yields the same results, although not as direct, is to first express $\hat{\mathbf{e}}_t$ and $\hat{\mathbf{e}}_n$, in terms of $\hat{\mathbf{e}}_r$, and $\mathbf{e}_\theta$ then apply

$$a_t = \mathbf{a}.\hat{e}_t = (\mathbf{a}_r + \mathbf{a}_\theta).\hat{\mathbf{e}}_t$$

$$a_n = \mathbf{a}.\hat{e}_n = (\mathbf{a}_r + \mathbf{a}_\theta).\hat{\mathbf{e}}_n$$

This procedure is left to be done as an exercise in Prob. 9-8.4.

It is Instructive to observe the geometric significance of the radial and transverse components of velocity and acceleration by discussing what happens if only **r** varies while θ is held constant, and then if only θ varies while **r** has a constant length. Permitting **r** to vary while θ is held constant is equivalent to rectilinear motion along the fixed direction of θ. This results in $v = \dot{r}$ and $a = \ddot{r}$. On the other hand, if **r** is of constant length while θ varies, we have a rotation along a circular path in which $v = r\dot{\theta} = r\omega$ and $a = a_n \nrightarrow a_t = (r\dot{\theta}^2 = r\omega^2) \nrightarrow \left(r\ddot{\theta} = r\alpha\right)$. In the equivalent expressions enclosed in the parentheses, $\omega = \dot{\theta}$ is the common symbol for angular velocity and $\alpha = \ddot{\theta}$ represents angular acceleration.

CYLINDRICAL COORDINATES

The extension of radial and transverse coordinates to three dimensional space is accomplished by adding a fixed Z coordinate axis to the polar coordinate axes, The position vector is then given by $\mathbf{r} = r\hat{\mathbf{e}}_r + z\hat{\mathbf{k}}$ where $\hat{\mathbf{k}}$ is a unit vector in the Z direction. Note that the positive Z direction is that which makes the unit vectors $\hat{\mathbf{e}}_r$, $\hat{\mathbf{e}}_\theta$, and $\hat{\mathbf{k}}$ form a right-handed triad so that $\hat{\mathbf{e}}_r \times \hat{\mathbf{e}}_\theta = \hat{\mathbf{k}}$. Since $\hat{\mathbf{k}}$ is constant in direction as well as magnitude, $\dot{\hat{\mathbf{k}}} = 0$. Hence, the successive derivatives of $z\hat{\mathbf{k}}$ are merely $\dot{z}\,\hat{\mathbf{k}}$ and $\ddot{z}\,\hat{\mathbf{k}}$. By adding these terms to Eqs. (9-8.2) and (9-8.3), we obtain the following expressions for velocity and acceleration in cylindrical coordinates:

$$\left.\begin{aligned} v &= \dot{r} = \dot{r}\hat{e}_r + r\dot{\theta}\hat{e}_\theta + \dot{z}\hat{\mathbf{k}} \\ a &= \ddot{r} = (\ddot{r} - r\dot{\theta}^2)\hat{e}_r + (r\ddot{\theta} + 2\dot{r}\dot{\theta})\hat{e}_\theta + \ddot{z}\hat{\mathbf{k}} \end{aligned}\right\} \quad (9\text{-}8.5)$$

ILLUSTRATIVE PROBLEMS

9-8.1 The plane curvilinear motion of a particle is defined in polar coordinates by $r = t^3/3 + 2t$ and $\theta = 0.3t^2$ where r is in metres, θ is in radians, and t is in seconds. At the instant when $t = 2$ sec, determine the magnitudes of velocity, acceleration, and the radius of curvature of the path.

Solution

Into successive time derivatives of r and θ, we substitute $t = 2$ sec to obtain

$$\left.\begin{aligned} r &= \frac{1}{3}t^3 + 2t & \theta &= 0.3t^2 \\ \dot{r} &= t^2 + 2 & \dot{\theta} &= 0.6t \\ \ddot{r} &= 2t & \ddot{\theta} &= 0.6 \end{aligned}\right]_{t=2} \qquad \begin{aligned} r &= \frac{20}{3}\text{ m} & \theta &= 1.2\text{ rad} = 68.8^\circ \\ \dot{r} &= 6\text{ m/s} & \dot{\theta} &= 1.2\text{ rad/sec} \\ \ddot{r} &= 4\text{ m/s}^2 & \ddot{\theta} &= 0.6\text{ rad/sec}^2 \end{aligned}$$

These values are now substituted into Eqs. (9-8.2) and (9-8.3) to obtain the following components of v and a:

$$v_r = \dot{r} = 6 \text{ m/sec}$$

$$v_\theta = r\dot{\theta} = \frac{20}{3} \times 1.2 = 8 \text{m/sec}$$

$$a_r = \ddot{r} - r\dot{\theta}^2 = 4 - \frac{20}{3}(1.2)^2 = -5.6 \text{m/sec}^2$$

$$a_\theta = r\ddot{\theta} + 2\dot{r}\dot{\theta}$$

$$= \frac{20}{3}(0.6) + 2 \times 6 \times 1.2 = 18.4 \text{ m/sec}^2$$

from which we find the velocity and acceleration to be

$[v = v_r \nrightarrow v_\theta]$ $\quad v = 6 \nrightarrow 8 = 10$ m/s **Ans.**

$[a = a_r \nrightarrow a_\theta]$ $\quad a = 5.6 \nrightarrow 18.4 = 19.25$ m/s^2 **Ans.**

The directions of the velocity and acceleration components, with due regard for sign, are shown in Fig. 9-8.3. By projecting the acceleration components upon the normal N to the path (i.e., perpendicular to the tangential direction defined by v), we find the normal component of acceleration to be

$$a_n = \Sigma N = 5.6 \cos\phi + 18.4 \sin\phi$$

$$= 5.6\,(0.8) + 18.4(0.6) = 15.52 \text{ m/sec}^2$$

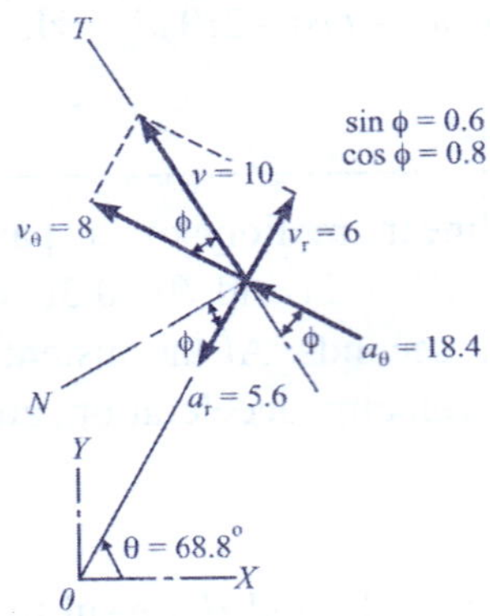

Figure 9-8.3

Using Eq. (9-7.1), we then find the instantaneous radius of curvature to be

$\left[a_n = \dfrac{v^2}{\rho}\right]$ $\quad 15.52 = \dfrac{(10)^2}{\rho}$ $\quad \rho = 6.44$ m **Ans.**

9.8.2 The 0.2 m crank OA of an oscillating arm quick-return mechanism is rotating counterclockwise at a constant rate of 10 rad/ sec. When the mechanism is in the position as shown in Fig. 9-8.4, determine the angular acceleration of arm *BD*.

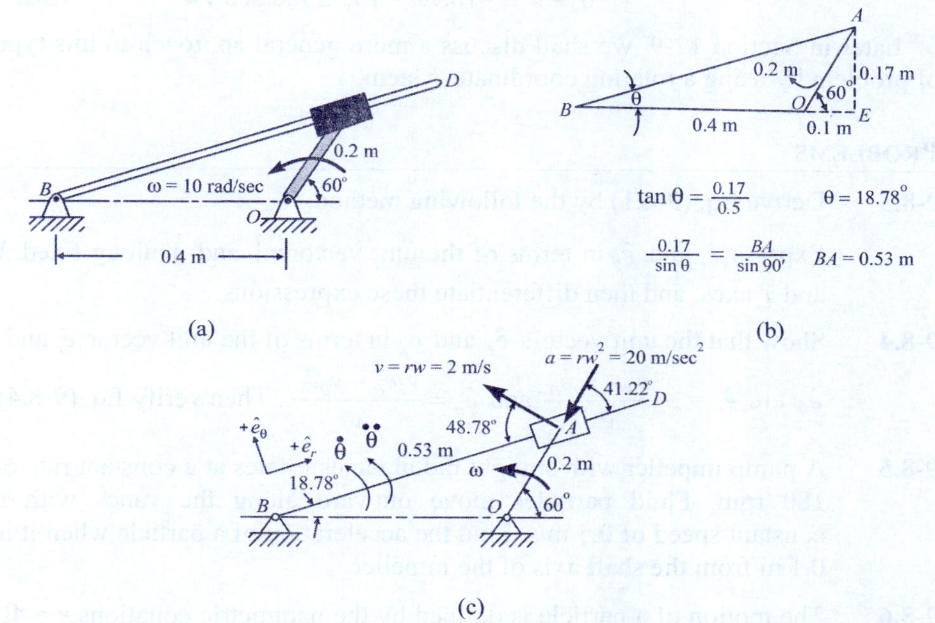

Figure 9-8.4

Solution

From the preliminary geometry indicated in part (b), we find $\theta = 18.78$ and $BA = 0.53$ m which we insert on the skeleton outline of the mechanism in part (c). The motion of *A* is common to both *OA* and *BD*. As part of *OA*, point *A* has pure rotation about *O* which causes the values of *v* and *a as* shown. With respect to *BD*, however, point *A* is sliding toward *B* as *BD* rotates about *B*. It is natural, therefore, to describe the motion of *A* on *BD* in terms of r-θ components with an origin at *B*, the positive senses being as indicated.

By projecting *v* onto the radial and transverse directions, we obtain

$$v_r = \dot{r} = -2\cos 48.78 = -1.32 \text{ m/s}$$

and

$$v_\theta = r\dot{\theta} = 2\sin 48.78°; \qquad 0.53\,\dot{\theta} = 2\sin 48.78; \qquad \dot{\theta} = 2.84 \text{ rad/sec} \circlearrowleft$$

Similarly, by projecting *a* upon the transverse direction, we obtain a_θ which enables us to find $\ddot{\theta}$ now that r, $\dot{r}$ and $\dot{\theta}$ are known. Thus, we get

$[a_\theta = r\ddot{\theta} + 2\dot{r}\dot{\theta}]$ $\quad -20 \sin 41.22 = 0.53\ddot{\theta} + 2(-1.32)(2.84)$

$$\ddot{\theta} = \alpha = -10.72 = 10.72 \text{ rad/sec}^2) \circlearrowleft \qquad \textbf{\textit{Ans.}}$$

Later in Section 12-9, we shall discuss a more general approach to this type of problem by using a rotating coordinate system.

PROBLEMS

9-8.3 Derive Eq. (9-8.1) by the following method:

Express $\hat{e}_r$ and $\hat{e}_\theta$ in terms of the unit vectors $\hat{\mathbf{i}}$ and $\hat{\mathbf{j}}$ along fixed X and Y axes, and then differentiate these expressions.

9-8.4 Show that the unit vectors $\hat{\mathbf{e}}_t$ and $\hat{\mathbf{e}}_n$ in terms of the unit vector $\hat{e}_r$ and $\hat{e}_\theta$ are $\hat{e}_t = \dfrac{v_r\hat{e}_r + v_\theta\hat{e}_\theta}{v}$ and $\hat{e}_n = \dfrac{v_r\hat{e}_\theta - v_\theta\hat{e}_r}{v}$. Then verify Eq. (9-8.4)

9-8.5 A pump impeller with straight radial vanes rotates at a constant rate of 180 rpm. Fluid particles move outward along the vanes with a constant speed of 0.5 m/s. Find the acceleration of a particle when it is 0.1 m from the shaft axis of the impeller.

9-8.6 The motion of a particle is defined by the parametric equations $r = 4t^2$ and $\theta = 2t$ where r is in metres and t is in seconds. Find the total acceleration of the particle at $t = 2$ sec.

9-8.7 The polar coordinates of a particle moving on a plane curve are given by $r = t^3 - 3t + 10$ and $r = 2\theta$ where r is in metres, θ is in radians, and t is in seconds. Determine the acceleration of the particle at $t = 2$ sec.

9-8.8 In a certain plane motion, it is known that when the radial distance from a fixed origin is $r = 0.5$ m, the r-θ components of velocity are $v_r = 1.5$ m/s and $v_\theta = 2$ m/s, while the components of acceleration are $a_r = -1.5$ m/s^2 and $a_\theta = 25$ m/s^2. At this instant, find the angular acceleration α (i.e., $\ddot{\theta}$) of the position vector and the component of acceleration normal to the path.

9-8.9 At the position shown in Fig. P-9-8.9, the block B is sliding outward along the straight rod with the given values of velocity and acceleration relative to the rod. Simultaneously, the rod has the given values of angular velocity ω and angular acceleration $\propto$. Find the total acceleration of the block. What is the component of this acceleration normal to the path described by the block?

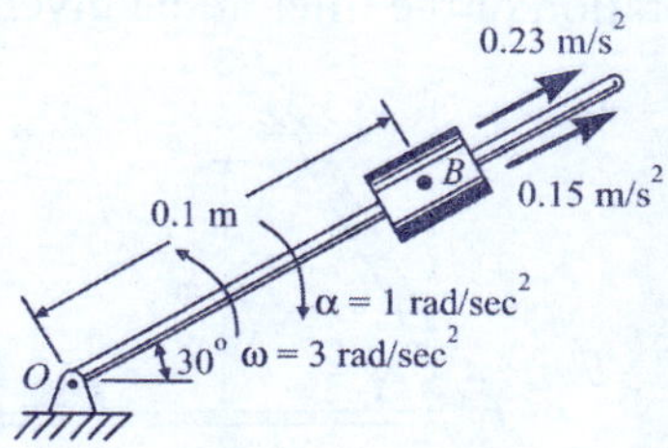

Figure P-9-8.9

9.8.10 A particle is constrained to move along the cardiod $r = 2(1 + \cos\theta)$ where r is in m. If its position vector has a constant angular speed of 2 rad/sec counterclockwise, at what rate is the particle changing its speed at $\theta = 60°$?

9-8.11 A rocket is fired vertically and tracked by radar as shown in Fig. P.9-8.11. At the instant $\theta = 60°$, it is known that $r = 9144$ m, $\dot{\theta} = 0.02$ rad/sec, and $\ddot{\theta} = 0.002$ rad/sec^2. Determine the velocity and acceleration of the rocket.

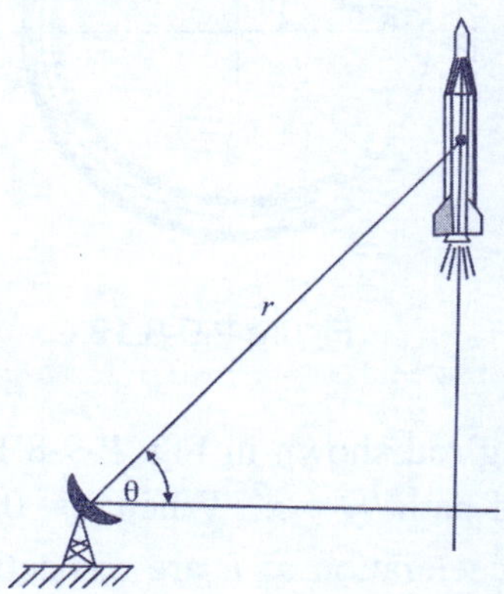

Figure P-9-8.11

9-8.12 A marine steering gear known as Rapson's slide is shown in Fig. P-9-

8.12. The tiller AB is turned by the link DE acting upon the slider D. If link DE is moving rightward at a constant speed of 1 m/min, find the angular acceleration of the tiller at the given position.

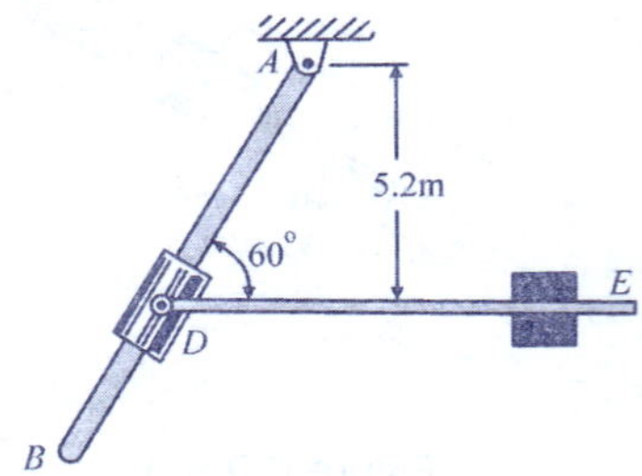

Figure P-9-8.12

9.8.13 In Fig. P-9-8.13, the motion of the pin P on a fixed circular path of radius r is controlled by the slotted rod OA. Between $\theta = \pm\ \pi/4$, the rod OA has a constant angular speed $\dot{\theta}$. Find the acceleration of P at any value of θ between the indicated limits.

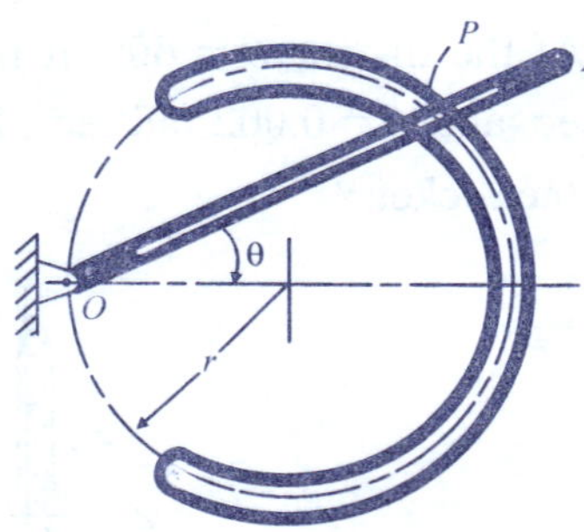

Figure P-9-8.13

9.8.14 The telescoping rod shown in Fig. P-9-8.14 forces the pin P to move along the fixed path $9y = x^2$. When $x = 0.15$ m., it is known that the velocity and acceleration of P are respectively $\mathrm{v} = 0.3\,\hat{\mathbf{i}} + 0.5\,\hat{\mathbf{j}}$ m/s and $\mathrm{a} = 0.25\,\hat{\mathbf{i}} + 0.5\,\hat{\mathbf{j}}$ m/s^2. What is then the angular acceleration of the rod?

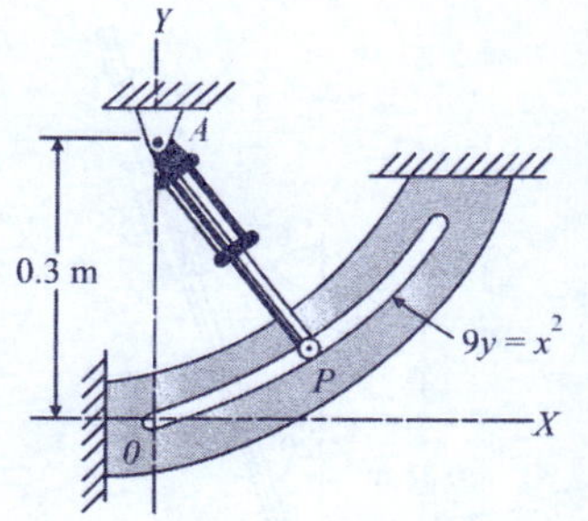

Figure P-9-8.14

9.8.15 The driving crank AB of the oscillating arm quick-return mechanism shown in Fig. P-9-8.15 rotates at a constant clockwise rate of 11.2 rad/sec. Compute the angular acceleration of the arm CD at the instant when crank AB is horizontal as shown.

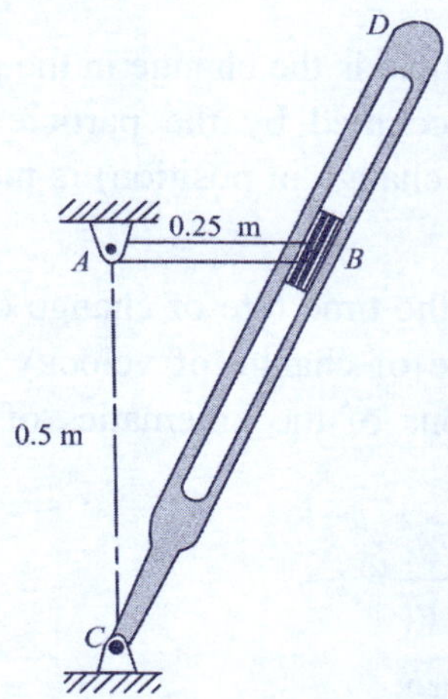

Figure P-9-8.15

9.8.16 The elements of a quick-return mechanism have the given configuration at the instant shown in Fig. P.9-8.16. Find the angular acceleration of arm CD if the driving crank AB rotates at a constant clockwise rate of 5 rad/sec.

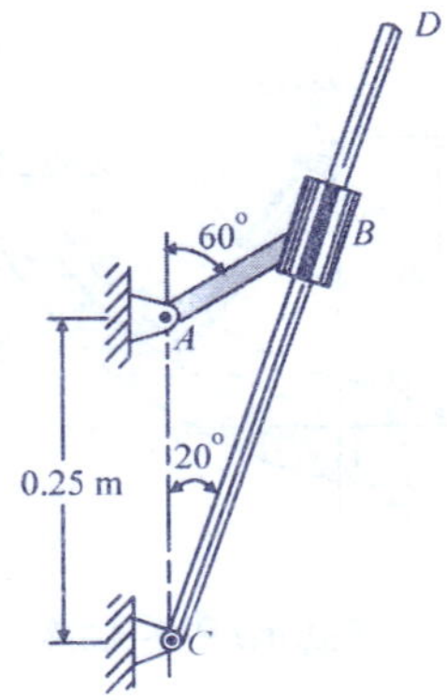

Figure P-9-8.16

SUMMARY

The displacement of a particle is the change in the position vectors from a fixed origin to the positions occupied by the particle on its path of travel. The absolute displacement (or change in position) is independent of the choice of a fixed origin.

Velocity is defined as the time rate of change of displacement; acceleration is defined as the time rate of change of velocity. From these definitions, the vector differential equations of the kinematics of a particle are given by the equations

$$v = \frac{dr}{dt} = \dot{r} \qquad (9\text{-}2.1)$$

and

$$a = \frac{dv}{dt} = \dot{v} = \ddot{r} \qquad (9\text{-}2.3)$$

For rectilinear motion in which displacement, velocity, and acceleration are unidirectional, the signed position coordinate s replaces the position vector **r** and the scalar differential equations of kinematics become

$$v = \frac{ds}{dt} \qquad (9\text{-}3.1)$$

$$a = \frac{dv}{dt} \qquad (9\text{-}3.2)$$

$$a\,ds = v\,dv \qquad (9\text{-}3.3)$$

In applying these equations, take the positive values of s, v, and a as corresponding to the positive sense of initial motion. When the motion can be expressed graphically (Section 9-4), the area under a v-t curve represents the change of displacement during the corresponding time interval, and the area under the a-t curve represents the corresponding change in velocity. The a-t curve can also be used to find the change in displacement during a time interval by means of the equation

$$\Delta s = v_1(\Delta t) + (\text{Area})_{a-t}(\bar{t}_2) \tag{9-4.3}$$

where v_1 is the velocity at the start of the time interval and the last term is the moment of the area under the a-t curve about an ordinate at the *end* of the time interval.

The curvilinear motion of a particle may be expressed in terms of rectangular components, normal and tangential components, or polar and cylindrical coordinates. In terms of rectangular components, we have

$$\mathbf{r} = x\hat{\mathbf{i}} + y\hat{\mathbf{j}} + z\hat{\mathbf{k}} \tag{9-6.1}$$

$$\mathbf{v} = v_x\hat{\mathbf{i}} + v_y\hat{\mathbf{j}} + v_z\hat{\mathbf{k}} \tag{9-6.2}$$

$$\mathbf{a} = a_x\hat{\mathbf{i}} + a_y\hat{\mathbf{j}} + a_z\hat{\mathbf{k}} \tag{9-6.3}$$

These formulations show that curvilinear motion may be considered equivalent to the simultaneous projection of the motion along fixed rectangular axes, and each projected motion can be treated as a rectilinear motion.

Tangential and normal components of acceleration separate and denote respectively the rate of change of magnitude and direction of velocity. Using unit vectors $\hat{\mathbf{e}}_t$ and $\hat{\mathbf{e}}_n$ to denote directions respectively tangent, and normal to the path, we have

$$\mathbf{a} = \frac{dv}{dt}\hat{\mathbf{e}}_t + \frac{v^2}{\rho}\hat{\mathbf{e}}_n = \mathbf{a}_t + \mathbf{a}_n \tag{9-7.1}$$

where the magnitudes of the acceleration components are $a_t = \dfrac{dv}{dt}$ and $a_n = \dfrac{v^2}{\rho}$

and p is the instantaneous radius of curvature of the path. Their directions are such that positive a_t is tangent to the path in the direction of increasing speed whereas a_n is always directed inward toward the center of curvature.

In polar coordinates, the position vector $\mathbf{r}$ is expressed in terms of its magnitude r and its angular coordinate θ measured from a fixed reference axis.

The radial and transverse components of velocity and acceleration are

$$\mathbf{v} = \dot{r}\,\hat{\mathbf{e}}_r + r\dot{\theta}\hat{\mathbf{e}}_\theta = \mathbf{v}_r + \mathbf{v}_\theta \tag{9-8.2}$$

and

$$\mathbf{a} = \left(\ddot{r} - r\dot{\theta}^2\right)\hat{\mathbf{e}}_r + \left(r\ddot{\theta} + 2\dot{r}\dot{\theta}\right)\hat{e}_\theta = \mathbf{a}_r + \mathbf{a}_\theta \tag{9-8.3}$$

where $\hat{e}_r$ and $\hat{e}_\theta$ are unit vectors acting in the positive directions of increasing r and θ, i.e., the unit vectors $\hat{e}_r$ and $\hat{e}_\theta$ are directed parallel to the positive senses of $\mathbf{v}_r$ and $\mathbf{v}_\theta$.

The extension of polar coordinates to cylindrical coordinates is accomplished by adding a fixed Z axis perpendicular to the polar coordinate axes. The position vector is then $\mathbf{r} = r\hat{e}_r + z\hat{\mathbf{k}}$ where $\hat{\mathbf{k}}$ is a unit vector in the positive Z direction. The positive Z direction is that which makes the unit vectors $\hat{e}_r, \hat{e}_\theta$ and $\hat{\mathbf{k}}$ form a right-handed triad so that $\hat{e}_r \times \hat{e}_\theta = \hat{\mathbf{k}}$.

CHAPTER 10

GENERAL PRINCIPLES OF DYNAMICS

10-1 INTRODUCTION

Our objective in this chapter is to derive the two basic equations of dynamics that we stated in the preview of dynamics presented in Section 9-1. After discussing several associated concepts, we shall use kinematic principles to extend Newton's laws of motion for a single particle to bodies consisting of systems of particles. We concentrate here on the general principles of dynamics and in subsequent chapters consider their application to various types of rigid-body motion. But first it is appropriate to present the following brief background.

Dynamics is the branch of mechanics which deals with the study of bodies in motion. Compared with statics, dynamics is relatively new[1]; it is generally considered to have been begun by Galileo (1564-1642). Its development was greatly retarded by the lack of precise methods for measuring time. The experiments which form the foundation of dynamics require the use of three kinds of units: force, length, and time. Precise methods for measuring force and length are relatively simple and account in part for the early development of statics in which only these units of measurements are required. No accurate time-measuring devices, such as the pendulum clock developed by Huygens in 1657 and the balance wheel watch developed by Robert Hooke around 1666, were devised until after Galileo's death.

An understanding of dynamics was also retarded by the principles of natural

[1]An excellent summary of the history of dynamics and those who contributed greatly to its development is given in Goodman and Warner's *Dynamics*, Wadsworth Publishing Co., pp. 1- 8 and in Fanger's *Engineering Mechanics*, Merrill Publishing Co., pp. 501- 512.

philosophy which were promulgated by Aristotle and in Galileo's time were regarded as infallible. Galileo's experimental turn of mind led him to doubt these dogmas of abstract thought. For example, he did not accept the notion that heavy weights fall more rapidly than light ones. His experiments with dropping weights from the Leaning Tower of Pisa exploded this theory, but precipitated such bitter conflicts that he was forced to leave Pisa.[2]

Galileo's experiments with blocks sliding down inclined planes led to a relation between force and acceleration which Sir Isaac Newton (1642-1727) generalized and incorporated into the laws governing the motion of a particle that are named after him. Newton's laws of motion are the basis for extending the kinetics of a particle to a body composed of a system of particles. This extension is discussed in Sections 10-5 and 10-6.

At this point, definitions of the terms particle and body are pertinent. The term *particle* usually denotes an object of point size. The term *body* denotes a system of particles which form an object of appreciable size. The criterion of size is only relative however; the terms particle and body may apply equally to the same object. For example, in astronomical calculations the earth may be assumed to be a particle in comparison with the size of its path, whereas to earthbound observers, it is a body of great size. More realistically, a particle is a body so small that any differences in the motions of its parts can be ignored with negligible error. We shall also learn that any rigid body, regardless of its size, may be considered to be a particle if all of its parts move in identical parallel paths.

10-2 NEWTON'S LAWS OF MOTION FOR A PARTICLE

From his study of falling bodies, Galileo discovered the first two of what are commonly called Newton's laws of motion for a particle. Newton's name is associated with the laws of motion, however, because it was he who generalized them and, together with the law of universal gravitation, demonstrated their truth by astronomical predictions based on them.

Newton's laws of motion have been stated in a variety of ways. For our purposes, we shall phrase them as follows:

1. A particle acted upon by a balanced force system has no acceleration.
2. A particle acted upon by an unbalanced force system has an acceleration in line with and directly proportional to the resultant of the force system.[3]

[2] The Star Gazer by Z. Harsanyi (Putnam) is an excellent account of Galileo's life.

[3] An alternate version states that the resultant force on a particle is in line with and proportional to the time rate of change of momentum of the particle. Either statement is correct when applied to a particle of constant mass.

3. Action and reaction forces between two particles are always equal and oppositely directed.

The reference frame or set of axes in which Newton's laws are valid is one having a fixed origin and fixed directions of the axes. It is called an *inertial, Newtonian*, or *Galilean frame of reference*. An inertial frame may also be a frame whose axes remain fixed in direction while its origin moves along a straight line with constant speed. Motion with respect to an inertial frame is known as absolute motion. The most accurate of such reference frames available to us is one having its origin at the mass center of the solar system and its axes directed toward the so-called fixed stars whose parallaxes are too small to be measured. Such a reference frame is suitable for celestial mechanics, but for orbiting satellites or long-range rockets, the inertial frame could be one fixed in the earth with its origin at the earth's center. With respect to most engineering applications, however, it is sufficiently accurate to use an inertial frame attached to the surface of the earth.

It should be observed that modern concepts of relativistic mechanics initiated by Albert Einstein (1879-1955) and of quantum mechanics developed by Max Planck (1858-1947), Niels Bohr (1885-1962), and Werner Heisenberg (1901-1976) establish the boundaries of Newtonian mechanics but do not invalidate it. Except for speeds approaching that of light or the motion of elements of the atom, Newton's laws will continue to be used by engineers and scientists. Their accuracy has been definitely proven by the Apollo space missions.

Another outstanding contribution to the study of mechanics is Newton's law of universal gravitation which states that any two particles of masses m_1 and m_2 separated by a distance r attract each other with equal, opposite forces that are directed from either particle toward the other. The magnitude of each of this pair of action and reaction forces is given by

$$F = G\frac{m_1 m_2}{r^2} \qquad (10\text{-}2.1)$$

where G is the universal constant of gravitation[4]. Its value has been shown to be $66.7 \times 10^{-12}\ \text{m}^3/\text{kg-sec}^2$ in the SI system.

Observe that Newton's law of gravitation does not define any of the physical quantities (force, mass, and length) contained in it. The defining relations among these quantities is expressed in Newton's second law. In Section 10-4 we shall discuss various systems of units that may be used. Of importance here

---

[4]Do not confuse G with the acceleration g due to gravity. Note that g denotes the acceleration of a freely falling body; its value varies with position but for most engineering applications has the value of $9.81\ \text{m/s}^2$.

is the concept of weight. The weight of a body is the gravitational force of attraction exerted on the body by the earth and depends on its position relative to the earth's center.

10-3 FUNDAMENTAL EQUATION OF KINETICS FOR A PARTICLE

Consider a particle of weight W acted upon by the forces shown in Fig. 10-3.1*a*. The resultant of these forces is obtained by a tip-to-tail addition as in Fig.10-3.1*b*. Because all the forces on a particle are assumed to be concurrent (the size of a particle is a point), the direction and position of the resultant is indicated by the dashed vector **R** in Fig.10-3.1*a*. From Newton's second law of motion, this resultant causes an acceleration **a** which is in line with **R** and directly proportional to it, or

$$\mathbf{R} = k\mathbf{a} \tag{a}$$

where k is some constant of proportionality.

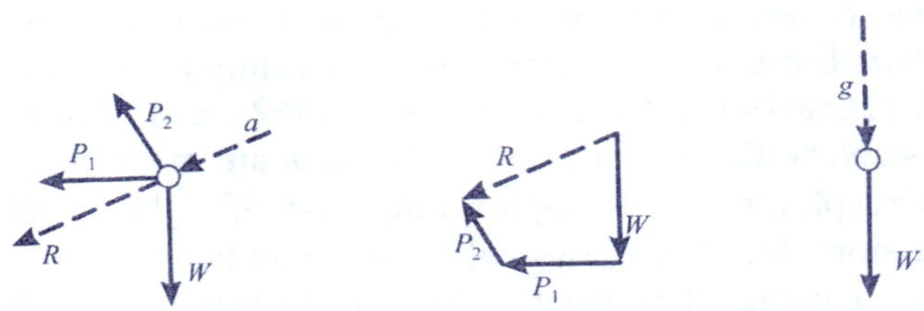

(a) (b) (c)

Figure 10.3.1 Resultant force on and acceleration of a particle.

If the same particle is now assumed to be in a vacuum, the resultant force acting upon it is its weight W. By experiment, the acceleration produced by W is found to be the value of the gravitational constant g which acts in the direction of W, as shown in Fig. 10-3.1c. Again applying Newton's second law and assuming the same constant of proportionality, we have

$$W = kg \tag{b}$$

Dividing Eq. (*a*) by Eq. (*b*) gives

$$\mathbf{R} = \frac{W}{g}\mathbf{a} = ma \tag{10-3.1}$$

Comparison with Eq. (*a*) discloses the value of the constant of proportionality to be W/g. This ratio of weight divided by the gravitational constant is often called the *gravitational mass m* of the particle since the weight

W and the associated gravitational acceleration g vary with position on the earth, whereas inertial mass m is an invariant absolute property of the particle which is independent of position. However, the most refined experiments have failed to show any difference between gravitational mass and inertial mass and they are therefore used interchangeably in this book. One advantage of using gravitational mass is that, when applying Eq. (10-3.1) it results in a consistent set of dimensional units by automatically including the conversion factor g. Perhaps as good a concept of mass as any is to consider that it represents the inertia of a body, i.e., the resistance a body offers to a change in its motion.

EFFECTIVE FORCE ON A PARTICLE

The effective force on a particle is defined as the resultant force on a particle. Since **R** and m**a** are numerically equivalent, either may be said to be the effective force on a particle. The use of ma to represent this force is especially convenient in cases when the acceleration of the particle is known but the actual force system producing this acceleration is not known. This concept is used in succeeding sections to extend the laws of motion for a single particle to the motion of a body composed of a system of particles.

According to Newton's third law, for every force there is an equal but opposite reaction. In the case of a particle accelerated by a resultant force, this reaction is called the *inertia force* of the particle. It is convenient to think of the inertia force as a force numerically equal to ma but directed oppositely to the acceleration.

If the inertia force is considered to act on a particle together with the resultant force, the particle will be in a state of equilibrium. This will be called *dynamic equilibrium* to distinguish it from static equilibrium in which the particle is at rest or is moving with constant velocity.

10-4 GRAVITATIONAL AND ABSOLUTE SYSTEM OF UNITS

In applying $\mathbf{R} = m\mathbf{a}$, any system of units can be used which will result in a force of one unit giving a unit acceleration to a unit mass. Four quantities are involved in this equation – force (F), mass (M), length (L), and time (T) – which are dimensionally related by

$$F = \frac{ML}{T^2}$$

Of these four quantities, L, T, and either F or M may be chosen as basic dimensions. Most engineers select F as a basic dimension and mass is the derived dimension expressed in terms of the others; namely, $M = FT^2/L$. This is known as a gravitational system since F is defined as the weight of a standard object. In the English system using foot-pound-second units, a unit mass (sometimes called a slug) has the dimensions of lb-sec^2/ft. In the metric system using meter-kilogram-second units, a unit mass has the dimensions of kg-sec^2/m.

In the absolute **MKS** system of units used by physicists and electrical engineers, inertial mass is a basic dimension and force is the derived dimension defined as $F = \frac{ML}{T^2}$. The unit of force, called a newton (nt), has the dimensions of kg-m/sec^2. The advantage to the electrical engineer of an absolute **MKS** system of units stems from the definition of a watt or volt-ampere as one newton-meter per second.

In general, either a gravitational or an absolute system of units may be used since each has its own advantage over the other. For example, in situations where the value of g may vary considerably as in rocket flight or space probes, it is preferable to consider mass as a basic unit since it is a property independent of position. However, most engineers are concerned with problems in which the variation in g is negligible; this makes the common concept of the weight of a standard body as the unit of force convenient to use. A comparison of various systems of units is shown in Table 10-4.1.

Table 10-4.1 Systems of Units.

System	Length	Time	Force	Mass	* Used by
Gravitational English (FPS)	foot (ft)	Sec	Pound (lb)	slug[a] (lb-sec^2/ft)	Engineers in English-speaking * countries
Gravitational Metric (MKS)	meter (m)	sec	kilogram (kg)	metric slug[a] (kg-sec^2/m)	Engineers elsewhere
Absolute Metric (CGS)	centimeter (cm)	sec	dyne[a] (gm-cm/sec^2)	gram (gm)	Physicists rarely
Absolute Metric (MKS)	Meter (m)	Sec	Newton[a] (kg-m/sec^2)	Kilogram (kg)	Physicists generally

[a]derived unit

10-5 D'ALEMBERT'S PRINCIPLE – MOTION OF THE MASS CENTER

D'Alembert's principle expresses the relation between the external forces applied to a system of particles and the effective force on each particle of the system. It is the basis for extending Newton's laws for a single particle to a system of particles. The principle may be stated as follows: *The resultant of the external forces applied to a body (rigid or nonrigid) composed of a system of particles is equal to the vector summation of the effective forces acting on all particles*. Because of the fundamental importance of this principle, we shall prove it both in terms of its physical meaning and again by a formal mathematical development.

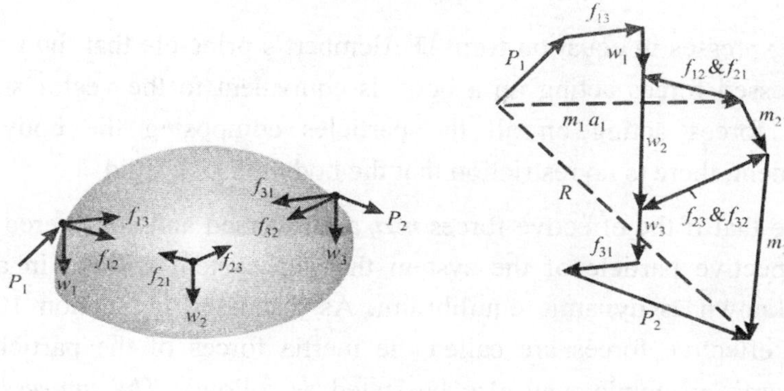

(a) (b)

$$R = W \nleftrightarrow p_1 \nleftrightarrow p_2 \nleftrightarrow \ldots = m_1a_1 \nleftrightarrow m_2a_2 \nleftrightarrow m_3a_3 \nleftrightarrow \ldots$$

Figure 10-5.1 Resultant of external forces is equivalent to that of the effective forces.

The physical significance of D'Alembert's principle is demonstrated in Fig.10-5.1. The heavy outline denotes the boundary of a body which consists of a system of particles. The body may be rigid so that the particles remain at fixed distances from one another, or the body may be nonrigid like a gas or a liquid. The external forces acting on this body are its weight W and several forces of which two are shown, P_1 and P_2. For convenience and also for clarity, assume the body to consist of only the three particles shown. Each of these particles is acted upon by its own weight and by the internal forces exerted upon it by the other particles. In addition, the first particle at the left boundary is acted upon by the external force P_1 while the third particle at the right boundary is acted upon by the external force P_2.

In part (b) of the Fig. 10-5.1 the effective force m_ia_i on each particle is shown as the tip-to-tail sum of the forces acting on each particle. Since the internal forces occur in equal opposite pairs, they cancel out of a vector summation of forces involving all particles.[5] Hence, as shown in the figure, we

[5]In order to draw vector polygons of the forces acting on each particle so that the effective force m_ia_i on each particle will be obvious, it is not possible to draw f_{13} and f_{31}

see that the resultant force **R** of the impressed forces P_1, P_2 and W (note that $W = w_1 + w_2 + w_3$) is equivalent to the vector sum of the effective forces. Hence, we obtain

$$\begin{aligned} R &= W +\!\!\!\!\rightarrow P_1 +\!\!\!\!\rightarrow P_2 +\!\!\!\!\rightarrow \ldots \\ &= m_1 a_1 +\!\!\!\!\rightarrow m_2 a_2 +\!\!\!\!\rightarrow m_3 a_3 +\!\!\!\!\rightarrow \ldots \end{aligned} \qquad (10\text{-}5.1)$$

This expresses in equation form D'Alembert's principle that the resultant of the impressed forces acting on a body is equivalent to the vector sum of the effective forces acting on all the particles composing the body. In this development, there is no restriction that the body must be rigid.

Notice that if the effective forces $m_i a_i$ are reversed and considered to act on each respective particle of the system that the system will be in a state of balance known as dynamic equilibrium. As discussed in Section 10-3, these reversed effective forces are called the inertia forces of the particles. Thus, D'Alembert's principle may also be stated as follows: *The impressed forces acting on any body are in dynamic equilibrium with the inertia forces of the particles of the body.*

The preceding discussion should make it simple to follow a formal mathematical development of D'Alembert's principle and its extension to the motion of the mass center of any body. In a system of n particles, let $\mathbf{F}_i$ be the resultant of external forces acting on the ith particle and $\sum_{j=1}^{n} \mathbf{f}_{ij}$ be the resultant of internal forces exerted on particle i by all other particles. (Note that $\mathbf{f}_{ii} = 0$ since a particle does not exert an internal force upon itself.)

Newton's second law for any particle may now be expressed by

$$\mathbf{F}_i + \sum_{j=1}^{n} \mathbf{f}_{ij} = m_i \mathbf{a}_i$$

Then the sum of the external and internal forces on all particles of a system is

$$\sum_{i=1}^{n} \mathbf{F}_i + \sum_{i=1}^{n} \sum_{j=1}^{n} \mathbf{f}_{ij} = \sum_{i=1}^{n} m_i \mathbf{a}_i$$

---

as the equal, opposite, collinear vectors they really are. However, it should be obvious that f_{13} and f_{31} will cancel each other when we consider a tip-to-tail summation of all the forces shown.

However, the internal forces occur in equal, opposite pairs ($\mathbf{f}_{ij} = -\mathbf{f}_{ji}$ by Newton's third law) so that the double sum is zero. Hence, we obtain

$$\sum_{i=1}^{n} \mathbf{F}_i = \sum_{i=1}^{n} m_i \mathbf{a}_i = \sum_{i=1}^{n} m_i \ddot{\mathbf{r}}_i \tag{a}$$

which is the mathematical statement of D'Alembert's principle.

To extend this discussion to determine the motion of the mass center of any system, we start by noting that the total moment of mass is the moment sum of the component masses, or

$$m\bar{\mathbf{r}} = \sum_{i=1}^{n} m_i \mathbf{r}_i$$

Note the use of the overbar in $\bar{\mathbf{r}}$ to denote the position vector to the mass center. When this relation is differentiated twice with respect to time, it becomes

$$m\ddot{\bar{r}} = \sum_{i=1}^{n} m_i \ddot{\mathbf{r}}_i \tag{b}$$

Comparison of the right-hand terms of Eqs. (a) and (b) shows them to be identical; hence, we conclude that the left-hand terms must be equal, or

$$\sum_{i=1}^{n} F_i = \mathbf{m}\ddot{\bar{\mathbf{r}}} \quad \text{equivalent to} \quad \mathbf{R} = m\bar{\mathbf{a}} = \frac{W}{g}\bar{\mathbf{a}} \tag{10-5.2}$$

This equation expresses the acceleration of the mass center of any body (rigid or nonrigid) in terms of the applied external forces. Since it is identical in form to Newton's second law for a single particle, we conclude that insofar as the relation between the resultant force and the acceleration of the mass center is concerned, *any system of particles may be considered equivalent to a single particle concentrated at the mass center and having the same mass as the mass of the system*. This concept as expressed by Eq. (10-5.2) is fundamental to the study of dynamics since it is valid for any body having any motion. We shall refer to it frequently in later chapters.

But there is an important difference between the system and the equivalent particle concept; the line of action of the resultant force on the system does not usually pass through the equivalent particle at the mass center of the system. The action line of **R** on a body or a system of particles must be found by applying the principle of moments; i.e., the moment of the resultant force is equal to the moment sum of its parts. We do this in the next section.

10-6 MOMENT EFFECT OF EXTERNAL FORCES

As we learned in statics, any force system can be reduced to a resultant force-couple system. Because the resultant force is independent of the choice of a reference center, we applied this in the previous section to show the equivalence of the force polygon of the applied forces on a body to the vector polygon of the effective forces. The magnitude of the resultant couple, however, depends on the position of a reference center. We now discuss the relation between the resultant couple and the motion of a system of particles. Of particular importance will be the rotational effect of the resultant couple on the angular motion of a rigid body.

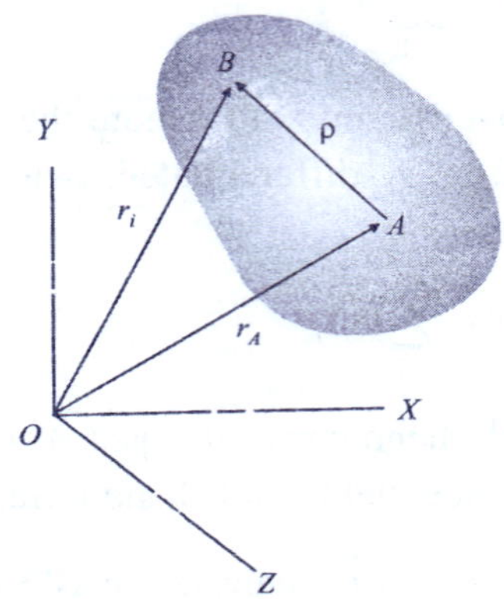

Figure 10-6.1

Consider a swarm of particles (enclosed by the boundary shown in Fig. 10-6.1) which is acted upon by any set of forces. The moment effect $\Sigma \boldsymbol{M}_A$ of these forces (not shown) about any center A in or outside the system is equal to the moment sum of the equivalent force distribution expressed in terms of the effective forces. Since Newton's second law is only valid in an inertial frame, the effective force on a typical particle B is given by $m_i\mathbf{a}_i = m_I\ddot{\boldsymbol{r}}_i$ where $\mathbf{r}_i$ is the position vector of B from the origin O of the inertial reference frame. The moment of this typical effective force about the moment center A depends on the position vector $\boldsymbol{\rho}$ of B from A. Thus, we have

$$\Sigma \mathbf{M_A} = \Sigma\, \boldsymbol{\rho} \times m_i\ddot{\boldsymbol{r}}_i \tag{a}$$

From the Fig. 10-6.1, we see that $\mathbf{r_i} = \mathbf{r_A} + \boldsymbol{\rho}$ so that two time differentiations result in $\ddot{\mathbf{r}}_\mathbf{i} = \ddot{\mathbf{r}}_\mathbf{A} + \ddot{\boldsymbol{\rho}}$ which we substitute into Eq. (a) to obtain

$$\Sigma \mathbf{M_A} = \Sigma\boldsymbol{\rho} \times m_i\ddot{\mathbf{r}}_A + \Sigma\boldsymbol{\rho} \times m_i\ddot{\boldsymbol{\rho}} \tag{b}$$

Now we simplify the first right-hand term by noting that $\ddot{\mathbf{r}}_A$ is independent of $\boldsymbol{\rho}$ so that we can take it out of the summation and rewrite it as

$$\Sigma\rho \times m_i\ddot{\mathbf{r}}_A = (\Sigma\rho m_i) \times \ddot{\mathbf{r}}_A = \overline{\boldsymbol{\rho}}_A m \times \ddot{r}_A = \overline{\boldsymbol{\rho}}_A \times m\ddot{r}_A \qquad (c)$$

In the second of these equivalent terms, observe that $\Sigma\boldsymbol{\rho} m_i$ is the total moment of mass which is equivalent in the third term to the total mass m multiplied by the position vector $\overline{\boldsymbol{\rho}}_A$ from A to the center of mass. Finally we reassociated m with $\ddot{\mathbf{r}}_A$.

Consider next the second right-hand term of Eq. (b). Let $m_i\dot{\boldsymbol{\rho}}$ be defined as the linear momentum of a particle relative to A so that $\rho \times m_i\dot{\boldsymbol{\rho}}$ is the moment of this relative momentum about A. Then using the symbol $\boldsymbol{H}_A$ to represent the total relative moment of momentum about A, we have

$$\mathbf{H}_A = \Sigma\rho \times m_i\dot{\boldsymbol{\rho}} \qquad (d)$$

whose time derivative is

$$\dot{\mathbf{H}}_A = \Sigma\dot{\rho} \times m_i\dot{\rho} + \Sigma\rho \times m_i\ddot{\rho} \qquad (e)$$

in which the first right-hand term is zero, since it is the cross product of a vector with itself. Therefore, Eq. (e) reduces to

$$\dot{\mathbf{H}}_A = \Sigma\rho \times m_i\ddot{\rho} \qquad (f)$$

We now return to Eq. (*b*) and replace its right-hand terms by their equivalents from Eqs. (*c*) and (*f*) to obtain the following fundamental moment equation for any system of particles:

$$\Sigma \boldsymbol{M}_A = \overline{\rho}_A \times m\ddot{r}_A + \dot{\mathbf{H}}_A \qquad (10\text{-}6.1)$$

Usually, we select a moment center A that is either a fixed point O in space so that $\ddot{r}_A = 0$, or else is the mass center G so that $\overline{\rho}_A = 0$. Either of these choices of a moment center reduces Eq. (10-6.1) to the more convenient form

$$\Sigma \boldsymbol{M}_o = \dot{\mathbf{H}}_o \text{ or } \Sigma M_o = \dot{H}_o \qquad (10\text{-}6.2)$$

This simpler form also applies to the special case[6] when the acceleration of the reference point (i.e., $\ddot{\mathbf{r}}_A$) passes through the mass center, since then $\overline{\boldsymbol{\rho}}_A$ and $\ddot{\mathbf{r}}_A$ coincide and their cross product is zero.

[6] This occurs in a free-rolling body (see Section 12-5).

In, general, by adding an inertial force $-m\ddot{\mathbf{r}}_A$ acting at the mass center of the system and considering it as an additional external force acting on the system, all of the results derived by applying Newton's laws of motion relative to an inertial frame apply equally well with respect to a *translating* frame having the acceleration of the base point A.

The physical interpretation of Eq. (10-6.2) as shown in Fig. 10-6.2 will greatly enhance its universal application. This Fig. 10-6.2 shows that at the mass center G, the resultant force-couple system of the forces acting on any system of particles is equivalent to a resultant effective force $m\bar{\mathbf{a}}$ at G plus the couple $\dot{\mathbf{H}}$ representing the time rate of change of the relative angular momentum $\mathbf{H}$ about G.

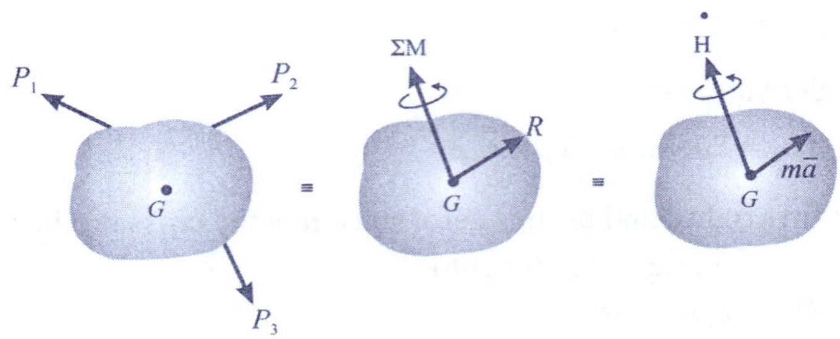

Figure 10-6.2 Equivalence of applied forces to their dynamic effects.

Some final observations are pertinent. We have now developed the two fundamental equations of dynamics; namely, Eqs. (10-5.2) and (10-6.2) which we merely stated in the preview of dynamics in Section 9-1. We shall discuss their application in subsequent chapters. At this time, though, consider the application of Eq. (10-6.2) and its physical interpretation in Fig. 10-6.2 to a *translating* rigid body in which all particles travel along identical parallel paths. In this case, no angular motion of the body is involved and both the relative angular momentum $\mathbf{H}$ and its time derivative $\dot{\mathbf{H}}$ will be zero. We conclude, from the equivalences in Fig. 10-6.2, that the resultant of all forces on a translating rigid body reduces to a single force $\mathbf{R} = m\bar{a}$ which passes through the mass center. This conclusion will be very useful when we apply it later in Section 11-4.

SUMMARY

The laws of motion for a particle (Section 10-2) are extended, by means of D'Alembert's principle (Section 10-5), to include a body (rigid or nonrigid) composed of a system of particles. D'Alembert's principle states that the resultant of the impressed forces acting on a body is equivalent to the vector sum of the effective forces acting on all the particles composing the body.

A useful variation of this principle states that the impressed forces acting on any body are in dynamic equilibrium with the inertia forces acting on the particles of the body.

By extending D'Alembert's principle to determine the motion of the mass center of any body, we find that the resultant of the applied external forces is equivalent to the product of the mass of the body and the acceleration of its mass center; it is expressed by the equation

$$\bar{R} = m\bar{a} = \frac{W}{g}\bar{a} \qquad (10\text{-}5.2)$$

This equation is fundamental to dynamics; it is valid for both nonrigid and rigid bodies. However, the location of the resultant force is not specified by this equation.

The effect of the resultant couple acting on any body is given by

$$\Sigma M_A = \bar{\boldsymbol{\rho}}_A \times m\ddot{\mathbf{r}}_A + \dot{\mathbf{H}}_A \qquad (10\text{-}6.1)$$

where A is any center of moments and $\dot{\mathbf{H}}_A$ is the time derivative of the relative angular momentum about A. When the moment center is either a fixed point O in space or else is the mass center G, it reduces to

$$\Sigma \mathbf{M} = \dot{\mathbf{H}} \qquad (10\text{-}6.2)$$

which is the second fundamental equation of dynamics. Its principal use is to determine the angular motion of rigid bodies which we do in subsequent chapters. In particular when applied to a *translating rigid body*, Eqs. (10-5.2) and (10-6.2) show that both the resultant of applied forces and the resultant of the effective forces passes through the mass center.

CHAPTER 11

KINETICS OF PARTICLES

11-1 INTRODUCTION

In this chapter, we consider only particle motion. This restriction does not mean that we are limited to bodies of point size; it only means that we consider bodies so small that any differences in the motions of their parts can be ignored or that the bodies have no angular motion. The translational motion of rigid bodies is such a type.

Translation is defined as the motion of a *rigid* body in which a straight line passing through any two of its particles always remains parallel to its initial position. The condition that the body be rigid means that the distance separating any two particles remains constant. This, coupled with the requirement that a straight line in the body remain parallel to its initial direction, means that all particles travel the same or parallel paths. Thus, the outstanding *kinematic* characteristic of a translating rigid body is that all the particles have the same values of displacement, velocity, and acceleration. The motion is completely described by the motion of *any* particle of the body. The particle usually selected is the one at the mass center of the body. Furthermore, as the resultant of all forces acting upon the translating rigid body passes through its mass center.

Translation may be either rectilinear or curvilinear, depending on whether the path described by any particle of the body is straight or curved. This is illustrated in Fig. 11-1.1 where the motion in part (a) is called *rectilinear translation* and the motion in part (b) is called *curvilinear translation.* In either case, the pertinent kinetic equation is $R = ma = \frac{W}{g}a$ obtained from

Eq. (10-5.2). Note that the bar sign over **a** to identify the acceleration of the mass center can be omitted since all particles have the same motion.

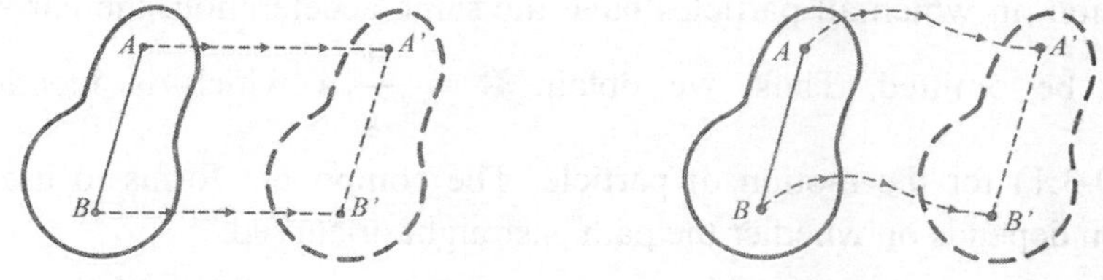

(a) Rectilinear translation (b) Curvilinear translation

Figure 11.1.1 Translation of a rigid body.

The general problem in dynamics is to determine the relation between the motion of a body and the force system acting upon it. When the motion of a particles specified, it is simple to determine the force causing the motion directly from **R** = *m***a**. The converse problem of determining the motion of a particle when the force is specified is more involved. This requires solving **R** = *ma* for the instantaneous acceleration ***a*** and then integrating one or more of the differential equations of kinematics (as in Section 9-3) to obtain the appropriate .*s-t*, *v-t*, or *v-s* relation desired.

An alternate procedure is to integrate **R** = *ma* directly, either in terms of the displacement or the time. The displacement integral leads to what is known as the work-energy relation, whereas the time integral yields the impulse-momentum relation. As we shall learn, the work-energy equation relates force, displacement, and velocity. Moreover, we shall see that it is a scalar equation which makes its use much simpler than the impulse-momentum relation which is a vector equation relating force, velocity, and time.

Each of these various methods – force-acceleration, work-energy, and impuls momentum-and their respective advantages are discussed in subsequent chapters. Our consideration here of the kinetics of particle motion allows us to neglect temporarily the complications due to the angular motion of a general rigid-body motion.

11-2 TRANSLATION – ANALYSIS AS A PARTICLE

The kinetic equations for translation as a particle are obtained from the general equation governing the motion of the mass center (see Section 10-5). It was shown that any translating body could be treated as though it were a particle which had the same mass as the body and the same motion as the mass center of the body, i.e. $\mathbf{R} = \dfrac{W}{g}\bar{a}$. When this equation is applied to the motion of translation, in which all particles have the same acceleration, the bar sign over $\bar{\mathbf{a}}$ can be omitted. Thus, we obtain $\mathbf{R} = \dfrac{W}{g}\mathbf{a}$ which is identical with Eq. (10-3.1) for the motion of particle. The component forms to use for this equation depends on whether the path is straight or curved.

For rectilinear motion, it is convenient to select the line of motion as the X axis. *Let this axis be considered positive in the initial direction of motion.* Using this convention, we consider displacement, velocity, acceleration and X components of forces as positive when directed in the initial direction of motion. Since the acceleration coincides with the X axis, we have $a_x = a$, and $a_y = a_z = 0$. Then the component forms of the kinetic equation for rectilinear motion become

$$\Sigma X = \frac{W}{g}a \qquad \Sigma Y = \Sigma Z = 0 \tag{11-2.1}$$

in which the force summations ΣX, ΣY, and ΣZ are the equivalent components of **R**.[1]

For curvilinear motion, the convenient components of acceleration to use are normal and tangential to the path, or on occasion the radial and transverse components. Calling the normal axis N and the tangential axis T, the corresponding scalar forms of $\mathbf{R} = \dfrac{W}{g}\mathbf{a}$ are

$$\Sigma N = \frac{W}{g}a_n = \frac{W}{g}\frac{v^2}{r} \qquad \Sigma T = \frac{W}{g}a_t \tag{11-2.2}$$

In applying these equations, the positive senses of ΣN and ΣT are in the directions of a_n and a_t.

---

[1]Sometimes the force summation are represented by ΣF_x, ΣF_y, etc. We recommend using ΣX, ΣY, etc., since this is both easier to write and to read.

In curvilinear motion, however, it is usually preferable to create dynamic equilibrium by applying inertia forces of magnitudes $\frac{W}{g}\frac{v^2}{r}$ and $\frac{W}{g}a_t$ acting through the mass center opposite in direction to the normal and tangential acceleration components respectively. These inertia forces are known as the centrifugal inertia force (or reversed normal effective force) and the tangential inertia force (or reversed tangential effective force). Recall that the normal component of acceleration is always directed inward toward the center of curvature of the path while the positive sense of the tangential component of acceleration will be in the direction of motion.

When motion along a curved path is defined by polar coordinates, it is convenient to use the radial and transverse components of acceleration as derived in Section 9-8. The corresponding components of $\mathbf{R} = \frac{W}{g}\mathbf{a}$ along radial and transverse direction then become

$$\Sigma F_r = \frac{W}{g}a_r = \frac{W}{g}\left(\ddot{r} - r\dot{\theta}^2\right) \qquad \text{(11-2.3)}$$

$$\Sigma F_\theta = \frac{W}{g}a_\theta = \frac{W}{g}\left(r\ddot{\theta} + 2\dot{r}\dot{\theta}\right)$$

In these equations, the positive senses are in the positive directions of increasing r and θ.

ILLUSTRATIVE PROBLEMS

11-2.1 Two blocks A and B are released from rest on a 30° incline when they are 18 m apart. The coefficient, of friction under the upper block A is 0.20 and that under the lower block B is 0.40. Compute the elapsed time until the blocks touch. After they touch and move as a unit, what will be the contact force between them?

Solution

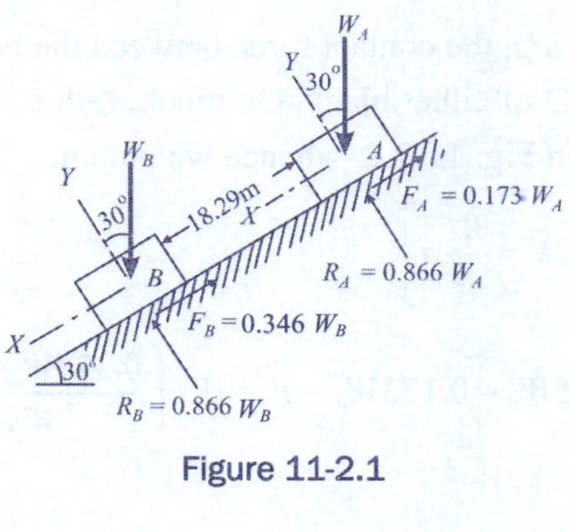

Figure 11-2.1

The FBD of each body is shown in Fig. 11-2.1. Select reference axes as shown with the X axis positive down the incline in the direction of motion. Applying $\Sigma Y = 0$ and $F = fR$, the normal and frictional components of the incline reaction are as shown. Applying $\Sigma X = (W/g)a$ to each block, we have

For A: $$0.5\,W_A - 0.173 W_A = \frac{W_A}{g}\,a_A, \qquad a_A = 3.21 \text{ m/sec}^2$$

For B: $$0.5 W_B - 0.346 W_B = \frac{W_B}{g}\,a_B, \qquad a_B = 1.51 \text{ m/sec}^2$$

Thus, the relative acceleration with which A overtakes B is $a = a_A - a_B = 1.7$ m/s^2. Therefore, the time to traverse the relative distance between them of 18.29 m at this constant acceleration is found from Eq. (9-3.5) to be

$$\left[s = v_o t + \frac{1}{2}at^2\right] \qquad 18 = 0 + \frac{1}{2}(1.7)\,t^2$$

from which

$$t^2 = 21.2 \text{ and } t = 4.6 \text{ sec} \qquad \textbf{\textit{Ans.}}$$

After the blocks touch, the common acceleration for the system is $\left[\Sigma X = \frac{W}{g}a\right]$

$$0.5 W_A - 0.173\,W_A + 0.5\,W_B - 0.346 W_B = \left(\frac{W_A + W_B}{g}\right)a$$

which reduces to

$$\frac{a}{g} = \frac{0.327 W_A + 0.154 W_B}{W_A + W_B}$$

With this value of a/g, the contact force between the bodies may now be obtained from the FED of either block. On block A, this contact force acts up the incline as shown in Fig. 11-2.2, whence we obtain.

$$\left[\Sigma X = \frac{W}{g}a\right]$$

$$0.5 W_A - 0.173 W_A - P = W_A\left(\frac{0.327 W_A + 0.154 W_B}{W_A + W_B}\right)$$

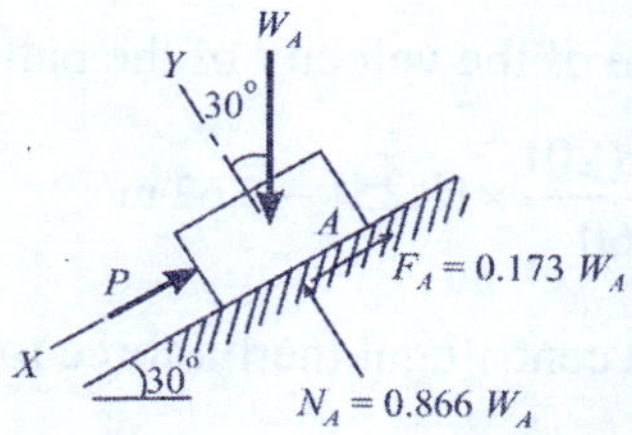

Figure 11-2.2 FBD of A after blocks touch.

from which we find

$$P = \frac{0.173 W_A W_B}{W_A + W_B} \qquad \textit{Ans.}$$

11-2.2 A rod *ABC* rotating at 20 rpm about a vertical axis through *A* supports a 900 N ball at its lower end as shown in Fig. ll-2.3 (a). It is fixed in position by the rod *BD*. Neglecting the weights of rods *AC* and *BD*, compute the force *F* in rod *BD*. Is the force tensile or compressive? At what rpm will this force be zero?

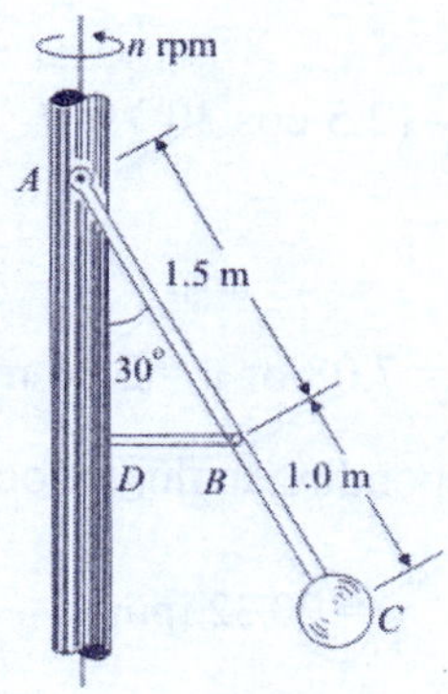

(a)

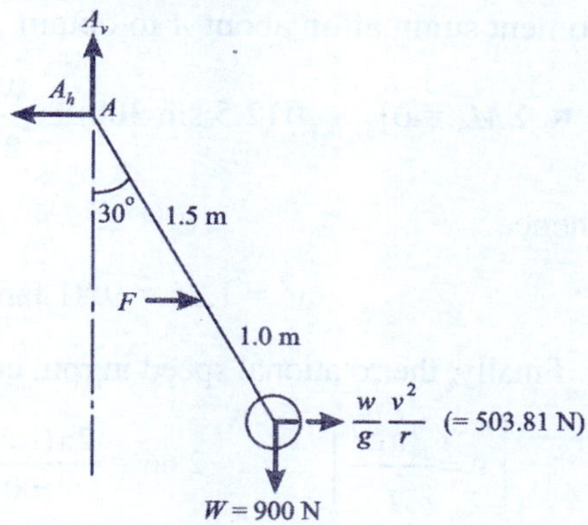

(b)

Figure 11-2.3

Solution

The FBD of rod *AC* is shown in Fig. 11-2.3(b). The ball may be considered as a particle moving in a horizontal circle of radius $r = 2.5 \sin 30° = 1.25$ m. In order to eliminate the components of the hinge reaction at *A* by using a moment summation about *A*, we create dynamic equilibrium by applying to the ball an inertia force $\frac{W}{g}\frac{v^2}{r}$ acting radially outward from the center of its path.

The constant magnitude of the velocity of the ball is found from

$$\left[v = \frac{s}{t} = \frac{2\pi rn}{t}\right] \quad v = \frac{2\pi(20)}{60} \times (1.25) = 2.62 \text{ m/s}$$

which then determines the centrifugal inertia force to be

$$\frac{W}{g}\frac{v^2}{r} = \frac{900 \times (2.62)^2}{9.81\ (1.25)} = 503.81 \text{ N}$$

Assuming the force F in BD to be compressive, it acts on AC as shown. Then a moment summation of dynamic equilibrium about A gives

$$[+\nwarrow \Sigma M_A = 0] \quad F(1.5 \cos 30^\circ) + 503.81(2.5) \cos 30^\circ) - 900\ (1.25) = 0$$

$$F = 26.34 \text{ N} \qquad \textbf{\textit{Ans.}}$$

Since F is positive, it acts as shown and confirms the assumption that BD is in compression.

To find the velocity v when $F = 0$ at the given position, we take another moment summation about A to obtain

$$[+\nwarrow \Sigma M_A = 0] \quad W(2.5 \sin 30^\circ) - \frac{W}{g}\frac{v^2}{1.25}(2.5 \cos 30^\circ) = 0$$

whence

$$v^2 = 1.25 \times 9.81 \tan 30^\circ = 7.08 \text{ or } v = 2.66 \text{ m/s}$$

Finally, the rotational speed in rpm corresponding to this velocity is found to be

$$\left[v = \frac{2\pi rn}{t}\right] \quad 2.66 = \frac{2\pi(1.25)n}{60} \quad n = 20.32 \text{ rpm} \qquad \textbf{\textit{Ans.}}$$

11.2.3 Determine the maximum speed v at which a car of weight W can make a horizontal turn around a highway curve of radius r that is banked (i.e., inclined) at θ° with the horizontal. The coefficient of static friction between the tires and the road is f and the corresponding angle of friction is ϕ.

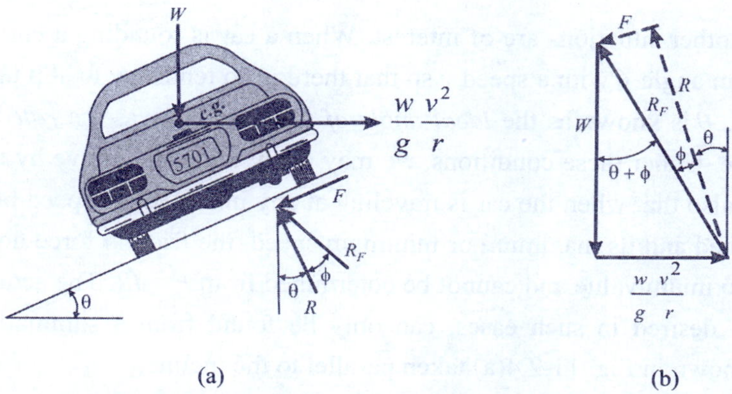

Figure 11.2.4 Maximum velocity, friction considered.

Solution

If the car was making a very sharp turn so that the dimensions of the car are not negligible compared with the radius of the turn, its motion would be an example of rigid-body rotation about the vertical axis of the turn. However, usually the radius of the turn is so large that no significant error is caused by analyzing the car as a particle.

A rear view of the FBD of the car is shown in Fig. 11-2.4(a). The centrifugal inertia force $\frac{W}{g}\frac{v^2}{r}$ is applied through the mass center to create dynamic equilibrium; it acts horizontally outward from the vertical axis of the turn. If its upplane component exceeds the downplane component of the weight of the car, a downplane friction force is necessary to prevent skidding up the incline. When the car is traveling at maximum speed around the turn, all available static friction is required to prevent such skidding. Then the resultant R of the total road reaction is inclined at the angle of friction ϕ with the normal pressure N. Observe that since the car is considered as a particle, the total reaction R is directed through the *c.g.* of the car. Consequently, the outer wheels carry a greater load than the inner set of wheels.

From the force polygon shown in Fig. 11-2.4(b), we obtain

$$\tan(\theta+\phi) = \frac{\frac{W}{g}\frac{v^2}{r}}{W} = \frac{v^2}{gr} \qquad (11\text{-}2.4)$$

Note that this relation is independent of the weight. If the car is on the verge of slipping down the incline (because of insufficient speed), we would have

$$\tan(\theta-\phi) = \frac{v^2}{gr} \qquad (11\text{-}2.5)$$

Two other situations are of interest. When a car is rounding a curve banked at such an angle θ with a speed v so that there is no tendency to slip up or down the road, θ is known as the *ideal angle of banking* and v as the *rated speed* of the curve. Under these conditions, we may apply the result above by taking $\phi = 0$. Note also that when the car is traveling at any intermediate speed between its rated speed and its maximum or minimum speed, the friction force does not act at its maximum value and cannot be determined from $F = fR$. The actual friction force, if desired in such cases, can only be found from a summation of the forces shown in Fig. 11-2.4(a) taken parallel to the incline.

As a specific application of this discussion, suppose the rated speed of a highway curve of 90 m radius is 18 m/s. If the coefficient of friction between the tires and the road is 0.6, what is the maximum speed, at which a car can round the curve without skidding? The angle of banking for the rated speed (i.e., $\phi = 0$) is found from

$$\left[\tan\theta = \frac{v^2}{gr}\right] \qquad \tan\theta = \frac{(18)^2}{9.81(90)} = 0.367, \qquad \theta = 20.3^\circ$$

From the definition of the angle of friction, we have

$$[\tan\phi = f] \qquad \tan\phi = 0.6 \qquad \phi = 31^\circ$$

We may now apply Eq. (11-2.4) to get

$$\left[\tan(\theta+\phi) = \frac{v^2}{gr}\right] \qquad \tan(20.3 + 31) = \frac{v^2}{9.81(90)}$$

hence

$$v_{max} = 33.2 \text{ m/sec}$$ ***Ans.***

PROBLEMS

11.2.4 A magnetic particle weighing 3.6 grams is pulled through a solenoid with an acceleration of 6 meters per sec^2. Compute the force in grams acting on the particle.

11.2.5 Determine the force P that will give the body in Fig. P.11-2.5 an acceleration of 0.2 m/s^2. The coefficient of kinetic friction is 0.20.

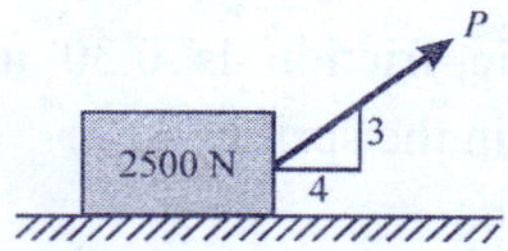

Figure P-11-2.5

11.2.6 In Fig. P.11-2.5, assume $P = 900$ N and the coefficient of kinetic friction is 0.5. Find the acceleration of the body under the following conditions: (a) the body is moving rightward; (b) the body is moving leftward; (c) the body starts from rest.

11-2.7 When a 3000 N boat is moving at 3 m/s, the motor conks out. How much farther will the boat glide, assuming its resistance to motion is $9v$ N where V is in m/s?

11.2.8 Referring to Fig. P-11-2.8, assume *A* weighs 900 N and *B* weighs 30 N. Determine the acceleration of the bodies if the coefficient of kinetic friction is 0.10 between the cable and the fixed drum.

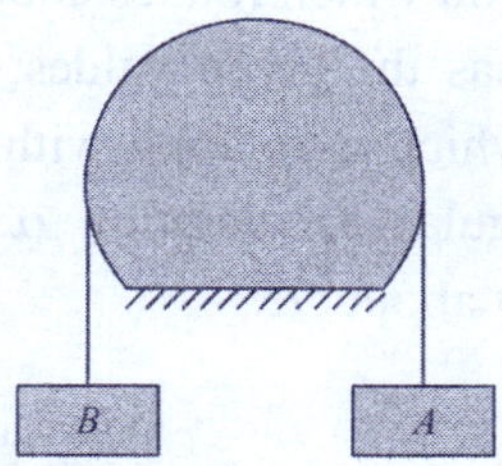

Figure P-11-2.8

11.2.9 Determine the acceleration of the bodies in Fig. P-11-2.9 if the coefficient of kinetic friction is 0.20 at all contact surfaces. Body *A* weighs 900 N and *B* weighs 1350 N.

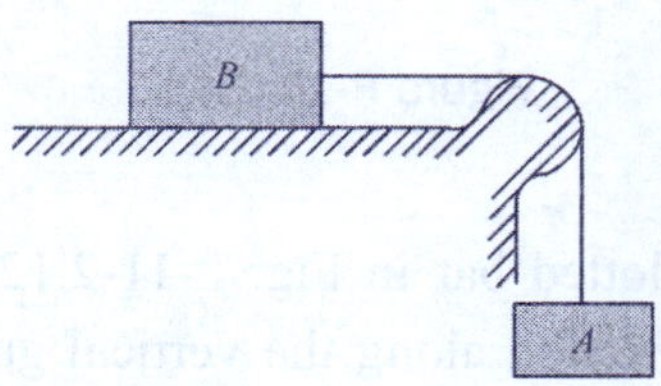

Figure P-11-2.9

11.2.10 The two bodies in Fig. P-11-2.10 are separated by a spring. Their motion down the incline is resisted by a force $P = 900$ N. The coefficient of kinetic friction is 0.30 under A and 0.10 under B. Determine the force in the spring.

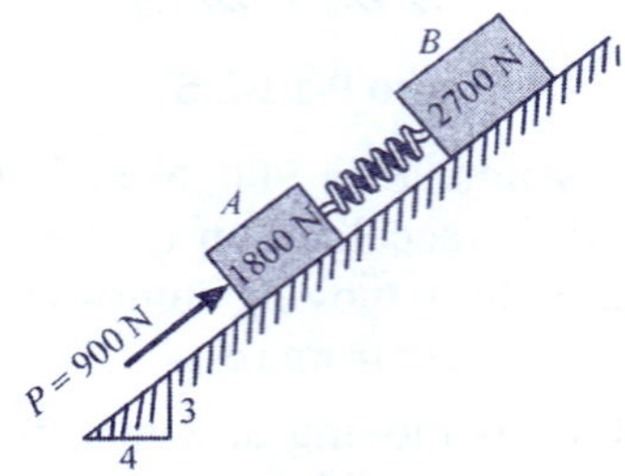

Figure P-11-2.10

11-2.11 In Fig. P-11-2.11, the motion of block B on a smooth horizontal table is controlled by the rod which rotates about a vertical axis at O. At the position shown, B has the given values of velocity and acceleration relative to the rod which is rotating with the given values of angular velocity ω and angular acceleration α. If B weighs 55 N, what moment does it exert about O?

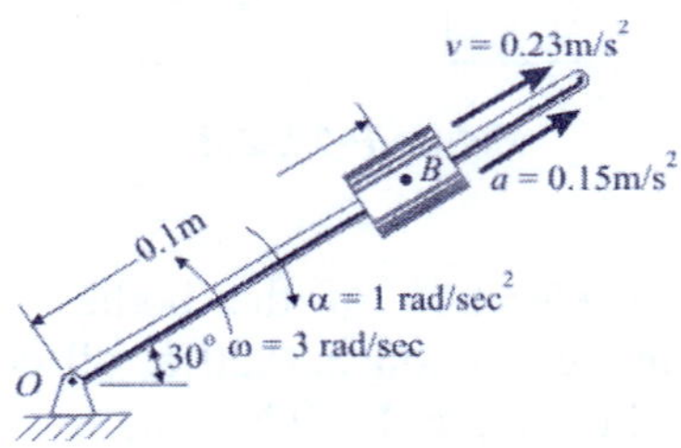

Figure P-11-2.11

11-2.12 The horizontal slotted bar in Fig. P-11-2.12 weighs 90 N and slides with negligible friction along the vertical guide. Determine the force exerted on it by pin P when y = 0.05 m, if AB is then rotating at a constant angular velocity of 20 rad/sec.

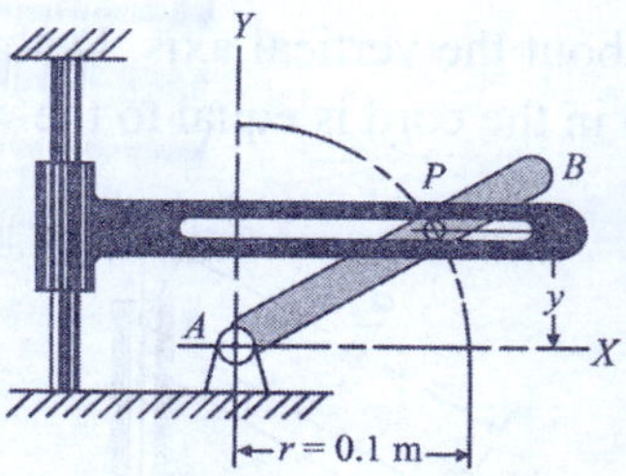

Figure P-11-2.12

11-2.13 A rocket fired vertically is tracked by radar as shown in Fig. P-11-2.13. Determine the thrust P on the 45 KN rocket if when $\theta = 60°$, it is known that r = 6 Km, $\dot{\theta}$ = 0.02 rad/sec, and $\ddot{\theta}$ = 0.003 rad/sec^2, Assume g remains constant.

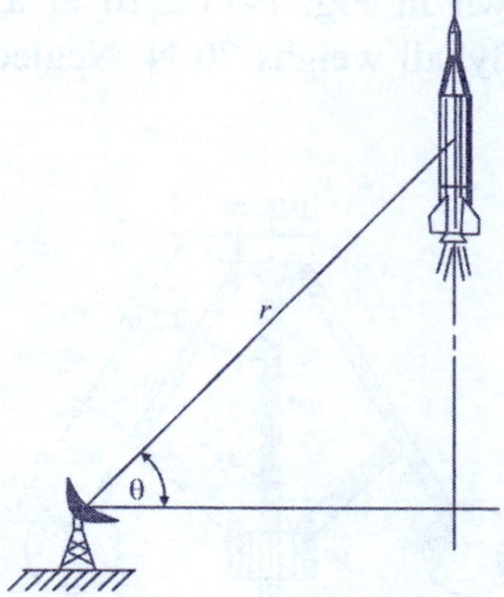

Figure P-11-2.13

11.2.14 In an amusement park ride, gondola cars of weight W are attached by rods 7.5 m long to a bull wheel 5 m in diameter. The bull wheel rotates in a horizontal plane about a vertical axis at its center. At how many rpm is it rotating when the supporting rods of the gondolas are inclined at 30° with the vertical? Consider each gondola to be a particle concentrated at the end of each 7.5 m rod.

11-2.15 A body of weight W rests on the smooth inclined surface of the frame shown in Fig. P-11-2.15. A peg attached to the frame forces the body to rotate with it about the vertical axis. Determine the speed in rpm at which the tension in the cord is equal to the weight of the body.

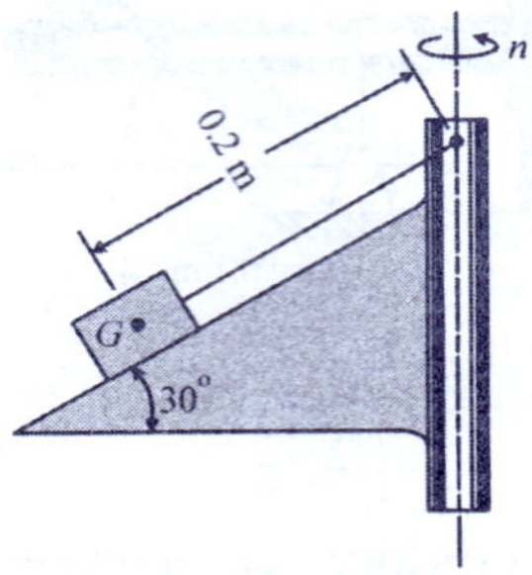

Figure P-11-2.15

11-2.16 What counterweight W will maintain the Corliss engine governor in the position shown in Fig. P-11-2.16 at a rotational speed of n = 120 rpm? Each flyball weighs 70 N. Neglect the weight of the other links.

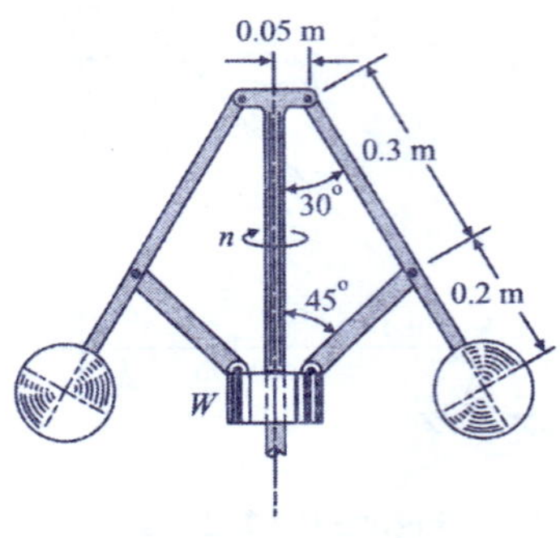

Figure P-11-2.16

11.2.17 Why are railroad curves laid out in the form of a spiral with a gradually decreasing radius of curvature instead of a circle of constant radius?

11-2.18 An airplane makes a turn in a horizontal plane without side slip at 140 m/s. At what angle must the plane be banked if the radius of the turn is 3300 m? If the pilot weighs 650 N, what force does he exert on his seat?

11-2.19 A stuntman drives a motorcycle around a circular vertical wall 30 m in diameter. The coefficient of friction between tires and wall is 0.60. What is the minimum speed that will prevent his sliding down the wall? At what angle will the motorcycle be inclined to the horizontal? What is the effect of traveling at a greater speed?

11.2.20 The superelevation of a railroad track is the number of mm that the outside rail is raised to prevent side thrust on the wheel flanges of cars rounding the curve at rated speed. Determine the superelevation e for a track having a gauge of 1.5 m of 600 m radius and a rated speed of 26 m/s. What is the flange pressure P on the wheels of a 445070 N car that rounds the curve at 35 m/s?

11-2.21 A car weighing 17803 N rounds a curve of 60 m radius banked at an angle of 30°. Find the friction force acting on the tires when the car is traveling at 26 m/s. The coefficient of friction between the tires and the road is 0.60.

11-2.22 Find the angle of banking for a highway curve of 90 m radius designed to accommodate cars traveling at 45 m/s if the coefficient of friction between the tires and the road is 0.60. What is the rated speed of the curve?

11-2.23 Transverse sections through the curves on an automobile speedway have the parabolic shape as shown in Fig. P-11-2.23. Determine the y coordinate of an auto so that when it is moving at a speed of v m/s there will be no tendency for side slip.

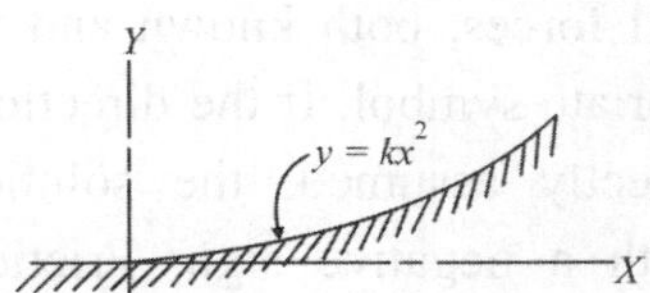

Figure P-11-2.23

11-2.24 The segment of road passing over the crest of a hill is defined by the parabolic curve $y = 0.4x - 0.01x^2$. A car weighing 14350 N travels along the road at a constant speed of 13.5 m/s. What is the normal pressure on the wheels of the car when it is at the crest of the hill where $y = 1.22$ m. At what speed will the road pressure be zero?

Hint: The radius of curvature is defined by

$$\rho = \frac{\left[1 + \left(\frac{dy}{dx}\right)^2\right]^{3/2}}{\frac{d^2y}{dx^2}}$$

11.2.25 As a car passes over the crest of the bill described in the previous problem, the brakes are locked and the wheels skid. The tangential retardation produced is 0.4 g. If the coefficient of friction between the tires and the road is 0.70, how fast is the car moving?

11-2.26 A satellite is put into a circular orbit at an altitude of 966 Km above the earth. The mean radius of the earth is 6375.6 Km. Determine the time to complete one orbit.

Hint: Refer to Eq. (10-2.1).

11-3 FURTHER DISCUSSION OF PARTICLE KINETICS

The equations and methods developed in the preceding section are extended here to more complex situations involving the kinetics of particles; particularly, connected systems of particles. A general plan for the solution of such problems consists of the following guide:

1. Draw a free-body diagram for each body involved in the problem. Indicate thereon all forces, both known and unknown, representing the latter by an appropriate symbol. If the direction of any unknown (except friction) is incorrectly assumed, the solution will yield its correct magnitude but with a negative sign. Friction forces, however, must *always* be directed to oppose relative motion.
2. In order to ensure that friction forces oppose the motion, determine the direction of motion if not evident or specified, and indicate it by a dashed arrow near each free-body diagram. This dashed arrow will be a convenient reminder of the positive direction of the equation of motion.
3. Determine the kinematic relations between the bodies involved in a connected system.

4. Select the X axis as positive in the direction of initial motion and apply $Y = 0$ and $\Sigma X = (W/g)a$ to each body. For curvilinear motion, create dynamic equilibrium.
5. Solve for the unknowns, using such additional equations of kinematics as may be required to determine relations among displacement, velocity, and time.

ILLUSTRATIVE PROBLEMS

11.3.1 A rocket-propelled sled weighing 1780 N is at rest on a horizontal track for which the coefficient of static friction is 0.30 and the coefficient of kinetic friction is 0.20. A horizontal thrust of $P = 66.75t$ is applied for 20 sec and then shut off. P is in Newton's and t is in seconds. What is the maximum velocity of the sled and how far will it be from its initial position when it comes to rest? Neglect the weight of expelled fuel.

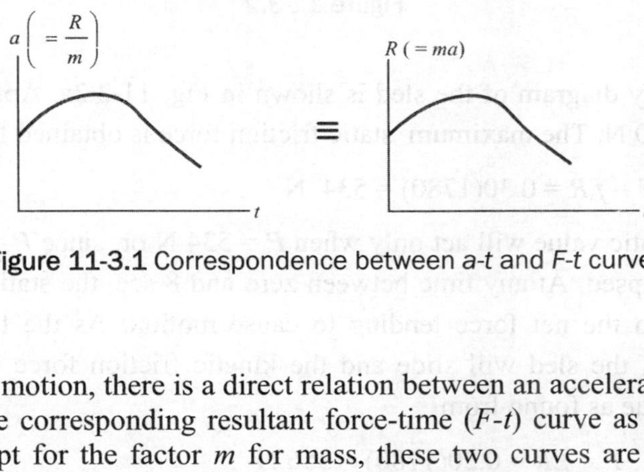

Figure 11-3.1 Correspondence between *a-t* and *F-t* curves.

Preliminary

In rectilinear motion, there is a direct relation between an acceleration-time (*a-t*) curve and the corresponding resultant force-time (*F-t*) curve as shown in Fig. 11-3.1. Except for the factor m for mass, these two curves are equivalent. In other words, the area under the *a-t* curve is $1/m$ times the area under the *F-t* curve. Hence, in place of using $(\text{Area})_{a\text{-}t}$ as discussed in Section 9-4 to find velocity and displacement, we can use $(\text{Area})_{F\text{-}t}$. The relations corresponding to Eqs. (9-4.1) and (9-4.3) then become

$$v_2 - v_1 = \Delta v = \frac{1}{m}(Area)_{F-t} \tag{11-3.1}$$

$$s_2 - s_1 = \Delta s = v_1(\Delta t) + \frac{1}{m}(Area)_{F-t}(\bar{t}_2) \tag{11-3.2}$$

In this problem we shall see the advantage of using these equations to determine the direct effect of forces on the motion. Actually, these equations are an elemental application of the impulse-momentum method mentioned in Section 11.1 which, by eliminating the acceleration, directly relates force, velocity, and time. A general discussion of the impulse-momentum method is presented later in Chapter 15.

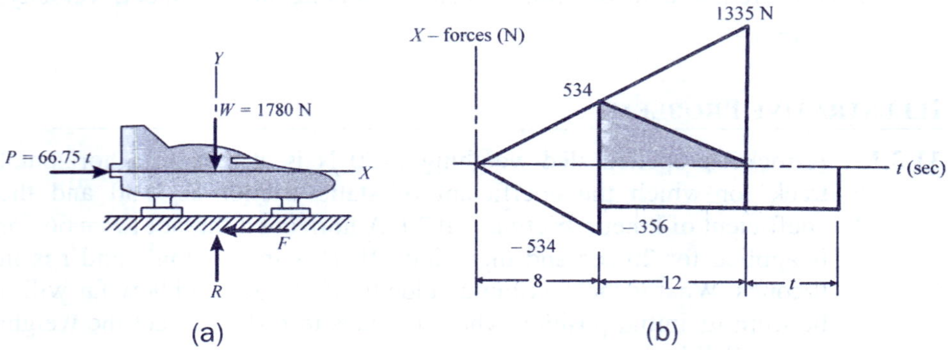

Figure 11-3.2

Solution

A free-body diagram of the sled is shown in Fig. 11-3.2a. Applying $\Sigma Y = 0$ gives $R = 1780$ N. The maximum static friction force is obtained from

$$\boldsymbol{F} = f_s R = 0.30(1780) = 534 \text{ N}$$

But this static value will act only when $P = 534$ N or, since $P = 66.75t$, after 8 sec have elapsed. At any time between zero and 8 sec, the static friction will adjust itself to the net force tending to cause motion. As the time increases beyond 8 sec, the sled will slide and the kinetic friction force will act at its maximum value as found from

$$F = f_k R = 0.20(1780) = 356 \text{ N}$$

We shall assume that the transition from static friction to kinetic friction takes place instantaneously. Although this is not strictly true, it involves negligible error. With this assumption, a graphic representation of the X components of the forces acting, plotted against time, will be shown in Fig. 11-3.2b. Notice that it is convenient to show separately the variation of positive and negative X components. Note that during the initial 8 sec, the resultant force is zero but attains its maximum positive value during the next 12 sec, at the end of which the sled has its maximum velocity. This value may now be found by applying Eq. (11-3.1) to the shaded area during this 12 sec interval to obtain

$$[\Delta v = \frac{1}{m}\ (\text{Area})_{F-t}]$$

$$\max\ v = \frac{1}{1780/9.81}\left[\left(\frac{1335+534}{2}\right)(12) - 356(12)\right] = 38.26 \text{ m/s} \qquad \textbf{\textit{Ans.}}$$

After the rocket thrust is shut off, the sled will coast to a stop during a time *t*. This time may be found by applying Eq. (11-3.1) to the coasting interval since we now know the velocity at the beginning of this coasting interval, but as an independent approach, we will apply Eq. (11-3.1) to the entire motion over which the velocity change is zero. This gives

$$\left[\Delta v = \frac{1}{m}(Area)_{F-t}\right] \quad 0 = \frac{1}{\text{m}}\left[\left(\frac{1335+890}{2}\right)(12) - 356(12+\text{t})\right]$$

$$t = 25.5 \text{ sec}$$

The required displacement can now be found by applying Eq. (11-3.2) to the entire shaded area of the *F-t* diagram. Remember to take the moment of each component area about the final time ordinate. Thus, we obtain

$$\left[\Delta s = v_1(\Delta t) + \frac{1}{m}(Area)_{F-t}(\bar{t}_2)\right]$$

$$s = 0 + \frac{1}{1780/9.81}\left[\left(\frac{1335\times 12}{2}\right)\left(\frac{12}{3}+25.5\right) + \left(\frac{534\times 12}{2}\right)\left(\frac{2}{3}\times 12 + 25.5\right) - 356(37.5)\left(\frac{37.5}{2}\right)\right]$$

which gives

$$s = 514.3 \text{ m}$$

11-3.2 In the system of connected blocks shown in Fig. 11-3.3 coefficient of kinetic friction under blocks *A* and *C* is 0.20. Compute the acceleration of each block and the tension in the connecting cable. The pulleys are assumed to be frictionless and of negligible weight.

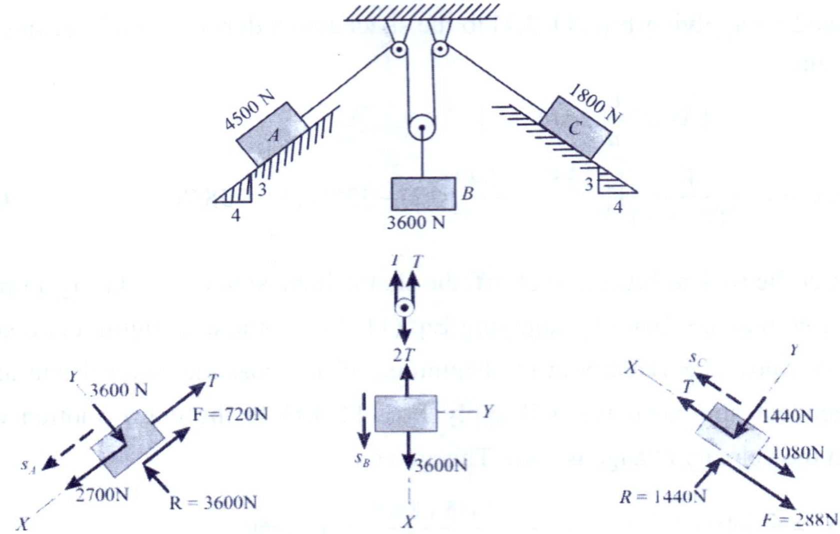

Figure 11-3.3

Preliminary

To simplify the computations, represent the weights of A and C by their components acting parallel and perpendicular to the inclines as shown on the free-body diagrams. To determine the direction of motion of the system, we begin by assuming that one part of the system does not move and then computing the tension necessary to keep it at rest. With this assumption, if any part of the system is found to be acted upon by an unbalanced force, that part will move in the direction of that force. When friction is involved, as in this problem, we must consider whether there is sufficient friction available to prevent motion.

For example, if we temporarily assume B is at rest, $2T = 3600$ N which gives $T = 1800$ N acting on A and C. Notice that since the floating pulley supporting B is of negligible weight, $\Sigma X = (W/g)a$, when applied to it reduces to $\Sigma X = 0$ as also is $\Sigma M = 0$. Hence, from the FBD of the floating pulley, the tension supporting B is $2T$ and the tension in the connecting cable is constant. From force summations parallel to the inclines and assuming $T = 1800$ N, the friction forces required to prevent motion of A and C are for A: 2700 – 1800 = 900 N and for C: 1800 – 1080 = 720 N. However, from $\Sigma Y = 0$ and $F = fR$, we find the available friction forces on A and C to be $F_A = 0.2(3600) = 720$ N and $F_C = 0.2(1440) = 288$ N. These are insufficient to keep A and C at rest and are the actual friction forces acting as shown to oppose the downward motion of A and the upward motion of C, the directions of motion being indicated by the dashed arrows adjacent to each FBD.

We do not yet know the motion of B because although the downplane motion of A tends to raise B, the upplane motion of C tends to lower B. We shall assume that B moves down. Since no friction acts on B, an incorrect assumption of its motion will be of no importance. The actual value of T will determine its motion. Notice that the FBD's of A and C require that T be less than 2700 – 720 = 1980 N to permit downward motion of A, but more than 1080 + 288 = 1368 N to cause upward motion of C. An estimated value of T will be the average of these limiting values or $\frac{1}{2}(1980 + 1368) = 1674$ N. Using this, approximation on the FBD of B gives an unbalanced downward force on B and justifies the assumption that B moves down.

We must next determine the kinematic relations among the motion of the blocks. The simplest way to do this is to use a concept from the method of virtual work and that we redevelop later in the work-energy method: The total work done by internal connecting forces on a system is zero. Noting that work is the product of force and distance, and that positive work is done by a force acting in the direction of motion, we have as the sum of the work done by T on the system of connected blocks

$$Ts_c - Ts_A - 2Ts_B = 0$$

whence, cancelling out T, the kinematic relation between the displacements is

$$s_c = s_A + 2s_B \tag{a}$$

on which successive differentiation with respect to time gives

$$v_c = v_a + 2v_b \tag{b}$$

and

$$a_c = a_A + 2a_B \tag{c}$$

An alternate approach to finding the kinematic relations among the blocks is the following method: The net motion of B is the difference between the partial motion s'_B of B caused by C while A is held at rest and the partial motion s''_B of B due to A while C is held at rest. Noting that these partial motions are only one-half the absolute motions of C and A, we obtain

with A at rest; $s'_B = \frac{1}{2}s_c$ (down)

with C at rest; $s''_B = \frac{1}{2}s_A$ (up)

Assuming that B moves down, its net motion is the difference of these partial motions or

$$s'_B - s''_B = s_B = \frac{1}{2}s_C - \frac{1}{2}s_A$$

which agrees with Eq. (a) above.

Solution

We are now ready to apply Eq. (11-2.1) to each body. Note carefully that in each case ΣX is applied as positive in the direction of motion indicated by the dashed displacement vectors adjacent to each FBD.

$$\left[\Sigma x = \frac{w}{g}\,a\right] \quad \text{For } A: \quad 2700 - 720 - T = \frac{4500}{g}\,a_A \qquad (d)$$

$$\text{For } B: \quad 3600 - 2T = \frac{3600}{g}\,a_B \qquad (e)$$

$$\text{For } C: \quad T - 288 - 1080 = \frac{1800}{g}\,a_C \qquad (f)$$

When a_c is replaced by $a_A + 2a_B$ from Eq. (*c*), these equations are easily solved for their common term T by multiplying Eq. (d) by $-\frac{4}{10}$, Eq. (*e*) by -1, and adding them, whence we obtain

$$T = 1694 \text{ N} \qquad \textbf{\textit{Ans.}}$$

and then by substitution,

$$a_A = 1.17 \text{ m/s}^2$$

$$a_B = 0.58 \text{ m/s}^2 \qquad \textbf{\textit{Ans.}}$$

$$a_c = 2.32 \text{ m/s}^2$$

The positive values of these accelerations confirm our assumptions of the direction of motion of each block.

Some final comments are pertinent. If the coefficient of friction were increased to 0.30, the available friction force under A and C would become F_A = 1080 N and F_C = 432 N. Under these conditions, enough friction is available to keep A at rest, but not sufficient to keep C at rest. It is essential to first determine which bodies can move and in which direction, or else much time will be wasted in getting impossible answers.

PROBLEMS

11.3.3 A 300 N block is resting on a smooth horizontal surface. It is acted upon by a horizontal force P which varies according to the relation $P = (12t - 3t^2)\,\hat{\mathbf{i}}$ where P is in Newton's and t is in seconds. Determine the maximum positive velocity of the block, and the time when it reverses its motion.

11.3.4 A 300 N body is free to slide on a horizontal smooth surface. It starts from rest and is struck three successive blows of 40 N, 225 N and

– 450 N at intervals of 2 sec. Assuming that each blow is along the line of motion and is constant for 0.1 sec, what is the velocity and displacement of the body after 8 sec?

Hint: Use an F-t diagram.

11.3.5 A 1400 N body moves under the action of a force given by the relation $P = (18 - 3t)\,\hat{\mathbf{i}}$ where P is in newtons and t is in seconds. If the body starts from rest, in how many seconds will it return to its initial position? What then is its velocity?

11.3.6 If the force on the 1400 N body of the previous problem is changed to $P = (10t - t^2)\,\hat{\mathbf{i}}$, how far will the body move from rest before it starts to reverse its direction?

11.3.7 Air resistance on a parachutist in free fall is given by kv^2. Find his terminal velocity if the parachute opens when his velocity is v_o m/s:

11.3.8 A rocket weighing 44500 N is launched vertically with a constant thrust of 17800 N. Assuming air resistance to be given by $R = 20v$ where R is in newtons and v in m/s, determine its height when its velocity is 300 m/s. Consider g and the weight of the rocket to be constant.

11.3.9 The frame shown in Fig. P-11-3.9 rotates about a vertical axis. The coefficient of friction under block A is 0.40. Compute the coefficient of friction at block B if B starts to rise when the frame rotates at 40 rpm.

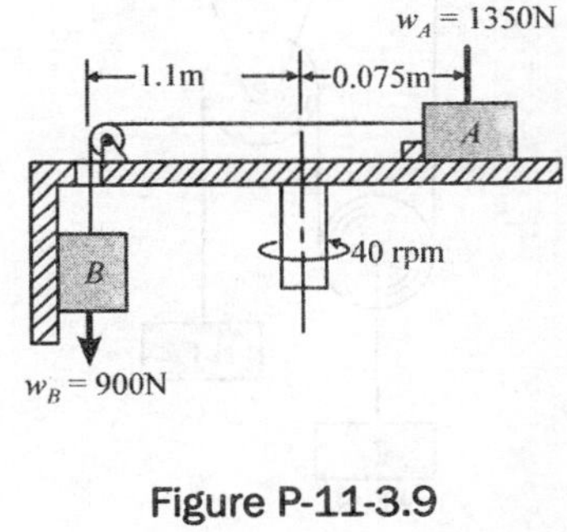

Figure P-11-3.9

11-3.10 In Prob. 11-3.9, block B starts to rise when the frame rotates at 38.2 rpm. Find the coefficient of friction under block A if the coefficient of friction at block B is 0.20.

11.3.11 Two blocks having the weights and positions shown in Fig. P-11-3.11 rest on a frame which rotates about its vertical axis at a constant speed. The coefficient of friction between the blocks and the frame is 0.20. Neglecting the weight and friction of the pulley, at what speed in rpm will the blocks start to slide? What is the tension in the cord at this instant?

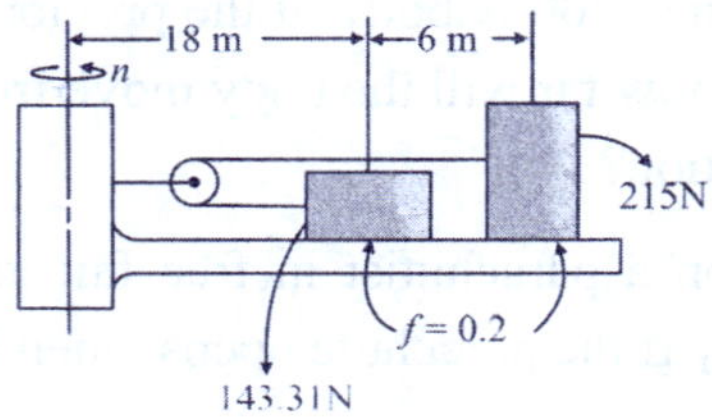

Figure P-11-3.11

11.3.12 Repeat Prob. 11-3.11 if the weights of the blocks are interchanged.

11.3.13 In the system of connected bodies shown in Fig. P-11-3.13, the pulleys are frictionless and of negligible weight. (a) Determine the weight of A to give B a downward acceleration of $0.6g$. (b) Can you find the weight of A to give B an upward acceleration of $0.6g$?

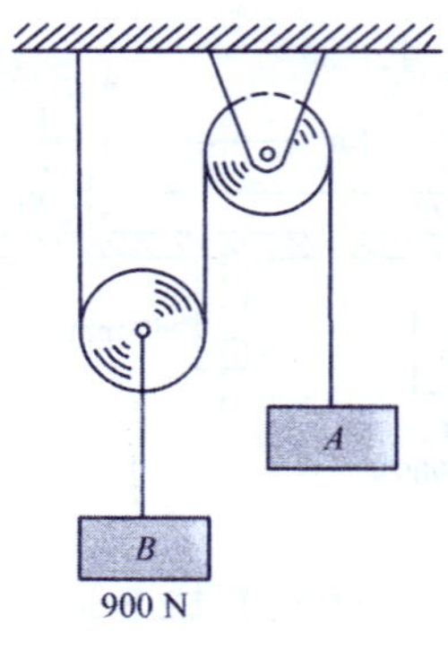

Figure P-11-3.13

11-3.14 Three bodies, each of weight W, are arranged as shown in Fig. P-11-3.14. The coefficient of kinetic friction between B and C is 0.30, and is 0.10 between C and the ground. How far does C move when B has moved s meters relative to C?

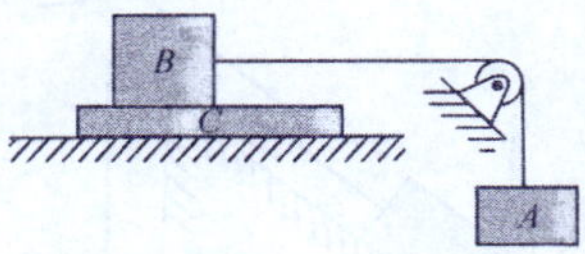

Figure P-11-3.14

11.3.15 Determine the acceleration of body A in Fig. P-11-3.15, assuming the pulleys to be frictionless and of negligible weight.

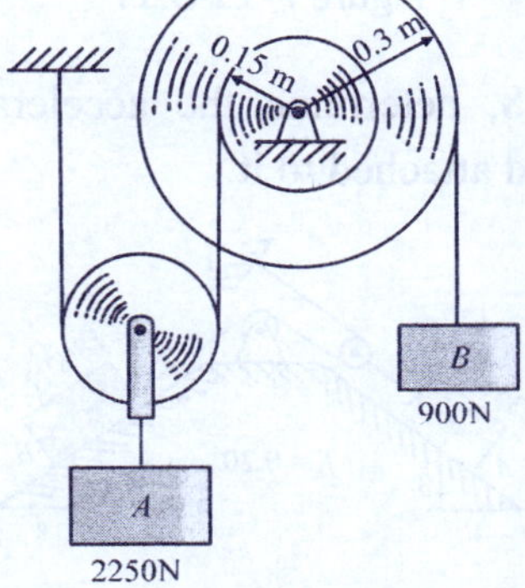

Figure P.11-3.15

11-3.16 Find the acceleration of each body in the system shown in Fig. P-11-3.16.

Hint: First determine the direction of motion.

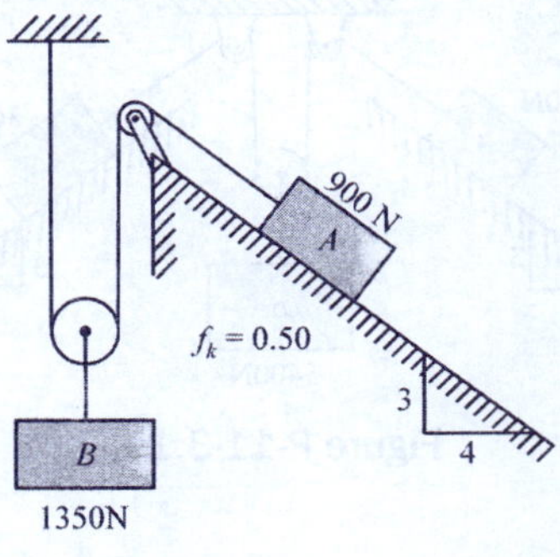

Figure P-11-3.16

11-3.17 Referring to Fig. P-11-3.17, compute the acceleration of body *B* and the tension in the cord supporting body *A*.

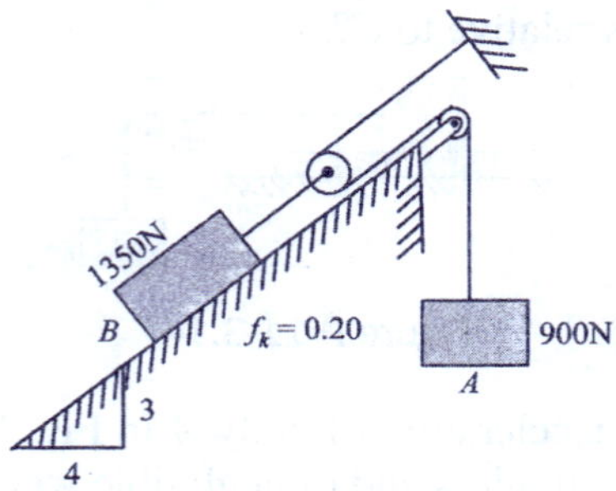

Figure P-11-3.17

11.3.18 In Fig. P-11-3.18, determine the acceleration of body *B* and the tension in the cord attached to it.

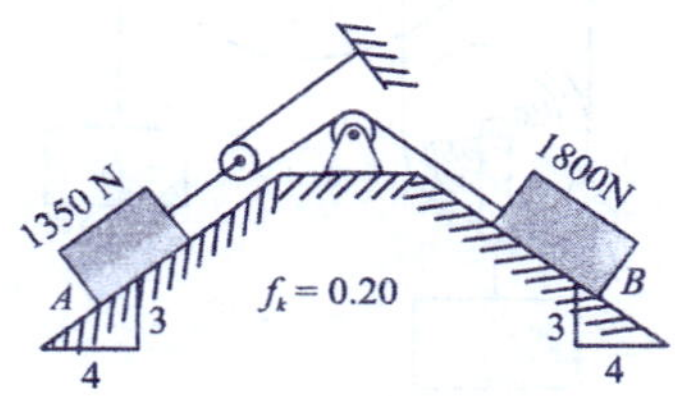

Figure P-11-3.18

11-3.19 If the coefficient of kinetic friction is 0.25 under each body in the system shown in Fig. P-11-3.19, how far and in what direction will body *B* move in 5 sec starting from rest?

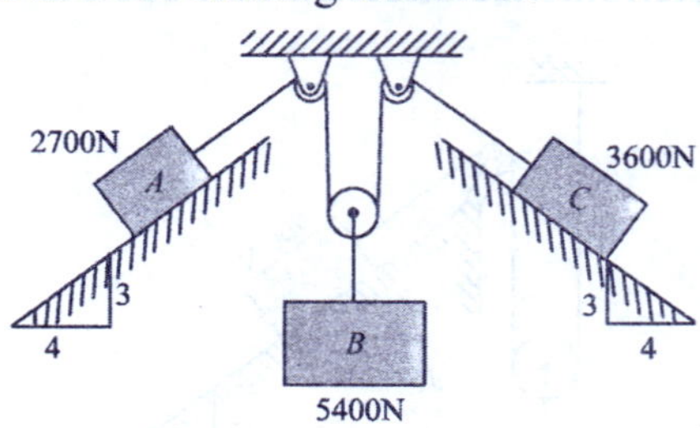

Figure P-11-3.19

11-3.20 Determine the acceleration of each body in Fig. P-11-3.20, assuming the pulleys to be frictionless and of negligible weight. The inclined plane is smooth.

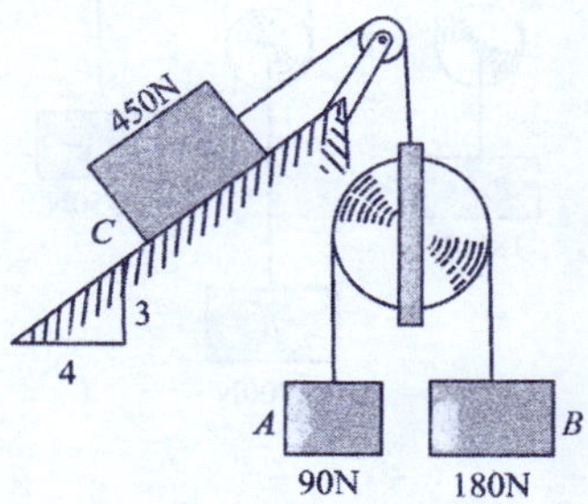

Figure P-11-3.20

11-3.21 Determine the tension in the cord supporting body C in Fig. P-11-3.21. The pulleys are frictionless and of negligible weight.

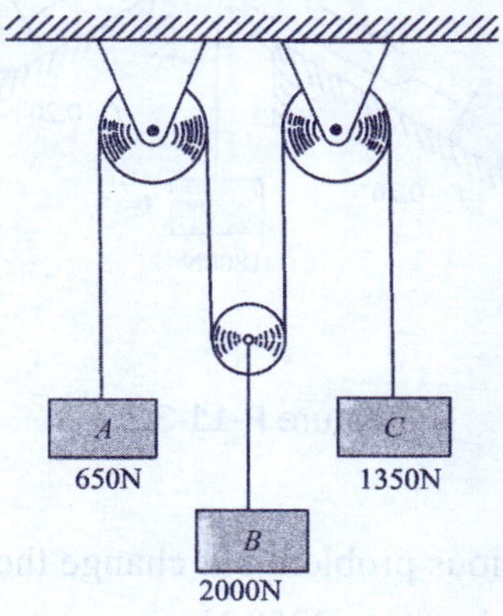

Figure P-11-3.21

11-3.22 The pulleys in Fig. P-11-3.22 are frictionless and of negligible weight. Determine the tension in the cable supporting body C.

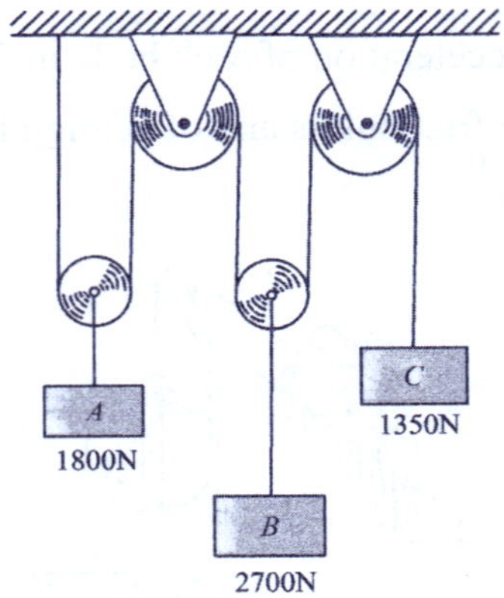

Figure P-11-3.22

11-3.23 In the system of connected bodies in Fig. P-11-3.23, the coefficient of kinetic friction is 0.20 under bodies *B* and *C*. Determine the acceleration of each body and the tension in the cord supporting *A*.

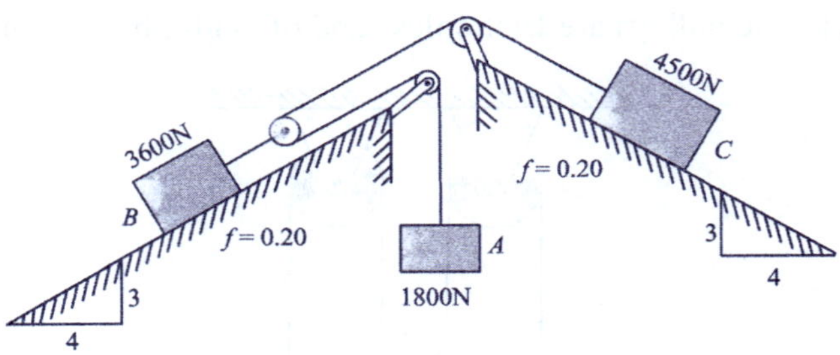

Figure P-11-3.23

11-3.24 Repeat the previous problem but change the weight of *A* to 2700 N, of *B* to 4500 N, and of *C* to 2250 N.

11-3.25 Two bodies *A* and *B*, weighing 1800 N and 1350 N respectively are connected by a rigid bar of negligible weight and move along the smooth surfaces as shown in Fig. P-11-3.25. If they start from rest at the given position, determine the acceleration of *B* at this instant.

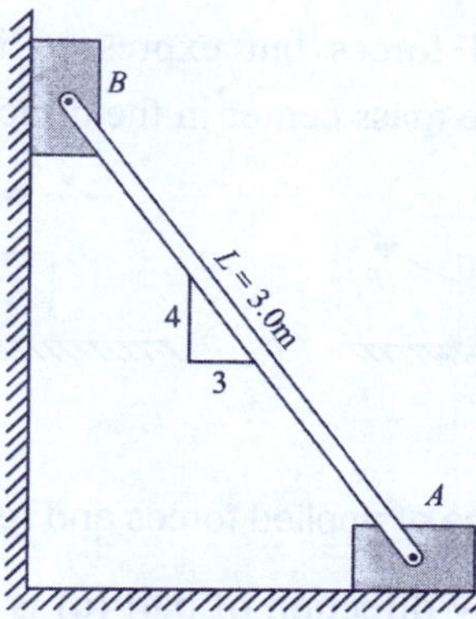

Figure P-11-3.25

11.3.26 Repeat the previous problem if a leftward horizontal force of 1.5 KN is applied to body *A*.

11.3.27 As shown in Fig. P-11-3.27, body *A* slides down the smooth surface of wedge *B* which is on a smooth horizontal floor. Their weights are W_A and W_B. Determine the acceleration of wedge *B*.
Hint: a_A equals the vector sum of a_B and the acceleration of *A* relative to *B*.

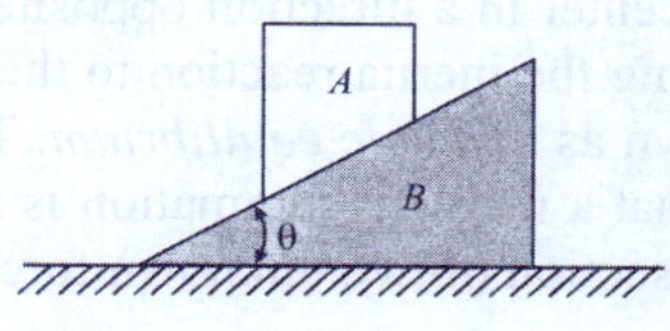

Figure P-11-3.27

11-4 TRANSLATION – ANALYSIS AS A RIGID BODY

In this section, we consider a group of problems which involve the magnitude and/or the position of forces that will constrain a body to be in translation. Our basic criterion is that, in a translating body, the resultant of all applied forces will pass through the mass center of the body (see Section 10-6). Then as shown in Fig. 11-4.1, the free-body diagram of the applied forces in part (a) is equivalent to part (b) where the resultant effective force $(W/g)a$ (equivalent to the resultant of the applied forces but expressed in terms of the mass and acceleration) acts through the mass center in the direction of the acceleration.

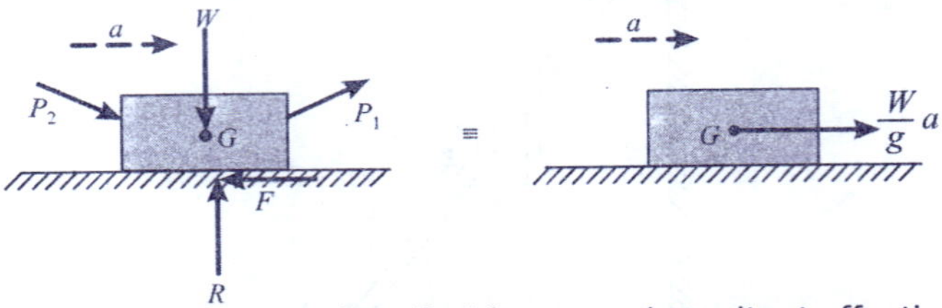

Figure 11-4.1 Equivalence of applied forces and resultant effective forces.

A force summation in any direction of part (a) is equal to the corresponding force summation of part (b); also, a moment summation about any center for part (a) is equal to the corresponding moment summation for part (b). Obviously moments about the mass center G are zero for either part, but both diagrams should be drawn when equating moment summations for both parts about centers other than G.

A useful stratagem that eliminates drawing two free-body diagrams is add to each part a force equal to (W/g) a reversed as shown in Fig. 11-4.2. There it is evident that part (b) then has a zero resultant; hence, part (a) also has a zero resultant. Consequently, in Fig. 11-4.2(a), force and moment summations are zero along any direction or about any axis. Thus, all the methods developed in statics may be applied to a single diagram upon which, in addition to the applied forces, there is included what may be called an inertia reaction[2] equal to $(W/g)a$ acting through the mass center in a direction opposite to the acceleration of the translating body. By adding the inertia reaction to the original force system, we create the condition known as *dynamic equilibrium*. The principal advantage of dynamic equilibrium is that a moment summation is zero about any convenient center such as at the intersection of two unknown forces.

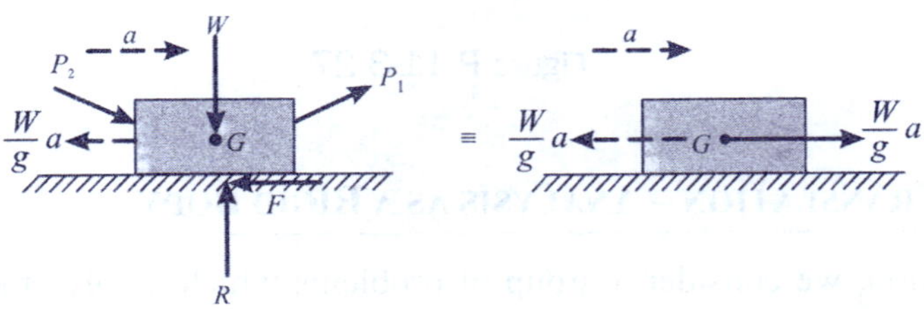

(a) (b)

Figure 11-4.2 Creation of dynamic equilibrium by adding the inertia reaction $(w/g)a$.

---

[2] The inertia reaction should be represented by a dashed vector to avoid any possible confusion with an applied force

ILLUSTRATIVE PROBLEMS

11-4.1 The frame of a certain machine accelerates rightward at 4/5 $g\ m/s^2$. As shown in Fig. 11-4.3, it carries a uniform bent bar *ABC* weighing 400 N/m pinned to it at *C* and braced by the uniform strut *DE* which weighs 500 N. Determine the components of the force at pin *D*.

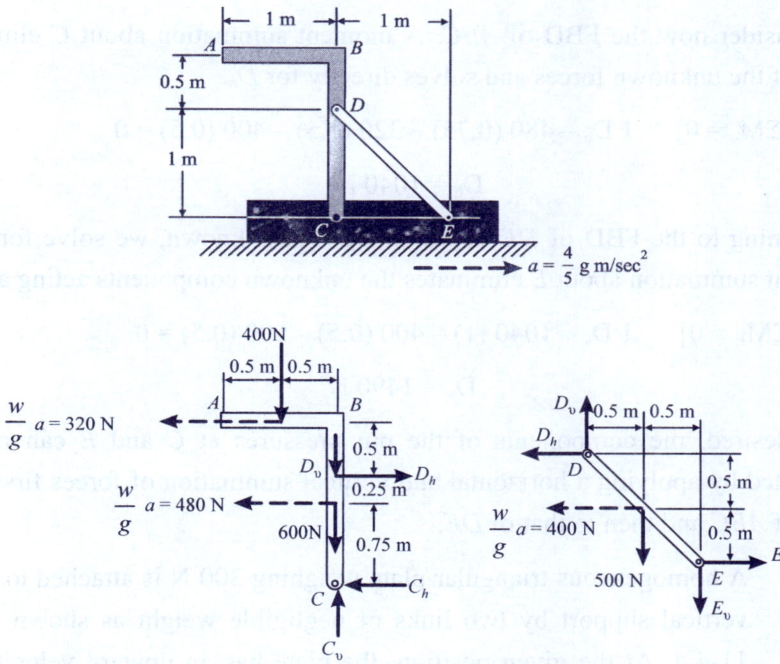

Figure 11-4.3

Solution

Except for the application of the inertia reactions, the analysis of this problem is exactly the same as for the method of members in Section 4-6. The free-body diagram of each member is placed in dynamic equilibrium by including the inertia reaction, acting through the gravity center of each part, directed opposite to the acceleration. Note that instead of locating the center of gravity of *ABC*, it is more convenient to apply the inertia reaction acting on each segment of its length. The inertia reactions are

For *AB*: $\frac{w}{g}a = \frac{400}{g}\left(\frac{4}{5}g\right) = 320$ N

For BC: $\frac{w}{g}a = \frac{600}{g}\left(\frac{4}{5}g\right) = 480$ N

For DE: $\frac{w}{g}a = \frac{500}{g}\left(\frac{4}{5}g\right) = 400$ N

Consider now the FBD of *ABC*. A moment summation about *C* eliminates three of the unknown forces and solves directly for D_h.

$[\curvearrowright + \Sigma M_c = 0] \quad 1\, D_h - 480\,(0.75) - 320\,(1.5) - 400\,(0.5) = 0$

$$D_h = 1040 \text{ N}$$ ***Ans.***

Turning to the FBD of *DE* in which D_h is now known, we solve for D_v. A moment summation about *E* eliminates the unknown components acting at *E*.

$[\curvearrowright + \Sigma M_E = 0] \quad 1\, D_v - 1040\,(1) - 400\,(0.5) - 500\,(0.5) = 0$

$$D_v = 1490 \text{ N}$$ ***Ans.***

If desired, the components of the pin pressures at *C* and *E* can now be computed by applying a horizontal and vertical summation of forces first to the FBD of *ABC* and then to that of *DE*.

11-4.2 A homogeneous triangular plate weighing 300 N is attached to a rigid vertical support by two links of negligible weight as shown in Fig. 11-4.4. At the given position, the plate has an upward velocity of 4 m/s and the hydraulic cylinder exerts a downward force at *E* of *P* = 600 N. Compute the total force acting on hinge *C*.

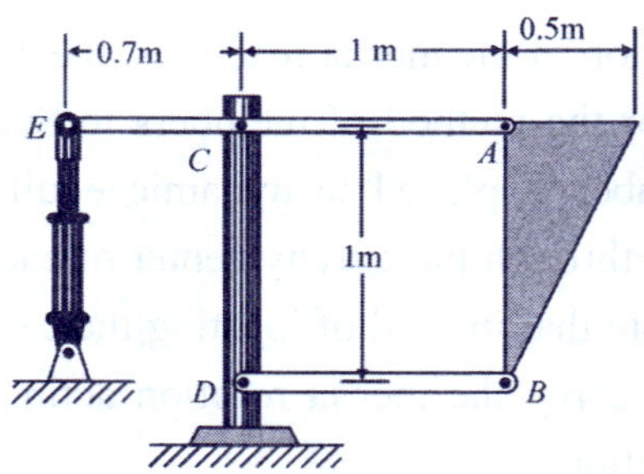

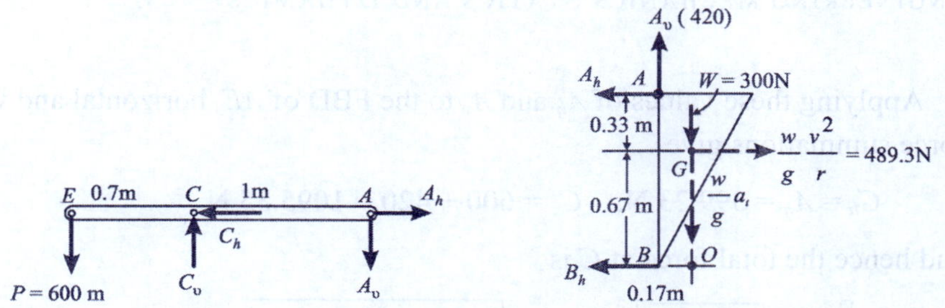

Figure 11-4.4

Solution

The parallel linkage constrains the plate to move without rotation; i.e., to be in curvilinear translation. The motion of its mass center is therefore the same as that of A or B which move in circular paths about C and D respectively. The acceleration of the mass center G thereby has the same normal and tangential components as A or B. These components determine the centrifugal inertia reaction and tangential inertia reaction which are applied as shown on the FBD of the plate to create dynamic equilibrium. The magnitude of the centrifugal inertia reaction is

$$\frac{w}{g}a_n = \frac{W}{g}\frac{v^2}{r} = \frac{300}{9.81}\frac{(4)^2}{1} = 489.3 \text{ N}$$

but the tangential inertia reaction $(W/g)a_t$ is as yet unknown.

Since link BD has negligible weight, it acts as an axial force member and exerts only B_h on the FBD of the plate. However, link AE is a multi-force member having the FBD shown. The common unknowns acting on the FBD's of the link AE and the plate are A_v and A_h. Considering first AE, we have

$[\curvearrowright + \Sigma M_C = 0] \qquad 1\,A_v - 600(0.7) = 0 \qquad A_v = 420 \text{ N}$

Consider next the FBD of the plate where, with A_v known, a convenient moment center is at the intersection O of B_h and $(W/g)a_t$ which thereby determines A_h, as follows:

$[\circlearrowleft + \Sigma M_O = 0] \qquad 1\,A_h - 420(0.17) - 489.3(0.67) = 0$

$$A_h = 399.23 \text{ N}$$

Applying these values of A_h and A_v to the FBD of AE, horizontal and vertical force summations give

$$C_h = A_h = 399.23 \text{ N}; \quad C_v = 600 + 420 = 1095.35 \text{ N}$$

and hence the total force at C is

$$\left[C = \sqrt{C_h^2 + C_v^2}\right] \qquad C = \sqrt{(399.23)^2 + (1020)^2} = 1095.35 \text{ N}$$

If desired, we may also apply horizontal and vertical force summations to the FBD of the plate to obtain B_h = 90.07 N and $(W/g)a_t$ = 120 N whence $a_t = 3.924 \text{ m/s}^2$.

PROBLEMS

11.4.3 A juggler places the lower end of a vertical rod upon his finger. As it starts to tip, explain how he keeps the rod in balance by moving his finger back and forth.

11.4.4 The cable of a cargo crane can support a maximum load of 2 tons. While the crane is lowering a 7000 N weight at uniform speed, the brake on the winch is applied too rapidly, thereby causing a sudden deceleration of the weight equal to 30 m/s^2. The cable snaps and the weight falls, badly injuring a workman. For the purpose of establishing liability in this accident, is it likely that failure of the cable was due to its being weaker than its test strength of 2 tons? Why?

11-4.5 A uniform box of weight W is 0.5 m square and 2 m high. It stands on end upon a truck with its sides parallel to the truck's motion. If the coefficient of friction between the box and the truck is 0.30, show whether the box will slide or tip first as the acceleration of the truck is increased. What happens if the truck is being braked?

11-4.6 The 2000 N body in Fig. P-11-4.6 is supported by wheels at B which roll freely without friction and by a skid at A under which the coefficient of friction is 0.50. Compute the value of P to cause an acceleration of $(\frac{g}{4}\ m/s^2)$

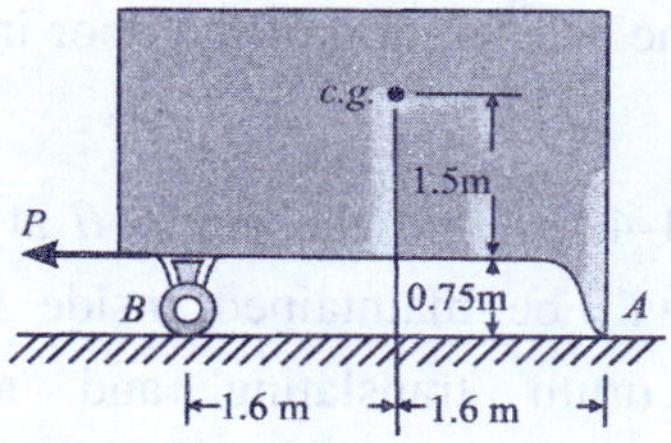

Figure P-11-4.6

11-4.7 If the value of P in Prob. 11-4.6 is 900 N, compute the acceleration. If this value of P were applied at a higher position on the body, would the acceleration be changed in any way? Explain your conclusion.

11-4.8 An auto has a wheel base of 3.5 m and its *c.g.* is 1 m above the ground and 1.5 m behind the front wheels. Which would give the greater acceleration on a level road - a front wheel or a rear wheel drive?

11-4.9 An auto has a wheel base of 3 m with its *c.g.* located 1.5 m ahead of the rear wheels and 0.8 m above the pavement. If $f = 0.80$ at the tires and only the front wheel brakes are operating, compute the minimum distance in which the auto can be brought to rest from a speed of 30 m/sec assuming the driver's reaction time before applying the brakes is $\frac{3}{4}$ sec.

11-4.10 The coefficient of friction under the sliding supports of the door in Fig. P-11-4.10 is 0.30 at A and 0.20 at B. What force P will give the 3000 N door a leftward acceleration of 2.5 m/s^2?

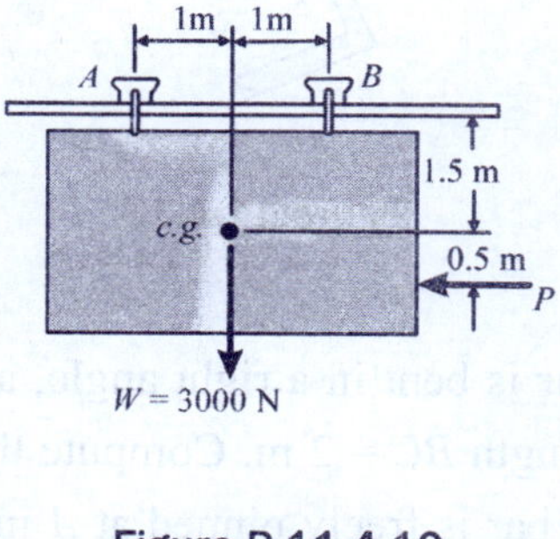

Figure P-11-4.10

11-4.11 Determine the acceleration of the door in the preceding problem if $P = 1150$ N.

11-4.12 In Fig. P-11-4.12, find the angle θ at which a uniform bar of weight W will be maintained inside the smooth surface of a cylindrical drum translating and accelerating leftward at $\frac{4}{5}g$ m/s^2?

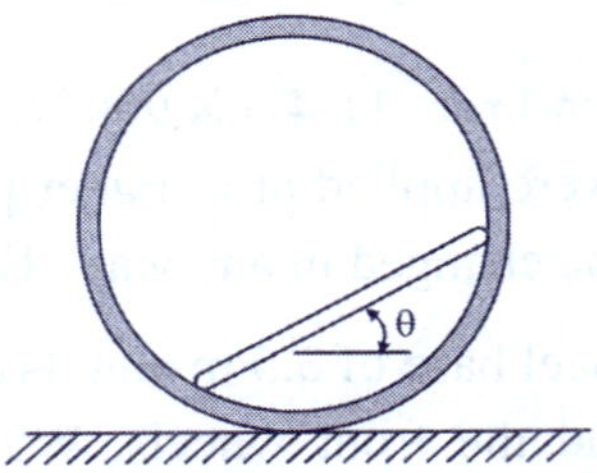

Figure P-11-4.12

11-4.13 A bar weighing 30 N/m is bent at right angles into segments 0.6 m and 0.3 m long. It takes the position shown in Fig. P-11-4.13 when the frame F to which it is pinned at A is accelerated horizontally. Determine this acceleration and the total reaction at A.

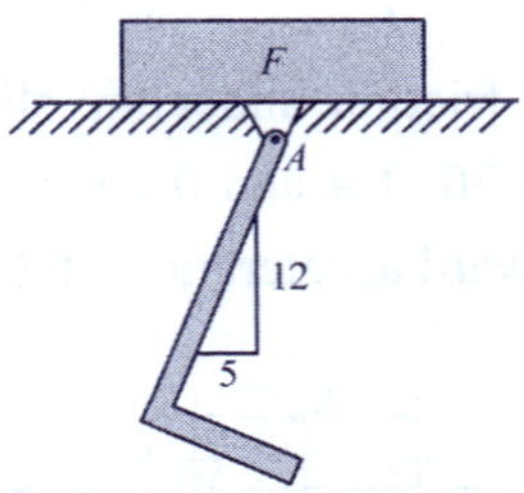

Figure 11-4.13

11-4.14 A 4.5 m/s bar is bent in a right angle, as shown in Fig. P-11-4.14, so that the length $BC = 2$ m. Compute the acceleration of the block to which the bar is freely pinned at A in order to maintain the bent bar in the given position. The bar weighs W N/m.

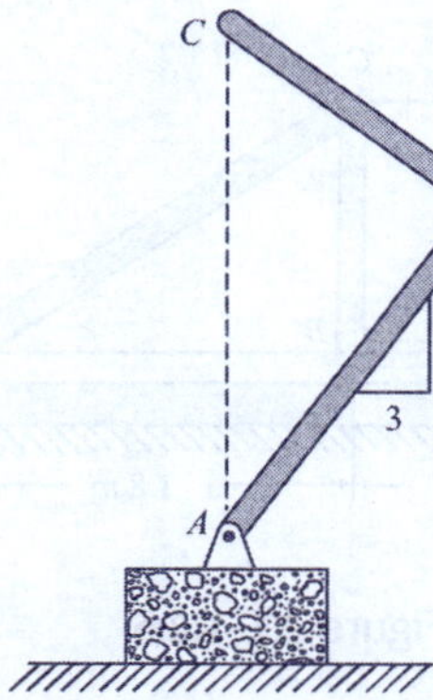

Figure 11-4.14

11-4.15 The uniform crate shown in Fig. P-11-4.15 weighs 900 N. It is pulled up the incline by a counterweight W of 1800 N. Find the maximum and minimum values of d so that the crate does not tip over as it slides up the incline.

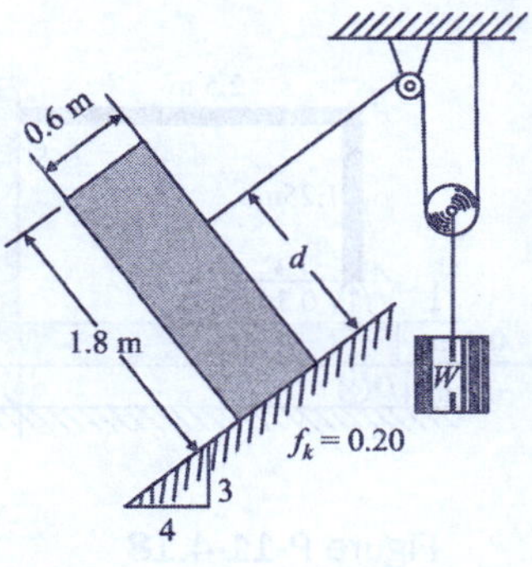

Figure 11-4.15

11-4.16 The frame of a machine is accelerated leftward at 0.6 g m/s^2. As shown in Fig. P-11-4.16, it carries a uniform angle ABC weighing 400 N which is braced by the uniform strut CD weighing 300 N. Determine the components of the hinge force at C upon CD.

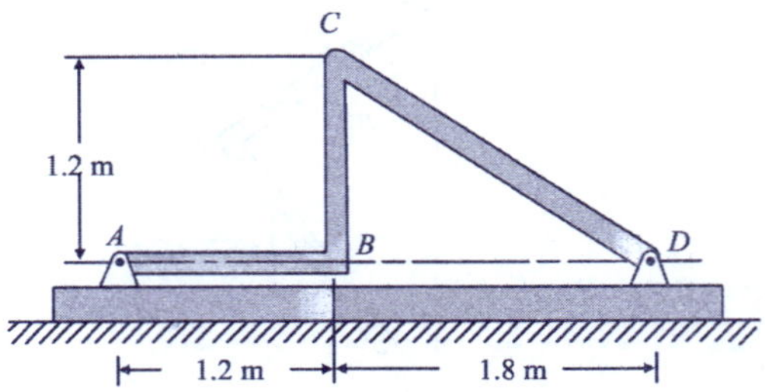

Figure 11-4.16

11.4.17 Solve the preceding problem if the frame of the machine is accelerated rightward at $\frac{4}{5}$ g m/s^2.

11.4.18 The bent bar *ABC* weighing 150 N/m is mounted as shown in Fig. P-11-4.18 upon a carriage weighing 1050 N. The center of gravity of the carriage is at *G* midway between the wheels. If $P = 500$ N and there is no frictional resistance at the wheels, find the wheel reactions and also the horizontal and vertical components of the hinge force at *A*.

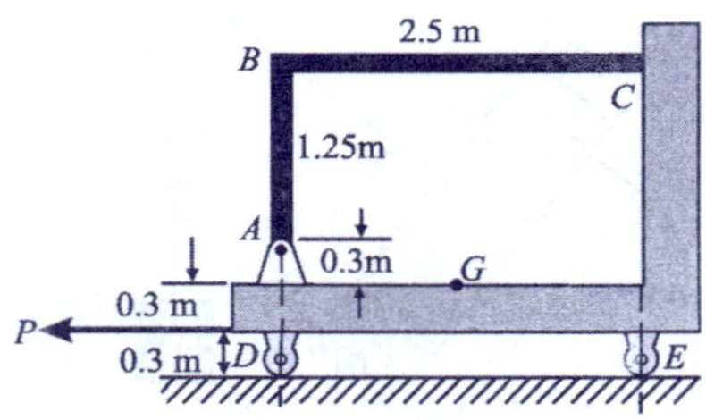

Figure P-11-4.18

11-4.19 Repeat the preceding problem if the magnitude and sense of *P* is such that the reaction of the carriage upon the bent bar at *C* is 200 N leftward.

11-4.20 Two bodies *A* and *B*, weighing 1350 N and 1800 N respectively, are connected by a rigid bar of negligible weight attached to them at their gravity centers as shown in Fig. P-11-4.20. The angle of friction at the wall and floor is 15°. If the bodies start from rest at the given position, determine the acceleration of *B* at this instant.

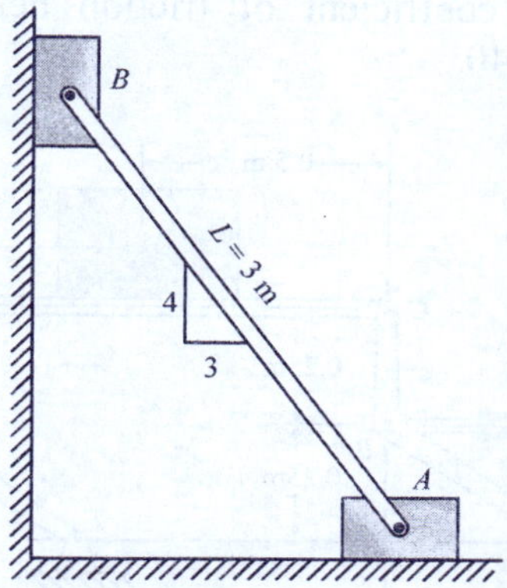

Figure 11-4.20

11-4.21 Repeat the preceding problem if a leftward horizontal force of 3150 N is applied to body A.

11-4.22 The coefficient of friction between the road and the tires of the car shown in Fig. P.11-4.22 is 0.60. The car weighs 1800 N. It is rounding a horizontal curve of 200 m radius at the maximum speed at which side slipping impends. What is the value of the friction force acting under each wheel? How high above the road must the center of gravity be to limit this maximum speed by the tendency to overturn?

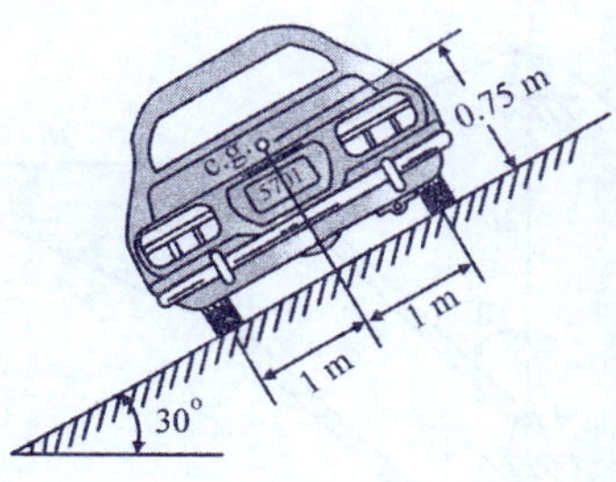

Figure P-11-4.22

11-4.23 Repeat the preceding problem if the road is banked at 20° instead of 30° as shown in Fig. P-11-4.22.

11-4.24 The loading platform shown in Fig. P-11-4.24 is raised from rest at the given position by a leftward pull of 11150 N exerted at *B* by the hydraulic cylinder. The weight of the platform and links are negligible compared with the 4500 N weight of the crate whose center of gravity is at *G*. Determine the force at the hinge *E* at this instant. The coefficient of friction between the crate and the platform is 0.40.

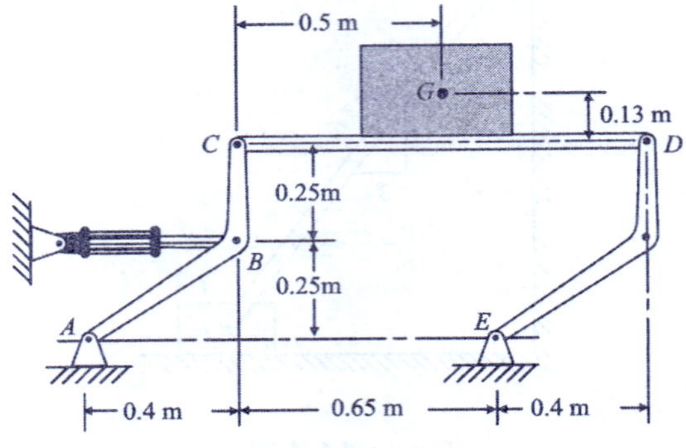

Figure P-11-4.24

11-4.25 Solve the preceding problem if, in place of the hydraulic cylinder, a counterclockwise torque of 700 N-m is applied to link *AC* at hinge *A*.

11-4.26 A homogeneous triangular plate weighing 450 N is supported by two parallel links of negligible weight as shown in Fig. P-11-4.26. It moves under the action of the cable attached to a 900 N weight. If the system starts from rest when $\theta = 90°$; find the force in link *CD*.

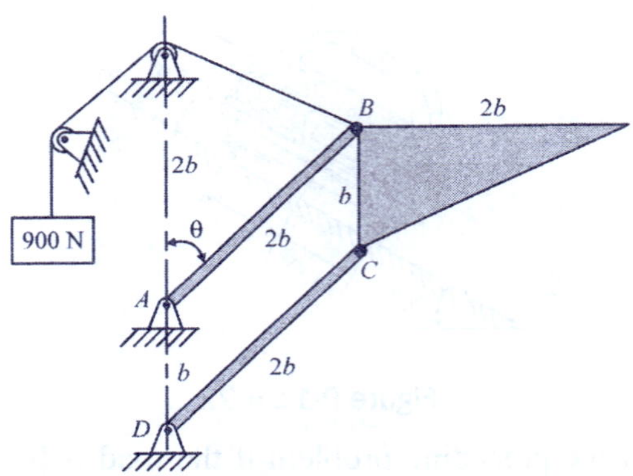

Figure P-11-4.26

11-4.27 Solve the preceding problem if the system starts from rest when $\theta = 60°$.

SUMMARY

The rigid-body motion of translation is defined as motion in which a straight line passing through any two particles of a body always remains parallel to its initial position. All particles travel along the same or parallel paths and therefore have identical values of displacement, velocity, and acceleration. If the path is straight, the motion is called rectilinear translation; if curved, the motion is called curvilinear translation. Insofar as Newton's laws of motion for a particle are concerned, translation of a rigid body is equivalent to that of a particle having the mass of the body and the motion of the mass center of the body.

For rectilinear motion, let the X axis be directed along and positive in the initial direction of motion. Then the component forms of the kinetic equation for rectilinear motion become

$$\Sigma X = \frac{W}{g} a \qquad \Sigma Y = \Sigma Z = 0 \tag{11-2.1}$$

For curvilinear motion, the convenient components of acceleration to use are those normal and tangential to the path or, occasionally, the radial and transverse components. When we call the normal axis N and the tangential axis T, the corresponding scalar forms of $\mathbf{R} = \frac{W}{g}\mathbf{a}$ are

$$\Sigma N = \frac{W}{g} a_n = \frac{W}{g}\frac{v^2}{r} \tag{11-2.2}$$

$$\Sigma T = \frac{W}{g} a_t$$

In applying these equations, be careful to take the positive senses of ΣN and ΣT in the positive directions of a_n and a_t. Usually, however, it is simpler to put curvilinear motion problems in dynamic equilibrium by applying inertia reactions of magnitudes $\frac{W}{g}\frac{v^2}{r}$ and $\frac{W}{g} a_t$ acting through the mass center which are directed opposite to the normal and tangential components of acceleration respectively.

When motion along a curved path is best defined by polar coordinates, the component forms of the kinetic equation along radial and transverse directions are

$$\Sigma F_r = \frac{W}{g} a_r = \frac{W}{g}\left(\ddot{r} - r\dot{\theta}^2\right) \qquad (11\text{-}2.3)$$

$$\Sigma F_\theta = \frac{W}{g} a_\theta = \frac{W}{g}\left(r\ddot{\theta} + 2\dot{r}\dot{\theta}\right)$$

In these equations, the positive senses are in the positive directions of increasing r and θ.

When considering the distribution of forces on a translating rigid body, it is often convenient to create dynamic equilibrium by applying the components of $\frac{W}{g}$a acting at the mass center of the body and directed opposite to the corresponding components of a. The advantage of dynamic equilibrium is that all the methods of statics may be applied to a single free-body diagram which includes the components of the inertial reaction $\frac{W}{g}$a. Especially useful is that with dynamic equilibrium a moment summation may be set equal to zero at a center which is at the intersection of two unknown forces.

CHAPTER 12

KINEMATICS OF RIGID BODIES

12-1 INTRODUCTION. TYPES OF RIGID-BODY MOTION

In the preview of dynamics presented in Section 9-1, we discussed the correlation between the unbalanced resultant force-couple system acting on a body and its consequent motion. Two basic equations were stated; namely, $\mathbf{R} = (W/g)\,\bar{a}$ and $\Sigma\mathbf{M} = \mathbf{H}$. These were later derived, in Sections 10-5 and 10-6. In addition to these basic equations, we noted the necessity for the correlation between acceleration and the motion which is the area known as kinematics.

For the same reasons that we began with a study of the kinematics of a particle as a prerequisite to the study of its kinetics, we similarly begin our study of rigid-body motion with the kinematics of rigid-body motion. The various types of motion to be discussed are:

1. Translation
2. Fixed-axis rotation
3. General plane motion
4. Rotation about a fixed point
5. General space motion

We have already shown in Section 11-2 that translation (either rectilinear or curvilinear) has the characteristic that all particles in the body have the same identical motion. This was the basis for treating a translating body as a particle having the mass of the body and the motion of any point of the body. More particularly, the motion of translation is characterized by the absence of any spin or angular motion of the body. Because of this, the kinematics of particle motion was sufficient to handle the translation of a rigid body, and therefore we included this type of motion in Chapter 11 as a simple extension of particle kinetics.

In the next section on fixed-axis rotation, we consider the concept whereby the position of a body is uniquely defined by its angular coordinate rather than by a linear coordinate which suffices for translation. In particular, we shall demonstrate an analogy between rectilinear motion and angular motion whereby much of our knowledge of rectilinear motion may be applied directly to rotational kinematics. Each of the remaining types of motion will be shown to be combinations of linear and angular motion for which our key objective will be to correlate the angular motion with the linear motion of the mass center. When this kinematic correlation is established, we will be ready to discuss the kinetics of the various types of rigid-body motions.

12-2 ANGULAR MOTION – FIXED-AXIS ROTATION

Fixed-axis rotation is defined as that motion of a rigid body in which the particles move in circular paths with their centers on a fixed straight line that is called the axis of *rotation*. The planes of the circles in which the particles move are perpendicular to the axis of rotation.

These concepts are illustrated in Fig.12-2.1 where the motion of a typical particle P is determined by the change in its position vector $\mathbf{r}$ drawn from an origin O anywhere on the fixed axis of rotation, The condition that the body be rigid requires that the length of $\mathbf{r}$ and its inclination ϕ be constant; hence, P describes a circular path, perpendicular to the axis of rotation, centered about its projection B on this axis and of radius $BP = r \sin \phi$.

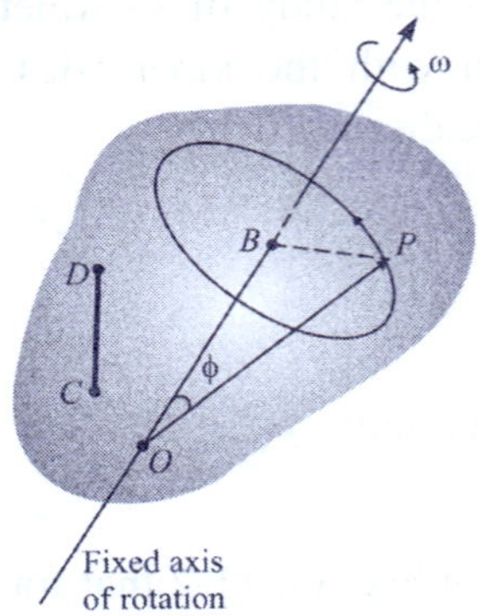

Figure 12-2.1 Motion of P in a fixed axis rotation of a rigid body.

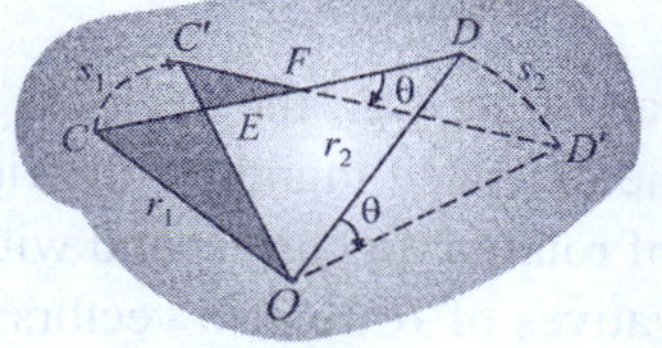

Figure 12-2.2 In fixed-axis rotation of a rigid body, all lines rotate through the same angular displacement.

This motion, and that of any line CD in the body, is most conveniently described by projecting the rotating body upon a plane perpendicular to the axis of rotation as shown in Fig. 12-2.2. There O represents the fixed axis and CD is the projection of any line such as CD in Fig.12-2.1. We see that if the radius to any point C is permitted to rotate through θ radians, point C moves through the arc distance $s_1 = r_1\theta$. Since the body is rigid, angle COD cannot change; hence, the radius to any other point D will also rotate through θ radians and point D will move through the arc distance $s_2 = r_2\theta$. From this we conclude that the radii to all particles of a rotating body have the same angular motion (i.e., the same θ) although their linear movements (i.e., s_1 and s_2) vary directly with their radial distances from the axis of rotation.

Of particular importance is the fact that the angle between the line CD and its subsequent position $C'D'$ is also equal to θ. We can easily prove this statement by noting that triangles CEO and $C'EF$ are similar because two of their angles are equal. Thus, angles CEO and $C'EF$ are equal vertical angles and angles ECO and $FC'E$ are the same apex angle of the congruent triangles COD and $C'OD'$. Consequently, the angles COE and $EF\ C'$ are equal to each other and have the same value θ as must angle DFD' which is the vertical angle of EFC'. Our conclusion is that the projections of all line segments in a rigid body, such as CD in Fig. 12-2.1, upon a plane perpendicular to the axis of rotation all rotate through the same angle. We define this angle as the *angular displacement* of the rigid-body rotation. The unit of angular displacement may be radians, degrees, or revolutions, but radian measurement is preferred in order to correlate angular motion with linear motion.

For the case of fixed-axis rotation, it is permissible to define angular displacement θ as a vector[1] directed along the axis of rotation, its sense being

[1]For other than plane motion, $\boldsymbol{\theta}$ is not a vector in that the sequence of combining its components causes different results; i.e., $\boldsymbol{\theta} = \boldsymbol{\theta}_z + \boldsymbol{\theta}_y$ is not equal to $\boldsymbol{\theta}_y + \boldsymbol{\theta}_z$. For example, consider a top initially in the position shown in Fig. 12-2.3. The sequence in which two separate 90° rotations of the top about the X and Y axes are done, will produce the two different dashed positions shown. However, although $\boldsymbol{\theta}$ is not a vector, its time derivatives $\boldsymbol{\omega}$ and $\boldsymbol{\alpha}$ are *always* vectors because any sequence of combining two *infinitesimal* rotations will produce the same result.

determined by the right-hand rule; e.g., the vector is directed along the axis of rotation as indicated by the extended thumb of the right hand when the fingers are curled about the axis of rotation to correspond with the direction of rotation. The successive time derivatives of θ are also vectors; i.e., angular velocity $\omega = \frac{d\theta}{dt}$ and angular acceleration $\alpha = \frac{d\omega}{dt} = \frac{d^2\theta}{dt^2}$. They too are directed along the axis of rotation in accordance with the right-hand rule. Since fixed-axis rotation, however, is usually specified in terms of its projection upon a plane perpendicular to the axis of rotation, it is more convenient to denote θ, ω, and α as clockwise or counterclockwise quantities, letting the initial direction of rotation denote the positive sense of θ, ω, and α. This then agrees with the previously established convention for rectilinear motion; i.e., the initial direction of motion determines the positive sense of s, v, and a.

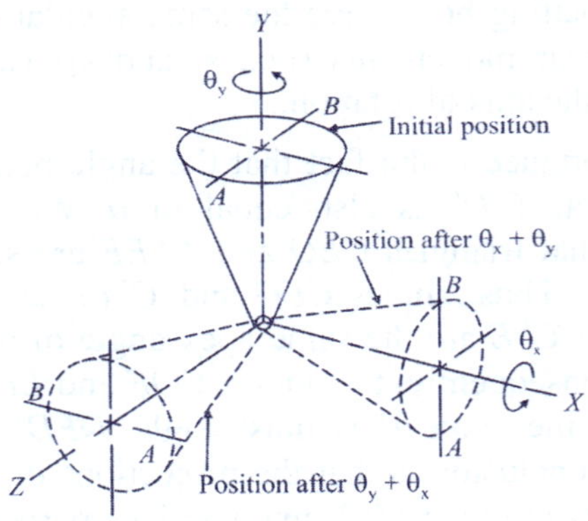

Figure 12.2.3 Effect of two different sequences of combining 90° rotations about X and Y axes.

The correlation between linear and angular displacement is presented in Fig.12-2.4 which shows a pulley free to rotate about an axle O under the action of a weight W suspended from a cord wound around the pulley. As the weight descends s m, a point B on the rim of the pulley moves through an equal arc length to position A. The corresponding angular displacement θ of the pulley is obviously subtended by radii to points B and A. Evidently, the relation between the linear displacement of the weight and the angular displacement (in radians) of the pulley is given by

$$s = r\theta \tag{a}$$

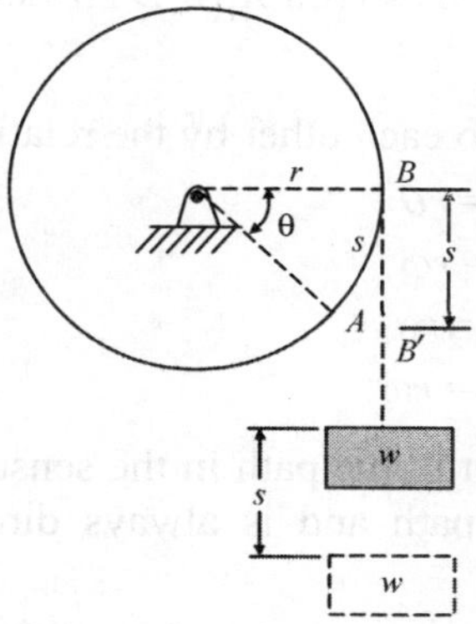

Figure 12.2.4 Relation between linear and angular displacement.

Since the radius r is constant, differentiating Eq. (a) with respect to time gives

$$\frac{ds}{dt} = r\frac{d\theta}{dt} \quad \text{or } v = r\omega \tag{b}$$

and a second differentiation of Eq. (b) gives

$$\frac{dv}{dt} = r\frac{d\omega}{dt} \quad \text{or } a = r\alpha \tag{c}$$

The expression $\frac{dv}{dt}$ in Eq. (c) is the linear acceleration of the weight, but it also represents the tangential acceleration a_t of a point on the rim of the pulley.

Recalling Eq. (9-7.2), we may use $v = r\omega$ to write the corresponding normal acceleration of any point on the rim of the pulley in any of the following forms:

$$a_n = \frac{v^2}{r} = r\omega^2 = v\omega \tag{d}$$

Summarizing this discussion, we see that the differential kinematic equations for rectilinear motion and for rotation are completely analogous in form, differing only in the symbols used; namely,

Rectilinear Motion	***Rotation***
$v = \frac{ds}{dt}$	$\omega = \frac{d\theta}{dt}$
$a = \frac{dv}{dt} = \frac{d^2s}{dt^2}$	$\alpha = \frac{d\omega}{dt} = \frac{d^2\theta}{dt^2}$
$a\,ds = v\,dv$	$\alpha\,d\theta = \omega\,d\omega$

They can be transformed into each other by the relations

$$
\begin{aligned}
s &= r\theta \\
v &= r\omega \\
a_t &= r\alpha \\
a_n &= r\omega^2
\end{aligned}
\qquad (12\text{-}2.1)
$$

Observe that a_t is tangent to the path in the sense that α rotates the radius r, while a_n is normal to the path and is always directed toward the center of rotation.

As an extension of the analogy between rectilinear motion and rotation, we observe that the procedures discussed previously in Section 9-3 for constant and variable acceleration will yield similar results in rotation that differ only in the symbols used. Thus, the equations of rotation with constant *angular acceleration* become

Rectilinear Motion	***(Related by)***	***Rotation***
$v = v_o + at$	$s = r\theta$	$\omega = \omega_o + \alpha t$
$s = v_o t + \frac{1}{2} at$	$v = r\omega$	$\theta = \omega_o t + \frac{1}{2} \alpha t^2$
$v^2 = v_o^2 + 2as$	$a = r\alpha$	$\omega^2 = \omega_o^2 + 2\alpha\theta$

ILLUSTRATIVE PROBLEMS

12-2.1 For the system of connected bodies shown in Fig. 12-2.5, the initial angular velocity of the compound pulley B is 6 rad per sec counterclockwise and weight D is decelerating at the constant rate of 1.22 m/s². What distance will weight A travel before coming to rest?

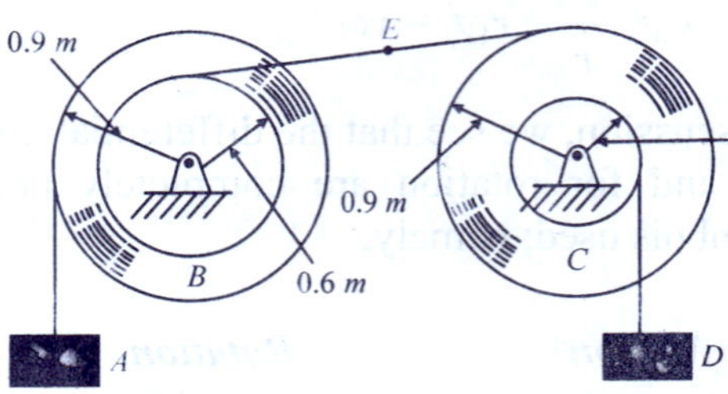

Figure 12-2.5

Solution

To correlate the given data, we start by finding the kinematic relations among the bodies. Using Eq. (12-2.1) and denoting by E any point on the cord connecting B and C, we obtain

$[s = r\theta]$ $\qquad s_A = 0.9\theta_B \quad , \quad s_E = 0.6\theta_B = 0.9\theta_C \quad , \quad s_D = 0.45\theta_C$

Combining these relations into one continuous equation, we have

$$s_A = 0.9\,\theta_B = 1.35\theta_C = 3s_D \qquad \text{(a)}$$

The velocity and acceleration relations among the bodies are obtained by taking two successive time derivatives of Eq. (a). The results are equivalent to merely changing the symbols in Eq. (a) since $v = r\omega$ and $a_t = r\alpha$ have the same mathematical form as $s = r\theta$. Therefore, we obtain

$$v_A = 0.9\,\omega_B = 1.35\omega_C = 3v_D \qquad \text{(b)}$$

$$a_A = 0.9\,\alpha_B = 1.35\alpha_c = 3.a_D \qquad \text{(c)}$$

Consider now the motion of A. From Eq. (*b*), $v_{oA} = 0.9\omega_{\bar{o}_B} = 0.9\,(6) = 5.4$ m/s and, from Eq. (*c*), $a_A = 3\ a_D = 3\,(-1.22) = -3.66$ m/s^2. Since the acceleration of A is constant, the distance it travels in coming to rest is

$[v^2 = v_o^2 + 2as]$ $\qquad 0 = (5.4)^2 - 2(3.66)s_A \qquad s_A = 4\ m$ $\qquad$ ***Ans.***

If desired, the displacement of the other bodies can now be found from Eq. (*a*).

12-2.2 A body rotates according to the relation $\alpha = 3t^2 + 4$, angular displacement being measured in radians and time in seconds. If its initial angular velocity is 4 rad per sec and the initial angular displacement is zero, compute the values of ω and θ for the instant when $t = 3$ sec. Solve analytically and graphically.

Solution

Rewriting $\alpha = d\omega/dt$ as $d\omega = \alpha\,dt$ and integrating between the given limits, we have

$$\int_4^{\omega} d\omega = \int_0^{t} \left(3t^2 + 4\right) dt$$

$$\omega - 4 = t^3 + 4t$$

$$\omega = t^3 + 4t + 4 \qquad (a)$$

Applying $\omega = d\theta/dt$ in the form $d\theta = \omega\,dt$, substituting for ω its value from Eq. (*a*), and integrating gives

$$\int_0^{\theta} d\theta = \int_0^{t} \left(t^3 + 4t + 4\right) dt$$

$$\theta = \frac{t^4}{4} + 2t^2 + 4t \qquad (b)$$

Substituting $t = 3$ sec in Eqs. (*a*) and (*b*), we have

$$\omega = (3)^3 + 4 \times 3 + 4 = 43 \text{ rad per sec} \qquad \textbf{\textit{Ans.}}$$

$$\theta = \frac{(3)^4}{4} + 2 \times (3)^2 + 4 \times 3 = 50.25 \text{ rad} \qquad \textbf{\textit{Ans.}}$$

Let us now check this solution by means of the motion curves shown in Fig. 12-2.6. Using an equation similar to Eq. (9-4.1), we find that the change in angular velocity is

$[\Delta\omega = (\text{Area})_{\alpha\text{-}t}]$ $\qquad \Delta\omega = 4 \times 3 + \frac{1}{3}(3)(27)$

$$= 12 + 27 = 39 \text{ rad per sec}$$

Adding this value to $\omega_0 = 4$ gives $\omega = 43$ rad per sec at $t = 3$ sec as before.

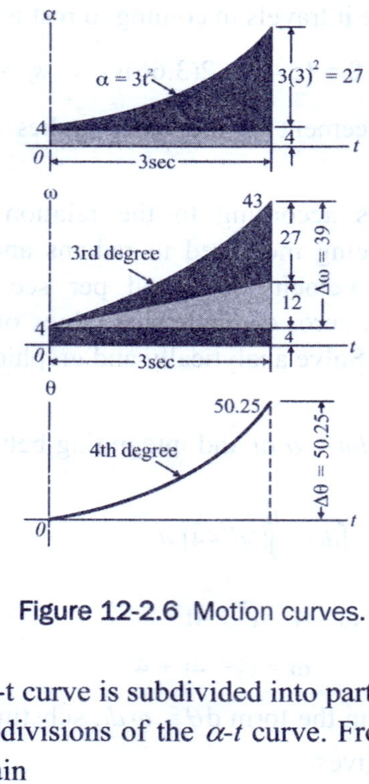

Figure 12-2.6 Motion curves.

The area under the ω-t curve is subdivided into parts shaded to correspond to the similarly shaded subdivisions of the α-t curve. From an equation similar to Eq. (9-4.2), we then obtain

$[\Delta\theta = (\text{Area})_{\omega-t}]$ $\qquad \Delta\theta = 4 \times 3 + \dfrac{1}{2} \times 3 \times 12 + \dfrac{1}{4} \times 3 \times 27$

$$= 50.25 \text{ rad} \qquad \textbf{\textit{Check}}$$

If preferred, $\Delta\theta$ may be computed by applying an equation similar to Eq. (9-4.3) to the area under the α-t curve. This gives

$[\Delta\theta = \omega_1(t_2 - t_1) + (\text{Area})_{\alpha-t}(\bar{t}_2)]$

$$\Delta\theta = 4\times3+(4\times3)\left(\tfrac{1}{2}\times3\right)+\left(\tfrac{1}{3}\times3\times27\right)\left(\tfrac{1}{4}\times3\right)$$

$$= 50.25 \text{ rad} \qquad \textit{Check}$$

which is identical term for term with the preceding computation although based on a different concept.

Finally, a comparison of these computations obtained from motion curves will be found to be identical term for term with those of the calculus solution.

PROBLEMS

12-2.3 A flywheel rotating at 3 rev per Sec has its speed increased at the constant rate of 45 rad per min each second during an interval of 10 sec. Through how many revolutions does it rotate in this time?

12-2.4 The rim of a 1.5 m, wheel on a brakeshoe testing machine has a speed of 27 m/s when the brake is applied. It comes to rest after the rim has traveled a linear distance of 185 m. What is the constant angular acceleration of the wheel and how many revolutions does it make after the brake is applied?

12-2.5 A gear is accelerated from rest to a speed of 1800 rpm and then immediately decelerated to a stop. If the total elapsed time is 12 sec, determine the total number of revolutions of the gear. Assume that both acceleration and deceleration are constant but not necessarily of the same magnitude.

12-2.6 When the angular velocity of a 1.2 m diameter pulley is 3 rad per sec, the total acceleration of a point on its rim is 9.0 m/s^2. Determine the angular acceleration of the pulley at this instant.

12-2.7 Determine the horizontal and vertical components of the acceleration of point B on the rim of the flywheel shown in Fig. P-12-2.7. At the given position, ω = 4 rad per sec and α = 12 rad per sec^2, both clockwise.

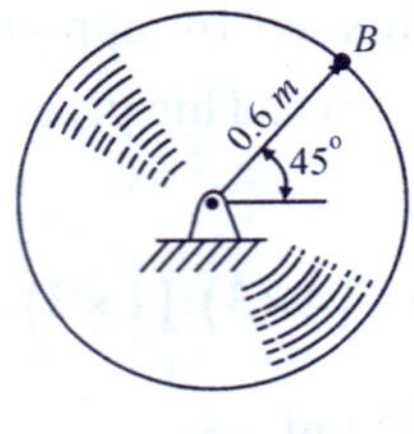

Figure P-12-2.7

12-2.8 Repeat the preceding problem if α is changed to 10 rad per sec^2 clockwise.

12-2.9 A pulley has a constant angular acceleration of 3 rad per sec^2. When the angular velocity is 2 rad per sec, the total acceleration of a point on the rim of the pulley is 3.0 m/s^2. Compute the diameter of the pulley?

12-2.10 The step pulleys shown in Fig. P-12-2.10 are connected by a crossed belt. If the angular acceleration of *C* is 2 rad per sec^2, what time is required for *A* to travel 20 m from rest? Through what distance will *D* move while *A* moves 32 m?

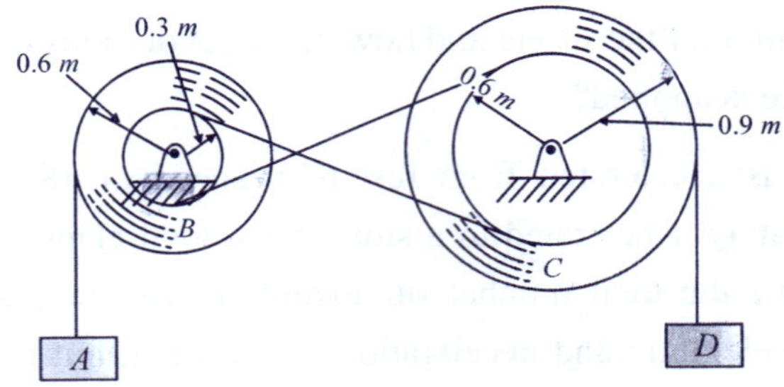

Figure P.12-2.10

12-2.11 Repeat Prob. 12-2.10 if the radii of pulley *B* are changed to 0.76 m. and 0.46 m.

12-2.12 At the position shown in Fig. P-12-2.12, the end *A* of the rod has a rightward component of velocity of 0.6 m/s and an upward component of acceleration of 1.2 m/s^2. Determine the angular acceleration of the rod at this position.

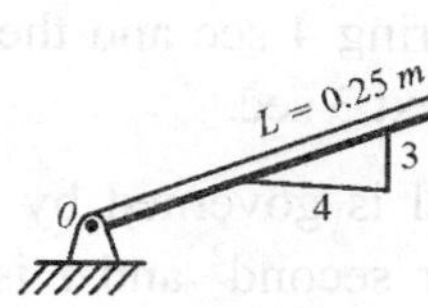

Figure P.12-2.12

12-2.13 The rod *BO* in Fig. P-12-2.13 rotates in a vertical plane about a horizontal axis at *O*. At the given position, end *B* has a downward vertical component of velocity of 2 m/s and also a downward vertical component of acceleration of 0.23 m/s^2. Compute the angular acceleration of the rod *BO* and the total acceleration of point *A*.

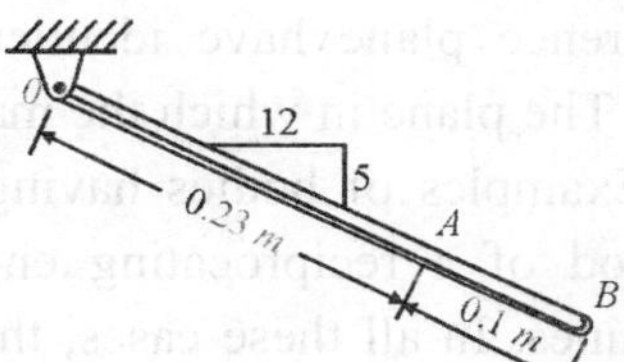

Figure P.12-2.13

12-2.14 The rotation of a flywheel is governed by the equation $\omega = 4\sqrt{t}$; ω is in radians per second and *t* is in seconds. If θ = 2 rad when *t* = 1 sec, determine the values of θ and α at the instant when *t* = 3 sec.

12-2.15 The angular acceleration of a flywheel is given by $\alpha = 8 - t$ where α is in radians per second^2 and *t* is in seconds. If the angular velocity of the flywheel is 42 rad per sec at the end of 6 see, determine its initial angular velocity and the number of revolutions made during the 6 sec.

12.2.16 The angular acceleration of a flywheel increases uniformly from 3 rad per sec^2 to 9 rad per sec^2 in an interval of 6 sec. If its angular velocity was 12 rad per sec at the start of the interval, compute the number of revolutions made during the 6 sec.

12-2.17 Determine the number of revolutions through which a pulley will rotate from rest if its angular acceleration is increased uniformly from zero to 12 rad per sec^2 during 4 sec and then uniformly decreased to 4 rad per sec^2 during the next 3 sec.

12-2.18 The rotation of a flywheel is governed by the relation $\alpha = 10t - t^2$ where α is in radians per second^2 and t is in seconds. How many revolutions will the flywheel make, starting from rest, before it momentarily stops prior to reversing its direction? Solve both graphically and analytically.

12-3 DEFINITION AND ANALYSIS OF PLANE MOTION

Plane motion is that motion of a rigid body in which all particles in the body remain at a constant distance from a fixed reference plane. Consequently, all particles move in parallel planes and all particles on the same straight line perpendicular to the reference plane have identical values of displacement velocity, and acceleration. The plane in which the mass center moves is defined as the *plane of motion*. Examples of bodies having plane motion are rolling wheels, the connecting rod of a reciprocating engine, or the links joining rotating elements of machines. In all these cases, the identifying characteristic of plane motion is that a reference line connecting any two points in the same plane of motion simultaneously undergoes an angular displacement and a linear displacement. Actually, a plane motion is a simultaneous combination of the motions of translation and rotation. The physical meaning of this characteristic will be amplified shortly.

We shall discuss three methods of relating the displacement, velocity, and acceleration of any two points in a body having plane motion. These methods are differing ways of solving the same problem; sometimes one or another will be simpler to apply. One method involves scalar calculus; another, vector calculus; and the third is their geometric application. An understanding of each method will reinforce comprehension of the others.

The most direct approach (but not necessarily the simplest to apply) is by means of scalar calculus. In Fig. 12-3.1 we represent a rigid body having plane motion by its projection upon the parallel fixed XY reference plane. Here A and B are the projections upon this plane of any two points in the body, r is the constant distance separating them, and θ is the angular coordinate of r measured from the X direction . Evidently the coordinates of B and A are related by

$$x_B = x_A + r\cos\theta \qquad y_B = y_A + r\sin\theta \tag{a}$$

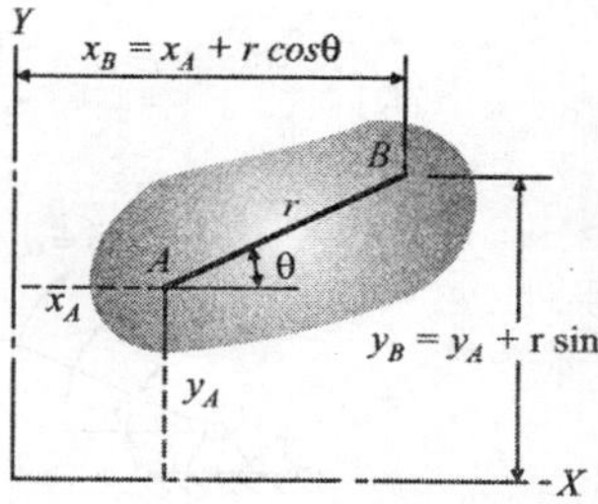

Figure 12-3.1

Assuming that these coordinates, including θ but not r (which is a constant length), vary with time, we may find the relations between the components of velocity and acceleration by successive differentiation of Eq. (*a*). Thus, we obtain

$$\dot{x}_B = \dot{x}_A - r\dot{\theta}\sin\theta \qquad \dot{y}_B = \dot{y}_A - r\dot{\theta}\cos\theta \qquad (b)$$

$$\ddot{x}_B = \ddot{x}_A - r\ddot{\theta}\sin\theta - r\dot{\theta}^2\cos\theta \qquad \ddot{y}_B = \ddot{y}_A + r\ddot{\theta}\cos\theta - r\dot{\theta}^2\sin\theta \qquad (c)$$

By combining these scalar components in the order shown in parts (*a*) and (*b*) of Fig. 12-3.2, we can visualize the geometric relations among the velocity and acceleration vectors of B and A. From part (*a*), we see that $V_{B/A}$ is the difference between v_B and v_A. Its magnitude is

$$v_{B/A} = \sqrt{(r\dot{\theta}\sin\theta)^2 + (r\dot{\theta}\cos\theta)^2} = r\dot{\theta}\sqrt{\sin^2\theta + \cos^2\theta} = r\dot{\theta}$$

and its direction with the vertical is

$$\tan\phi = \frac{r\dot{\theta}\sin\theta}{r\dot{\theta}\cos\theta} = \tan\theta$$

whence

$$\phi = \theta$$

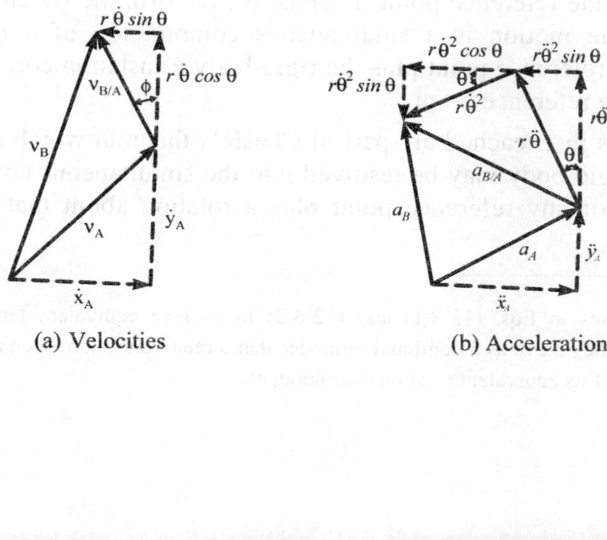

(a) Velocities (b) Accelerations

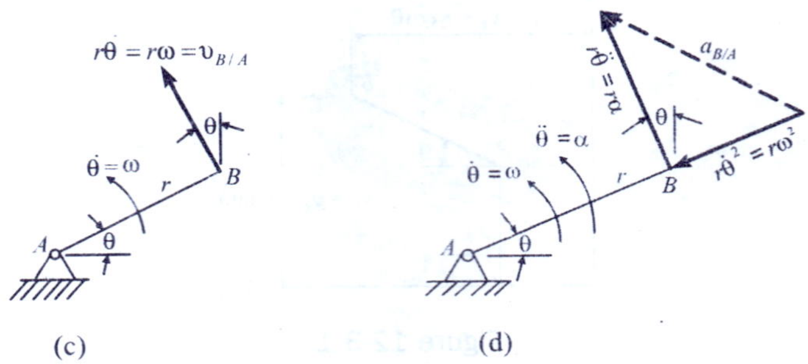

Figure 12-3.2 Interpretation of rectangular components of velocity and acceleration.

A simple physical meaning of these results is obtained by comparing them with the velocity of B as found from a fixed-axis rotation of the body about A as represented in part (c). Since they are identical, we conclude that $v_{B/A}$ is equivalent to the velocity of B as found from a fixed-axis rotation of B about A. For this reason it is convenient to designate $v_{B/A}$ as the relative velocity of B rotating about A.

Similarly, from a comparison of parts (b) and (d), we see that the difference $a_{B/A}$ between the accelerations a_B and a_A is composed of terms which denote the normal and tangential components of acceleration of B rotating about A.

This discussion may be summarized by the equations[2]

$$v_B = v_A \nrightarrow (v_{B/A} = r\omega) \tag{12-3.1}$$

$$a_B = a_A \nrightarrow (a_{B/A} = r\omega^2 \nrightarrow r\alpha) \tag{12-3.2}$$

which define the instantaneous velocity and acceleration of B in terms of the velocity and acceleration of any reference point A plus vectorially the velocity and acceleration of a body point B assumed simultaneously rotating about a fixed axis through the reference point A. Thus, we confirm the statement made earlier that a plane motion is a simultaneous combination of a rigid-body rotation about any reference point plus the rigid-body translation corresponding to the motion of the reference point.

The conclusions just reached are part of Chasle's theorem which states that any motion of a rigid body may be resolved into the simultaneous combination of the translation of any reference point plus a rotation about that reference point.

---

[2]The use of parentheses in Eqs. (12-3.1) and (12-3.2) to enclose equivalent terms may be confusing at first, but they serve as a continual reminder that a relative motion, such as $a_{B/A}$, may be expressed in terms of its equivalent rotational components.

Using a geometric approach to the application of Chasle's theorem, consider the rigid body represented in Fig, 12-3.3 by its projection upon a fixed reference plane coinciding with the plane of motion. Assume that the body changes its position from the full to the dashed outline during the time Δt. Assume the motion of two points A and B in the body to be as shown and to have the respective movements s_A and s_B.

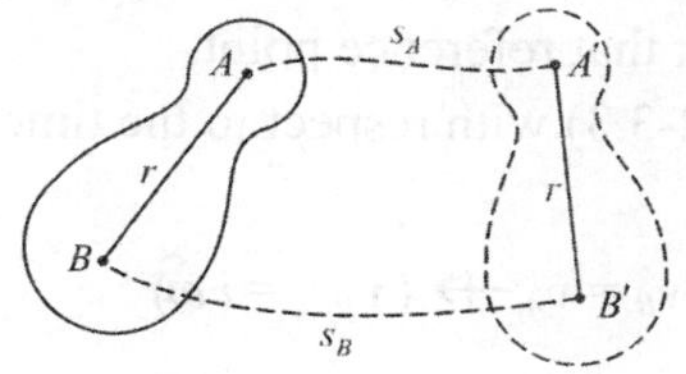

Figure 12-3.3 Plane motion.

To determine a relation between the movements of points A and B, assume that the body is rotated through an angle θ about axis passing through A until line AB is parallel to line $A'B'$ (Fig.12-3.4). The body may now be given the curvilinear translation defined by the motion of A, thereby causing the body to coincide with its final position. The motion of B thereby is the geometric sum of $s_{B/A}$ and s_A and may be expressed by the vector equation

$$s_B = s_A \;+\!\!\!\!\!\!\rightarrow\; (s_{B/A} = r\theta) \qquad (12\text{-}3.3)$$

The symbol $s_{B/A}$ is to read as the motion of B assumed rotating about A; it is equal to rθ where r is the distance between A and B.

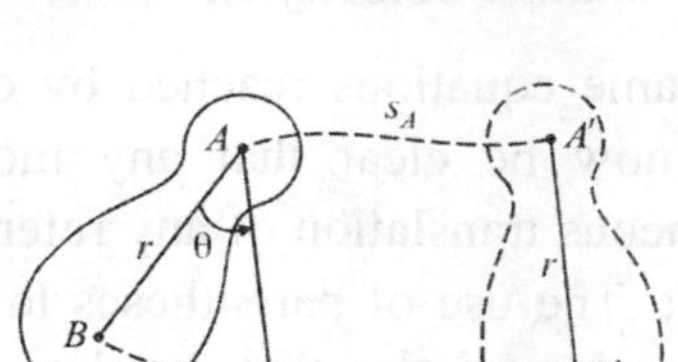

Figure 12-3.4 Plane motion resolved into a combined rotation about A and translation of A.

In a finite time interval, the actual movement of point B will only approximately equal $s_A \;+\!\!\!\!\!\!\rightarrow\; s_{B/A}$ since this equation assumes that the translation and rotation of the body take place independently whereas they actually occur simultaneously. For purposes of analysis, however, there is the valuable concept that plane motion is equivalent to a rigid-body rotation about *any* point plus the rigid-body translation of that point. *This assumption is exact when the time interval between any two successive positions approaches zero as a limit.* Then the actual motion is equivalent to successive infinitesimal displacements from one instantaneous position to the next. Each such infinitesimal displacement consists of the vector sum of the translation of an arbitrarily selected reference point plus a rotation about that reference point.

Differentiating Eq. (12-3.3) with respect to the time (i.e., assuming $\Delta t \to 0$), we have

$$v_B = v_A \nrightarrow (v_{B/A} = r\omega) \tag{12-3.1}$$

The symbol $v_{B/A}$ is read as the velocity of B assumed to be rotating about A. It is equal to $r\omega$ where the angular velocity ω is defined as $d\theta/dt = \dot{\theta}$ i.e., the time rate of change of the body rotation θ and r is the distance between A and B. In like manner, differentiating the velocity relation with respect to time yields

$$a_B = a_A \nrightarrow (a_{B/A} = r\omega^2 \nrightarrow r\alpha) \tag{12-3.2}$$

where the symbol $a_{B/A}$ is to be read as the acceleration of B assumed to be rotating about A. This acceleration may in turn be resolved into a normal component $r\omega^2$ directed from B toward the center of rotation A and a tangential component $r\alpha$ directed perpendicular to r in the sense that α moves the tip of r. (See Section 12-2). The angular acceleration α is defined by dω/dt, i.e., the time rate of change of the body angular velocity ω.

Thus we obtain the same equations reached by our first purely analytical development. It should now be clear that any motion of a rigid body is equivalent to the simultaneous translation of any reference point plus a rotation about that reference point. The use of parentheses to enclose equivalent terms should serve as a continual reminder that a relative motion, such as $a_{B/A}$, should be expressed in terms of its equivalent rotational components. It should be noted that the equations we have developed will also determine the relative motion between any two points whose absolute motion is known.

Finally, it is very important to understand that the rotational components of plane motion are properties of the body, and are independent of the choice of a reference point. Our discussion of rotation proved that θ, ω and α represent the angular properties of any line in the rotating body. Similarly, in plane motion,

θ, ω, and α are defined in terms of the angular motion of any line in the body. It is easy to demonstrate that such a reference line has the same angular motion regardless of the choice of any arbitrarily selected reference point. For example, if the body in Fig. 12-3.5 were rotating about B instead of A as previously assumed in Fig.12-3.4, its motion would be equivalent to a rotation about B, plus the translation of B. The reference line AB would then rotate about B to the position BA'' parallel to the final position $B'A'$, after which the body would be translated along the path $BB''B'$ to its final position. Obviously the angle θ defined by ABA'' is counterclockwise about B and equal to the angle BAB'' which defines the same counterclockwise rotation θ about A as a reference point. Hence, in plane motion, the values of θ, ω, and α refer to the body and have the same values and directional sense regardless of the reference point.

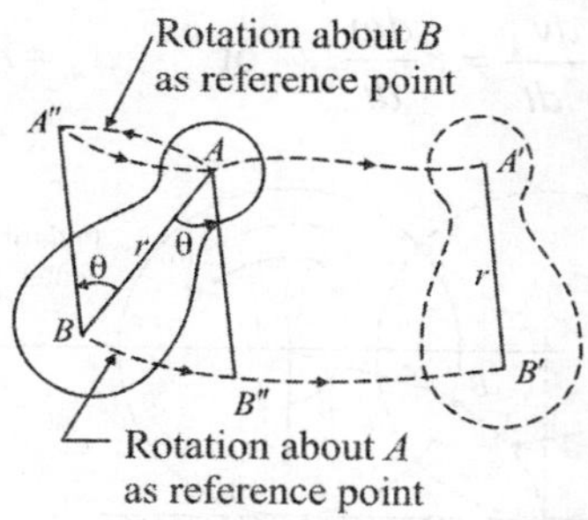

Figure 12-3.5 Angular motion is independent of reference point.

12-4 APPLICATION OF KINEMATIC EQUATIONS

As we shall see when we study Chapter 13 which relates applied forces to the motion of rigid bodies, it is necessary to determine the relation between the angular acceleration of a body and the linear acceleration of its center of gravity. This objective is achieved through the proper application of the kinematic equations. Since the acceleration equation involves the angular velocity ω of the body, it is usually necessary to first determine by a velocity analysis.

In the majority of cases involving plane motion, the linear motion $A - A'$ in Fig. 12-3.3 of any point A in the body can be related to the angular motion only by careful application of the kinematic equations. However, in the special case of a disk or sphere rolling freely, (i.e., without slipping at the point of contact) along a plane as in Fig. 12-4.1, we can obtain a direct relation between the linear motion of its geometric center and its angular motion in the position shown in Fig. 12-4.1, the disk has rolled to the right without slipping from a position with B originally coincident with B' i.e., from the dashed outline to the solid outline. This motion is equivalent to translating the disk through the distance $s_A = A'A$ and then rotating it about A through the angle θ. If there is no slipping the distance $B'C$ equals the arc length $BC = r\theta$. But $B'C$ is equal to the linear motion of the geometric center A. Hence, we have

$$s_A = r\theta \tag{12-4.1}$$

and two successive time differentiations yield the following results which are true at any particular instant:

$$\frac{ds_A}{dt} = r\frac{d\theta}{dt} \quad \text{or} \quad v_A = r\omega \tag{12-4.2}$$

$$\frac{dv_A}{dt} = r\frac{d\omega}{dt} \quad \text{or} \quad a_A = r\alpha \tag{12-4.3}$$

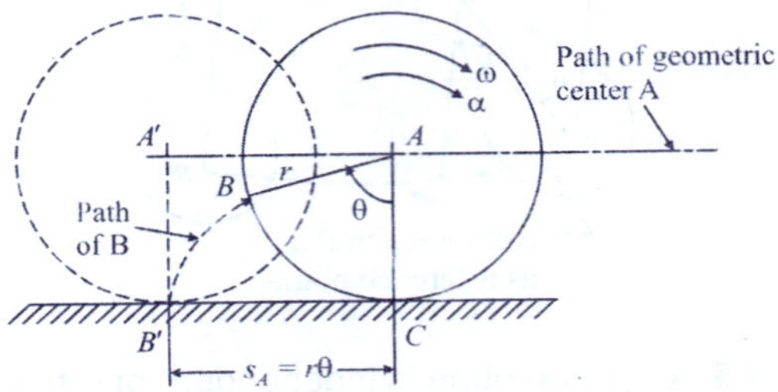

Figure 12.4.1 Free-rolling disk or sphere.

Remember that these relations between the linear motion of the geometric center of a disk or sphere and its angular motion are correct only when the disk or sphere rolls without slipping. If slipping occurs, these relations are not valid. Note also that positive ω and α have the same direction as positive θ, i.e., clockwise for a rightward motion of the disk and counterclockwise for a leftward motion. These directions are illustrated in Fig. 12-4.1. Observe also that when the plane motion of the disk is resolved into a series of simultaneous infinitesimal translations and rotations, the actual path of B will be a cycloid.

ILLUSTRATIVE PROBLEMS

12-4.1 The wheel of 0.9 m radius shown in Fig. 12-4.2 rolls freely to the right. At the given position, $\omega = 3$ rad/sec and $\alpha = 5$ rad/sec^2, both clockwise. Compute the velocity and acceleration of point B which is 0.6 m from the center A of the wheel.

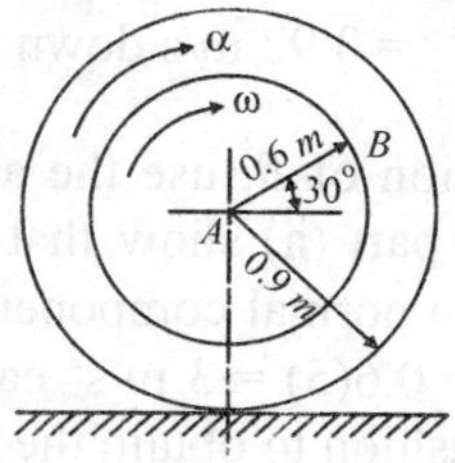

Figure 12-4.2

Solution

Applying the relations just developed for a free-rolling wheel, the velocity and acceleration of the geometric center are respectively $v_A = r\omega = 0.9 \times (3) = 2.7$ m/s and $a_A = r\alpha = (0.9)(5) = 4.5$ m/s^2, both right ward.

Since the motion of A is known, we use it as a reference point to resolve the plane motion of the wheel into a simultaneous rotation about A plus the translation of A. Then the velocity of B is determined as shown in part (a) of Fig. 12-4.3. The sketches below the velocity equation illustrate the magnitude and sense of each known term. Note that A is not actually stationary but is shown so only to emphasize the relative rotation of B about A.

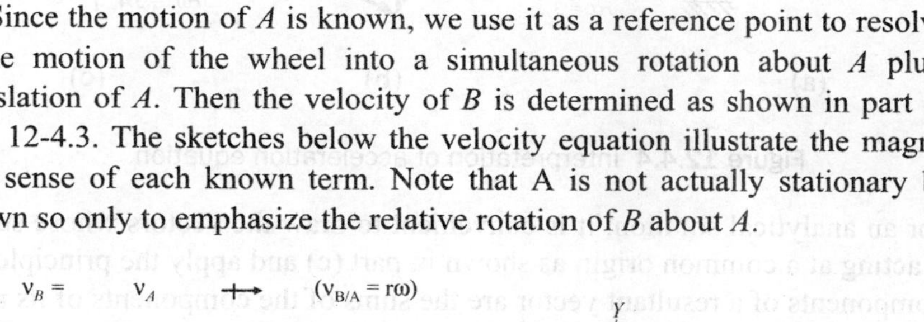

Figure 12.4.3 Interpretation of velocity equation.

If the terms in this vector equation are now plotted to scale as in part (b), the values of v_B and θ_x are scaled off as 4 m/s and 23.4°, respectively. Analytically, the vector polygon can be used to compute these results either by the cosine law or by projection upon conveniently selected reference axes. Using this latter method, we find the components of v_B are

$$v_{B_x} = \Sigma v_x = 2.7 + 1.8 \cos 60° = 3.6 \text{ m/s}$$

$$v_{B_y} = \Sigma v_y = -1.8 \sin 60° = -1.56 \text{ m/s}$$

whence

$$v_B = \sqrt{(v_{Bx})^2 + (v_{B_y})^2} = 3.92 \text{ m/s down to the right at } \theta_x = 23.4^\circ$$

To determine the acceleration of B, use the acceleration equation shown in Fig. 12-4.4a. The sketches in part (a) show that the relative acceleration of B rotating about A consists of the normal component $r\omega^2 = 0.6(3)^2 = 5.4$ m/s^2 and the tangential component $r\alpha = 0.6(5) = 3$ m/s^2 each directed as shown. Plot the vector equation in tip-to-tail fashion to obtain the acceleration polygon shown in part (b) from which we scale the values $a_B = 5.4$ m/s^2 and $\theta_x = 76°$

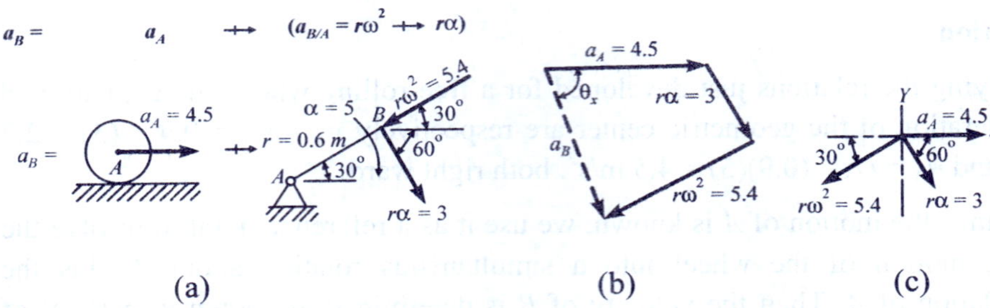

Figure 12.4.4 Interpretation of acceleration equation.

For an analytical solution, it is convenient to draw the vectors whose sum is a_B as acting at a common origin as shown in part (c) and apply the principle that the components of a resultant vector are the sums of the components of its parts. Then by taking summations with respect to the reference axes shown, we find the components of a_B are

$$\overset{+}{\rightarrow} a_{B_x} = \Sigma a_x = 4.5 + 3 \cos 60 - 5.4 \cos 30 = 1.324 \text{ m/s}^2$$

$$+\uparrow a_{B_y} = \Sigma a_y = -3 \sin 60° - 5.4 \sin 30° = -5.3 \text{ m/s}^2$$

whence

$$a_B = \sqrt{(a_{Bx})^2 + (a_{B_y})^2}$$

$$= 5.46 \text{ m/s}^2 \text{ down to the right at } \theta_x = 76° \quad \textbf{\textit{Ans.}}$$

12-4.2 A slender rod 0.25 m long moves with its ends in contact with a horizontal floor and an inclined surface as shown in Fig. 12-4.5. At the given position, end A has a leftward velocity of 0.9 m/s and a rightward acceleration of 0.1 m/s^2. Determine the velocity and acceleration of end B.

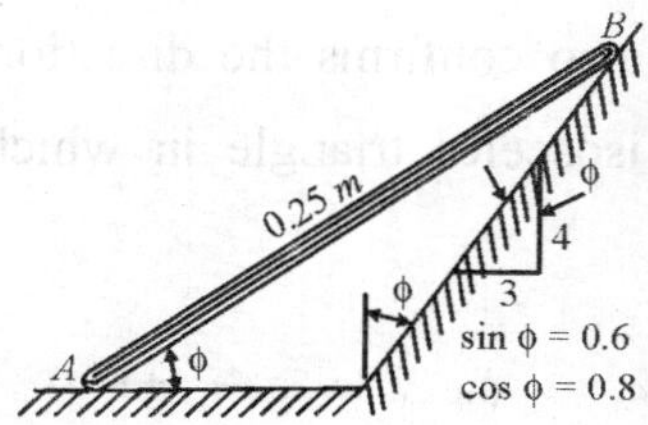

Figure 12.4.5

Solution

Since the motion of A is known, resolve the plane motion of the rod into a rotation about A plus the translation of A.

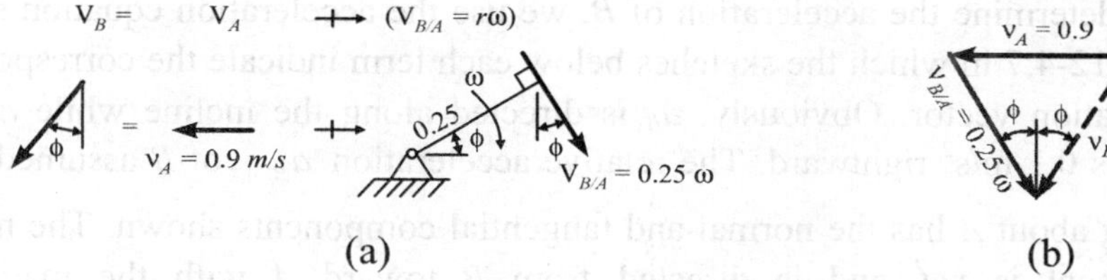

Figure 12-4.6 Interpretation of velocity equation.

To determine the velocity of B, use the vector equation shown in Fig. 12-4.6 in which the sketches below each velocity term specify its corresponding direction. The linear velocity of B is evidently constrained to be along the incline while the relative velocity of B rotating about A is perpendicular to the rod AB in the assumed clockwise sense of ω. Combining these terms gives the velocity polygon shown in part (b) where the tip-to-tail sum of v_A and $v_{B/A}$ equals the resultant v_B extending from the tail of v_A to the tip of $v_{B/A}$. Since the components of the resultant velocity equal the summations of the components of its parts, we have with respect to horizontal and vertical axes

$$[\overset{+}{\rightarrow} v_{B_h} = \Sigma v_h] \qquad v_B \sin\phi = 0.9 - 0.25\omega \sin\phi$$

$$[+\downarrow v_{B_v} = \Sigma v_v] \qquad v_B \cos\phi = 0.25\omega \cos\phi$$

from which, on substituting $\sin\phi = 0.6$ and $\cos\phi = 0.8$, we obtain

$\omega = 3$ rad/sec (clockwise) and $v_B = 0.75$ m/s **Ans.**

The positive value of ω confirms the assumption in Fig. 12-4.6a that B is rotating in a clockwise sense about A.

The velocity polygon also confirms the directions of v_B and $v_{B/A}$, and in addition is evidently an isosceles triangle in which $v_B = v_{B/A} = 0.25\omega$ by inspection.

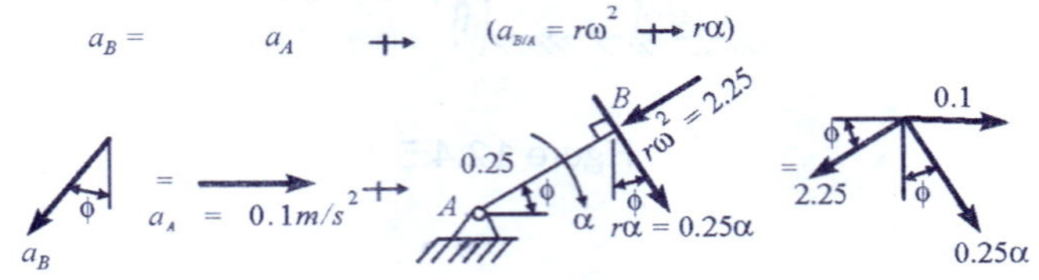

Figure 12-4.7 Interpretation of acceleration equation.

To determine the acceleration of B, we use the acceleration equation shown in Fig. 12-4.7 in which the sketches below each term indicate the corresponding acceleration vector. Obviously, a_B is directed along the incline while a_A was given as 0.1 m/s^2 rightward. The relative acceleration $a_{B/A}$ of B assumed to be rotating about A has the normal and tangential components shown. The normal component is $r\omega^2$ and is directed from B toward A with the magnitude $0.25(3)^2 = 2.25$ m/s^2. Observe that in order to evaluate $r\omega^2$ it was first necessary to determine ω from a velocity analysis. The tangential component $r\alpha$ is perpendicular to AB as shown with α assumed to be clockwise. Finally, the right-hand terms of the acceleration equation are shown acting at a common origin so that the components of the resultant acceleration a_B may be easily correlated with the summations of the components of its parts. Equating horizontal and vertical summations, we obtain

$$[\overset{+}{\leftarrow} a_{B_h} = \Sigma a_h] \qquad a_B \sin\phi = 2.25\cos\phi - 0.1 - 0.25\alpha\sin\phi$$

$$[+\downarrow a_{B_v} = \Sigma a_v] \qquad a_B \cos\phi = 2.25\sin\phi + 0.25\alpha\cos\phi$$

from which, on substituting $\sin\phi = 0.6$ and $\cos\phi = 0.8$, we obtain

$$\alpha = 2.3 \text{ rad/sec}^2 \text{ (clockwise) and } a_B = 2.26 \text{ m/s}^2 \qquad \textbf{\textit{Ans.}}$$

The positive values of α and a_B confirm our assumptions of their directions.

Let us consider now two variations of the problem we have completed. The first variation is that of a slider-crank chain as shown in Fig.12-4.8 in which the motion of B is controlled by the crank BC instead of the inclined plane. For ease of comparison, assume the 0.1 m crank BC to have the angular position shown.

Then the instantaneous velocity of B, acting perpendicular to BC, would be equivalent to the previous motion of sliding down the inclined plane. Hence, a velocity analysis of AB would result in $v_B = 0.75$ m/s as before, but the angular velocity of crank BC would be

$$[v = r\omega] \qquad 0.75 = 0.1\omega_{BC} \qquad \omega_{BC} = 7.5 \text{ rad/sec}$$

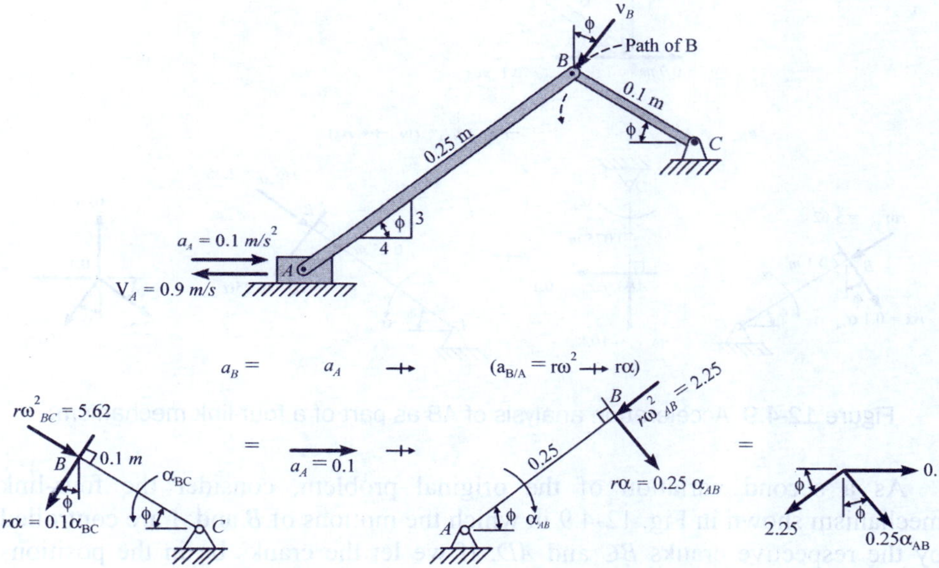

Figure 12-4.8 Effect on a_B caused by replacing incline by crank BC.

The absolute acceleration of B, instead of being along the incline, now consists of a normal component $r\omega_{BC}^2 = 0.1 \times 7.5^2 = 5.62$ m/sec^2 as well as a tangential component $r\alpha = 0.1\ \alpha_{BC}$ directed perpendicular to the crank BC. These components of a_B are shown below the term a_B in the acceleration equation shown in Fig. 12-4.8. The components of the acceleration equation become

$$[\overset{+}{\leftarrow} a_{B_h} = \Sigma a_h]$$

$$0.1\alpha_{BC} \sin\phi - 5.62 \cos\phi = 2.25 \cos\phi - 0.1 - 0.25\ \alpha_{AB} \sin\phi$$

$$[+\downarrow\ a_{B_v} = \Sigma a_v]$$

$$0.1\alpha_{BC} \cos\phi + 5.62 \sin\phi = 2.25 \sin\phi + 0.25\alpha_{AB} \cos\phi$$

from which, on substituting $\sin\phi = 0.6$ and $\cos\phi = 0.8$, we obtain

$\alpha_{AB} = 26$ rad/sec^2 and $\alpha_{BC} = 40$ rad/sec^2 ***Ans.***

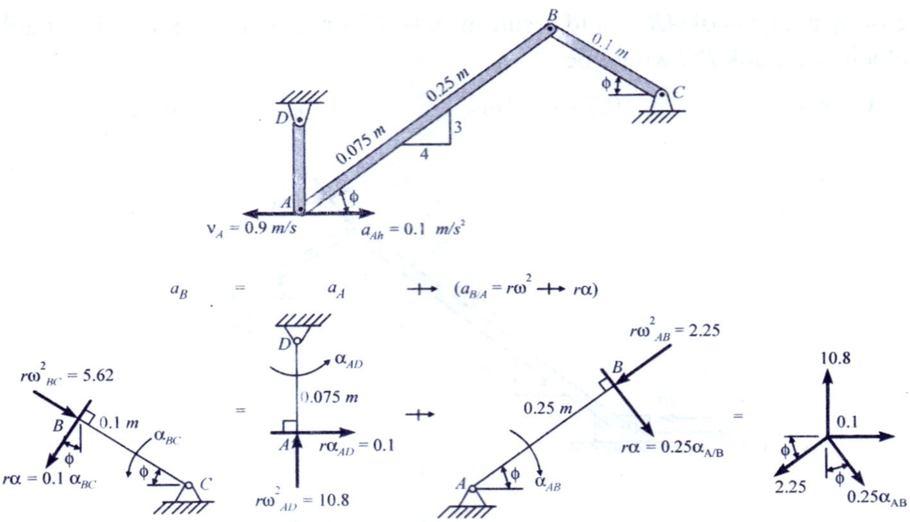

Figure 12-4.9 Acceleration analysis of *AB* as part of a four-link mechanism.

As a second variation of the original problem, consider the four-link mechanism shown in Fig. 12-4.9 in which the motions of *B* and *A* are controlled by the respective cranks *BC* and *AD*. If we let the cranks be in the positions shown and $v_A = 0.9$ m/s leftward, the velocity analysis of rod *AB* is identical with the original problem resulting in $\omega_{AB} = 3$ rad/sec and $v_B = 0.75$ m/s. When we apply the acceleration equation in this variation where both *B* and *A* now travel in circular paths, we must include the normal and tangential components of acceleration of both points *B* and *A*. These components are shown below the various terms of the acceleration equation in Fig. 12-4.9. Note that $r\omega_{AB}^2$ and $r\omega_{BC}^2$ for *B* are the same as in the slider-crank variation while the normal component of *A* rotating about *D* is $r\omega_{AD}^2 = \dfrac{v^2}{r} = \dfrac{(0.9)^2}{0.075} = 10.8\,\text{m/s}^2$.

Now the components of the acceleration equation are

$[\overset{+}{\leftarrow} a_{B_h} = \Sigma a_h]$

$$0.1\alpha_{BC} \sin\phi - 5.62 \cos\phi = 2.25 \cos\phi - 0.1 - 0.25\, a_{AB} \sin\phi$$

$[+\!\downarrow a_{B_v} = \Sigma a_v]$

$$0.1\, \alpha_{BC} \cos\phi + 5.62 \sin\phi = 2.25 \sin\phi - 10.8 + 0.25\alpha_{AB} \cos\phi$$

On substituting sin $\phi = 0.6$ and cos $\phi = 0.8$ in these equations, we find

$$\alpha_{AB} = 53 \text{ rad/sec}^2 \curvearrowright$$

and $$\alpha_{BC} = -27 \text{ rad/sec}^2 = 27 \text{ rad/sec}^2 \curvearrowright$$ ***Ans.***

The minus sign for α_{BC} indicates that it acts oppositely to the counterclockwise sense originally assumed.

Notice that the velocity analysis of AB produced the same results in all three variations of the original problem since we were careful to ensure that the velocity directions of A and B were unchanged. Similarly, except for the addition of the cranks which introduced normal components of acceleration in a_B and a_A, the acceleration analysis of AB in all three variations followed the same general procedure.

12-4.3 A link of length L is constrained to move as shown in Fig. 12-4.10. Its mass center G is at a distance b from end A. Use the scalar calculus approach to express the horizontal and vertical components of the acceleration of G in terms of the accelerations a_A and a_B of the ends of the link.

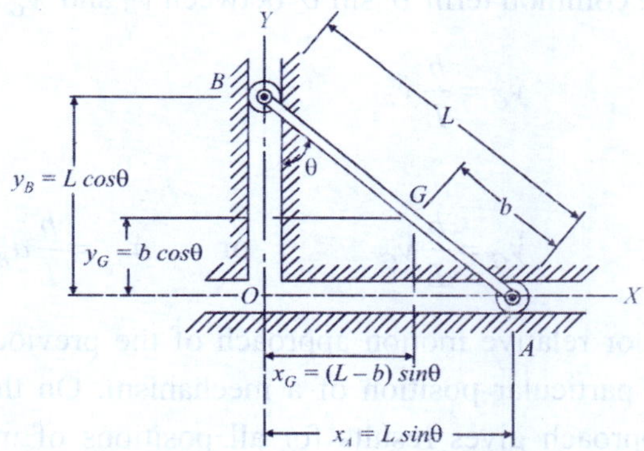

Figure 12-4.10 Analysis of motion by scalar calculus.

Solution

We first take successive time differentiations of the x coordinates of A and G.

x components of A	x components of G
$x_A = L \sin\theta$	$x_G = (L-b)\sin\theta$
$\dot{x}_A = L\dot{\theta}\cos\theta$	$\dot{x}_G = (L-b)\dot{\theta}\cos\theta$

By eliminating the common term $\dot{\theta}\cos\theta$ between $\dot{x}_A$ and $\dot{x}_G$, we obtain

$$\dot{x}_G = \left(\frac{L-b}{L}\right)\dot{x}_A$$

whence another time differentiation gives

$$\ddot{x}_G = \left(\frac{L-b}{L}\right)\ddot{x}_A \quad \text{or} \quad a_{G_h} = \left(\frac{L-b}{L}\right)a_A \qquad \textbf{\textit{Ans.}}$$

Similarly, successive time differentiations of the y coordinates of B and G yield.

y components of B	y components of G
$y_B = L\sin\theta$	$y_G = b\cos\theta$
$\dot{y}_B = -L\dot{\theta}\sin\theta$	$\dot{y}_G = -b\dot{\theta}\sin\theta$

Eliminating the common term $\dot{\theta}\sin\theta$ between y_B and $\dot{y}_G$ gives

$$\dot{y}_G = \frac{b}{L}\dot{y}_B$$

whence

$$\ddot{y}_G = \frac{b}{L}\ddot{y}_G \qquad \text{or} \qquad a_{G_v} = \frac{b}{L}a_B \qquad \textbf{\textit{Ans.}}$$

The geometric or relative motion approach of the previous problems gives results only for a particular position of a mechanism. On the other hand, the scalar calculus approach gives results for all positions of mechanisms which have repetitive cycles such as the slider-crank and four-link mechanisms, Such general results can then be adapted to compute analysis of a mechanism,

PROBLEMS

12-4.4 The compound wheel shown in Fig. P-12-4.4 rolls without slipping. At the given position, the velocity of A is 2 m/s and the acceleration of A is 6 m/s^2, both directed to the right. Compute the acceleration of points B and C.

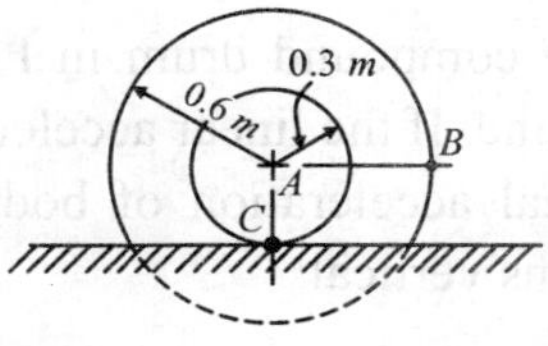

Figure P-12-4.4

12.4.5 The force P causes the unbalanced wheel shown in Fig. P.12-4.5 to have at the given instant $\omega = 3$ rad per sec and $\alpha = 6$ rad per sec^2 both clockwise. If the wheel does not slip, determine the acceleration of the mass center G.

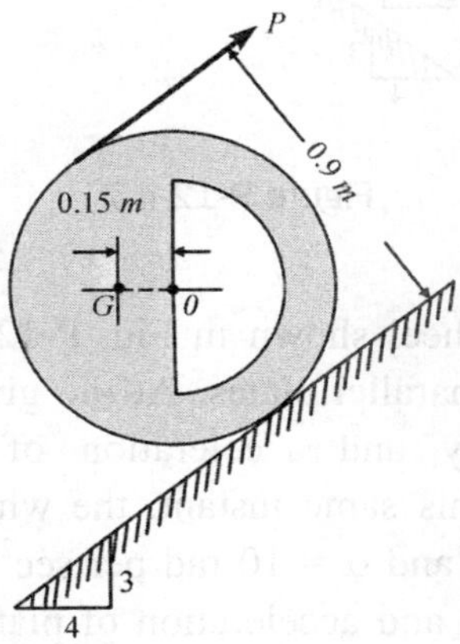

Figure P-12-4.5

12-4.6 As shown in Fig. P-12-4.6, a cord being unwrapped from the rim of a solid cylinder A of radius r passes over a fixed pulley P and back to the center of the cylinder. Determine the relation between the angular acceleration of the pulley and the cylinder.

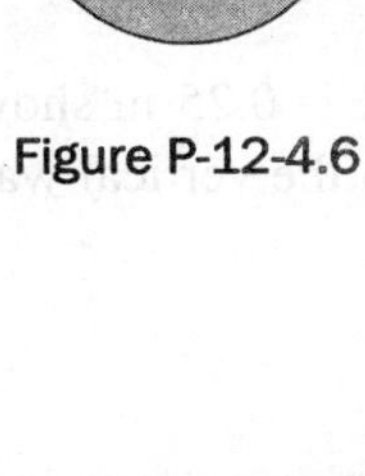

Figure P-12-4.6

12-4.7 Body B causes the compound drum in Fig. P-12-4.7 to roll without slipping up the incline. If the linear acceleration of B is 0.6 m/s^2 down, compute the vertical acceleration of body A. Assume that the cord supporting A remains vertical.

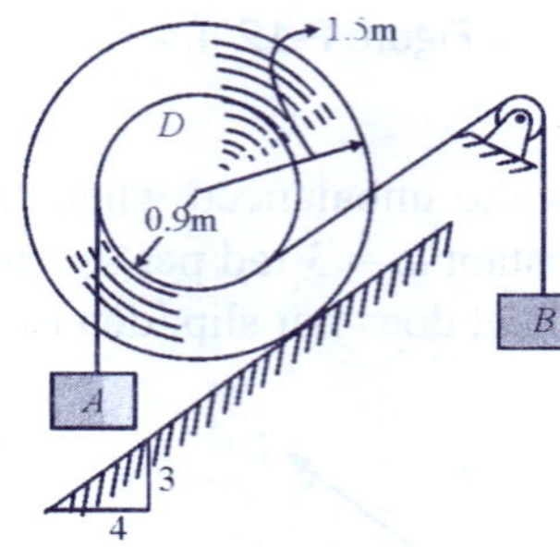

Figure P-12-4.7

12-4.8 The compound wheel shown in Fig. P-12-4.8 rolls without slipping between the two parallel plates. At the given position, plate A has a rightward velocity and acceleration of 2.4 m/s and 3.6 m/s^2, respectively. At this same instant, the wheel has absolute values of $\omega = 4$ rad per sec and $\alpha = 10$ rad per sec^2, both clockwise. Compute the linear velocity and acceleration of plate B and of point D on the wheel.

Hint: First find v_G and a_G.

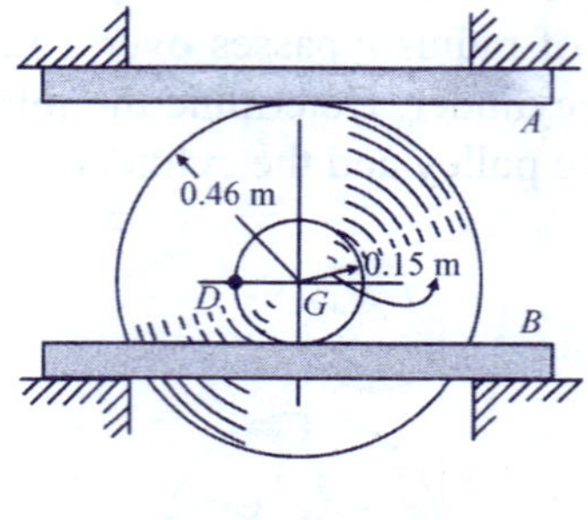

Figure P-12-4.8

12-4.9 The bar AB of length $L = 0.25$ m shown in Fig. P-12-4.9 moves with its ends in contact with the vertical wall and the horizontal floor. Use the method of scalar calculus to find the velocity and acceleration of points A and B if, at the instant $\theta = 60°$, $\omega = 4$ rad per sec and $\alpha = 6$ red per sec^2 both counterclockwise.

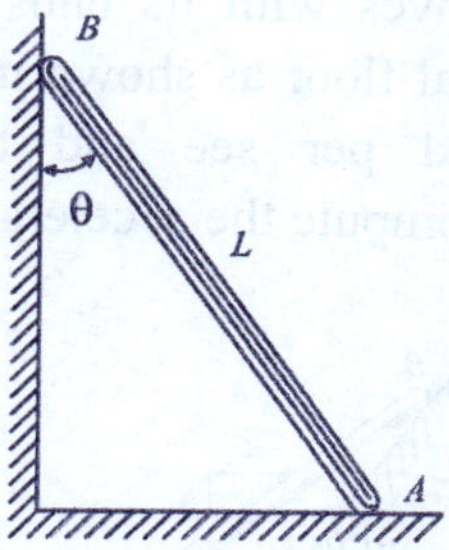

Figure P-12-4.9

12-4.10 Referring to Fig. P-12-4.9 with $L = 0.25$ m, suppose that when $\tan \theta = \dfrac{3}{4}$, the velocity and acceleration of A is 0.8 m/s rightward and 1.0 m/s^2 leftward, respectively. Using scalar calculus, compute the acceleration of B.

12-4.11 In Fig. P-12-4.9, assume that A is moving rightward with a constant velocity v_A. Use scalar calculus to find ω and α in terms of v_A and θ. What are these values expressed in terms of v_A and time t? Assume $t = 0$ when A is vertically below B.

12-4.12 The 2 m rod in Fig. P-12-4.12 moves with its ends in contact with the floor and the inclined surface. At the given instant when $\omega = \sqrt{2}$ rad per sec and $\alpha = 3$ rad per sec^2 both clockwise, determine the velocity and acceleration of point B.

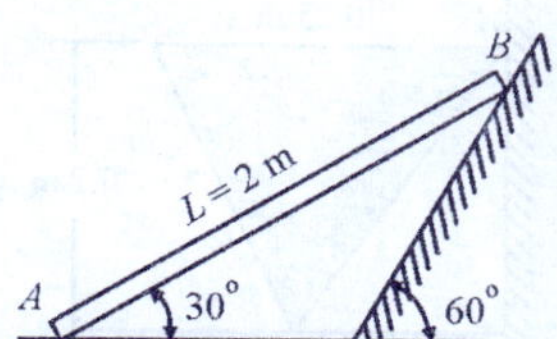

Figure P-12-4.12

12-4.13 In Fig. P-12-4.12, assume that the velocity of A is 5.5 m/s leftward and the acceleration of A is 6 m/s^2 rightward. Compute the velocity and acceleration of point B.

12-4.14 A bar 1 m long moves with its ends in contact with the inclined surface and horizontal floor as shown in Fig. P-12-4.14. At the given instant, $\omega = 2$ rad per see and $\alpha = 3$ rad per sec^2, both counterclockwise. Compute the acceleration of the midpoint M of AB.

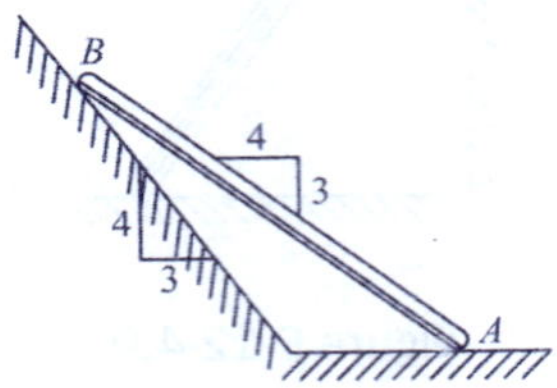

Figure P-12-4.14

12-4.15 In a rigid body having plane motion, let A and B be any two points in the plane of motion and let M be a point midway between A and B. Prove that the acceleration of M is one-half the vector sum of the accelerations of points A and B; i.e., $a_M = \frac{1}{2}(a_A +\!\!\!\!\rightarrow a_B)$. Use this result to check the midpoint acceleration of bar AB in Prob. 12-4.14.

12-4.16 The triangular plate ABC in Fig. P-12-4.16 moves with its vertices A and B in contact with a vertical wall and a horizontal floor. If $\omega = \sqrt{5}$ rad per sec and $\alpha = 3$ rad per sec^2, both counterclockwise at the given position, compute the acceleration of points A, B, and C.

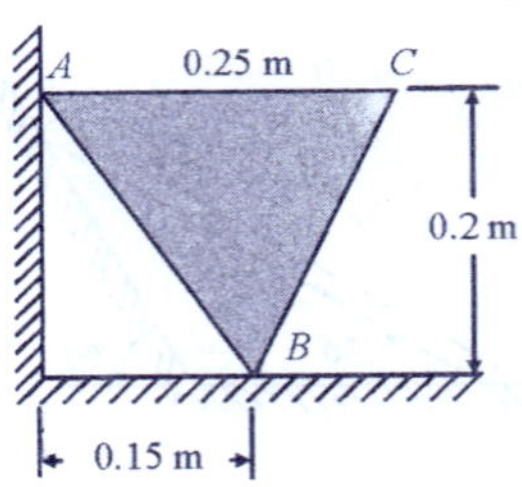

Figure P.12.4.16

12-4.17 A plate *ABC* moves with its ends *A* and *B* on the horizontal and inclined guides shown in Fig. P-12-4.17. At the given position, $\omega = 4$ rad per sec and $\alpha = 5$ rad per sec^2, both clockwise. Find the accelerations of points *A*, *B*, and *C*.

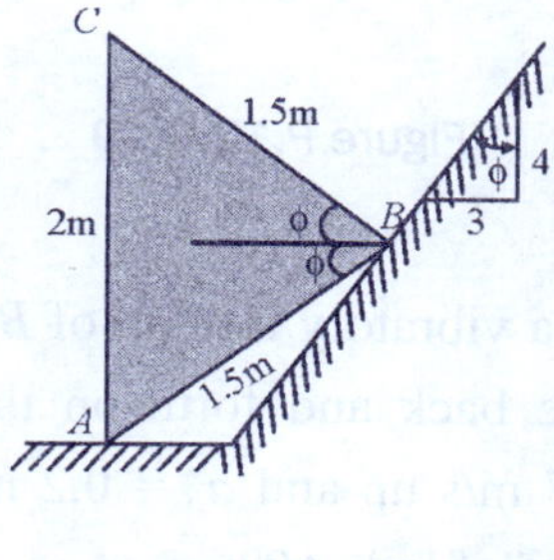

Figure P.12.4.17

12.4.18 The motion of the rigid link *ABC* in Fig. p-12-4.18 is governed by the control rods *AD* and *BE*. At the given instant when *AB* is horizontal, the angular velocity and acceleration of the link is $\omega = \sqrt{10}$ rad per sec and $\alpha = 5$ rad. per sec^2, both clockwise. Determine the acceleration of point *C*.

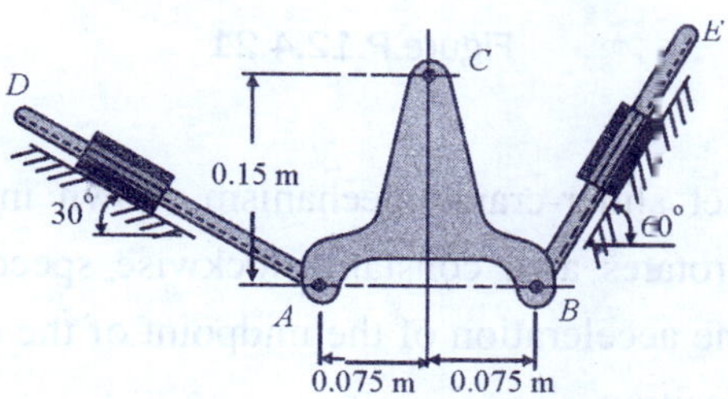

Figure P.12.4.18

12-4.19 The 2 m diameter disk shown in Fig. P-12-4.19 rolls without slipping back and forth over a short distance. A 1.5 m bar *BC* is pinned to the rim of the disk at *B* and its other end *C* drags on the ground. If the center of the disk has a constant rightward velocity $v_A = 3.6$ m/sec at the given position compute v_C and a_C.

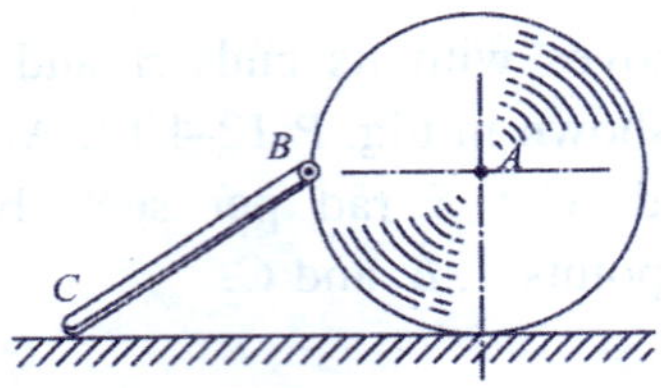

Figure P.12.4.19

12-4.20 In Fig. P-12-4.20, a vibratory motion of B in the vertical guide causes gear A to oscillate back and forth on the fixed rack. At the given position, $v_B = 0.67$ m/s up and $a_B = 0.2$ m/s^2 down. What is then the angular acceleration of link AB?

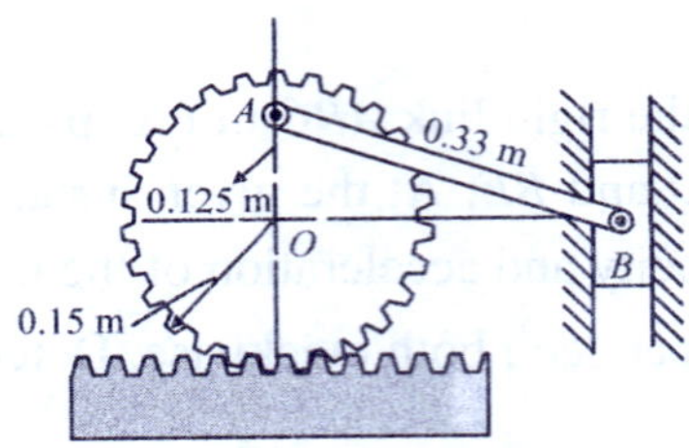

Figure P.12.4.21

12-4.21 In the offset slider-crank mechanism shown in Fig. P-12-4.21, the crank OA rotates at a constant clockwise speed of 10 rad per sec. Compute the acceleration of the midpoint of the connecting rod AB at the given position.

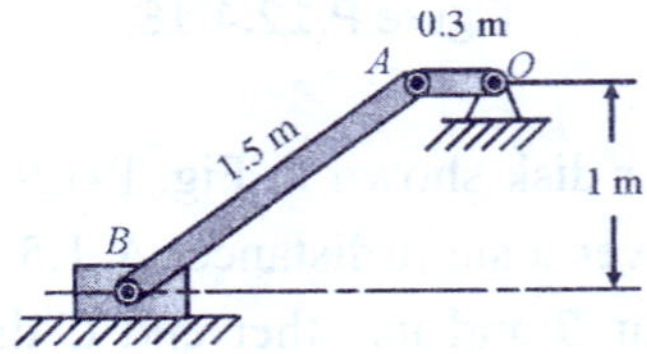

Figure P.12.4.21

12.4.22 When a slider-crank mechanism is in the position shown in Fig. P-12-4.22, the velocity and acceleration of slider C are $v_C = 5$ m/s and $a_C = 0.85$ m/s^2, both vertically downward. What is then the angular acceleration of crank AB?

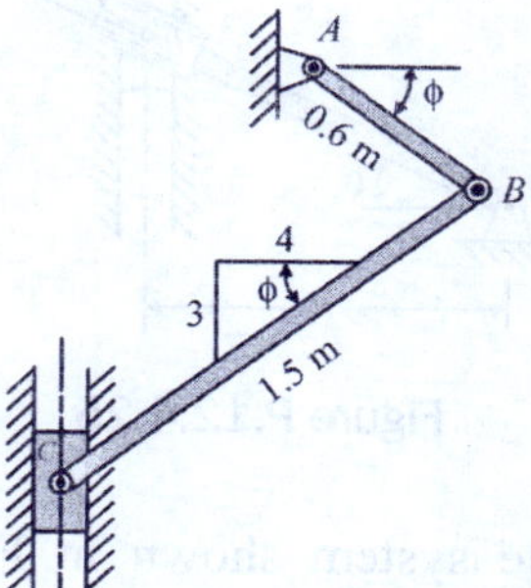

Figure P.12.4.22

12-4.23 At the instant shown in Fig. P-12-4.23, the driving crank AB of the four-link mechanism has a constant clockwise speed of 4 rad per sec. Find the angular acceleration of the link CD.

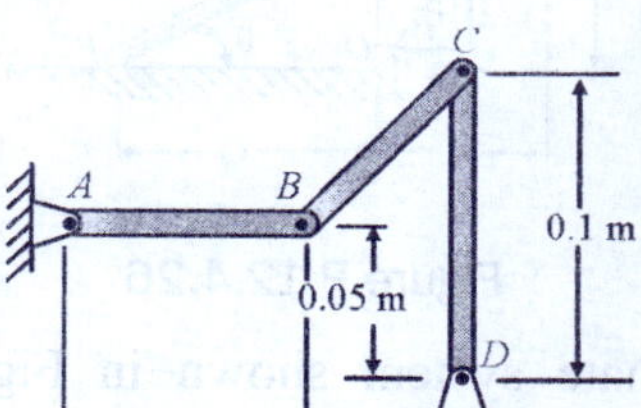

Figure P.12.4.23

12-4.24 In Fig. P-12-4.23, find the angular acceleration of link CD if at the instant shown the driving crank AB has a clockwise angular velocity of 4 rad per sec and a counterclockwise angular acceleration of 2 rad per sec^2.

12-4.25 A slider S is pinned at B to a collar C which is free to slide along AB as shown in Fig. P-12-4.25. If the slider S has a constant upward velocity v, use scalar calculus to find the angular velocity and angular acceleration of rod AB.

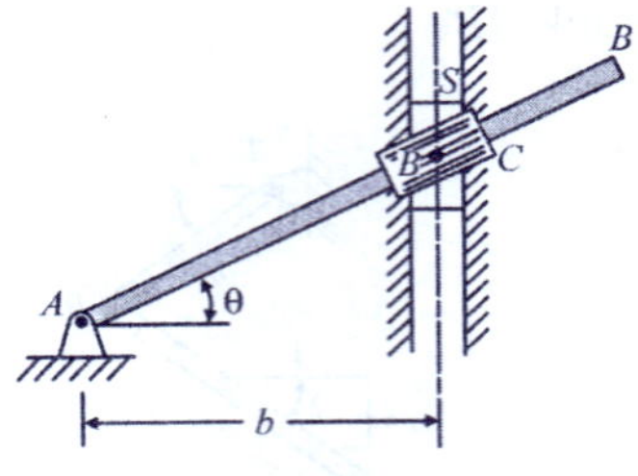

Figure P.12.4.25

12-4.26 Use the coordinate system shown in Fig. P-12-4.26. to relate the velocity and acceleration of points A and B. Then apply these results to find the acceleration of B if $r = 1.5$ m, $\tan\theta = \dfrac{3}{4}$, and A is moving down the incline with $v_A = 6$ m/s and $a_A = 4.5$ m/s^2.

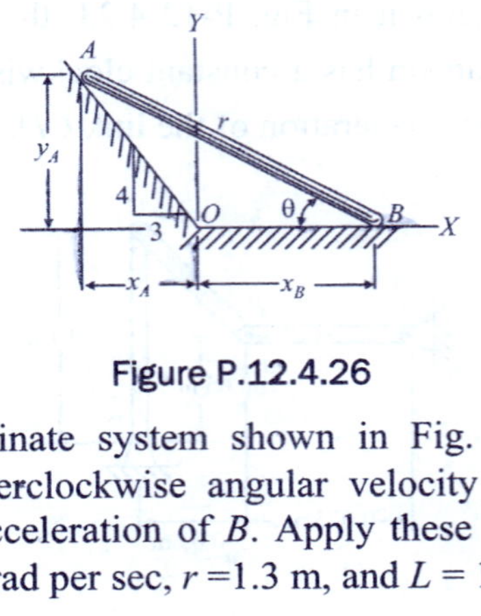

Figure P.12.4.26

12-4.27 Use the coordinate system shown in Fig. P.12-4.27 to relate the constant counterclockwise angular velocity ω of crank AO to the velocity and acceleration of B. Apply these results to the case when $\theta = 45°$, $\omega = 4$ rad per sec, $r = 1.3$ m, and $L = 1.5$ m.

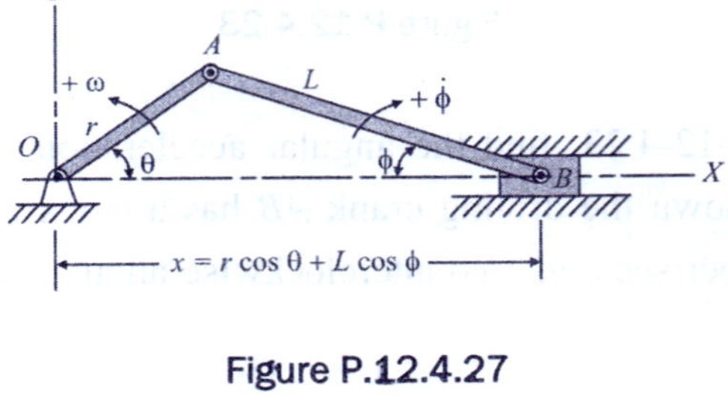

Figure P.12.4.27

12-4.28 Use the data specified in Prob. 12-4.27 to find the velocity and acceleration of *B* using the geometric approach.

12-4.29 An inversion of the slider-crank mechanism has the slider *B* fixed as shown in Fig. *P*-12-4.29. The rod *OD* slides back and forth through the fixed block *B* as the link *AB* oscillates about *B* and causes *AO* to describe plane motion. At the instant shown, the rod *OD* has a constant leftward velocity of 2.2 m/s. Determine graphically using velocity and acceleration polygons the angular acceleration of link *AO* for this position.

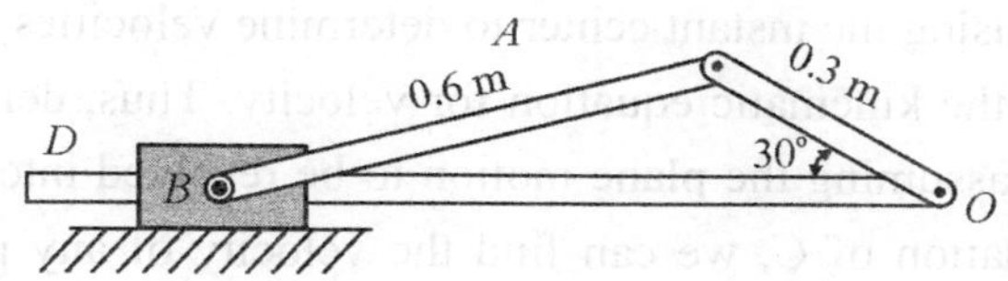

Figure P.12.4.29

12-5 INSTANT CENTER AND INSTANTANEOUS AXIS OF ROTATION

In the preceding sections, we considered the plane motion of a rigid body as equivalent to successive infinitesimal displacements from one instantaneous position to the next. Each such infinitesimal displacement consisted of the vector sum of the translation of an arbitrarily selected reference point plus a rotation about that reference point. If we could find a reference point that is momentarily at rest, however, each such successive displacement would consist only of a rotation about that reference point.

Consider now the body shown in Fig. 12-5.1 in which, at a particular instant, the velocities of points *A* and *B* have the given directions. If there is a center momentarily at rest, about which the body rotates, it will be located at the intersection *C* of two lines drawn through *A* and *B* perpendicular to v_A and v_B respectively. The point *C* is on the instantaneous axis of rotation. This is the line, perpendicular to the plane of motion, which joins those points in the body which are instantaneously at rest and have zero velocity. The intersection of this instantaneous axis with the plane of motion is called the instantaneous center of rotation or, more briefly, the instant center.

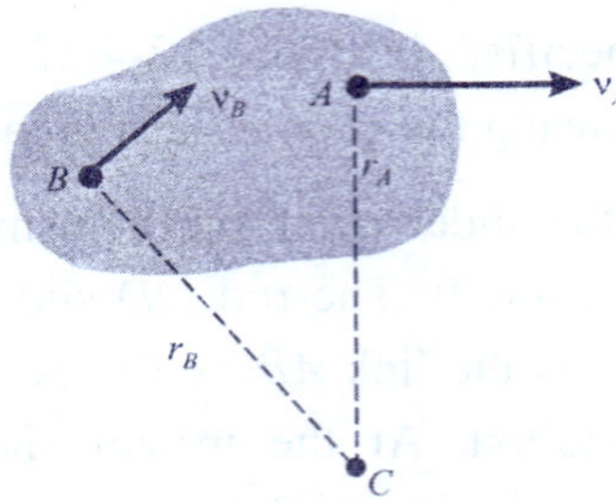

Figure 12-5.1 Graphic determination of instant center.

The value of using the instant center to determine velocities is evident from a consideration of the kinematic equation for velocity. Thus, denoting the instant center by C and assuming the plane motion to be resolved into a rotation about C plus the translation of C, we can find the velocity of any point B from the equation $v_B = v_c \nrightarrow (v_{B/C} = r\omega)$. The velocity of C, however, must be zero by the definition of an instant center. Hence the kinematic equation for determining velocities reduces to $v_B = r\omega$, which is equivalent to the kinematic equation of rotation. During the instantaneous rotation, all points in the body have the same angular velocity about the instant center. Hence the direction of the instantaneous velocity of any point is perpendicular to the line joining the point with the instant center.

Returning to Fig. 12-5.1, if the velocity of one of the points, say v_A, is known, we can use the following equation to determine the angular velocity of the body and the linear velocity of any other point B in the body:

$$\omega = \frac{v_A}{r_A} = \frac{v_B}{r_B}$$

If the two points A and B have parallel but unequal velocities, shown in Fig. 12-5.2a or (b), the instant center C lies on the common perpendicular to these velocities and is located by direct proportion.

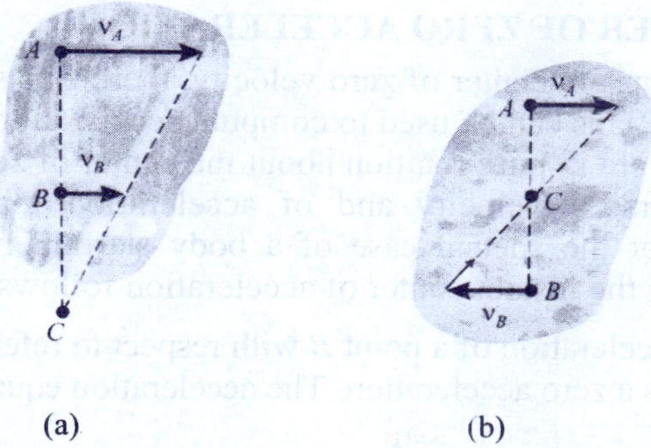

Figure 12.5.2 Determination of instant center when velocities are parallel, but unequal.

The instant center can also be located if the linear velocity of one point and the angular velocity of the body are both known any instant. Thus, if point A in Fig.12-5.3 has a known rightward velocity v_A in a body with a known clockwise value of ω, the instant center C is located along the line passing through A perpendicular to v_A and lying a distance r below A as given by the relation.

$$r = \frac{v_A}{\omega}$$

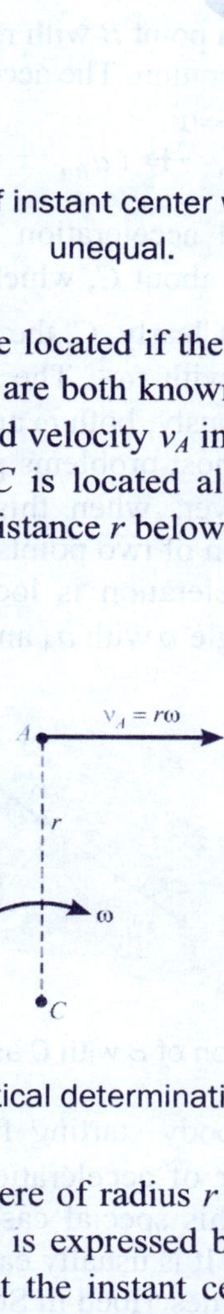

Figure 12.5.3 Analytical determination of instant center.

In the case of a wheel or sphere of radius r rolling freely along a plane, the velocity of the geometric center is expressed by $v_A = r\omega$, as shown in Section 12-4.1. Hence, we conclude that the instant center of a wheel is located at a distance r from the geometric center and, from the above discussion, that it is at the point of contact with the plane. There is visual evidence to confirm this, as for example the clear imprint left by a tire after it has run over a patch of wet pavement.

INSTANT CENTER OF ZERO ACCELERATION

Analogous to the instant center of zero velocity, there exists an instant center of zero acceleration[3]. This can be used to compute accelerations in plane motion as though the body were in pure rotation about the center of zero acceleration. The two instant centers of velocity and of acceleration, unfortunately, do not coincide except for the special case of a body starting from rest. A general method of locating the instant center of acceleration follows.

Consider the acceleration of a point *B* with respect to reference center *C* which instantaneously has a zero acceleration. The acceleration equation becomes

$$a_B = \overset{=0}{a_C} \nrightarrow (a_{B/C} = r\omega^2 \nrightarrow r\alpha)$$

so that with $a_C = 0$, the total acceleration a_B consists only of the relative acceleration $a_{B/C}$ of *B* rotating about *C*, which has the components $r\omega^2$ and $r\alpha$ acting as shown in Fig. 12-5.4. Clearly, *C*, the instant center of acceleration, lies on a line through *B* collinear with $r\omega^2$. The angle this line makes with a_B is defined by $\tan \phi = \alpha/\omega^2$. Obviously, both ω and α must be known before ϕ can be determined, but usually in most problems prior knowledge of either or both ω and α is unknown. However, when this information is given, and the direction of the total acceleration of two points *A* and *B* in Fig, 12-5.5 is known, then the instant center of acceleration is located at the intersection of lines through *A* and *B* making the angle ϕ with a_A and a_B as is shown.

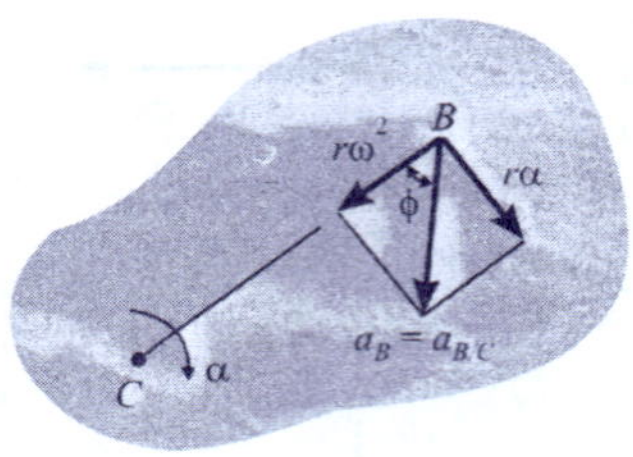

Figure 12-5.4 Acceleration of *B* with *C* as instant center of acceleration.

For the special case of a body starting from rest (i.e., $\omega = 0$), the angle $\phi = 90°$ and the instant center of acceleration will coincide with the instant center of velocity. Except in this special case, however, the instant center of acceleration is of limited value. It is usually easier to determine accelerations by using the fundamental methods described in Sections 12-4 and 12-7.

---

[3]Hereafter, the unmodified expression "instant center" will always mean the instant center of zero velocity. If there is any possibility of confusion, the explicit terms "instant center of velocity" or "...nter of acceleration" will be used.

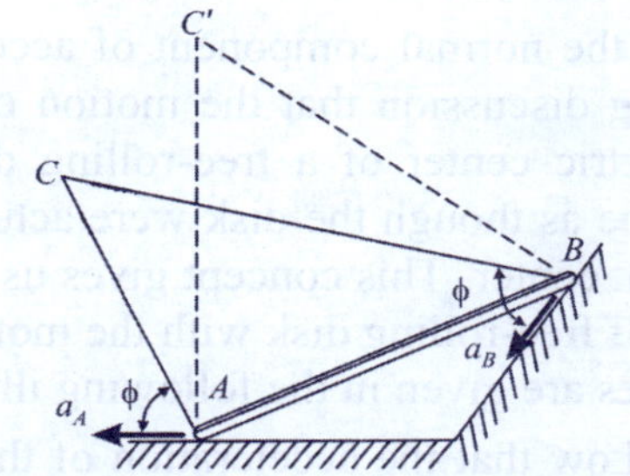

Figure 12-5.5 Location of instant center of acceleration C when $\phi = \tan^{-1}\alpha/\omega^2$ is known. When ω = 0, instant center of acceleration is at C′ and coincides instantaneously with the instant center of velocity.

There is another important special case, that of a free-rolling disk, in which the tangential acceleration of certain points can be computed as though the disk were actually rotating about the instant center of zero velocity. These points lie on the line joining the geometric center of the disk with its instant center of zero velocity. Thus, consider the free-rolling disk in Fig. 12-5.6a whose geometric center has the acceleration $a_o = r\alpha$ as shown in Section 12-4. If the plane motion is resolved into a rotation about *O* plus the translation of *O*, the acceleration of a point *B* is given by the equation

$$a_B = (a_o = r\alpha) \mapsto (a_{B/O} = b\omega^2 \mapsto b\alpha)$$

The graphic interpretation of this equation is shown in Fig. 12-5.6b. Obviously, the tangential acceleration of *B* (direced perpendicular to the line *BOC*) is given by

$$a_{B_t} = r\alpha + b\alpha = (r + b)\alpha = r\,\alpha$$

where r_1 is the distance from *B* to the instant center *C*.

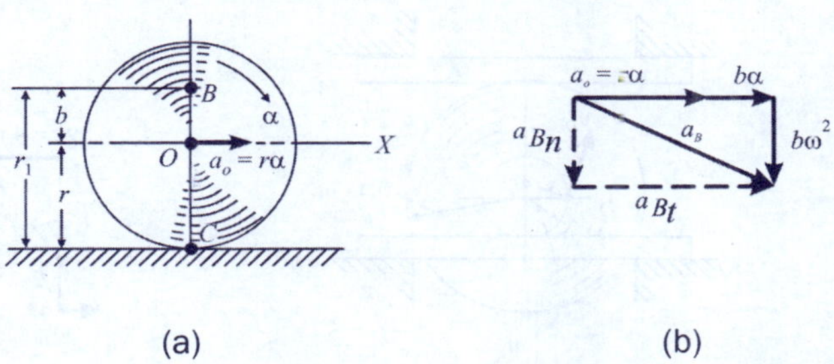

Figure 12-5.6 Acceleration of any point *B* on the line joining the geometric center O with the instant center *C*.

If we disregard the normal component of acceleration, we can conclude from the preceding discussion that the motion of any point along the line joining the geometric center of a free-rolling disk with instant center of velocity is the same as though the disk were actually rotating about an axis through this instant center. This concept gives us a convenient way to relate the plane motion of free-rolling disk with the motion of a body connected to it. Several examples are given in the following illustrative problems.

Finally, let us show that the acceleration of the velocity instant center C of the disk is not zero. Figure 12-5.7 shows the graphic interpretation of the equation

$$a_C = (a_o = r\alpha) \twoheadrightarrow (a_{C/O} = r\omega^2 \twoheadrightarrow r\alpha)$$

from which it is evident that the acceleration of C is equal to $r\omega^2$ and is directed from C toward O. Observe that the spatial position of the instant center of velocity is never stationary but always moves with some acceleration. Were the instant center of velocity to have no acceleration, it would remain fixed in position and plane motion would reduce to rotation about a fixed axis.

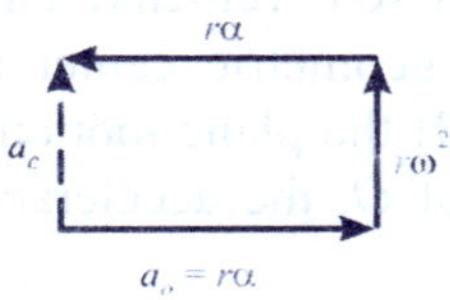

Figure 12-5.7 Acceleration of the instant center C.

ILLUSTRATIVE PROBLEMS

12-5.1 The compound wheel shown in Fig. 12-5.8a rolls without slipping between two parallel plates. The velocity of the upper plate A is $v_A = 4$ m/s to the right and that of the lower plate B is $v_B = 2$ m/s to the left. Find the angular velocity of the wheel and the linear velocity of point D on it using the instant center method.

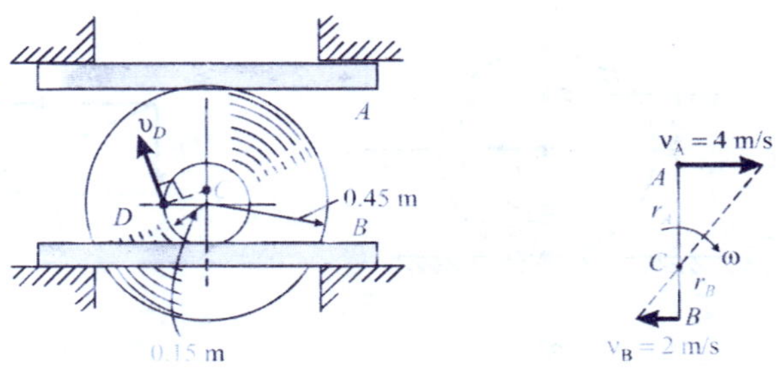

(a) (b)

Figure 12-5.8

Solution

The points on the vertical centerline of the wheel which are in contact with the plates have the velocities shown in part (b) of the Fig. 12.5.8. The instant center C lies on the line AB which is the common perpendicular to v_A and v_B. Its location from points A and B is determined from

$$\left[\frac{v_A}{r_A}=\frac{v_B}{r_B}\right] \qquad \frac{4}{r_A}=\frac{2}{r_B} \quad \text{or} \quad r_A = 2r_B$$

which, substituted into $r_A + r_B = 0.6$ m obtained from the dimensions of the wheel, gives $r_A = 0.4$ m and $r_B = 0.2$ m. Therefore, the angular velocity is

$$\omega = \frac{v_B}{r_B} = \frac{2}{0.2} = 10 \text{ rad per sec clockwise} \qquad \textbf{\textit{Ans.}}$$

Returning to Fig. 12-5.8a, we find the velocity of D is perpendicular to the length CD drawn from the instant center and its magnitude is

$$[v = r\omega] \qquad v_D = \left(\sqrt{(0.05)^2+(0.15)^2}\right)(10) = 1.58 \text{ m/sec} \qquad \textbf{\textit{Ans.}}$$

It is useful to observe that the horizontal and vertical components of v_D are equal to the vertical and horizontal components of CD respectively multiplied by ω ; i.e.,

$$v_{D_h} = 0.05\,(10) = 0.5 \text{ m/sec left and } v_{D_v} = 0.15\,(10) = 1.5 \text{ m/sec up}$$

12-5.2 The disks in Fig. 12-5.9 are free to roll on the inclined planes. Disk A is fastened to pulley P by cord Q, which is wrapped around the drum of the disk and around the smaller pulley. Disk B is similarly attached to the outer wheel of the pulley. The linear acceleration of the center of disk A is constant at 3.0 m/sec^2 down the incline. Determine the linear displacement, velocity, and acceleration of the center of disk B after 4 sec, starting from rest.

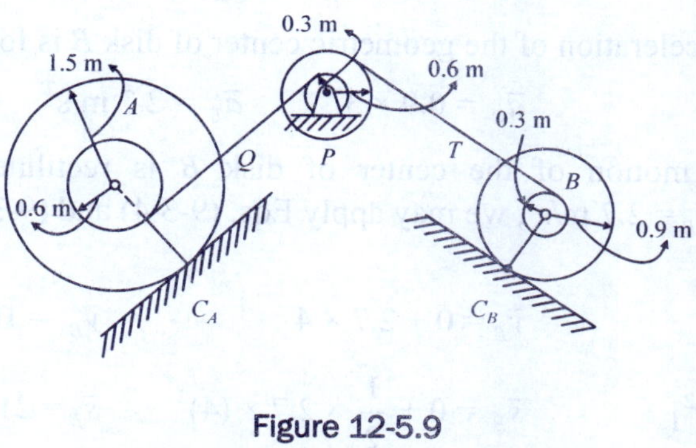

Figure 12-5.9

Solution

The motion of pulley P is related to the motions of disks A and B by cords Q and T. All the bodies can be considered as in a state of rotation – P about its axle, A about its instant center C_A, and B about its instant center C_B. Applying the relation between linear and angular displacement in rotation, we may express the linear displacements of points on cords Q and T in terms of the angular displacements of the bodies by

$$[s = r\theta] \qquad s_Q = 0.9\,\theta_A = 0.3 \times \theta_P \qquad (a)$$

$$s_T = 0.6\,\theta_P = 1.2\,\theta_B \qquad (b)$$

From Eqs. (*a*) and (*b*) we obtain

$$3\,\theta_A = \theta_P = 2\,\theta_B \qquad (c)$$

Velocity and acceleration relations between the disks and the pulley have similar forms because $v = r\omega$ and $a_t = r\alpha$ have the same mathematical form as $s = r\theta$. Therefore we may also write

$$3\omega_A = \omega_P = 2\omega_B \qquad (d)$$

$$3\alpha_A = \alpha_P = 2\alpha_B \qquad (e)$$

Not all of the above discussion is really necessary to solve this problem; however, it will be found useful in applying the work-energy method in Section 14-8.

From the given data the angular acceleration of disk A can be expressed in terms of its linear acceleration by

$$[\bar{a}_t = \bar{r}\alpha] \qquad 3.0 = 1.5\alpha_A \qquad \alpha_A = 2 \text{ rad per sec}^2$$

Applying the relation (*e*), we now compute the angular acceleration of disk B.

$$[3\alpha_A = 2\alpha_B] \qquad 3 \times 2 = 2 \times \alpha_B \qquad \alpha_B = 3 \text{ rad per sec}^2$$

whence the acceleration of the geometric center of disk B is found from

$$[\bar{a}_t = \bar{r}\alpha] \qquad \bar{a}_B = 0.9 \times 3 \qquad \bar{a}_B = 2.7 \text{ m/s}^2$$

Since the motion of the center of disk B is rectilinear with constant acceleration $\bar{a}_B = 2.7$ m/s^2, we may apply Eqs. (9-3.4) and (9-3.5) of Section 9-3 to obtain

$$[v = v_o + at] \qquad \bar{v}_B = 0 + 2.7 \times 4 \qquad \bar{v}_B = 10.8 \text{ m/s} \quad \textbf{Ans.}$$

$$[s = v_o t + \frac{1}{2}at^2] \qquad \bar{s}_B = 0 + \frac{1}{2} \times 2.7 \times (4)^2 \qquad \bar{s}_B = 21.6 \text{ m} \quad \textbf{Ans.}$$

In this problem as disk A rolls down the incline, its center travels the distance $\bar{s}_A = 1.5\ \theta_A$ while the cord Q travels $s_Q = 0.9\theta_A$ Hence for each radian of angular displacement of A, 0.6 m of cord are unwound from the drum of A. Actually the unwinding of $0.6\theta_A$ m of cord from A is the result of the rotation of the drum about the geometric center of A. Similarly we may show that for each radian of angular displacement of B, 3.0 m of cord is unwound from B as it rolls up the incline. If the motion of the system were reversed with A rolling up and B rolling down their respective inclines, cords Q and T would wind up these amounts on A and B respectively. In either case, the geometric center of A would still move faster than cord Q while cord T would still move faster than the geometric center of B.

12.5.3 At a certain instant a four-link mechanism occupies the position shown in Fig. 12-5.10. The dimensions are as shown. The angular velocity of AB is $\omega_{AB} = 11$ rad per sec. Locate the instant center of BC and use it to determine the linear velocity of C and the angular velocity of CD.

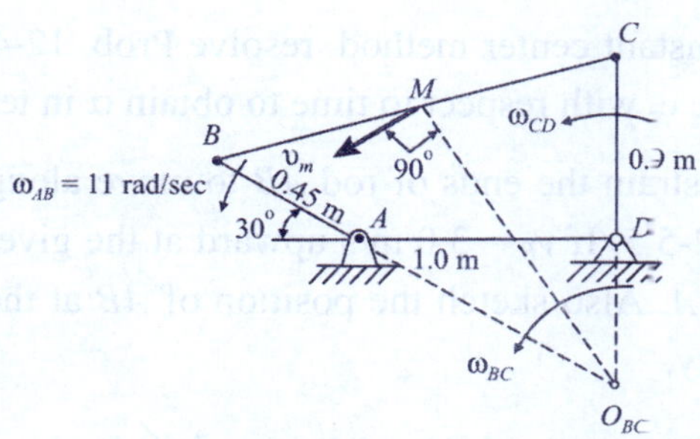

Figure 12-5.10

Solution

Points B and C are constrained to rotate about the fixed bearings A and D respectively. Since the velocity of B is therefore perpendicular to AB and the velocity of C is perpendicular to CD, the intersection of AB and CD prolonged determines the instant center O_{BC}, From the diagram, we have

$$DO = AD \tan 30^\circ = 1.0 \times 0.58 = 0.58 \text{ m}$$

$$AO = \frac{AD}{\cos 30^\circ} = \frac{1.0}{0.866} = 1.155 \text{ m}$$

These distances could have been obtained graphically by drawing the mechanism to scale. The graphic method is generally used in this type of problem.

Applying the principle that the velocity of any point is equivalent to rotation about the instant center, we have

$[v_B = r_{AB}\,\omega_{AB} = r_{BO}\,\omega_{BC}]$ $0.45 \times 11 = 1.604\,\omega_{BC}$ $\omega_{BC} = 3.08$ rad per sec

$[v_C = r_{CO}\,\omega_{BC}]$ $v_c = (0.9 + 0.58) \times 3.08 = 4.56$ m/s ***Ans.***

Also

$[v_C = r_{CD}\omega_{CD}]$ $4.56 = 0.9 \times \omega_{CD}$ $\omega_{CD} = 5$ rad per sec ***Ans.***

The velocity of any other point *M* is found by scaling the distance *MO* from a scale diagram of the mechanism and applying $v_M = r_{MO}\,\omega_{BC}$. The direction of v_M is perpendicular to *MO* as shown.

PROBLEMS

12-5.4 Using the instant center method, resolve Prob. 12-4.11 for ω and then differentiate ω with respect to time to obtain α in terms of ω.

12-5.5 Rollers constrain the ends of rod *AB* to move along the guides shown in Fig. P-12-5.5, If $v_B = 3.0$ m/s upward at the given position, find the velocity of *A*. Also sketch the position of *AB* at the instant end *B* has zero velocity.

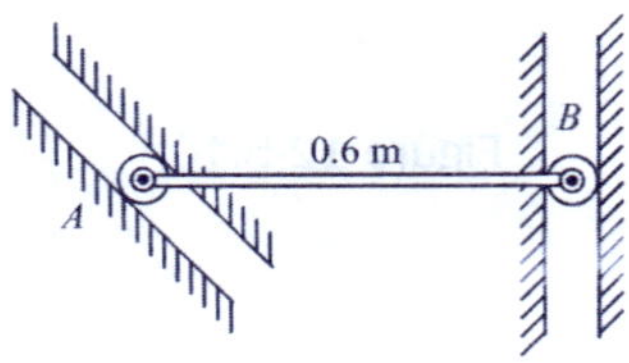

Figure P-12-5.5

12.5.6 The compound disk in Fig. P-12-5.6 rolls and slips on the horizontal floor. The linear velocity of the cord leaving *A* is 3.6 m/s right ward and the disk rotates counterclockwise at 6 rad per sec. Use the instant center method to find the velocity of points *O* and *B*.

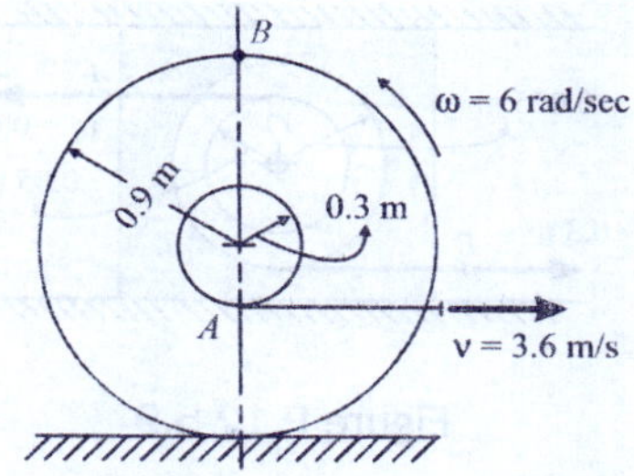

Figure P-12-5.6

12-5.7 If the center of the free-rolling disk in Fig. P-12-5.7 travels 2.7 m to the right, how far does body *A* move?

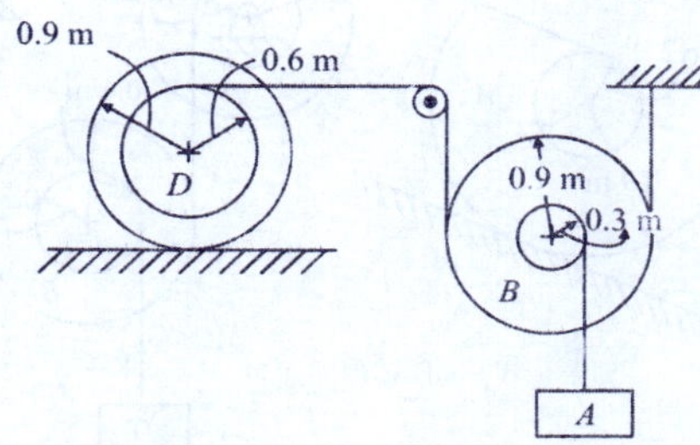

Figure P-12-5.7

12-5.8 Make the following changes in Illus. Prob. 12-5.2: The radii of disk *A* are changed to 0.9 m and 1.5 m; of disk *B* to 0.6 m and 1.2 m; of pulley *P* to 0.23 m and 0.45 m. If the angular acceleration of disk *A* is 3 rad per sec^2 clockwise, determine the time for the center of disk *B* to travel 18 m from rest.

12-5.9 The compound disk in Fig. P-12-5.9 is pinned to the sliding block at *O*. Cords attached to the disk have the constant velocities shown. Determine the velocity of the sliding block and the number of meters of cords *A* and *B* wound on or unwound from the disk each second.

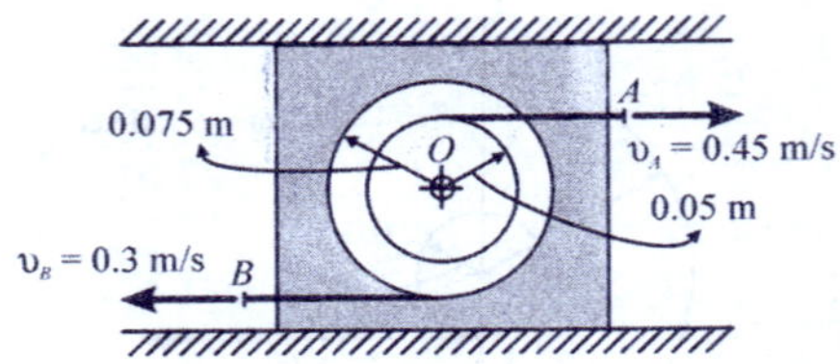

Figure P.12-5.9

12-5.10 In Fig. P.12-5.10, the initial downward velocity of body *D* is 0.30 m/s and its acceleration is constant at 0.05 m/s^2. Determine the linear displacement of the center of the free-rolling disk A after 2 sec.

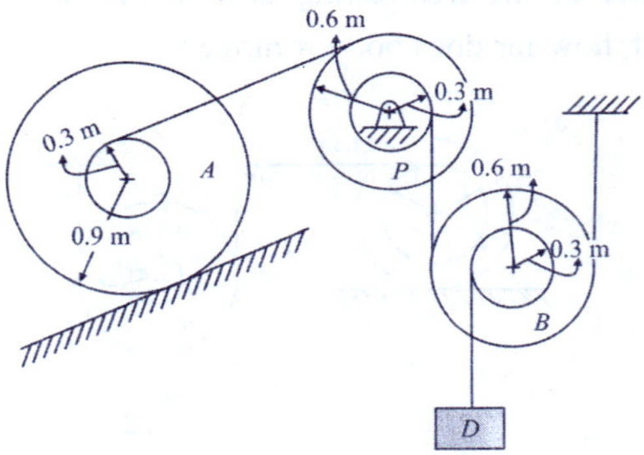

Figure P.12-5.10

12-5.11 The wheel in Fig. P.12-5.11 rolls freely on the circular arc. Show that $v_A = r\omega$ and $a_{At} = r\alpha$.

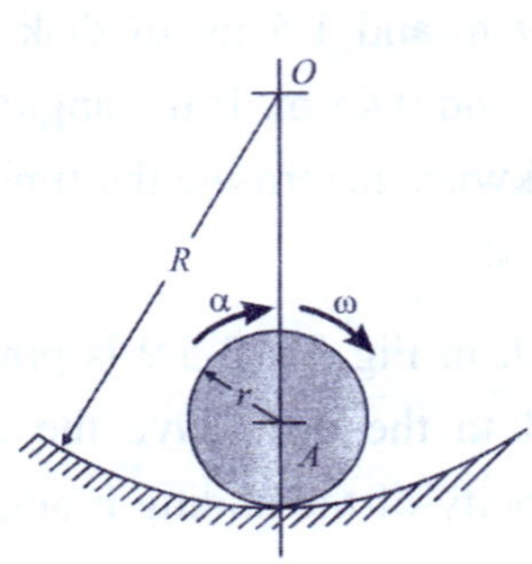

Figure P.12-5.11

12-5.12 Show that the acceleration of the instant center of the wheel rolling freely on the circular arc in Fig. P.12-5.11. as

$$a = \left(\frac{Rr}{R-r}\right)\omega^2 = \frac{Rv_A^2}{(R-r)r}$$

12-5.13 At the position shown in Fig. P-12-5.13, crank AB has a clockwise angular velocity of 3 rad per sec. Use the instant center method to find the linear velocity of point D.

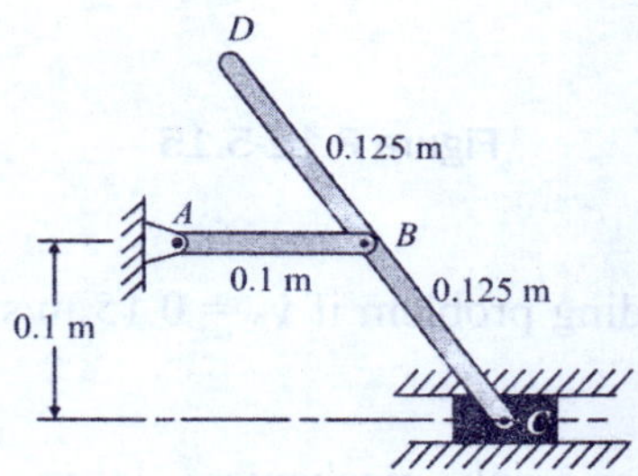

Figure P.12-5.13

12-5.14 At the given position of the mechanism shown in Fig. P.12-5.14, the velocity of the center of the free-rolling wheel is v_A = 0.8 m/s rightward. Determine the corresponding velocity of slider D.

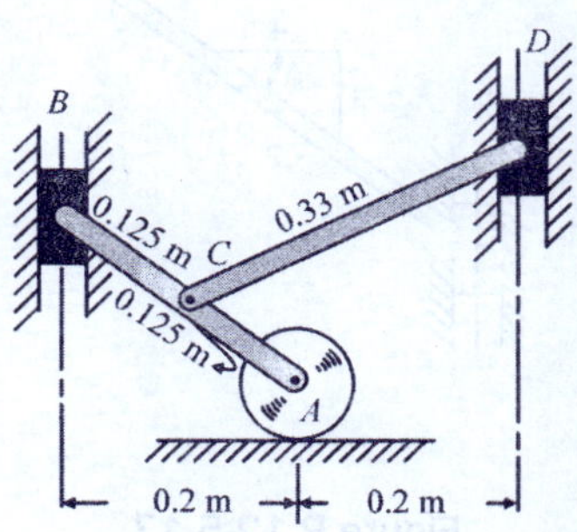

Figure P.12-5.14

12-5.15 The rigid-slab in Fig. P-12-5.15 moves in the XY plane with $v_{A_y} = 0.15$ m/s, $v_{B_x} = 0.45$ m/s, and $v_{C_y} = -0.3$ m/s. Determine the coordinates of the instant center.

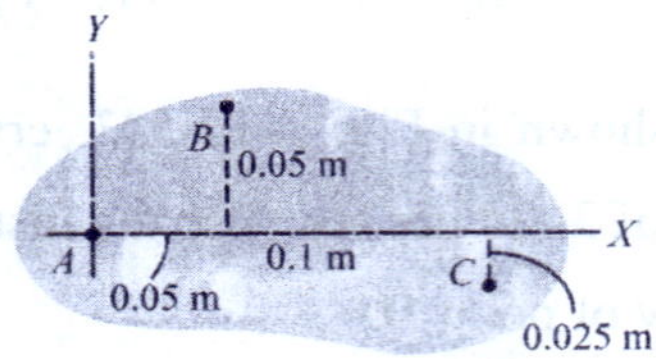

Figure P.12-5.15

12-5.16 Solve the preceding problem if $v_A = 0.15$ m/s, $v_B = 0.05$ m/s and $v_C = 0.1$ m/s.

12-5.17 When the slider-crank mechanism is in the position shown in Fig. P-12-5.17, the slider C has a downward velocity of 5 m/s. Determine the velocity of the midpoint M of BC and the angular velocity of crank AB.

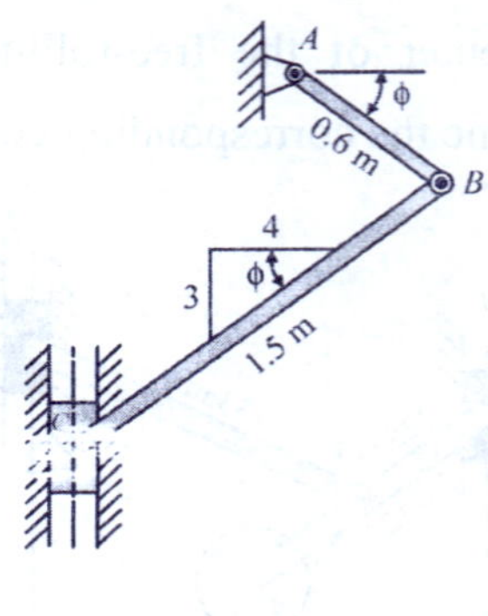

Figure P.12-5.17

12-5.18 At the instant shown in Fig. P-12-5.18, point B has an upward velocity of 0.6 m/s. Use the instant center method to determine the velocity of point C and the angular velocity of crank AO.

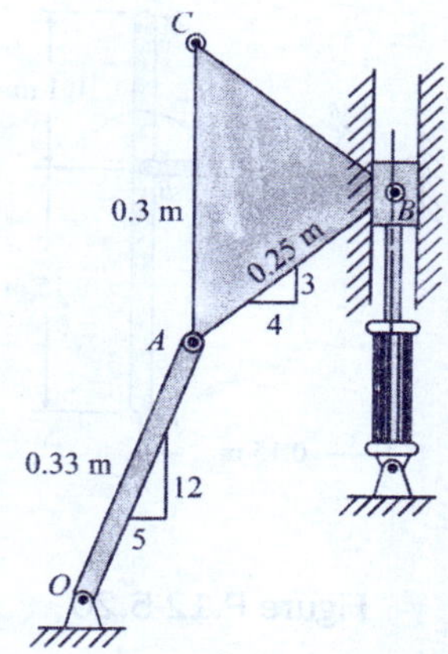

Figure P.12-5.18

12-5.19 In the four-link mechanism shown in Fig. P- 12-5.19, the crank *AB* has a constant angular speed of ω_{AB} = 3 rad per sec. Use the instant center method to determine ω_{BC} and ω_{CD}. Compare your results with those in Illus. Prob. 12-7.2.

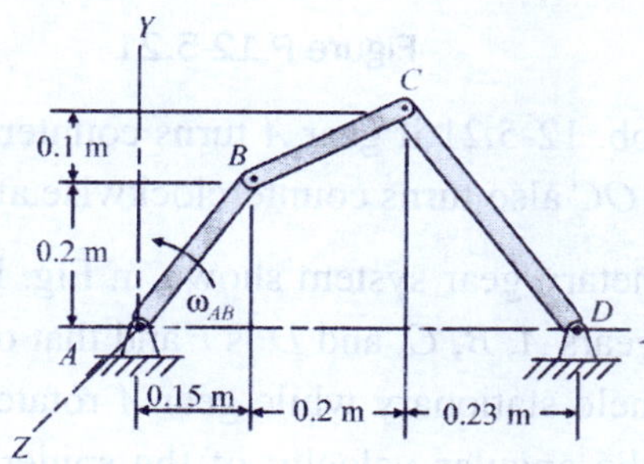

Figure P.12-5.19

12.5.20 Rods *AB* and *CD* are pinned together at *B* as shown in Fig. P-12-5.20 and move in a vertical plane with the absolute angular velocities ω_{AB} = 4 rad per sec and ω_{CD} = 3 rad per sec. Determine the linear velocities of points *C* and *D*.

12.5.21 Fig. P-12-5.21 shows the pitch circles of two gears *A* and *B* connected by the arm *OC*. Gear *A* turns counterclockwise at 12 rad per sec while arm *OC* turns clockwise at 6 rad per sec, thereby causing gear *B* to roll around gear *A*. Locate the instant center of gear *B* and compute its angular velocity.

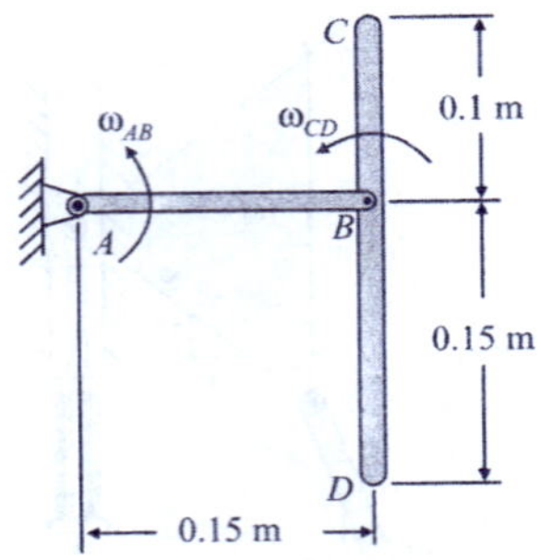

Figure P.12-5.20

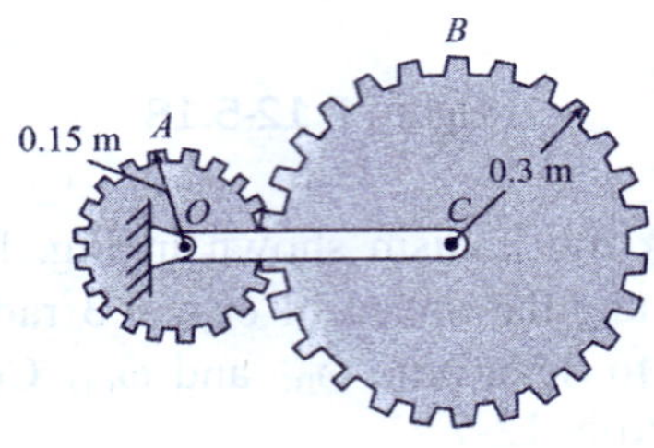

Figure P.12-5.21

12-5.22 Repeat Prob. 12-5.21 if gear *A* turns counterclockwise at 9 rad per sec while arm *OC* also turns counterclockwise at 1 rad per sec.

12-3.23 In the planetary gear system shown in Fig. P-12-5.23, the pitch circle radius of gears *A*, *B*, *C*, and *D* is *r* and that of gear *E* is 3*r*. If the outer gear *E* is held stationary while gear *A* rotates clockwise at ω_A rad per sec, find the angular velocity of the spider connecting the planetary gears.

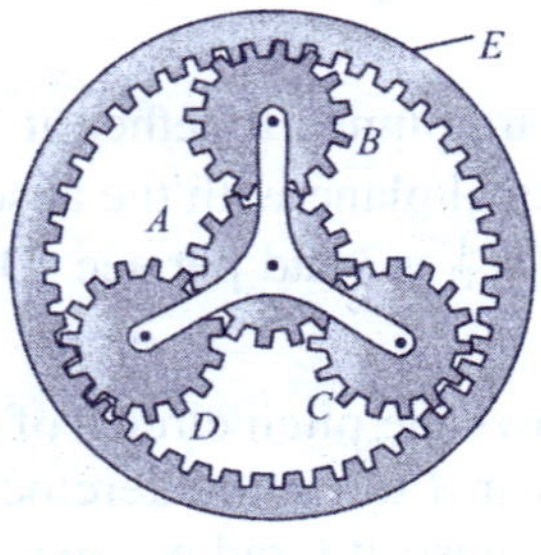

Figure P.12-5.23

12.5.24 For the planetary gear system described in the preceding problem, assume that gear *A* is held fixed while the spider rotates clockwise at ω_{AB} rad per sec. What will then be the angular velocity of the ring gear *E*?

12-6 THE OMEGA THEOREM

We consider now an application of the rotation of a rigid body to a concept known as the omega theorem. This theorem shows how to find the time derivative of any vector of constant magnitude, but whose direction changes as a result of rotating with the body in which it is fixed.

In Fig. 12-6.1 we show a body rotating about a fixed point *O* with an absolute angular velocity ω coinciding with the axis of rotation. If the axis of rotation were fixed in space, any point *P*, located by the position vector *r* from *O*, would describe a circular path perpendicular to the axis of rotation. However, with only one point of a body fixed in space as in a spinning top, the axis of rotation will move or precess, and the path of *P* will be circular only for a differential time *dt*. During this time *dt*, the radius of this circular path about the instantaneous axis of rotation is $r \sin \phi$ and it sweeps through the angle $\omega\, dt$, thereby causing the displacement of *P* to be

$$d\mathbf{r} = (\omega dt)(r \sin \phi)\, \hat{e}_t$$

The velocity of *P* is therefore

$$\mathbf{v} = \frac{d\mathbf{r}}{dt}(\omega r \sin \phi)\hat{e}_t$$

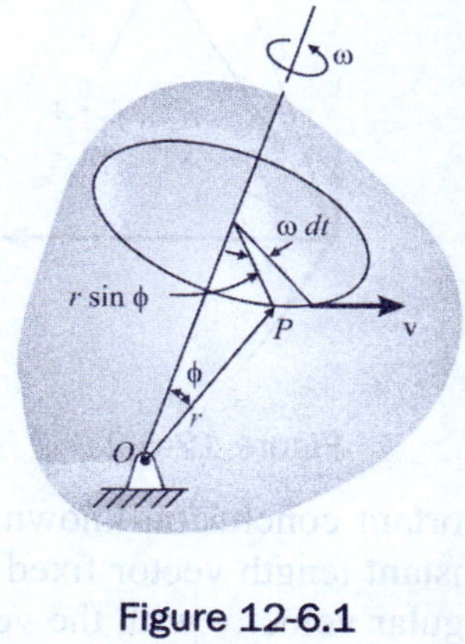

Figure 12-6.1

It is evident that **v** is perpendicular to the plane formed by the vectors **ω** and **r** so that the right side of this result represents the definition of the cross product, ω × r. Hence, the velocity of any point in a rotating body is expressed by

$$\mathbf{v} = \dot{\mathbf{r}} = \boldsymbol{\omega} \times \mathbf{r} \tag{12-6.1}$$

where **r** may be the position vector to *P* from *any* origin on the axis of rotation regardless of whether the axis of rotation is fixed or moving. If the axis has a constant spatial orientation, ω can change only in magnitude so that then $\boldsymbol{\alpha} = \dot{\boldsymbol{\omega}}$ will be collinear with **ω**. With a moving axis, however, **ω**, will always change direction in addition to possibly changing its magnitude which will cause α to have a different direction from **ω**. This is analogous to the differing directions of **v** and **a** in curvilinear motion.

But more significantly, Eq. (12-6.1) represents the rate at which a constant length vector fixed in a rotating body changes its direction. Thus, consider a vector ρ, fixed in a rotating body, joining any two particles *A* and *B* as in Fig. 12-6.2. The time derivative of ρ can be found by differentiating the relation $\boldsymbol{\rho} = \mathbf{r}_B - \mathbf{r}_A$ to obtain

$$\dot{\boldsymbol{\rho}} = \dot{\mathbf{r}}_B - \dot{\mathbf{r}}_A \tag{a}$$

But from Eq. (12-6.1), we have $\dot{\mathbf{r}}_B = \boldsymbol{\omega} \times \mathbf{r}_B$ and $\dot{\mathbf{r}}_A = \boldsymbol{\omega} \times \mathbf{r}_A$ which we substitute into Eq. (*a*) to obtain

$$\dot{\boldsymbol{\rho}} = \boldsymbol{\omega} \times \mathbf{r}_B - \boldsymbol{\omega} \times \mathbf{r}_A = \boldsymbol{\omega} \times (\mathbf{r}_B - \mathbf{r}_A) = \boldsymbol{\omega} \times \boldsymbol{\rho} \tag{b}$$

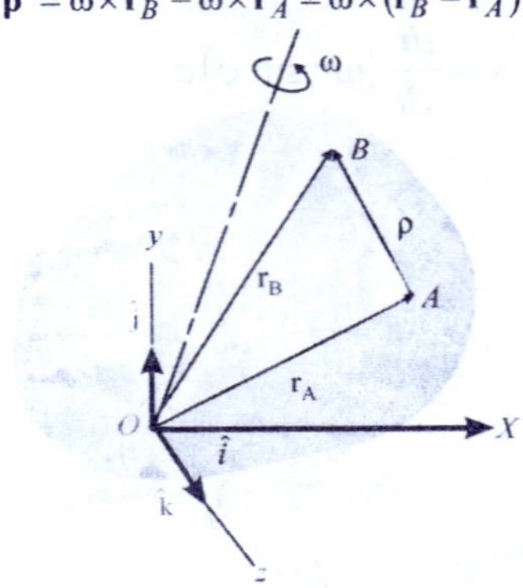

Figure 12-6.2

Thus we reach the important conclusion known as the omega theorem that the time derivative of a constant length vector fixed in a rotating body is simply the cross product of the angular velocity with the vector. We shall see later that the omega theorem can be extended to include the *directional* rate of change of any type of vector quantity.

Since the angular velocity ω is the time rate of change of the angular displacement of *any* line in the body as was proved, we may consider ω to be a free vector. Consequently, we may define $\dot{\boldsymbol{\rho}}$ as the velocity of *B* relative to A, or

$$\mathbf{v}_{B/A} = \dot{\boldsymbol{\rho}} = \boldsymbol{\omega} \times \boldsymbol{\rho} \tag{12-6.2}$$

We can now apply the omega theorem to determine the time derivatives of the unit vectors in a coordinate system rotating with the body. In Fig. 12-6.2, we also show a reference frame *xyz* attached to the body and rotating with it at the angular velocity **ω**. Here the unit vectors $\hat{\mathbf{i}}, \hat{\mathbf{j}}$ and $\hat{\mathbf{k}}$, although of constant length, change their directions as a result of rotating with the body. Using the omega theorem, their time derivatives are

$$\begin{aligned} \dot{\hat{\mathbf{i}}} &= \boldsymbol{\omega} \times \hat{\mathbf{i}} \\ \dot{\hat{\mathbf{j}}} &= \boldsymbol{\omega} \times \hat{\mathbf{j}} \\ \dot{\hat{\mathbf{k}}} &= \boldsymbol{\omega} \times \hat{\mathbf{k}} \end{aligned} \tag{12-6.3}$$

and, in general, for any unit vector $\hat{\mathbf{e}}$ changing its direction at a rate ω,

$$\dot{\hat{\mathbf{e}}} = \boldsymbol{\omega} \times \hat{\mathbf{e}} \tag{12-6.4}$$

As an application of Eq. (12-6.4), consider the rates of change of the unit vectors $\hat{\mathbf{e}}_r$, $\hat{\mathbf{e}}_\theta$, in polar or cylindrical coordinates which we derived previously by other methods in Section 9-8. Using = $\omega = \dot{\theta}\hat{\mathbf{k}}$ and the right-hand system of unit vectors $\hat{\mathbf{e}}_r$, $\hat{\mathbf{e}}_\theta$, $\hat{\mathbf{k}}$ we obtain at once

$$\dot{\hat{\mathbf{e}}}_r = \omega \times \hat{\mathbf{e}}_r = \dot{\theta}\,\hat{\mathbf{k}} \times \hat{\mathbf{e}}_r = \dot{\theta}\hat{\mathbf{e}}_\theta$$

and

$$\dot{\hat{\mathbf{e}}}_\theta = \omega \times \hat{\mathbf{e}}_\theta = \dot{\theta}\,\hat{\mathbf{k}} \times \hat{\mathbf{e}}_\theta = -\dot{\theta}\hat{\mathbf{e}}_r$$

Finally, let us re-examine the case of fixed-axis rotation using the omega theorem. The velocity of any point at a distance **r** from the axis of rotation is

$$\mathbf{v} = \dot{\mathbf{r}} = \boldsymbol{\omega} \times \mathbf{r} \tag{12.6.5}$$

and its acceleration

$$\begin{aligned} \mathbf{a} = \dot{\mathbf{v}} &= \dot{\omega} \times \mathbf{r} + \omega \times (\dot{\mathbf{r}} = \mathbf{v}) \\ &= \propto \times \mathbf{r} + \boldsymbol{\omega} \times (\boldsymbol{\omega} \times \mathbf{r}) \end{aligned} \tag{12.6.6}$$

These vector equations are readily correlated with our previous discussion of rotation by referring to Fig. 12-6.3. There, the rotating body is represented by its projection upon a plane perpendicular to the axis of rotation. Noting that **ω** and **α** are perpendicular to **r**, the definition of a cross product gives the magnitude of

v as **v** = |ω × **r**| = $r\omega$, and the magnitude of the components of a as a_t = |**a** × **r**| = $r\alpha$ and a_n = |ω × **v**| = $r\omega^2$. Further, in agreement with Section 12-2, **v** is perpendicular to **r** in the sense that ω rotates r; while a_t is perpendicular to r in the sense of α, and a_n is directed inward toward the center of rotation. For plane motion problems, therefore, a convenient way of denoting the vector component of normal acceleration is the reduction

$$a_n = \omega \times (\omega \times \mathbf{r}) = -\omega^2 \mathbf{r} \tag{12-6.7}$$

which we may use whenever **ω** and **r** are mutually perpendicular. (See Illus. Prob. 12-7.1.) Be warned, however, that this reduction does not apply to spatial motion where **ω** and **r** generally are *not* perpendicular.

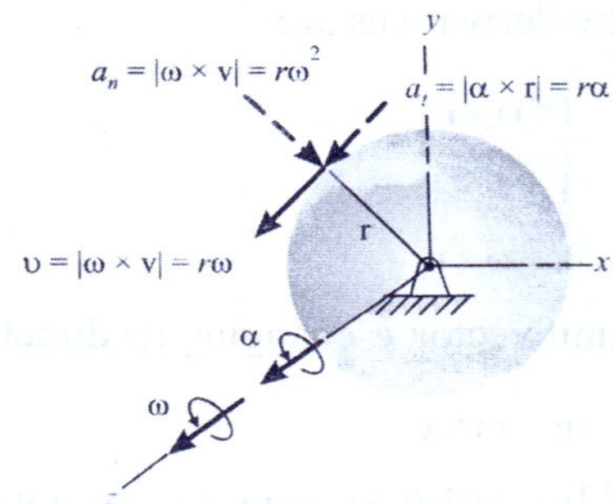

Figure 12-6.3 Interpretation of velocity and acc eleration components.0

CHAPTER 13

KINETICS OF RIGID BODIES

13-1 INTRODUCTION

By now we have covered so many topics in such detail that you may have forgotten the overall preview of dynamics that was presented in Section 9-1. *Please read this section again.* It will enable you to see the "forest" of dynamics rather than the trees.

In this chapter we shall consider only the plane motion of rigid bodies that are symmetrical with respect to the plane in which the mass center moves. Since most engineering applications are of this type, they are separated here from the motion of unsymmetrical bodies and general spatial kinetics. The latter are discussed in later chapters and involve complications which, although important, are not essential to the plane motion of symmetrical bodies.

Only two basic equations are involved: $\mathbf{R} = m\bar{\mathrm{a}} = (W/g)\,\bar{\mathrm{a}}$ and $\Sigma\mathbf{M} = \dot{\mathbf{H}}$. These were derived in Sections 10-5 and 10-6. The first relates the resultant of applied forces to the acceleration of the mass center while the second relates the moment of applied forces to the angular motion. In addition, these equations are coupled by the correlation between linear and angular motion as presented in the preceding chapter on kinematics.

13-2 EQUATIONS OF PLANE MOTION

Consider a symmetrical rigid body constrained to move in the *XY* plane. Because of this constraint, all forces affecting the motion can be reduced to a resultant force **R** acting through the mass center *G*, and a resultant couple $\Sigma\mathbf{M}_G$ representing the moment sum of the applied forces about a *Z* axis through the mass center perpendicular to the plane of motion.

We know from the principle of the motion of the mass center that the relation between **R** and the linear acceleration of the mass center is expressed by $\mathbf{R} = (W/g)\,\bar{\mathbf{a}}$. The resultant couple $\sum \mathbf{M}_G$ causes an angular acceleration α, and the kinematics of constraints determine the relation between and $\bar{a}$ and $\boldsymbol{\alpha}$. What we do not yet know is the explicit relation between $\sum \mathbf{M}_G$ and α. We shall soon show that, this is $\sum \mathbf{M}_G = I_G\boldsymbol{\alpha}$ by using several approaches which complement one another.

But first, let us review the possible effects of the resultant force-couple system upon the motion of a body starting from rest. If the resultant is a single force which passes through the mass center, we have learned (see Section 11-2) that a motion of translation ensues: rectilinear translation if the direction of the resultant force is constant, or curvilinear translation if the direction of the resultant force varies. A force system whose resultant is a couple will cause the body on which it acts to have a centroidal rotation even if there is no fixed axis through the mass center. Conversely, if motion occurs about a fixed centroidal axis, the resultant must be a couple. When the resultant is both a force and a couple, plane motion ensues, consisting of both a linear motion of the mass center and an angular acceleration. The angular acceleration is a property of the body and is independent of the actual or assumed axis of rotation (see Sections 12-2 and 12-3). The important cases of plane motion are noncentroidal rotation (i.e., rotation about a fixed axis which does not pass through the mass center), rolling bodies, and general plane motion. In all these cases, kinematic analyses must be used to relate the linear acceleration $\bar{a}$ of the mass center with the angular acceleration $\boldsymbol{\alpha}$ of the body.

Let us now derive the relation between the moment sum of the applied forces and the angular acceleration of the body, and then later discuss its application to each of the various motions just described. To obtain this relation from the basic equation $\sum \mathbf{M} = \dot{\mathbf{H}}$, we must first determine the angular momentum **H** for the rigid-body motion.[1]

Consider in Fig. 13-2.1 the outline of a symmetrical rigid body constrained to move in the *XY* plane through its mass center. Because of the plane motion restriction, the angular velocity and acceleration vectors of the body are perpendicular to the plane of motion and are expressed by $\boldsymbol{\omega} = \omega\hat{\mathbf{k}}$ and $\boldsymbol{\alpha} = \dot{\boldsymbol{\omega}} = \alpha\hat{\mathbf{k}}$. Fix *xyz* axes in the body as shown so that the *z* axis is always perpendicular

---

[1] *Later in Chapter 16, we shall develop a general expression for* **H** applicable to any body having any motion.

to the *XY* plane of motion while the *xy* axes rotate with the body at the angular rates ω and α. Let the origin of these axes be the mass center *G*, although, it may also be a center *A* which is fixed in space, or else one whose acceleration is directed through the mass center. Recall from Section 10-6 that with any of these conditions, the moment equation has the form $\sum \mathbf{M} = \dot{\mathbf{H}}$.

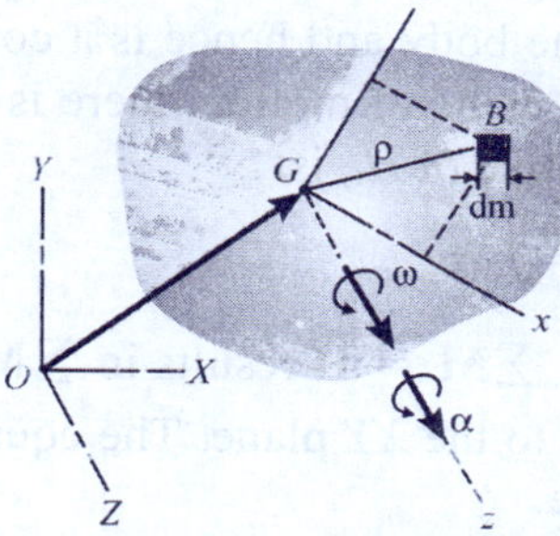

Figure 13-2.1 G and the *xy* axes move in the XY plane of motion.

Because of symmetry, the mass center of any pair of symmetrically placed elements lies in the *XY* plane. Denote their mass by *dm* and their location at *B* by the relative position vector from *G* having the constant length $\rho = x\hat{\mathbf{i}} + y\hat{\mathbf{j}}$. The moment of their relative momentum about *G* is then $\boldsymbol{\rho} \times v_{B/G}\, dm = \boldsymbol{\rho} \times (\boldsymbol{\omega} \times \boldsymbol{\rho})\, dm$ and the total relative moment of momentum of the body (or angular momentum) is then

$$H_G = \int \rho \times (\omega \times \rho)\, dm \qquad (a)$$

To evaluate the vector triple product of the integrand, we can use the identity $\mathbf{a} \times (\mathbf{b} \times \mathbf{c}) = (\mathbf{a} \cdot \mathbf{c})\mathbf{b} - (\mathbf{a} \cdot \mathbf{b})\mathbf{c}$ [Eq. (2-8.6)] to yield $\rho^2\omega$ directly[2] or proceed fundamentally as follows:

$$\boldsymbol{\omega} \times \boldsymbol{\rho} = \omega\hat{\mathbf{k}} \times (x\hat{\mathbf{i}} + y\hat{\mathbf{j}}) = \omega x\hat{\mathbf{j}} - \omega y\hat{\mathbf{i}}$$

whence

$$\boldsymbol{\rho} \times (\boldsymbol{\omega} \times \boldsymbol{\rho}) = (x\hat{\mathbf{i}} + y\hat{\mathbf{j}}) \times (\omega x\hat{\mathbf{j}} - \omega y\hat{\mathbf{i}})$$
$$= (x^2 + y^2)\,\omega\hat{\mathbf{k}} = \rho^2\omega$$

[2]Thus, $\boldsymbol{\rho} \times (\omega \times \rho) = (\rho \cdot \rho)\omega - (\rho \cdot \omega)\boldsymbol{\rho} = \boldsymbol{\rho}^2\omega$ since $\boldsymbol{\rho} \cdot \omega = 0$ because $\boldsymbol{\rho}$ and are ω are mutually perpendicular.

On substituting this evaluation of the integrand into Eq. (*a*) and placing $\boldsymbol{\omega}$ outside the integral since it is a property of the body and thus independent of the position of *B*, we obtain

$$H_G = \boldsymbol{\omega} \int \rho^2 \, dm = I_G \boldsymbol{\omega} \qquad (b)$$

Note that the inertia integral is defined with respect to a *z* axis through *G* fixed in and moving with the body and hence is a constant quantity. Since $\boldsymbol{\omega}$ is always perpendicular to the plane of motion, there is no change in its direction, and the absolute time derivative of H_G is

$$\dot{H}_G = I_G \dot{\omega} = I_G \boldsymbol{\alpha} \qquad (c)$$

Thus, the application of $\sum \mathbf{M} = \dot{\mathbf{H}}$ results in $\sum \mathbf{M}_G = \mathbf{I}_G \boldsymbol{\alpha}$ in which each of the vectors is perpendicular to the *XY* plane. The equivalent scalar forms of this equation and of $\mathbf{R} = m\bar{a}$ are

$$\sum X = \frac{W}{g}\bar{a}_x ; \qquad \sum Y = \frac{W}{g}\bar{a}_y ; \qquad \sum \bar{M} = \bar{I}\alpha \qquad (13\text{-}2.1)$$

where the subscript *G* has been replaced by bar superscripts as is customary when referring to centroidal axes. These equations are valid for any plane motion of a rigid body. Their physical interpretation is shown in Fig. 13-2.2 which is the plane motion equivalent of Fig. 10-6.2.

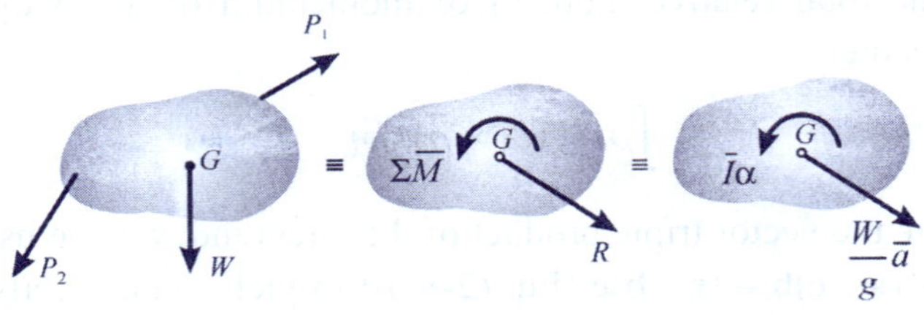

Figure 13-2.2 Equivalence of applied forces to their dynamic effects.

Observe that these three independent scalar equations are the coplanar kinetic equivalents of the three independent scalar equations of coplanar statics. Later in Chapter 16, we shall expand $\mathbf{R} = m\bar{a}$ and $\sum \mathbf{M} = \dot{\mathbf{H}}$ to obtain six independent scalar equations for spatial motion. It is recommended that you clearly identify the number of unknowns in any problem to ensure that there are a corresponding number of independent equations available to determine them. In this regard, the suggestions made in Chapters 3 and may be useful.

Before proceeding with the application of these equations, it may be informative to consider other methods of obtaining $\sum \overline{\mathbf{M}} = \bar{\mathbf{I}}\alpha$. A direct evaluation of the *magnitude* of $\mathbf{H}_G$ may be found, without using vector analysis in this plane motion, by summing up the moments of relative momentum $v_{B}/_{G}$ dm of all typical particles such as B of mass dm in Fig. 13-2.3. Clearly, the relative velocity $v_{B/G}$ of B about G is $v_{B/G} = \rho\omega$ whence the total relative angular momentum is

$$H_G = \int \rho(v_{B/G}\, dm) = \omega \int p^2\, dm = \bar{I}\omega$$

and, from differentiating this expression for H_G, we obtain again

$$\left[\sum M = \dot{H}\right] \qquad \sum M = \bar{I}\dot{\omega} = \bar{I}\alpha$$

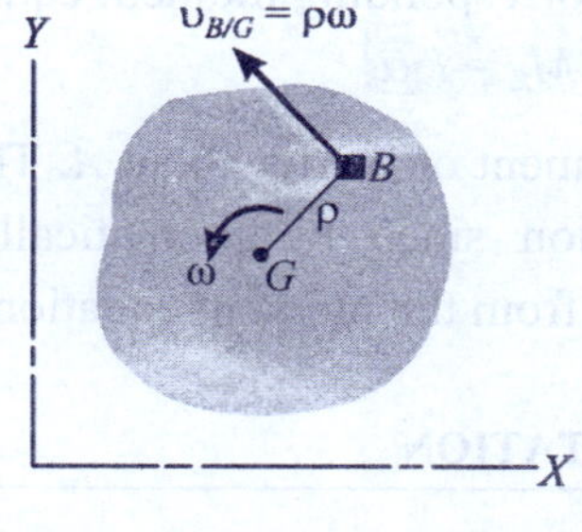

Figure 13-2.3 Moment of momentum of B about G is $p(v_{B/G}\, dm)$.

Another fundamental way of obtaining this result is to equate the moments of the applied forces about the mass center G to the moment sum about G of the effective forces on each particle. Thus, a typical particle at B is subjected to the effective force system shown in Fig. 13-2.4 found by multiplying each term of $a_B = a_G \nrightarrow (a_{B/G} = r\omega^2 \nrightarrow r\alpha)$ by dm. Summing their moments about G eliminates all normal effective forces like $r\omega^2\, dm$ which pass through G and also the effects of the uniformly distributed parallel force system of a_G dm acting on all particles, since their resultant is $a_G \int dm = m a_G$ which passes through G. We are left with only the moment sum $\int r(r\alpha\, dm)$ of the tangential effective forces, resulting once more in

$$\sum \bar{M} = \alpha \int r^2 dm = \bar{I}\alpha$$

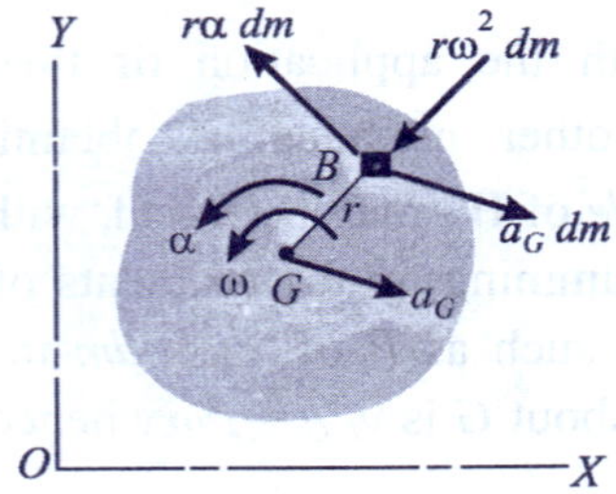

Figure 13-2.4 Effective forces on particle *B*.

Finally, since $\Sigma\mathbf{M} = \dot{\mathbf{H}}$ is also valid about a center *A* which is fixed in space, it should be observed that if such a point were substituted for *G* in the preceding derivations, the corresponding moment equation becomes

$$\Sigma M_A = I_A\alpha \tag{13-2.2}$$

where I_A is the mass moment of inertia about *A*. This equation is particularly useful in fixed-axis rotation since it automatically eliminates the usually unknown axis reaction at *A* from the moment equation.

13-3 FIXED-AXIS ROTATION

A common approach to all problems in plane motion consists of the following method:

1. Determine the kinematic relation between the acceleration $\bar{a}$ of the mass center and the angular acceleration α of the body.
2. Apply Eq. (13-2.1) diagramatically by equating a free body diagram of the body to a diagram of the equivalent dynamic effects.
3. An alternative to item (2) is to use a single diagram showing both the applied forces and the *reversed dynamic* effects, thereby creating dynamic equilibrium. The principal advantage of dynamic equilibrium is that only one diagram need be drawn, and no special sign conventions need be observed since force and moment summations are zero with respect to any axis or moment center.

We now illustrate how this common approach is applied to fixed-axis rotation.

CENTROIDAL ROTATION

One of the commonest types of fixed-axis rotation is centroidal rotation in which the axis of rotation passes through the mass center.

Since $\bar{a} = 0$, the equivalent diagrams of applied forces and dynamic effects are as shown in Fig. 13-3.1 from which we deduce the following equations:

$$\sum X = 0; \qquad \sum Y = 0; \qquad \sum \overline{M} = \bar{I}\alpha;$$

The bar sign is used to indicate that both $\sum \overline{M}$ and $\bar{I}$ are taken with respect to a centroidal axis of rotation.

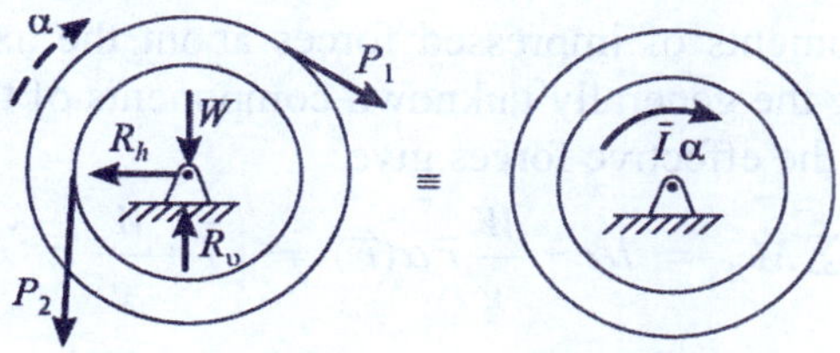

Figure 13-3.1 Applied forces and equivalent dynamic effect in centroidal rotation.

Since both $\sum X$ and $\sum Y$ equal zero for a centroidal rotation, the resultant of the impressed forces always is a couple of magnitude $\sum \overline{M}$. The converse of this observation is that if the force system applied to a body reduces to a couple, the body will undergo a centroidal rotation. The resultant couple will create centroidal rotation even if there is no fixed axis of rotation as we mentioned initially in Section 9-1.

NONCENTROIDAL ROTATION

Consider now the rigid body of weight W in Fig. 13-3.2 which is constrained to rotate about a noncentroidal axle at A. The action of the applied forces W and P, as well as the components of the bearing reaction at A, give the body the instantaneous values of the angular velocity ω and angular acceleration α as shown. The mass center G moves in a circle of radius $\bar{r}$ from the axis of rotation at A and has an acceleration $\bar{a}$ whose components are $\bar{a}_n = \bar{r}\omega^2$ directed toward A and $\bar{a}_t = \bar{r}\alpha$ directed perpendicular to $\bar{r}$ in the sense of α. The corresponding components of the resultant force $R = \left(\dfrac{W}{g}\right)\bar{a}$ therefore act as shown and, together with $\bar{I}\alpha$ in the sense of α, compose a dynamic system equivalent to the impressed forces.

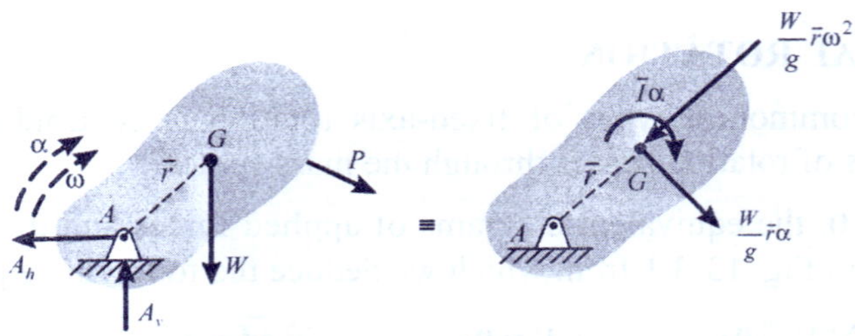

Figure 13-3.2 Equivalence of applied forces and dynamic effects in non-centroidal rotation.

Equating the moments of impressed forces about the axis of rotation at A (thereby eliminating the generally unknown components of the axle reaction) to the moment sum of the effective forces give

$$\sum M_A = \bar{I}\alpha + \frac{W}{g}\bar{r}\alpha(\bar{r}) = \left(\bar{I} + \frac{W}{g}\bar{r}^{\,2}\right)\alpha$$

in which we recognize that the term in parentheses means that the centroidal moment of inertia is transferred from G through the distance $\bar{r}$ to the axle at A. Consequently, we obtain

$$\sum M_A = I_A\alpha$$

which corresponds with Eq. (13-2.2) as obtained directly from $\sum \mathbf{M} = \dot{\mathbf{H}}$.

In summary, the general equations of noncentroidal rotation of symmetrical body are

$$\sum X = \frac{W}{g}\bar{r}\omega^2; \qquad \sum Y = \frac{W}{g}\bar{r}\alpha$$
$$\sum \overline{M} = \bar{I}\alpha \quad \text{or} \quad \sum M_A = \bar{I}_A\alpha \tag{13-3.2}$$

where the X and Y axes through the mass center are positive in the directions of $\bar{a}_n$ and $\bar{a}_t$, respectively. The direction of $\bar{a}_n$ will always be toward the center of rotation; that of $\bar{a}_t$ is determined from the sense of α. The sense of $\sum \overline{M}$ and $\sum M_A$ is positive in the initial direction of rotation.

In the practical application of Eqs. (13-3.2), it is usually best apply $\sum M_A = I_A\,\alpha$ directly to a free-body diagram of the impressed forces, thereby eliminating the unknown axle reaction. Force summations can then be obtained from equating the free-body diagram to its equivalent dynamic effects, or else from creating dynamic equilibrium by adding the reversed dynamic effects to the free-body diagram. Both procedures are discussed in the illustrative problems.

ILLUSTRATIVE PROBLEMS

13-3.1 The pulley assembly shown in Fig. 13-3.3 weighs 675 N and has a centroidal radius of gyration of 0.6 m. The blocks are attached to the assembly by cords wrapped around the pulleys. Determine the acceleration of each body and the tension in each cord.

Solution

For any connected system, apply the procedure outlined in Section 11-3. Thus, after drawing a FBD of each body and determining the direction of motion, we next determine the kinematic relations among the bodies, and finally apply the kinetic equation appropriate to the motion of each body.

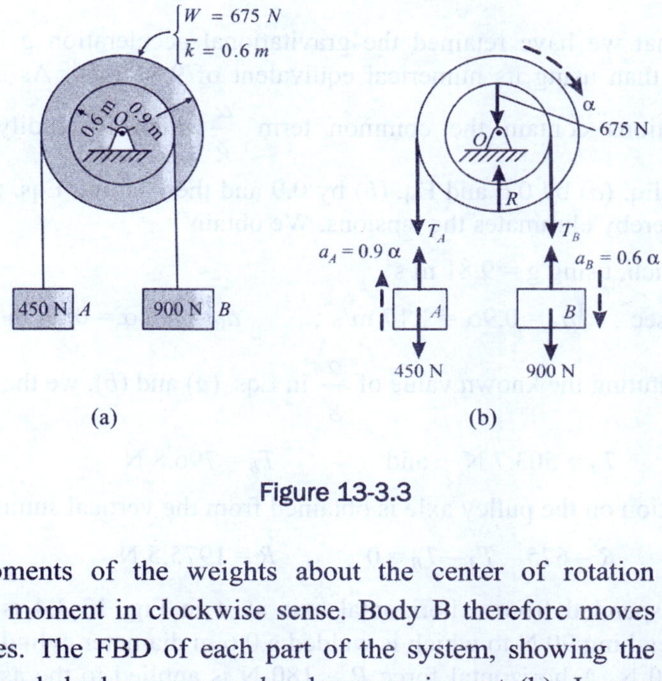

Figure 13-3.3

The moments of the weights about the center of rotation O give an unbalanced moment in clockwise sense. Body B therefore moves down while body A rises. The FBD of each part of the system, showing the direction of motion of each body, can now be drawn as in part (b). In accordance with Section 12-2, the linear accelerations of A and B are each expressed in terms of the angular acceleration of the pulley by $a_t = r\alpha$, or in this case by $a_A = 0.9\alpha$ and $a_B = 0.6\alpha$ which we draw as dashed vectors adjacent to each body. Their directions also indicate the positive sense of force and moment summations.

The kinetic equations for the translating bodies and the rotating pulley are respectively $\Sigma X = \left(\frac{W}{g}\right) a$ and $\Sigma \overline{M} = \overline{I}\alpha$. We recall that the moment of inertia is given by $\overline{I} = \left(\frac{W}{g}\right)\overline{k}^2$ and the application of these equations gives

For *B*: $$900 - T_B = \frac{900}{g} a_B = \frac{900}{g}(0.6\alpha) \qquad (a)$$

For *A*: $$T_A - 450 = \frac{450}{g} a_A = \frac{450}{g}(0.9\alpha) \qquad (b)$$

For pulley: $$0.6\,T_B - 0.9T_A = \frac{675}{g}(0.6)^2\alpha = \frac{243}{g}\alpha \qquad (c)$$

Notice that we have retained the gravitational acceleration g in symbolic form rather than using its numerical equivalent of 9.81 m/s^2. As a result, the above equations contain the common term $\frac{\alpha}{g}$ which is readily found by multiplying Eq. (*a*) by 0.6 and Eq. (*b*) by 0.9 and then adding Eqs. (a), (b), and (c) which thereby eliminates the tensions. We obtain

From which, using g = 9.81 m/s^2

$\alpha = 1.3$ rad/sec^2; $a_A = 0.9\alpha = 1.17$ m/s^2; $a_B = 0.6\,\alpha = 0.78$ m/s^2

By substituting the known value of $\frac{\alpha}{g}$ in Eqs. (*a*) and (*b*), we then find

$T_A = 503.7$ N and $T_B = 796.8$ N ***Ans.***

The reaction on the pulley axle is obtained from the vertical summation

$[\Sigma V = 0]$ $R - 675 - T_A - T_B = 0$ $R = 1975.5$ N ***Ans.***

13-3.2 Suspended from a horizontal axis at *A* in Fig. 13-3.4 is a 2 m rod weighing 90 N to which is welded a 0.6 m diameter cylinder weighing 140 N. A horizontal force $P = 180$ N is applied to the assembly at a distance *d* from *A*. (a) Find the horizontal component A_h of the bearing reaction at *A* when $d = 0.6$ m. (b) Find the distance *d* at which *P* should act so that A_h will be zero. This position of *P* is called the *center of percussion*.

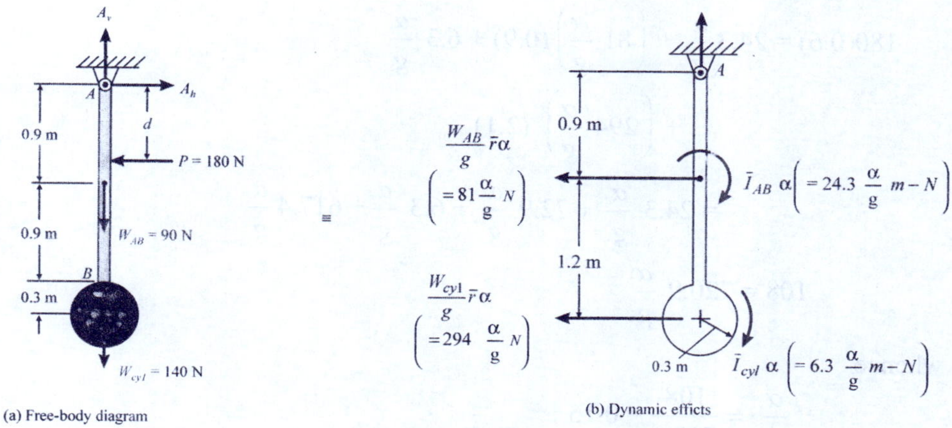

(a) Free-body diagram (b) Dynamic effects

Figure 13-3.4

Solution: Part (a)

The FBD of impressed forces in (a) is equivalent to the dynamic effects in (b). For each part of the assembly, these consist of the effective force $\left(\frac{W}{g}\right)\bar{a}$ (having only a component $\left(\frac{W}{g}\right)\bar{r}\alpha$ since $\omega = 0$) and a couple $\bar{I}\alpha$. Their values are

$$\text{Rod:}\quad \frac{W}{g}\bar{r}\alpha = \frac{90}{g}(0.9)\alpha = 81\frac{\alpha}{g}\ \text{N}; \qquad \bar{I}\alpha = \frac{1}{12}\frac{W}{g}L^2\alpha = \frac{90}{12\,g}(1.8)^2\alpha = 24.3\frac{\alpha}{g}\ \text{N-m}$$

$$\text{Cyl.:}\quad \frac{W}{g}\bar{r}\alpha = \frac{140}{g}(2.1)\alpha = 294\frac{\alpha}{g}\ \text{N}; \qquad \bar{I}\alpha = \frac{1}{2}\frac{W}{g}r^2\alpha = \frac{140}{2g}(0.3)^2\alpha = 6.3\frac{\alpha}{g}\ \text{N-m}$$

To eliminate the unknown components of the bearing reaction, we equate moments about A of the impressed forces and the equivalent dynamic effects. Noting that the moment of the effective couples are independent of any moment center, we obtain

$$[\circlearrowleft + (\Sigma M_A)_{\text{FBD}} = (\Sigma M_A)_{\text{dyn}}]$$

$$180(0.6) = 24.3\frac{\alpha}{g} + \left(81\frac{\alpha}{g}\right)(0.9) + 6.3\frac{\alpha}{g} + \left(294\frac{\alpha}{g}\right)(2.1)$$

$$= 24.3\frac{\alpha}{g} + 72.9\frac{\alpha}{g} + 6.3\frac{\alpha}{g} + 617.4\frac{\alpha}{g}$$

$$108 = 720.9\frac{\alpha}{g}$$

whence

$$\frac{\alpha}{g} = \frac{108}{720.9} = 0.15$$

Equating horizontal force summations for each system, we get

$$[\overset{+}{\leftarrow}(\Sigma F_h)_{FBD} = (\Sigma F_h)_{dyn}]$$

$$180 - A_h = 81\frac{\alpha}{g} + 294\frac{\alpha}{g} = 375\frac{\alpha}{g}\left(\frac{\alpha}{g} = 0.15\right)$$

$A_h = 123.75$ N **Ans.**

A few comments are pertinent. First, numerical simplicity is achieved by solving for $\frac{\alpha}{g}$ from which α can be found if desired. Second, if only α were to be found, we could have applied $\Sigma M_A = I_A\,\alpha$ directly to the FBD in (a), but to determine I_A requires transferring each centroidal moment of inertia to A as follows:

$$[I = \Sigma(\bar{I} + M\,d^2)] \qquad I_A = \frac{24.3}{g} + \frac{90}{g}(0.9)^2 + \frac{6.3}{g} + \frac{140}{g}(2.1)^2$$

$$= \frac{720.9}{g} \text{ m-N-sec}^2$$

whence

$$[\Sigma M_A = I_A\alpha] \qquad 180\,(0.6) = \frac{720.9}{g}\alpha \quad \text{or} \quad \frac{\alpha}{g} = \frac{108}{720.6} = 0.15$$

as before. Notice that the previous solution automatically included the transfer terms for moments of inertia.

Part (b)

Here both the location of P and the value of $\dfrac{\alpha}{g}$ are unkrown, but $A_h = 0$. Hence, equating force summations (with $A_h = 0$), we get

$$[\overset{+}{\leftarrow}(\Sigma F_h)_{FBD} = (\Sigma F_h)_{dyn}] \qquad P = 81\frac{\alpha}{g} + 294\frac{\alpha}{g}, \qquad \frac{\alpha}{g} = \frac{P}{375}$$

With this value of $\dfrac{\alpha}{g}$ the moment summations obtained previously gives

$$[\circlearrowleft + (\Sigma M_A)_{FBD} = (\Sigma M_A)_{dyn}] \qquad Pd = 720.9\frac{\alpha}{g} = 720.9\left(\frac{P}{375}\right)$$

$$d = \frac{720.9}{375} = 1.93 \text{ m} \qquad \textbf{\textit{Ans.}}$$

Note that the value of d is independent of the value of P. In other words, the center of percussion is a unique property of the body which depends only on the distribution of its mass.

13-3.3 A turntable rotating in a horizontal plane about a vertical axis O carries a bent bar weighing 225 N/m pinnec to it at A and forced to rotate with it by a smooth peg at C. At the instant shown in Fig. 13-3.5, $\omega = 4$ rad/sec and $\alpha = 6$ rad/sec^2 both clockwise. Determine the forces acting on the bar at A and C.

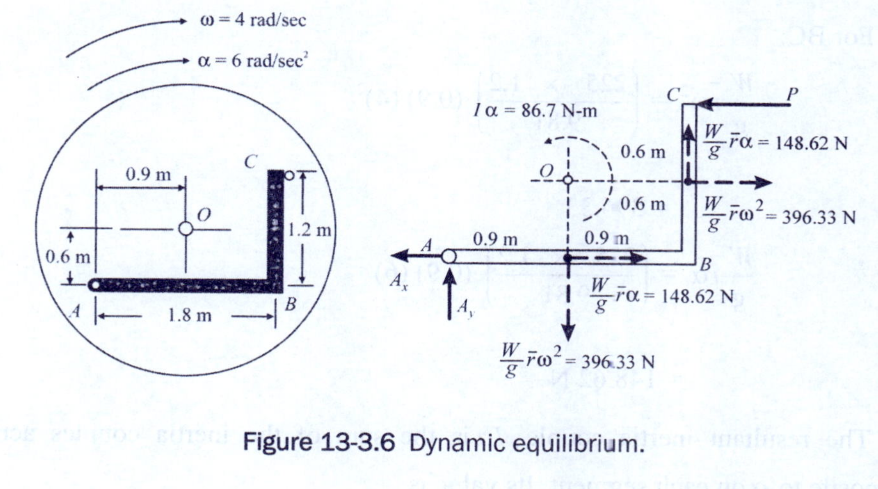

Figure 13-3.6 Dynamic equilibrium.

Solution

Instead of drawing two separate diagrams of impressed forces and equivalent dynamic effects, consider here the method of dynamic equilibrium which permits a free choice of axes and moment centers. Rather than locating the mass center of the entire bent bar, it is more convenient to apply the inertia components acting as shown in Fig. 13-3.6 at the mass center of each segment. Each of these inertia components act respectively opposite to $\bar{a}_n$ and $\bar{a}_t$ of each segment.

The values of the inertia components are

For AB:

$$\frac{W}{g}\bar{r}\omega^2 = \left(\frac{225 \times 1.8}{9.81}\right)(0.6)(4)^2$$

$$= 396.33 \text{ N}$$

$$\frac{W}{g}\bar{r}\alpha = \left(\frac{225\times 1.8}{9.81}\right)(0.6)(6)$$

$$= 148.62 \text{ N}$$

For BC:

$$\frac{W}{g}\bar{r}\omega^2 = \left(\frac{225 \times 1.2}{9.81}\right)(0.9)(4)^2$$

$$= 396.33$$

$$\frac{W}{g}\bar{r}\alpha = \left(\frac{225 \times 1.2}{9.81}\right)(0.9)(6)$$

$$= 148.62 \text{ N}$$

The resultant inertia couple $\bar{I}\alpha$ is the sum of the inertia couples acting opposite to α on each segment. Its value is

$$\bar{I}\alpha = \Sigma\left(\frac{1}{12}\frac{W}{g}L^2\right)\alpha = \left[\frac{1}{12}\left(\frac{405}{9.81}\right)(1.8)^2\right](6) +$$

$$\left[\frac{1}{12}\left(\frac{270}{9.81}\right)(1.2)^2\right](6)$$

$$\bar{I}\alpha = 86.7 \text{ N-m}$$

We now apply the equations of dynamic equilibrium. The value of P is determined from a moment summation about A.

$[+\curvearrowleft \Sigma M_A = 0]$ $\quad 1.2P + 148.62(1.8) - 396.33\ (0.6)$

$$- 396.33\ (0.9) + 86.7 = 0$$

$P = 200$ N **_Ans._**

Using force summations directed along the perpendicular components of the reaction at A, we obtain

$[\overset{+}{\leftarrow} \Sigma X = 0]$ $\quad A_x + (P = 200) - 396.33 - 148.62 = 0$ $\quad A_x = 345$ N

$[+\uparrow \Sigma Y = 0]$ $\quad A_y - 396.33 + 148.62 = 0$ $\quad A_y = 247.71$ N

from which the total reaction at A is found to be

$$A = \sqrt{[345]^2 + (24771)^2} = 424.71 \text{ N}$$

PROBLEMS

13-3.4 A weight of 430 N is suspended from a cord wrapped around a solid cylinder of 0.9 m radius weighing 1425 N. The cylinder rotates about a horizontal axle through its mass center. Compute the total bearing reaction on this axle.

13.3.5 What torque applied to the cylinder of the previous problem will raise the weight with an acceleration of 3.6 m/s^2? What will be the total bearing reaction?

13.3.6 During the operation of a punch press, its flywheel decelerates uniformly from 600 rpm to 400 rpm in 1 sec. The rim of the flywheel weighs 5725 N, its inside and outside diameters are 1.4 m, and 1.5 m, and it is attached to its hub by 6 spokes. What average shearing force is developed between the rim and each spoke during the 1-sec interval?

13.3.7 The train of gears in Fig. P-13.3.7 is represented by their pitch circles. Their centrodial moments of inertia, in units of N-m-sec^2, are $\bar{I}_A = 0.15$, $\bar{I}_B = 0.4$, and $\bar{I}_C = 0.12$. Determine the angular acceleration of gear A if the difference in torques applied at *A* and removed at *C* is 0.2 N-m.

Hint: Create dynamic equilibrium applying inertia couples to each gear.

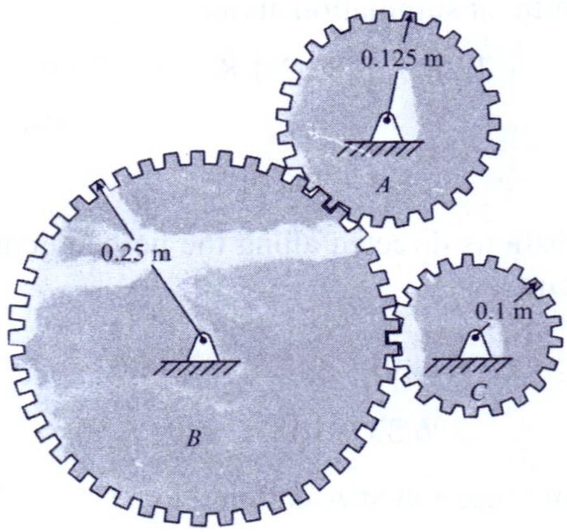

Figure P-13-3.7

13.3.8 A 14 KN flywheel is mounted at the midpoint of a horizontal shaft 1.8 m long. The mass center of the flywheel is 2.5 mm from the centerline of the shaft. When the flywheel is rotating at a constant speed of 1800 rpm, determine the maximum and minimum values of the bearing reactions at each end of the shaft.

13-3.9 A uniform bar, 3 m long and weighing 20 N/m, is rotating in a horizontal plane about a vertical axis at one end. If the angular velocity is constant at 4 rad/sec, compute the axial force on a transverse section at its midpoint.

13-3.10 A 1.8 m rod weighing 450 N is rigidly fastened to a 900 N cylinder as shown in Fig. P-13-3.10. Find the linear acceleration of the 1350 N block at the given position.

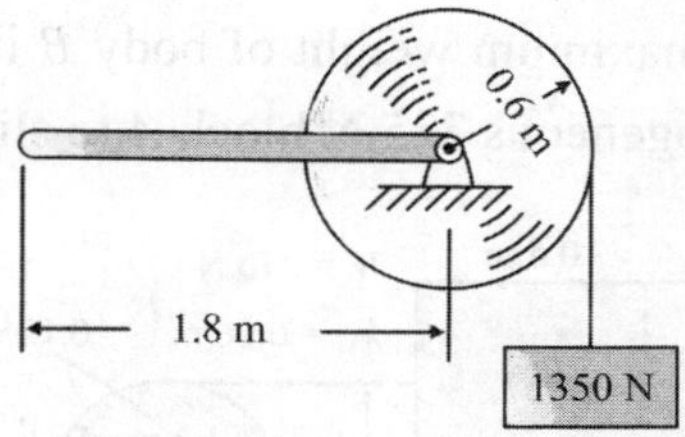

Figure P-13-3.10

13-3.11 If the coefficient of kinetic friction is 0.25 under each block in Fig. P-13-3.11, compute the total reaction at the axle of the compound cylinder.

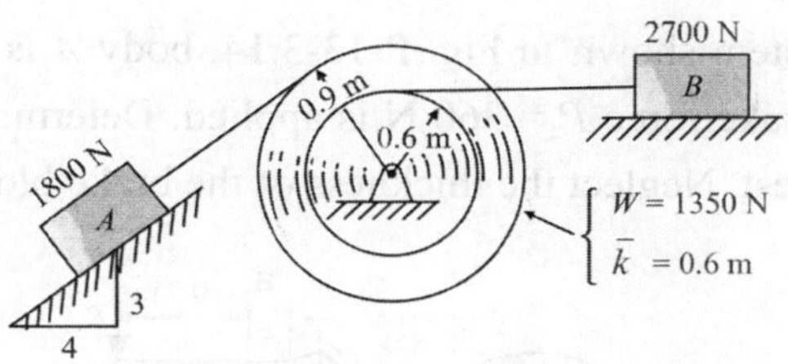

Figure P-13-3.11

13-3.12 For the system shown in Fig. P-13-3.12, determine the minimum and maximum weight of body *C* that will keep it at rest. For each incline, the coefficient of friction is 0.25.

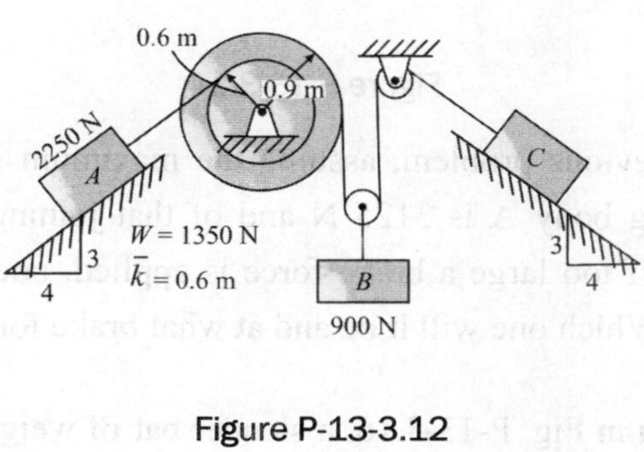

Figure P-13-3.12

13-3.13 Determine the maximum weight of body B in Fig. P-13-3.13 that will permit the homogeneous 225 N block A to slide without tipping over.

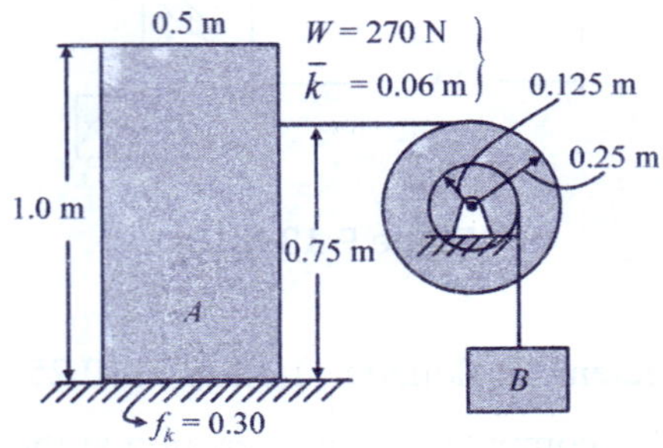

Figure P-13-3.13

13-3.14 In the system shown in Fig. P-13-3.14, body A is dropping at 12 m/s when a brake force P = 360 N is applied. Determine the time for A to come to rest. Neglect the thickness of the brake block.

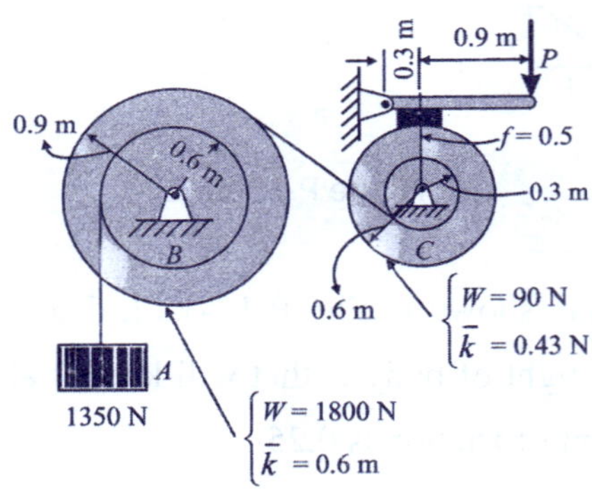

Figure P-13-3.14

13-3.15 In the previous problem, assume the maximum strength of the cord supporting body A is 3125 N and of that joining drums B and C is 3600 N. If too large a brake force is applied, one of these cords will rupture. Which one will it be and at what brake force P?

13-3.16 As shown in Fig. P-13-3.16, a slender bar of weight W and length L is supported at one end and at a distance b from its mass center G.

Determine the reaction at A the instant after the support at B is suddenly removed.

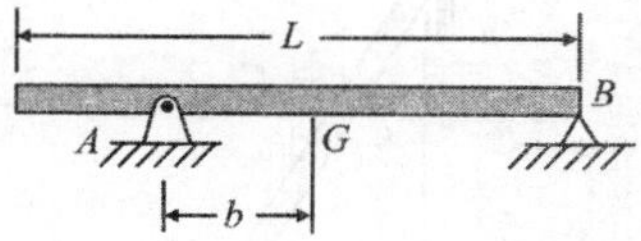

Figure P-13-3.16

13-3.17 A cylinder weighing 900 N is welded to the end of a 450 N bar. The assembly is supported by a horizontal axis at A and by a vertical cable as shown in Fig. P-13-3.17. Compute the reaction at A an instant after cutting the cable.

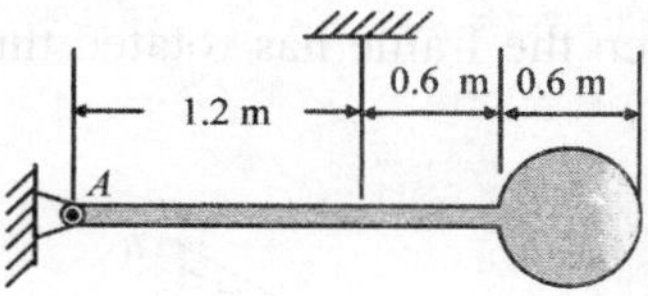

Figure P-13-3.17

13-3.18 At the instant the assembly of the previous problem has rotated through an angle $\theta = \tan^{-1}\frac{3}{4}$ from the horizontal, its angular velocity is known to be $\omega = \sqrt{0.2g}$ rad/sec. What is then the magnitude of the axle reaction at A?

13-3.19 A uniform slender rod of length L is hinged to a frame rotating about a vertical axis as in Fig. P-13-3.19. Show that the angle between the rod and the axis is defined by $\cos\theta = 3g/2L\omega^2$.

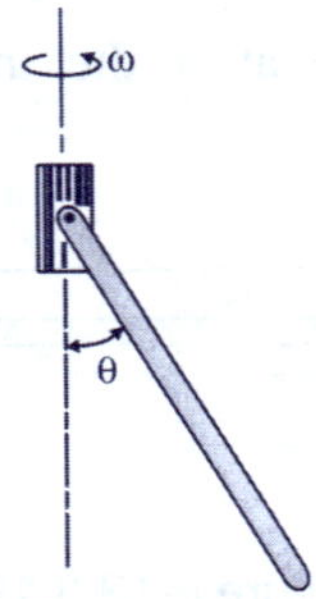

Figure P-13-3.19

13-3.20 Three bars, each 0.6 m long and weighing 43 N, are pinned together to form the equilateral frame shown in Fig. P-13-3.20. They rotate in a horizontal plane about a vertical axis at A. What torque M is required to cause a constant angular acceleration of 12 rad/sec^2? What is the reaction at A when the frame has rotated through $\frac{2}{3}$ rad starting from rest?

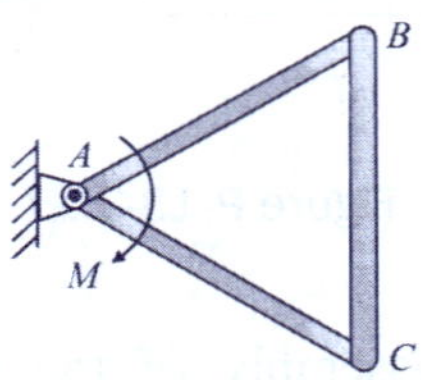

Figure P-13-3.20

13-3.21 Two bars are welded together as shown in Fig. P-13-3.21. The assembly rotates about a horizontal axis at A. Segment AB weighs 300 N and segment CD weighs 150 N. At the given position it is known that $\omega = \sqrt{0.3g}$ rad/sec. Determine the horizontal and vertical components of the axle reaction and the moment transmitted at B by AB upon CD.

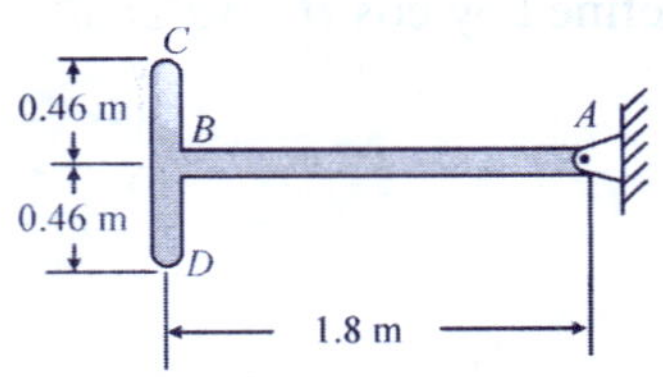

Figure P-13-3.21

13-3.22 A turntable, rotating in a horizontal plane about a vertical axis O, carries a bent bar pinned to it at A and forced to rotate with it by a smooth peg at C. Each segment of the bar weighs 10 N. At the instant shown in Fig. P-13-3.22, $\omega = \sqrt{20}$ rad/sec and $\alpha = 10$ rad/sec^2 both clockwise. Determine the forces acting at A and C.

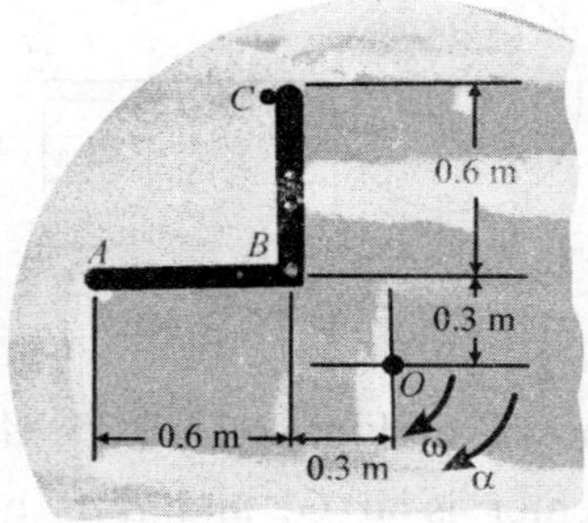

Figure P-13-3.22

$A = 63.28$ N; $C = 71.21$N *Ans.*

13-3.23 The bent bar shown in Fig. P-13-3.23 weighs 150 N/m and is free to rotate in a vertical plane about a horizontal axis at A. Determine the moment at B exerted by AB on BC an instant after the bar is released from rest at the given position.

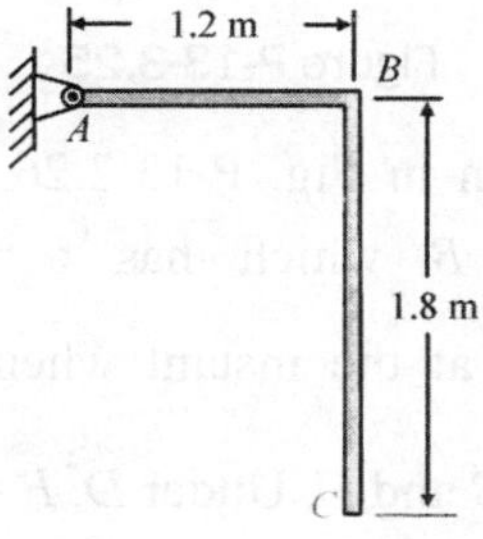

Figure P-13-3.23

13-3.24 A 0.3 m, bar weighing 15 N is welded to the rim of an 360 N cylinder as shown in Fig. P-13-3.24. Find the moment exerted on the weld at B when the system is released from rest at the given position where $\tan\theta = \dfrac{3}{4}$.

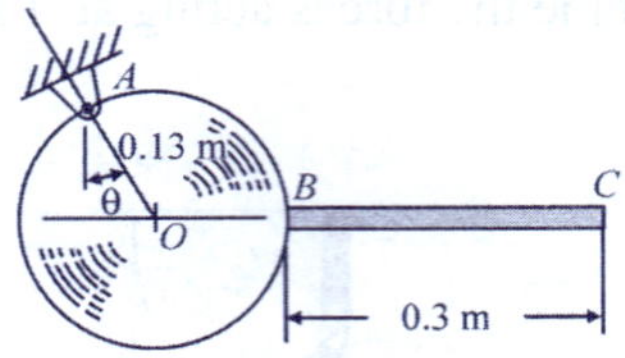

Figure P-13-3.24

13.3.25 In the system shown in Fig. P-13-3.25, A is an unbalanced wheel whose gravity center is at G, and B is a solid cylinder. Find the components of each axle reaction if $\omega_A = \sqrt{g}$ rad/sec at the instant shown.

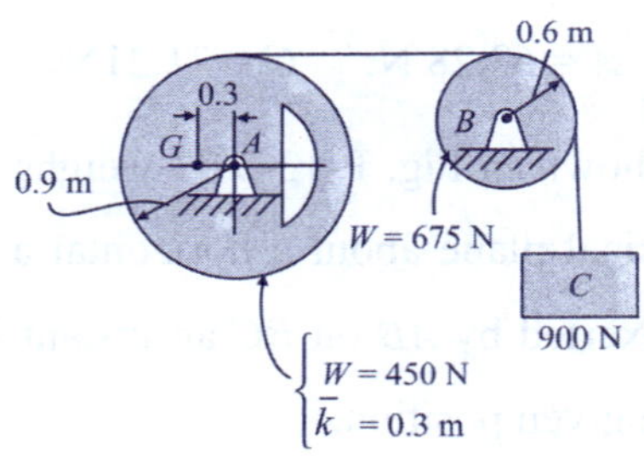

Figure P-13-3.25

13-3.26 In the system shown in Fig. P-13-3.26, a 225 N bar 1.8 m long is attached to drum B which has a clockwise angular velocity $\omega = \sqrt{g/3}$ rad/sec at the instant when $\tan\theta = \dfrac{3}{4}$. Find the axle reactions on drums B and C. Under D, $F = 0.225$.

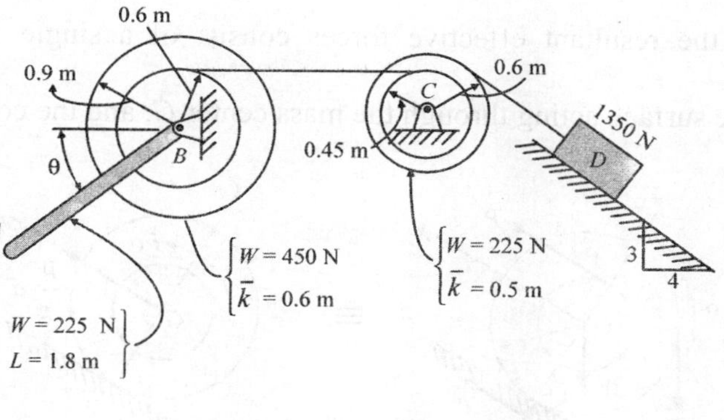

Figure P-13-3.26

13.3.27 Two eccentric weights $W_1 = 450$ N and $W_2 = 900$ N are fastened to the rotating horizontal shaft as shown in Fig. P-13-3.27. Compute the values of balance weights concentrated 0.3048 m from the shaft and rotating in vertical planes through *A* and *B* that will balance the dynamic effects of W_1 and W_2. What are the angular positions of the balance weights measured from the plane containing W_1 and axis *AB*?

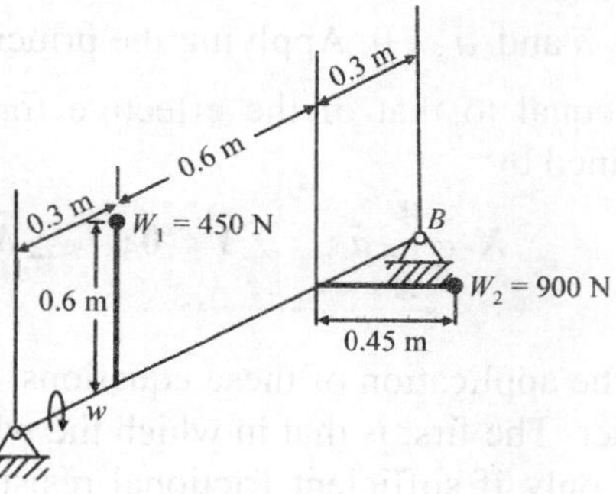

Figure P-13-3.27

13-4 ROLLING BODIES

Let us consider next the application of Eq. (13-2.1) to rolling bodies, postponing the discussion of general plane motion to the next section. The case of the *homogeneous* rolling wheel is simplified by the fact that its mass center has a rectilinear motion parallel to the flat surface on which it rolls. Then, as shown in

Fig. 13-4.1, the resultant effective forces consist of a single force $\left(\frac{W}{g}\right)\bar{a}$ parallel to the surface acting through the mass center G, and the couple $\bar{I}\alpha$.

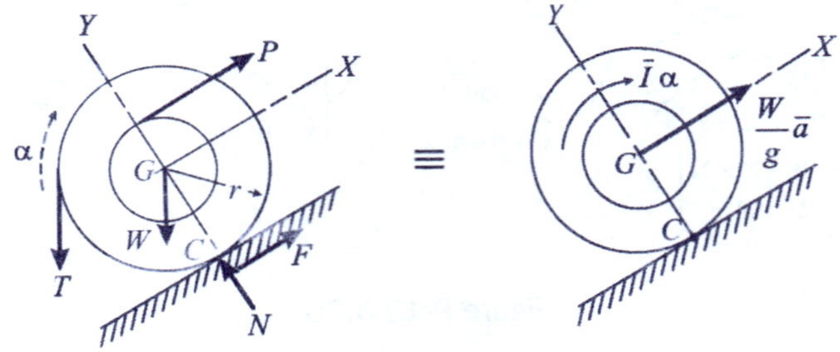

Figure 13-4.1 Equivalence of applied forces and resultant effective forces. With a free-rolling wheel, $\bar{a} = r\alpha$ and $\frac{\omega}{g}\bar{a}$ becomes $\frac{\omega}{g}rd$.

The best reference axes to choose are those which pass through the mass center with the X axis directed parallel to the surface on which the body rolls and considered positive in the initial direction of motion. The components of $\bar{a}$ are then given by $\bar{a}_x = \bar{a}$ and $\bar{a}_y = 0$. Applying the principle that the resultant of the applied forces is equal to that of the effective forces, the motion of the rolling body is determined by

$$\Sigma X = \frac{W}{g}\bar{a}; \quad \Sigma Y = 0; \quad \Sigma \bar{M} = \bar{I}\alpha \qquad (13\text{-}4.1)$$

Before discussing the application of these equations, there are two important special cases to consider. The first is that in which the wheel rolls freely without slipping. It can do so only if sufficient frictional resistance acts at the instant center C to hold that point instantaneously at rest. The value of this static frictional resistance is generally unknown since it may have any value ranging from zero up to the limiting static friction force $f_s N$. It is therefore convenient to eliminate this unknown static frictional force by taking moments about the instant center C. Thus, by equating the moment sum about C of the applied forces (i.e., ΣM_C) to the moment sum of the effective forces, we obtain

$$\Sigma M_C = \bar{I}_C\alpha + \left(\frac{W}{g}\bar{a}\right)r \qquad (a)$$

When free-rolling exists, then $\bar{a} = r\alpha$, so that Eq. (a) becomes

$$\Sigma M_C = \bar{I}\alpha + \left(\frac{W}{g} r\alpha\right) r = \left(\bar{I} + \frac{W}{g} r^2\right)\alpha$$

in which we recognize that the sum in parentheses is an application of the transfer formula and represents the moment of inertia I_C about the instant center C. Consequently, for free-rolling wheels, we obtain

$$\Sigma M_C = \bar{I}_C\,\alpha \tag{13-4.2}$$

If slipping occurs, the position of the instant center is unknown and the relation $\Sigma M_C = I_C\alpha$ cannot be used. However, Eqs. (13-4.1) are always valid, whether or not free rolling exists. Observe also that for a homogeneous free-rolling wheel, we could obtain $\Sigma M_C = I_C\alpha$ directly from $\Sigma \mathbf{M} = \dot{\mathbf{H}}$ since the acceleration of the instant center is directed through the mass center (see Section 12-5) which is an other condition for which the general moment equation is valid.

The second special case is that of the unbalanced free-rolling wheel or cylinder in Fig. 13-4.2 whose mass center G does *not* coincide with the geometric center O. The mass center in this case does *not* have a rectilinear motion and consequently Eq. (13-4.2) *does not apply*. Nevertheless, $\Sigma M_C = I_C\alpha$ is still valid at the two instants when G crosses the line containing the geometric center O and the instant center C because at such positions the acceleration of the instant center is directed toward the mass center.

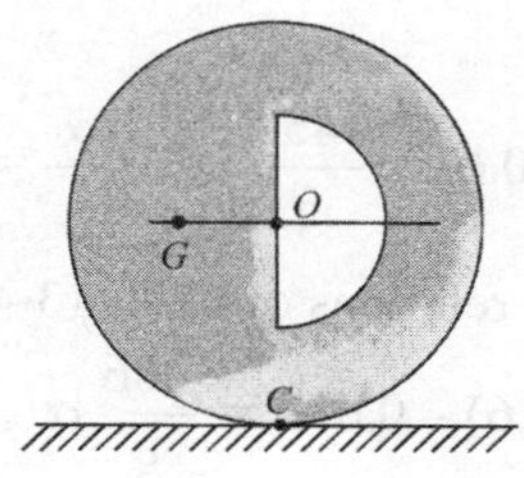

Figure 13-4.2 Unbalanced wheel.

ILLUSTRATIVE PROBLEMS

13-4.1 The solid cylinder in Fig. 13-4.3 is 1.2 m in diameter and weighs 2700 N. It is acted upon by an upward force of 450 N applied by a cord wrapped around it. Find the coefficient of friction required to prevent slipping.

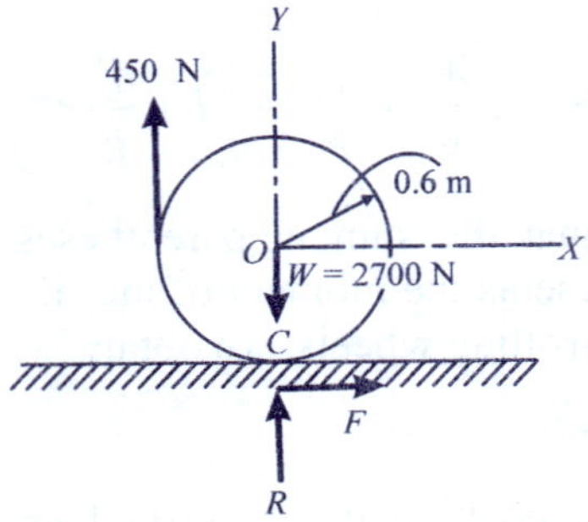

Figure 13-4.3

Solution: Part (a)
Since free rolling of the cylinder is implied in the statement of the problem, the instant center C is at the point of contact with the ground, and Eq. (13-4.2) may be used to obtain α. The centroidal mass moment of inertia of the cylinder is

$$\bar{I} = \frac{1}{2}\frac{W}{g}r^2 = \frac{1}{2}\left(\frac{2700}{g}\right)(0.6)^2 = \frac{486}{g} \text{ m-N-sec}^2$$

and the transfer formula gives the mass moment of inertia about the instant center as

$$[I = \bar{I} + m\,d^2] \qquad I_C = \frac{486}{g} + \left(\frac{2700}{g}\right)(0.6)^2 = \frac{1458}{g} \text{ m-N-sec}^2$$

We now have

$$[\curvearrowleft + \Sigma M_C = I_C\,\alpha] \quad 450\,(0.6) = \frac{1458}{g}\alpha \qquad \frac{\alpha}{g} = \frac{270}{1458} = \frac{1}{5.4}$$

whence, applying the general relations of Eqs. (13-4.1), we have

$$[\curvearrowleft + \Sigma \bar{M} = \bar{I}\alpha] \quad 450\,(0.6) - 0.6\;F = \frac{486}{g}\alpha$$

$$= 486\left(\frac{1}{5.4}\right) = 90$$

$$0.6\text{ F} = 270 - 90$$

$$\text{F} = 300\text{ N}$$

$$[+\downarrow \Sigma Y = 0] \quad R + 450 - 2700 = 0 \qquad R = 2250\text{ N}$$

Applying the friction relationship gives

$$\left[f_s = \frac{F}{R}\right] \qquad f_s = \frac{300}{2250} = 0.1334$$ ***Ans.***

Part (b)

If the coefficient of static friction were specified as 0.20, what would be the friction force? The available friction force would be found from the relation $F = f_s R = 0.2(2250 \text{ N}) = 450$ N. But not all of this is required to prevent slipping; only 300 N would be acting as before. This illustrates that the actual static friction force cannot be found from the relation defining the available static friction force.

Part (c)

If $f_s = 0.12$ and $f_k = 0.11$, what would happen? What would be the values of $\bar{a}$ and a? Under these conditions, the available static friction of 0.12(2250) = 270 N would be less than required to prevent slipping. Therefore, slipping would occur at the ground and a kinetic friction force $F_k = f_k.R = 0.11(2250) =$ 247.5 N would act. $\Sigma M_c = I_c \propto$ cannot then be used because it applies only to free-rolling bodies. Therefore, we must use the general relations of Eqs. (13-4.1) to obtain

$$[\circlearrowleft + \Sigma \overline{M} = \bar{I}\alpha] \quad 450\,(0.6) - 247.5\,(0.6) = \frac{486}{g}\alpha$$

$$\frac{486}{g} \times \alpha = 121.5$$

$$\alpha = \frac{121.5 \times 9.81}{486} = 2.45 \text{ rad/sec.}$$

$$\left[\overset{+}{\rightarrow} \Sigma X = \frac{W}{g}\bar{a}\right] \quad 247.5 = \frac{2700}{g}\bar{a} \qquad \bar{a} = 0.899 \text{ m/s}^2$$

Note that when slipping occurs, the relation between $\bar{a}$ and α for free rolling is not valid; i.e.,

$$[\bar{a} \neq r\alpha] \qquad 0.899 \neq 0.6(2.5)$$

13-4.2 In the system shown in Fig. 13-4.4, the floating pulley *D* is supported by a cord wound around the drum and a second cord wound around its outer radius. The second cord, after passing over a frictionless pulley of negligible weight, is wound around the drum of disk *A*. Determine the angular accelerations and the tensions *T* and *P* if both bodies roll without slipping.

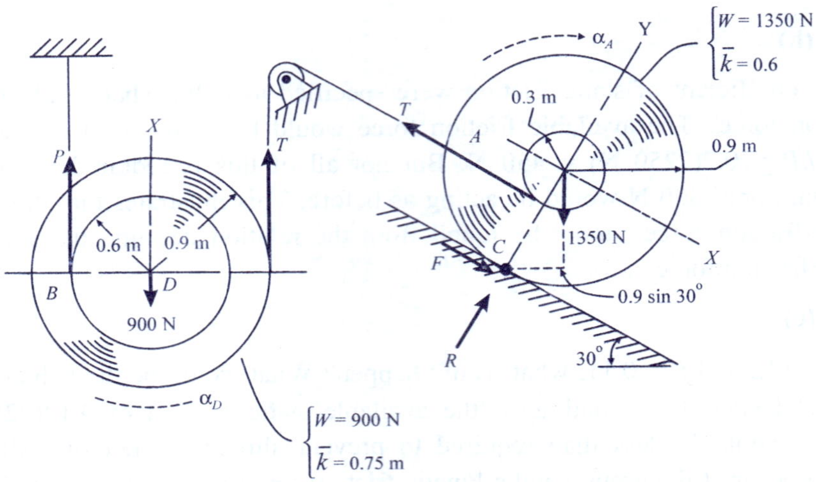

Figure 13-4.4

Solution

To determine whether motion occurs and in which direction, temporarily assume disk A to be at rest. Setting a moment summation about A's instant center C equal to zero, we find 0.6 T = 1350 (0.9 sin 30°) from which a preliminary value of T is 1012.5 N. Consider now disk D whose instant center is at B. If $\Sigma M_B = 0$, there will be equilibrium; but if $\Sigma M_B \neq 0$, the disk D will move in the direction of the unbalanced moment. Thus $\Sigma M_B = 1012.5(1.5) - 900(0.6)$ indicates an unbalanced counterclockwise moment acting on disk D. It therefore moves up, climbing along cord P which is wound upon its drum. Simultaneously, cord T, moving up, is unwound from disk D. The absolute motion of T is also *down* the incline, its point of attachment with disk A moving slower than the mass center of A, so that T also unwinds from A as disk A rolls down the incline. These directions of motion specify the positive directions of moments and forces.

The kinematic relations between disks A and D are found as described in Illus. Prob. 12-5.2. The bodies may be assumed to be instantaneously rotating about their instant centers B and C so that the motion of the cord which connects them is given by

$$[s = r\theta] \qquad s_T = 1.5\ \theta_D = 0.6\theta_A$$

whence successive differentiation yields the following relation between their accelerations:

$$1.5\alpha_D = 0.6\ \alpha_A$$

The centroidal mass moments of inertia are

$[\bar{I} = m\bar{k}^2]$ For A: $\bar{I} = \dfrac{1350}{g}(0.6)^2 = \dfrac{486}{g}$ N-m-sec^2

For D: $\bar{I} = \dfrac{900}{g}(0.75)^2 = \dfrac{506.25}{g}$ N-m-sec^2

whence, using the transfer formula, the values with respect to the instant centers are

$[I = \bar{I} + md^2]$ For A: $I_C = \dfrac{486}{g} + \dfrac{1350}{g}(0.9)^2 = \dfrac{1579.5}{g}$ N-m-sec^2

For D: $I_B = \dfrac{506.25}{g} + \dfrac{900}{g}(0.6)^2 = \dfrac{830.25}{g}$ N-m-sec^2

We are now ready to consider the kinetics of each disk. The friction force F acting on disk A is a static friction holding point C momentarily at rest. It is directed so as to resist rotation about the mass center. Since a static friction may vary from zero to its maximum available value, F will be unknown eventhough f_s may be specified. Only when slipping impends will F be determined by f_sR. However, this unknown friction force can be easily eliminated by taking advantage of the condition of free rolling and applying $\Sigma M_C = I_C\alpha$ which permits a moment summation about the instant center C through which F acts. Similarly, the unknown tension P on disk B is eliminated by a moment summation about its instant center B. Taking moments positive in the initial directions of motion, we obtain

$[\Sigma M_C = I_C\alpha]$ $1350\,(0.9 \sin 30^\circ) - 0.6\text{T} = \dfrac{1579.5}{g}\alpha_A$

$[\Sigma M_B = I_B\alpha]$ $1.5\text{ T} - 900\,(0.6) = \dfrac{830.25}{g}\alpha_D = \dfrac{830.25}{g}(0.4\,\alpha_A)$

The tension T is easily eliminated from these equations by multiplying Eq. (c) by $\dfrac{2}{5}$ and adding them, whence

$$\frac{\alpha_A}{g} = 0.23$$

and

$$\alpha_A = 2.26 \text{ rad/sec}^2; \quad \alpha_D = \frac{0.6}{1.5}\,\alpha_A = 0.9 \text{ rad/sec}^2 \qquad \textit{Ans.}$$

after which either value of α may be substituted in (*b*) or (*c*) to determine

$$T = 406 \text{ N} \qquad \textbf{\textit{Ans.}}$$

To determine *P*, we apply the general relations of Eq. (13-4.1) to disk *D*, using Σ*M*. We obtain

$$[+\uparrow \Sigma \overline{M} = \bar{I}\alpha] \qquad 406\,(0.9) - 0.6P = \frac{506.25}{9.81}(0.9)$$

$$P = 531.6 \text{ N} \qquad \textbf{\textit{Ans.}}$$

PROBLEMS

13-4.3 A solid cylinder and a homogeneous sphere, each of weight *W* and radius *r*, roll without slipping down a plane inclined at θ° with the horizontal. For each body, determine the minimum coefficient of friction to prevent slipping and the acceleration of the mass center.

13-4.4 In Fig. P-13-4.4, at what height h above the table should a billiard ball of radius *r* be stroked so that it will roll without slipping on a frictionless surface?

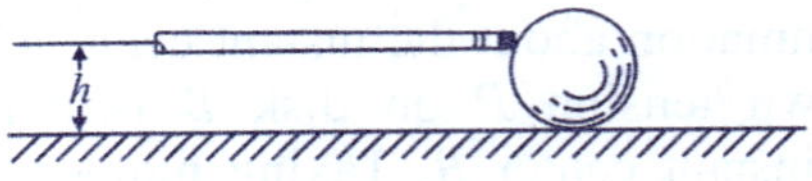

Figure P-13-4.4

13-4.5 A spherical shell and a circular hoop of the same weight and mean radius are side by side on an incline. If they are simultaneously released from rest, will the shell or the hoop reach the bottom of the incline first, or do they arrive together?

13-4.6 The body shown in Fig. P-13-4.6 consists of two solid cylinders each weighing 450 N. They are separated by a hub of negligible weight. Find the acceleration of the mass center, the friction force acting, the angular acceleration, and the direction of motion.

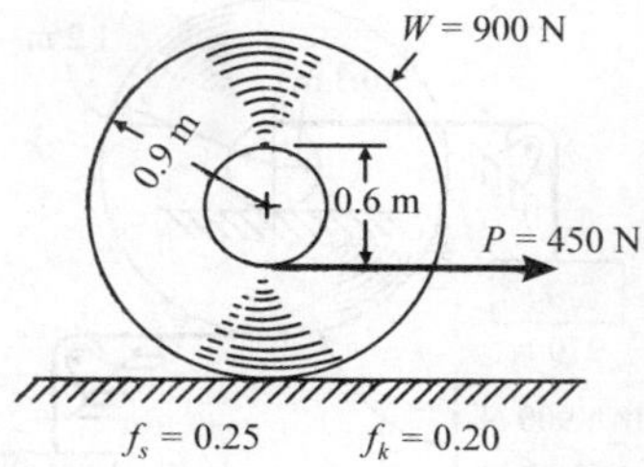

Figure P-13-4.6

13-4.7 The compound disk and drum shown in Fig. P-13-4.7 is acted upon by a force P = 450 N which always remains horizontal. Assuming free rolling, determine $\bar{a}$ and the required friction force.

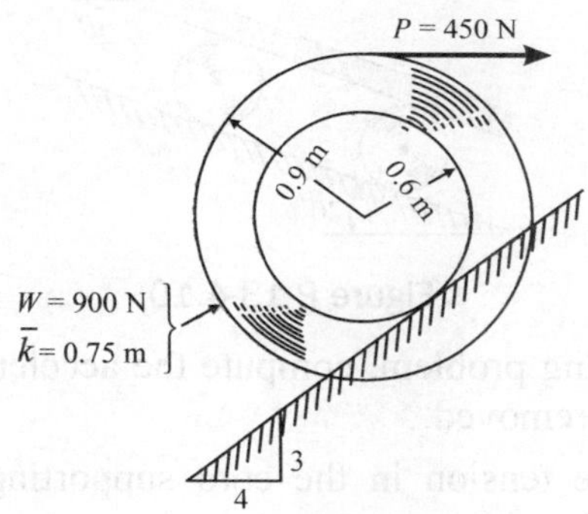

Figure P-13-4.7

13-4.8 A bowling ball is sent down an alley with an initial velocity v_o and without any spin. Assuming the coefficient of friction is f, how far does it skid before pure rolling occurs.

13-4.9 If the compound disk shown in Fig. P-43-4.9 rolls without slipping, determine the acceleration of its mass center. Assume the fixed pulleys to be frictionless and of negligible weight.

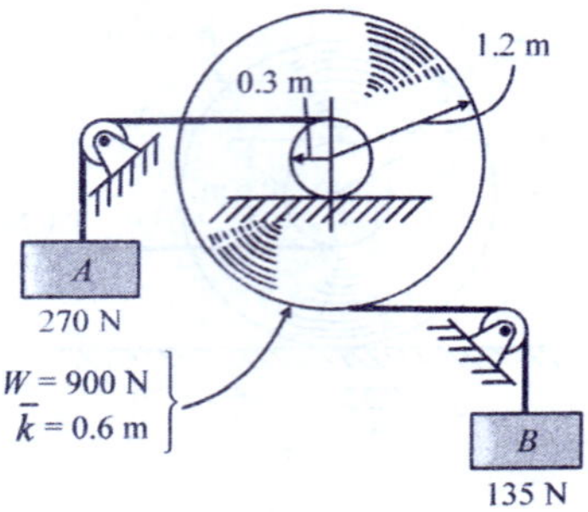

Figure P-13-4.9

13-4.10 The 360 N plank shown in Fig. P-13-4.10 rests on two cylindrical rollers of radius r each weighing 135 N. Determine the force P that will accelerate the plank up the incline at 1.8 m/s^2. Assume no slipping occurs either at the plank or at the incline.

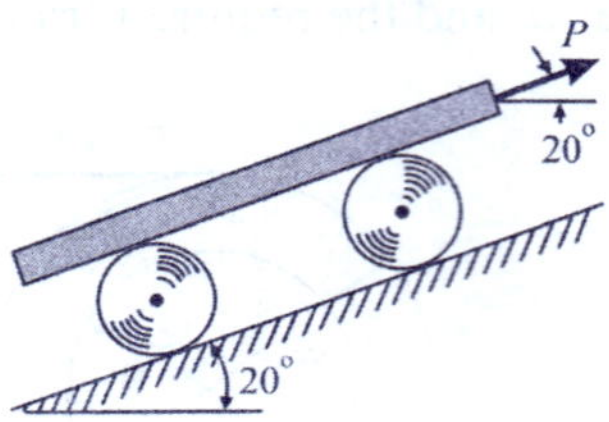

Figure P-13-4.10

13-4.11 In the preceding problem, compute the acceleration of the plank after the force P is removed.

13-4.12 Determine the tension in the cord supporting the 300 N shown in Fig. P-13-4.12.

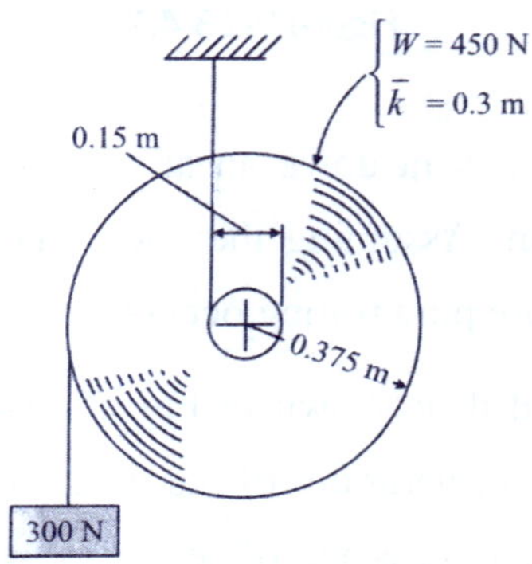

Figure P-13-4.12

13-4.13 Determine the tensions in the cords supporting the floating pulley shown in Fig. P-13-4.13.

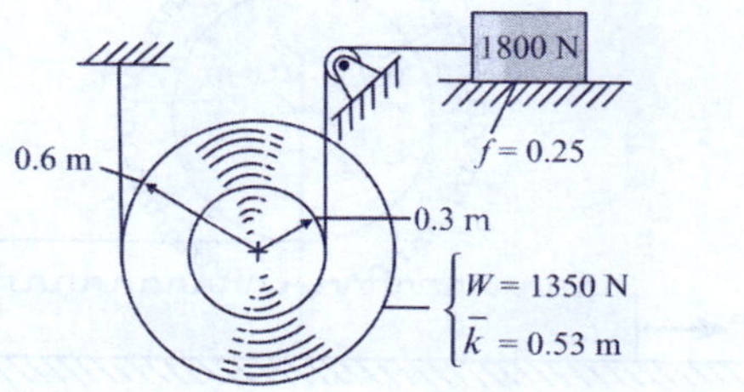

Figure P-13-4.13

13-4.14 The 450 N weight causes the compound disk shown in Fig. P.13-4.14 to roll and slip on the horizontal floor. If the coefficient of friction between the disk and the floor is 0.40, determine the acceleration $\bar{a}$ of the center of the disk.

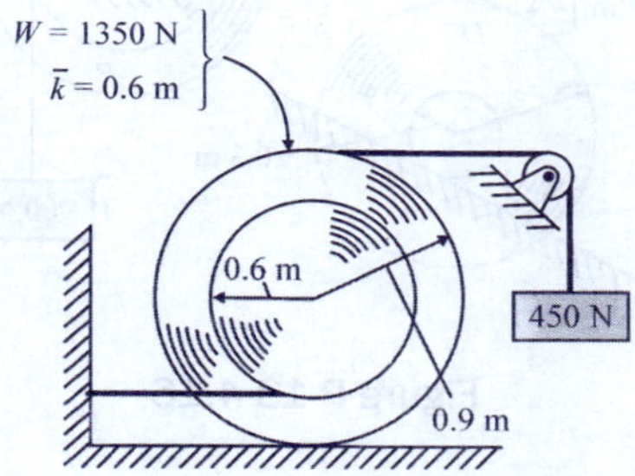

Figure P-13-4.14

13-4.15 The gear and concentric drum shown in Fig. P-13-4.15 rests on a rack *B*. The rack, which weighs 900 N is free to slide on smooth horizontal guides. A force $P = 1250$ N applied, to the rack causes gear *A* to roll along the rack. Determine the tension in the cable and how much cable is wound on or off the drum in 2 sec starting from rest.

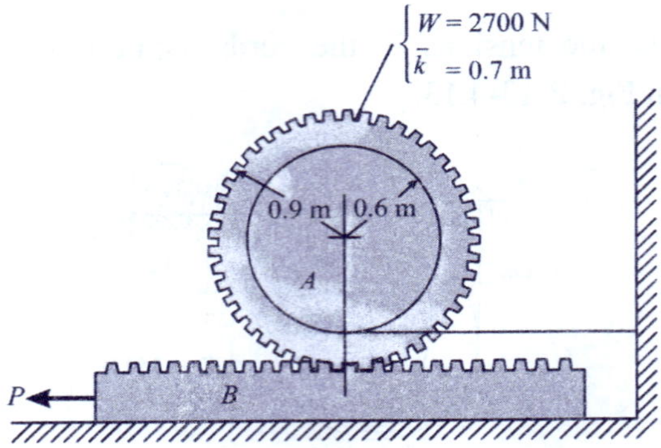

Figure P-13-4.15

13-4.16 Determine the value of the static friction force acting on disk D in Fig. P-13-4.16 if the disk rolls without slipping upon the incline.

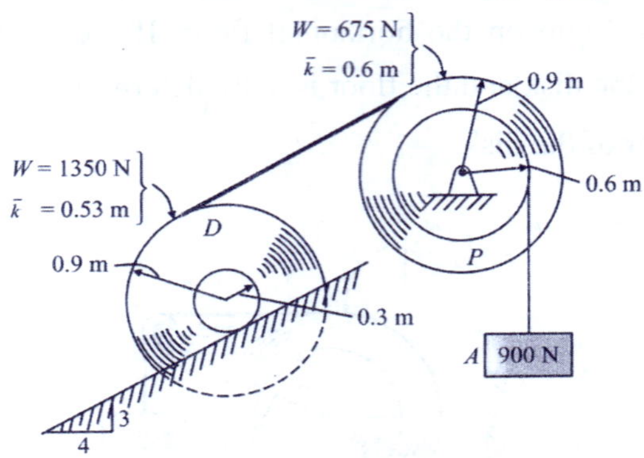

Figure P-13-4.16

13.4.17 Find the tension in the cord supporting body A in Fig. P.13-4.17. Assume that W_A = 4500 N; for disk D, W = 1.34 N and $\bar{k}$ = 0.53 m; for floating disk B, W = 900 N and $\bar{k}$ = 0.52 m.

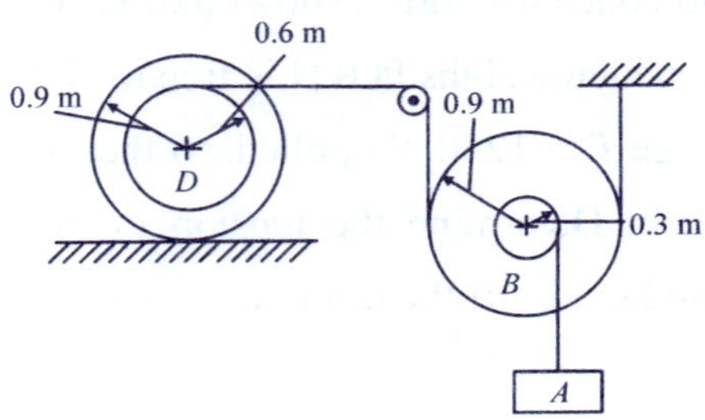

Figure P-13-4.17

13-4.18 Determine the weight of body B that will make the compound wheel shown in Fig. P-13-4.18 roll freely up the 30° incline at 4 rad/sec^2. Neglect the backward inclination of B caused by inertia.

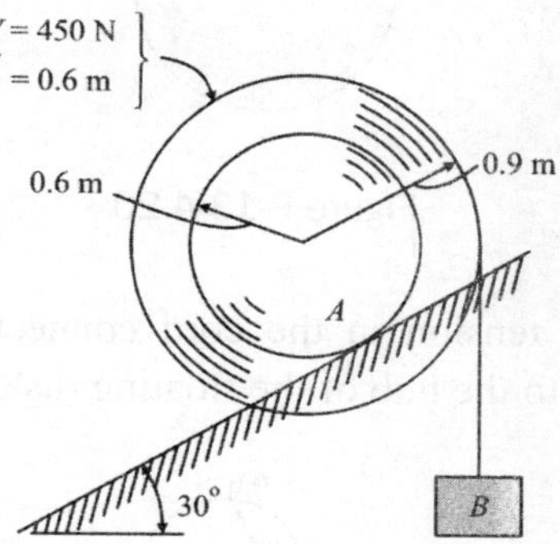

Figure P-13-4.18

13-4.19 In Fig P.13-4.19 is shown the pitch circles of two gears A and B connected by the arm OC. If gear A is fixed and the system is released from rest at the given position, find the angular acceleration of gear B. Assume gear B is equivalent to a solid cylinder weighing 135 N and that arm OC weighs 90 N.

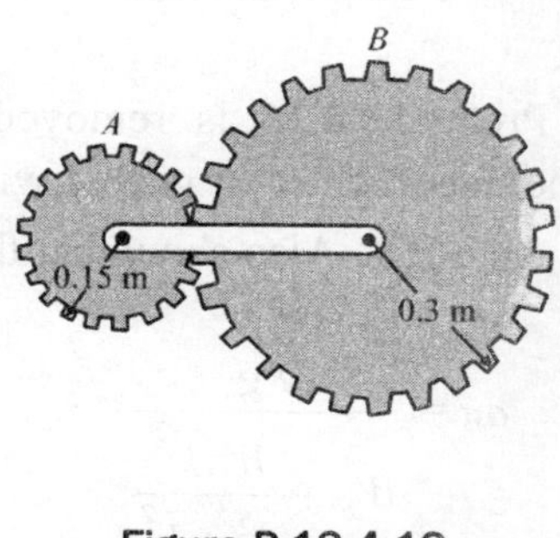

Figure P-13-4.19

13-4.20 As shown in Fig. P-13-4.20, a cord passes from the rim of a solid cylinder of weight W and radius r over a frictionless pulley of negligible weight and back to the center of the cylinder. Determine the tension in the cord.

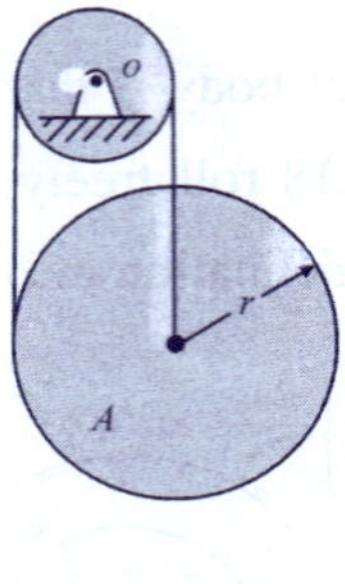

Figure P-13-4.20

13-4.21 Determine the tension in the cord connecting the solid cylinder in Fig. P.13-4.21 to the hub of the floating disk.

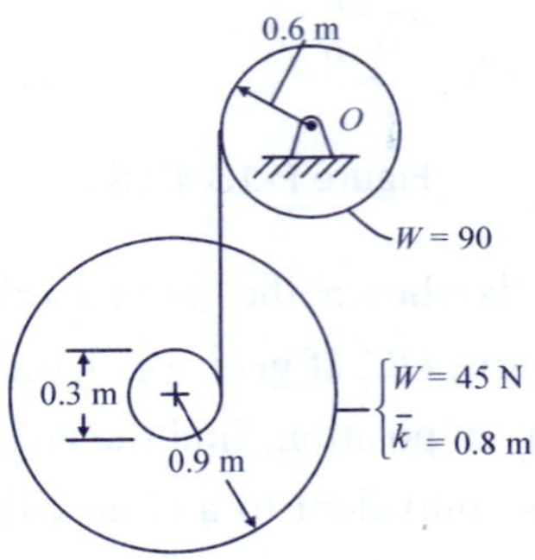

Figure P-13-4.21

13-4.22 If the cable in Prob. 13-4.15 is removed, show that the relative acceleration of the mass center of gear A with respect to the rack B is given by $a_r = r^2 a_B / r^2 + \bar{k}^2$. Also show that the acceleration of the rack will be

$$a_B = \frac{Pg}{W_B + \dfrac{W_A \bar{k}^2}{r^2 + \bar{k}^2}}$$

13-4.23 In the compound gear and rack assembly shown schematically in Fig. P-13-4.23, a clockwise torque of 1600 N-m is applied to the gear. If the racks slide in frictionless guides, compute the acceleration $\bar{a}$ of the center of the compound gear.

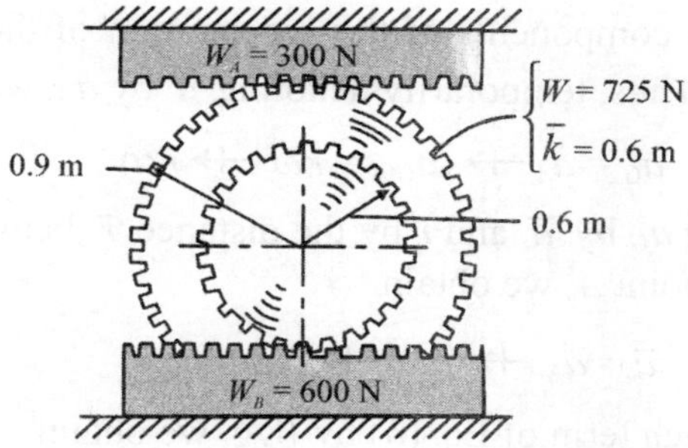

Figure P-13-4.23

13-4.24 As shown in Fig. P-13-4.24, a solid cylinder *A* rolls without slipping along wedge *B* which is on a smooth horizontal floor. If the cylinder and wedge each weigh *W* N, determine the acceleration of wedge *B*.

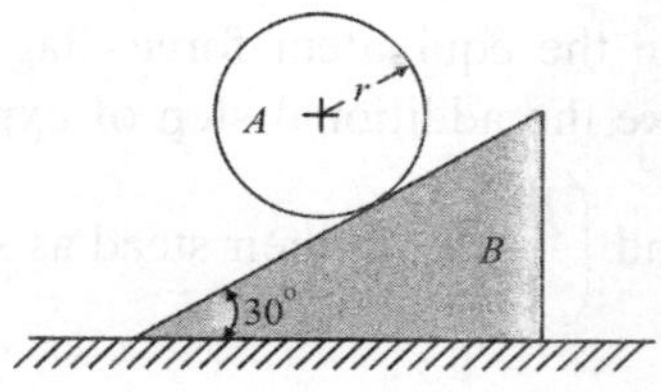

Figure P-13-4.24

13-5 GENERAL PLANE MOTION

For general plane motion, we use the same approach discussed previously for fixed-axis rotation and for rolling bodies. Conceptually there is nothing new; we still equate a free-body diagram of the impressed forces to its dynamic equivalent of the effective force $\left(\dfrac{W}{g}\right)\bar{a}$ at the mass center and an effective couple $\bar{I}\alpha$, or we may apply these dynamic effects reversed on the free-body diagram and use the method of dynamic equilibrium.

There is an additional complication, however, in that generally no simple relation exists between and $\bar{a}$ and α. Usually, a kinematic analysis of the type discussed in Section 12-4 is necessary in which the acceleration $\bar{a}$ of the mass center is found by combining the known acceleration a_A of any reference point

A with the rotational components of the acceleration of the mass center about that reference point. Thus, temporarily denoting $\bar{a}$ by a_G, we apply

$$a_G = a_A \mathrel{+\!\!\!\!\!\rightarrow} (a_{G/A} = r\omega^2 \mathrel{+\!\!\!\!\!\rightarrow} r\alpha)$$

whence, on replacing a_G by $\bar{a}$ and r by the distance $\bar{r}$ between the mass center G and the reference point A, we obtain

$$\bar{a} = a_A \mathrel{+\!\!\!\!\!\rightarrow} \bar{r}\omega^2 \mathrel{+\!\!\!\!\!\rightarrow} \bar{r}\alpha \qquad (a)$$

On multiplying each term of Eq. (a) by W/g, we obtain

$$\frac{W}{g}\bar{a} = \frac{W}{g}a_A \mathrel{+\!\!\!\!\!\rightarrow} \frac{W}{g}\bar{r}\omega^2 \mathrel{+\!\!\!\!\!\rightarrow} \frac{W}{g}\bar{r}\alpha \qquad (b)$$

in which the right-side terms represent the components of the resultant effective force $\left(\frac{W}{g}\right)\bar{a}$. Although these components are not mutually perpendicular, they may be applied directly on the equivalent force diagram shown in part (b) of Fig. 13-5.1, or we may take the additional step of expressing $\bar{a}$ in terms of $\bar{a}_x$ and $\bar{a}_y$ and use $\left(\frac{W}{g}\right)\bar{a}_x$ and $\left(\frac{W}{g}\right)\bar{a}_y$ in their stead as shown in part (c).

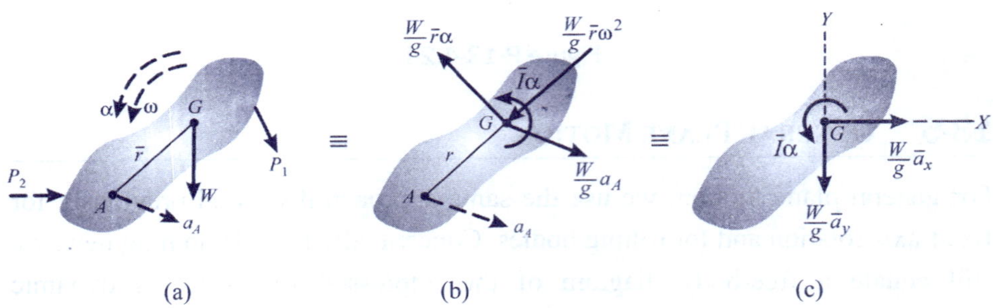

Figure 13-5.1 Equivalent sets of dynamic effects.

As a study of Fig. 13-3.2 will show, part (b) of Fig. 13-5.1 actually combines the effective forces of a noncentroidal rotation about A with an effective force caused by the acceleration of the reference point A. Sometimes this set of effective forces may be more convenient than the set in part (c), particularly if A is the instant center of acceleration because then the moment relation takes the simple form $\Sigma M_A = I_A\alpha$. Unfortunately, it is difficult to locate the instant center of acceleration except in the special case when a body starts from rest. Then, as shown in Section 12-5, the instant center of acceleration coincides with the instant center of velocity which is usually easy to locate. Notice also that $\Sigma M_A = I_A\alpha$ also applies if the acceleration a_A is directed through G. Of course, this is really another example of applying $\Sigma \mathbf{M} = \dot{\mathbf{H}}$ when the acceleration of the reference point acts through the mass center.

ILLUSTRATIVE PROBLEMS

13.5.1 A uniform slender bar, 2 m long and weighing 450 N moves with its ends in contact with the smooth surfaces shown in Fig. 13-5.2. If $\tan\theta = \frac{3}{4}$, determine its angular acceleration and the normal reactions R_A and R_B the instant after it starts from rest.

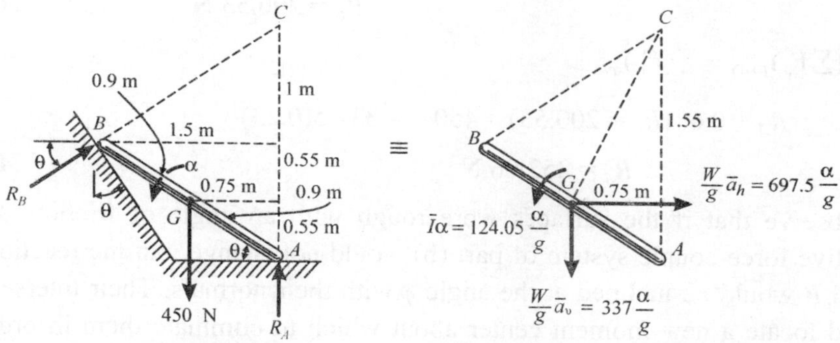

(a) Free-body diagram (b) Equivalent dynamic effects

Figure 13-5.2

Solution

Since the bar starts from rest, the instant centers of acceleration and of velocity coincide at C. Hence, in terms of α assumed counterclockwise, the components of $\bar{a} = a_G$, with G considered rotating about C, are $\bar{a}_h = 1.55\,\alpha \rightarrow$ and $\bar{a}_v = 0.75\,\alpha$. The corresponding effective forces, shown in part (b) are $\frac{W}{g}\bar{a}_h = \frac{450}{g}(1.55\alpha) = 697.5\,\frac{\alpha}{g}$ N and $\frac{W}{g}\bar{a}_v = \frac{450}{g}(0.75\,\alpha) = 337.5\frac{\alpha}{g}$ N. The effective couple

$$\bar{I}\alpha = \frac{1}{12}\frac{W}{g}l^2\alpha = \frac{1}{12}\left(\frac{450}{g}\right)(1.8)^2\alpha = 121.5\frac{\alpha}{g}\ \text{Nm}$$

Taking moments about C, which eliminates the normal reactions R_A and R_B, we obtain

$[+\circlearrowleft (\Sigma M_C)_{FBD} = \Sigma (M_C)_{dyn}]$

$$450(0.75) = 121.5\frac{\alpha}{g} + 337.5\frac{\alpha}{g}(0.75) + 697.5\frac{\alpha}{g}(1.55)$$

whence

$$\frac{\alpha}{g} = 0.23 \text{ and } \alpha = 2.27 \text{ rad/sec}^2 \circlearrowleft$$ ***Ans.***

Force summations now give

$$\left[\xrightarrow{+}(\Sigma F_h)_{FBD} = \Sigma(F_h)_{dyn}\right] \qquad 0.8\, R_B = 697.5\,(0.23)$$

$$R_B = 200.53 \text{ N}$$ ***Ans.***

$$\left[+\uparrow(\Sigma F_v)_{FBD} = \Sigma(F_v)_{dyn}\right]$$

$$R_A + 0.6\ \ (R_B = 200.53) - 450 = -337.5(0.23)$$

$$R_A = 252.06 \text{ N}$$ ***Ans.***

Observe that if the surfaces were rough with an angle of friction ϕ, the effective force-couple system of part (b) would not change, but the reactions at A and B would be inclined at the angle ϕ with their normals. Their intersection would locate a new moment center about which to eliminate them in order to find α directly.

13-5.2 One cylinder of an internal combustion engine is shown in Fig. 13-5.3. At the given position of the power stoke, a force $P = 4450.7$ N acts on the 7.165 N piston. Assuming the connecting rod AB to be a uniform bar 0.1524 m long and weighing 8.901 N, determine the forces on the piston pin B and the crankpin A for a constant clockwise crank speed of 200 rad/sec (equivalent to 1910 rpm).

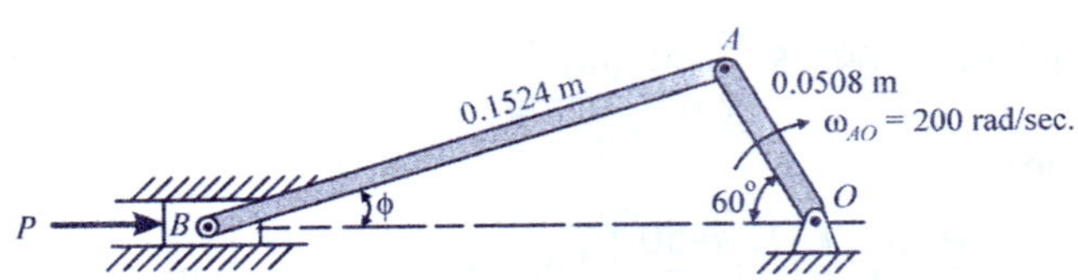

(a)

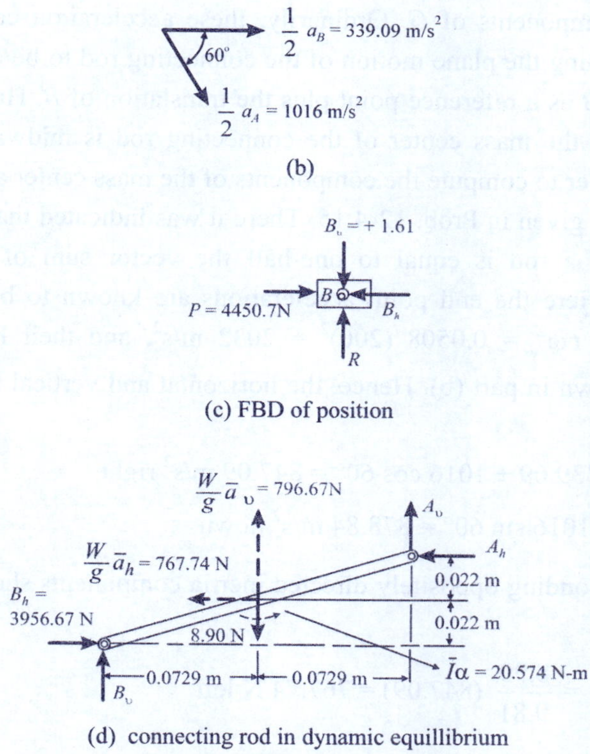

(b)

(c) FBD of position

(d) connecting rod in dynamic equillibrium

Figure 13-5.3

Preliminary

The kinematic properties of the connecting rod must first be determined. Using the methods discussed in Section 12-4, you will find them to be ω = 34.8 rad/sec counterclockwise, α = 11,700 rad/sec^2 clockwise, and a_B = 678.18 m/s^2. Since $AB \sin \phi = AO \sin 60^\circ$, we also have

$$\sin \phi = \frac{0.044}{0.1524} = 0.289 \quad \text{and} \quad \cos \phi = \frac{0.1458}{0.1524} = 0.957$$

which are used to compute the dimensions shown in part (d).

Solution

Here we illustrate the method of dynamic equilibrium in which the inertia components are inserted directly on the free-body diagram of the connecting rod. These inertia components act through the mass center G opposite to the acceleration components of G. Ordinarily, these acceleration components are found by assuming the plane motion of the connecting rod to be equivalent to a rotation about B as a reference point plus the translation of B. However, in this example where the mass center of the connecting rod is midway between its ends, it is simpler to compute the components of the mass center acceleration by using the result given in Prob. 12-4.15. There it was indicated that the midpoint acceleration of a rod is equal to one-half the vector sum of its end point accelerations. Here the end point accelerations are known to be a_B = 678.18 m/s^2 and $a_A = r\omega_{A0}^2 = 0.0508\ (200)^2 = 2032$ m/s^2, and their half-values are directed as shown in part (b). Hence, the horizontal and vertical components of $\bar{a}$ are

$$\bar{a}_h = 339.09 + 1016 \cos 60° = 847.09 \text{ m/s}^2 \text{ right}$$

$$\bar{a}_v = 1016 \sin 60° = 878.84 \text{ m/s}^2 \text{ down}$$

and the corresponding oppositely directed inertia components shown in part (d) are

$$\frac{W}{g}\bar{a}_h = \frac{8.90}{9.81}(847.09) = 767.74 \text{ N left}$$

$$\frac{W}{g}\bar{a}_v = \frac{8.90}{9.81}(878.84) = 797.44 \text{ N up}$$

The inertia couple $\bar{I}\alpha$, directed opposite in sense to α, is

$$\bar{I}\alpha = \left(\frac{1}{12}\frac{W}{g}L^2\right)\alpha = \frac{1}{12}\left(\frac{8.90}{9.81}\right)(0.1524)^2\ (11,700) = 20.57 \text{ N-m}$$

The kinetic equation applied to the piston gives B_h as follows:

$$\left[\overset{+}{\rightarrow}\Sigma X = \frac{W}{g}a\right] \qquad 4450.7 - B_h = \frac{7.165}{9.81}(678.18) = 495.33$$

$$B_h = 3955.37 \text{ N} \qquad \textbf{\textit{Ans.}}$$

When we apply the conditions of dynamic equilibrium to the connecting rod, a moment summation about A yields B_v.

$$[\circlearrowleft \Sigma M_A = 0] \qquad 0.1458B_v - 3956.67(0.044) - 8.9014\ (0.0728) + 767.74\ (0.022) + 796.67\ (0.0728) - 20.574 = 0$$

$$B_v = 825.98 \text{ N} \qquad \textbf{\textit{Ans.}}$$

Using this value of B_v in a vertical summation of forces we have

$$[+\uparrow \Sigma V = 0] \qquad 825.98 + 796.67 - 8.90 + A_v = 0$$

$$A_v = -1613.75 \text{ N} = 1613.75 \text{ N}\downarrow \qquad \textbf{Ans.}$$

while from a horizontal force summation, we obtain

$$\left[\overset{+}{\rightarrow} \Sigma H = 0\right] \qquad 3956.67 - 767.74 - A_h = 0$$

$$A_h = 3188.92 \text{ N} \leftarrow \qquad \textbf{Ans.}$$

By combining their components, the total forces on the piston pin B and crankpin A are found to be

$$B = 4045.68 \text{ N} \quad \text{and} \quad A = 3565.01 \text{ N}$$

Supplementary

An independent check of A_v may be made by taking a moment summation about the intersection of A_h and B_v. Also, the crank effort (i.e., torque) exerted on the crankshaft is equal to the moment sum about O of the components of the crankpin force and equals 1613.75 (0.0254) + 3188.92(0.044) = 181.3 N-m.

13-5.3 In Fig. 13-5.4, a rod AB 1.8 m long and weighing 300 N hangs from the center of a solid cylinder of 0.6 m radius and weighing 450 N. If the cylinder can roll without slipping, determine the acceleration of its mass center A directly after a horizontal force P = 300 N is applied at B.

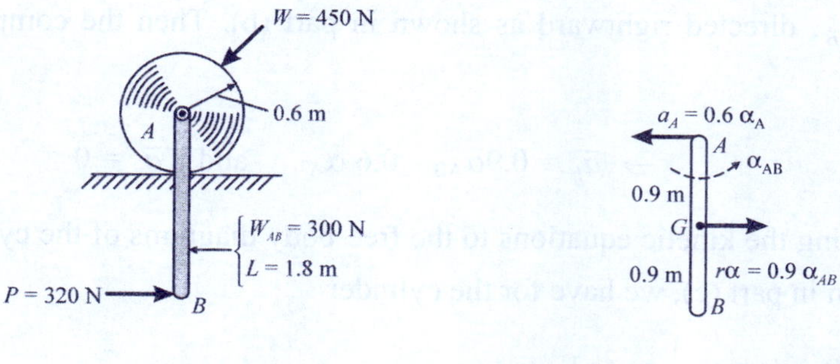

(a) (b)

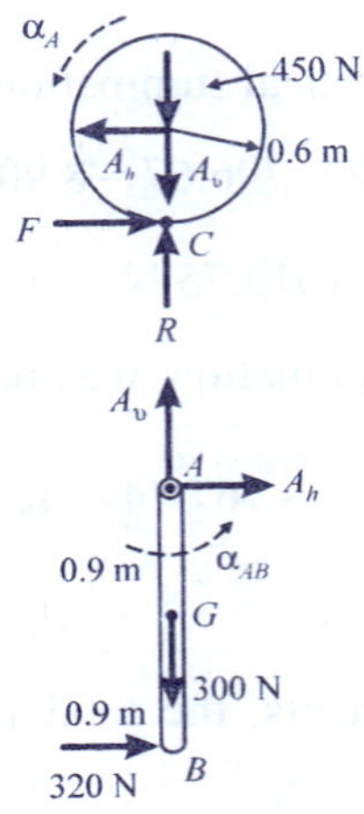

(c)

Figure 13.5.4

Solution

The rod *AB* simultaneously translates with and rotates about hinge *A* which, as part of the free-rolling cylinder, has a leftward acceleration $a_A = 0.6\,\alpha_A$ as shown in part (b). The acceleration of the mass center *G* of the rod is related to the acceleration of *A* by

$$\bar{a} = a_G = a_A \nrightarrow (a_{G/A} = \bar{r}\,\overset{=0}{\omega^2} \nrightarrow \bar{r}\,\alpha)$$

Note that the assumed counterclockwise rotation of *G* about *A* causes $\bar{r}\alpha$ to be $0.9\,\alpha_{AB}$, directed rightward as shown in part (b). Then the components of $\bar{a}$ are

$$\overset{+}{\rightarrow}\ \bar{a}_h = 0.9\alpha_{AB} - 0.6\,\alpha_A \qquad \text{and} \quad \bar{a}_v = 0$$

Applying the kinetic equations to the free-body diagrams of the cylinder and rod shown in part (c), we have for the cylinder

$$[\circlearrowright + \Sigma M_C = I_c\,\alpha = \frac{3}{2}\,mr^2\alpha]$$

$$0.6\,A_h = \frac{3}{2}\left(\frac{450}{g}\right)(0.6)^2\alpha_A = \frac{243}{g}\,\alpha_A \qquad (a)$$

and for the rod *AB*

$$[\circlearrowright + \Sigma M = \bar{I}\alpha = \frac{1}{12} mL^2 \alpha]$$

$$0.9(320) - 0.9A_h = \frac{300}{12g}(1.8)^2\alpha_{AB} = \frac{81}{g}\alpha_{AB} \qquad (b)$$

$$\left[\xrightarrow{+} \Sigma H = \frac{W}{g}\bar{a}_h\right] \qquad 320 + A_h = \frac{300}{g}(0.9\alpha_{AB} - 0.6\alpha_A) \qquad (c)$$

Solve for the common unknown A_h from the equations (*a*) (*b*) and (*c*) to obtain

$A_h = 144.5$ N

after which, substituting A_h in Eq. (*a*), we get

$\alpha_A = 3.5$

whence

$a_A = 0.6\alpha_A \ 0.6 \times 3.5 = 2.1 \text{ m/s}^2$ ***Ans.***

PROBLEMS

13-5.4 A 3.5 m girder weighing 4500 N is being lifted from rest by two separate cables as shown in Fig. P-13-5.4. If cable *A* is accelerated at 3 m/s^2 and cable *B* at 1 m/s^2, find the tension in each cable. Assume the girder is equivalent to a long slender bar.

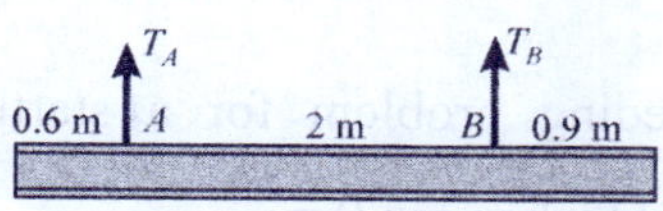

Figure P-13.5.4

13-5.5 In the previous problem, suppose both cables are being accelerated upward at the same rate of *a* m/s^2 when cable *A* suddenly breaks. What will then be the tension in cable *B*?

13-5.6 A force P is applied as shown in Fig. P-13-5.6 to a bar of weight W and length L resting on a smooth horizontal surface. Determine the acceleration of end B.

Figure P-13.5.6

13-5.7 A uniform slender bar, 1.8 m long and weighing 450 N, slides along the smooth wall and floor as shown in Fig. P-13-5.7. Find the force P that will give end A a starting acceleration of 3 m/s^2 rightward when $\tan\theta = \frac{4}{3}$. Also find the reactions at A and B.

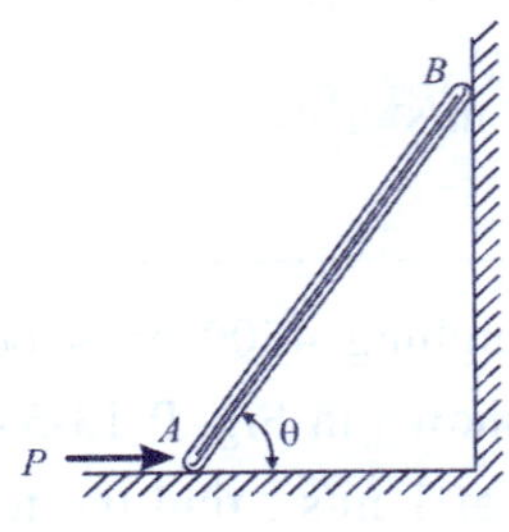

Figure P-13.5.7

13-5.8 Solve the preceding problem for a starting acceleration of a_A = 3.3 m/s^2 rightward when $\tan\theta = \frac{3}{4}$.

13-5.9 The system shown in Fig. P-13-5.9 starts from rest on the smooth surfaces at the given position. The bar AB is freely pinned to the mass centers of bodies A and B. Determine the angular acceleration of AB if AB weighs 450 N and bodies A and B are of negligible weight.

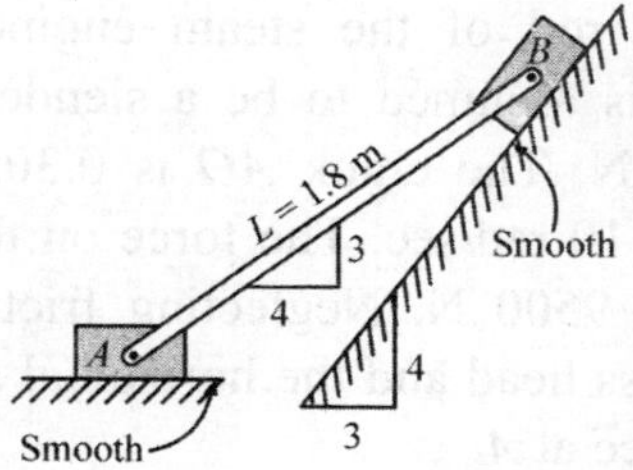

Figure P-13.5.9

13.5.10 Determine the initial acceleration of bodies A and B in the preceding problem if $W_{AB} = 450$ N, $W_A = 225$ N, and $W_B = 700$ N.

13-5.11 If the system shown in Fig. P-13-5.11 is released from rest, determine the initial acceleration of body C. Neglect friction at all surfaces. The uniform bar AB is 1.83 m long and weighs 450 N while body C weighs 700 N.

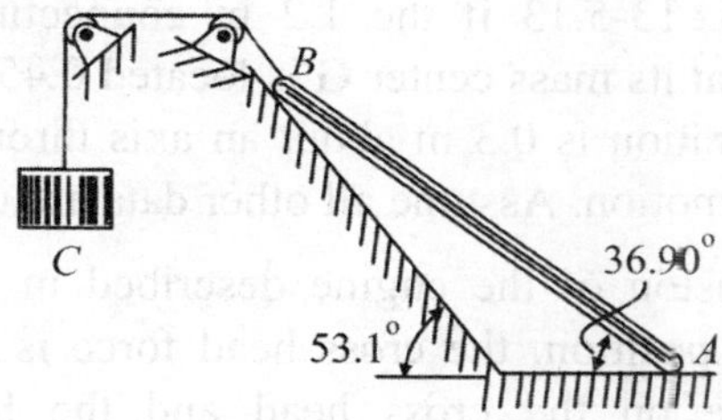

Figure P-13.5.11

13.5.12 The 1.8 m bar AB, weighing 450 N, is released from rest on the smooth surfaces as shown in Fig. P-13-5.12. Compute the normal reaction at A.

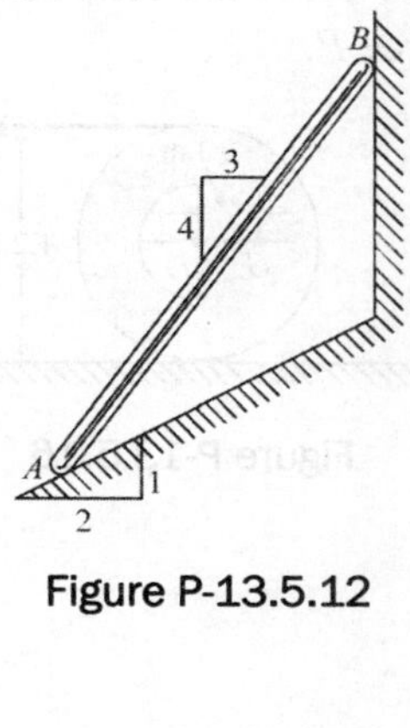

Figure P-13.5.12

13.5.13 The connecting rod of the steam engine shown schematically in Fig. P-13-5.13 is assumed to be a slender uniform rod 1.2 m long weighing 1450 N. The crank *AO* is 0.305 m long and rotates at a constant rate of 10 rad/sec. The force on the 300 N cross head at the given instant is 9500 N. Neglecting friction, determine the normal force on the cross head and the horizontal and vertical components of the crankpin force at *A*.

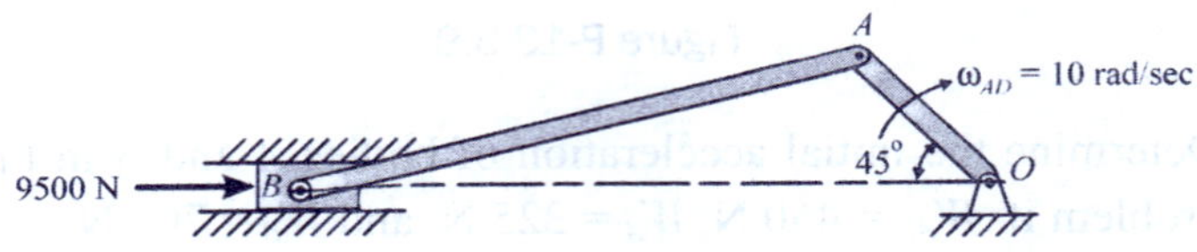

Figure P-13.5.13

$R = 1789.18$ N; $A_h = 5741.40$ N; $A_v = 1637.85$ N **Ans.**

13-5.14 Repeat Prob. 13-5.13 if the 1.2 m connecting rod varies in cross section so that its mass center G is located 0.45 m from *A* and its mass radius of gyration is 0.3 m about an axis through *G* perpendicular to the plane of motion. Assume all other data are unchanged.

13-5.15 When the piston of the engine described in Prob. 13-5.13 is at its extreme left position, the cross head force is zero. What is then the normal force on the cross head and the horizontal and vertical components of the crankpin force at *A*?

13.5.16 The circular disk shown in Fig. P-13-5.16 weighs 700 N after the semicircular hole is cut out of it. Its radius of gyration about its mass center *G* is 0.3 m. At the given instant, ω = 2 rad/sec and α = 2 rad/sec^2, both clockwise. If the disk does not slip, determine the value of the horizontal force *P*.

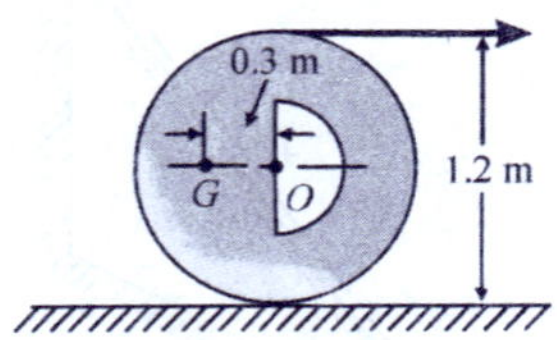

Figure P-13.5.16

13-5.17 A thin hoop of negligible weight has a 300 N bar welded to its rim as shown in Fig. P-13-5.17. At the instant the bar is horizontal, the angular velocity and acceleration of the hoop are $\omega = \sqrt{0.4g}$ rad/sec and $\alpha = 0.1\ g$ rad/sec^2, both clockwise. Find the value of P at this instant. P is parallel to the incline and the hoop rolls without slipping.

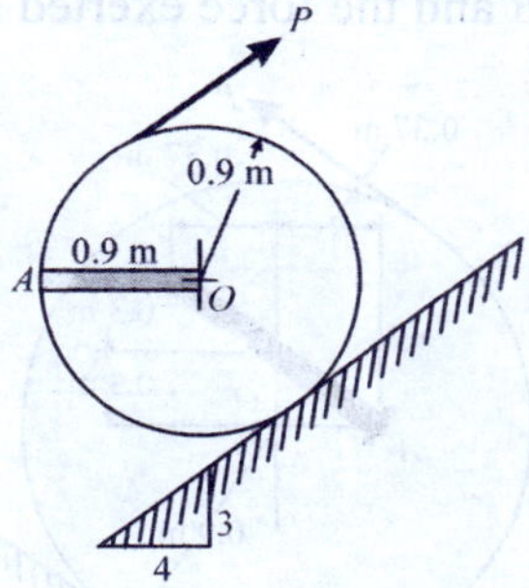

Figure P-13.5.17

13-5.18 In the preceding problem, determine the angular acceleration of the hoop and the moment exerted by the weld at A upon the bar at the given position when $P = 300$ N and $\omega = \sqrt{0.4g}$ rad/sec.

13-5.19 An unbalanced wheel rolls without slipping on the cylindrical surface as shown in Fig. P-13-5.19. It weighs 900 N and has a centroidal radius of gyration $\bar{k} = 0.43$ m. Find the normal and friction forces on the wheel at the given position when its angular velocity is $\omega = \sqrt{0.4g}$ rad/sec.

Hint: Some useful results will be found in Prob. 12-5.11.

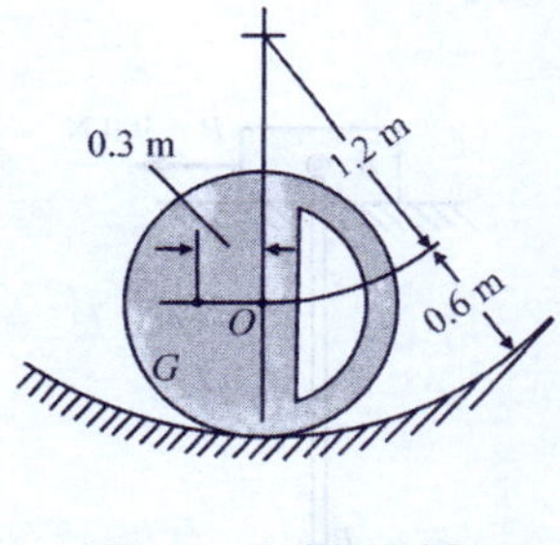

Figure P-13.5.19

13-5.20 The homogeneous wheel shown in Fig. P-13-5.20 rolls without slipping under the action of the pull P parallel to the incline. At the given position, $\omega = 3$ rad/sec and $\alpha = 4$ rad/sec^2 both clockwise. The wheel carries a uniform slender bar weighing 143.31 N which is hinged to the wheel at B and rests between two smooth pins at A. At the given position, compute the horizontal and vertical components of the hinge force at B and the force exerted by the active pin at A.

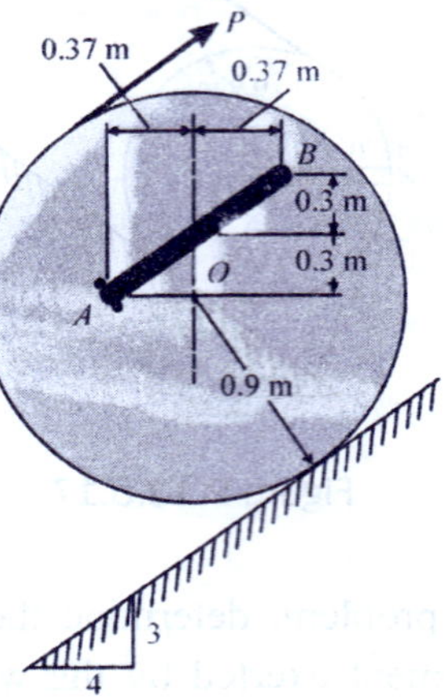

Figure P-13.5.20

13-5.21 Set up dynamic equilibrium for the combined system of wheel and bar described in the preceding problem and determine the value of the pull P at the given position. The wheel weighs 300 N.

13-5.22 In Fig. P-13-5.22, a 1.8 m rod weighing 450 N hangs vertically from a smooth hinge at A on a 900 N block at rest on a smooth horizontal surface. Find the acceleration of the block directly after a horizontal force $P = 300$ N is applied to it. What will be the acceleration of end B of the rod?

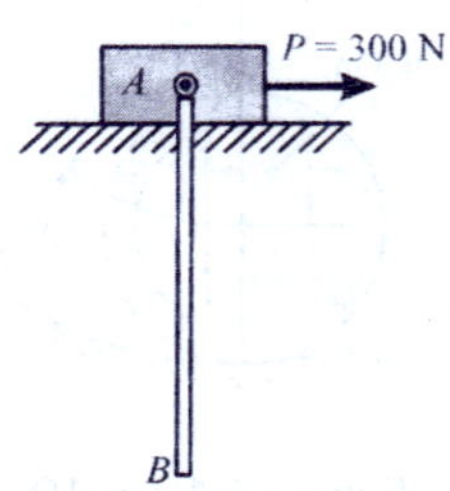

Figure P-13.5.22

13-5.23 In the system shown in Fig. P-13.23, a uniform rod *AB* is suspended from a frictionless hinge at the rim of a solid cylinder. Determine the acceleration of point *A* directly after a horizontal force $P = 250$ N is applied at *B*. If *P* were applied at the mass center of the rod, what change occurs in the acceleration of *A*?

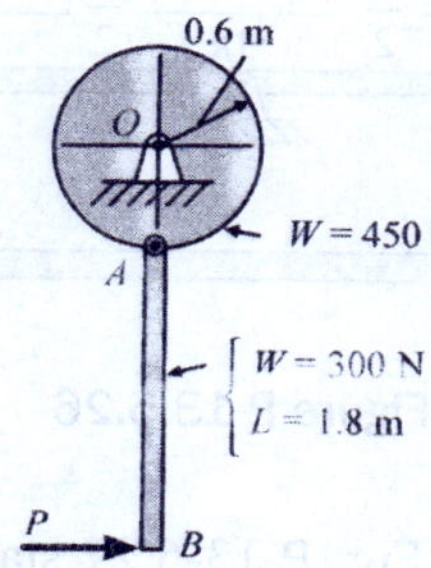

Figure P-13.5.23

13-5.24 In Fig. P-13-5.24 is shown a double pendulum consisting of two equal rods, each of weight *W* and length *L*. If a rightward horizontal force *P* is applied at the midpoint of *BC*, determine the angular acceleration of each rod.

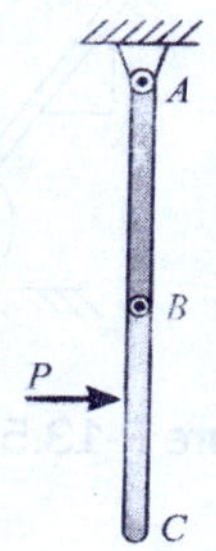

Figure P-13.5.24

13-5.25 Solve the preceding problem if the rightward horizontal force *P* is applied at the midpoint of *AB*.

13-5.26 Two identical slender bars, each of weight W and length L, are connected by two cords to form the assembly supported by a frictionless hinge at B as shown in Fig. P-13-5.26. Find the tension T in cord CE directly after cord AD is cut.

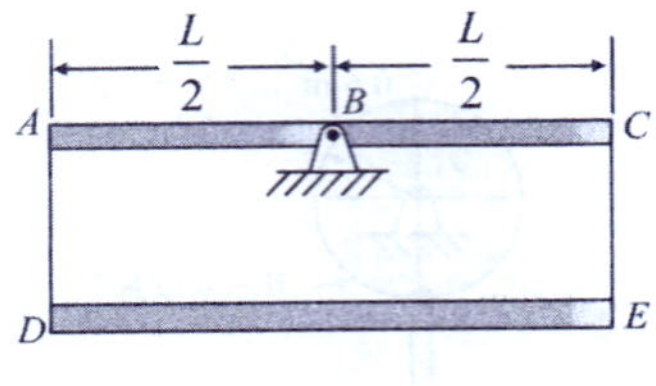

Figure P-13.5.26

13-5.27 The system shown in Fig. P-13-5.27 starts from rest. Body A weighs 225 N; bar AB is 1.5 m long and weighs 300 N; and the solid cylinder B, which rolls freely, has a radius of 0.3 m and weighs 450 N. Determine the initial acceleration of body A.

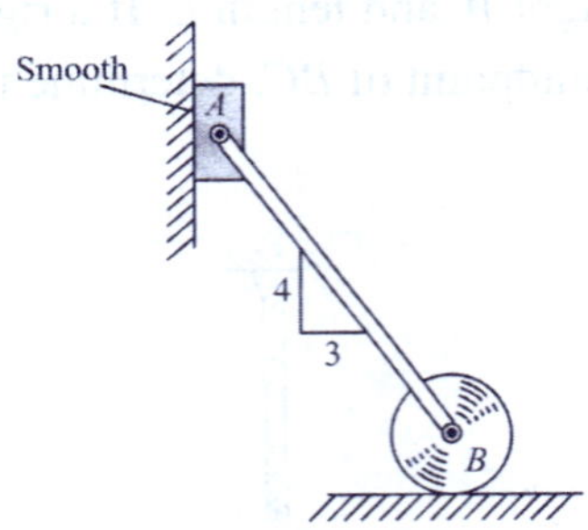

Figure P-13.5.27

13-5.28 In Fig. P-13-5.28, determine the force P that will give body C an initial leftward acceleration of 3.6 m/s^2. The various parts of the system have the following weights: $W_C = 450$ N; $W_{AB} = 300$ N; $W_{CD} = 550$ N.

Hint: Apply the virtual work method to diagrams of active forces and dynamic effects.

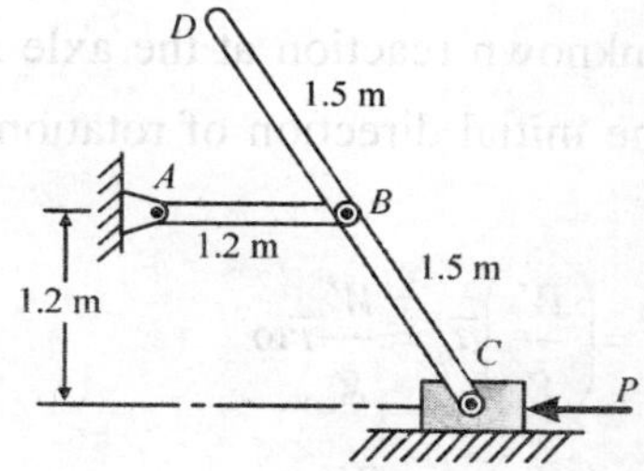

Figure P-13.5.28

P = 901.266 N **Ans.**

SUMMARY

A common approach to all problems involving bodies symmetrical about their plane of motion is to apply the equations **R** = (W/g)$\bar{a}$ and ΣM_G = $\mathbf{H}_G$ in the scalar forms

$$\Sigma X = \frac{W}{g}\bar{a}_x; \quad \Sigma Y = \frac{W}{g}\bar{a}_y; \quad \Sigma \overline{M} = \bar{I}\alpha \qquad (13\text{-}2.1)$$

where the bar superscripts denote reference to centroidal axes.

Two methods of applying these equations may be used. Either equate a free-body diagram of the impressed forces to its dynamic equivalent of the effective force $\left(\frac{W}{g}\right)\bar{a}$ at the mass center plus an effective couple $\bar{I}\alpha$, or apply these dynamic effects reversed on the free-body diagram and use dynamic equilibrium. Usually a kinematic analysis to relate $\bar{a}$ and α must also be used except in the case of fixed-axis rotation and homogeneous bodies rolling without slipping. Since a direct relation between $\bar{a}$ and α is known for these motions, convenient supplemental equations can be deduced and used as discussed in Sections 13-3 and 13-4.

Thus, for noncentroidal rotation, it is convenient to select X and Y axes through the mass center which are positive in the directions of $\bar{a}_n$ and $\bar{a}_t$ respectively. Also it is useful to supplement $\Sigma \overline{M} = \bar{I}\alpha$ with $\Sigma M_A = I_A\alpha$ which eliminates the generally unknown reaction at the axle A. The positive sense of $\Sigma\overline{M}$ and ΣM_A is that of the initial direction of rotation. Then Eqs. (13-2.1) are transformed into

$$\Sigma X = \left(\frac{W}{g}\right)\bar{a}_n = \frac{W}{g}\bar{r}\omega^2$$

$$\Sigma Y = \left(\frac{W}{g}\right)\bar{a}_t = \frac{W}{g}\bar{r}\alpha \qquad (13\text{-}3.2)$$

$$\Sigma\overline{M} = \bar{I}\alpha \quad \text{or} \quad \Sigma M_A = I_A\alpha$$

In the case of centroidal rotation where A coincides with the mass center, $\bar{r} = 0$ and the above equations reduce to

$$\Sigma X = 0;\ \Sigma Y = 0;\ \Sigma\overline{M} = \bar{I}_a \qquad (13\text{-}3.1)$$

For homogeneous rolling bodies (Section 3-4) in which the mass centre has a rectilinear motion parallel to the surface on which it rolls, it is convenient to select reference axes at the mass center with the X axis parallel to the surface and positive in the initial direction of motion. Equations (13-2.1) then become.

$$\Sigma X = \frac{W}{g}\bar{a};\ \Sigma Y = 0;\ \Sigma\overline{M} = \bar{I}_a \qquad (13\text{-}4.1)$$

If the body is known to roll without slipping, an additional equation that can be used is

$$\Sigma M_C = I_C\alpha \qquad (13\text{-}4.2)$$

which eliminates both the unknown static frictional resistance and the normal force acting at the instant center C.

CHAPTER **14**

WORK-ENERGY METHOD

14-1 INTRODUCTION

In this chapter we present an alternate approach to dynamics which bypasses accelerations and directly relates forces, displacements, and velocities. This approach, known as the wok-energy method, will usually be faster and easier to apply than the force-inertia method discussed in preceding chapters. As we shall see, the work-energy method involves only scalar quantities which are combined arithmetically and thereby eliminates the directional aspects of the vector quantities inherent in the force-inertia method. Moreover, the work-energy method can be applied directly to a complete system of connected bodies thereby eliminating forces which are internal to the system, and dispensing with drawing separate free-body diagrams of each part of the system. Finally, we deal with displacements and velocities which are much easier to visualize than are accelerations.

The work-energy method is based on integrating the fundamental equations $\mathbf{R} = m\bar{\mathbf{a}}$ and $\Sigma\mathbf{M} = \mathbf{H}$ directly in terms of the displacement. Another approach, to be considered in the next chapter, is to integrate these equations with respect to time, thereby leading to the impulse-momentum method. Frequently, a combination of work-energy and impulse-momentum methods will solve problems that otherwise would be very difficult. After deriving the basic work-energy equations, we shall first apply them to rigid-body translation (both rectilinear and curvilinear) and then extend them to all other cases of rigid-body plane motion. As in the preceding chapter, we discuss only the plane of motion of rigid bodies which are symmetrical with respect to the plane of motion through the mass center of the body. Other cases, including those of spatial motions, are presented in Chapter 16. In a sense, the work-energy method is an adjunct and introduction to Lagrangian and Hamiltonian mechanics.

14-2 WORK-ENERGY EQUATION FOR TRANSLATION

The general problem in dynamics is to determine the relation between the motion of a body (or a system of bodies) and the force system acting upon it. For a particle or a translating rigid body, this may be accomplished by solving **R** = *ma* for the acceleration a and then integrating one of the differential kinematic equations ($a_t = dv/dt,\ v = ds/dt,\ or\ a_t\ ds = v\ dv$) to find the appropriate *s-t*, *v-t*, or *v-s* relation desired.

An alternate procedure is to integrate **R** = *m***a** directly either in terms of the time or in terms of the displacement. The time integral leads to the impulse-momentum method which we shall discuss in Chapter 15. The basis of the work-energy method is the displacement integral which is obtained by multiplying both sides of **R** = *m***a** by *d***r**. Since vector multiplication can either be the dot or the cross product, our choice is determined by expressing **a** as $a_t + a_n$ and nothing that *dr* is tangent to the path. Since $\mathbf{a}_n$ is normal to the path, its dot product with *dr* will be zero. Therefore, using the dot product, we obtain

$$\mathbf{R} \cdot d\mathbf{r} = m\mathbf{a} \cdot d\mathbf{r} = ma_t\, ds = mv\, dv \qquad (a)$$

in which we have for convenience replaced the magnitude of *d***r** by *ds* measured along the path and then substituted *v dv* for $a_t\, ds$. The dot product **R** . *d***r** defines the differential work done by the resultant force **R** during the displacement *dr*. Similarly **F** . *d***r** defines the differential work done by a component of **R**.

Integrating both sides of Eq. (a), we obtain

$$\int_{r_o}^{r} \mathbf{R} \cdot dr = m \int_{v_o}^{v} v\, dv = \frac{1}{2} m\,(v^2 - v_o^2) \qquad \text{(b)}$$

Of particular importance is the fact that those terms are scalar quantities which means that they can be combined arithmetically with similar equations without considering directional effects. The left-hand integral is known as resultant work (abbreviated as **RW**) done upon the particle while the term $\frac{1}{2}mv^2$ is called kinetic energy (abbreviated as KE). On replacing *m* in Eq. (*b*) by its equivalent W/g, we obtain

$$RW = \frac{W}{2g}(v^2 - v_o^2) = \Delta(KE) \qquad (14\text{-}2.1)$$

which is the work-energy equation for a particle. Expressed in words, it states that the resultant work done upon a particle is equal to its change in kinetic energy.

An alternate derivation of the work-energy equation which is somewhat more direct is the following: Assume an *X* axis to be tangent to the path of the particle during each distance ds measured along the path. Then, by eliminating a_t between $\Sigma X = \frac{W}{g} a_t$ and $a_t\, ds = v\, dv$, we obtain

$$\Sigma X(ds) = \frac{W}{g} v\, dv$$

which is integrated between an initial velocity v_o at the initial position to the final velocity v at a distance s measured along the path to give

$$\int_o^s \Sigma X(ds) = \frac{W}{2g}(v^2 - v_o^2) \tag{14-2.2}$$

The left-hand integral is an alternate expression for resultant work in which ΣX represents the component of the resultant force along the path at each instant.

The unit of work is the dimensional product of force and distance; i.e., N-mm, dyne-cm, etc. However, since we usually use the N-m-s system of units in dynamics, work should be correspondingly expressed in N-m units to make it dimensionally consistent with kinetic energy whose unit is seen to be dimensionally.

$$\text{KE} = \frac{1}{2}\frac{W}{g}v^2 = \frac{N}{m/\sec^2}\left(\frac{m^2}{\sec^2}\right) = N - m$$

Then the work-energy equation will be dimensionally homogeneous which is required for the validity of any equation. Before considering the application of the work-energy equation, we examine next the meaning of work and how it is computed.

14-3 INTERPRETATION AND COMPUTATION OF WORK

Recalling that the dot product of two vectors is a scalar quantity equal to the product of the magnitudes of the two vectors by the cosine of their included angle, resultant work can be written as

$$\int \mathbf{R} \,.\, dr = \int (R\cos\theta)\, ds = \int R\,(ds\cos\theta)$$

In which we have replaced for convenience the magnitude of dr by ds measured along the path of motion.

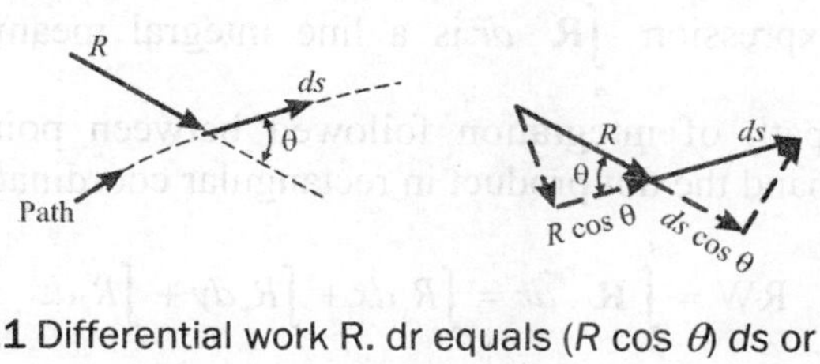

Figure 14-3.1 Differential work R. dr equals (R cos θ) ds or R(ds cos θ).

By referring to Fig. 14-3.1, we see that differential work may be interpreted as the product of the component $R \cos \theta$ of the resultant force in the direction of the displacement multiplied by the displacement ds. Also if differential work is rewritten as the product $R(ds \cos \theta)$, we see that resultant work may also be interpreted as the product of the resultant force multiplied by the component of displacement in the direction of the resultant force. Work done by a force may be positive or negative depending on whether the force component is directed respectively along or opposite to the direction of displacement.

Let us apply these concepts to the case of a body of weight W moving up an incline under the action of a constant force P as in Fig. 14-3.2. Interpreting work as the component of force in the direction of displacement multiplied by the displacement, the work done by P is $(P \cos \theta)s$. Note that in accordance with this concept, forces perpendicular to the displacement, such as N, do no work because they have no component in the direction of displacement.

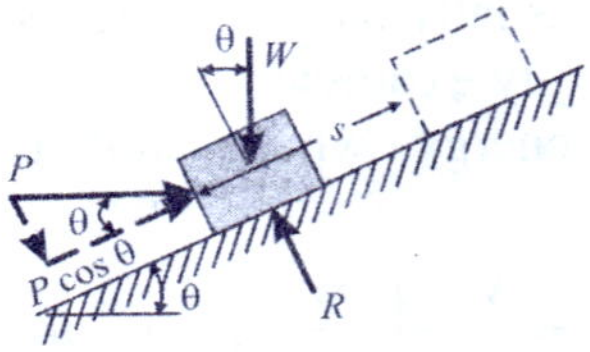

(a) Work = $(P \cos \theta)s$

Work = $(-W \sin \theta)s$

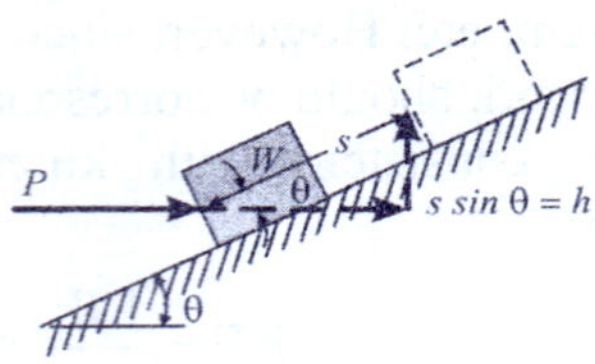

(b) Work = $P(s \cos \theta)$

Work = $-W(s \sin \theta) = -Wh$

Figure 14-3.2 Two interpretations of work done by a force.

The scalar product $(P \cos \theta)s$ can also be written as $P(s \cos \theta)$ which expresses work as the product of a force multiplied by the component of displacement in the direction of the force. This concept is especially useful in defining the work done by gravity forces as the product of the weight multiplied by the change in vertical displacement, here being $-W(s \sin \theta) = -Wh$ where h is the vertical rise of the body. It is important to recognize that work done by gravity forces depends only on a change in elevation of a body, being negative (as in this case) when the direction of W is opposite to the vertical displacement and positive when they are both downward.

The general expression $\int_A^B \mathbf{R} \cdot dr$ is a line integral meaning that its value depends on the path of integration followed between points A and B. To evaluate it, we expand the dot product in rectangular coordinates to obtain

$$RW = \int_A^B \mathbf{R} \cdot dr = \int R_x dx + \int R_y dy + \int R_z dz \tag{14-3.1}$$

A suitable selection of the reference axes will simplify the application of this result. Thus, for rectilinear motion, the X axis is chosen as positive in the initial direction of motion so that the components of the resultant force are $R_x = \Sigma X$ and $R_y = R_z = 0$. Also dx may be replaced by ds (equal to the magnitude of $d\mathbf{r}$). Then the resultant work in rectilinear motion reduces to

$$\text{RW} = \int_o^s \Sigma X(ds) \qquad (14\text{-}3.2)$$

For constant forces, this becomes $\Sigma X(s)$ while for variable forces, ΣX must be expressed in terms of s before resultant work can be evaluated.[1]

For cases in which the resultant force is not constant, and where a mathematical relation between ΣX and s is difficult to obtain (as for the variation of pressure with the position of a piston in an engine), experimental methods may be available to express graphically the variation of force with displacement. Figure 14-3.3 shows such a *force-displacement diagram.* Since the area of the shaded strip represents the term $\Sigma X(ds)$, it follows that the resultant work $\int_o^s \Sigma X(ds)$ is represented by the area under the force-displacement diagram.

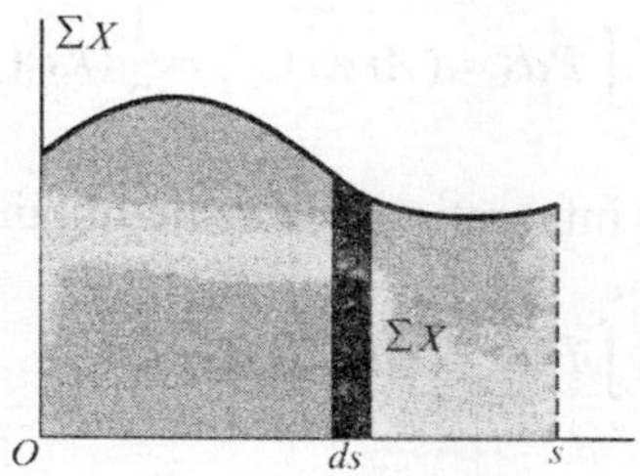

Figure 14-3.3 Force-displacement diagram.

This concept also provides a simple method of computing the work done by forces which are proportional to displacement, as in springs. The force-deformation diagram for a linear spring[2] (Fig. 14-3.4) is a straight line as determined from Hooke's law and expressed by the equation

$$P = ks$$

[1]Refer also to the alternate derivation of the work-energy equation.

[2]There also are nonlinear springs classified as "hard" or "soft," depending on whether their stiffness increases or decreases with deformation.

in which k is known as the spring modulus. This modulus or spring constant represents the force required to deform a given spring through a unit distance. The spring deformation s is the difference between the original or free length of the spring and its deformed length.

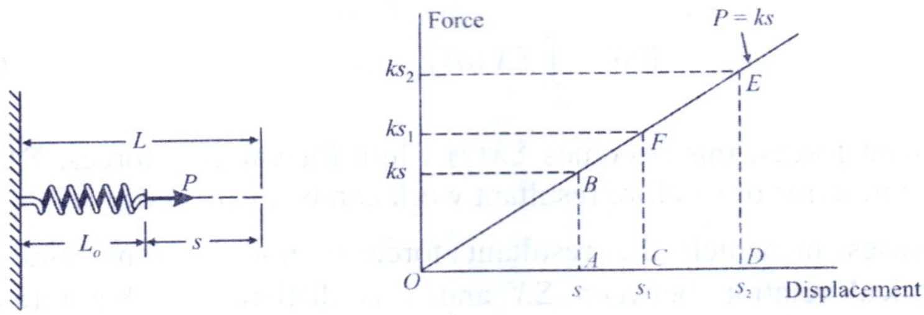

Figure 14-3.4 Force-displacement diagram for a spring.

Since resultant work is represented by the area under a force-displacement diagram, we see from Fig. 14-3.4 that the work done in deforming a spring from its free or unloaded length to an extension (or compression) of s units is the area of triangle OAB or

$$\int_o^S P ds = (Area)_{F-s} = \frac{1}{2}(ks)(s) = \frac{1}{2}ks^2 \qquad \text{(a)}$$

or, alternatively, by direct integration, we again obtain

$$\int_o^S P ds = \int_o^S ks\, ds = \frac{1}{2}ks^2 \qquad \text{(b)}$$

The force-displacement diagram is especially useful in determining the work required to stretch a spring from an initial deformation s_1 to a larger deformation s_2. In this case the work done is the area of the trapezoid $CDEF$, which is equivalent to the average force multiplied by the change in deformation, or

$$\text{RW} = \int_{S_1}^{S_2} ks\, ds = \frac{ks_1 + ks_2}{2}(s_2 - s_1) \qquad \text{(c)}$$

Since the force exerted by a spring depends only on the spring constant and the magnitude of spring deformation, the work done on a spring is due only to the change in length of the spring and is independent of any rotation of the spring.[3] For example, consider the spring in Fig. 14-3.5 which has the deformations s_1 and s_2 at positions A and C. As one end of this spring is moved from the dashed position A to the solid position C, imagine that it is first moved to B and then rotated to C. During this rotation, the spring force is normal to the circular path from B to C and therefore does no work during the rotation. This concept is much simpler than attempting an evaluaticn of the work integral in terms of the variable component of the spring force over the distance AC, and agrees with Eq. (c).

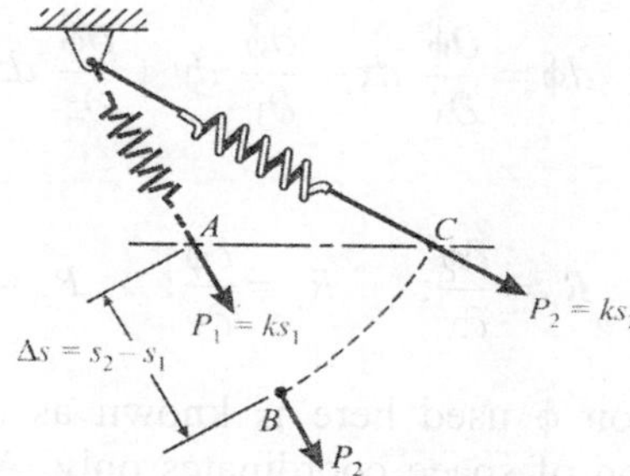

Figure 14-3.5 Work = $P_{ave}(\Delta S) = \dfrac{1}{2}(P_1 + P_2)(\Delta S)$.

A similar situation is shown in Fig. 14-3.6. There a weight W at one end of an inextensible cord is raised as the other end moves along a curved path from A to B. Here the change in elevation of W is equal to the difference ΔL between the lengths L_1 and L_2.

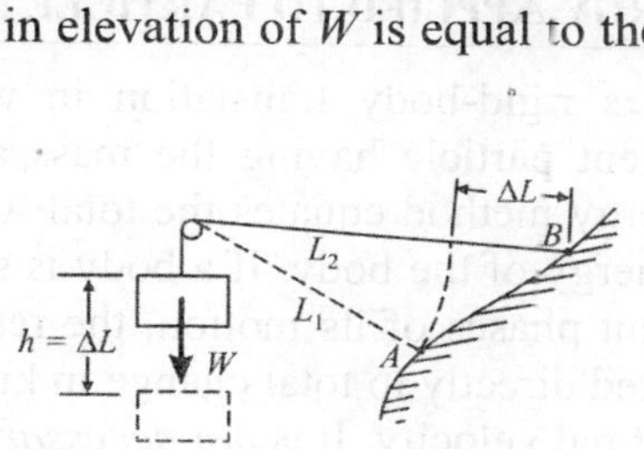

Figure 14-3.6 Work done on weight is $Wh = W(L_2 - L_1)$.

Figures 14-3.5 and 14-3.6 illustrate cases in which the line integral $\int_A^B \mathbf{R} \cdot dr$ is independent of the path of integration and depends only on the limits of

---

[3]Any work done by the weight of the spring is neglected since its weight and change in elevation are usually negligible compared with other data.

integration. For this to be generally true, **R**. dr must be an exact differential of some function ϕ such that $d\phi = \mathbf{R} \cdot dr$, and hence

$$\int_A^B \mathbf{R} \cdot \mathrm{dr} = \int_A^B d\phi = \phi\Big]_A^B = \phi(B) - \phi(A) \qquad (14\text{-}3.3)$$

If the function ϕ exists, the force may be derived from it as the two following expressions show:

$$\mathbf{R} \cdot \mathrm{dr} = R_x\, dx + R_y\, dy + R_z\, dz$$

$$d\phi = \frac{\partial \phi}{\partial x} dx + \frac{\partial \phi}{\partial y} dy + \frac{\partial \phi}{\partial z} dz$$

and hence

$$R_x = \frac{\partial \phi}{\partial x}; \qquad R_y = \frac{\partial \phi}{\partial y}; \qquad R_z = \frac{\partial \phi}{\partial z} \qquad (14\text{-}3.4)$$

The potential function ϕ used here is known as a force function and is assumed to be a function of space coordinates only. A spring is one example where the force is a function of position. A gravity field is another which we will discuss in detail later in Section 15-7 dealing with the motion of satellites. However, if the force is a function of velocity, it is not derivable from a potential and the line integral will not be independent of the path of integration.

14-4 WORK-ENERGY APPLIED TO PARTICLE MOTION

Particle motion includes rigid-body translation in which the body may be replaced by an equivalent particle having the mass and motion of the body. Basically, the work-energy method equates the total work done upon a body to the change in kinetic energy of the body. If a body is subjected to different sets of forces during different phases of its motion, the resultant work done during all these phases is equated directly to total change in kinetic energy expressed in terms of the initial and final velocity. It is *not necessary* to compute the velocity at the end of any phase so that it may be used as the initial velocity for the next phase.

The preceding section showed how to compute the work done on a body under various conditions. Specific details are explained in the illustrative problems. Note especially that the work done by gravity[4] depends only on the

[4]Actually correct only when the gravitational acceleration g is essentially constant. See Section 15-7 on satellite motion for work done by gravitational attraction when g varies.

change in vertical position. Also note that a spring does negative work upon a body whose motion stretches the spring. Sometimes the work absorbed by a spring is called potential energy since the work done in deforming the spring can be recovered if the spring is permitted to return to its undeformed position.

ILLUSTRATIVE PROBLEMS

14-4.1 As shown in Fig. 14-4.1, a 450 N body moves along the two inclines for which the coefficient of friction is 0.20. If the body starts from rest at *A* and slides 60 m down the 30° incline, how far will it then move along the other incline? What will be its velocity when it return to *B*?

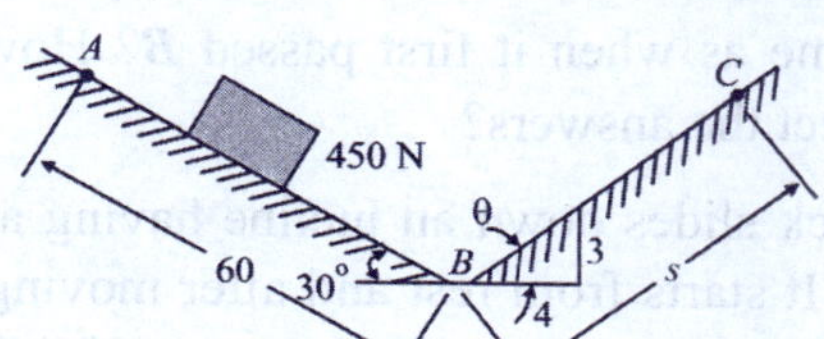

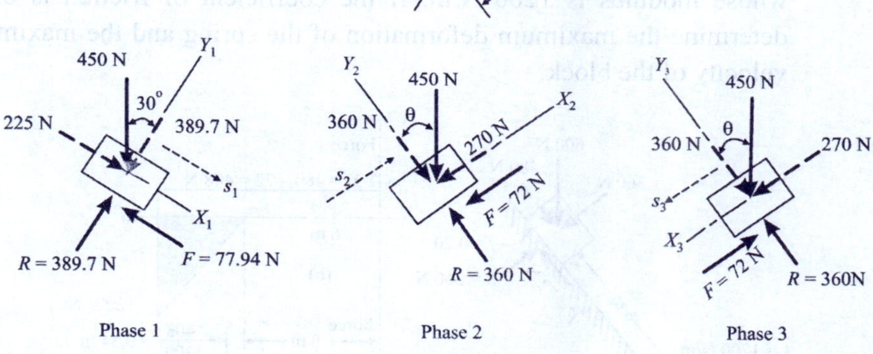

Figure 14-4.1

Solution

The FBD of the body during each phase of its motion is as shown. Computing the normal and frictional forces in the usual manner, we apply the work-energy equation to phases 1 and 2. Between *A* and *C*, the change in kinetic energy will be zero since the initial and final velocities are zero. We obtain

$$\left[\Sigma X_1(s_1) + \Sigma X_2(s_2) = \frac{W}{2g}(v^2 - v_o^2)\right]$$

$$(225 - 77.94)(60) - (270 + 72)s = 0$$

from which

$$s = 25.8 \text{ m}$$

Note how simply the W-E method obtains this result directly even though two different accelerations are involved during the displacements s_1 and s_2.

On applying the work-energy equation to the motion from C back to B during phase 3, we have

$$\left[\Sigma X_3(s_3) = \frac{W}{2g}(v^2 - v_o^2)\right] \qquad (270-72)(25.8) = \frac{450}{2\times 9.81}v_B^2$$

from which

$$v_B = 14.92 \text{ m/s} \qquad \textit{Ans.}$$

Is this velocity the same as when it first passed ***B***? How would changing the weight of the body affect the answers?

14-4.2 A 600 N block slides down an incline having a slope of 4 vertical to 3 horizontal. It starts from rest and after moving 6 m, strikes a spring whose modulus is 1200 N/m. If the coefficient of friction is 0.20, determine the maximum deformation of the spring and the maximum velocity of the block.

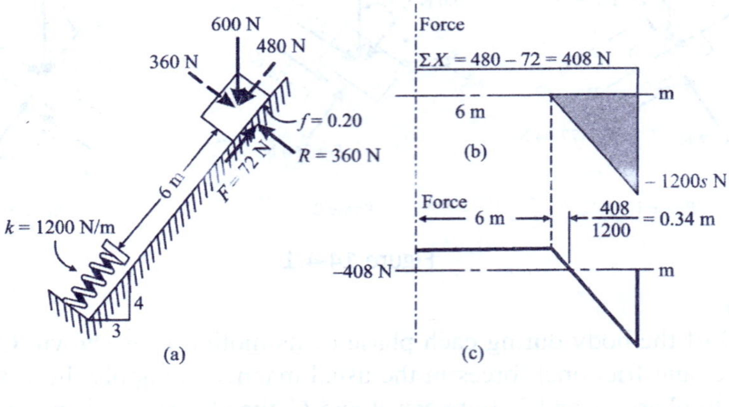

Figure 14-4.2

Solution

Since the block starts from rest and again has zero velocity at the instant of maximum spring deformation, its change in kinetic energy is zero. Referring to Fig. 14-4.2, we obtain the resultant work from the area under a force-displacement diagram in which the effects of the X components of the constant forces and of the variable spring force have been plotted separately. The negative force exerted by the spring on the body reaches its maximum value of

$-1200s$ N at the instant the spring has reached its maximum deformation of s m. We now, obtain

$$[\text{RW} = \Delta\text{KE}] \qquad 408(6 + s) - \frac{1}{2}(1200s)(s) = 0$$

from which

$$s = 2.39 \text{ m} \qquad \textbf{\textit{Ans.}}$$

The block will reach its maximum velocity when the net force on it is zero; not when it first contacts the spring which may have been your first thought. At the instant of maximum velocity, the acceleration of the block must be zero according to maxima-minima theory, or $a = dv/dt = 0$. You may understand this more clearly, however, by examining the net force-displacement diagram in part (c). This shows that positive work is being done on the block up to the position of zero net force, thereby increasing the KE and the velocity of the block; thereafter, negative work slows the body down. Since the spring force increases at 1200 N/m, the critical position occurs after a spring deformation of 408/1200 = 0.34 m. Now applying the W-E equation, we have

$$\left[\text{RW} = \frac{W}{2g}(v^2 - v_o^2)\right] \qquad 408(6 + 0.34) - \frac{1}{2}(408)(0.34) = \frac{600}{2 \times 9.81} v^2$$

from which

$$v^2 = 82.32 \text{ and } v = 9.1 \text{ m/s} \qquad \textbf{\textit{Ans.}}$$

14-4.3 A 450 N weight is swung in a vertical circle at the end of a 1.2 m cord. The topmost velocity of the weight is 3.6 m/s. (a) Find the tension in the cord at 120° past top position. (b) What minimum velocity at the bottom will keep the weight in the circular path at the top?

Solution

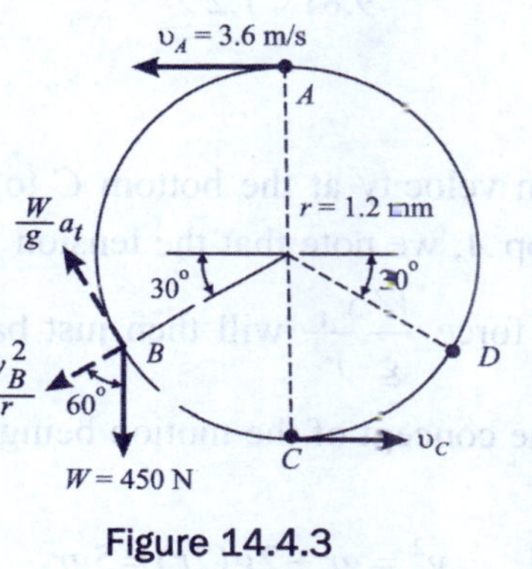

Figure 14.4.3

Part (a)

The tension is found by creating dynamic equilibrium at the 120° position at B as shown in Fig. 14-4.3. But first the centrifugal inertia force involving the velocity at B must be found. We could do this by computing the acceleration at any general position between A and B and then integrating to find v_B, but the work-energy method has the advantage of finding v_B directly. Note that the tension T is everywhere normal to the path and hence does no work. Only the weight W does work by dropping a vertical distance h = 1.2 + 1.2 sin 30°. We obtain

$$\left[Wh = \frac{W}{2g}(v_B^2 - v_A^2)\right] \qquad 450(1.2 + 1.2\sin 30^\circ) = \frac{450}{2 \times 9.81}(v_B^2 - 3.6^2)$$

$$v_B^2 = 48.3$$

It may be observed that if W is cancelled out of this equation, the terms may be rearranged to give $v_B^2 = v_A^2 + 2gh$ which is equivalent to the equation for free-falling bodies. This result will be obtained for any *frictionless translation* in which only the weight does work. Hence the velocity at D, which is at the same level as B, will be the same as at B. However, the time elapsed will, in general, differ from that required for free fall.

Now taking a summation of forces along the normal to the path at B gives

$$[\Sigma N = 0] \qquad T - \frac{W}{g}\frac{v_B^2}{r} - 450\cos\ 60^\circ = 0$$

$$T = \frac{450}{9.81}\left(\frac{48.3}{1.2}\right) + 225 = 2071.33N \qquad \textbf{\textit{Ans.}}$$

Part (b)

To find the minimum velocity at the bottom C to just keep the weight in the circular path at the top A, we note that the tension in the cord will be zero at A since the centrifugal force $\frac{W}{g}\frac{v_A^2}{r}$ will then just balance the weight W. Hence $v_A^2 = gr$. Applying the concept of the motion being equivalent to a free fall, we obtain

$$[v_c^2 = v_A^2 + 2gh] \qquad v_c^2 = gr + 2g(2r) = 5gr$$

from which

$$v_c = \sqrt{5(9.81)(1.2)} = 7.67 \text{ m/s} \qquad \textbf{Ans.}$$

PROBLEMS

14-4.4 The expression $\int ma \cdot dr$ may be rewritten as $\int m \frac{dv}{dt} \cdot dr$ to obtain $\int mv \cdot dv$. Show that this integral is always equivalent to $\int mv\, dv$ eventhough, as in curvilinear motion, dv is not necessarily collinear with v.

14-4.5 After the block in Fig. P-14-4.5 has moved 3 m from rest, the constant force P is removed. Find the velocity of the block when it returns to its initial position.

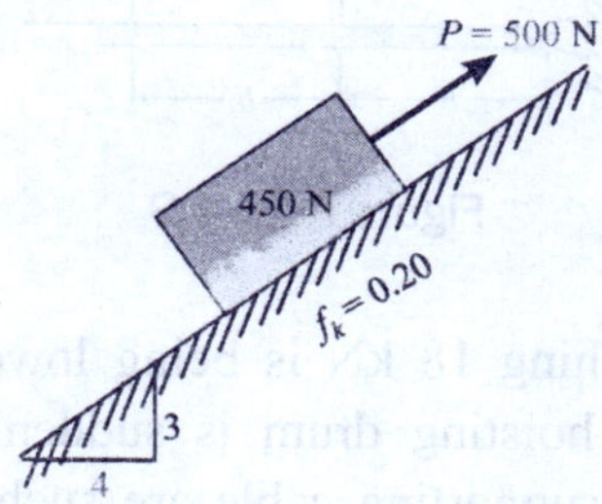

Figure P.14-4.5

14-4.6 A weight of W N is suspended from a vertical spring (Fig. P-14-4.6) whose modulus is k N/m. The weight is pulled down S m from its equilibrium position and then released. Determine its velocity when it returns to its equilibrium position.

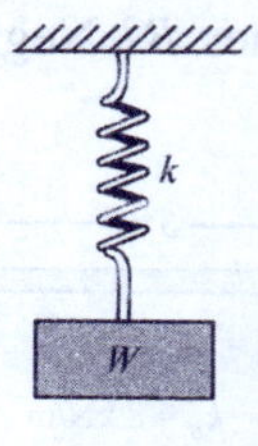

Figure P-14-4.6

14-4.7 Through what distance must the 2500 N block of Illus. Prob.14-4.2 slide from rest before touching the spring if its velocity is 5 m/s at the instant the spring is deformed 0.15 m? Assume the spring constant is changed to 35 kN/m.

14-4.8 A weight is dropped from a position just above, but not touching, a spring. Show that the maximum deformation produced will be twice that if the same weight were gradually lowered upon the spring.

14-4.9 A weight W is dropped from a height h upon a spring-supported bar as shown in Fig. P-14-4.9. Determine h so that the maximum spring force is three times its final static equilibrium value. Neglect the weight of the bar.

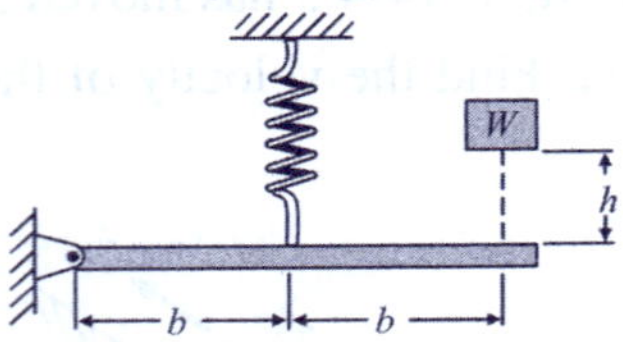

Figure P-14-4.9

14-4.10 An elevator weighing 18 kN is being lowered at a constant rate of 3 *m/s* when the hoisting drum is suddenly stopped. If the elastic properties of the supporting cable are such that it is equivalent to a spring with a modulus of 350 kN/m, determine by how many times the sudden stop of the hoisting drum momentarily increases the cable tension. (This problem illustrates another insight to situations like that in Prob. 11-4.4.)

14-4.11 The rigid horizontal bar in Fig. P-14-4.11 is supported by two springs. A 1350 N weight is then placed upon the bar at A. A sudden blow projects the weight toward B with an initial velocity of 1.8 *m/s*. What is its velocity when it reaches B? Neglect friction and the weight of the bar.

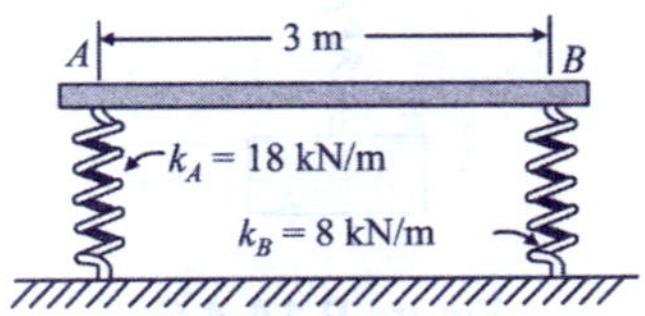

Figure P-14-4.11

14-4.12 A flexible rope, 24 m long and weighing 8 N/m, passes over two smooth pegs as shown in Fig. P-14.4.12. The rope starts from rest when $d = 3$ m. Determine the velocity of the rope at the instant when d = 12 m.

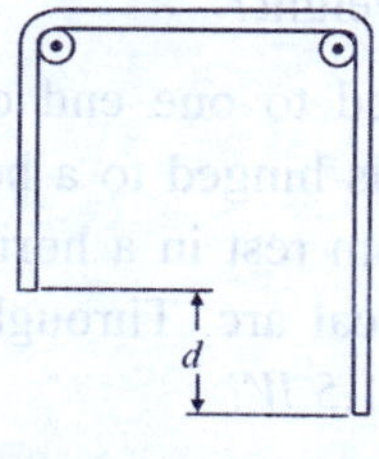

Figure P-14-4.12

14-4.13 A chain of length L and weighing w N/unit length is released from rest on a smooth table when in the position as shown in Fig P-14-4.13. Determine the velocity of the chain as the last link leaves the table.

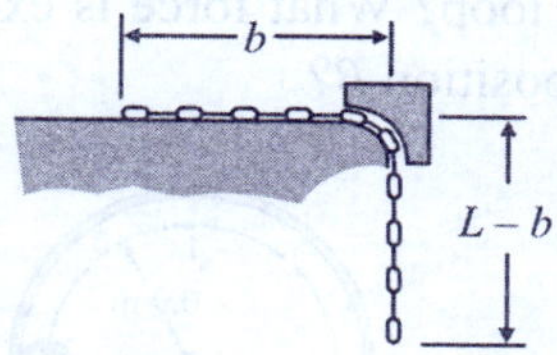

Figure P-14-4.13

14-4.14 A 445 N weight moves along the smooth rigid guide from A to B as shown in Fig. P-14-4.14. Find its velocity at B if it starts from rest at A. The free length of the spring is 600 mm and its modulus is 0.9 N/mm.

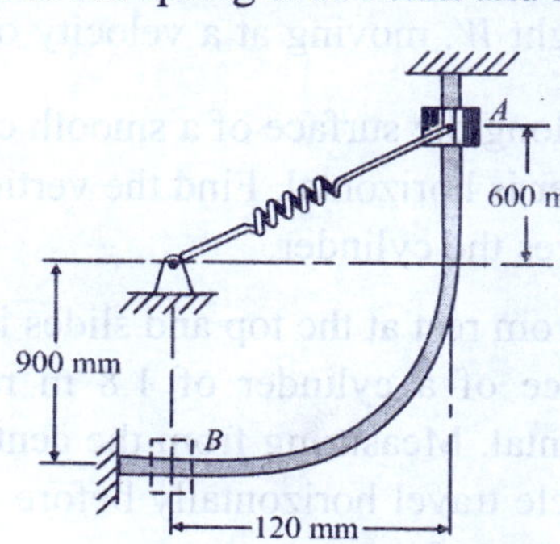

Figure P-14-4.14

14-4.15 A weight of 90 N is swung in a vertical circle at the end of a 0.9 m cord, At the lowest position of the weight, the tension in the cord is 360 N. (a) How high above the lowest position will the weight rise on the circular path? (b) What would be this result if the cord is replaced by a stiff rod of negligible weight?

14.4.16 A weight W is attached to one end of a stiff rod of length L and negligible weight that is hinged to a horizontal axis at the other end. The rod is released from rest in a horizontal position and allowed to swing freely in a vertical arc. Through what angle must it swing to cause a tension in it of 1.5 W?

14-4.17 The rod in the previous problem is displaced an angle θ from its lowest position and released from rest. Find θ so that the tension in the rod at the lowest position is four times that just after release.

14-4.18 In Fig. P-14-4.18, by how much should the spring be compressed so that it will cause the 4.5 N pellet to travel completely around the frictionless vertical loop? What force is exerted by the track upon the pellet when it is at position B?

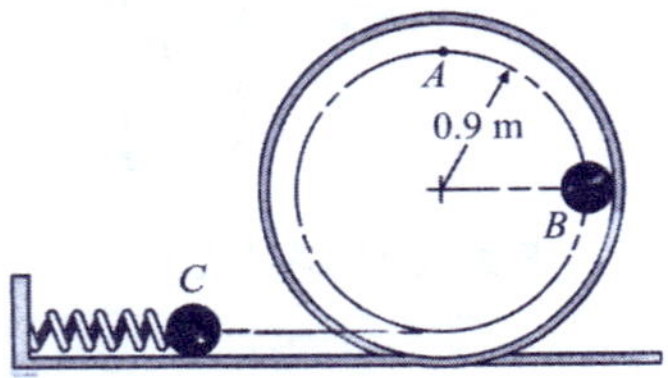

Figure P-14-4.18

14-4.19 A particle of weight W, moving at a velocity of $\frac{1}{2}\sqrt{gr}$ m/s at the top, slides vertically along the surface of a smooth cylinder of radius r, The axis of the cylinder is horizontal. Find the vertical distance the particle falls before it leaves the cylinder.

14-4.20 A particle starts from rest at the top and slides in a vertical plane along the smooth surface of a cylinder of 1.8 m radius. The axis of the cylinder is horizontal. Measuring from the center of the cylinder, how far does the particle travel horizontally before it strikes the ground on which the cylinder rests?

14-4.21 Neglect friction of the 275 N collar against its vertical guide and compute the velocity of the collar after it has fallen 2 m, starting from rest in the position shown in Fig. P-14-4.21. The unstretched length of the spring is 0.6 m.

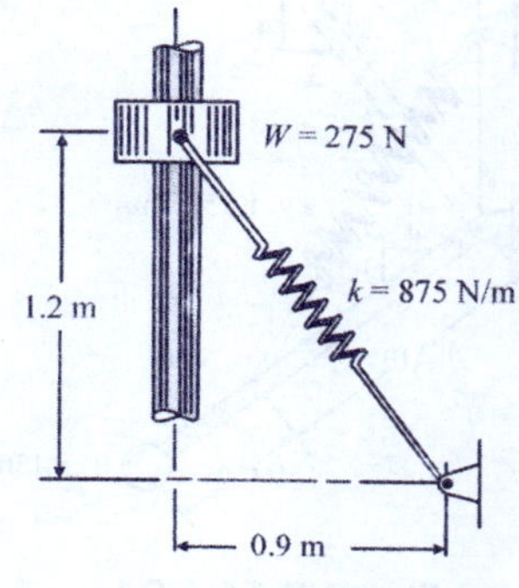

Figure P-14-4.21

14.4.22 The car in Fig. P-14-4.22 is moving toward the bumper spring and has a kinetic energy of 12 kN-m. The main bumper shield (*a-a*) is connected to the main spring which has a modulus of 175 N/mm. The two auxiliary bumper shields (b) are 300 mm behind *a-a* and are attached to secondary springs, each of which has a modulus of 90 N/mm. Determine the greatest movement of *a-a*. What percentage of the energy will then be absorbed by the main spring?

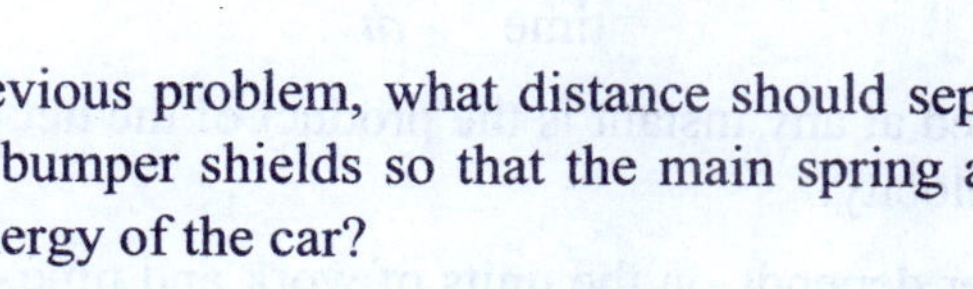

Figure P-14-4.22

14-4.23 In the previous problem, what distance should separate the main and auxiliary bumper shields so that the main spring absorbs 80% of the kinetic energy of the car?

14-4.24 A 450 N weight rotates in a vertical plane at the end of a 1.8 m rod of negligible weight as shown in Fig. P-14-4.24. The spring, whose modulus is 1.75 N/mm, does not act on the rod until it exceeds its free length of 0.9 m. Determine the velocity of the weight after it has swung from rest at the dashed position to the given position. Can the weight reach a position vertically below the pivot? If so, what will be its velocity then?

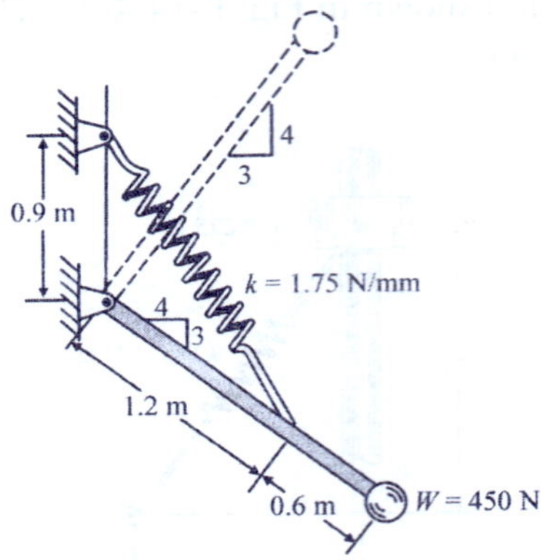

Figure P-14-4.24

14-4.25 Solve Prob. 14-4.24 if the spring is a "hard" spring in which the spring force in Newtons is given by $P = 60s^2$ in which s is the spring deformation in meter.

14-5 POWER EFFICIENCY

Power is the time rate at which work is done on a body. For example, if a train is being pulled by a locomotive, the work done on the train may be measured by the work done by the drawbar pull. Expressed mathematically,

$$\text{power} = \frac{\text{work}}{\text{time}}$$

This gives the average power. If F is the net force doing the work, then the work during a differential time is given by $F\,ds$, and the power exerted at any instant is

$$\text{power} = \frac{\text{work}}{\text{time}} = \frac{Fds}{dt} = Fv \qquad (14\text{-}5.1)$$

i.e., the power exerted at any instant is the product of the net force multiplied by the instantaneous velocity.

The unit of power depends on the units of work and time. Common units are N-m per sec in the gravitational system, the joule per sec in the absolute system. These units are usually too small for use in engineering. The units commonly used here are the watt or kilowatt (kW). The watt equals 10^7 ergs per sec, and the kilowatt equals 1000 watts.

The watt and kilowatt are used in electrical engineering. Relations between horsepower and kilowatts are as follows:

$$1 \text{ hp} = 0.746 \text{ kW}$$

$$1 \text{ kW} = 1.34 \text{ hp}$$

For large quantities of work, the units are the horsepower-hour (hp-hr) and the kilowatt-hour (kW-hr). These units indicate the amount of *work* done in one hour at the constant rate of 1 hp or 1 kW.

Because of losses resulting from friction and other causes, the power delivered from a machine or other device is never equal to the power put into it. Efficiency is the ratio of power output to power input. This ratio is usually multiplied by 100 so that efficiency may be given as a percentage. Since power output and power input are measured during the same time, efficiency may also be defined as the ratio of energy output to energy input or of work output to work input. The *overall efficiency* of a number of machines placed in series is equal to the product of their individual efficiencies.

ILLUSTRATIVE PROBLEMS

14-5.1 A 1000 ton train is accelerated at a constant rate up a 2% grade. The train resistance is constant at 45 N per ton. The velocity increases from 9 m/s to 17 m/s in a distance of 600 m. Determine the maximum power developed by the locomotive.

Preliminary

The inclination of a grade is expressed as the rise in meters measured in a horizontal distance of 100 m. Thus, a 2% grade means a rise of 2 m in a horizontal distance of 100 m. For grades up to 10%, the length of the incline is practically equal to its horizontal projection so that the cosine of the angle of the grade is practically equal to unity. Therefore the normal pressure may be considered equal to the weight of the train; hence, the train resistance remains constant. Further more, for small angles the sine may be considered equal to the tangent, so that in this case of a 2% grade, the sine will be equal to $\frac{2}{100}$.

Solution

The constant drawbar pull exerted by the locomotive is found by applying the work-energy equation. From the point diagram of the forces acting on the train (Fig. 14-5.1) we have

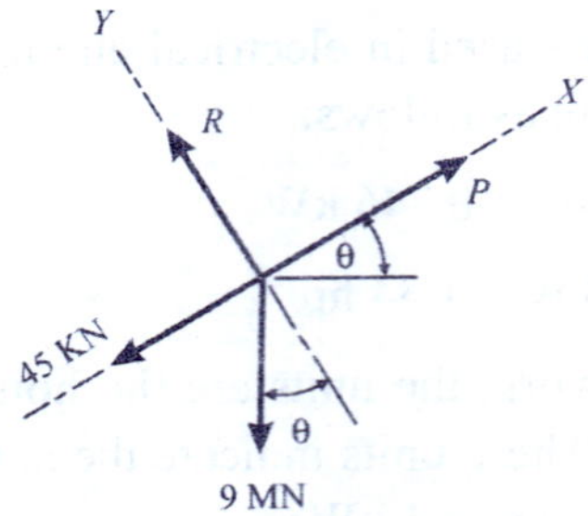

Figure 14-5.1

$$\left(P - 45 \times 10^3 - 9 \times 10^6 \times \frac{2}{100}\right) \times 600$$

$$= \frac{9 \times 10^6}{2 \times 9.81} \times (17^2 - 9^2)$$

$$P = 384 \times 10^3 \text{N}$$

The maximum power is developed when the train is moving at maximum speed. Therefore

$$1\text{kW} = \frac{F \times V}{746} = \frac{384 \times 10^3 \times 17}{746} \qquad \text{power} = 8750 \text{ Watts}$$

PROBLEMS

14-5.2 A train weighs 1200 tons. The train resistance is constant at 53.4 N/ton. If 4476 kW are available to pull this train up a 3% grade, find its maximum speed in m/s.

14.5.3 A train weighing 1200 tons is pulled up a 2% grade at 4.5 m/s. If train resistance is constant at 45 N/ton and the power exerted by the locomotive remains constant, determine the speed of the train on a level track. What horsepower does the locomotive develop?

14-5.4 A train weighing 1000 tons is being pulled up a 2% grade. The train resistance is constant at 45 N/ton. The speed of the train is increased at a constant rate from 6.0 m/s to 12 m/s in a distance of 300 m. Find the maximum power developed by the locomotive, in kw.

14-5.5 A conveyor belt moving at 1.5 m/s transports 100 tons of material per hour through a vertical height of 46 m. Assuming the mechanical efficiency of the system to be 85% and the drive motor to be 94% efficient, determine the power required in kilowatts.

14-5.6 Water flows through a nozzle of 25 mm diameter under a head of 125 m to drive a turbine. The turbine is 90% efficient and drives a generator which is 94% efficient. Compute the power output in kilowatts.

14-5.7 Water enters a hydraulic reaction turbine with a velocity of 3.75 m/s and leaves it 1.25 m lower with a velocity of 0.915 m/s. If 450 KN of water flow through the turbine each second, compute the power output. Assume the turbine is 90% efficient and water weighs 9.8 KN/m^3.

14-5.8 Determine the constant horsepower required to change the speed of a 18 KN auto from 9 m/sec to 17 m/s in a distance of 60 m along a level road. What is the elapsed time? Assume road and air resistance are negligible.

Hint: With constant power, the acceleration is not constant.

14-6 WORK-ENERGY APPLIED TO CONNECTED SYSTEMS

An outstanding advantage of the work-energy method is that it may be applied directly to connected systems of bodies without the necessity for considering forces which are internal to the system, such as those in inextensible connecting bars or cords. This is particularly useful in cases where the internal forces may vary (as in Illus. Prob. 14-6.2). To understand why internal forces are automatically eliminated, consider the system of two bodies in Fig. 14-6.1 which are connected by a rigid link of negligible weight. From Newton's second and third laws, the reactions of the bodies on the link are equal and oppositely directed. If the length of the link does not change, each end has the same component of displacement along the link and hence the positive work done by the connecting link on one of the bodies is equal to the negative work done on the other body. Consequently, the total change in the kinetic energy of a connected system of bodies is caused only by the resultant work of the external forces, provided that the length of the connecting members does not change and they are of negligible weight.

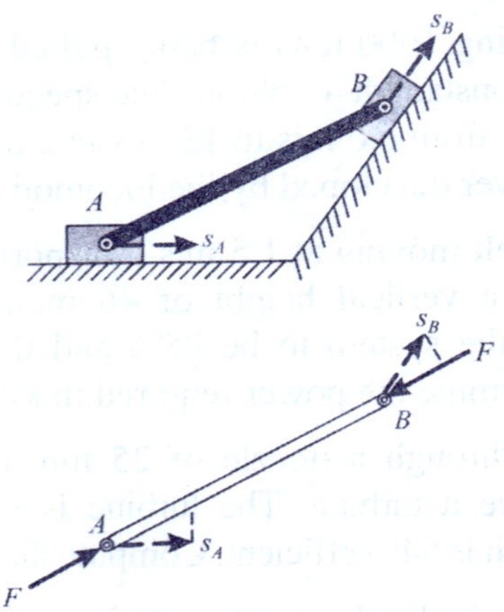

Figure 14-6.1

As an illustration of this principle, consider the system in Fig. 14-6.2. The work done by T on A is Ts_A, being positive because T is in the direction of s_A. On B, the work done by $2T$ is $-2Ts_B$, being negative because $2T$ is opposite to the direction of s_B. Since the work done by the internal tensions must be zero, we have

$$Ts_A - 2Ts_B = 0 \quad \text{or} \quad S_A = 2s_B$$

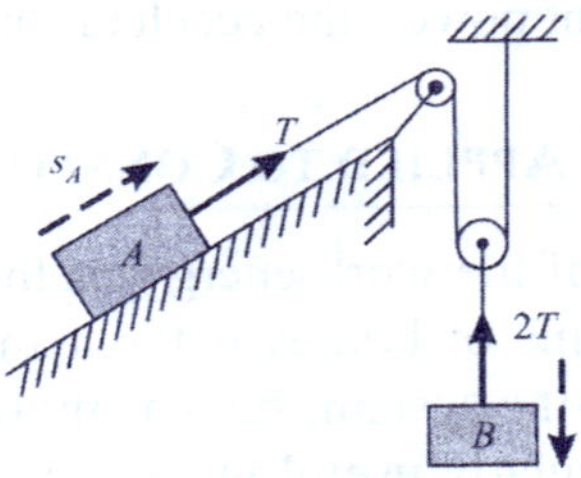

Figure 14-6.2

whence successive differentiation gives $v_A = 2v_B$ and $a_A = 2a_B$. We have already "borrowed" this principle for use in Illus. Prob. 11-3.2, but now we can also confirm these relations from our knowledge of kinematics. Thus, using the instant center of the floating pulley, it is evident that $s_A = 2s_B$, etc.

A general plan for applying the work-energy method consists of the following steps:

1. Determine the direction of motion. Alternatively, assume a direction of motion, confirming this assumption by noting that the resultant work must be positive to speed up a system, and vice versa.

2. Draw an *active-force diagram* for the system, showing only those forces that do work upon the system.
3. Determine the kinematic relations among the bodies corresponding to the system.
4. Apply the work-energy equation by equating the resultant work on the system to the sum of the changes in kinetic energy of the bodies in the system.
5. If the internal force in a connecting member is desired, apply the work-energy equation to a free-body diagram of that part of the system on which this force is then exposed as an external force. If the internal force is not constant, however, this step will determine only its average value. The instantaneous value of a variable force must be found by the force-inertia method.

ILLUSTRATIVE PROBLEMS

14-6.1 If the system in Fig. 14-6.3 starts from rest in the given position, how much further will A move up the incline after B hits the ground? The coefficient of kinetic friction is 0.20 and the pulleys are frictionless and of negligible weight.

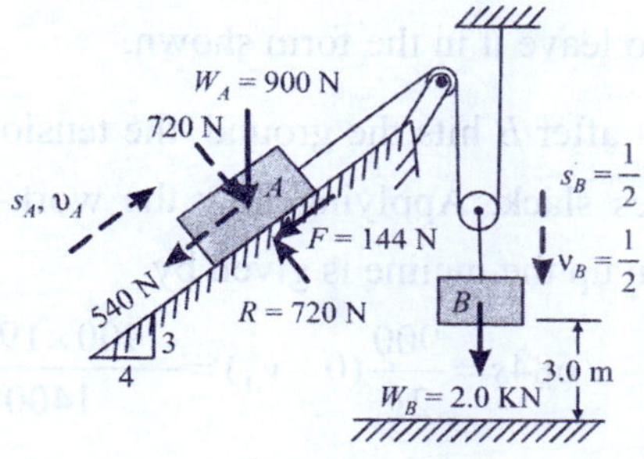

Figure 14-6.3

Solution

Since the initial and final velocities are zero, there is a strong temptation to likewise equate the total resultant work to zero. However, this approach would not account for the kinetic energy that is lost when B strikes the ground. Instead, we must consider two phases of the motion: (1) when A and B move as a connected system and (2) when A continues to move alone.

Phase (1). Only the external forces acting on the system which do work are shown. The internal forces (tensions in the cords) do no work on the system and may be used to show that the displacement and velocity of B are one-half the corresponding values of A.

For convenience, we represent the normal and downplane components of the 900 N weight of A by the dashed components, whence the friction force on A is $F = 0.20(720) = 144$ N. If we now apply the work-energy equation to the system and note that the resultant work RW = $(\Sigma X\ s)_B + (\Sigma X\ s)_A$ where the X summations are positive in the indicated directions of motion, we obtain

$$\left[\text{RW} = \Sigma \frac{W}{2g}(v^2 - v_o^2)\right]$$

$$2 \times 10^3 \times (3) - (540 + 144)\ 6.0 = \frac{2 \times 10^3}{2g}\left(\frac{1}{4}v_A^2\right) + \frac{900}{2g}v_A^2$$

$$6 \times 10^3 - 4.1 \times 10^3 = \frac{v_A^2}{2g}\ (500 + 900)$$

whence

$$\frac{v_A^2}{2g} = \frac{1900}{1400}$$

This result could be used to determine the initial velocity of A in phase (2); however, it is simpler to leave it in the form shown.

Phase (2). Note that after B hits the ground, the tension in the cord becomes zero; i.e., the cord goes slack. Applying now the work-energy equation to A alone, its further motion up the incline is given by

$$\left[\text{RW} = \frac{W}{2g}(v^2 - v_o^2)\right] \quad -\ 684s = \frac{900}{2g}(0 - v_A^2) = -\frac{900 \times 1900}{1400}$$

whence

$$s = \frac{900 \times 1900}{1400 \times 684} = 1.786 \text{ m} \qquad \textbf{\textit{Ans.}}$$

Some additional comments may be worthy of note. The tension in the cord supporting B during its descent can be found by applying the work-energy equation to a FBD of B as follows:

$$\left[\text{RW} = \frac{W}{2g}(v^2 - v_o^2)\right] \quad (2000 - T_B)(3) = \frac{2000}{2g}\left(\frac{1}{4}v_A^2\right) = 500\left(\frac{1900}{1400}\right)$$

$$T_B = 2000 - 226.2 = 1773.8 \text{ N} \qquad \textbf{\textit{Ans.}}$$

Also, since the forces and accelerations are constant, we may substitute the known value of v_A^2 in Eq. (9-3.6) to compute (a_A during phases (1) and (2). Doing this gives

$$[v^2 = v_o^2 + 2as] \qquad \frac{1900}{1400}(2g) = 2(a_A)_1(6) \qquad (a_A)_1 = 2.2 \text{ m/sec}^2$$

and

$$0 = \frac{1900}{1400}(2g) + 2(a_A)_2(1.786) \quad (a_A)_2 = -7.45 \text{ m/sec}^2$$

14-6.2 Body A is pulled along the frictionless level surface in Fig. 14-6.4 by a flexible, inextensible cord passing over smooth fixed pegs at C and D to the weight B. If A starts from rest at the given position, determine its velocity when it is in the dashed position.

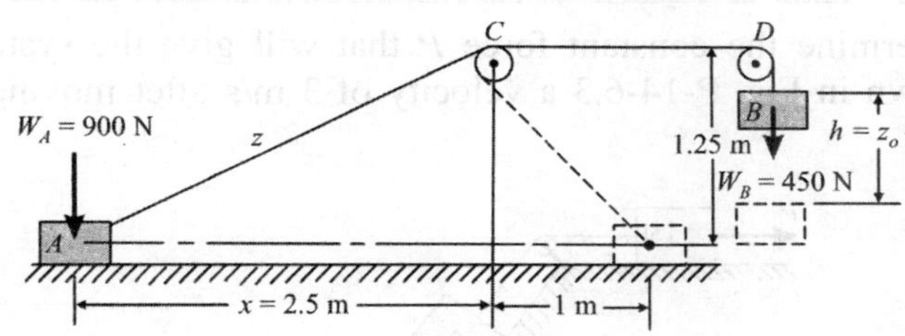

Figure 14-6.4

Solution
The only active force doing work on the system is W_B. The vertical movement of B is equal to the difference in length of AC as it moves between the two specified positions. Designating this variable length by z, and noting that the velocity of B is equal to the rate of change of this length $\left(i.e., v_B = \frac{dz}{dt}\right)$, we have

$$z^2 = x^2 + 1.25^2 \quad \text{hence} \quad z_o = \sqrt{2.5^2 + 1.25^2} = 2.8 \text{ m}$$

$$\text{and} \quad z_f = \sqrt{1^2 + 1.25^2} = 1.6 \text{ m}$$

To relate v_A to v_B, we differentiate the preceding expression with respect to the time. This gives

$$2z\frac{dz}{dt} = 2x\frac{dx}{dt} \quad \text{or} \quad zv_B = xv_A$$

hence for the dashed position

$$1.6\ v_B = 1\ v_A$$

Applying the work-energy equation, we obtain

$$\left[\text{RW} = \Sigma \frac{W}{2g}(v^2 - v_o^2) \right]$$

$$450(2.8 - 1.6) = \frac{900}{2 \times 9.81} v_A^2 + \frac{450}{2 \times 9.81}\left(v_B^2 = \frac{1}{2.56} v_A^2 \right)$$

where $v_B = \dfrac{1}{1.6} v_A$

from which

$$v_A^2 = 9.85 \quad \text{and} \quad v_A = 3.14 \text{ m/sec} \qquad \textit{Ans.}$$

PROBLEMS

14-6.3 Determine the constant force P that will give the system of bodies shown in Fig. P-14-6.3 a velocity of 3 m/s after moving 4.6 m from rest.

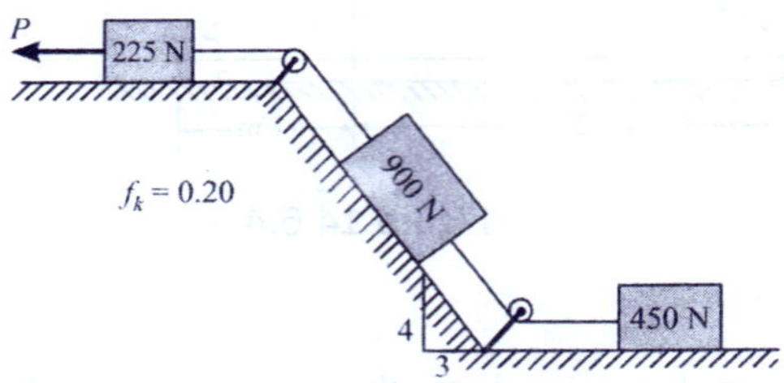

Figure P-14-6.3

14-6.4 Through what distance will body A in Fig. P-14-6.4 move in changing its velocity from 2 m/s to 4 m/s? Assume the pulleys to be friction less and of negligible weight.

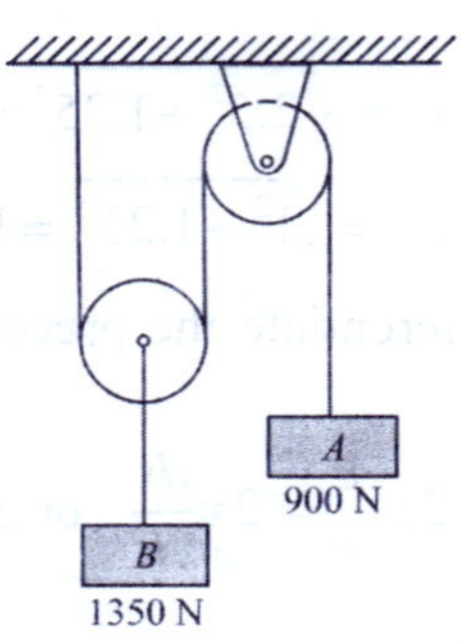

Figure P-14-6.4

14-6.5 In what distance will body A of Fig. P-14-6.5 attain a velocity of 3 m/s, starting from rest?

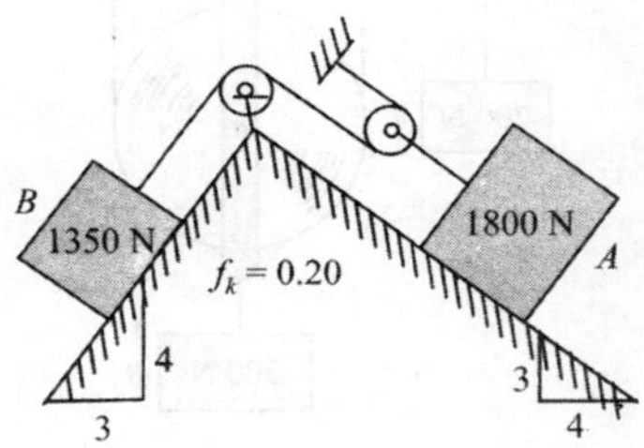

Figure P-14-6.5

14-6.6 Determine the velocity of body A in Fig. P-14-6.6 after it has moved 4 m starting from rest. Assume the pulleys to be frictionless and of negligible weight.

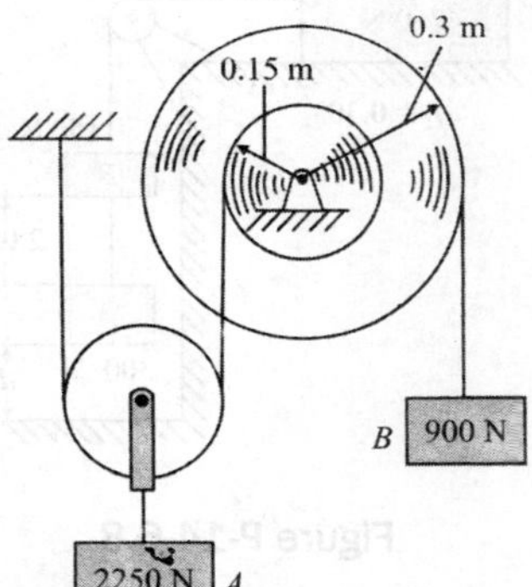

Figure P-14-6.6

14-6.7 Assume the pulleys in Fig. P-14-6.7 to be frictionless and of negligible weight. Find the velocity of body B after it has moved 3 m from rest. Then use this result to determine the acceleration of body B.

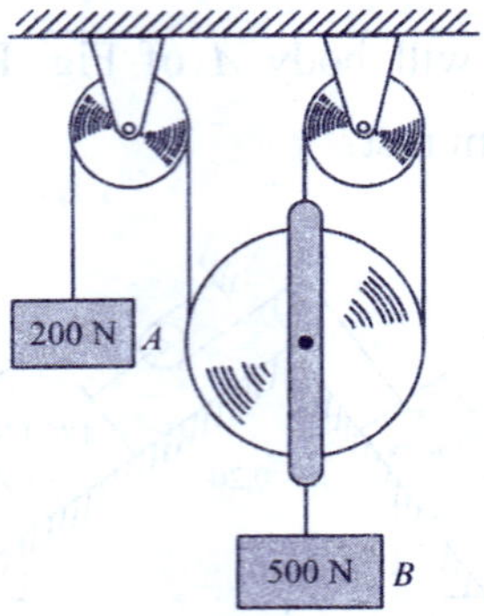

Figure P-14-6.7

14-6.8 The system shown in Fig. P-14-6.8 is connected by flexible, inextensible cords. If the system starts from rest, find the distance d between A and the ground so that the system comes to rest with body B just touching A.

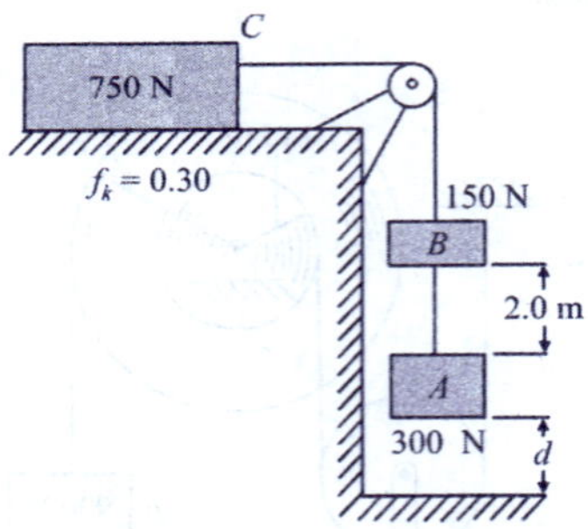

Figure P-14-6.8

14-6.9 If d = 1.5 m in the preceding problem, determine the total distance moved by body C.

14-6.10 In Prob. 14-6.8, what should be the weight of body C so that for d = 1.5 m the system stops with B just touching A?

14-6.11 Body A starts from rest in the position shown in Fig. P-14-6.11. Determine its velocity after it has moved 5 m along the frictionless surface.

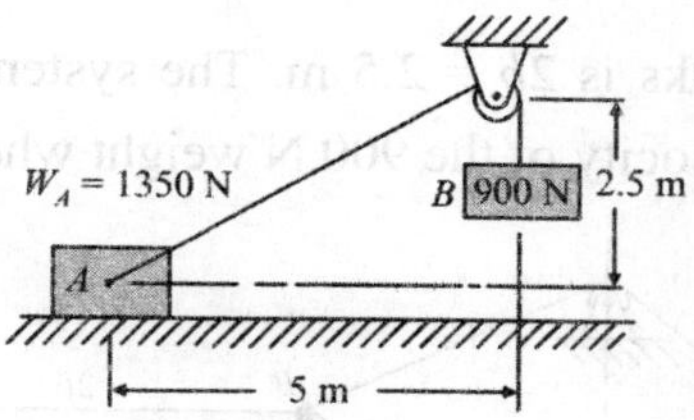

Figure P-14-6.11

14-6.12 Find the velocity of body A in Prob. 14-6.11 after it has moved, starting from rest at the given position, for 2.75 m along the frictionless surface.

14-6.13 Determine the tension in the cord connecting bodies A and B in Prob. 14-6.11 at the instant A has moved 2.75 m starting from rest in the given position.

Hint: Use the result of Prob. 14-6.12.

14-6.14 Two sliders, connected by a light rigid link 3 m long, move in the frictionless guides as shown in Fig. P-14-6.14. If B starts from rest when it is vertically below A, determine the velocity of B when x = 1.8 m. Assume $W_A = W_B$ = 900 N and W_c = 450 N.

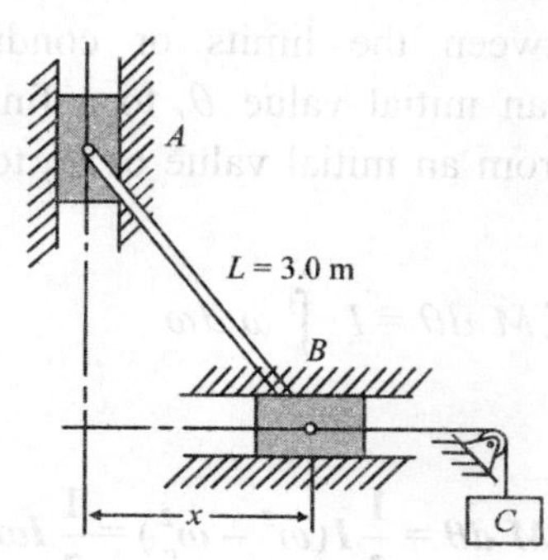

Figure P-14-6.14

14-6.15 Repeat Prob. 14-6.14 when x = 2.5 m

14-6.16 A homogeneous triangular plate weighing 450 N is supported by two parallel links of negligible weight as shown in Fig. P-14-6.16. The length of the links is $2b = 2.5$ m. The system starts from rest at $\theta = 90°$. Find the velocity of the 900 N weight when $\theta = 60°$.

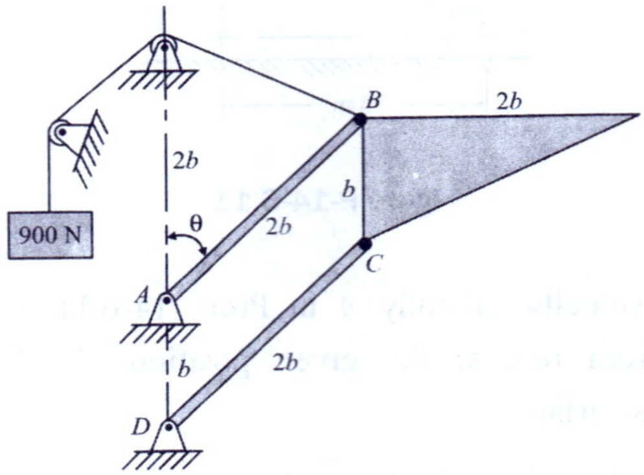

Figure P-14-6.16

14-6.17 Repeat Prob. 14-6.16 when $\theta = 30°$.

14-7 WORK-ENERGY METHOD APPLIED TO FIXED-AXIS ROTATION

The simplest and most direct way to obtain the work-energy equation for fixed-axis rotation is to eliminate α, between the equations $\Sigma M = I\alpha$ and $\alpha\, d\theta = \omega\, d\omega$, to obtain

$$\Sigma M\, d\theta = I\omega\, d\omega \qquad (a)$$

This is integrated between the limits or conditions that the angular displacement varies from an initial value θ_o to a final value of θ while the angular velocity changes from an initial value of ω_o to a final value of ω. We thus obtain

$$\int_{\theta_o}^{\theta} \Sigma M\, d\theta = I \int_{\omega_o}^{\omega} \omega\, d\omega$$

or

$$\int_{\theta_o}^{\theta} \Sigma M\, d\theta = \frac{1}{2} I(\omega^2 - \omega_o^2) = \frac{1}{2} I\omega^2 - \frac{1}{2} I\omega_o^2 \qquad (14\text{-}7.1)$$

This is the fundamental work-energy equation for fixed-axis rotation. In it, both ΣM and I are computed about the axis of rotation. Hence, for centroidal rotation, these terms become $\Sigma \overline{M}$ and $\overline{I}$, respectively.

As in translation, this is a scalar equation.[5] The left-hand term represents the resultant work of rotation during an angular displacement θ, expressed in Newton-meter. The right-hand terms represent the change in kinetic energy of the rotating body, also expressed in Newton-meter, during this angular displacement.

The expression for kinetic energy of a rotating body can also be developed from the fact that the total kinetic energy is the scalar summation of the kinetic energies of all particles of the body. Thus, in Fig. 14-7.1, if B is a typical particle, its kinetic energy is

$$\text{KE} = \frac{1}{2} dm\, v_B^2 = \frac{1}{2} dm\, r^2 \omega^2$$

For the entire body, therefore,

$$\text{Total KE} = \int \frac{1}{2} dm\, r^2 \omega^2 = \frac{1}{2} \omega^2 \int r^2 dm$$

or, as before

$$\text{KE} = \frac{1}{2} I \omega^2$$

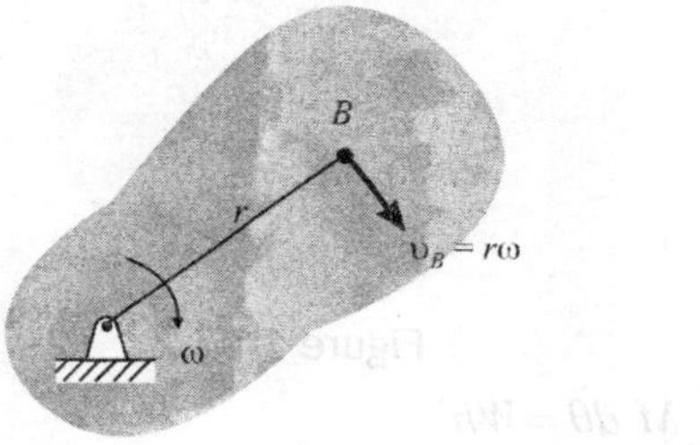

Figure 14-7.1 Velocity of a typical particle.

If ΣM is variable, the resultant work in rotation can be found only if ΣM can be expressed in terms of θ, or from the area under a ΣM- θ diagram. If ΣM is constant, the resultant work may be expressed as $\Sigma M(\theta)$, in which case the work-energy equation becomes

$$\boldsymbol{\Sigma M(\theta) = \frac{1}{2} I(\omega^2 - \omega_o^2)} \qquad (14\text{-}7.2)$$

---

[5]Thus, by taking the dot product of $d\theta$ with each side of $\Sigma \mathbf{M} = \dot{\mathbf{H}}$, we obtain

$$\Sigma \mathbf{M} \cdot d\theta = \dot{\mathbf{H}} \cdot d\theta = \frac{d\mathbf{H}}{dt} \cdot d\theta = \omega \cdot d\mathbf{H}$$

Since all these vectors are perpendicular to the plane of motion, and $\mathbf{H} = I\omega$ so that $d\mathbf{H} = I\, d\omega$, this reduces as before to Eq. (*a*).

Notice that, as in translation, moments in the initial direction of rotation do positive work on the body, and vice versa.

The work done by gravity forces on a rotating body will be the product of the weight W by its change in elevation h as discussed in Section 14-3. It is worthwhile to check this statement by considering Fig. 14-7.2 in which the body rotates through θ radians from the horizontal. The work done by W can be found by applying

$$\int_o^\theta \Sigma M \, d\theta \text{ which gives}$$

$$\int_o^\theta \Sigma M \, d\theta = \int_0^\theta W\bar{r}\cos\theta \, d\theta = W\bar{r}[\sin\theta]_o^\theta = W\bar{r}\sin\theta$$

and, as $\bar{r}\sin\theta = h$, we obtain Wh as stated above.

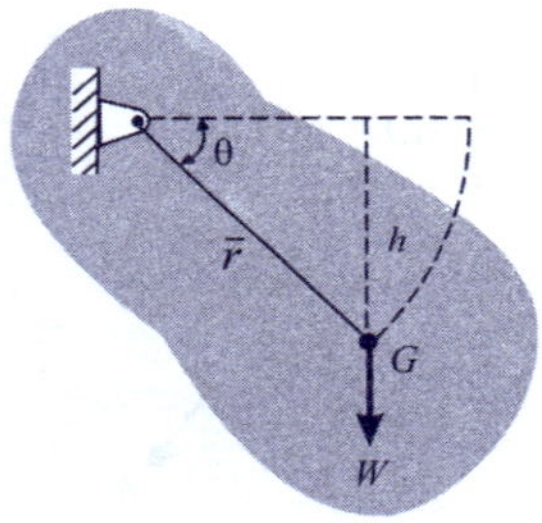

Figure 14-7.2

$$\text{RW} = \int M \, d\theta = Wh.$$

As in the case of translation, the work-energy equation may be applied directly to a connected system of translating and rotating bodies without considering any of the internal forces in the connecting elements. The resultant work (however computed) is always equal to the scalar changes in kinetic energy of the system, or

$$(\text{RW})_{sys} = \Sigma\frac{W}{2g}(v^2 - v_o^2) + \Sigma\frac{1}{2}I(\omega^2 - \omega_o^2) \qquad (14\text{-}73)$$

Then after the relation between forces, displacements, and velocities for the original system has been determined, we may isolate any part of the system and reapply the work-energy equation to that part to determine constant internal forces.

The illustrative problems demonstrate several advantages of the work-energy method. In the first one, a single work-energy equation replaces a simultaneous solution of four equations resulting from applying a kinetic equation to each of four bodies, plus an additional kinematic equation. In the second problem, the work-energy equation leads directly to a velocity which otherwise would require integrating a kinematic equation involving a variable acceleration. In the third problem, we illustrate a solution combining both work-energy and force-inertia methods.

ILLUSTRATIVE PROBLEMS

14-7.1 Determine the distance that body D in Fig. 14-7.3 must move in order to reach a velocity of 12 m/sec starting from rest. What tension is acting in the cord joining the step pulleys B and C during this movement?

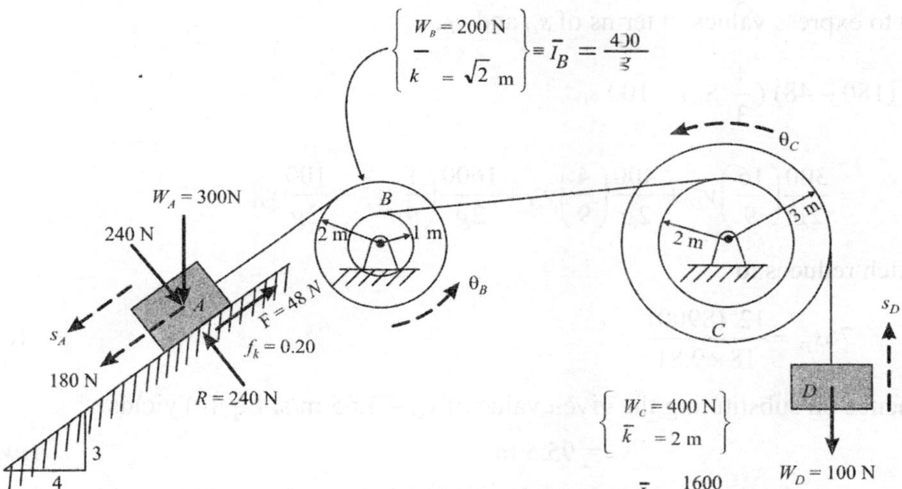

Figure 14-7.3

Solution

A preliminary calculation of the tensions in the various cords, assuming the system to be at rest, discloses an unbalanced upward force to act on D which therefore rises while the rest of the system moves as shown by the dashed displacement arrows. Using $s = r\theta$, the kinematic relations among the bodies are

$$S_A = 2\ \theta_B; \qquad \theta_B = 2\theta_C; \qquad 3\theta_C = S_D$$

Combining these relations gives

$$s_A = 2\theta_B = 4\theta_C = \frac{1}{3}\ s_D \tag{a}$$

and since $v = r\omega$ is of the same mathematical form as $s = r\theta$, the velocities are similarly related as follows:

$$v_A = 2\omega_B = 4\omega_C = \frac{4}{3}\ v_D \tag{b}$$

We are now ready to equate the resultant work of external forces on the system to the total change in kinetic energy of the system. We apply $\Sigma X(s)$ to the external forces on each translating body and $\Sigma M(\theta)$ to the moments of external forces on the rotating bodies (of which there are none in this case). Remember that ΣX and ΣM are positive in the initial direction of motion. The kinetic energy change in the system is $\Sigma \frac{W}{2g}(v^2 - v_o^2)$ for the translating bodies and is $\Sigma \frac{1}{2}\bar{I}(\omega^2 - \omega_o^2)$ for the rotating bodies. We then have, using Eqs. (*a*) and (*b*) to express values in terms of s_D and v_D,

$$(180 - 48)\left(\frac{4}{3}\,S_D\right) - 100\,s_D$$

$$= \frac{300}{2g}\left(\frac{16}{9}\right)v_D^2 + \frac{400}{2g}\left(\frac{4}{9}\right)v_D^2 + \frac{1600}{2g}\left(\frac{1}{9}\right)v_D^2 + \frac{100}{2g}v_D^2$$

which reduces to

$$76 s_D = \frac{12^2(8900)}{18 \times 9.81} \qquad (c)$$

whence on substituting the given value of $v_D = 3.65$ m/s, Eq. (*c*) yields

$$s_D = 95.5 \text{ m} \qquad \textbf{\textit{Ans.}}$$

The advantage of using general symbols for s_D and v_D will be evident if we now differentiate Eq. (*c*) with respect to the time. Doing this gives

$$76\,v_D = \frac{8900}{18g}(2v_D)\frac{dv_D}{dt}$$

whence, cancelling the common term v_D, the acceleration of D is[6]

$$a_D = \frac{dv_D}{dt} = \frac{76(18)\,(9.81)}{8900(2)} = 0.754 \text{ m/sec}^2$$

Now, if desired, relations similar to Eqs. (*a*) or (*b*) can be used to find the accelerations of the other bodies. Similarly, the computed value of s_D can be used to determine the displacements of the other bodies.

---

[6]This method of finding instantaneous acceleration is a general procedure applicable to constant or variable accelerations. When the acceleration is constant, as is the case here when forces and moments are constant, the procedure outlined in Illus. Prob. 14-6.1 is more direct.

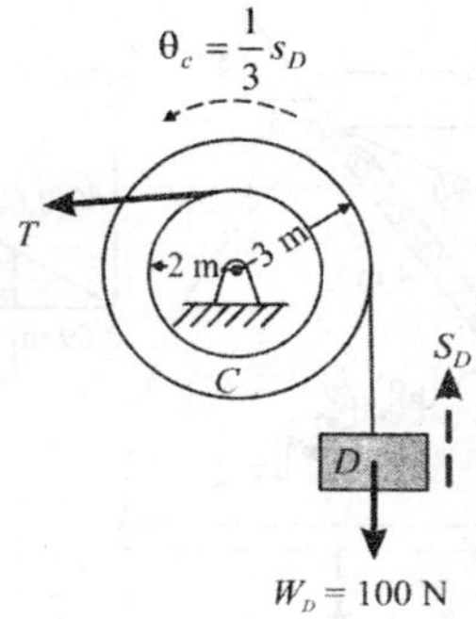

Figure 14-7.4

To determine the tension in the cord between B and C, we cut this cord and isolate the system of bodies C and D as shown in Fig. 14-7.4. As this system moves through the distance s_D, its change in KE is the sum of the last two terms enclosed in the parentheses of Eq. (c). The tension T is an external force on this system and its work is computed from $\Sigma M(\theta_C)$ where the moment of T is $2T$ and $\theta_C = \frac{1}{3}s_D$. An equivalent way of finding the work done by T is to multiply T by the length of cord unwrapped from C which is $s = r\theta_C =$ 0.6096 (1.1 s_D). Applying the work-energy equation to this system and using the value of $s_D = 95.5$ m corresponding to a velocity of $v_D = 12$ m/s, we obtain

$$[\text{RW} = \Delta\text{KE}] \qquad 2T\left(\frac{1}{3}\right)95.5 - 100(95.5) = \frac{12^2}{18g}(2500)$$

$$T = 182 \text{ N}$$ ***Ans.***

14-7.2 At the position shown in Fig. 14-7.5, the uniform bar, 4 m long and weighing 60 N, has an angular velocity ω_o. The spring has a modulus of 5 N/mm and a free length of 2 m. Determine the value of ω_o so that the bar just reaches the horizontal dashed position. Is it possible for the bar to rebound from this horizontal position past its topmost position? If so, what will then be its angular velocity?

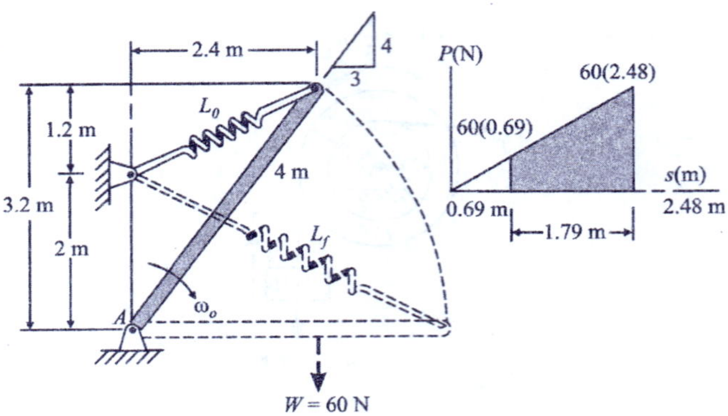

Figure 14-7.5

Solution

We first compute the initial and final lengths of the stretched spring to be respectively $L_o = 2.69$ m and $L_f = 4.48$ m. The corresponding deformation of the spring, found by subtracting the free length of 2 m, is shown in the spring force versus deformation diagram. The corresponding forces, after converting the spring modulus to 3901 N/m are also shown. The shaded area represents the negative work done by the spring (stored in it as elastic potential energy).

We now apply the work-energy equation for rotation. Noting that

$$I_A = \frac{1}{3}\frac{W}{g}L^2 = \frac{1}{3}\left(\frac{60}{g}\right)(4)^2 = \frac{320}{g} \text{ m-N-sec}^2$$

and that the work done by gravity is Wh, we obtain

$$\left[\text{RW} = \frac{1}{2}I_A(\omega^2 - \omega_o^2)\right]$$

$$60\left(\frac{3.2}{2}\right) - \frac{60(0.69 + 2.48)}{2}(1.79) = \frac{1}{2}\left(\frac{320}{g}\right)(0 - \omega_o^2)$$

from which

$$\omega_o^2 = 4.55 \quad \text{and} \quad \omega_o = 2.14 \text{ rad/sec} \qquad \textbf{\textit{Ans.}}$$

To rebound past the topmost position, the potential energy stored in the spring must be larger than the negative work done by gravity. To check this and also determine the possible ω_2 at the topmost position, the work-energy equation gives

$$\left[RW = \frac{1}{2} I_A(\omega_2^2 - \omega_1^2) \right]$$

$$\frac{1}{2}(60 \times 2.48)(2.48) - 60(2) = \frac{1}{2}\left(\frac{320}{g}\right)(\omega_2^2 - 0)$$

$$184.6 - 120 = \frac{320}{2 \times 9.81}\omega_2^2$$

Hence, there is a sufficient excess of potential energy over the work of gravity to give the bar an angular velocity at its topmost position of

$$\omega_2 = 1.99 \text{ rad/sec}$$ ***Ans.***

If you are surprised at the small difference between ω_2 and ω_o, compute the resultant work done between the given position and the topmost position. You should obtain a value of – 9.4 N-m.

14-7.3 A 150 N uniform rod 6 m long which carries a 30 N weight at its end is bolted to a 200 N solid cylinder as shown in Fig. 14-7.6. A cable wrapped around the cylinder passes horizontally over a small pulley to support a 400 N body. The system starts from rest when the rod is in its lowest vertical position. At the instant the rod has rotated through 60° to the given position, determine the values of ω, α, and the tension T in the cable. Then find the instantaneous bearing reaction.

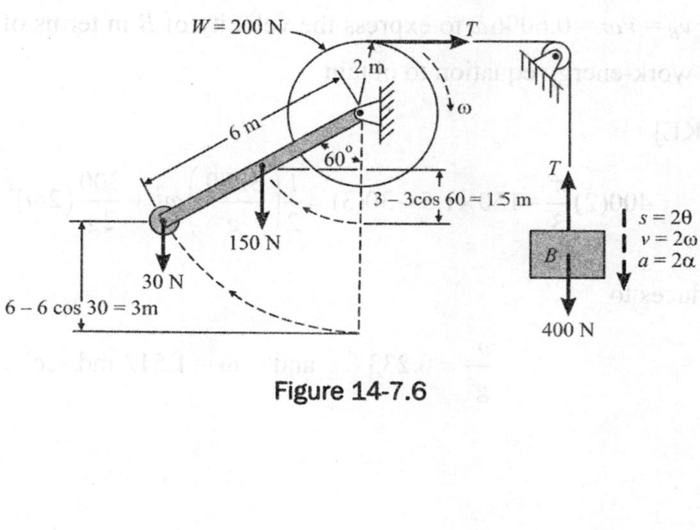

Figure 14-7.6

Solution

The angular velocity is found by equating the resultant work of external forces to the total change in kinetic energy of the system.

The cylinder rotates through 60(π/180) = π/3 rad; hence, the vertical distance that body *B* falls is given by $s = r\theta = 0.61\ (\pi/3)$ m. The vertical components of displacement of the gravity centers of the rod and attached weight are respectively 0.457 m and 0.914 m, computed as indicated in the figure.

The mass moment of inertia of the rotating elements about the axis *A* is the sum of the following values:

$$I_{cyl} = \frac{1}{2}mr^2 = \frac{200}{2g}(2)^2 = \frac{400}{g} \quad m\text{-}N\text{-}sec^2$$

$$I_{rod} = \frac{1}{3}mL^2 = \frac{150}{3g}(6)^2 = \frac{1800}{g} \quad m\text{-}N\text{-}sec^2$$

$$I_{\omega} = mk^2 = \frac{30}{g}(6)^2 = \frac{1080}{g} \quad m\text{-}N\text{-}sec^2$$

whence

$$I_A = I_{cyl} + I_{rod} + I_{\omega} = \frac{3280}{g} m\text{-}N\text{-}sec^2$$

Using $v_B = r\omega = 0.6096\omega$ to express the velocity of *B* in terms of ω, we now apply the work-energy equation to obtain

[RW = ΔKE]

$$400(2)\frac{\pi}{3} - 150 \times 1.5 - 30(3) = \frac{1}{2}\left(\frac{3280}{g}\right)\omega^2 + \frac{300}{2g}(2\omega)^2$$

which reduces to

$$\frac{\omega^2}{g} = 0.233 \qquad \text{and} \qquad \omega = 1.512 \text{ rad/sec} \qquad \textbf{\textit{Ans.}}$$

Note that by solving first for ω^2/g from which we found ω, we have the common term in the following expression for centrifugal inertia force on the rotating elements:

$$\text{CIF} = \Sigma\frac{W}{g}\bar{r}\omega^2 = \frac{\omega^2}{g}\Sigma W\,\bar{r} = 0.233(150\times3+30\times6) = 147\,\text{N}$$

which acts radially outward from A.

If only the instantaneous angular acceleration were required, we could have used the method discussed in Illus. Prob. 14-7.1. However, since the instantaneous values of both α and T are asked for, it is simpler and more direct to apply the kinetic equations of translation and rotation respectively to the body B and the rotating elements. This gives

$$\left[+\downarrow \Sigma X = \frac{W}{g}a\right] \qquad 400 - T = \frac{400}{g}(2\alpha)$$

$[\curvearrowright + \Sigma M_A = I_A\alpha]$

$$2T - 150(3\sin 60^\circ) - 30(6\sin 60^\circ) = \frac{3280}{g}\alpha$$

which are solved to yield

$$\frac{\alpha}{g} = 0.052; \quad \alpha = 0.51\ \text{rad/sec}^2 \curvearrowleft; \quad T = 358\ \text{N} \qquad \textbf{\textit{Ans.}}$$

Note that α/g is the common term in the following expression for the total tangential inertia force on the rotating elements:

$$\text{TIF} = \Sigma\frac{W}{g}\bar{r}\alpha = \frac{\alpha}{g}\Sigma W\,\bar{r} = 0.052(150\times3+30\times6) = 32.8\ \text{N}$$

Its direction is opposite to the tangential accelerations of the gravity centers of the rod and attached weight.

Finally, the components of the bearing reaction at A can be found by using these inertia effects to place the FBD of the rotating elements in dynamic equilibrium. However, the point diagram of dynamic equilibrium (shown in Fig. 14-7.7) is simpler to draw and just as effective since only force summations are needed to find the components of the bearing reaction. Note that the value of the resultant couple $\Sigma\bar{I}\alpha$ need not be computed because it has no effect on the following vertical and horizontal force summations of dynamic equilibrium:

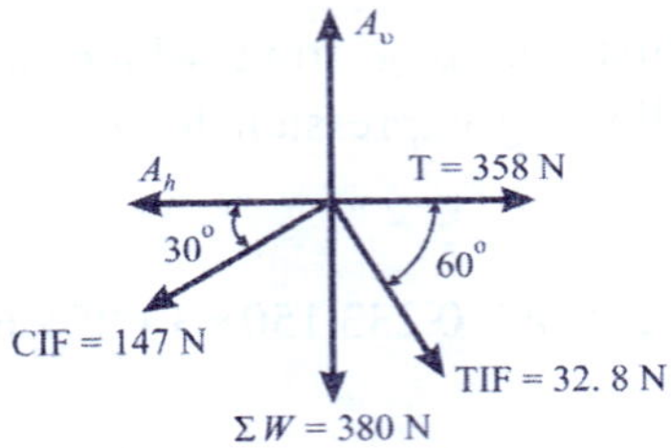

Figure 14-7.7

$$[+\uparrow \Sigma V = 0] \qquad A_v - 380 - 32.8 \sin 60^\circ - 147 \sin 30^\circ = 0$$

$$A_v = 482 \text{ N} \uparrow$$

$$[\overset{+}{\leftarrow} \Sigma H = 0] \qquad A_h + 147 \cos 30^\circ - 32.8 \cos 60^\circ - 358 = 0$$

$$A_h = 247 \text{ N} \leftarrow$$

When we combine these components, the instantaneous bearing reaction is

$$\left[A = \sqrt{A_h^2 + A_v^2}\right] \qquad \text{A} = 541 \text{ N} \measuredangle 62.9^\circ$$ ***Ans.***

PROBLEMS

14-7.4 Body A in Fig. P-14-7.4 drops 3.65 m from rest before striking the ground. Pulley B is mounted on an axle 0.15 m in diameter, and the axle friction is constant at 178 N. How many turns will B make after A stops? Also compute the tension in the cord before A hits the ground.

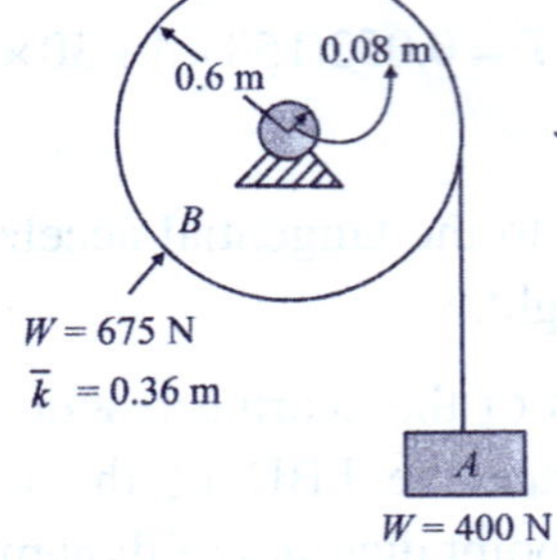

Figure P-14-7.4

14-7.5 Determine the distance moved by body A in Fig. P-14-7.5 in changing its velocity from 4 to 8 m/s.

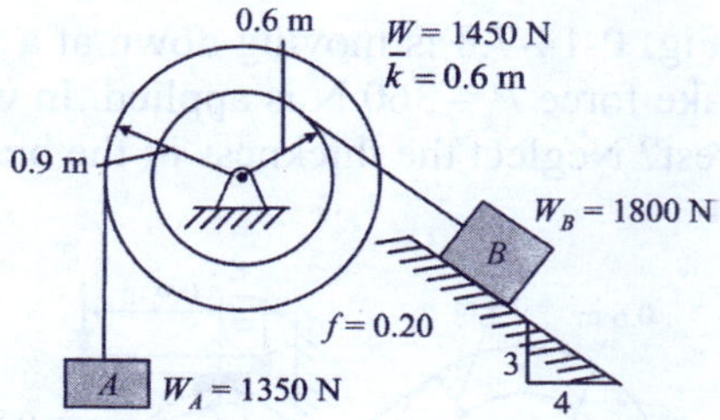

Figure P-14-7.5

14-7.6 For the system shown in Fig. P-14-7.6, what velocity will body D have after it has moved 3.0 m starting from rest? What is the tension in the cord joining pulleys B and C?

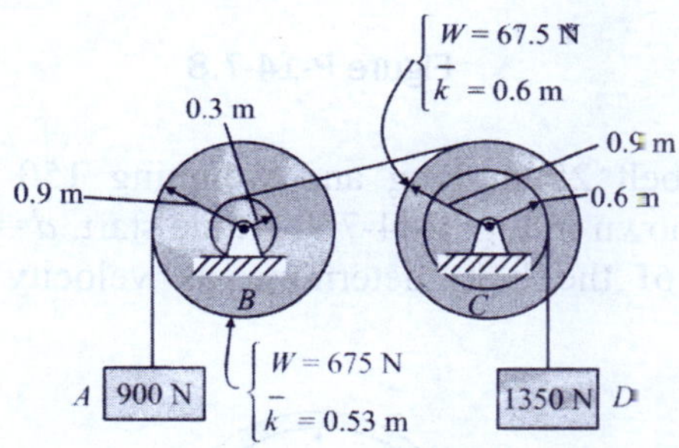

Figure P-14-7.6

14-7.7 The compound pulleys in Fig. P-14-7.7 are connected by a crossed belt. Determine the velocity attained by body A after it has moved 4 m from rest, and the tension in the cord supporting it.

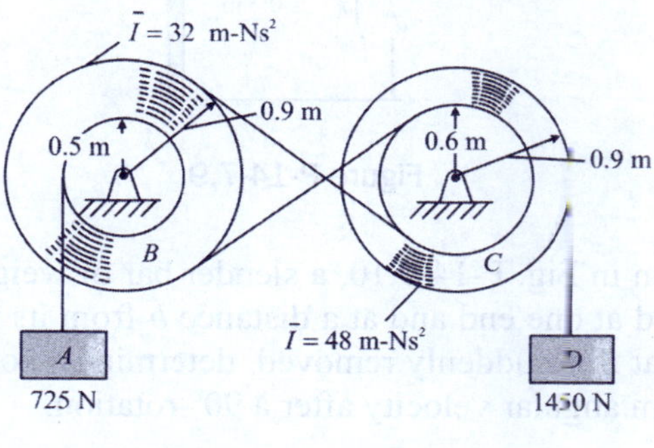

Figure P-14-7.7

14-7.8 Body A in Fig. P-14-7.8 is moving down at a velocity of 9 m/s at the instant a brake force P = 360 N is applied. In what distance does body A come to rest? Neglect the thickness of the brake block.

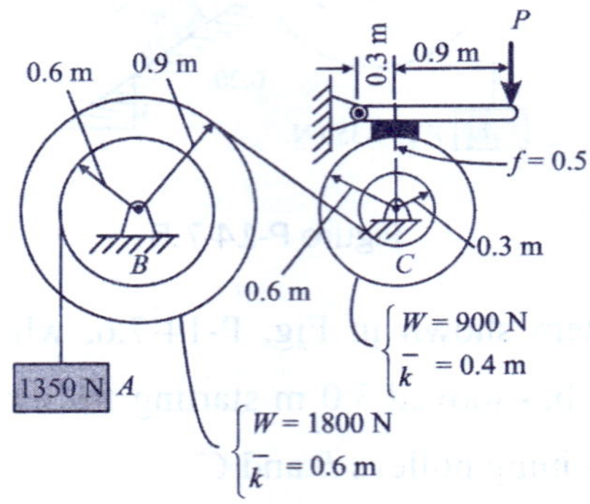

Figure P-14-7.8

14-7.9 A heavy belt 25 m long and weighing 150 N/m passes over the cylinder shown in Fig. P-14-7.9. At the start, d = 3.0 m. Neglecting the thickness of the belt, determine its velocity at the instant when d = 15 m.

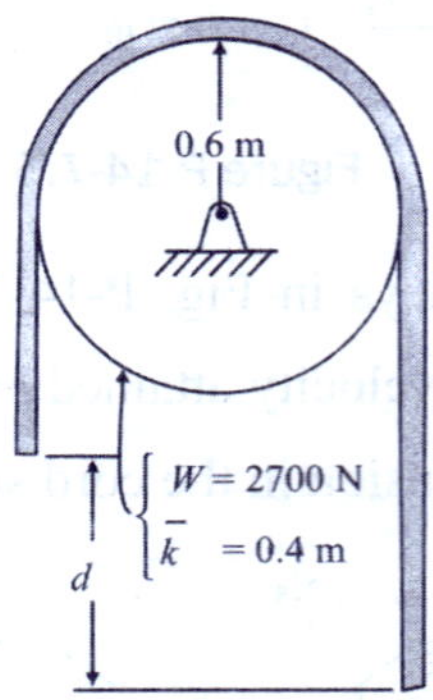

Figure P-14-7.9

14-7.10 As shown in Fig. P-14-7.10, a slender bar of weight W and length L is supported at one end and at a distance b from its mass center G. If the support at B is suddenly removed, determine b so that the bar attains a maximum angular velocity after a 90° rotation.

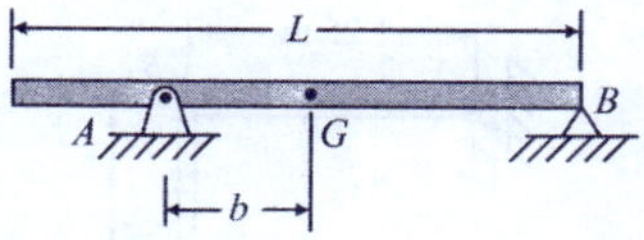

Figure P-14-7.10

14-7.11 Two bars are welded together as shown in Fig. P-14.7.ll. Segment AB weighs 300 N and segment CD weighs 150 N. The assembly starts from rest at the given position and rotates in a vertical plane about a horizontal axis at A. Compute the value of the bearing reaction at A after a 90° counterclockwise rotation.

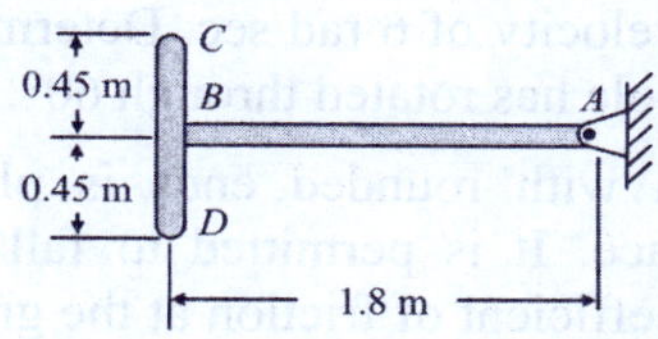

Figure P-14-7.11

14-7.12 A system, consisting of a 200 N disk attached to a 150 N rod AB as shown in Fig. P-14-7.12, is released from rest when AB is horizontal. Compute the angular velocity of AB after rotating 90° clockwise if (a) the disk is welded to the rod, (b) the disk is fastened to the rod by a smooth pin at B.

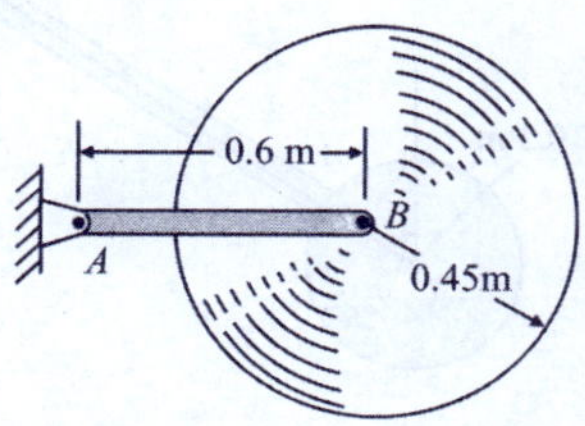

Figure P-14-7.12

14-7.13 The bent bar shown in Fig. P-14-7.13 weighs 150 N/m and is free to rotate in a vertical plane about a horizontal axis at A. If the bar is released from rest at the given position, determine its angular velocity at the instant it has rotated through 90°.

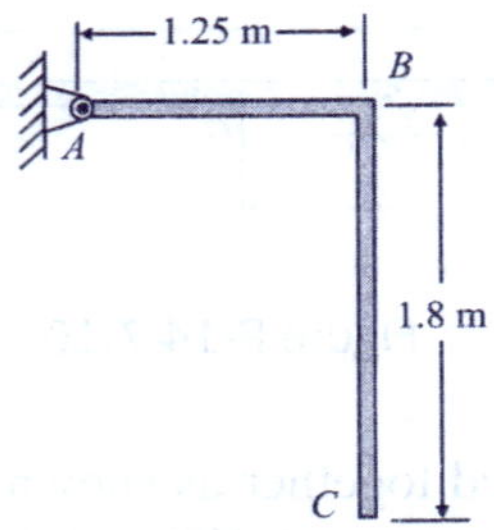

Figure P-14-7.13

14-7.14 A pole 2 m long and weighing 450 N hangs vertically from a horizontal axis through its upper end. A sudden blow gives the pole an initial angular velocity of 6 rad/sec. Determine the bearing reaction at the instant the pole has rotated through 60°.

14-7.15 A uniform rod with rounded ends is placed upright on a rough horizontal surface. It is permitted to fall from rest to the ground. Compute the coefficient of friction at the ground if the lower end slips when the rod has a slope of 4 vertical to 3 horizontal.

14-7.16 A rod *AB*, 2 m long and weighing 30 N is welded to the surface of a 900 N cylinder as shown in Fig. P-14-7.16. If the system starts from rest when *AB* is vertically above the axle *C*, find the axial force (directed along *AB*) acting on the weld at *A* when $\tan\theta = \dfrac{4}{3}$.

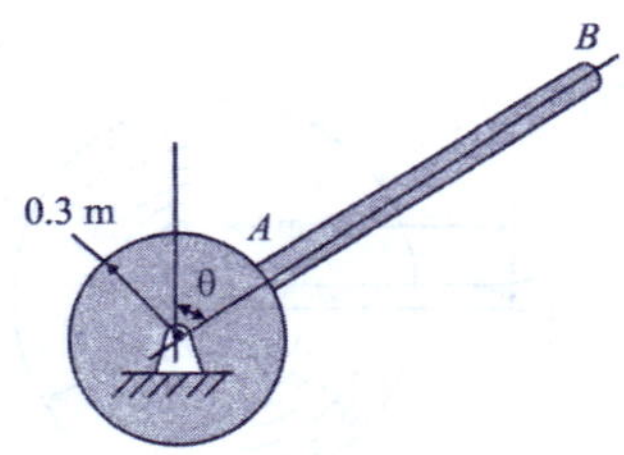

Figure P-14-7.16

14-7.17 A rod *AB* weighs 90 N and carries a 45 N weight at *B* as shown in Fig. P-14-7.17, It rotates in a vertical plane about a horizontal axis at *A*. If the rod is released from rest in the given position, find the velocity of end *B* when the rod is horizontal.

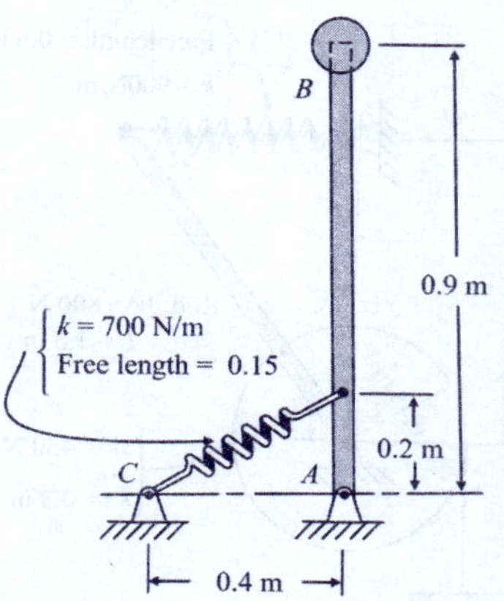

Figure P-14-7.17

14-7.18 A 30 N weight rotates in a vertical plane at one end of a 1.8 m rod which weighs 400 N. As shown in Fig. P-14-7.18, a spring, having a constant of 1460 N/m, only acts on the rod after the spring exceeds its free length of 1 m. If the rod starts from rest when it is vertically above the pivot *A*, find the angular velocity and angular acceleration of the rod when it has a downward slope of 3 vertical to 4 horizontal as shown.

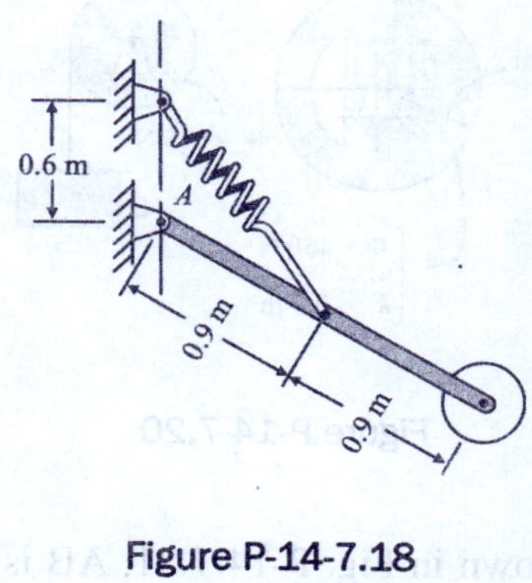

Figure P-14-7.18

14-7.19 At the instant shown in Fig. P-14-7.19, the spring is horizontal. Determine the clockwise angular velocity of the rod at the given position so that it will just reach a horizontal position.

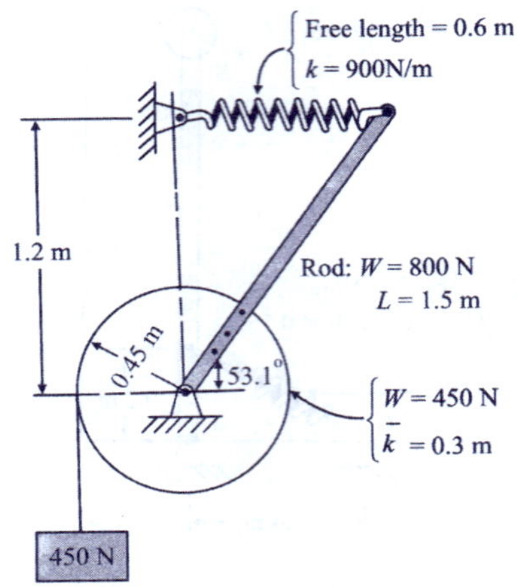

Figure P-14-7.19

14-7.20 In the system shown in Fig. P-14-7.20, A is an unbalanced wheel whose gravity center is at G, and B is a solid cylinder. If the system starts from rest at the given position, find the angular velocity of *A* after it has rotated 270° clockwise.

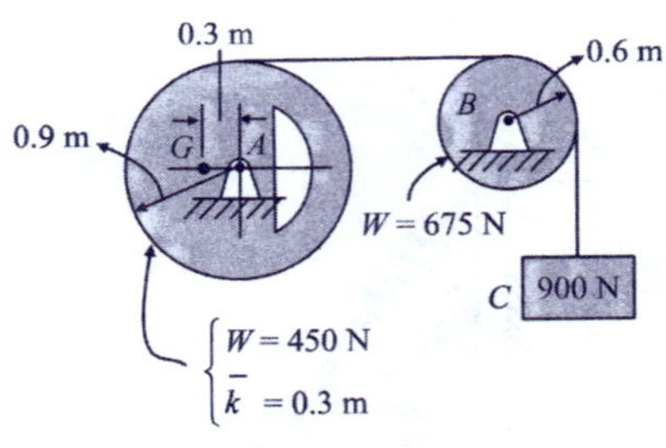

Figure P-14-7.20

14-7.21 In the system shown in Fig. P-14-7.21, AB is a 225 N bar 1.8 m long which is firmly attached to drum B. The system starts from rest when AB is in its lowest position. Determine the angular velocity of *B* at the instant $\theta = 30^\circ$.

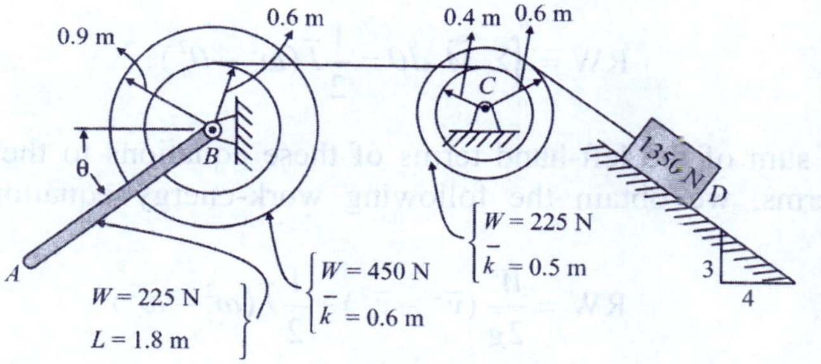

Figure P-14-7.21

14-8 WORK-ENERGY APPLIED TO PLANE MOTION

In the force analysis of plane motion problems, the dynamic effects are expressed in terms of a resultant force $\frac{W}{g}\bar{a}$ acting through the mass center and a resultant couple $\bar{I}\alpha$. In many problems the instantaneous angular velocity ω must be known as a preliminary step to finding the acceleration $\bar{a}$ of the mass center. Thus, recall from Section 13-5 that $\bar{a} = a_G = a_A \mathrel{+\!\!\!\!\rightarrow} (a_{G/A} = \bar{r}\omega^2 \mathrel{+\!\!\!\!\rightarrow} \bar{r}\alpha)$ where a_A is the acceleration of any reference point and $a_{G/A}$ is the acceleration of the mass center rotating about *A*. When α varies, as in noncentroidal rotation and general plane motion, it can be very laborious to use the differential equations of kinematics to determine an instantaneous value of ω. This is where work-energy methods become very useful since ω can be found directly without involving α.

In Section 13-2 we showed that a plane motion is equivalent to a combination of pure rotation about its centroidal axis and a translation of its mass center. We shall now show that the work-energy equation for plane motion is the sum of the work-energy expressions previously developed for the motions of translation and of centroidal rotation.

For the translational component of plane motion, taking the *X* axis tangent to the path of the gravity center during each displacement $d\bar{s}$ along the path, we have

$$\text{RW} = \int_0^s \Sigma X \, d\bar{s} = \frac{W}{2g}(\bar{v}^2 - \bar{v}_o^2) \qquad \text{(a)}$$

and for the rotational component,

$$\text{RW} = \int_0^\theta \Sigma \bar{M}\, d\theta = \frac{1}{2}\bar{I}(\omega^2 - \theta_o^2) \qquad \text{(b)}$$

Equating the sum of the left-hand terms of these equations to the sum of the right-hand terms, we obtain the following work-energy equation for plane motion:

$$\text{RW} = \frac{W}{2g}(\bar{v}^2 - \bar{v}_o^2) + \frac{1}{2}\bar{I}(\omega^2 - \omega_o^2) \qquad \text{(14-8.1)}$$

This arithmetical addition is possible because both work and kinetic energy are scalar quantities expressed in the same units. The resultant work **(RW)** of constant or variable forces and moments is computed as explained in previous sections. The kinetic energy is expressed in terms of the velocity $\bar{v}$ of the gravity center and $\bar{I}$ which is the centroidal mass moment of inertia.

If the instant center of plane motion can be located easily, it is usually more convenient to express the kinetic energy in terms of an instantaneous rotation about the instant center. Thus expressing the kinetic energy at any instant as

$$\text{KE} = \frac{1}{2}\frac{W}{g}\bar{v}^2 + \frac{1}{2}\bar{I}\omega^2 \qquad \text{(c)}$$

we may replace $\bar{v}$ by $\bar{r}\omega$ where $\bar{r}$ is the distance from the instant center to the center of gravity. Doing this gives

$$\text{KE} = \frac{1}{2}\frac{W}{g}\bar{r}^2\omega^2 + \frac{1}{2}\bar{I}\omega^2 = \frac{1}{2}\omega^2\left(\bar{I} + \frac{W}{g}\bar{r}^2\right)$$

which reduces to

$$\text{KE} = \frac{1}{2}I_C\omega^2 \qquad \text{(d)}$$

since, by the transfer formula $\bar{I} + \left(\frac{W}{g}\right)\bar{r}^2 = I_C$ which is the mass moment of inertia about the instant center C.

Except for homogeneous free-rolling wheels, the position of the instant center relative to the gravity center of the body usually changes, so that the change in kinetic energy must be expressed as

$$\Delta\text{KE} = \frac{1}{2}I_C\omega^2 - \frac{1}{2}I_{C_o}\omega_o^2 \qquad \text{(14-8.2)}$$

in which I_C and I_{C_o} represent, respectively the moments of inertia about the final and initial positions of the instant center C.

This last expression suggests a special but convenient form of the work-energy equation for *homogeneous* free-rolling wheels in which I_C is constant. Considering the plane motion as analogous to rotation about the instant center, we may write

$$\Sigma M_C(\theta) = \frac{1}{2} I_C(\omega^2 - \omega_o^2) \qquad (14\text{-}8.3)$$

in which ΣM_C is the moment about C of the constant external forces acting during an angular displacement θ.[7] This moment summation about C automatically eliminates the static frictional force holding the instant center momentarily at rest and indicates that such friction forces (as the term "static" indicates) do no work on free-rolling wheels. Notice that the static friction force F at the instant center C may be replaced by an equal force at the mass center G plus a couple. See Fig. 14-8.1. If we do this, the sum of the translational work done by F at the mass center will cancel the rotational work of its couple, i.e., $F\bar{s} + (-F\bar{r})\theta = 0$, thereby verifying that the static friction does no work on free-rolling wheels. *If the wheel slips, Eq.* (14-8.3) *will not be valid*; we shall then have to use Eq. (14-8.1) and also consider the work done by kinetic friction forces.

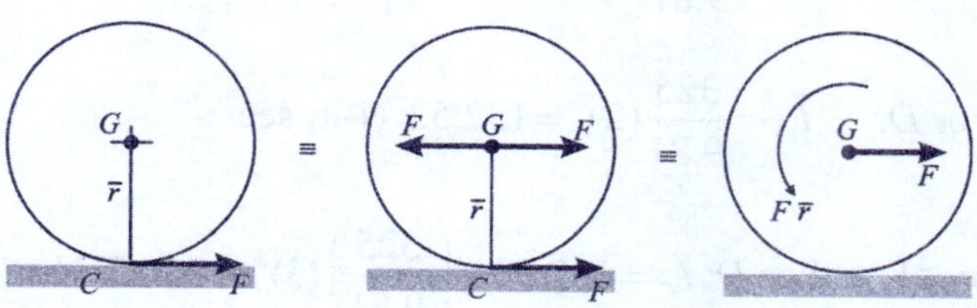

Figure 14-8.1

ILLUSTRATIVE PROBLEMS

14-8.1 The system in Fig. 14-8.2 has the properties shown. Assuming disk D to roll without slipping, determine the linear velocity of its center after body B has moved 12 m starting from rest. Also determine the static friction force acting on disk D.

---

[7]Equation (14-8.3) can also be obtained by eliminating α from between $\Sigma M_c = I_c\alpha$ and $\alpha\, d\theta = \omega\, d\omega$ and then integrating according to the procedure outlined in Section 14-7. Naturally, the ensuing result is valid only for homogeneous free-rolling wheels to which $\Sigma M_c = I_c\alpha$ may be applied.

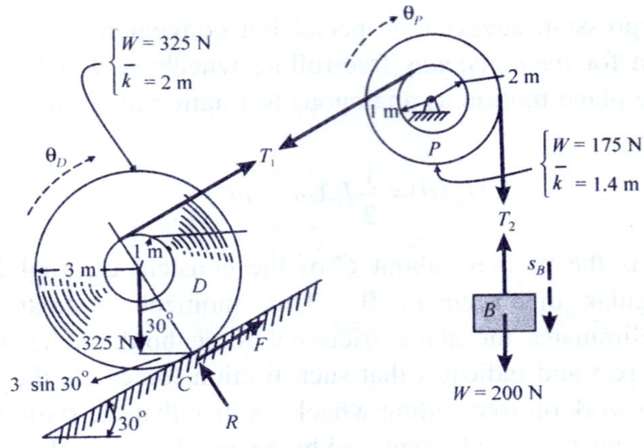

Figure 14-8.2

Solution

We first determine the mass moments of inertia.

$$\left[\bar{I} = m\bar{k}^2\right] \quad \text{For } P: \quad \bar{I}_p = \frac{175}{9.81}(1.4)^2 = 34.96 \text{ N-m-sec}^2$$

$$\text{For } D: \quad \bar{I}_D = \frac{325}{9.81}(2)^2 = 132.52 \text{ N-m-sec}^2$$

$$[I_C = \bar{I} + md^2] \qquad \text{For } D: I_C = 132.52 + \left(\frac{325}{9.81}\right)(3)^2 = 430.7 \text{ N-m-sec}^2$$

The directions of motion may be found by the method described in Illus. Prob. 13-4.2. These directions are shown by the dashed arrows, adjacent to each body, which specify the positive directions of forces and moments. Alternatively, the directions of motion may be assumed and then confirmed by observing that, for a system starting from rest, the resultant work based on these assumptions must be positive to speed up the system.

The kinematic relations among the bodies are

$$[s = r\theta] \qquad s_{T_1} = 4\theta_D = 1(\theta_P) \qquad \text{also} \qquad s_{T_2} = S_B = 2\theta_p$$

which are combined into the following relation:

$$4\theta_D = \theta_P = \frac{1}{2}s_B \tag{a}$$

whence, by differentiation, the velocities are similarly related by

$$4\omega_D = \omega_P = \frac{1}{2} v_B \qquad \text{(b)}$$

We now apply the work-energy equation to the system, considering only the work of active external forces. There are several alternate methods of computing the work done by the weight W of disk D. Since the center of D moves the distance $\bar{s}_D = -3\theta_D$, the work done by W is either $-W \sin 30^\circ(\bar{s}_D)$ or $-Wh = -W(\bar{s}_D \sin 30^\circ)$, both reducing to $-W(3 \sin 30^\circ)\vartheta_D$. But probably the best method is to apply $\Sigma Mc(\theta_D)$ to obtain $-W(3 \sin 30^\circ)\theta_D$, giving the same result directly. The value of θ_D, using Eq. (a) and the given value of $s_B = 12$ m, is $\theta_D = \frac{12}{8} = 1.5$ rad. The resultant work done on the system is the sum of the left-hand terms in Eqs. (14-2.2), (14-7.2), and (14-8.3) used respectively for the translational, rotational, and plane motion parts of the system. The total change in kinetic energy is the sum of the right-hand terms of these equations. Note that although no forces external to the system do work upon the pulley, it nevertheless gains kinetic energy. Thus we obtain

$$\left[\Sigma X(s)\right]_B + \left[\Sigma \bar{M}(\theta)\right]_P + \left[\Sigma M_C(\theta)\right]_D$$

$$= \left[\frac{W}{2g}(v^2 - v_o^2)\right]_B + \left[\frac{1}{2}\bar{I}(\omega^2 - \omega_o^2)\right]_P + \left[\frac{1}{2}I_C(\omega^2 - \omega_o^2)\right]_D$$

$$200\,(12) - 325\,(3 \sin 30)\,(1.5)$$

$$= \frac{200}{2 \times 9.81} v_B^2 + \frac{1}{2}(34.96)\,\omega_p^2 + \frac{1}{2}(430.7)\,\omega_D^2 \qquad \text{(c)}$$

Substituting $v_B = 8\,\omega_D$ and $\omega_P = 4\,\omega_D$ from Eq. (*b*) in Eq. (*c*) gives

$$2400 - 731.25 = (652.4\ \omega_D^2) + (279.68\ \omega_D^2) + 215.35\,\omega_D^2$$

whence

$$\omega_D^2 = 1.45 \text{ and } \omega_D = 1.2 \text{ rad/sec}$$

The velocity of the center of disk D therefore is

$[\bar{v} = \bar{r}\omega]$ $\qquad \bar{v}_D = 3\,(1.2) = 3.6$ m/sec $\qquad$ ***Ans.***

The friction force acting on D (and also the tension T_1) is found by applying the following work-energy relations to the FBD of the disk:

$$[\curvearrowright +\Sigma M_c(\theta)] = \frac{1}{2} I_C\left(\omega^2 - \omega_o^2\right)$$

$$[4T_1 - 325\ (3 \sin 30)]\ (1.5) = \frac{1}{2}\ (430.7)\ (1.45) \qquad \text{(d)}$$

$$[\curvearrowright +\Sigma \bar{M}(\theta) = \frac{1}{2}\bar{I}(\omega^2 - \omega_o^2)$$

$$[1(T_1) - 3\ F]\ (1.5) = \frac{1}{2}\ (132.52)\ (1.45) \qquad \text{(e)}$$

which are solved to yield

$T_1 = 173.92$ N and $F = 36.62$ N ***Ans.***

In place of Eq. (*e*) representing the rotational component of the work-energy equation for the disk, we could have used the translational component in terms of the motion of its mass center, i.e.,

$$\Sigma X(\bar{s}) = \frac{W}{2g}(\bar{v}^2 - \bar{v}_o^2)$$. Try this to see which is easier.

14-8.2 The circular disk shown in Fig. 14-8.3 weighs 200 N after the semicircular hole is cut out of it. Its radius of gyration about the gravity center G is 0.36 m. Determine the angular velocity of the disk after it has rolled without slipping for 450° or $1\frac{1}{4}$ turns clockwise, starting with an initial clockwise angular velocity of 4 rad/sec. What is then the normal reaction N?

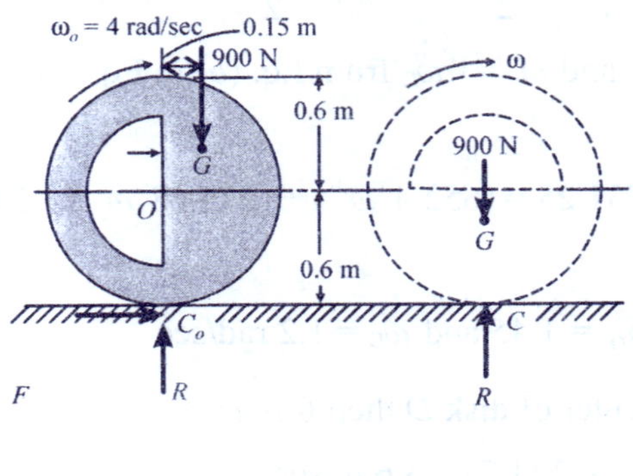

Figure 14-8.3

Solution

During the movement from the initial to the dashed final position, the work done is caused only by the center of gravity dropping a net vertical distance of 0.15m. Since free-rolling is assumed, the static friction F does no work. Before equating the resultant work to the change in kinetic energy given by Eq. (14-8.2), let us compute the values of I_{C_o} and I_c. Using the transfer formula, we obtain

$[I = \bar{I} + md^2]$

$$Ic_o = \frac{900}{g}(0.36)^2 + \frac{900}{g}(0.15^2 + 0.6^2) = \frac{460.89}{g} \text{N-m-sec}^2$$

$$I_c = \frac{900}{g}(0.36)^2 + \frac{900}{g}(0.45)^2 = \frac{298.89}{g} \text{N-m-sec}^2$$

The work-energy equation now gives

$$\left[\text{RW} = \frac{1}{2}I_c\omega^2 - \frac{1}{2}I_{C_o}\omega_o^2\right]$$

$$900(0.15) = \frac{1}{2}\left(\frac{298.89}{g}\right)\omega^2 - \frac{1}{2}\left(\frac{460.89}{g}\right)(1.2)^2$$

from which we get

$\omega^2 = 11.08$ and $\omega = 3.33$ rad/sec ***Ans.***

The value of N at the final position will be found by using $R = (W/g)\ \bar{a}$. At this instant $\Sigma \bar{M} = \bar{I}\alpha$ gives $\alpha = 0$ and hence the acceleration of the geometric center O is $a_o = r\alpha = 0$. Then, if we use O as a reference point, the acceleration of the gravity center is $\bar{a} = a_G = a_o \nrightarrow (a_{G/o} = r\omega^2 \nrightarrow r\alpha)$ which reduces to $\bar{a} = r\omega^2 = 0.1525(32.9)\text{m/s}^2\uparrow$. Hence, we obtain

$$\left[+\uparrow R = \frac{W}{g}\bar{a}\right] \qquad R - 900 = \frac{900}{g}(0.15)\ (11.08) \qquad R = 1052.48 \text{ N} \qquad \textbf{\textit{Ans.}}$$

14-8.3 In Fig. 14-8.4, the slider A, weighing 900 N, moves in the vertical frictionless guide. It is connected to the center of a solid cylinder B, 1.2 m in diameter and weighing 1350 N, by a uniform slender rod AB which is 3 m long and weighs 800 N. If the system starts with negligible velocity when B is vertically below A, and B rolls freely, compute the velocity of A in the given position.

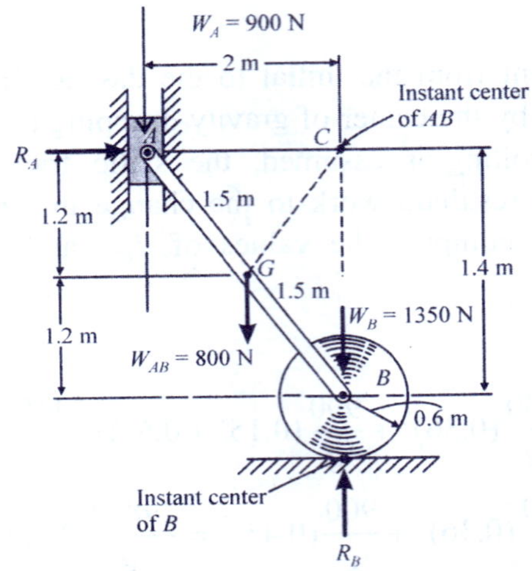

Figure 14-8.4

Solution

The reactive forces R_A and R_B, which do no work, may be omitted from the active-force diagram. The resultant work done on the system is due only to W_A and W_{AB}, dropping respectively through 3 – 1.4 = 1.6 m and 1.5 – 1.2 = 0.3 m.

If we use the instant center C of AB, the kinematic relations between the bodies at the instant shown are

$$[v = r\omega] \qquad v_A = 2\,\omega_{AB} \quad \text{and} \quad \bar{v}_B = 1.4\,\omega_{AB} = 0.6\,\omega_B \tag{a}$$

Since the system starts with negligible velocity, the change in kinetic energy is the sum of $\dfrac{W}{2g}v^2$ for the translating slider A and $\Sigma\dfrac{1}{2}I_c\omega^2$ for the plane motions of bar AB and cylinder B. About their respective instant centers, their moments of inertia are

$$[I = \bar{I} + md^2]$$

$$\text{For } AB\text{:}\quad I_{C_{AB}} = \frac{1}{12}\left(\frac{800}{g}\right)(3)^2 + \frac{800}{g}(1.5)^2 = \frac{2400}{g}\ \text{N-m-sec}^2$$

$$\text{For } B\text{:}\quad I_{C_B} = \frac{1}{2}\left(\frac{1350}{g}\right)(0.6)^2 + \frac{1350}{g}(0.6)^2 = \frac{729}{g}\ \text{N-m-sec}^2$$

Applying the work-energy equation, and using the kinematic relations (a) to express v_A and ω_B in terms of ω_{AB}, we have

[RW = ΔKE]

$$900(2) + 800(1) = \frac{900}{2g}(2\omega_{AB})^2 + \frac{1}{2}\left(\frac{2400}{g}\right)\omega_{AB}^2 + \frac{1}{2}\left(\frac{729}{g}\right)(1.2\omega_{AB})^2$$

whence

$$2600 = \frac{\omega_{AB}^2}{2g}[3600 + 2400 + 1049.76]$$

$$= \frac{\omega_{AB}^2}{2g}(7049.76)$$

$$\omega^2_{AB} = 7.24 \quad \text{and} \quad \omega_{AB} = 2.7 \text{ rad/sec}$$

Hence, the required velocity of slider A is

$[v_A = 2\omega_{AB}]$ $\qquad v_A = 2(2.7) = 5.4$ m/sec **Ans.**

PROBLEMS

14-8.4 As shown in Fig. P-14-8.4, a cord from a fixed support is wrapped around a solid homogeneous cylinder of radius r and weight W. Find the velocity of the mass center of the cylinder after it has dropped h m, starting from rest.

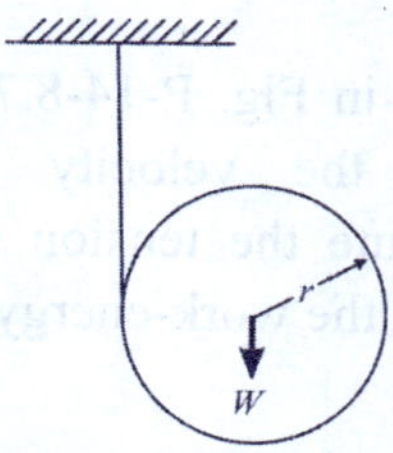

Figure P-14-8.4

14-8.5 The disk shown in Fig. P-14.8.5 rolls freely on a horizontal track. Compute the angular velocity of the disk after its center has moved 4 m from rest. Neglect the backward inclination of the block caused by its inertia.

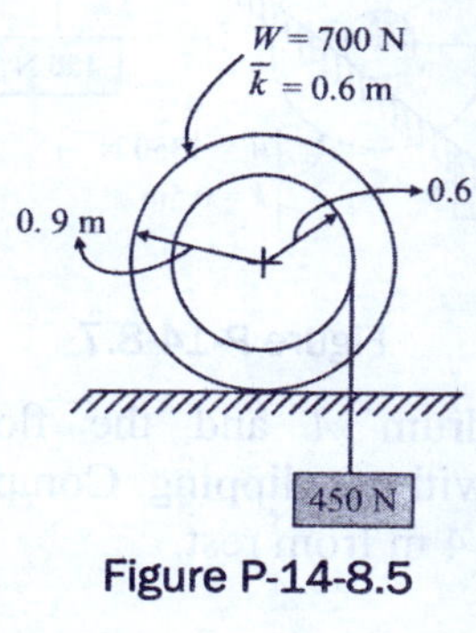

Figure P-14-8.5

14-8.6 Disk A is composed of a 0.9 m radius cylinder sandwiched between two cylinders each of 0.6 m radius. It rolls without slipping between two parallel rails as shown in Fig. P-14-8.6. Determine the angular velocity of disk A after its center has moved 4 m, starting from rest. Assume body B weighs 700 N and neglect the effect of inertia on its inclination with the vertical.

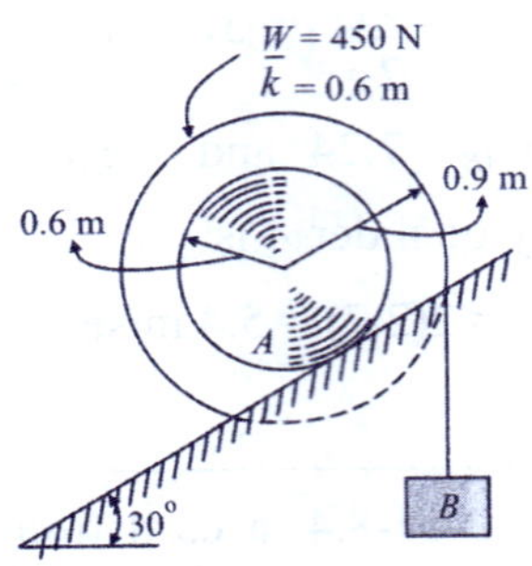

Figure P-14-8.6

14-8.7 Assume the body A in Fig. P-14-8.7 moves 4 m, starting from rest. After determining the velocity of A corresponding to this displacement, compute the tension in the cord joining disk D and pulley P by applying the work-energy method to a FBD of disk D.

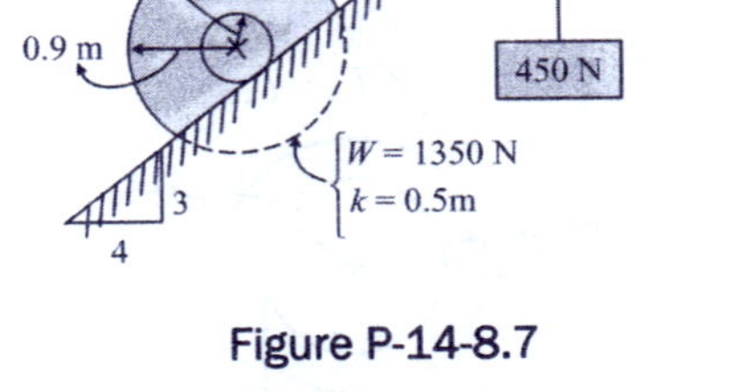

Figure P-14-8.7

14-8.8 The compound drum A and the floating pulley B shown in Fig. P-14-8.8 roll without slipping. Compute the velocity of the center of A after it moves 4 m from rest.

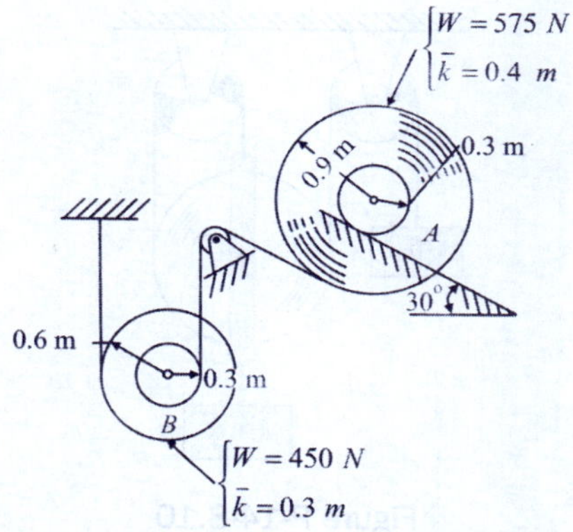

Figure P-14-8.8

14-8.9 The floating disk D in Fig. P-14-8.9 rolls without slipping on the cord passing from the fixed support and around it to pulley P. Compute the velocity of body A after A has moved 4 m from rest. What is the tension in the cord connecting pulley P with disk D?

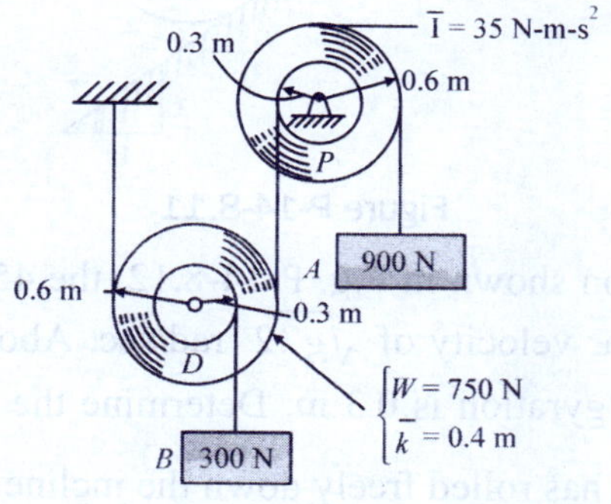

Figure P-14-8.9

14-8.10 The floating pulley in Fig. P-14-8.10 is equivalent to a 0.3 m diameter solid cylinder weighing 350 N. The other pulleys have negligible weight. Determine the velocity of body B after it has moved 3 m from rest.

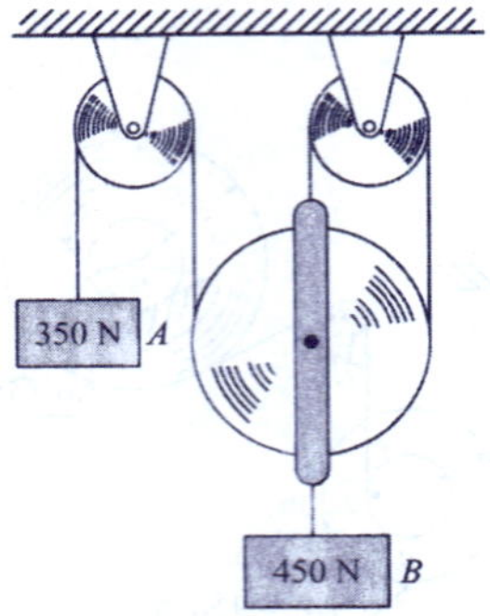

Figure P-14-8.10

14-8.11 Disk B rolls without slipping on the incline shown in Fig. P-14-8.11. Compute the velocity of its center after the rod attached to cylinder A has rotated from rest at its lowest position to the horizontal position shown.

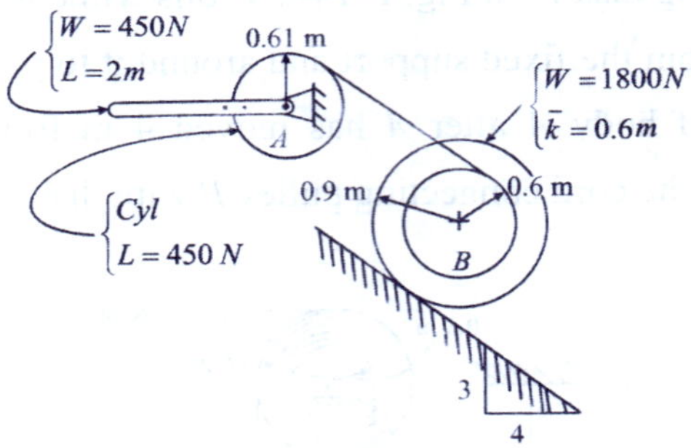

Figure P-14-8.11

14-8.12 At the position shown in Fig. P-14-8.12, the 450 N unbalanced wheel has an angular velocity of $\sqrt{g/2}$ rad/sec. About its gravity center G, the radius of gyration is 0.3 m. Determine the angular velocity of the wheel after it has rolled freely down the incline for $\frac{1}{2}$ turn.

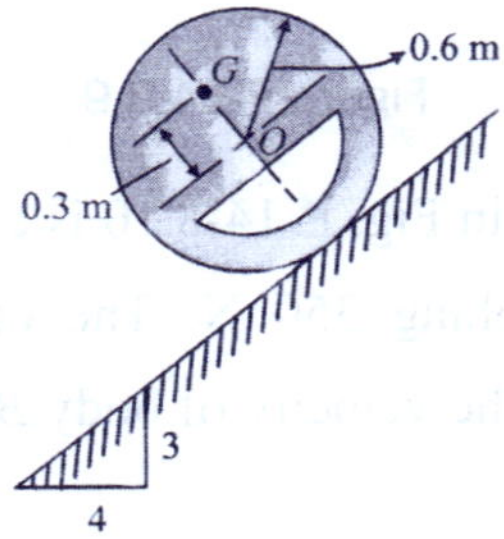

Figure P-14-8.12

14-8.13 For the system shown in Fig. P-14-8.13, find the angular velocity of the unbalanced wheel after it has rolled freely for $\frac{1}{4}$ turn, starting from rest at the given position.

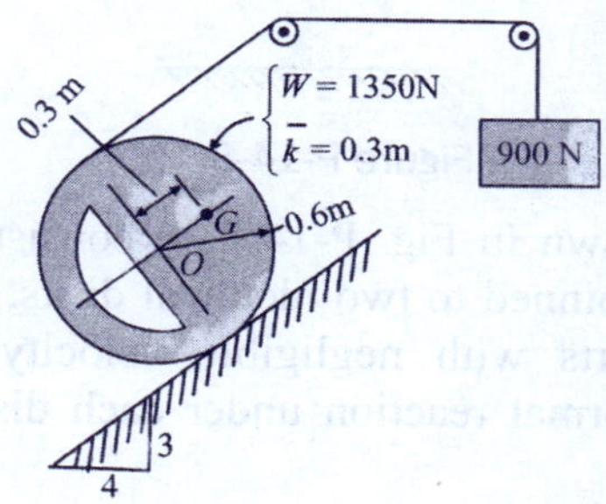

Figure P-14-8.13

14-8.14 A homogeneous slender rod L m long and weighing W N slides down a smooth vertical wall and along a smooth horizontal floor as shown in Fig. P-14-8.14. Assume that the ends of the rod remain in contact with the wall and the floor. If the rod is at rest when $\theta = 0^\circ$, determine the value of the instantaneous angular velocity ω in terms of the angular displacement θ. Then by differentiating ω, obtain the value of α in terms of θ. What is the effect upon α if the lower end of the rod is not at rest when $\theta = 0^\circ$?

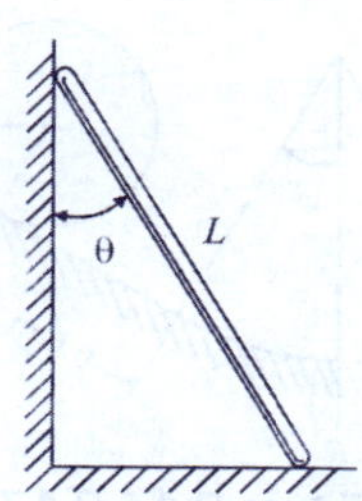

Figure P-14-8.14

14-8.15 An elliptical plate has semi axes a and b. The plate is released from rest with the major axis vertical as in Fig. P-14-8.15, and rolls freely on the horizontal surface shown. Determine the angular velocity at the instant the major axis is horizontal.

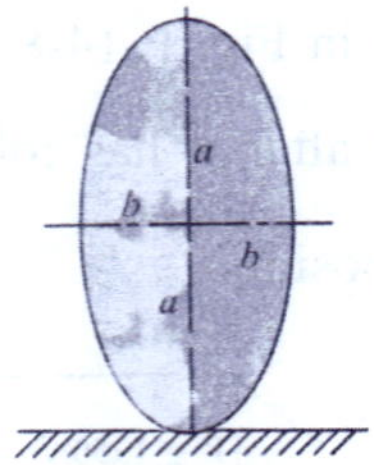

Figure P-14-8.15

14-8.16 The system shown in Fig. P-14-8.16 consists of a 300 N bar 1.5 m long smoothly pinned to two identical disks, each weighing 200 N. If the system starts with negligible velocity at the given position, compute the normal reaction under each disk when the bar is at its lowest position.

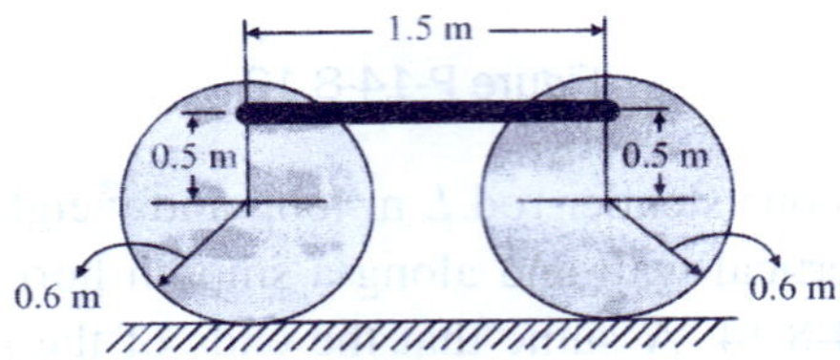

Figure P-14-8.16

14-8.17 A solid cylinder weighing 450 N starts from rest at the position shown in Fig. P-14-8.17 and rolls without slipping down the incline and on to the circular path. Determine the vertical reaction upon the cylinder at its lowest position.

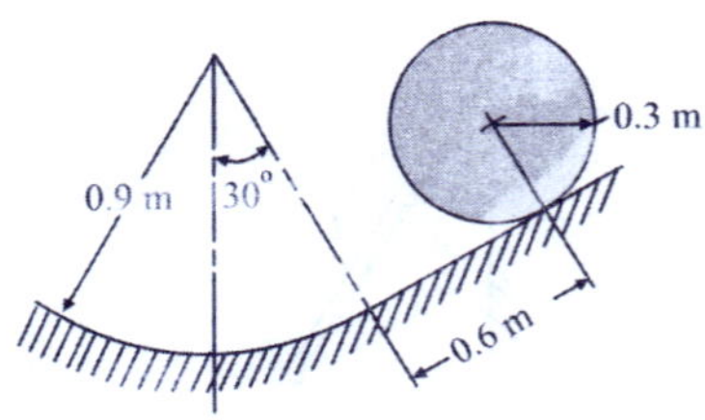

Figure P-14-8.17

14-8.18 Bar AB, 3 m long and weighing 900 N. is fastened at A to a roller of negligible weight moving in a vertical slot as shown in Fig. P-14-8.18. The other end B is smoothly pinned to a cylinder weighing 1350 N and of 1.5 m radius. The spring has a free length of 0.6 m and a modulus of 900 N/m. If the system starts from rest when B is vertically below A, compute the linear velocity of point B when in the given position.

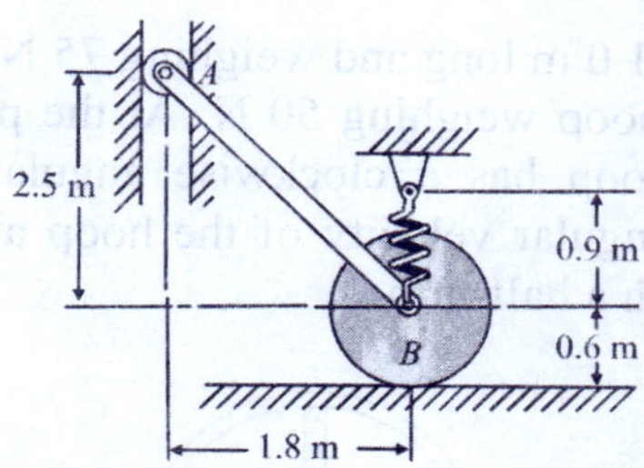

Figure P-14-8.18

14-8.19 At the position shown in Fig. P-14-8.19, the given system is released from rest on the smooth surfaces. The uniform bar *AB*, weighing 450 N, is smoothly pinned to the mass centers of bodies A and B which weigh respectively 200 N and 600 N. Determine the velocity of body *B* at the instant when bar *AB* is horizontal.

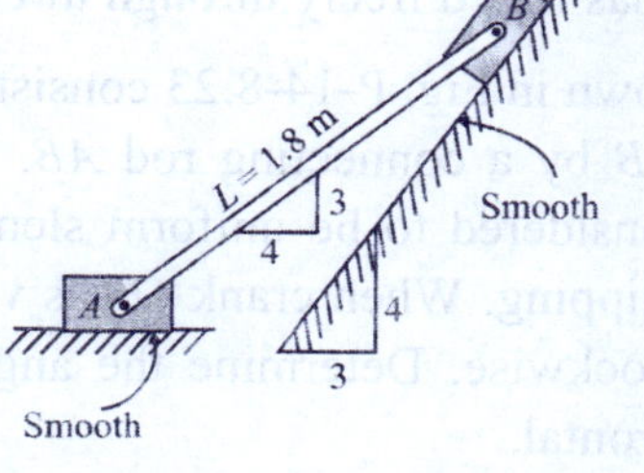

Figure P-14-8.19

14-8.20 The pitch circles of two gears, *A* and *B*, connected by the link *OC* are shown in Fig. P-14-8.20. If gear *A* is held stationary and the system is released from rest at the given position, find the angular velocity of gear *B* when it is at its lowest position. Assume gear *B* is equivalent to a solid cylinder weighing 150 N and that link *OC* weighs 100 N.

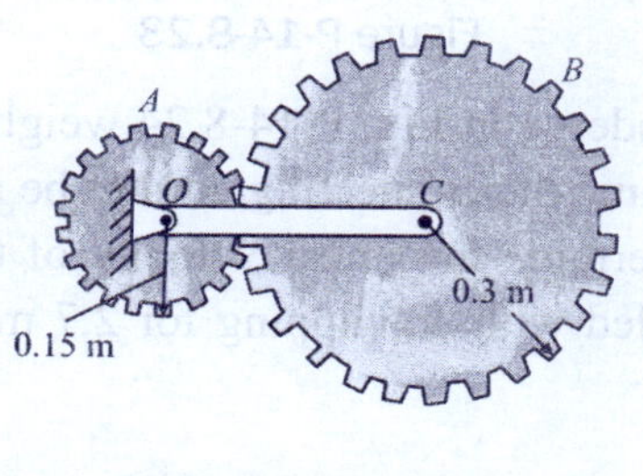

Figure P-14-8.20

14-8.21 A uniform rod 1.0 m long and weighing 75 N is welded to the rim of a 2 m diameter hoop weighing 50 N. At the position shown in Fig. P-14-8.21, the hoop has a clockwise angular velocity of 4 rad/sec. Compute the angular velocity of the hoop after it has rolled without slipping through a half-turn.

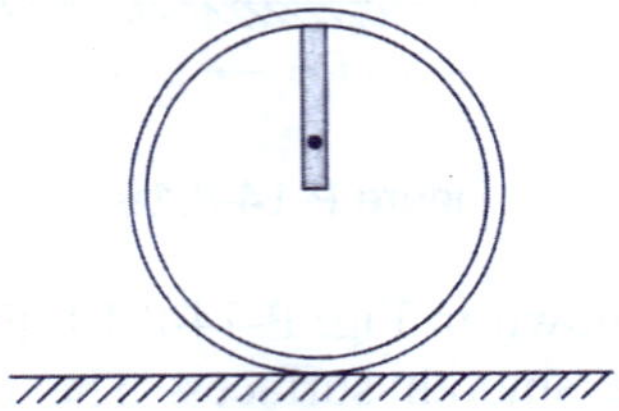

Figure P-14-8.21

14-8.22 Determine the angular velocity of the hoop in the preceding problem after the hoop has rolled freely through three-quarters of a turn.

14-8.23 The system shown in Fig. P-14-8.23 consists of a crank *OA* joined to a solid cylinder *B* by a connecting rod *AB*. The crank and connecting rod may be considered to be uniform slender rods, and the cylinder rolls without slipping. When crank *OA* is vertical, its angular velocity is 2 rad/sec clockwise. Determine the angular velocity of the crank when it is horizontal.

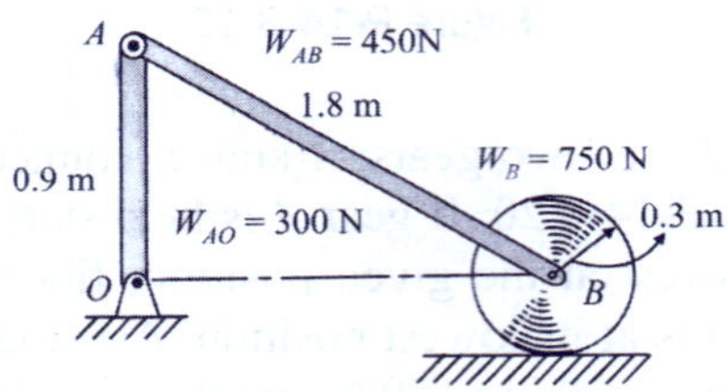

Figure P-14-8.23

14-8.24 The solid cylinder *A* in Fig. P-14-8.24 weighs 1450 N and is 0.9 m in diameter. Assume the connecting cord to be attached to the cylinder at its center. Determine the linear velocity of the center of the cylinder after it has rolled without slipping for 2.7 m, starting from rest at the given position.

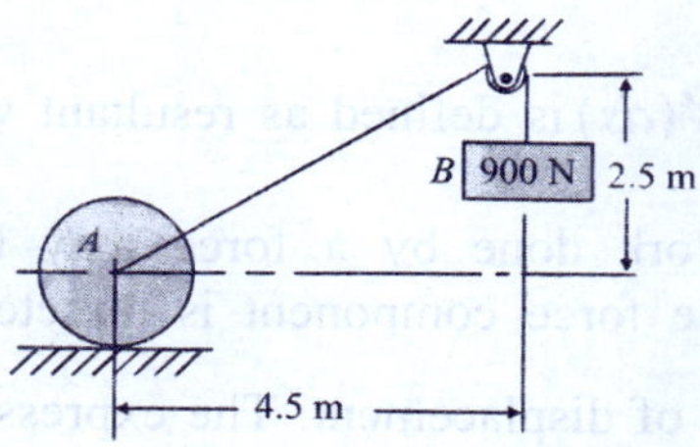

Figure P-14-8.24

14-8.25 The linkage shown in Fig. P-14-8-25 consists of four identical bars, each of weight W and length L, connected by smooth pins. The linkage is released from rest when D coincides with A. Compute the velocity of D at the given position.

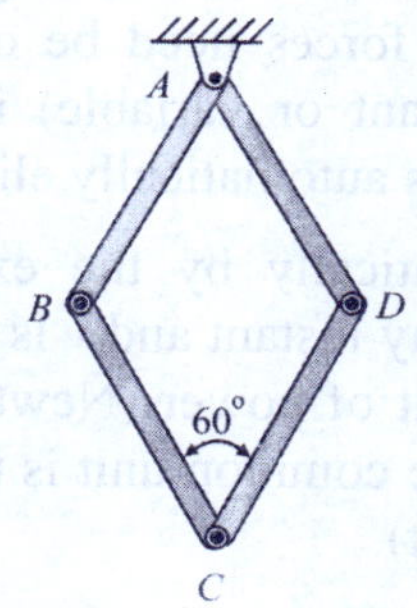

Figure P-14-8.25

SUMMARY

By eliminating the acceleration from the equations $\Sigma X = \frac{W}{g} a_t$ and $a_t\, ds = v\, dv$, we obtain the relation

$$\int_o^s \Sigma X(ds) = \frac{W}{2g}(v^2 - v_o^2) \qquad (14\text{-}2.2)$$

which is the fundamental work-energy equation for a particle or a translating rigid body. The X axis is assumed to be tangent to the path of the particle during each distance ds measured along the path. Hence ΣX represents the component of the resultant force along the path at each instant.

The expression $\int_o^s \Sigma X(ds)$ is defined as resultant work [or as $\Sigma X(s)$ if the forces are constant]. Work done by a force may be positive or negative according to whether the force component is directed respectively along or opposite to the direction of displacement. The expression $\frac{W}{2g}v^2$ is defined as kinetic energy. The usual unit for work and energy is the Newton-meter.

In Section 14-3 it is shown that the geometric significance of work is the product of the component of force in the direction of displacement multiplied by the displacement, or the product of displacement in the direction of the force multiplied by the force. Forces perpendicular to the displacement do no work.

In the application of the work-energy method, the work done during all phases of a motion is equated to the total change in kinetic energy. Only the work done by *active* external forces need be considered; the work done by internal forces (whether constant or variable) in inextensible links or cords connecting a system of bodies is automatically eliminated.

Power is defined mathematically by the expression Fv where F is the resultant force doing work at any instant and v is the corresponding value of the instantaneous velocity. The unit of power (Newton-meter-second) is too small for use in engineering; the more common unit is the horsepower (hp) defined as equivalent to 746 N-m/sec (watt).

The work-energy equation for rotation and the methods employed are analogous to those applied to translation. The fundamental equation is obtained by eliminating the angular acceleration from $\Sigma M = I\alpha$ and $\alpha\, d\theta = \omega\, d\omega$; this gives

$$\int_o^\theta \Sigma M(d\theta) = \frac{1}{2} I(\omega^2 - \omega_o^2) \tag{14-7.2}$$

The expression $\int_o^\theta M(d\theta)$ [or $\Sigma M(\theta)$ if the moments are constant] is defined as resultant work in rotation. Moments in the sense of initial angular motion do positive work, and vice versa. The expression $\frac{1}{2}I\omega^2$ is defined as kinetic energy in rotation. The unit for both expressions is usually the Newton-meter.

As in the case of translation, the work-energy method may be applied directly to a system of connected bodies, some of which translate and others rotate, without considering any of the internal forces in the connecting elements.

The resultant work done by active external forces on the system is equal to the sum of changes in kinetic energy of each body of the system, or

$$(RW)_{sys} = \Sigma \frac{W}{2g}(v^2 - v_o^2) + \Sigma \frac{1}{2} I(\omega^2 - \omega_o^2) \quad (14\text{-}7.3)$$

The work-energy equation applied to plane motion is the sum of the work-energy expressions for translation of the mass center plus that of centroidal rotation, namely,

$$RW = \frac{W}{2g}(\bar{v}^2 - \bar{v}_o^2) + \frac{1}{2}\bar{I}(\omega^2 - \omega_o^2) \quad (14\text{-}8.1)$$

If the instant center of zero velocity can easily be located, it is usually more convenient to express the change in kinetic energy as

$$\Delta KE = \frac{1}{2} I_C \omega^2 - \frac{1}{2} I_{C_o} \omega_o^2 \quad (14\text{-}8.2)$$

in which I_c and I_{C_o} are the moments of inertia about the final and initial positions of the instant center C.

For homogeneous free-rolling bodies, the most convenient form of the work-energy equation is

$$\Sigma M_C(\theta) = \frac{1}{2} I_C(\omega^2 - \omega_o^2) \quad (14\text{-}8.3)$$

in which ΣM_C is the moment about the instant center C of the constant external forces acting during an angular displacement θ.

CHAPTER 15

IMPULSE AND MOMENTUM

15-1 INTRODUCTION

We have seen that the work-energy method was based on integrating the kinetic equations of motion with respect to the displacement. The resulting displacement integrals were scalar quantities which were particularly useful in problems relating force, velocity, and displacement. In this chapter we present another approach to dynamics, known as the impulse-momentum method, which directly relates force, velocity, and time. It is based on integrating the kinetic equations of motion with respect to the time.

The impulse-momentum method is particularly convenient in situations when forces act for very small time intervals during which the forces may not vary, as in an impact or sudden blow. The method is also useful for solving problems in which a system gains or loses mass. Still other cases will involve a combination of work-energy and impulse-momentum methods, as in satellite motion.

We start with the application of the impulse-momentum method to particle motion or the equivalent rigid-body motion of translation, and then extend its application in subsequent sections to fixed-axis rotation and general plane motion.

15-2 LINEAR IMPULSE-MOMENTUM

As was shown in Section 10-5, the motion of the mass center of any body, rigid or nonrigid, is governed by $\mathbf{R} = m\bar{a}$. On replacing $\bar{a}$ by its equivalent $d\,\bar{v}/dt$, we obtain

$$\mathbf{R}\,dt = m\,d\bar{v} \tag{a}$$

which is the differential form of the linear impulse-momentum equation for all systems which neither gain nor lose mass. If we assume that the velocity has the value $\bar{v}_o$ when t is zero and the value $\bar{v}$ at any other value t, then integration of Eq. (a) between these limits gives the following general form of the linear impulse- momentum equation:

$$\int_o^t \mathbf{R}\,dt = m\bar{\mathbf{v}} - m\bar{\mathbf{v}}_o \qquad (15\text{-}2.1)$$

Note that this equation refers to the motion of the mass center of any system of constant mass. When it is applied to the motion of a single particle or a translating rigid body, there is no need to retain the bar sign over the velocity terms.

Resultant Linear Impulse

The expression $\int_o^t \mathbf{R}\,dt$ is known as resultant linear impulse. Examining it dimensionally shows the unit of linear impulse to be Newton-sec. Obviously, linear impulse is a vector quantity since it represents a vector force multiplied by scalar time. In analytical solutions, linear impulse is usually evaluated in terms of its rectangular components, namely,

$$\int_o^t \mathbf{R}\,dt = \hat{\mathbf{i}}\int_o^t \Sigma X\,dt + \hat{\mathbf{j}}\int_o^t \Sigma Y\,dt + \hat{\mathbf{k}}\int_o^t \Sigma Z\,dt \qquad (15\text{-}2.2)$$

Clearly, these integrals can only be evaluated if the components of **R** are constant or known functions of time, or are available experimentally from the areas under curves representing the variation of these components with time. Review now the discussion concerning the use of force-time curves to determine velocity and displacement.

Linear Momentum

The quantity $m\bar{v} = (W/g)\bar{v}$ is known as the linear momentum of a body at any instant and may be symbolized by $\bar{\mathbf{p}}$, and the quantity $m\,d\,\bar{v}$ may be represented by $d\,\bar{\mathbf{p}}$. Substituting dimensional equivalents in the expression for linear momentum gives

$$\frac{W}{g}\bar{v} = \frac{\text{N}}{\text{m/sec}^2}\left(\frac{\text{m}}{\text{sec}}\right) = \frac{\text{N(sec}^2)}{\text{m}}\left(\frac{\text{m}}{\text{sec}}\right) = \text{N.sec}$$

This shows the dimensional unit of linear momentum to be identical with that of linear impulse.

As in the case of impulse, it is important to note that linear momentum consists of a vector $\overline{v}$ multiplied by a scalar coefficient; this results in a vector having the same sense and inclination as $\overline{v}$ but a different magnitude. In analytical solutions, the change in linear momentum is $\overline{v}$ best expressed as the difference between the components of final and initial momentum.

Equation (15-2.1) may now be summarized in the following principle: *The component of the resultant linear impulse along any axis if equal to the change in the components of the linear impulse along that axis.* The explicit form of this statement is

$$\int_0^t \Sigma X\, dt = \frac{W}{g}(v_x - v_{o_x}), \text{ etc.} \tag{15-2.3}$$

In applying the component form of the impulse-momentum principle, it is convenient to assume that the positive sense of each term is directed along the components of the initial direction of motion. Alternatively, we may rearrange Eq. (15-2.1) to $m\overline{v}_o + \int \mathbf{R}\, dt = m\overline{v}$ and denote it diagrammatically as in Fig. 15-2.1. Then choosing any convenient direction along which to equate the components of each term in this diagrammed equation, we revert to Eq. (15-2.3). The advantage of this approach is to emphasize the vector nature of the impulse-momentum equation and to ensure that all terms are properly included.

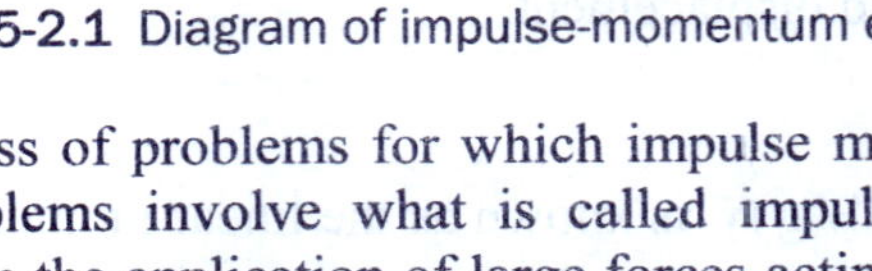

Figure 15-2.1 Diagram of impulse-momentum equation.

But there is a class of problems for which impulse momentum is uniquely suitable. These problems involve what is called impulsive motion; that is, motion resulting from the application of large forces acting for short periods of time due to a sudden impact. Such forces and their moments are known as impulsive forces and impulsive moments and are so much larger than ordinary forces such as weight that impulsive effects of the latter can be considered negligibly small in comparison. In situations involving impact, significant changes in velocity occur eventhough the time interval involved is too short to cause significant changes in position. Observe that to produce an impulse, a force need only exist for a time interval.

ILLUSTRATIVE PROBLEMS

15-2.1 Determine the velocity of body B in Fig. 15-2.2 after moving for 5 sec (a) starting from rest and (b) starting with a downward velocity of 2 m/s.

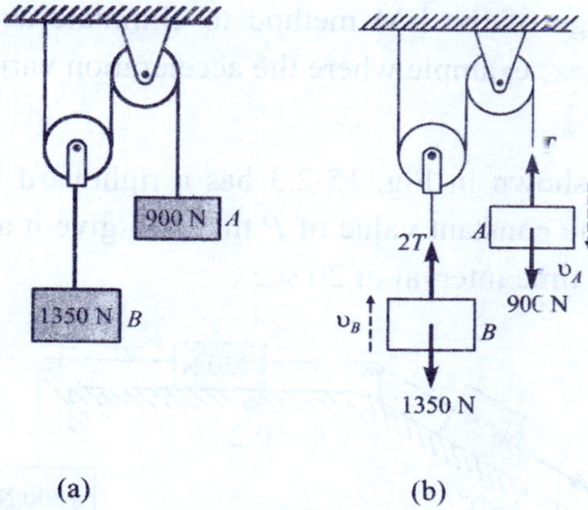

Figure 15-2.2

Solution

The FBD of each body is shown in part (b) of the figure. Noting that $v_A = 2V_B$ and applying the impulse-momentum equation (15-2.3) (with $\sum X$ constant) to each body gives

$$\left[\sum X(t) = \frac{W}{g}(v - v_o)\right] \qquad (900 - T)(5) = \frac{900}{g}(2)(v_B - v_{B_o})$$

$$(2T - 1350)(5) = \frac{1350}{g}(v_B - v_{B_o})$$

Multiplying the first of these equations by 2 and adding them eliminates the tension T and gives

$$450 = \frac{990}{g}(v_B - v_{B_o})$$

whence, for part (a) with $v_{B_o} = 0$, we obtain $v_B = 4.46$ m/s up and for part (b) with $v_{B_o} = -2$ m/s (minus because down), we obtain $v_B = 2.46$ m/sec up.

Note that T is not automatically eliminated by adding the I-M equations for each body as it would be in a W-E solution. It would have been just as simple in this problem to apply $\Sigma X = \left(\frac{W}{g}\right) a$ to each body to find the acceleration of B, and then substitute this in the kinematic relation $v = v_0 + at$ to find the desired velocity. The advantage of the I-M method to eliminate acceleration will be more apparent in the next example where the acceleration varies during the time interval considered.

15-2.2 The system shown in Fig. 15-2.3 has a rightward velocity of 3 m/s. Determine the constant value of P that will give it a leftward velocity of 6 m/s in a time interval of 20 sec.

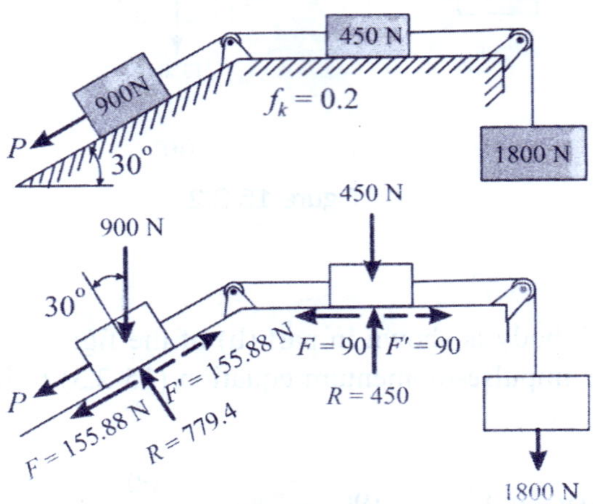

Figure 15-2.3

Solution

Since all bodies have the same displacement in the same time, the impulses of internal tensions in the connecting cords will be eliminated. Hence, using a free-body diagram of the system showing only the external forces, we apply

$$(\Sigma X)_{ext}(t) = \Sigma \frac{W}{g}(v - v_o)$$

Two sets of equal friction forces are shown: F (drawn solid) which acts during rightward motion and equal but reversed values F' (shown dashed) which act during leftward motion. Using the friction values F which resist the rightward motion during the time t to come to rest, we have

$$(1800 - 90 - 155.88 - 900 \sin 30° - P)t_1 = \frac{3150}{9.81}(0 - 3)$$

which reduces to

$$(P - 1104.12)t_1 = 963.3 \qquad (a)$$

Using the dashed friction values F' during the time t_2 to move leftward from rest to a velocity of 20 m/s, we have

$$(P + 900 \sin 30° - 155.88 - 90 - 1800)t_2 = \frac{3150}{9.81}(6 - 0)$$

which reduces to

$$(P - 1595.88)t_2 = 1926.6 \qquad (b)$$

Also, from the given data, we have

$$t_1 + t_2 = 20 \qquad (c)$$

These equations may be solved simultaneously, but it is easier and faster to use the following method of successive approximations. From Eq. (b), P must be larger than 1595.88 N. Let us try $P = 1800$ N as a start. Then from Eq. (a), we obtain

$$t_1 = \frac{963.3}{(1800 - 1104.12)} = 1.38 \text{ sec}$$

which is substituted into Eq. (c) to give $t_2 = 20 - 1.38 = 18.62$ sec. Using this value of t_2 in Eq. (b) yields

$$P = 1595.88 + \frac{1926.6}{18.62} = 1699.35 \text{ N}$$

Repeat the cycle of operations. From Eq. (a),

$$t_1 = \frac{963.3}{(1699.35 - 1104.12)} = 1.62 \text{ sec}, \quad \therefore t_2 = 18.38 \text{ sec}$$

From Eq. (b),

$$P = 1595.88 + \frac{1926.6}{18.38} = 1700.7 \text{ N} \qquad \textit{Ans.}$$

This result may be accepted as the answer since there has been no appreciable change from the previous value.

PROBLEMS

15-2.3 A 1.12 N baseball is thrown with a velocity of 60 m/s toward a batter. After being struck by the bat B, the ball has a velocity of 160 m/s directed as shown in Fig. P-15-2.3. Find the average force exerted on the ball if the impact lasts for 0.02 sec.

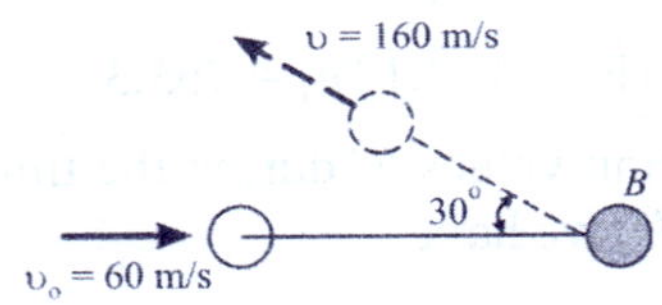

Figure P-15-2.3

15-2.4 A 40 ton railroad car moving at 0.9 m/sec along a level track strikes and is coupled to a stationary 60 ton car in 0.50 sec. Determine the average force acting on each car during the coupling.

15-2.5 During an Apollo mission to the moon, the lunar excursion module (LEM) is separated from the command module by an average thrust P lasting for 1 sec. As a result, the LEM has a velocity 6 m/s slower than the command module. Determine P assuming the mass of the LEM and the command module to be respectively 1750 N-sec^2/m and 4375 N-sec^2/m.

15-2.6 The system shown in Fig. P-15-2.6 is moving rightward at a velocity of 5 m/sec when a constant horizontal force P is applied as shown. Determine the value of P that will give the system a leftward velocity of 10 m/sec in a time interval of 10 sec.

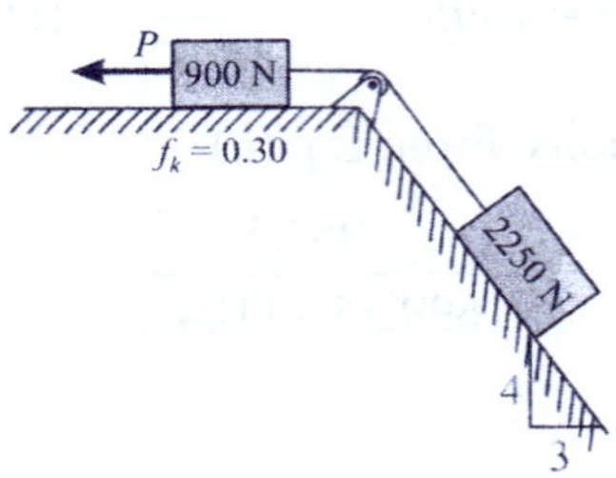

Figure P-15-2.6

15-2.7 Repeat the previous problem if the velocity of the system is to be changed from a rightward velocity of 10 m/sec to a leftward velocity of 5 m/sec in a time interval of 10 sec.

15-2.8 Oil with a specific gravity of 0.9 is flowing at 3 m/sec in a straight pipe 0.6 m in diameter and 600 m long. How long will it take to close a valve at the discharge end without raising the average pressure more than 300 KN/m^2

15-3 DYNAMIC ACTION OF JET STREAMS

The force produced by a jet of water or other system of particles impinging upon a stationary or moving object is easily obtained by applying the concept that the component of resultant impulse is equal to the change in the corresponding components of momentum. Details of application are discussed in the illustrative problems.

Impulse-momentum methods are also useful to solve situations in which a moving body absorbs or ejects material. These cases are often called variable mass problems[1] since the mass of the moving body does not stay constant. Consider Fig. 15-3.1 showing a thruster ejecting an elemental mass Δm in the time Δt at a relative velocity u from a body of mass m moving with an absolute velocity v. The velocity of Δm changes in the time Δt from an initial velocity **v** to the absolute final value $\mathbf{v}_A$, equal to the sum of **v** and **u** as shown. The force **P** exerted on the moving body is equal and opposite to the force acting on the ejected mass. This force is easily found by equating the impulse of P to the change in momentum of the ejected mass, thereby giving

$$\mathrm{P}\,\Delta t = \Delta m\,(\mathbf{v}_A - \mathbf{v}) = \Delta m(\mathbf{u})$$

or

$$\mathbf{P} = \frac{\Delta m}{\Delta t}\mathrm{u} \qquad (a)$$

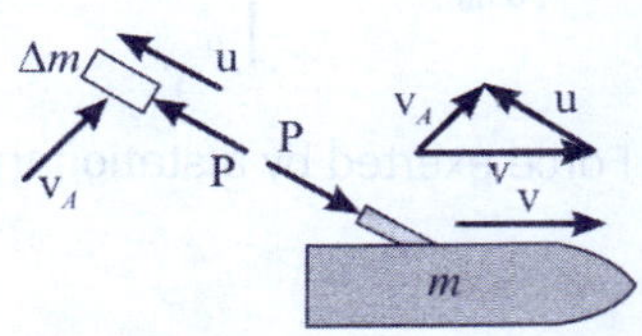

Figure 15-3.1 Jet thrust P on a moving body.

[1]We exclude problems of relativistic mechanics in which the mass is a function of velocity.

If mass is ejected continuously, although not necessarily at the same rate, the instantaneous value of **P** is found by letting Δt approach zero as a limit. We thus obtain

$$\mathbf{P} = \mathbf{u}\frac{dm}{dt} \tag{15-3.1}$$

Observe that **P** and **u** are collinear. Usually, both **u** and dm/dt are constant so that **P** remains at a constant value with respect to the moving body. A similar analysis gives the same result for the retarding force exerted on a moving body absorbing material at a relative velocity **u**.

The subsequent motion of the moving body under the influence of **P** and all other external forces that may be acting is determined by applying the kinetic equations as we have done previously. However, there is one important difference. The mass of the moving body is now time-dependent and may change appreciably as in the case of a rocket in which the amount of ejected fuel is a significant portion of its original total weight.

ILLUSTRATIVE PROBLEMS

15-3.1 A jet of water 50 mm in diameter issues from a nozzle with a velocity of 30 m/sec and impinges tangentially upon a perfectly smooth stationary vane which deflects it through an angle of 30° without loss of velocity as shown in Fig. 15-3.2. Determine the total force exerted by the jet upon the vane.

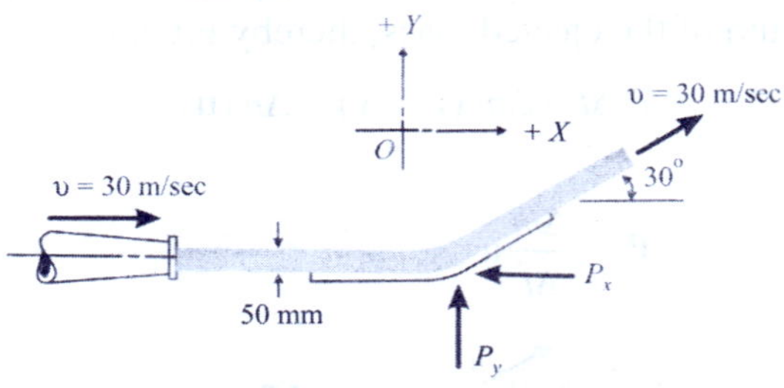

Figure 15-3.2 Force exerted by a stationary vane on a jet.

Solution

Since it is more convenient to treat the jet of water as the free body, we represent the effect of the stationary vane upon the jet by means of its components P_x and P_y. These forces are equal but opposite to the effect of the jet upon the vane and are constant for a steady flow of water. The rate of flow in Newtons/sec is $Q = wAv$, where w is the specific weight of water (taken as 9810 N/m^3 for fresh water), A is the cross-sectional area of the jet in square meter and

v is the velocity of the jet in m/s. In an interval of t sec, the weight of water whose momentum is changed thus becomes

$$[W = Qt] \qquad W = 9810\left(\frac{\pi \times (0.05)^2}{4}\right)(30)t = 577.86t\ N$$

The weight of water in contact with the vane at any instant is relatively small compared with the impulsive forces P_x and P_y and may be neglected in applying the impulse-momentum relation.

Taking X and Y axes as shown (with the X axis positive in the initial direction of motion), and applying the condition that the X component of resultant impulse equals the change in X components of momentum, we have

$$\left[\int \Sigma X\ dt = \frac{W}{g}(v_x - v_{o_x})\right] \qquad -P_x \text{x}t = \frac{577.86\ t}{9.81}(30 \cos 30^\circ - 30)$$

$$P_x = 236.75\text{ N}$$

Similarly, considering the Y components of the impulse-momentum equation, we have

$$\left[\int \Sigma Y\ dt = \frac{W}{g}(v_y - v_{o_y})\right] \qquad P_y \text{x}t = \frac{577.86\ t}{9.81}(30 \sin 30^\circ - 0)$$

$$P_y = 883.58\text{ N}$$

Combining these components, the resultant force on the vane is

$$\left[P = \sqrt{P_x^2 + P_y^2}\right] \qquad P = \sqrt{236^2 + 883.58^2} \qquad P = 914.6\text{ N} \qquad \textit{Ans.}$$

15-3.2 Assume that the vane in the previous problem is moving to the right at a velocity $u = 12$ m/s; all other data remain unchanged.

Solution

In this problem, the total weight of water whose momentum is changed during a time interval t is expressed by $W' = wA(v - u)t$ where $(v - u)$ is the relative velocity with which the water approaches the vane. In terms of W, the total weight of water issuing from the nozzle in this time, we also have $W' = W\left(\frac{v - u}{v}\right)$ so that the weight of water whose momentum is being changed may be found from either

$$W' = 9810\left(\frac{\pi \times (0.05)^2}{4}\right)(30 - 12)t = 346.72t\ N$$

or, using the result of the preceding problem,

$$W' = W\left(\frac{v-u}{v}\right) = 577.86t\left(\frac{30-12}{30}\right) = 346.72tN \qquad \textbf{\textit{Check}}$$

In the case of a vane moving with constant velocity, it is convenient to use reference axes which are fixed in and moving with the vane. With respect to these axes, we proceed as in the previous problem to equate the components of impulse to the corresponding changes in the components of momentum. However, the velocities to be used now are the relative velocities of approach to and departure from the vane, each having the magnitude $v - u$ = 18 m/s and directed as shown in Fig. 15-3.3.

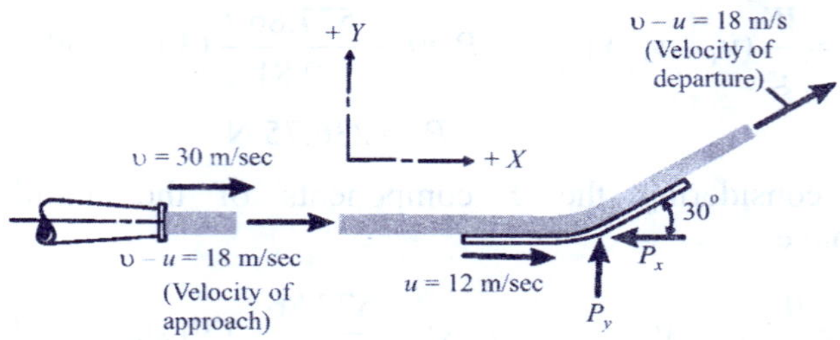

Figure 15-3.3 Jet impinging on a moving vane.

From the X component of the impulse-momentum equation we obtain

$$\left[\int_o^t \Sigma X\, dt = \frac{W'}{g}(v_x - v_{o_x})\right] \qquad -P_x x(t) = \frac{346.72t}{9.81}\ (18 \cos 30^\circ - 18)$$

$$P_x = 85.23 \text{ N}$$

Similarly, the Y component of the impulse-momentum equation gives

$$\left[\int_o^t \Sigma Y\, dt = \frac{W'}{g}(v_y - v_{o_y})\right] \qquad P_y \times (t) = \frac{346.72t}{9.81}(18 \sin 30 - 0)$$

$$P_x = 318.1 \text{ N}$$

As before, the resultant force on the vane is found from

$$[P = \sqrt{P_x^2 + P_y^2}] \qquad P = \sqrt{(85.23)^2 + (318.1)^2} \qquad \text{P} = 329.31 \text{ N} \qquad \textbf{\textit{Ans.}}$$

Power is developed by the X component of P which acts in the direction of motion of the vane. The power developed is

$$= \frac{85.23 \times 12}{1000} = 1.0 \text{ KW} \qquad \textbf{\textit{Ans.}}$$

This result is the power developed by one vane. If a series of moving vanes are used in which the next vane (as in a rotating Pelton turbine) intercepts the jet just after the first vane has moved out of the way, all the water from the jet is utilized, and we should use $wAvt$ instead of $wA(v - u)t$ in the preceding computations. Doing this changes the power developed to

$$= \frac{wAvt}{wA(v-u)t} \times 1 = \frac{30}{18} \times 1 = 1.67\text{KW}$$

15-3.3 The total weight of a rocket including its fuel is W N. The fuel is ejected at a constant rate of w N/sec with a constant velocity u m/s relative to the longitudinal axis of the rocket. Determine the velocity and height of the rocket after t sec, assuming vertical flight and that t is less than t_b at burnout of fuel. Neglect air resistance and the variation in g with altitude.

Solution

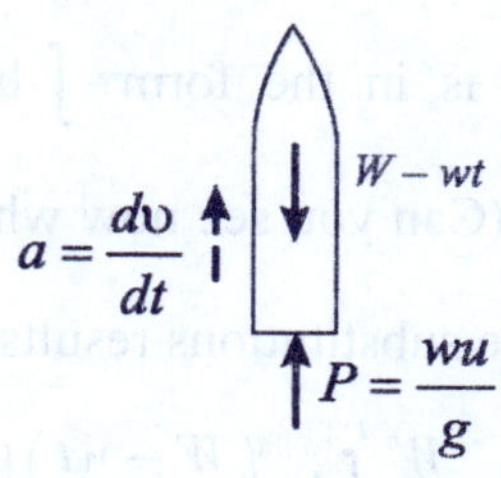

Figure 15-3.4

In Fig. 15-3.4 is shown the free-body diagram of the rocket. At any instant before burnout, the net weight of the rocket is $W - wt$ where wt is the weight of fuel expelled at time t. From Eq. (15-3.1), the jet thrust $P = wu/g$. The acceleration of the rocket at this instant is

$$\left[+\uparrow \Sigma Y = \frac{W}{g} a \right] \qquad \frac{wu}{g} - (W - wt) = \frac{W - wt}{g} a$$

whence replacing a by dv/dt and multiplying by $g/(W - wt)$, we obtain

$$a = \frac{dv}{dt} = \frac{wu}{W - wt} - g$$

Now separate the variables v and t, and integrate between appropriate limits. This procedure determines v as follows:

$$\int_o^v dv = \int_o^t \frac{wu}{W - wt} dt - g \int_o^t dt$$

$$v = -u \ln (W - wt)\Big]_o^t - gt$$

$$v = u \ln \frac{W}{W - wt} - gt \qquad \textbf{\textit{Ans.}}$$

To find the height h at time t, now replace v by dy/dt and integrate between appropriate limits to obtain

$$v = \frac{dy}{dt} = u \ln \frac{W}{W - wt} - gt$$

$$\int_o^h dy = u \int_o^t \ln \frac{W}{W - wt} dt - g \int_o^t t\, dt$$

$$h = -u \int_o^t \ln \left(\frac{W - wt}{W}\right) dt - \frac{1}{2} gt^2$$

The integral in this relation is in the form $\int \ln x \; dx = x \ln x - x$ where $x = \frac{W - wt}{W}$ and $dx = -\frac{w\, dt}{W}$ (Can you see now why the integrand was inverted in the last step?). Making these substitutions results in

$$h = u \frac{W}{w} \int_o^t \ln \left(\frac{W - wt}{W}\right)\left(-\frac{w}{W} dt\right) - \frac{1}{2} gt^2$$

which reduces to

$$h = \frac{uW}{w}\left[\frac{W - wt}{W} \ln \frac{W - wt}{W} - \frac{W - wt}{W} + 1\right] - \frac{1}{2} gt^2 \qquad \textbf{\textit{Ans.}}$$

Some supplement remarks on rocket motion are pertinent. Observe that liftoff will occur only if $P = wu/g$ is greater than the "wet" weight W. For example, liftoff of a 10-ton rocket with an exhaust velocity of 1830 m/s would require a rate of fuel consumption of at least 3.22 tons/min. Consequently, "exotic" fuels are required which have both a high combustion rate and a large exhaust velocity.

PROBLEMS

15-3.4 What force would a 50 mm diameter jet of water flowing at 0.06 m^3/sec exert on a stationary plate placed as in parts (a) and (b) of Fig. P-15-3.4?

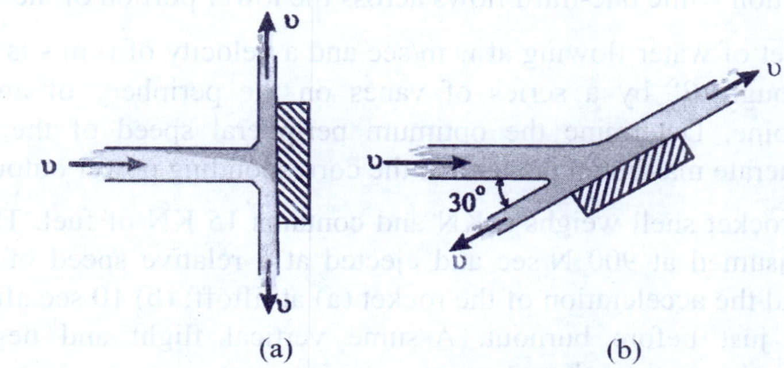

Figure P-15-3.4

15-3.5 A jet of water issuing from a nozzle at the rate of 90 N/sec at a velocity of 30 m/s strikes a series of vanes and is deflected through 120°. If the speed of the vanes is 18 m/s, compute the power delivered to the vanes.

15-3.6 A tank weighing 5 KN rests on platform scales. It is being filled with water from a vertical jet having a velocity of 45 m/s and a cross-sectional area of 0.02 m^2. What will be the total scale reading after 5 sec?

15-3.7 A 300 mm pipe carries water with a velocity of 3 m/s around a 60° elbow. The pressure intensity at both ends of the elbow is 280 KN/m^2. Compute the resultant force exerted by the water on the elbow. Neglect the weight of water in the elbow.

15-3.8 A stream of water flowing at 450 N/sec and moving rightward at 30 m/sec strikes the vane as shown in Fig. P-15-3.8. If half of the water flows across each part of the vane, determine the force required to hold it in place.

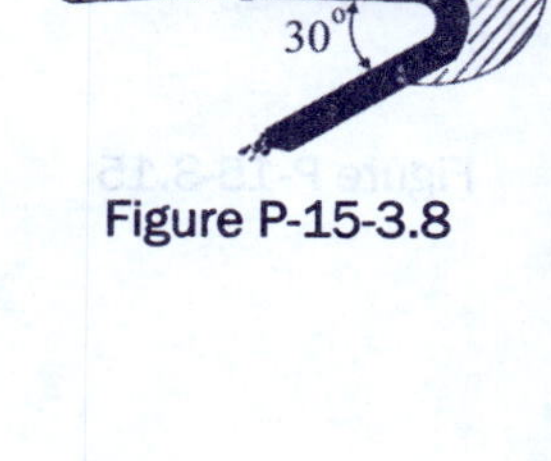

Figure P-15-3.8

15-3.9 If the vane of Prob. 15-3.8 is moving rightward at 12 m/s and one-third of the water flows over the upper portion while two-thirds flow across the lower portion, determine the power developed.

15-3.10 Solve Prob. 15-3.9 if two-thirds of the water flow across the upper portion while one-third flows across the lower portion of the vane.

15-3.11 A jet of water flowing at w m/sec and a velocity of υ m/s is deflected through θ° by a series of vanes on the periphery of an impulse turbine. Determine the optimum peripheral speed of the vanes to generate maximum power and the corresponding power output.

15.3.12 A rocket shell weighs 5 KN and contains 15 KN of fuel. The fuel is consumed at 900 N/sec and ejected at a relative speed of 490 m/s. Find the acceleration of the rocket (a) at liftoff; (b) 10 sec after liftoff; (c) just before burnout. Assume vertical flight and neglect any variation in the value of g.

15.3.13 A chain 24 m long that weighs 150 N/m is suspended vertically with the lover end just touching a platform. If the chain is dropped, determine the reaction of the platform at the instant the upper end of the chain has dropped 18 m

15-3.14 A 0.3 N ball is placed in a vertical jet of water flowing from a nozzle at the rate of 1.5 N/s with a velocity of 5 m/s. Determinet the height h above the nozzle at which the ball will stay suspended in the jet.

15-3.15 After a jet plane has touched down, reverse thrust to decelerate it is obtained by means of exhaust vanes which partially reverse the exhaust stream as shown in Fig. P-15-3.15. For a two-engine plane weighing 10 tons and moving at a ground speed of 70 m/s, determine the maximum deceleration if each engine scoops in air at a rate of 450 N/s and discharges it at a relative velocity of 600 m/s while consuming 9 N of fuel per sec.

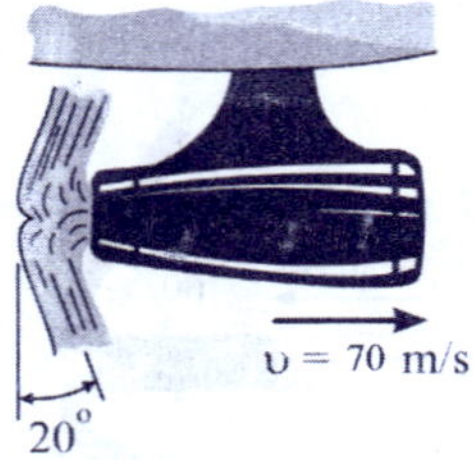

Figure P-15-3.15

15.3.16 A snowplow, moving at a constant speed of 9 m/s clears a 3 m path from an airport runway. The average depth of snow is 0.5 m and its density is 1100 N/m^3. The snow is diverted sideways at a 120° angle from the forward motion of the plow and, because the snow compacts somewhat, at a relative speed of 7 m/s. Determine the power required beyond that needed to move the snowplow itself.

15-3.17 The total weight of a rocket including its fuel is 10 tons. The fuel is consumed at the rate of 550 N/sec and ejected with a constant rearward relative velocity of 1850 m/s. If burnout occurs at 150 sec after launch, what will then be the velocity and height of the rocket? Assume vertical flight and neglect both air resistance and any variation in g.

15-3.18 The rocket of the previous problem is redesigned as a two-stage rocket. Each stage weighs 45 KN including 40 KN of fuel. When the first stage reaches burnout, it is detached and the second stage fires. If the fuel is again consumed at the rate of 550 N/sec and ejected at a constant relative velocity of 1850 m/sec, determine the velocity when the first stage is released and the maximum velocity reached by the second stage.

15-3.19 A 6 ton airplane is climbing at an angle of 10° at a constant airspeed of 135 m/s. Its jet engine scoops in air at the rate of 450 N/sec and discharges it at a relative velocity of 500 m/s while burning 9 N of fuel per sec. (a) Determine the lift and drag forces perpendicular and parallel to the flight path. (b) What will be the acceleration of the plane as it levels off?

15-4 Conservation of Linear Momentum

The integral form of the linear impulse-momentum equation for a body or a system of particles is

$$\int \mathbf{R}\, dt = \int m\, d\,\bar{\mathbf{v}} = \int d\mathbf{P}$$

in which $d\mathbf{p}$ represents differential momentum. Observe that if the resultant external force $\mathbf{R}$ is zero during any time interval, this equation reduces to $d\mathbf{p} = 0$ and therefore the total momentum $\mathbf{p}$ must be constant during this time interval.

The condition that $\mathbf{p}$ (the momentum of a system of particles) be constant applies either to static equilibrium of a system moving with constant velocity, or, more importantly, when **R** is zero because of mutual action and reaction between the particles composing the system. Although the total resultant external force may be zero, this does not mean that the resultant force on any single particle is zero. For example, when a shell is discharged from a gun, the force acting on the shell is always equal and opposite to the force acting on the gun. The resultant force acting on the system composed of shell and gun will be zero although a propulsive force is acting on both shell and gun if either is considered as a free body. From another viewpoint, for a system subjected to a zero resultant force, the motion of the mass center of the system remains unchanged although there may be changes in the motion of its individual parts. *The condition that the momentum be constant applies only to the system as a whole, never to its component parts. This is known as the principle of conservation of momentum.*

Therefore, if a system is composed of particles of weight W_1, W_2, etc., having velocities v_1, v_2, etc. and after a mutual reaction between the particles they possess new velocities v_1', v_2', etc., the condition that the momentum of the system be constant is expressed by means of the following equation:

$$\frac{W_1}{g}v_1 + \frac{W_2}{g}v_2 + \ldots\ldots = \frac{W_1}{g}v_1' + \frac{W_2}{g}v_2' + \ldots\ldots \qquad (15\text{-}4.1)$$

Note that it is possible to have a zero component of impulse along one coordinate direction, thereby resulting in conserving the components of linear momentum along that direction eventhough the components of momentum along other directions are not conserved,

Another approach which may give more meaning to conservation of momentum is the example shown in Fig. 15-4.1. Here we have the FBDs of two bodies moving in the same direction which have collided because it is assumed that $v_1 > v_2$. Assume the bodies to be either perfectly or partially elastic and to continue in the same direction with the primed velocities after the impact. Obviously $v_2' > v_1'$ for separation to occur.

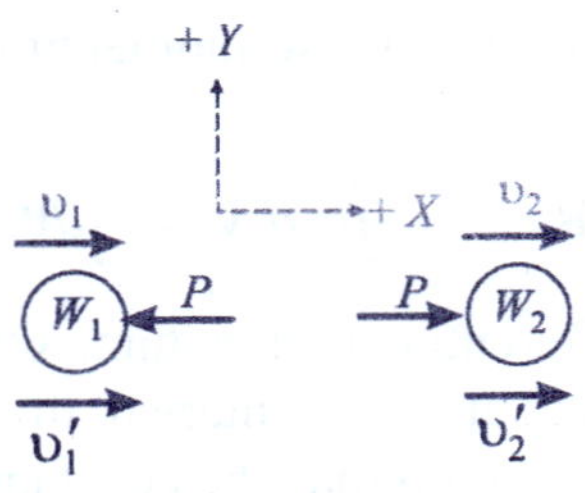

Figure 15-4.1 Free-body diagrams during impact.

The velocity changes are caused by the pair of equal and opposite reaction forces P which vary in an unknown manner during the impact. Significant information about the velocities can be obtained by writing the impulse-momentum equation for each body eventhough the variation of P with time is unknown. Using the same coordinate system shown for each body, and neglecting the impulse of the weights or of friction as negligibly small compared with that of P, we obtain

For W_1: $$-\int P\,dt = \frac{W_1}{g}(v_1' - v_1)$$

For W_2: $$\int P\,dt = \frac{W_2}{g}(v_2' - v_2)$$

Adding these equations eliminates the impulse $\int P\,dt$ and gives

$$0 = \frac{W_1}{g}(v_1' - v_1) + \frac{W_2}{g}(v_2' - v_2)$$

Rearranging the terms gives the following result which represents the conservation of the components of momentum along the X direction:

$$\frac{W_1}{g}v_1 + \frac{W_2}{g}v_2 = \frac{W_1}{g}v_1' + \frac{W_2}{g}v_2'$$

ILLUSTRATIVE PROBLEMS

15-4.1 A 750 N man, moving horizontally with a velocity of 3 m/s jumps off the end of a pier into a 2850 N boat. Determine the horizontal velocity of the boat (a) if it had no initial velocity and (b) if it was approaching the pier with an initial velocity of 0.9 m/s.

Solution: Part (a)

Since the boat and the man exert mutual forces on each other, the principle of conservation of momentum may be applied to the system of boat and man. The final horizontal velocity of the boat and man will be the same; hence, applying the horizontal components of Eq. (15-4.1) with the common factor g cancelled out, we have for part (a)

$$[W_1 v_1 + W_2 v_2 = (W_1 + W_2)v]$$

$$750(3) + 2850(0) = (750 + 2850)v$$

$$v = 0.625 \text{ m/s} \qquad \textit{Ans.}$$

The positive value of the final velocity indicates that the boat and man move in the direction of the man's initial velocity. If the vertical component of the man's velocity were known, a similar analysis would yield the common initial vertical velocity with which the boat and man would start to bob up and down in the water.

Part (b)

In this case the initial velocity of the boat is opposite to that of the man and hence is considered negative when we consider the horizontal components of momentum.

$[W_1 v_1 + W_2 v_2 = (W_1 + W_2)v]$

$$750(3) + 2850(-0.9) = (750 + 2850)v$$

$$v = -0.0875 \text{ m/s} \qquad \textbf{\textit{Ans.}}$$

The minus sign indicates a final velocity directed toward the pier.

15-4.2 A ballistic pendulum consists of a sand box weighing 260 N that is suspended from a cord 3 m long. A 5 N shell is fired horizontally into the box and remains embedded in it. Because of impact, the sand box swings through a maximum angle of 30°, as shown in Fig. 15-4.2. Determine the velocity with which the shell strikes the box and the loss in energy of the system.

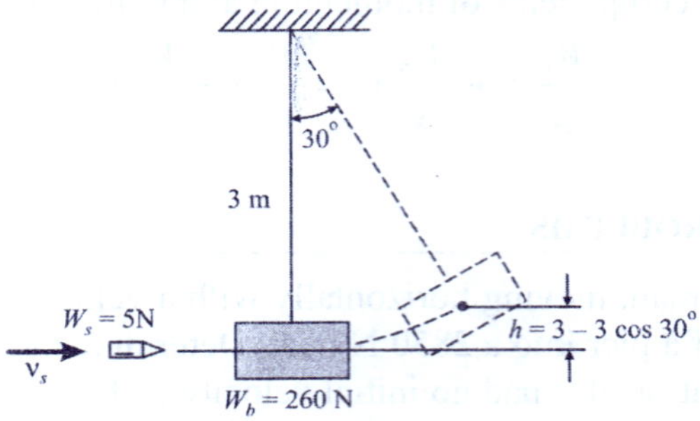

Figure 15-4.2

Solution

The initial velocity of the sand box with the embedded shell is found by the work-energy method. The work done is negative because the gravity force acts opposite to the upward rise.

$$\left[-Wh = \frac{W}{2g}(v^2 - v_0^2)\right] \quad -W(3 - 3\cos 30°) = \frac{W}{2 \times 9.81}(0 - v_0^2)$$

$$v_0 = 2.81 \text{ m/s} \qquad \textbf{\textit{Ans.}}$$

This value of velocity represents the common velocity of the shell and box directly after impact. The velocity of the shell before impact can be found now by applying the principle of conservation of momentum.

$[W_1 v_1 + W_2 v_2 = (W_1 + W_2)v]$

$$5(v_s) + 0 = (5 + 260)(2.81)$$

$$v_s = 147.12 \text{ m/s} \qquad \textit{Ans.}$$

To determine the loss in energy of the system, we have

$$\text{initial energy of shell} = \frac{1}{2}mv^2 = \frac{1}{2}\left(\frac{5}{9.81}\right)(147.1)^2 = 5515.88 \text{ Nm}$$

$$\text{final energy of system } \frac{1}{2}mv^2 = \frac{1}{2}\left(\frac{265}{9.81}\right)(2.81)^2 = 106.65 \text{ Nm.}$$

Their difference represents an energy loss of 5409.23 Nm which is dissipated in heat.

15-4.3 In Fig. 15-4.3, bodies A and B are connected by a flexible, inextensible cord. Body A is lifted vertically and then dropped from rest. If it acquires a velocity v_o just before the cord becomes taut, find its velocity right after body B is jerked into motion.

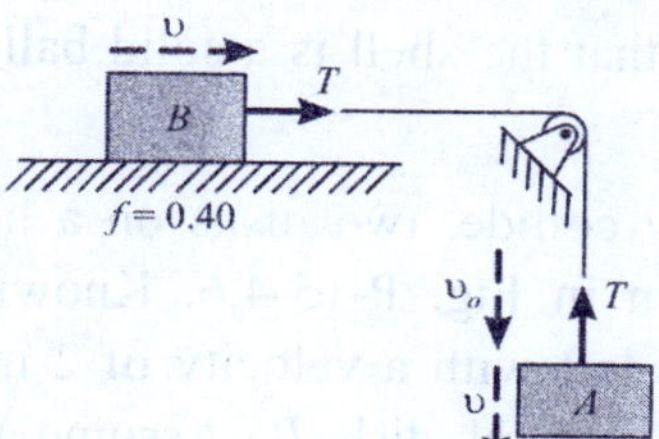

Figure 15-4.3

Solution

In this problem, neglect the impulse of the weights and friction as negligibly small compared to that of T. Since equal and opposite impulses act on each body due to the tension T, momentum of the system is conserved. We therefore obtain for the common final velocity v of both bodies,

$$[W_1 v_1 + W_2 v_2 = (W_1 + W_2)v] \qquad W_A v_o + 0 = (W_A + W_B)v$$

$$v = \frac{W_A v_0}{W_A + W_B} \qquad \textit{Ans.}$$

Alternatively, the same result can be found by a direct application of the impulse-momentum equation to each body as follows:

On A: $$-\int T\,dt = m_A v - m_A v_o$$

On B: $$\int T\,dt = m_B v - 0$$

Adding these equations eliminates the impulsive forces and, after converting mass to weight by cancelling out the common factor g, we again obtain

$$v = \frac{W_A v_o}{W_A + W_B}$$ ***Ans.***

PROBLEMS

15-4.4 A 4.5 KN shell is fired from a 900 KN cannon with a velocity of 600 m/s. Find the modulus of a nest of springs that will limit the recoil of the cannon to 0.9 m.

15-4.5 A cannon is mounted on one end of a railroad car and a rigid target is mounted on the other end. The car is free to roll without resistance on a level track. A shell is fired from the gun when the car is at rest. Discuss the motion of the car and what happens when the shell hits the target. Assume that the shell is a solid ball which remains embedded in the target.

15-4.6 Just before they collide, two disks on a horizontal surface have the velocities shown in Fig. P-15-4.6. Knowing that the 90 N disk A rebounds to the left with a velocity of 2 m/s, determine the rebound velocity of the 150 N disk B. Assume the colliding surfaces are smooth.

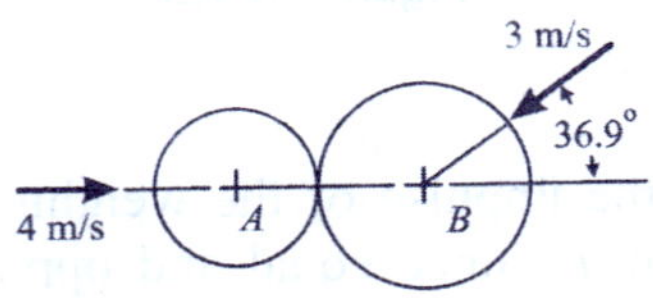

Figure P-15-4.6

15-4.7 Solve the previous problem if the coefficient of friction at the colliding surfaces is 0.40 and the collision lasts for 0.01 sec.

15-4.8 Two moving balls of putty collide and cling together. Determine the magnitude of their common final velocity and the loss in kinetic energy. The first ball A weighs 4.5 N and has a velocity $v_A = 3\,\hat{\mathbf{i}} + 5\,\hat{\mathbf{j}}$ m/s. The second ball B weighs 8.9 N and has a velocity $v_B = 1.5\,\hat{\mathbf{i}} - 1\,\hat{\mathbf{j}} + 2\,\hat{\mathbf{k}}$ m/s.

15-4.9 A bullet weighing 0.3 N and moving at 670 m/s penetrates the 45 N body as shown in Fig. P-15-4.9 and emerges with a velocity of 180 m/s. How far and how long does the body then move?

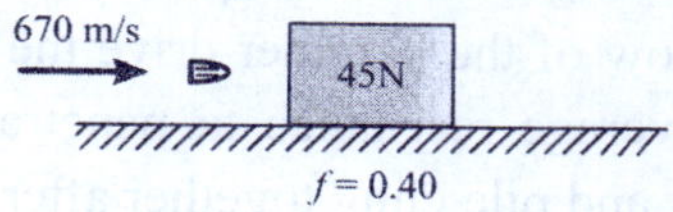

Figure P-15-4.9

15-4.10 The spring shown in Fig. P.15-4.10 has a free length of 0.30 m. It is compressed to half its length and the blocks are suddenly released from rest. Determine the velocity of each block when the spring is again 0.3 m long.

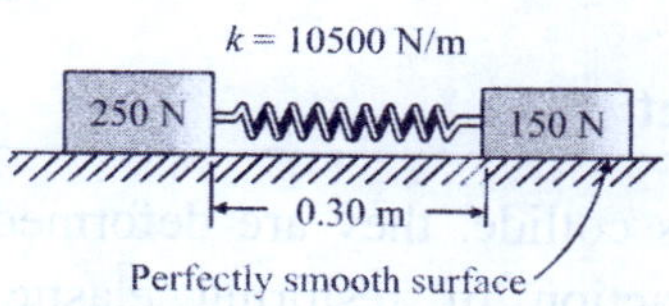

Figure P-15-4.10

15.4.11 Determine the velocity of each block in the preceding problem at the instant the spring is 0.3 m long.

15-4.12 After body A in Fig. P-15-4.12 moves 2 m from rest, it picks up body C. How much further does A move before reversing its direction? Assume $W_A = 45$ N, $W_B = 90$ N, $W_C = 65$ N.

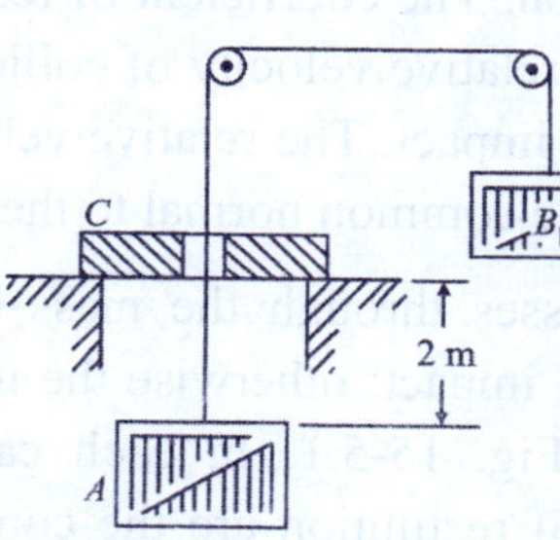

Figure P-15-4.12

15.4.13 A 900 N man and a 450 N boy jump to a pier from a 1350 N rowboat with a horizontal velocity of 3 m/s relative to the rowboat. If the rowboat is initially motionless in the water, under which of the following conditions will the rowboat acquire the largest velocity: (a) man and boy jump simultaneously; (b) man jumps first, followed by boy; (c) boy jumps first, followed by man.

15-4.14 A wooden pile that weighs 2500 N is driven vertically into the ground by successive blows from a piledriver whose 4500 N hammer falls freely through a height of 1.2 m upon the head of the pile. How deep does a single blow of the hammer drive the pile when the ground has developed an average resistance to penetration of 3000 N? Assume that the hammer and pile cling together after impact.

Hint: Resultant work done equals kinetic energy lost in impact.

15-4.15 A 5500 N hammer falling freely through 0.9 m drives a 2500 N pile 0.15 m vertically into the ground. Assuming the hammer and pile to cling together after impact, determine the average ground resistance to penetration by the pile.

15-5 Elastic Impact

When two elastic bodies collide, they are deformed at first; then they spring apart because of the action of restoring elastic forces. Throughout this elastic impact, there exist mutual action and reaction forces. (Refer back to Fig. 15-4.1.) The magnitude of these impact forces during the short time they exist is so large that nonimpulsive forces are negligible in comparison. Hence, the resultant force on the system is zero and conservation of momentum may be applied to express one relation between the unknown final velocities of the bodies. Another relation between these velocities is obtained from the definition of the coefficient of restitution. The coefficient of restitution (symbolized by e) is defined as the ratio of the relative velocity of colliding bodies after impact to their relative velocity before impact. The relative velocities are measured along the line of impact which is the common normal to the colliding surfaces.

If the line of impact passes through the mass centers of the bodies, the collision is known as central impact; otherwise the impact is eccentric. Several possibilities are shown in Fig. 15-5.1. In each case, the relative velocities involved in the coefficient of restitution are the components of velocity along the line of impact. Since the velocity components involved in both the definition of the coefficient of restitution and in the component of momentum along the line of impact are all along the same line, we may use their scalar values.

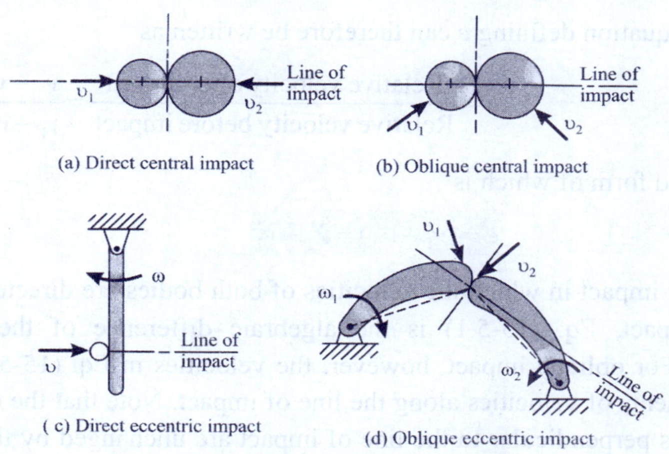

Figure 15-5.1 Various types of impact.

Now consider direct central impact of the bodies shown in Fig. 15-5.2a where the unprimed values are initial velocities and the primed values are their velocities after impact. Body W_1 is overtaking body W_2 hence, v_1 must be larger than υ_2 and the relative velocity before impact is $v_1 - v_2$. After impact, assume that the bodies continue moving in their original directions. In order for them to separate, body W_2 must have a larger final velocity than body W_1; therefore, the relative velocity after impact is $v_2' - v_1'$. Fig. 15-5.2b, shows another extreme case of direct central impact in which the bodies approach each other before impact and have opposite velocities after impact. In this case, the relative velocity again is the algebraic difference $v_1 - v_2$ which corresponds to the arithmetic sum of their magnitudes. Thus, since the sign of v_2 is minus (i.e., opposite to v_1 assumed as plus)

$$v_1 - v_2 = v_1 - (-v_2) = v_1 + v_2$$

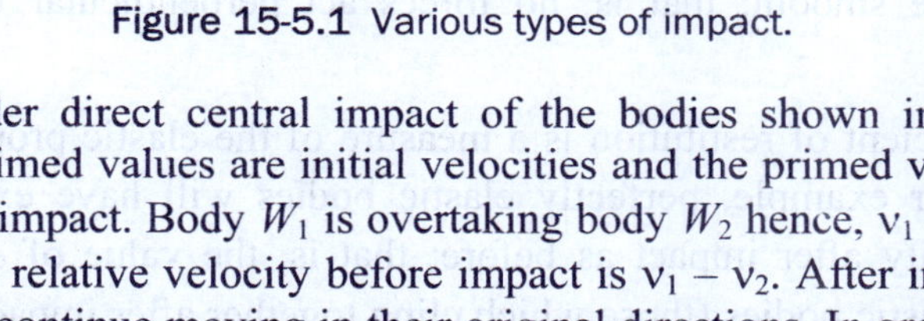

Figure 15-5.2 Two extreme cases of direct central impact.

Similarly, the relative velocity after impact is also given by

$$v_2' - v_1' = v_2' - (-v_1') = v_2' + v_1'$$

A general equation defining e can therefore be written as

$$e = \frac{\text{Relative velocity after impact}}{\text{Relative velocity before impact}} = \frac{v_2' - v_1'}{v_1 - v_2}$$

the preferred form of which is

$$e = (v_1 - v_2) = v_2' - v_1' \qquad (15\text{-}5.1)$$

In direct impact in which the velocities of both bodies are directed along the line of impact, Eq. (15-5.1) is the algebraic difference of the respective velocities. For oblique impact, however, the velocities in Eq. (15-5.1) must be the components of velocities along the line of impact. Note that the components of velocities perpendicular to the line of impact are unchanged by the impact if the bodies are smooth; that is, no forces act perpendicular to the common normal.

The coefficient of restitution is a measure of the elastic properties of a pair of bodies. For example, perfectly elastic bodies will have exactly the same relative velocity after impact as before; that is, the value of e will be unity. Perfectly inelastic bodies (those which cling together after impact) will have the same final velocity and the value of e is obviously zero. The coefficient of restitution will therefore always lie somewhere between the limits of zero and unity.

Impact problems usually imply point contact as in colliding spherical bodies. The value of e in any particular case depends not only on the two materials involved, but also on the impact velocity and the size and shape of the bodies. Handbook values of e for different pairs of materials usually refer to spherical bodies and should be used with prudence for other shapes.

Imperfect elastic action causes a loss in kinetic energy which can be found by taking the difference between the kinetic energies before and after impact. The kinetic energy lost is partially dissipated by heat generated at the contact surfaces. The rise in temperature of a steel hammer repeatedly struck against an anvil demonstrates this. Energy is also dissipated by vibrations within the bodies, hysteresis effects, and sound waves produced in the air.

ILLUSTRATIVE PROBLEMS

15-5.1 The 45 N and 90 N bodies in Fig. 15-5.3 are approaching each other with the velocities shown. If e = 0.60, determine the velocity of each body directly after impact.

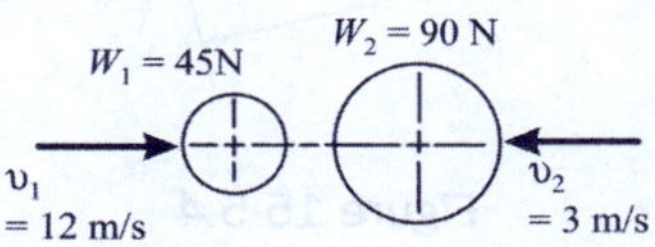

Figure 15-5.3

Solution

Applying the equation of conservation of momentum with the factor g canceled out gives

$$\left[W_1 v_1 + W_2 v_2 = W_1 v_1' + W_2 v_2'\right]$$

$$45(12) + 90(-3.0) = 45\, v_1' + 90\, v_2'$$

or

$$6 = v_1' + 2\, v_2' \qquad (a)$$

Using Eq. (15-5.1) to obtain another relation between these velocities, we have

$$[e(v_1 - v_2) = -\; v_2' - v_1'] \qquad 0.6[12 - (-3)] = v_2' - v_1'$$

or

$$9 = v_2' - v_1' \qquad (b)$$

Solving Eqs. (*a*) and (*b*) yields

$$v_1' = -\,4 \text{ m/s}$$

$$v_2' = +\,5 \text{ m/s} \qquad \textbf{\textit{Ans.}}$$

The minus sign before v_1' indicates that the 45 N body rebounds to the left after impact.

15-5.2 A ball is dropped from a height of 2.5 m upon a 15° incline and rebounds as shown in Fig. 15-5.4. If *e* = 0.80, find the distance *x* at which the ball again strikes the incline.

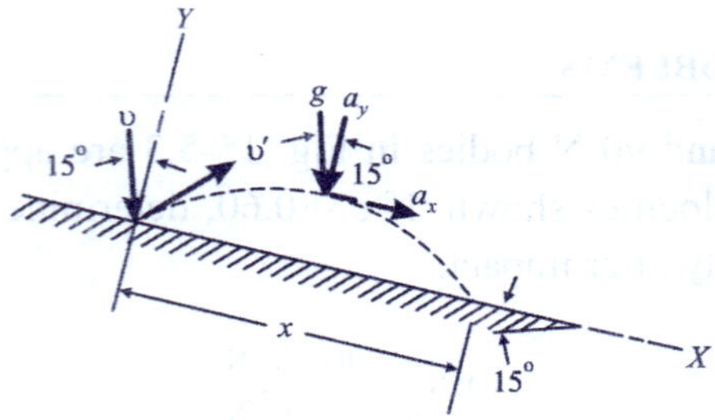

Figure 15-5.4

Solution

This problem is an example of oblique central impact. The initial velocity with which the ball strikes the incline is.

$$\left[v^2 = v_0^2 + 2gh\right] \qquad v = \sqrt{2 \times 9.81\ (2.5)} = 7 \text{ m/s}$$

and its components parallel and perpendicular to the incline are

$$v_x = v \sin 15° = 1.81 \text{ m/s} \qquad v_y = v \cos 15° = 6.76 \text{ m/s}$$

Assuming smooth contact surfaces, the X component of the rebound velocity is $v'_x = v_x = 1.81$ m/s. The value of the rebound component v'_y, as found from Eq: (15-5.1), is

$$[e(v_1 - v_2) = v'_2 - v'_1] \qquad 0.8(6.76 - 0) = 0 - (-v'_y)$$

$$v'_y = 5.41 \text{ m/s}$$

The rest of the problem involves the trajectory of the ball under the influence of a constant gravitational acceleration g m/s^2. Using the X-Y axes shown, the constant components of acceleration which affect the rebound components v'_x and v'_y are

$$a_x = g \sin 15° = 2.54 \text{ m/s}^2 \quad \text{and} \quad a_y = -g \cos 15° = -9.48 \text{ m/s}^2$$

Since these components of acceleration are constant, we may use $s = v_o t + \frac{1}{2}at^2$ to determine the corresponding rectangular components of displacement. Since $y = 0$ when the ball again strikes the incline, the time of flight is found to be

$$\left[y = v_{oy}t + \frac{1}{2}a_y t^2\right] \qquad 0 = 5.41\ t - \frac{1}{2}(9.48)\ t^2 \qquad t = 1.14 \text{ sec}$$

When we use this value of t, the distance x is

$$\left[x = v_{ox}t + \frac{1}{2}a_x t^2\right] \qquad x = 1.81(1.14) + \frac{1}{2}(2.54)(1.14)^2$$

$$x = 3.71 \text{ m} \qquad \textit{Ans.}$$

Notice the relative ease of this approach to the trajectory problem as compared with that of Illus. Prob. 9-6.3. To use horizontal and vertical components of acceleration as we did there would here require first transforming v_x and v_y', into corresponding horizontal and vertical components of velocity, and then solving a pair of simultaneous equations.

PROBLEMS

15-5.3 A ball is thrown at an angle θ with the normal to a smooth floor as shown in Fig. P-15-5.3. It rebounds at an angle θ' with the normal. Show that the coefficient of restitution is expressed by

$$e = \tan\theta / \tan\theta'$$

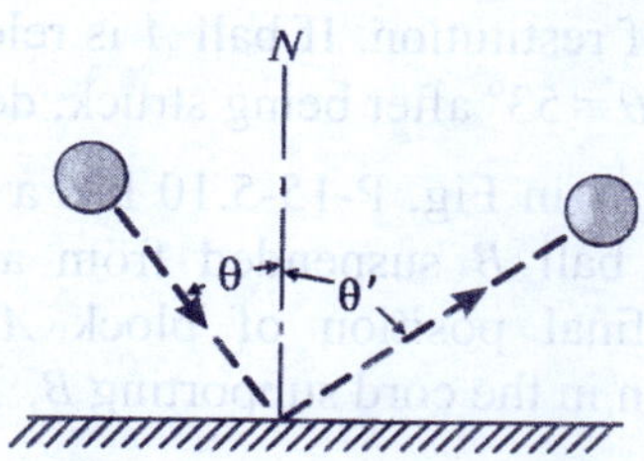

Figure P-15-5.3

15-5.4 Direct central impact occurs between a 300 N body moving rightward at 3 m/s and a 150 N body moving leftward at 6 m/s. If the coefficient of restitution is $e = 0.6$, determine the average impact force for a time of impact lasting 0.02 sec.

15-5.5 Direct central impact occurs between a 150 N body moving to the right at 2 m/s and a body of weight W moving to the left at 1.5 m/s. The coefficient of restitution is $e = 0.5$. After impact the 150 N body has a leftward velocity of 0.9 m/s. (a) Find the value of W. (b) If the impact lasts for 0.02 sec find the average impact force.

15-5.6 A golf ball is dropped from a height of 3 m upon a concrete floor. The coefficient of restitution is 0.984. Find the height to which the ball rebounds on the first, second, and third bounces.

15-5.7 Show that a general expression for the answer to Illus. Prob. 15-5.2 is $x = 4eh(1 + e)\sin\theta$.

15-5.8 The balls A and B in Fig. P-15-5.8 are attached to stiff rods of negligible weight. Ball A is released from rest and allowed to strike B. if $e = 0.6$, determine the maximum angle θ through which ball B will swing. What is the maximum and minimum tension in the rod attached to B? If the impact lasts for 0.01 sec, also find the average impact force.

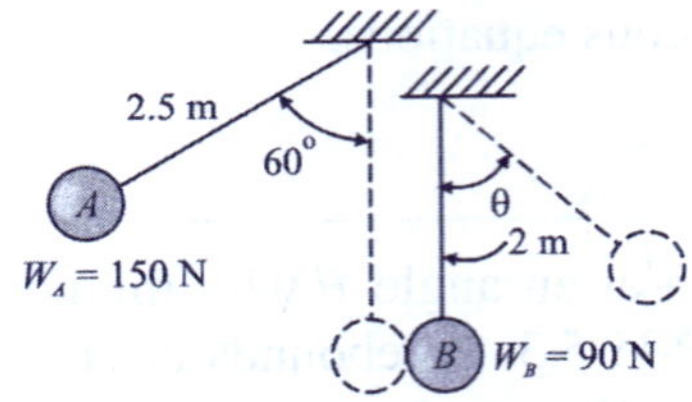

Figure P-15-5.8

15-5.9 The system shown in g. P-15-5.8 is used to determine experimentally the coefficient of restitution. If ball A is released from rest and ball B swings through $\theta = 53°$ after being struck, determine e.

15.5.10 The 45 N block A in Fig. P-15-5.10 has a velocity of 3 m/s when it strikes a 90 N ball B suspended from a 2 m cord. If $e = 0.80$, determine the final position of block A and the maximum and minimum tension in the cord supporting B.

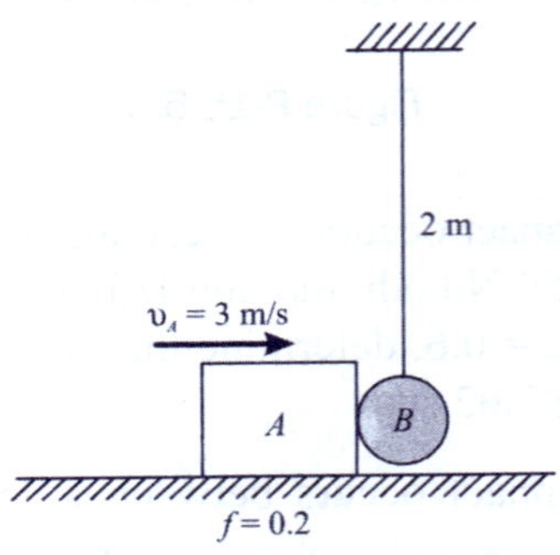

Figure P-15-5.10

15-5.11 As shown in Fig. P-15-5.11, a ball is thrown against a smooth vertical wall. The ball is released from a position 12 m from the wall and 2 m above the level ground and travels in a plane perpendicular to the wall. Its initial velocity is 15 m/s upward at an angle of 60° with the horizontal. The coefficient of restitution between ball and wall is 0.6. How far from the wall does the ball hit the ground?

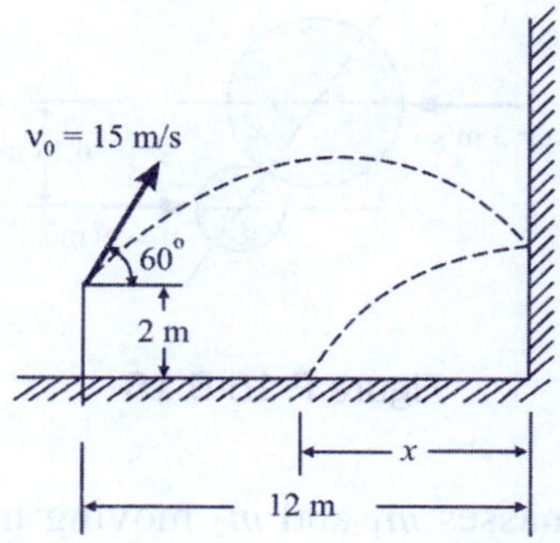

Figure 15-5.11

15-5.12 Solve Prob. 15-5.11 if the ball travels in a vertical plane which makes an angle of 45° with the wall.

15-5.13 A 30 N ball moving at 3 m/s up to the right at 60° to the horizontal collides with a 45 N ball moving horizontally leftward at 3 m/s. At the moment of impact, the line joining the centers of the balls is horizontal. If $e = 0.6$, determine the amount and direction of the velocity of each ball directly after impact.

15-5.14 In the oblique central impact shown in Fig. P-15-5.14, the coefficient of restitution is 0.6. The flat disks slide on a smooth horizontal surface. Compute the final velocity of each disk directly after impact.

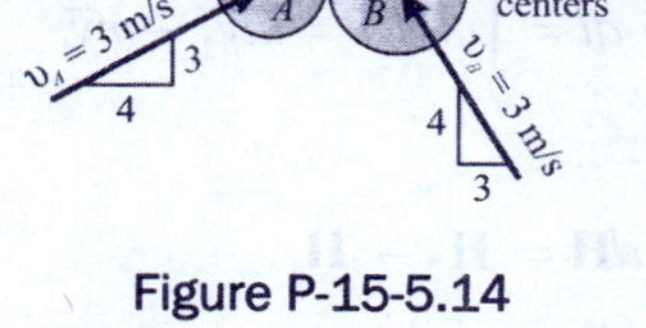

Figure P-15-5.14

15-5.15 If the surfaces of contact between the colliding disks of Prob. 15-5.14 were rough with a coefficient of friction of 0.40, determine the final velocity of each disk.

15.5.16 Two spheres are moving along parallel paths with the velocities shown in Fig. P-15-5.16. Sphere A has a weight of 300 N and a radius of 0.1 m, whereas sphere B has a weight of 90 N and a radius of

0.06 m. If $e = 0.6$, determine the velocity of each sphere directly after impact.

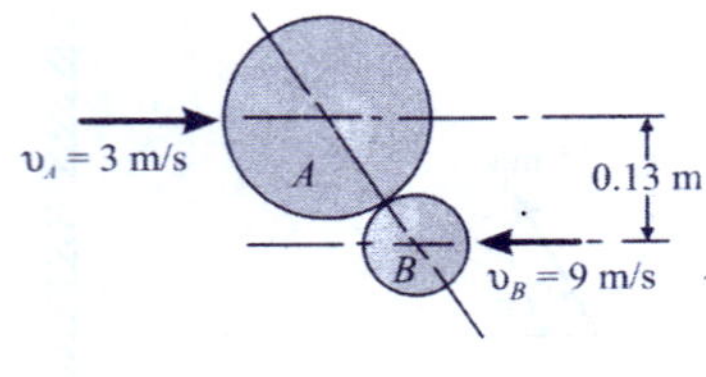

Figure P-15-5.16

15.5.17 Two bodies of masses m_1 and m_2 moving in opposite directions with velocities v_1 and v_2 collide. If e is the coefficient of restitution, show that the energy loss due to direct central impact is given by

$$\text{Loss in KE} = \frac{1}{2}\frac{m_1 m_2}{m_1 + m_2}(1 - e^2)(v_1 - v_2)^2$$

15-6 IMPULSE MOMENTUM IN PLANE MOTION

The impulse-momentum equations directly relating force, velocity, and time for any body are obtained by taking the time integrals of the fundamental equations $\mathbf{R} = m\bar{a}$ and $\Sigma\mathbf{M} = \dot{\mathbf{H}} = d\mathbf{H}/dt$. Recall that the moment equation is valid with respect to either the mass center G or a center A which is fixed in space. Thus, by multiplying each side of the fundamental equations by dt and integrating, we obtain

$$\int_{t_1}^{t_2} \mathbf{R}\, dt = \int_{t_1}^{t_2} m\bar{a}\, dt = \int_{\bar{v}_1}^{\bar{v}_2} m\, d\,\bar{v} = m\bar{v}_2 - m\bar{v}_1 \qquad (a)$$

$$\int_{t_1}^{t_2} \Sigma\mathbf{M}\, dt = \int_{\mathbf{H}_1}^{\mathbf{H}_2} d\mathbf{H} = \mathbf{H}_2 - \mathbf{H}_1 \qquad (b)$$

For the plane motion of bodies which are symmetrical about the plane of motion, we showed in Section 13-2 that the value of $\mathbf{H}$ about the mass center is $\overline{\mathbf{H}} = \bar{I}\omega$ while for rotation about a fixed axis A, the value of H is $\mathbf{H}_A = I_A\omega$. Hence, the scalar form of the impulse-momentum equations for the plane motion of symmetrical bodies becomes.

$$\left.\begin{aligned}
\int_{t_1}^{t_2} \Sigma X\, dt &= m\bar{v}_{2_x} - m\bar{v}_{1_x} = \mathrm{m}(\bar{v}_{2_x} - \bar{v}_{1_x}) \\
\int_{t_1}^{t_2} \Sigma Y\, dt &= m\bar{v}_{2_y} - m\bar{v}_{1_y} = \mathrm{m}(\bar{v}_{2_y} - \bar{v}_{1_y}) \\
\int_{t_1}^{t_2} \Sigma M\, dt &= \bar{I}\,\omega_2 - \bar{I}\,\omega_1 = \bar{I}(\omega_2 - \omega_1) \\
\int_{t_1}^{t_2} \Sigma M_A\, dt &= I_A\omega_2 - I_A\omega_1 = I_A(\omega_2 - \omega_1)
\end{aligned}\right\} \quad (15\text{-}6.1)$$

Resultant Angular Impulse

The expression $\int_{t_1}^{t_2} \Sigma M\, dt$ is called resultant angular impulse. In terms of the N-m-sec system of units, the dimensional unit of this expression obviously is N-m-sec. if $\Sigma\mathbf{M}$ is variable, the resultant angular impulse can be found only if $\Sigma\mathbf{M}$ can be expressed mathematically in terms of time t, or if a graph can be obtained showing its variation with time. In the latter case, the resultant angular impulse is equivalent to the area under the proper limits of the $\Sigma\mathbf{M}$ -t diagram.[2] In many cases, however, $\Sigma\mathbf{M}$ is constant and the resultant angular impulse is expressed simply as $\Sigma\mathbf{M}\,(t)$. It is important to note that resultant angular impulse is a vector, quantity having the same sense and inclination as $\Sigma\mathbf{M}$.

Angular Momentum

In plane motion of symmetrical bodies, the angular momentum **H** about either the mass center or a fixed axis is $I\omega$ and is also expressed in units of N-m-sec. The right-hand terms of the last two of Eqs. (15-6.1) represent the difference between the final and initial angular momentum and is called the change in angular momentum. The change in angular momentum depends only on the final and initial angular velocities since I is constant with respect to axes fixed in the body. Likewise, as in angular impulse, it is important to note that angular momentum is a vector quantity which, in plane motion, has the same sense and inclination as ω.

---

[2]Except for a change in symbols, this concept may be used to determine angular velocity and angular displacement analogous to the corresponding use of force-time curves as discussed.

The vector properties of angular impulse and angular momentum are usually of little importance in plane motion since they have a constant direction perpendicular to the plane of motion. Hence, in plane motion, they are usually written in scalar form as in Eqs. (15-6.1). Although scalar quantities are involved in Eqs. (15-6.1) the linear and angular terms have different units and directions. This fact discloses an important difference between work-energy and impulse-momentum methods. It was possible to apply the work-energy equations directly to a connected system of bodies having different motions because the terms involved were scalar quantities expressed in the same units which could be added arithmetically. In order to combine *linear* impulse-momentum terms with *angular* impulse-momentum terms, the linear terms must be transformed into angular terms by multiplying them by the appropriate moment arms. Usually this transformation is accomplished automatically if the impulse-momentum equations [Eqs. (15-6.1)] are applied to each separate body of the system. Notice also that all external forces on a system, including unknown constraint forces, have impulses which affect the momenta, whereas in work-energy the constraint forces may be neglected since they generally do no work on the system.

An alternate development which emphasizes the physical application of the impulse-momentum equations in plane motion is to start with the equivalence between applied forces and their dynamic effects as discussed in Section 13-2 and shown again here as Fig. 15-6.1. Multiplying both sides of these equivalents by *dt* gives Fig. 15-6.2 which relates the differential impulses to the differential momenta. This, after integrating over the time interval involved, then becomes Fig. 15-6.3 which is finally rearranged to give Fig. 15-6.4. Although the diagrams show only a single body, this body may also represent a system of bodies. Expressed in words, Fig. 15-6.4 states the impulse-momentum principle as follows:

Initial system momenta + Resultant ext. impulses

= Final system moments (15-6.2)

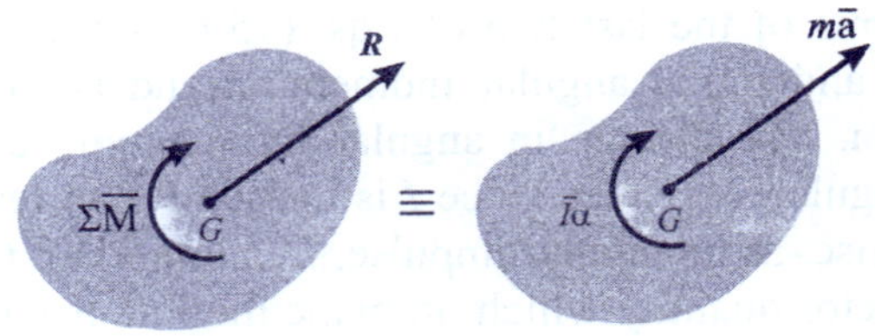

Figure 15-6.1 Equivalence of applied forces to their dynamic effects.

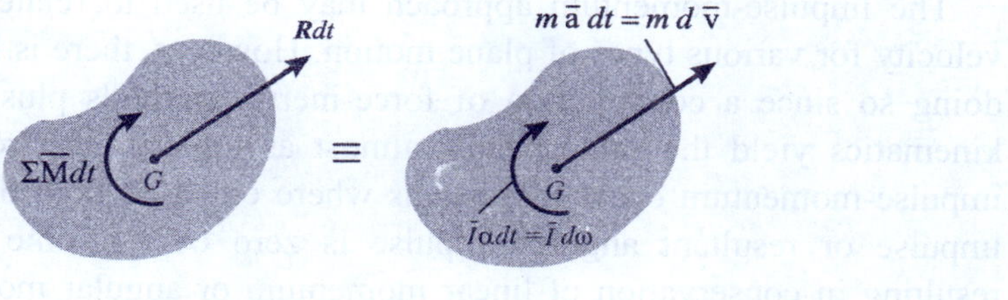

Figure 15-6.2 Equivalence of differential impulses to different momenta.

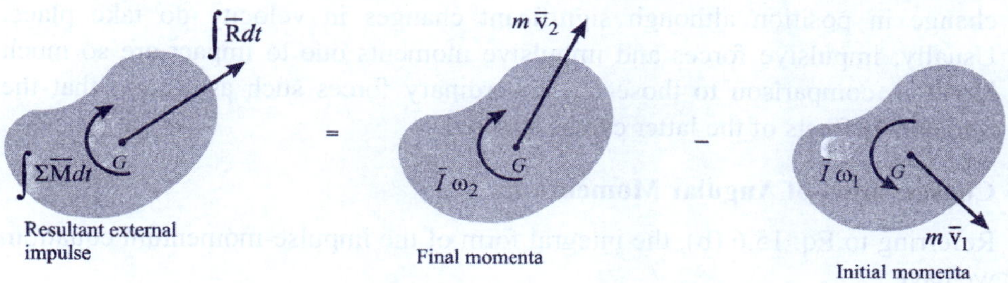

Figure 15-6.3 Effect of integrating differential impulse-momentum relations.

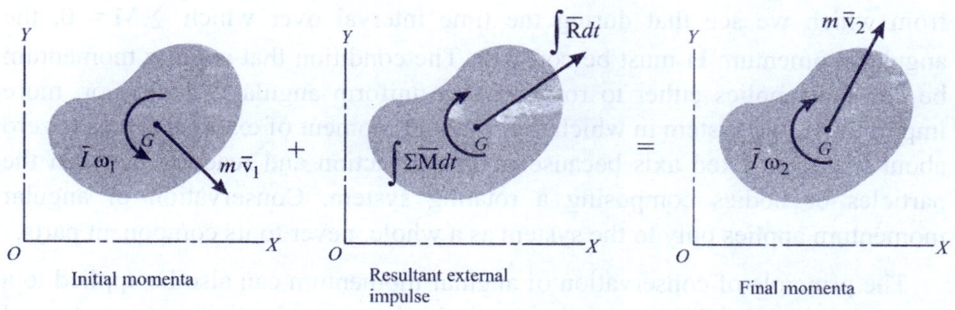

Figure 15-6.4 Rearranged impulse-momentum relations.

The advantage of this diagrammed approach is that angular momenta are seen to be couples, while the vector nature of the linear momenta is clearly shown by the different directions that $\bar{v}_1$ and $\bar{v}_2$ have after the elapsed time interval. Equations (15-6.1) may be obtained directly from Fig. 15-6.4 by summing and equating the X and Y components of the linear momenta and impulses, and by equating the moments of these vectors and the angular momentum couples about the same point in each part of the diagrammed equivalents. The coordinate axes may be fixed in space or may move with the mass center while maintaining fixed directions.

The impulse-momentum approach may be used to relate force, time, and velocity for various types of plane motion. However, there is little advantage in doing so since a combination of force-inertia methods plus the equations of kinematics yield the same results almost as rapidly. The real advantages of impulse-momentum occur in systems where either or both the resultant linear impulse or resultant angular impulse is zero over a time interval, thereby resulting in conservation of linear momentum or angular momentum, or both. Important also are situations involving impact where impulsive forces or impulsive moments act over so short a time interval that there is no significant change in position although significant changes in velocity do take place. Usually, impulsive forces and impulsive moments due to impact are so much larger in comparison to those due to ordinary forces such as weight that the impulsive effects of the latter can be ignored.

Conservation of Angular Momentum

Referring to Eq. 15.6 (b), the integral form of the impulse-momentum equation, we have

$$\int \Sigma \mathbf{M}\, dt = \int d\, \mathbf{H}$$

from which we see that during the time interval over which $\Sigma \mathbf{M} = 0$, the angular momentum **H** must be constant. The condition that angular momentum be constant applies either to rotation at a uniform angular velocity, or, more importantly, to a system in which the resultant moment of external forces is zero about the *same* fixed axis because of mutual action and reaction between the particles or bodies composing a rotating system. Conservation of angular momentum applies only to the system as a whole, never to its component parts.

The principle of conservation of angular momentum can also be applied to a translating body striking a rotating body. In this case, the linear momentum of the translating body must be multiplied by its moment arm about the axis of rotation to convert it into moment of linear momentum; i.e., its "angular momentum." To avoid the error of neglecting to convert linear momentum into moment of momentum, it is recommended the diagrammatic approach outlined in Fig. 15-6.4 and summarized by Eq. (15-6.2) be used in all cases. Recall that the momenta for each body consists of the linear momentum $m\bar{\mathbf{v}}$ acting through the mass center plus the moment couple $\bar{I}\omega$. This diagrammatic approach is very useful to obtain moment summations about a center that eliminates resultant angular impulse, thereby leading automatically to conservation of angular momentum.

ILLUSTRATIVE PROBLEMS

15-6.1 As indicated in Fig. 15-6.5, a 0.3 N bullet moving at 450 m/s strikes and pierces a 2 m rod weighing 45 N which is hanging motionless from a frictionless pivot at *A*. If the bullet emerges with a speed of 200 m/s, determine the maximum angle through which the rod will swing.

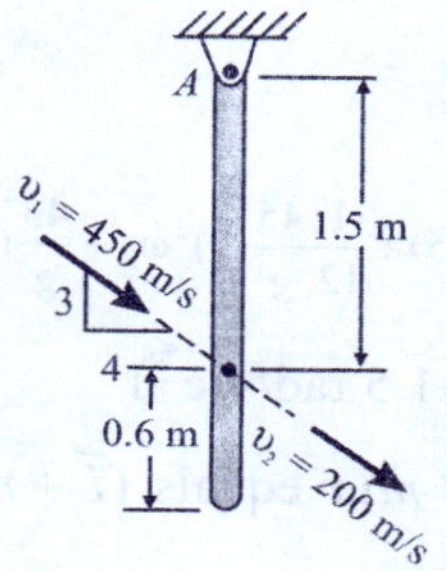

Figure 15-6.5

Solution

In Fig. 15-6.6, we equate the sum of the initial momenta and resultant external impulse to the final momenta. The initial linear momentum $m_b\mathbf{v}_l$ of the bullet is represented by its components. The only external impulse is that acting at the pivot *A* since those caused by mutual action and reaction of the bullet and rod cancel each other and the impulsive effect of the weights are negligibly small. The final momenta consist of the components of the final linear momentum $m_b\mathbf{v}_2$ of the bullet and the momenta $m\bar{\mathbf{v}}$ and $\bar{I}\omega_2$ acquired by the rod.

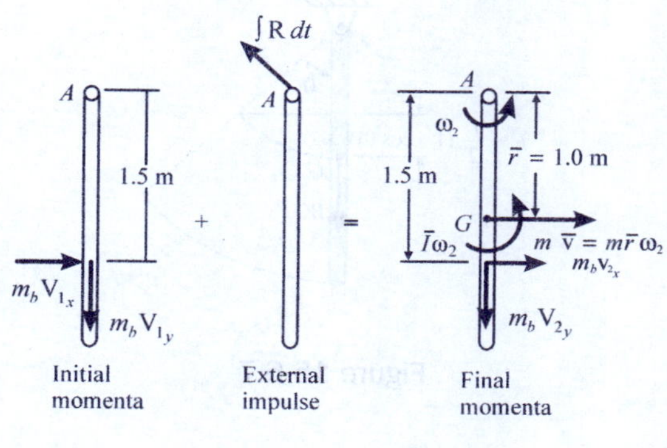

Figure 15-6.6

Equating moment summations about A eliminates the impulse at A and yields

$$m_b v_{1x} r = m_b v_{2x} r + \bar{I}\omega_2 + m\bar{v}\ \bar{\mathbf{r}}$$

which represents conservation of angular momentum about the fixed pivot A. Notice how the moment summation of the diagrammed impulse-momentum terms automatically converted linear momentum of the bullet into moment of linear momentum about A. Now substituting numerical data and noting that $\bar{I} = \frac{1}{12}\frac{W}{g}L^2$, we have

$$\frac{0.3}{g}(450)(0.8)(1.5) = \frac{0.3}{g}(200)(0.8)(1.5) + \frac{1}{12}\frac{45}{g}(2)^2\omega_2 + \frac{45}{g}(1.0\omega_2)(1.0)$$ from which

$$\omega_2 = 1.5 \text{ rad/sec} \circlearrowleft$$

Notice that the sum of $\bar{I}\omega$ and $m\bar{v}\bar{r}$ equals $(\bar{I} + m\bar{r}^2)\omega_2 = I_A\omega_2$.

We now apply the work-energy relation **RW** = **ΔKE** to determine the maximum swing angle of the rod. At the final dashed position in Fig. 15-6.7, the initial **KE** due to impact has been dissipated by the negative work done by gravity so that

$$\left[-Wh = \frac{1}{2}I_A(\omega^2 - \omega_0^2)\right] - \frac{WL}{2}(1 - \cos\,\theta) = \frac{1}{2}\left(\frac{W}{3g}L^2\right)(0 - \omega_2^2)$$

which reduces to

$$\cos\theta = 1 - \frac{L\omega_2^2}{3g} = 1 - \frac{2\,(1.5)^2}{3g} = 0.85 \qquad \theta = 31.8^{\circ} \qquad \textit{Ans.}$$

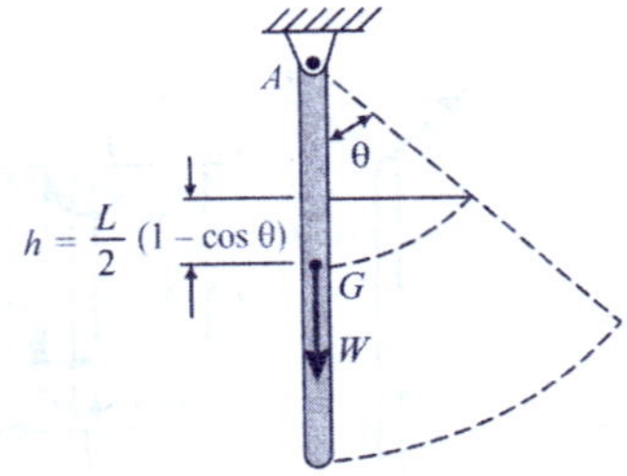

Figure 15-6.7

If the passage of the bullet occurs in a time assumed to be 0.002 sec, the average reaction R at A can be found by equating the X and Y components of linear momenta in Fig. 15-6.6. Thus,

$$\left[m_b v_{1x} - R_x \Delta t = m_b v_{2x} + m\bar{r}\omega_2\right]$$

$$\frac{0.3}{g}(450)(0.8) - R_x(0.002) = \frac{0.3}{g}(200)(0.8) + \frac{45}{g}(1.0)(1.5)$$

$$R_x = 0$$

$$\left[-m_b v_{1y} + R_y \Delta t = -m_b v_{2y}\right]$$

$$-\frac{0.3}{g}(450)(0.6) + R_y(0.002) = -\frac{0.3}{g}(200)(0.6)$$

$$R_y = 2293.58 \text{ N}\uparrow$$

$$\therefore \qquad R = \sqrt{R_x^2 + R_y^2} = 2293.58 \text{ N}\uparrow$$

Observe that while angular momentum about A is conserved, there is not conservation of linear momentum in the X or Y directions. If desired, the loss in KE is readily found by taking the difference between the initial and final **KE** as follows:

$$\text{Loss in KE} = \frac{1}{2}mv_1^2 - \frac{1}{2}mv_2^2 - \frac{1}{2}I_A\omega_2^2$$

$$= \frac{1}{2}\left(\frac{0.3}{g}\right)(450)^2 - \frac{1}{2}\left(\frac{0.3}{g}\right)(200)^2 - \frac{1}{2}\left(\frac{45}{3g}\right)(2)^2(1.5)^2$$

$$= 3096.33 - 611.62 - 6.88$$

$$= 2477.83 \text{ N-m}$$

Notice how small is the **KE** absorbed by the rod; the rest is largely dissipated by heat.

15-6.2 A rod 1.5 m long and weighing 40 N is rotating in a vertical plane about one end. At the instant the rod is in the vertical position as shown in Fig. 15-6.8, it has a clockwise angular velocity ω_1 = 2 rad/sec when it is hit by a 2 N ball moving horizontally at 60 m/s. If the coefficient of restitution is e = 0.6, determine the angular velocity of the rod directly after impact. Assuming the impact lasts for 0.005 sec, find the average impact force P between the ball and the rod.

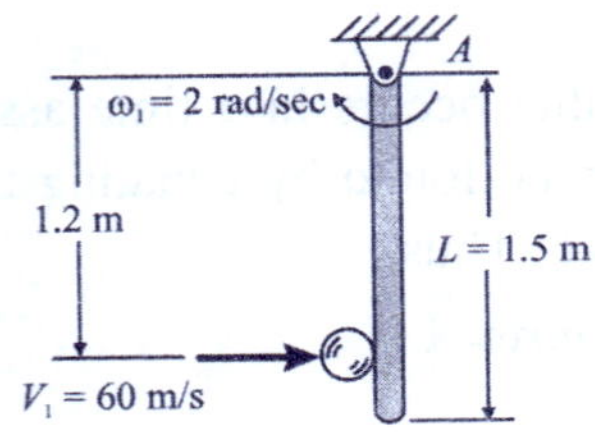

Figure 15-6.8

Solution

This problem illustrates direct eccentric impact in which we assume the ball and the rod rebound in opposite directions as shown in Fig. 15-6.9 by the diagrammed impulse-momentum equation. Noting that $\bar{I}\omega + m\bar{v}\bar{r} = I_A\omega$ as shown in the previous problem, we equate moment summations about A to eliminate the impulse at A, thereby giving the following equation of conservation of angular momentum:

$$m v_1 r - I_A \omega_1 = I_A \omega_2 - m v_2 r$$

Substituting numerical data and noting that

$$I_A = \frac{1}{3}\frac{W}{g}L^2 = \frac{40\,(1.5)^2}{3g} = \frac{30}{g}\ \text{N-m-sec}^2$$

we have

$$\frac{2}{g}(60)(1.2) - \frac{30}{g}(2) = \frac{30}{g}\omega_2 - \frac{2}{g}(\upsilon_2)(1.2)$$

which reduces to

$$84 = 30\,\omega_2 - 2.4\,\upsilon_2 \qquad (a)$$

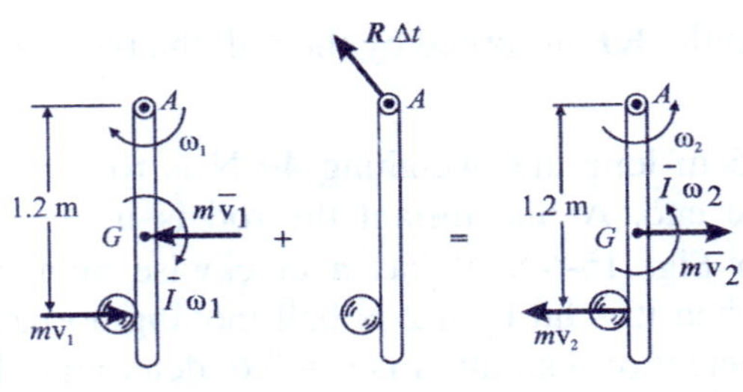

Figure 15-6.9

Another equation relating ω_2 and v_2 is available from the definition of the coefficient of restitution, Eq. (15-5.1). Along the line of impact, the original velocity of the impact point on the rod is $r\omega = 1.2(2) = 2.4$ m/s leftward and its final velocity is $1.2\omega_2$ rightward. Hence, taking the original rightward velocity of the ball as the positive sense of velocity, we obtain

$$\left[e(v_1 - v_2) = v_2' - v_1'\right] \qquad 0.6[60 - (-2.4)] = 1.2\,\omega_2 - (-v_2)$$

which reduces to

$$37.44 = 1.2\omega_2 + v_2 \qquad (b)$$

Solving Eqs. (*a*) and (*b*) simultaneously, we obtain

$$\omega_2 = 5.28 \text{ rad/sec} \;;\; v_2 = 31.08 \text{ m/s} \leftarrow$$

The positive results confirm the assumed directions to be correct.

The contact force P can be found either by applying angular impulse momentum to the rod or applying linear impulse momentum to the ball. Using both methods will serve as a check. For the rod, we obtain

$$[+\uparrow \Sigma M_A \Delta t = I_A(\omega - \omega_o)] \qquad 1.2P(0.005) = \frac{30}{g}[5.28 - (-2)]$$

$$P = 3714.36 \text{ N} \qquad \textbf{Ans.}$$

while for the ball, we have

$$[\overset{+}{\rightarrow} \Sigma X \Delta t = m(v_2 - v_1)] \qquad -P(0.005) = \frac{2}{g}(-31.08 - 60)$$

$$P = 3713.8 \text{ N} \qquad \textbf{Check}$$

15.6.3 A dumbbell has a mass m and a centroidal radius of gyration k about an axis perpendicular to its axis of symmetry. At the instant before one end of the dumbbell strikes the corner of a table as in Fig. 15-6.10, its angular velocity is zero but its mass center has a downward velocity $\bar{v}_o$. Determine its angular velocity directly after (*a*) plastic impact; (*b*) partially elastic impact with a coefficient of restitution *e*.

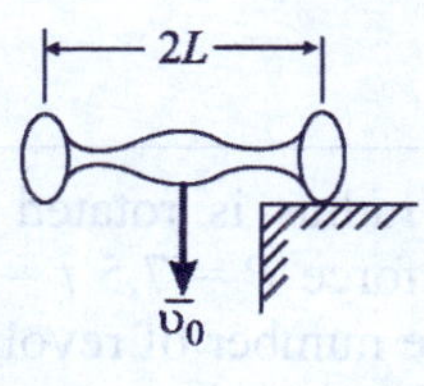

Figure 15-6.10

Solution: Part (a)

Applying the impulse-momentum equation shown in Fig. 15-6.11 and equating moments about A, we have

$$m\bar{v}_o L = m\bar{v}L + mk^2\omega \qquad (a)$$

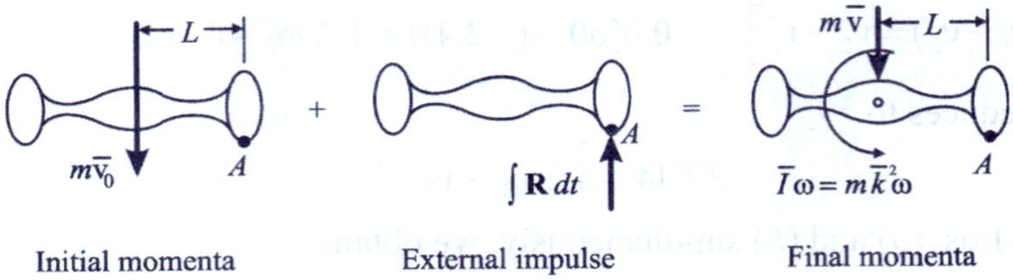

Figure 15-6.11

After plastic impact, the body is instantaneously rotating about A and the kinematic relation between $\bar{v}$ and ω is

$$[v = r\omega] \qquad \bar{v} = L\omega \qquad (b)$$

Solving Eqs. (a) and (b) then results in

$$\omega = \frac{\bar{v}_o L}{L^2 + k^2} \circlearrowright \qquad \textit{Ans.}$$

Part (b)

With partially elastic impact, end A rebounds upward with a velocity $e\bar{v}_o$. Using kinematics, the velocity $\bar{v}$ of the mass center now is

$$[v_G = v_A \twoheadrightarrow (v_{G/A} = rw)] \qquad \bar{v} = e\bar{v}_o \uparrow + L\omega \downarrow = (L\omega - e\bar{v}_o) \downarrow$$

Substituting this value of $\bar{v}$ in Eq. (a) above now results in

$$\omega = \frac{\bar{v}_o L\,(1+e)}{L^2 + k^2} \circlearrowright \qquad \textit{Ans.}$$

Observe that for perfectly elastic impact (e = 1), the final value of w is twice that for plastic impact.

PROBLEMS

15-6.4 A pulley of 0.8 m radius is rotated about its centroidal axis by a tangentially applied force $P = 7.5\,t - 1.5\,t^2$ where P is in N and t is 20 N-m-sec^2, find the number of revolutions through which the pulley rotates from rest before starting to reverse its direction. Also find its angular velocity when it returns to its original position.

15-6.5 Repeat the previous problem but change the radius of the pulley to 1.0 m and the value of the tangentially applied force to $P = 2t^2 - 1.5\,t^3$ where P is in N and t is in seconds.

15.6.6 A wheel 2 m in diameter is rotated about its centroidal axis by a tangentially applied force $P = 12 + 2t^2$ where P is in N and t is in seconds. The rotation is resisted by a brake (f 0.40) which exerts a constant normal force of 450 N on the rim. If P acts for 6 sec and is then removed, through how many revolutions will the wheel turn before stopping? Assume the mass moment of inertia is 2 N-m-sec^2.

15.6.7 As a golf ball of radius r hits a green as in Fig. P-15-6.7, it has a horizontal component of velocity v_h and a backspin angular velocity ω. Determine the relation between v_h and ω that will cause the golf ball to stop dead upon hitting the green. Assume plastic impact.

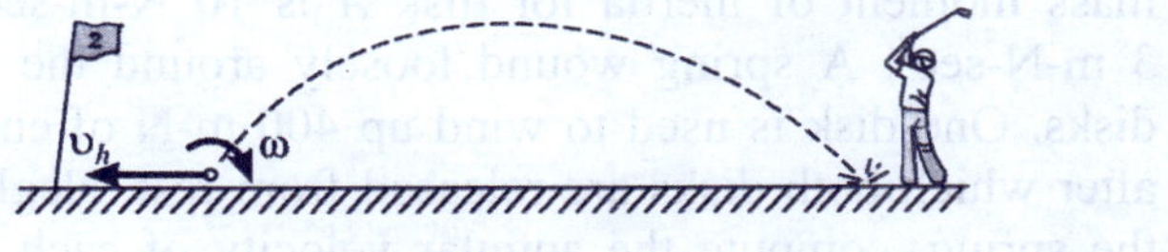

Figure P-15-6.7

15-6.8 A 10 N weight is attached to one end of a cord as shown in Fig. P-15-8.8. At the given position, the weight is rotating about the vertical axis of the hollow support in a circular path of 1.5 m radius at 9 rad/sec. If the cord is slowly drawn into the support so that the radius of rotation is 1.0 m, determine the new angular velocity of the weight and the tension in the cord. Further, determine and explain the increase in kinetic energy.

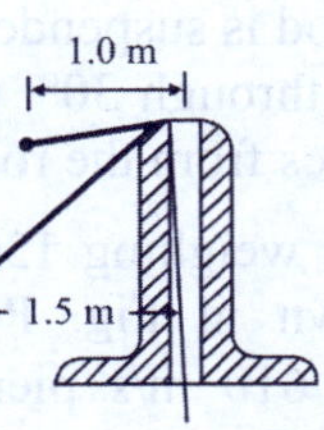

Figure P-15-6.8

15.6.9 When the speed becomes critical, the balls in a certain type of governor are released by a catch, and they move to the positions shown by the dashed outlines in Fig. P-15-6.9. Each ball weighs 90 N and $\bar{I}$ for the rod is 4 N-m-sec^2. If the speed of the assembly is 20 rad/sec when the governor operates, determine the final speed of the system. What would be the final speed if only one ball moved to the dashed position?

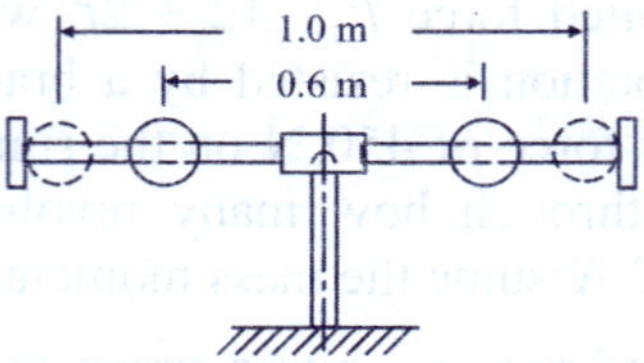

Figure P-15-6.9

15-6.10 Two disks A and B are mounted loosely on a horizontal shaft. The mass moment of inertia for disk A is 10 N-m-sec^2, and for B it is 3 m-N-sec^2. A spring wound loosely around the shaft connects the disks. One disk is used to wind up 400 m-N of energy in the spring, after which both disks are released from rest. Neglecting the mass of the spring, compute the angular velocity of each disk at the instant when the spring has released all its stored energy to the disks.

15-6.11 A uniform bar, 1.0 m long and weighing 90 N is hanging motionless from a horizontal axis at its upper end. It is struck at 0.6 m below this axis by a 10 N ball of putty moving horizontally at 20 m/s. If all the putty clings to the bar, compute the maximum angle through which the bar will swing and the loss in kinetic energy.

15-6.12 A uniform rod, 2 m long and weighing 45 N, is suspended vertically from a horizontal axis at its upper end. A 30 g bullet is fired at 300 m/s along a horizontal line 1.5 m below and perpendicular to the axis from which the rod is suspended. The bullet pierces the rod which is observed to swing through 30°. Compute the velocity of the bullet directly after it emerges from the rod.

15-6.13 A uniform 1.5 m bar weighing 150 N is hanging motionless from a 0.6 m cord as shown in Fig. P-15-6.13. A 30 g bullet moving horizontally at v_0 = 610 m/s pierces the bar and emerges with a velocity of 250 m/s. Determine the velocity of end B of the bar directly after impact.

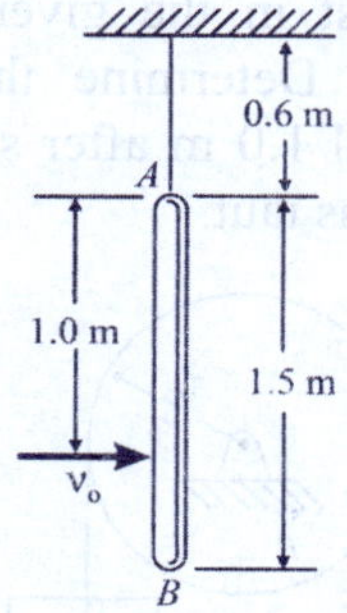

Figure P-15-6.13

15.6.14 Determine the velocity of end A of the previous problem if the path of the bullet is downward at a slope of 3 vertical to 4 horizontal.

15.6.15 A friction drive consists of two disks keyed to parallel shafts. They can be separated or brought into contact by moving disk B in the horizontal slot as shown in Fig. P-15-6.15. Initially, they are separated and rotate in opposite directions at $\omega_A = 6$ rad/sec and $\omega_B = 30$ rad/sec. Their radii and moments of inertia are $r_4 = 0.3$ m, $r_B = 0.15$ m, $I_A = 1.6$ m-N-see^2, and $I_B = 0.6$ m-N-sec^2. Find the angular velocity of the disks after they are brought into contact and slipping stops.

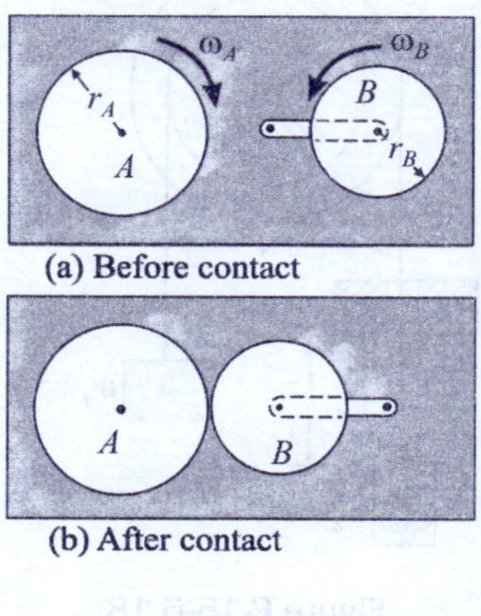

Figure P-15-6.15

15-6.16 Solve the previous problem if both disks are initially rotating clockwise at $\omega_A = 8$ rad/sec and $\omega_B = 30$ rad/sec.

15.6.17 In Fig. P-15-6.17, block A weighs 450 N and the homogeneous cylinder B weighs 900 N. An inextensible cord passes through a hole in the block and is wrapped around the cylinder which is at rest. Block

A is released from rest in the given position and drops onto the weightless stopper C. Determine the velocity of A after it has descended an additional 1.0 m after striking C. Neglect any rebound and assume cord remains taut.

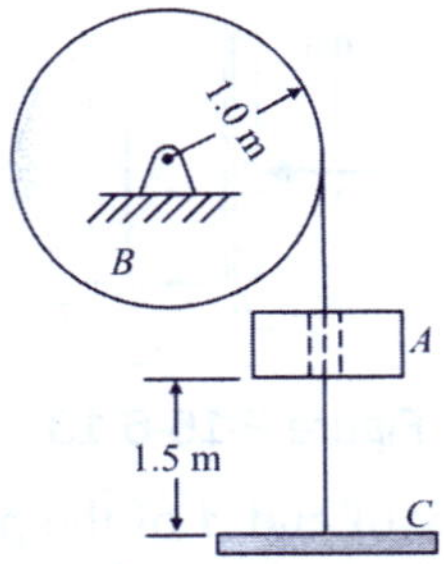

Figure P-15-6.17

15-6.18 The system shown in Fig. P-15-6.18 starts from rest in the given position. After rising h = 1.5 m, body B picks up body C. How much higher will B move?

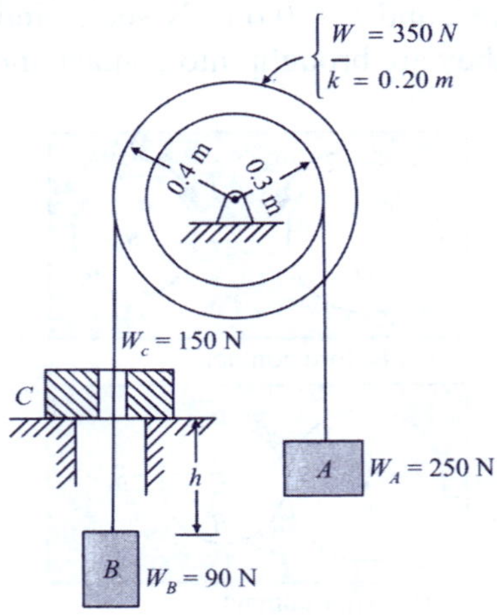

Figure P-15-6.18

15-6.19 The uniform sphere in Fig. P-15-6.19 rolls without slipping down the incline. Its center has a velocity $\mathbf{v}_o$ just before it touches the level floor. Determine the velocity of the center of the sphere as it starts rolling without slipping on the floor. Assume there is no rebound.

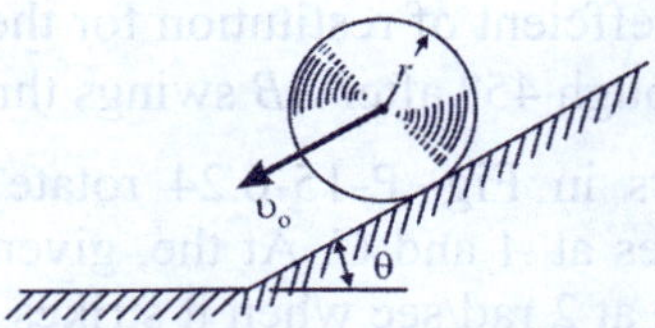

Figure P-15-6.19

15-6.20 A uniform sphere 0.8 m in diameter and weighing 45 N is rolling without slipping on a level floor when it hits a step 0.15 m high as shown in Fig. P45-6.20. Determine the smallest velocity $\bar{v}$ of the sphere so that it will roll over the step, assuming plastic impact.

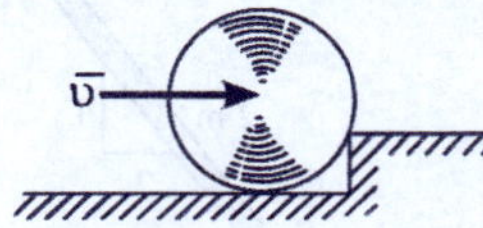

Figure P-15-6.20

15-6.21 As shown in Fig. P-15-6.21, a box weighing W N is moving with a velocity v_o along a level floor when it hits a cleat at A. Determine the angular velocity of the box directly after impact if (a) $e = 0$ and (b) for any value of e. Ignore the height of the cleat.

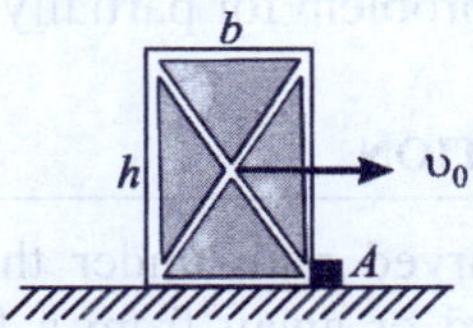

Figure P-15-6.21

15-6.22 In Fig. P-15-6.22, the uniform rods AB and CD are free to swing in a vertical plane about A and C. Each rod is 1.5 m long and weighs 150N. Rod AB is released from rest at $\theta = 90°$ and a hard rubber knob at its end strikes CD. If $e = 0.60$, through what angle will CD swing, starting from rest?

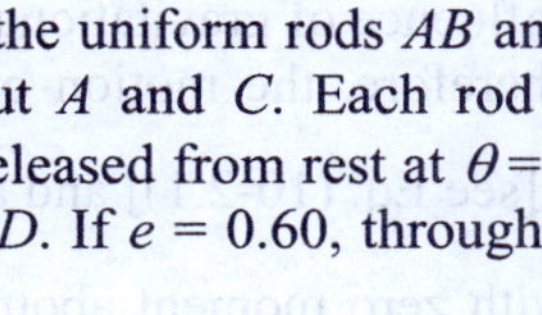

Figure P-15-6.22

15-6.23 Determine the coeffcient of restitution for the system of Prob. 15-6.22 if *CD* swings through 45° after *AB* swings through $\theta = 60°$.

15-6.24 The uniform bars in Fig. P-15-6.24 rotate on a horizontal surface about vertical axes at *A* and *C*. At the, given position, *AB* is rotating, counterclockwise at 2 rad/sec when it strikes *CD* which is at rest. Each bar is 2 m long and weighs 150 N. Assuming plastic impact, determine the angular velocity of each bar directly after impact and the loss in **KE**.

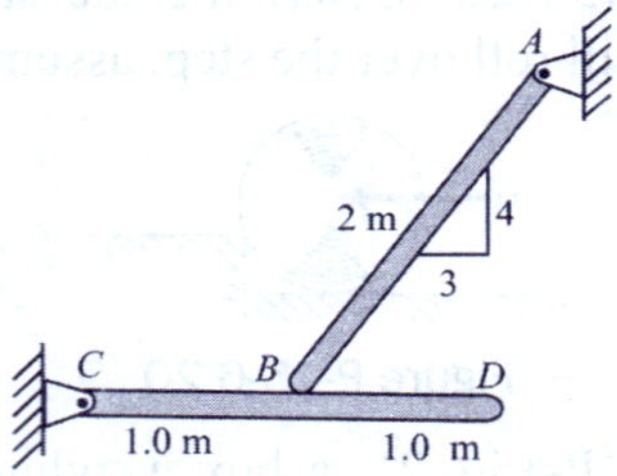

Figure P-15-6.24

15-6.25 Solve the preceding problem for partially elastic impact with $e = 0.60$.

*15-7 SATELLITE MOTION

When a body moves in a curved path under the action of a force directed through its mass center toward or away from a fixed point, we have what is known as central force motion. This is the condition which applies to satellites orbiting the earth under the influence of gravitational attraction.[3] With respect to the center *O* of the earth, therefore, the motion of satellite is governed by a gravitational pull $F = \dfrac{GMm}{r^2}$ [see Eq. (10-2.1)] and a moment $\mathbf{M}_o = 0$. Hence, as discussed in Section 15-6, with zero moment about *O*, the angular momentum $\mathbf{H}_o$ about *O* must be constant.

Refer now to the orbiting satellite shown in Fig. 15-7.1. The angular momentum at any instant is the moment about O of the linear momentum m v and hence, in terms of the position vector **r** from *O*,

$$\mathbf{H}_o = \mathbf{r} \times m\mathbf{v}$$

[3]By using appropriate gravitational pulls, the discussion applies also to satellites orbiting other planets. For an extended discussion of satellite motion, see W. T. Thomson, *Introduction to Space Dynamics*, John Wiley & Sons (1961).

from which we conclude that, since $\mathbf{H}_o$ is constant in magnitude and direction, the satellite moves in a fixed plane, defined by $\mathbf{r} \times m\mathbf{v}$, which always is perpendicular to the constant direction of $\mathbf{H}_o$.

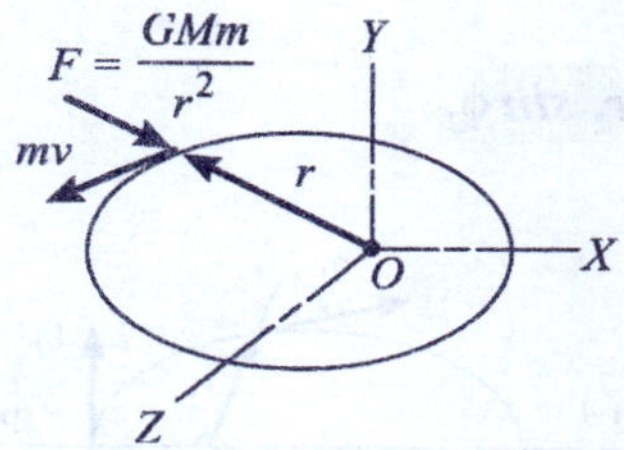

Figure 15-7.1 Satellite orbiting about the center O of the earth.

We now discuss how a combination of angular impulse-momentum and work-energy principles will solve various problems associated with satellite motion.

Assume that the path of an orbiting satellite is an ellipse as shown in Fig. 15-7.2. Other possible paths will be discussed later. The velocity of the satellite varies with changes in its position vector from O because of the work done on it by gravitational attraction. From Eq. (10-2.1) we recall that the magnitude of the gravitational force on the satellite is $F = GMm/r^2$. Noting that $\mathbf{F}$ is always directed toward O and hence opposite to $\mathbf{r}$, the resultant work done on the satellite as it moves from any position 1 to any other position 2 is

$$RW = \int F.\,dr = -\int F\,dr = -\int_{r_1}^{r_2} \frac{GMm}{r^2}\,dr = \mathrm{GMm}\left[\frac{1}{\mathrm{r}}\right]_{r_1}^{r_2}$$

which may be rewritten in either of the following forms:

$$RW = GMm\left(\frac{1}{\mathrm{r}_2} - \frac{1}{r_1}\right) = \frac{GMm}{r_2}\left(1 - \frac{r_2}{r_1}\right)$$

On equating the change in kinetic energy to this resultant work, we obtain

$$\frac{1}{2}m(v_2^2 - v_1^2) = GMm\left(\frac{1}{r_2} - \frac{1}{r_1}\right) = \frac{GMm}{r_2}\left(1 - \frac{r_2}{r_1}\right)$$

or

$$v_2^2 - v_1^2 = 2GM\left(\frac{1}{r_2} - \frac{1}{r_1}\right) = \frac{2GM}{r_2}\left(1 - \frac{r_2}{r_1}\right) \tag{15-7.1}$$

Another relation between these velocities is obtained by recalling that the angular momentum is constant, thereby giving

$$mv_1r_1 \sin \phi_1 = mv_2r_2 \sin \phi_2$$

or

$$v_1r_1 \sin \phi_1 = v_2r_2 \sin \phi_2 \qquad (15\text{-}7.2)$$

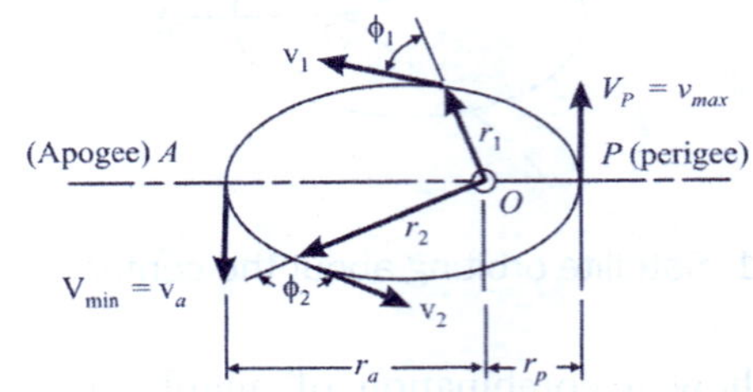

Figure 15-7.2 Velocity variations of a satellite in an elliptical orbit.

Observe also in Fig. 15-7.2 that sin $\phi = 1$ at both the apogee A, the position furthest from O where the velocity is a minimum, and at the perigee P, the position closest to O where the velocity is a maximum.

With some additional information, Eqs. (15-7.1) and (15-7.2) are sufficient to solve most problems of satellite motion. The value of GM is readily found by noting that the gravitational pull on a body at the earth's surface is its weight W. Denoting the radius of the earth by R, we then have

$$W = mg = \frac{GMm}{R^2}$$

whence substituting $R = 637.56 \times 10^4$ m we obtain

$$GM = gR^2 = 9.81(637.56 \times 10^4)^2$$

$$= 3.987 \times 10^{14} \text{ m}^3/\text{sec}^2 \qquad (15\text{-}7.3)$$

In order for a satellite to move in a circular orbit of radius r at the constant velocity υ, the gravitational pull at every position will be balanced by the centrifugal inertia force, resulting in the following relation:

$$\frac{GMm}{r^2} = m\frac{v^2}{r} \quad \text{or} \quad v^2 = \frac{GM}{r} \qquad (15\text{-}7.4)$$

For smaller or larger velocities, the various orbits possible as shown in Fig. 15-7.3. It is evident that multiplying a circular orbiting velocity by $\sqrt{2}$ will determine the velocity v_e to escape from the earth's gravitational field. At the surface of the earth, this value is

$$v_e^2 = \frac{2GM}{R} = \frac{2(3.987\times10^{14})}{637.56\times10^4} \quad \text{giving} \quad v_e = 11186.16 \text{ m/s}$$

while at larger radii where the gravitational pull is smaller, the escape velocity will decrease.

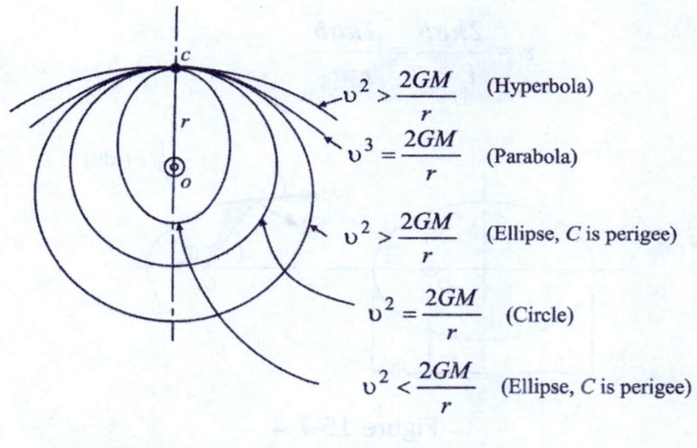

Figure 15-7.3 Various possible orbital paths.

We conclude our discussion of satellite motion by showing how to determine the major and minor semiaxes of an elliptical orbit and how they are related to the period of the orbit; i.e., the time τ required to complete an orbit. If we refer to Fig. 15-7.4 in which the earth is at one focus of the elliptical path, we find that analytic geometry expresses the major and minor semiaxes, a and b, in terms of the apogee and perigee distances, r_a and r_p, as follows:

$$\boldsymbol{a = \frac{1}{2}(r_a + r_p) \text{ and } b = \sqrt{r_a r_p}} \tag{15-7.5}$$

As a preliminary to determining the period of the orbit, we first define *areal velocity* as the time rate of the area swept out by the position vector. Thus, in Fig. 15-7.4, the shaded area dA generated by r in the time dt is $\frac{1}{2}r(r\,d\theta)$ from which

$$\text{areal velocity} = \frac{dA}{dt} = \frac{1}{2}r\left(r\frac{d\theta}{dt}\right) \tag{a}$$

Recalling from Eq. (15-7.2) that $v \sin\phi$ is a constant and that $v\sin\phi$ is the transverse component of velocity $v_\theta = r\dot{\theta}$, we see that Eq. (*a*) can be rewritten as

$$\frac{dA}{dt} = \frac{1}{2} rv \sin\phi \tag{b}$$

whence separating the variables in Eq. (*b*) and integrating, the period is found to be

$$\tau = \frac{2A}{rv \sin\phi}$$

which, in terms of the area πab of the ellipse and the identities $rv\sin\phi = r_a v_a = r_p v_p$, becomes

$$\tau = \frac{2\pi ab}{r_a v_a} = \frac{2\pi ab}{r_p v_p} \tag{15-7.6}$$

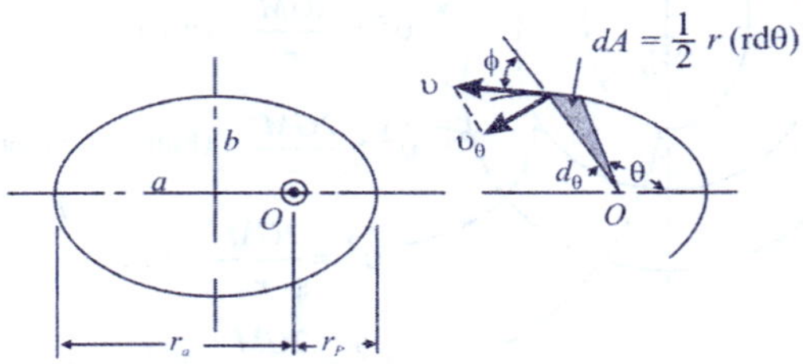

Figure 15-7.4

It is remarkable that from only astronomical observations, Johann Kepler (1571 –1630) was able to deduce his famous laws of planetary motion, two of which we have verified by Newtonian mechanics. Kepler's laws may be stated as:

1. Each planet describes an ellipse about the sun as a focus.
2. The position vector from the sun to a planet sweeps out equal areas in equal times.
3. For each planet, the square of its periodic time is proportional to the cube of the major semiaxis of its orbit. [The precise relation is $(\tau/2\pi)^2 = a^3/GM$ whose derivation constitutes Prob. 15-7.7.]

ILLUSTRATIVE PROBLEMS

15-7.1 The distances from the center of the earth to the apogee and perigee of an orbiting satellite are respectively r_a and r_p. Show that the corresponding velocities υ_a and υ_p at apogee and perigee are given by

$$v_a^2 = \frac{2GM}{r_a + r_p}\left(\frac{r_p}{r_a}\right) \quad \text{and} \quad v_p^2 = \frac{2GM}{r_a + r_p}\left(\frac{r_a}{r_p}\right)$$

Compute v_a and v_p in km/sec if the maximum and minimum altitudes of the orbit are respectively 34.45×10^5m and 38.64×10^4 m, and the radius of the earth is $R = 63.75 \times 10^5$m.

Solution

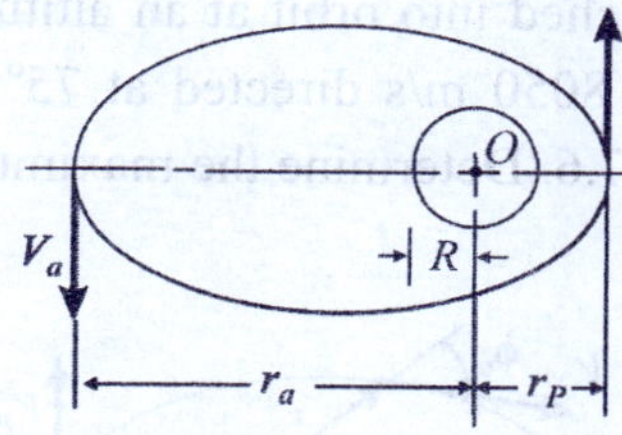

Figure 15-7.5

At apogee and perigee, the velocities are perpendicular to their position vectors as shown in Fig. 15-7.5. Applying conservation of angular momentum, we have

$$mv_a r_a = mv_p r_p \qquad \text{or} \qquad v_p = \frac{r_a}{r_p} v_a \tag{a}$$

From Eq. (15-7.1) we obtain

$$v_a^2 - v_p^2 = 2GM\left(\frac{1}{r_a} - \frac{1}{r_p}\right) = \frac{2GM}{r_a}\left(1 - \frac{r_a}{r_p}\right)$$

which, on replacing v_p by its value from Eq. (a), becomes

$$v_a^2\left(1 - \frac{r_a^2}{r_p^2}\right) = v_a\left(1 + \frac{r_a}{r_p}\right)\left(1 - \frac{r_a}{r_p}\right) = \frac{2GM}{r_a}\left(1 - \frac{r_a}{r_p}\right)$$

whence

$$v_a^2 = \frac{2GM}{r_a + r_p}\left(\frac{r_p}{r_a}\right)$$

and similarly, (15-7.7)

$$v_p^2 = \frac{2GM}{r_a + r_p}\left(\frac{r_a}{r_p}\right)$$

Noting from Eq. (15-7.3) that $GM = 4 \times 10^{14}$ m^3/sec^2 and that the apogee and perigee distances are $r_a = 6375000 + 3445000 = 9820000$ m and $r_p = 6375000 + 386400 = 6761400$ m, we obtain

$$v_a^2 = \frac{2(4\times10^{14})}{9820000+6761400}\times\left(\frac{6761400}{9820000}\right) = 33.22\times10^6, \qquad v_a = 5763 \text{ m/s.}$$

$$v_p^2 = \frac{2(4\times10^{14})}{9820000+6761400}\times\left(\frac{9820000}{6761400}\right) = 70.07\times10^6, \qquad v_a = 8370 \text{ m/s.}$$

15.7.2 A satellite is launched into orbit at an altitude of 2801400 m when it has a velocity of 8050 m/s directed at 75° with the local vertical as shown in Fig. 15-7.6. Determine the maximum and minimum altitudes of the orbit.

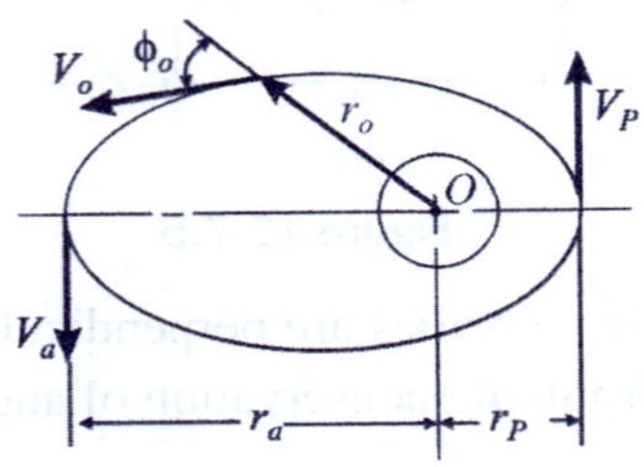

Figure 15-7.6

Solution

Using Eqs. (15-7.2) and (15-7.1), we have

$$r_p v_p = r_a v_a = r_o v_o \sin\phi_o \tag{a}$$

and

$$v_a^2 - v_o^2 = 2GM\left(\frac{1}{r_a} - \frac{1}{r_o}\right) = \frac{2GM}{r_o}\left(\frac{r_o}{r_a} - 1\right) \tag{b}$$

After replacing v_a by its value from Eq. (a), Eq. (b) becomes

$$\frac{r_o^2}{r_a^2}\, v_o^2 \sin\phi_o - v_o^2 = v_o^2\left(\frac{r_o^2}{r_a^2}\sin^2\phi_o - 1\right) = \frac{2GM}{r_o}\left(\frac{r_o}{r_a} - 1\right)$$

which, when divided by v_o^2, may be rearranged into the following quadratic equation in terms of (r_o/r_a):

$$\sin^2 \phi_o \left(\frac{r_o}{r_a}\right)^2 - \frac{2GM}{r_o v_o^2}\left(\frac{r_o}{r_a}\right) + \frac{2GM}{r_o v_o^2} - 1 = 0 \qquad (c)$$

Using the given data, we have

$$\frac{2GM}{r_o v_o^2} = \frac{2(4\times10^{14})}{(6375000+2801400)\times(8050)^2} = 1.343$$

and $$\sin^2 75^\circ = 0.933$$

which is substituted into Eq. (c) to yield

$$0.933\left(\frac{r_o}{r_a}\right)^2 - 1.343\left(\frac{r_o}{r_a}\right) + 0.343 = 0$$

of which the two roots are found to be

$$\frac{r_o}{r_a} = 0.332 \quad \text{and} \quad \frac{r_o}{r_a} = 1.107$$

The corresponding values of r_a are
(with $r_a = 6375000 + 2801400 = 91.77 \times 10^5$ meters)

$$r_a = \frac{9177000}{0.332} = 27.64\times10^6 \quad \text{and} \quad r_a = \frac{9177000}{1.107} = 83\times10^5 \text{ meters}$$

of which, in view of Eq. (a), the larger value is the apogee distance while the smaller value is the perigee distance. The corresponding maximum and minimum altitudes of the orbit are

Max. altitude = $27.64 \times 10^6 - 6375000 = 21.26 \times 10^6$ meters

Min. altitude = $8300000 - 6375000 = 1925000$ meters

PROBLEMS

15-7.3 Determine the period of a satellite moving at 6708.33 m/s in a circular orbit about the earth. If air resistance is neglected, what is the minimum time in which a satellite can complete a circular orbit about the earth?

15.7.4 Determine the velocity and altitude of a communications satellite that will occupy a fixed position relative to the earth's equator.

15-7.5 A satellite is in a circular polar orbit at an altitude of 547400 meters. Find the change in longitude of its orbit during each pass as observed at the equator. What result would be observed at an latitude of 40° North?

15-7.6 Compute the time required for the command module of an Apollo mission to complete one orbit of the moon at a constant altitude of 128800 meters. The radius of the moon is 17.38×10^5 m and its mass is 0.0123 that of the earth.

15-7.7 Derive Kepler's third law: $\left(\frac{\tau}{2\pi}\right)^2 = \frac{a^3}{GM}$.

15-7.8 Compute the maximum and minimum velocities of a satellite whose period is 7200 sec if its altitude at perigee is known to be 386400 m.

15-7.9 A lunar excursion module (LEM) is launched at an altitude of 161000 m with a velocity v parallel to the moon's surface. Determine the minimum value of v so that the LEM will go into orbit.

15-7.10 A satellite orbiting the earth has maximum and minimum altitudes of 1030400 meters and 386400 meters. If it is launched at an altitude of 547400 m, determine its launch velocity v and its angle ϕ with the local vertical. What is the period of the satellite?

15-7.11 Show that the kinetic energy required to change a circular orbit of radius r to a circular orbit of radius $3r$ is $GMm/3r$

15-7.12 A satellite whose earth weight is 4299 N is circling the earth at a constant altitude of 708400 m. Its thruster engines can exert a force of 445 N. In order to place the satellite into a new orbit at a constant altitude of 32.9×10^5 m, for what time intervals should its engines be fired? Neglect mass of fuel consumed and the change in altitude during thrust.

15-7.13 As it orbits the moon, the command module of an Apollo mission is moving at 1162.77 m/s at its apogee altitude of 96.6×10^4 m. Determine the velocity change at its perigee to put it into a circular orbit of 96600 m above the moon. The radius of the moon is 29×10^5 m and its mass is 0.0123 that of the earth.

15-7.14 As shown in Fig. P-15-7.14, a space shuttle at an altitude of 16.74×10^5 m has a velocity $v_A = 7602$ m/s directed at $\phi = 70°$. At B where its altitude is 24.8×10^5 m, a course correction θ is made so that the space shuttle at its apogee C will be tangent to the orbit of a skylab circling the earth at an altitude of 57×10^5 m. Determine the course correction angle θ.

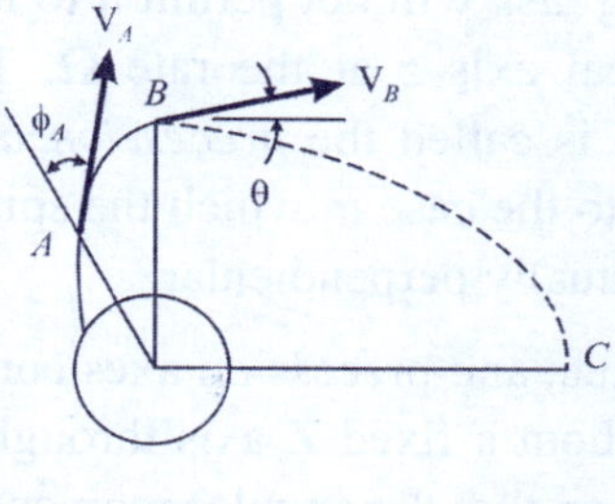

Figure P-15-7.14

*15-8 INTRODUCTION TO GYROSCOPIC ACTION

We now apply angular impulse-momentum principles to some elementary but important aspects of rigid-body rotation about a fixed point as contrasted to fixed-axis rotation. A more general discussion of rotation about a fixed point appears in the next chapter on spatial dynamics. Here we concentrate on the gyroscopic effect that is present whenever a body that rotates about one axis also rotates about another axis that is perpendicular to the first one. This effect appears, for example, in the wheels of an automobile which is rounding a curve or in the rotor of a jet engine of an airplane making a turn.

The usual concept of a gyroscope is a solid of revolution that rotates rapidly about its axis of summetry. Let us consider the case in which the gyroscope is a disk on an axle supported at point O by a ball-and-socket joint as in Fig. 15-8.1. The disk is spinning with a large angular velocity ω about its axis of symmetry. This axis will henceforth be called the spin axis. The angular momentum of the disk about this axis is $\mathbf{H}_O = \bar{I}\omega\hat{\mathbf{i}}$.

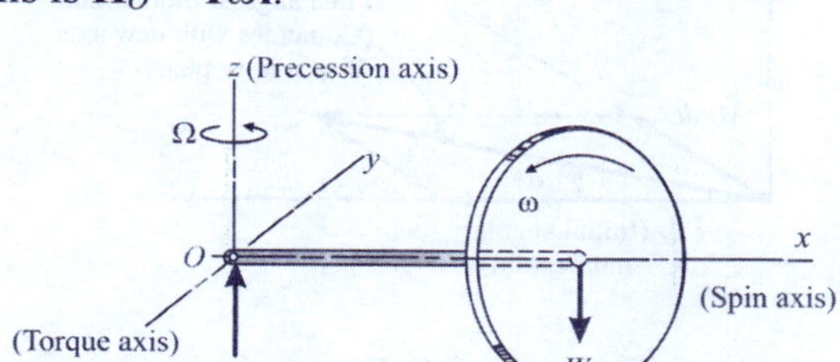

Figure 15-8.1 Gyroscope.

The external forces, consisting of the weight W of the disk and the vertical reaction R at O, create a torque $\mathbf{M} = \mathbf{M}\,\hat{\mathbf{j}}$ about the y axis. This axis is called the *torque axis*. Since the axle is free to rotate in any direction about O, it seems as though the disk will fall. Strangely enough, as we shall soon show, the angular momentum of the spinning disk will not permit it to fall but will cause it instead to rotate about the vertical axis z at the rate $\mathbf{\Omega}$. This rotation is known as *precession* and the z axis is called the *precession axis*. Our discussion of the gyroscope is limited here to the case in which the spin axis, the torque axis, and the precession axis are mutually perpendicular.

Note that the spin, torque, and precession axes constitute a rotating reference frame xyz which rotates about a fixed $\mathbf{Z}$ axis through O at the rate $\mathbf{\Omega}$. Usually, the spin rate ω is constant so that the angular momentum $\mathbf{H}_O$ may be considered to be a vector fixed in this rotating reference frame. Then since O is a fixed point about which the equation $\mathbf{M}_O = \dot{\mathbf{H}}_O$ is valid, and using the omega theorem (Section 12-6) to obtain the time derivative of $\mathbf{H}_O$, we have

$$[\mathbf{M}_O = \dot{\mathbf{H}}_O] \qquad \left[M_o\hat{\mathbf{j}} = \Omega \times \mathbf{H} = \Omega\hat{\mathbf{k}} \times \bar{I}\omega\hat{\mathbf{i}} = \bar{I}\omega\Omega\hat{\mathbf{j}}\right]$$

which in scalar form is

$$M_o = \bar{I}\omega\Omega \tag{15-8.1}$$

Generally the rate of precession Ω is small compared with ω so that angular momentum associated with Ω may be neglected.

An alternate development which gives the physical significance of this result is obtained by expressing the angular impulse-momentum relation in the following form:

Final angular momentum = Angular impulse + Initial angular momentum

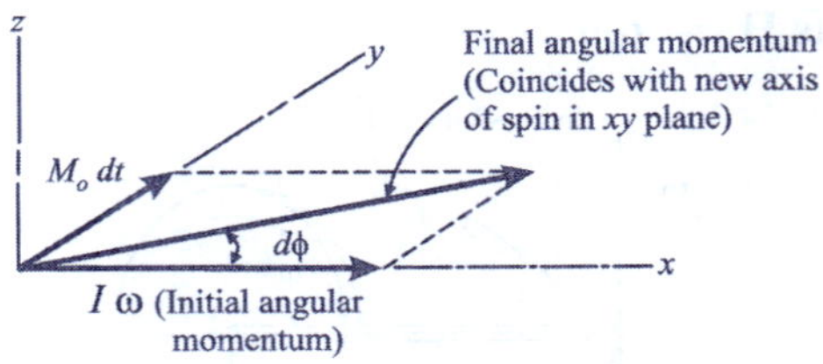

Figure 15-8.2 Precession.

These vector quantities, represented by the right-hand rule, are combined vectorially as shown in Fig. 15-8.2. In the differential time dt, the angular impulse caused by the external torque is $\mathbf{M}_O \, dt$. Since the final angular momentum vector coincides with the axis about which the disk tries to spin, the spin axis processes through the angle $d\phi$. Noting that $d\phi$ is a small angle whose tangent is equal to the angle in radians, we have

$$M_o \, dt = I\omega \, d\phi$$

whence dividing by dt and noting that $\Omega = d\phi / dt$ is the rate of change of angular displacement of the spin axis, we obtain, as before,

$$M_o = \bar{I}\omega\Omega$$

If we attempt to hurry the rate of precession by applying a force in the xy plane, we shall merely make the rotating disk and axle rise. This rise is caused by the angular impulse of the hurrying torque represented by a vector directed along the z axis as shown in Fig. 15-8.3. Similarly, a force in the xy plane exerted to retard the precession will make the axle descend.

We have shown that a torque about an axis perpendicular to the spin axis causes a precession about a third axis perpendicular to the other two. Conversely, a forced precession about one axis will induce a torque about a third axis perpendicular to the spin and precession axes. For example, the wheels of an automobile going around a horizontal curve are forced to precess about a vertical axis through the center of curvature. This forced precession causes a torque about a third axis whose direction is along the tangent to the curve. Because of this torque, the reaction on the outer wheel is increased and that on the inner wheel is decreased.

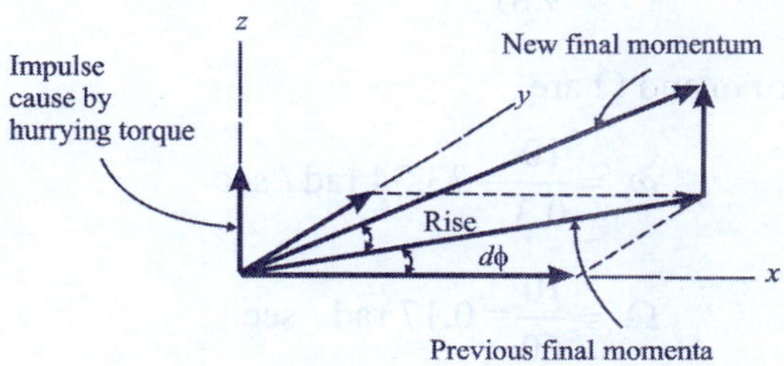

Figure 15-8.3 Effect of hurrying torque.

ILLUSTRATIVE PROBLEMS

15-8.1 A car is rounding a horizontal curve of 60 m radius at a speed of 10 m/s. Each wheel weighs 300 N is 0.8 m in diameter, and has a radius of gyration of 0.3 m. If the distance separating the wheels on the axles is 1.5 m, compute the change in ground reaction on each pair of wheels caused by the gyroscopic effect.

Solution

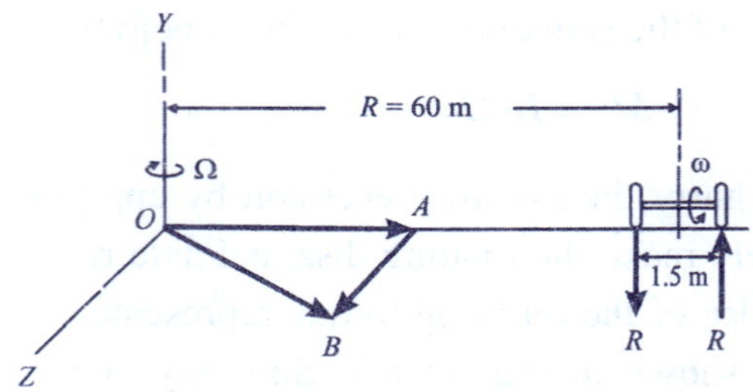

Figure 15-8.4

The vector *OA* in Fig. 15-8.4 represents the angular momentum of a pair of wheels as the car moves in the *OZ* direction. As the car rounds the curve, this vector rotates at the rate Ω about the vertical *Y* axis to position *OB*. The corresponding angular impulse *AB* is caused by an induced torque acting on the car wheels. If we apply the right-hand rule and extend our thumb in the direction *AB*, our fingers will curl about the vector *AB* in a counterclockwise sense, indicating that the gyroscopic reactions on the wheels must act as shown.

The moment of inertia of a pair of wheels is

$$\left[\bar{I} = \frac{W}{g}\bar{k}^2\right] \qquad \bar{I} = \frac{2\times 300}{9.81}(0.3)^2 = 5.5 \text{ N-m-sec}^2$$

and, the values of ω and Ω are

$$\left[\omega = \frac{v}{r}\right] \qquad \omega = \frac{10}{0.3} = 33.34 \text{ rad / sec}$$

$$\left[\Omega = \frac{v}{R}\right] \qquad \Omega = \frac{10}{60} = 0.17 \text{ rad / sec}$$

Applying Eq. (15-8.1), we now obtain

$$[M_o = \bar{I}\omega\Omega] \qquad 1.5R = 5.5(33.34)(0.17) \qquad R = 20.78 \text{ N} \quad \textbf{\textit{Ans.}}$$

PROBLEMS

15-8.2 A solid disk 0.3 m diameter weighing 1500 N rotates at 1800 rpm. It is keyed to a shaft 250 mm long which is supported by a friction less pivot attached to the outer race of a ball bearing, as shown in Fig. P-15-8.2. If the disk rotates in a clockwise sense when viewed from the right, determine the angular velocity and sense of the precession.

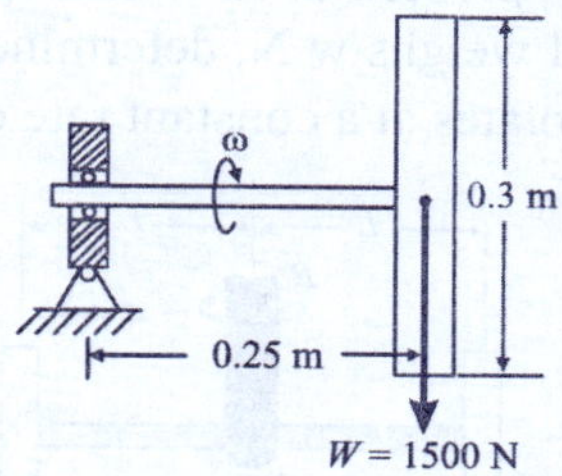

Figure P-15-8.2

15-8.3 A pair of locomotive driving wheels and their axle weigh 21500 N. Their diameter is 2 m and their radius of gyration is 0.8 m. Determine the changes in wheel reactions caused by the gyroscopic effect when the locomotive is rounding a curve of 900 m radius at 26.83 m/s. Assume the rails to be 1.5 m apart:

15.8.4 The wheel in Fig. P-15-8.4 weighs 1500 N and has a radius of gyration of 0.5 m. It rotates about its horizontal axis *AB* with a constant speed of 200 rad per sec. Axis *AB* is mounted in a yoke which when viewed from above rotates at 1 rad per sec clockwise about the *Y* axis passing through the center of gravity of the wheel. Compute the forces on the bearings at *A* and *B*.

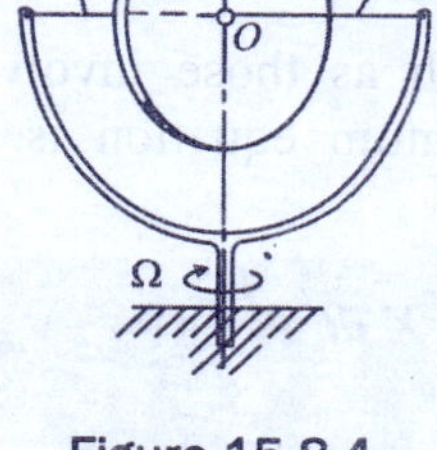

Figure 15-8.4

15-8.5 Compute the magnitude of the force P acting 0.6 m to the right of *O* in Fig. P-15-8.4 that is required to increase the rate of precession in Prob. 15-8.4 to 2 rad per sec. In what direction must it be applied?

15.8.6 A steam turbine in a ship is mounted with its axis parallel to the propeller Shaft. The turbine weighs 14500 N, rotates at 1800 rpm, and is mounted in bearings 1.2 m apart. If it has a radius of gyration of 0.9 m, compute the changes in bearing reactions caused by the gyroscopic effect of the ship making a turn of 600 m radius at 10 m/sec.

15-8.7 A grinding mill consists of two wheels free to spin about axles which are independently pivoted at A to the vertical shaft *AB* in Fig. P-15-8.7. If each wheel weighs w N, determine the ground reaction on each wheel when *AB* rotates at a constant rate of Ω rad/sec.

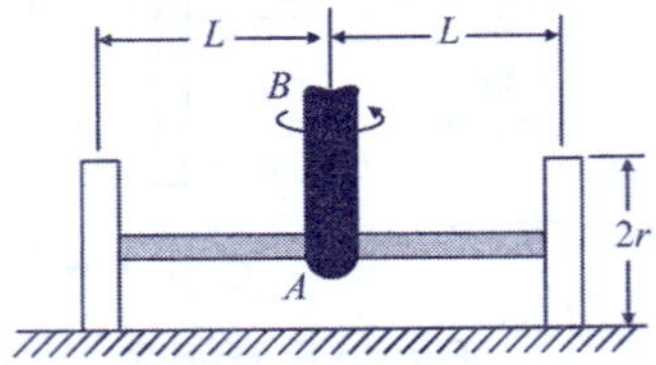

Figure

SUMMARY

By taking the time integral of $\mathbf{R} = m\bar{\mathbf{a}}$, we obtain

$$\int_o^t \mathbf{R}\,dt = m\bar{\mathbf{v}} - m\bar{\mathbf{v}}_o \qquad (15\text{-}2.1)$$

The equation refers to the motion of the mass center of a body of constant mass, rigid or nonrigid. When applied to a particle or to a translating rigid body, the bar sign can be omitted from the velocity terms. The expression $\int_0^t \mathbf{R}\,dt$ is known as resultant linear impulse. Its unit is N-sec. The quantity $m\bar{v} = (W/g)\,\bar{v}$ is defined as linear momentum and also has the dimensional unit of N-sec.

In most applications such as those involving the dynamic action of jet streams, the impulse-momentum equation is applied in terms of its scalar components, namely,

$$\int_o^t \Sigma X\,dt = \frac{W}{g}(v_x - v_{ox}), \text{etc.} \qquad (15\text{-}2.3)$$

which state that the component of resultant impulse along any axis is equal to the change in the components of linear momentum along that axis.

For variable mass situations in which a moving body absorbs or ejects material, the force **P** exerted on the moving body is given by

$$P = \mathbf{u}\frac{dm}{dt} \tag{15-3.1}$$

where **u** is the velocity of the absorbed or ejected mass relative to the moving body and $\frac{dm}{dt}$ is the rate at which the moving body gains or loses mass. Note that the mass of the moving body is time-dependent and may change appreciably in the case of a rocket in which the amount of ejected fuel is a significant portion of its original total weight.

The principle of conservation of linear momentum may be applied to any system of particles acted upon only by mutual reactive forces. This principle is expressed by the equation

$$\frac{W_1}{g}\mathbf{v}_1 + \frac{W_2}{g}\mathbf{v}_2 + \ldots = \frac{W_1}{g}\mathbf{v}_1' + \frac{W_2}{g}\mathbf{v}_2' + \ldots \tag{15-4.1}$$

Problems involving elastic impact between particles are solved by means of the principle of conservation of linear momentum in combination with the definition of the coefficient of restitution *e*. The coefficient of restitution is defined by the equation

$$e = \frac{\text{Relative velocity after impact}}{\text{Relative velocity before impact}}$$

and applied in the form

$$e(v_1 - v_2) = v_2' - v_1' \tag{15-5.1}$$

in which the velocities are the components of velocities along the common normal to the impacting bodies.

For rigid bodies in fixed-axis rotation or general plane motion, impulse-momentum methods are the only practical way of treating bodies which react mutually upon each other, or are subjected to impact. Starting with $\mathbf{R} = m\bar{a}$ *and* $\Sigma\mathbf{M}_G = \dot{\mathbf{H}}_G$ which relate the resultant force-couple system at the mass center *G* to their equivalent dynamic effects on a body, the time-integrated forms of these equations become

$$\int_{t_1}^{t_2} \mathbf{R}\, dt = m\bar{\mathbf{v}}_2 - m\bar{\mathbf{v}}_1$$

$$\int_{t_1}^{t_2} \Sigma \overline{M}\, dt = \overline{\mathbf{H}}_2 - \overline{\mathbf{H}}_1 = \overline{I}(\omega_2 - \omega_1)$$

Both of these equations are combined diagrammatically in Fig. 15-6.4 which expresses the impulse-momentum principle as

$$\text{Initial system momenta} + \text{Resultant ext. impulses} = \text{Final system momenta} \qquad (15\text{-}6.2)$$

The expression $\int_{t_1}^{t_2} \Sigma \mathbf{M}\, dt$ is called resultant angular impulse and has the dimensional unit of m-N-sec. The expression $I\omega$ represents angular momentum of plane motion and is also dimensionally in units of m-N-sec. Notice that both angular impulse and angular momentum correspond to couples whose vectors are perpendicular to the plane of motion whereas linear impulse and linear momentum correspond to forces in the plane of motion.

The advantage of the diagrammed approach of Eq. (15-6.2) is that it clearly distinguishes between linear and angular impulses and momenta. Moreover, a moment center which eliminates impulsive forces is readily apparent. Moment summations about such a center leads automatically to conservation of angular momentum. From Fig. 15-6.4, there also may be obtained the following scalar forms of the impulse-momentum equations:

$$\int_{t_1}^{t_2} \Sigma X\, dt = m\left(\overline{v}_{2_x} - \overline{v}_{1_x}\right)$$

$$\int_{t_1}^{t_2} \Sigma Y\, dt = m\left(\overline{v}_{2_y} - \overline{v}_{1_y}\right) \qquad (15\text{-}6.1)$$

$$\int_{t_1}^{t_2} \Sigma \overline{M}\, dt = \overline{I}\left(\omega_2 - \omega_1\right)$$

The following form of the angular impulse-momentum equation for fixed-axis rotation about an axle A is also useful:

$$\int_{t_1}^{t_2} \Sigma M_A\, dt = I_A\left(\omega_2 - \omega_1\right)$$

Observe that here the moment sum about A of $m\overline{v}$ and $\overline{I}\omega$ reduces to $I_A\omega$.

A combination of conservation of angular momentum and work-energy will solve many problems involving the motion of satellites. The specific equations used are

$$v_1 r_1 \sin\phi_1 = v_2 r_2 \sin\phi_2 \qquad (15\text{-}7.2)$$

and

$$v_2^2 - v_1^2 = 2GM\left(\frac{1}{r_2} - \frac{1}{r_1}\right) \qquad (15\text{-}7.1)$$

where

$$GM = 3.995 \times 10^{14}\ \text{m}^3/\text{sec}^2 \qquad (15\text{-}7.3)$$

The velocity at which a satellite will remain in a circular orbit is found from

$$v^2 = \frac{GM}{r} \qquad (15\text{-}7.4)$$

In terms of the apogee and perigee distances r_a and r_p the major and minor semiaxes of the elliptical path of an orbiting satellite are

$$a = \frac{1}{2}(r_a + r_p) \text{ and } b = \sqrt{r_a r_p} \qquad (15\text{-}7.5)$$

which are used to determine the period of the orbiting satellite from

$$\tau = \frac{2\pi ab}{r_a v_a} = \frac{2\pi ab}{r_p v_p} = \frac{2\pi a^{3/2}}{\sqrt{GM}} \qquad (15\text{-}7.6)$$

The introduction to gyroscopic action is limited to the case in which the spin axis, the torque axis, and the precession axis are mutually perpendicular. The applied torque about a fixed point O is given by

$$M_o = \bar{I}\omega\Omega \qquad (15\text{-}8.1)$$

where $\bar{I}$ is the centroidal moment of inertia about the spin axis of the rotor spinning at the rate ω and precessing at the rate Ω. A more extended discussion of gyroscopic action appears in Chapter 16.

CHAPTER 16

MECHANICAL VIBRATIONS

16-1 INTRODUCTION, DEFINITIONS AND CONCEPTS

Vibrations in an elastic structure are caused by disturbing forces which create a displacement of the structure from its position of static equilibrium. Such displacements create elastic forces which tend to restore the body to its original condition of equilibrium. When the disturbing force is removed, the elastic forces cause the body to accelerate back to its equilibrium position. However, the body will now possess some velocity as it passes through its equilibrium position, and this will cause it to overshoot. Thus, vibrations of it are set up which may or may not diminish, depending on whether any resisting forces are present.

When these vibrations are caused by an initially applied force which is then removed from the body, the result what are known as free vibrations. If resisting forces also act on the vibrating body, the motion is known as a damped free vibration. When the disturbing force continues to act at periodic intervals upon the body, the result is a forced vibration which may or may not be damped.

Generally speaking, a *vibration* is a periodic motion which repeats itself, after a definite interval of time. This time interval is called the *period* of the vibration and will be designated by the symbol τ, usually measured in seconds. Each repetition of the motion is called a *cycle*. The *frequency* (f) of the vibration is the reciprocal of the period, i.e., $f = 1/\tau$ and is measured in cycles per second (also called Hertz and abbreviated as **Hz**). The maximum displacement of the body from its equilibrium position is known as the *amplitude* of the vibration.

16-2 SIMPLE HARMONIC MOTION. FREE VIBRATIONS

Simple harmonic motion (shm) is the name used to describe the straight-line motion of a body whose acceleration is proportional to the displacement from a fixed origin and is always directed toward the origin. This motion is a special case of rectilinear motion with variable acceleration. The motion is essentially a vibratory displacement such as that described by a weight which is attached to one end of a spring and is allowed to vibrate freely.

The mathematical description of the motion is given by the equation

$$a = -Ks$$

where K is the constant of proportionality and s is the displacement from the origin. It is preferable, however, to denote this constant by ω^2, thus making

$$\mathbf{a = -\omega^2 s} \tag{16-2.1}$$

the mathematical statement of simple harmonic motion. The negative sign indicates that the direction of the acceleration is always opposite to that of the displacement.

The equations for simple harmonic motion may be derived from the kinematic differential equations by means of the method developed in Section 9-3. Another simpler method is to describe the motion graphically by considering a point that moves with constant speed around a circular path. The projection of this moving point upon a diameter of the circle possesses simple harmonic motion.

Consider, for example, a circle whose radius r is rotating at a constant rate of ω radians per second. Choose a diameter of the circle as the X axis (Fig. 16-2.1), and let the initial position of a point on the circle be at A. The speed of this point will be constant at the value $v = r\omega$. Its acceleration is a normal one due only to the change in the direction of the velocity because the magnitude of the latter is constant. In Section 12-2 this acceleration was given by $a_n = r\omega^2$. After a time interval of t sec, point A reaches position C. The velocity and acceleration of the point in this position are $r\omega$ and $r\omega^2$, directed as shown.

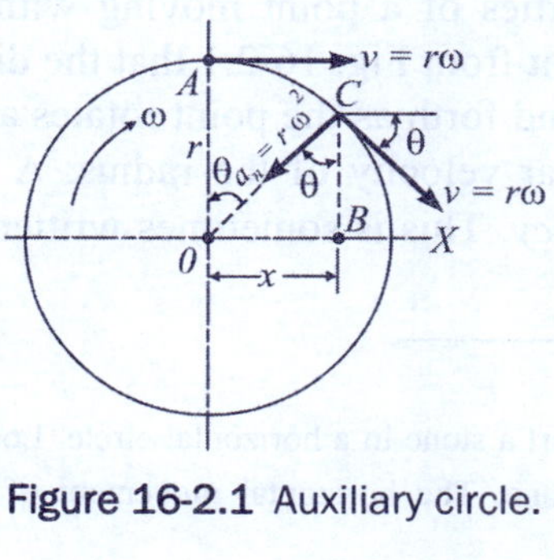

Figure 16-2.1 Auxiliary circle.

During this time the projection of the point on the X axis moves from O to B through a displacement x. From the figure arc AC projected upon the diameter is

$$x = r \sin \theta \tag{a}$$

If Eq. (a) is differentiated with respect to time, we obtain

$$\frac{dx}{dt} = r \cos \theta \frac{d\theta}{dt}$$

or

$$v_x = r\omega \cos \theta \tag{b}$$

since $\omega = d\theta/dt$. Observe that $r\omega \cos \theta$ is the diametral projection of the velocity of point C.

If Eq. (b) is differentiated with respect to time, the following expression for acceleration results:

$$\frac{dv_x}{dt} = - r\omega \sin \theta \frac{d\theta}{dt}$$

or

$$a_x = - r\omega^2 \sin \theta \tag{c}$$

The minus sign indicates that the acceleration is directed toward the origin O. Notice that $r\omega^2 \sin\theta$ is the diametral projection of the acceleration of point C. This projection is also directed toward the origin.

However, $r \sin \theta = x$ from Eq. (a); hence, Eq. (c) may be re written as

$$a_x = - \omega^2 r \sin \theta$$

or

$$a_x = - \omega^2 x$$

This is the mathematical description of simple harmonic motion given in Eq. (16-2.1). Hence, we may conclude that the displacement, velocity, and acceleration of a simple harmonic motion are described by the diametral projection of the properties of a point moving with constant speed around a circular path. It is evident from Fig. 16-2.1 that the diametral projection of such a point oscillates back and forth as the point rotates around the circle;[1] also that ω is the constant angular velocity of the radius. A convenient name for ω is *natural circular frequency*. This is sometimes written as ω_n the subscript being

---

[1]A physical analogy is to whirl a stone in a horizontal circle. Look at the edge view of the plane described by the rotating stone. The horizontal movement of the stone is describing simple harmonic motion.

added to avoid confusion when ω is used in its usual meaning of angular velocity.

The circle in this discussion is known as the auxiliary circle. Its radius is called the *amplitude* of the motion. The time required to complete one oscillation backward and forward is called the *period* of the motion. The period is determined by the time required for the radius of the auxiliary circle to complete one revolution, or

$$\tau = \frac{2\pi}{\omega} \tag{16-2.2}$$

The *frequency* is the number of oscillations per second, or

$$f = \frac{1}{\tau} = \frac{\omega}{2\pi} \tag{16-2.3}$$

ILLUSTRATIVE PROBLEMS

16-2.1 If a particle in simple harmonic motion has an amplitude of 300 mm and a period of 1 sec, determine the displacement, velocity, and acceleration after 0.4 sec from when the particle was at the right end of its path.

Solution

Construct an auxiliary circle having a radius equal to the amplitude, as shown in Fig. 16-2.2. Choose diameter *AOB* as the path of the simple harmonic motion. The natural circular frequency is determined from Eq. (16-2.2).

$$\left[\tau = \frac{2\pi}{\omega}\right] \qquad 1 = \frac{2\pi}{\omega} \qquad \omega = 2\pi \text{ rad per sec}$$

After 0.4 sec the radius will have swept through an angular distance

$$[\theta = \omega t] \qquad \theta = 2\,\pi \times (0.4) = 0.8\pi \text{ radians}$$

$$= 0.8\pi \times 57.3 = 144°$$

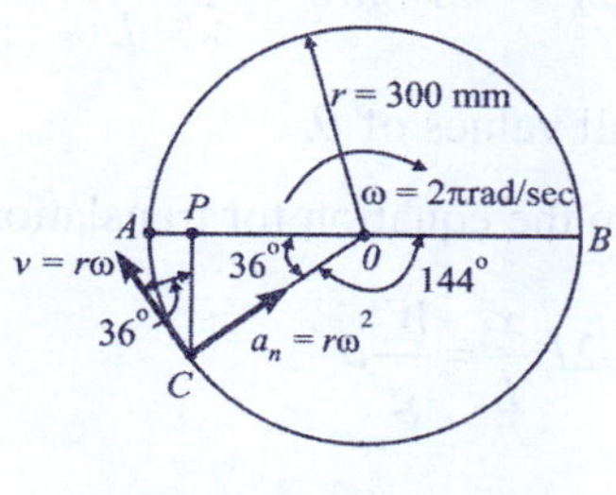

Figure 16-2.2

The moving point will then be at position C on the auxiliary circle. The properties of its diametral projection P are given by

$$[s = r\cos\theta] \qquad s = -300 \times \cos 36° = -242.7 \text{ mm}$$

$$[v = r\omega\sin\theta] \qquad v = -300 \times 2\pi\sin 36° = -1107.95 \text{ mm/sec}$$

$$[a = r\omega^2\cos\theta] \qquad a = 300 \times (2\pi)^2\cos 36° = 9581.6 \text{ mm/sec}^2$$

The signs are determined by inspection of Fig. 16-2.2, rightward values being plus and leftward minus.

16-2.2 An elastic string of length $2L$ tightly stretched between two rigid supports, as in Fig. 16-2.3, carries a small ball of weight W at its midpoint. Show that for small displacements, the ball will have a simple harmonic motion; compute the period.

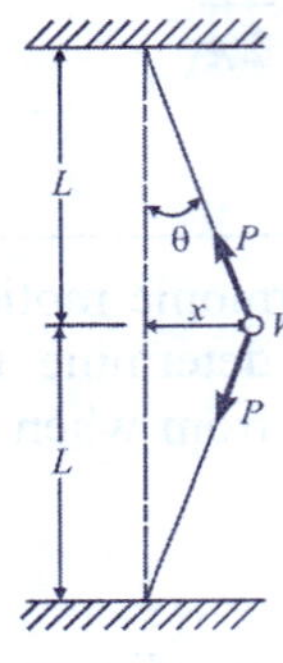

Figure 16-2.3

Solution

We shall assume that the initial tension P in the string is so large that we can neglect any increase in it caused by the small displacement of the ball. If the rightward direction from the equilibrium position is considered positive, the unbalanced force acting on the ball is

$$\Sigma X = -2P\sin\theta \approx -2P\frac{x}{L}$$

since $\sin\theta \approx \tan\theta$ for small values of θ.

Substituting this value in the equation for translation, Eq. (11-2.1), we obtain

$$\left[\Sigma X = \frac{W}{g}a\right] \qquad -2P\frac{x}{L} = \frac{W}{g}a$$

or

$$a = \frac{2Pg}{WL} x$$

which is in the form of a simple harmonic motion ($a = -\omega^2 s$) where $\omega^2 = \dfrac{2Pg}{WL}$.
Hence, from Eq. (16-2.2) the period is

$$\left[\tau = \frac{2\pi}{\omega}\right] \qquad \tau = 2\pi\sqrt{\frac{WL}{2Pg}} \qquad \textit{Ans.}$$

PROBLEMS

16-2.3 A simple harmonic motion is defined by the relation $a = -36\ s$. Determine its period and frequency.

16-2.4 The amplitude of a simple harmonic motion is 610 mm and the period is 1 sec. Determine the maximum velocity and the maximum acceleration.

16-2.5 A particle in simple harmonic motion has an amplitude of 400 mm and a period of $\pi/2$ sec. Find the velocity and acceleration of the particle when it has traveled 250 mm to the right of the center of its path. What time is required for this displacement?

16-2.6 A particle has a simple harmonic motion defined by $a = -9s$ and an amplitude of 250 mm. Find the velocity and acceleration of the particle when it is 150 mm away from the center of its path. What is its period?

16-2.7 A particle moving with simple harmonic motion has a maximum velocity of 6 m/s and a maximum acceleration of 15 m/s^2. Determine the velocity and acceleration of the particle when it is midway between the center and the right end of its path. How long does it take to move from the center to the specified position?

16-2.8 The amplitude of a particle moving with simple harmonic motion is 500 mm. When the particle is 250 mm from the extreme left position, its acceleration is 4 m/s^2. What is its velocity at that position? How many seconds are required to move 250 mm from the extreme left position?

16-2.9 A particle moving with simple harmonic motion is observed to have a velocity of 400 mm/sec when it is 150 mm, from the center of its path and a velocity of 300 mm/s when it is 200 mm from the center.

Determine the maximum velocity, maximum acceleration, and the frequency of vibration.

16-2.10 A weight of W N is attached to one end of a vertical spring (Fig. P-16-2.10) whose scale is k N per m. The weight is displaced vertically from its equilibrium position by an external force. It is then released and vibrates freely. By means of the equation $\Sigma X = (W/g)a$, show that the vibration is in the form of a simple harmonic motion and that the period of vibration is given by $\tau = 2\pi\sqrt{W/kg}$.

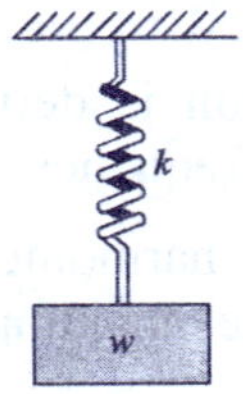

Figure P-16-2.10

16-2.11 A 750 N weight is suspended vertically from the end of a spring whose modulus is 4 N/mm. The weight is pulled down 150 mm. below its equilibrium position and then released. Compute the velocity of the weight as it passes through the equilibrium position and also its frequency of vibration. Check the velocity using the work-energy method.

16.2.12 The average depth of immersion of a ship is h and its area at the water line is A. Using the equation $\Sigma X = (W/g)a$, show that if the ship is displaced slightly downward and then released, the buoyant force of the water will cause vertical oscillations that have a simple harmonic motion. Determine the period of oscillation.

16-3 SIMPLE PENDULUM

A simple pendulum is defined as a particle at the end of a weightless cord that is allowed to oscillate in a vertical arc of a circle under the influence of gravity and the tension in the cord. This motion is approximated by suspending a small weight at the end of a thread, as shown in Fig. 16-3.1.

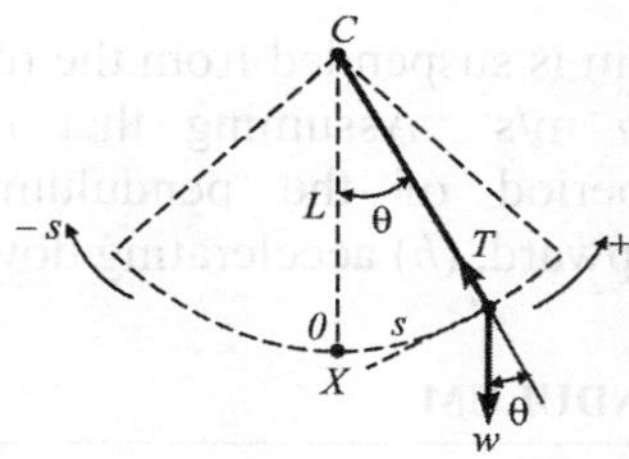

Figure 16-3.1 Simple pendulum.

Taking the displacement as positive in the direction shown, we have

$$\left[\Sigma X = \frac{W}{g}a\right] \qquad -W\sin\theta = \frac{W}{g}a \qquad (a)$$

If the oscillation is restricted to small angles of vibration, sin θ may be considered practically equivalent to θ measured in radians. From the figure, $\theta = s/L$, therefore

$$-W\frac{s}{L} = \frac{W}{g}a \qquad (b)$$

so that

$$a = -\frac{g}{L}s \qquad (c)$$

This is the equation of simple harmonic motion ($a = -\omega^2 s$) in which $\omega^2 = \dfrac{g}{L}$. From Section 16-2, the period is given by

$$\tau = \frac{2\pi}{\omega} = 2\pi\sqrt{\frac{L}{g}} \qquad (16\text{-}3.1)$$

and the frequency by

$$f = \frac{\omega}{2\pi} = \frac{1}{2\pi}\sqrt{\frac{g}{L}} \qquad (16\text{-}3.2)$$

These results show that for small angles of vibration, the period and frequency depend only on the length of the cord of the simple pendulum.

PROBLEMS

16-3.1 Determine the length of a simple pendulum whose period is 1 sec.

16.3.2 Find the period of a simple pendulum whose length is 1.3 m.

16-3.3 A simple pendulum is suspended from the roof of an elevator which is accelerating at a m/s^2. Assuming that the vibrations are small, determine the period of the pendulum when the elevator is (a) accelerating upward, (b) accelerating downward, (c) falling freely.

16-4 COMPOUND PENDULUM

A compound pendulum is a rigid body that is free to rotate about a horizontal axis and that oscillates under the influence of gravity. Such a body is represented in Fig. 16-4.1. Point O is called the *center of suspension.* The position of the pendulum is defined by its angular displacement θ from the equilibrium position.

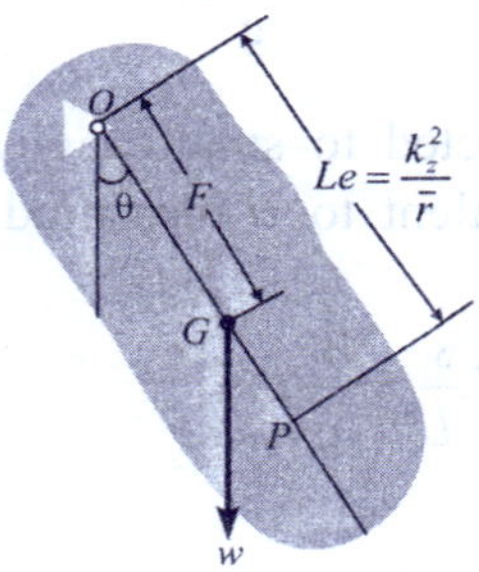

Figure 16-4.1 Compound pendulum.

Applying the equation of rotation about a fixed axis and noting that moments are positive in the sense of angular displacement, we obtain (neglecting friction at the axis)

$$[\Sigma M_z = I_z\alpha] \qquad -W\,\bar{r}\sin\theta = I_z\alpha \tag{a}$$

Expressing I_z in terms of its mass W/g and radius of gyration k_z gives

$$-W\,\bar{r}\sin\theta = \frac{W}{g}k_z^2\alpha \tag{b}$$

If the oscillations are restricted to small angles for which sin θ is practically equal to θ, Eq. (b) may be written in the form

$$\alpha = -\frac{g\bar{r}}{k_z^2}\theta \tag{c}$$

which is similar to a simple harmonic motion in which $\omega^2 = \frac{g\bar{r}}{k_z^2}$.

Therefore, the period is given by $\tau = 2\,\pi/\omega$ (Section 16-2), or

$$\tau = 2\pi\sqrt{\frac{k_z^2}{g\bar{r}}} = 2\pi\sqrt{\frac{L_e}{g}} \tag{16-4.1}$$

Observe that the period of a compound pendulum is the same as that of a simple pendulum (Section 16-3) that has an equivalent length L_e equal to $\frac{k_z^2}{\bar{r}}$.

From the transfer formula relating radii of gyration,

$$k_z^2 = \bar{k}^2 + \bar{r}^2$$

Hence the equivalent length of a compound pendulum may be written as

$$L_e = \frac{\bar{k}^2}{\bar{r}} + \bar{r} \tag{d}$$

This may be interpreted as meaning that the period is the same for all parallel axes of suspension located at the same distance from the center of gravity.

Point P on line OG at the distance $L_e = \frac{k_z^2}{\bar{r}}$ from O is called the *center of oscillation.* This point is also the center of percussion (Illus. Prob. 13-3.2,). Huygens showed that if the center of oscillation is made the center of rotation, the period will be unchanged, i.e., *the centers of suspension and oscillation are interchangeable.* Thus since $GO = \bar{r}$ and $GP = L_e - \bar{r}$, Eq. (*d*) gives

$$(GO)(GP) = \bar{k}^2 \tag{e}$$

If the pendulum is suspended from a horizontal axis through P and the new center of oscillation is denoted by O', then from Eq. (*e*) we have

$$(GO')(GP) = \bar{k}^2 \tag{f}$$

or O' and O must coincide. We conclude that if P becomes the center of suspension, O becomes the center of oscillation.

Equation (16-4.1) is sometimes used in experiments to determine the radius of gyration or moment of inertia of a body that can be swung as a pendulum. For example, if a connecting rod is swung from a knife edge that passes through the hole for the wrist pin or crank pin and the duration of the period is observed, the radius of gyration and hence the moment of inertia about the axis of suspension can be computed. The center of gravity can be determined by balancing the connecting rod on a knife edge, and the moment of inertia about any parallel axis can then be found by applying the transfer formula.

PROBLEMS

16-4.1 A circular ring is suspended from a knife edge as shown in Fig. P-16-4.1. The ring weighs 300 N, its outer radius is 900 mm, and its inner radius is 600 mm. Determine the period for small oscillations.

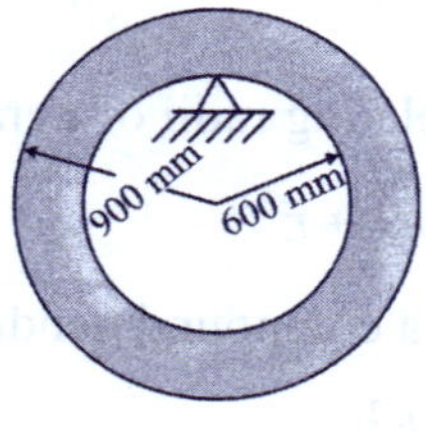

Figure P-16-4.1

16-4.2 The compound pendulum in Fig. P-16-4.2 consists of a slender rod 600 mm long weighing 30 N to which is attached a solid circular disk of 300 mm diameter that weighs 40 N. Compute the period of small oscillations.

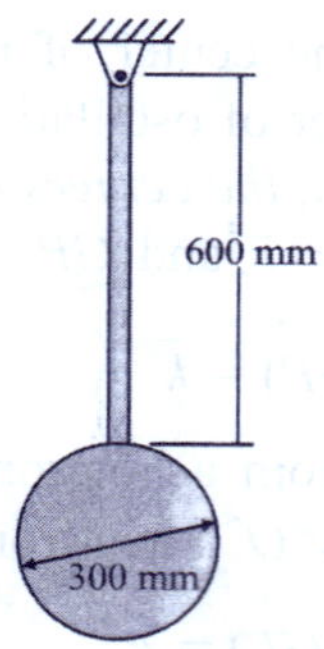

Figure P-16-4.2

$\tau = 1.67$ sec ***Ans.***

16-4.3 A connecting rod weighing 45 N is swung as a pendulum from a horizontal knife edge resting on the inner face of the wrist pin bearing. The dimensions are given in Fig. P-16-4.3. If the observed period of oscillation is 1 sec, compute the moment of inertia with respect to the axis of the crank bearing.

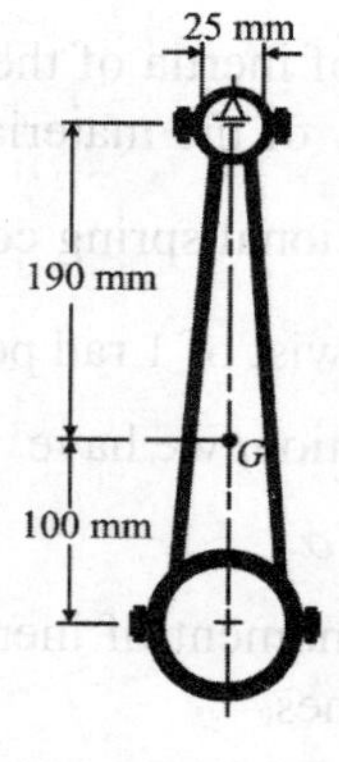

Figure P-16-4.3

16-5 TORSION PENDULUM

A torsion pendulum consists of a body that is fastened to the lower end of an elastic wire or rod whose upper end is clamped. The body is assumed to be fastened so that the axis of the rod passes through the center of gravity of the body, as in Fig. 16-5.1.

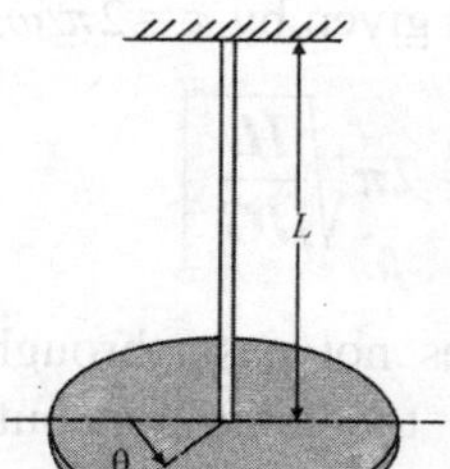

Figure 16-5.1 Torsion pendulum

Let an external moment be applied to the body that rotates it (and the lower end of the rod) through an angular displacement θ from the equilibrium position. When this moment is removed, the body undergoes an oscillatory rotation known as torsional vibration. During it, the only external moment acting on the body is supplied by the resisting moment M exerted by the rod. From strength of materials, it is known that the relation between the twisting moment and the angle of twist for a circular cylindrical rod is

$$M = \frac{JG}{L}\theta \qquad (a)$$

in which J is the polar moment of inertia of the cross-sectional area of the rod,[2] G the shear modulus of elasticity of the material, and L the length. The product $\frac{JG}{L}$ is sometimes called the torsional spring constant of the rod, i.e., the torque required to produce an angle of twist of 1 rad per unit length,

Applying the equation of rotation, we have

$$[\Sigma M = I\alpha] \qquad -M = I\alpha$$

in which I represents the mass moment of inertia of the body about the axis of rotation. From Eq. (a) this becomes

$$-\frac{JG}{L}\theta = I\alpha$$

or

$$\alpha = \frac{-JG}{IL}\theta \tag{b}$$

This is in the form of a simple harmonic motion in which $\omega^2 = \frac{JG^3}{IL}$.

In Section 16-2 the period was given by $\tau = 2\pi/\omega$, or

$$\left[\tau = 2\pi\sqrt{\frac{IL}{JG}}\right] \tag{16.5.1}$$

If the axis of the rod does not pass through the center of gravity of the attached body, the rod tends to be thrown out of plumb. However, if it is supported in bearings at the lower end to prevent this, the attached body will undergo torsional vibrations whose period is defined by Eq. (16-5.1). Under these conditions, observing the period of vibration is an easy method of determining the mass moment of inertia of any body with respect to an axis coinciding with that of the rod.

Many practical applications of torsional vibration occur when the load or power transmitted along a shaft between two heavy masses is suddenly released. An example is the shaft connecting the engine and propeller of a vessel that is pitching in a heavy sea. If Fig. 16-5.2 represents two heavy masses having moments of inertia I_1 and I_2 between which a transmitted torque is suddenly released, the elastic energy stored in the shaft will cause the masses to rotate in

---

[2]For a rod of diameter d, $J = \pi d^4/32$.

opposite directions (see Prob. 15-6.10). One section *n-n* of the shaft, called the nodal section, will remain stationary. The motion of each mass may be considered equivalent to a torsion pendulum on a shaft fixed at the nodal section.

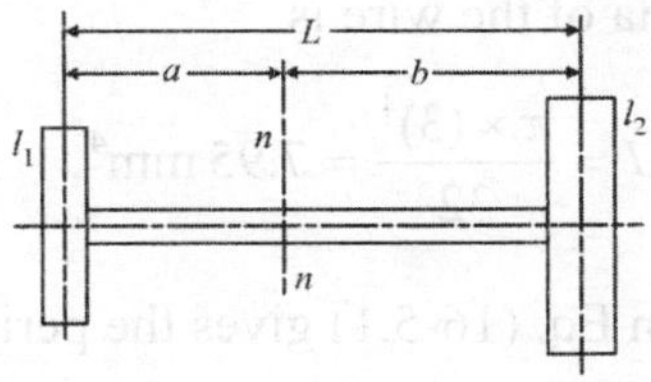

Figure 16-5.2 Nodal section

The position of the nodal section is located from the condition that each segment of the shaft has the same period of oscillation. Hence by applying Eq. (16-5.1) we obtain

$$I_1 a = I_2 b \qquad (c)$$

from which, since $a + b = L$

$$a = \frac{I_2 L}{I_1 + I_2} \qquad b = \frac{I_1 L}{I_1 + I_2} \qquad (d)$$

Since each segment has the same period, we may substitute either of these lengths in Eq. (16-5.1) to obtain the following expression for the period of free torsional vibration for the system:

$$\left[\tau = 2\pi\sqrt{\frac{IL}{JG}}\right] \qquad \tau = 2\pi\sqrt{\frac{I_1 I_2 L}{JG(I_1 + I_2)}} \qquad (e)$$

ILLUSTRATIVE PROBLEMS

16-5.1 A solid circular disk weighing 150 N and having a diameter of 900 mm is suspended at its midpoint horizontally from the end of a vertical steel wire 900 mm long and 3 mm in diameter. The shear modulus of elasticity for the wire is $G = 0.83 \times 10^5$ N/mm^2. Determine the period of torsional vibration.

Solution

The moment of inertia of the disk with respect to the axis of the wire is

$$\left[I = \frac{1}{2}\frac{W}{g}r^2\right] \qquad I = \frac{1}{2} \times \frac{150}{9.81} \times (0.45)^2 = 1.55 \text{ N-m-s}^2$$

$$= 1.55 \times 1000 = 1550 \text{ N-mm-Sec}^2$$

The polar moment of inertia of the wire is

$$\left[J = \frac{\pi d^4}{32}\right] \qquad J = \frac{\pi \times (3)^4}{32} = 7.95 \text{ mm}^4$$

Substituting these values in Eq. (16-5.1) gives the period

$$\left[\tau = 2\pi\sqrt{\frac{IL}{JG}}\right] \qquad \tau = 2\pi\sqrt{\frac{1550 \times 900}{7.95 \times 0.83 \times 10^5}} = 9.14 \text{ sec} \qquad \textbf{Ans.}$$

PROBLEMS

16-5.2 A horizontal bar 300 mm long that weighs 70 N is suspended at its midpoint by a vertical steel rod 1250 mm long and 6 mm in diameter. If the shear modulus of elasticity is 0.83×10^5 N/mm^2, compute the frequency of torsional vibration.

16-5.3 A certain body has a moment of inertia of 0.68 N-m-sec^2 with respect to an axis coincident with the center line of a wire from which the body is suspended. The period of torsional vibration is 0.6 sec. When another body is suspended from the same wire in place of the first body, the period is 0.9 sec. Determine the moment of inertia of the second body.

16-5.4 A shaft 80 mm in diameter and 3000 mm long connects the rotor of a steam turbine with the armature of a generator. The rotor weighs 10 KN and has a radius of gyration of 600 mm. The armature weighs 4 KN and has a radius of gyration of 900 mm. If G = 0.83×10^5 N/mm^2, compute the frequency of free torsional vibration of the system.

16-6 GRAPHICAL REPRESENTATION OF SIMPLE HARMONIC MOTION

As discussed in Section 16-2, the displacement-time relation defining a simple harmonic motion may be written in the form

$$x = r \sin \omega t \qquad (a)$$

Differentiating this equation with respect to the time determines the velocity-time relation

$$v = r\omega \cos \omega t \tag{b}$$

The maximum value of velocity is $v_m = r\omega$. The acceleration-time relation is obtained by another differentiation with respect to the time; this gives

$$a = -r\omega^2 \sin \omega t \tag{c}$$

from which the maximum value of acceleration is found to be $a_m = r\omega^2$.

One method of visualizing the variations in these equations is by means of the auxiliary circle described in Section 16-2. Another method, to be described, permits the displacement-time, velocity-time, and acceleration-time curves to be plotted easily. This technique is useful when analyzing mechanical vibrations, especially when combining vibrations of the same frequency.

Consider a point that moves with constant speed around a circle of radius r, as shown in Fig. 16-6.1. Let the position of this point be represented by the position vector r which is rotating at the constant angular velocity ω. Taking time as zero when the position vector is horizontal, we can write the vertical projection of r as $x = r \sin \omega t$, which is the simple harmonic motion expressed by Eq. (a). Similarly the velocity can be represented by the vertical projection of a vector of length $r\omega$ that rotates with the same angular velocity ω as the position vector but is always 90° *ahead* of it. The acceleration is represented by the vertical projection of a vector of length $r\omega^2$ that rotates with the same angular velocity ω as the position and velocity vectors but is 180° *ahead* of the position vector and 90° *ahead* of the velocity vector. Figure 16-6.1 indicates how the projections of these rotating vectors are used to plot the values of displacement, velocity, and acceleration against time.

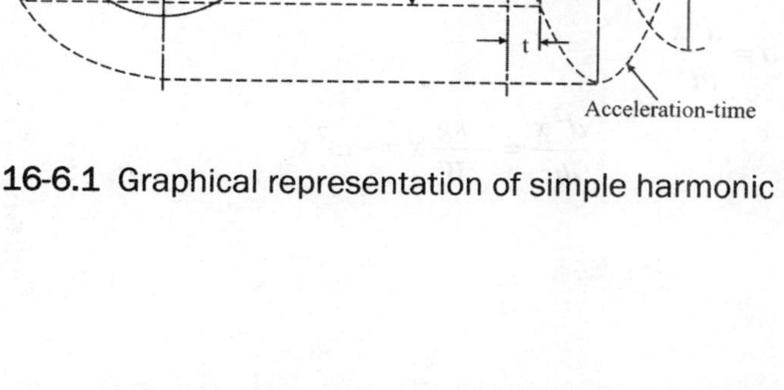

Figure 16-6.1 Graphical representation of simple harmonic motion.

16-7 FREE VIBRATIONS WITHOUT DAMPING. GENERAL CASE

Let us consider a spring-suspended weight which is displaced from its equilibrium position and allowed to vibrate freely. This is described in Fig. 16-7.1. It is assumed that all resistances such as air resistance and internal friction in the spring are neglected. This means that the spring will vibrate indefinitely under the action of the variable spring force exerted upon the weight. This condition is called free vibration.

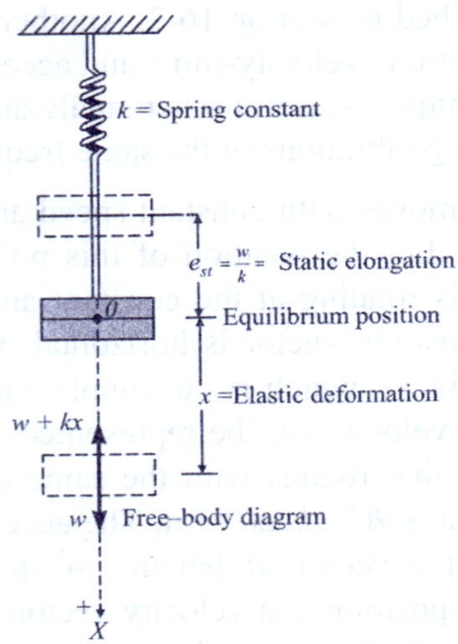

Figure 16-7.1 Free vibration of a spring-suspended weight.

The differential equation of motion for this case is found by applying the relation between the unbalanced force and the resulting acceleration. Considering downward forces and displacements as positive, we obtain

$$\left[\Sigma X = \frac{W}{g}a\right] \qquad -kx = \frac{W}{g}a$$

or, since $a = \dfrac{d^2x}{dt^2}$

$$\frac{d^2x}{dt^2} = -\frac{kg}{W}x = -\omega^2 x \qquad (a)$$

where

$$\omega = \sqrt{\frac{kg}{W}} \qquad (b)$$

Equation (*a*) is the differential equation of a free vibration. It indicates that the displacement x is a function of time such that, upon being differentiated twice with respect to time, it equals itself multiplied by a negative constant of value ω^2. The sine and cosine functions repeat themselves in this manner. Substitution of the relations $x = \sin \omega t$ and $x = \cos \omega t$ in Eq. (*a*) shows that they are solutions of that equation. A more general solution is obtained by multiplying these solutions by arbitrary constants C_1 and C_2 which are evaluated to fit the actual condition of the vibration. With this procedure, the complete solution of Eq. (*a*) is given by the sum of these solutions, or

$$x = C_1 \sin \omega t + C_2 \cos \omega t \qquad (c)$$

To determine the constants C_1 and C_2 it is necessary to understand that Eq. (*c*) describes all the possible motions that the system of weight and spring shown in Fig. 16-7.1 can have. If we now specify that the weight has a velocity v_o when given an initial displacement x_o, there result two conditions which can be used to determine the constants of integration. Measuring the time as zero under the given conditions, at $t = 0$ we obtain

$$x = x_o \qquad \text{and} \qquad v = v_o$$

Substituting the first condition in Eq. (*c*) yields

$$x_o = C_1(0) + C_2(1) \text{ or } C_2 = x_o \qquad (d)$$

The second condition is substituted in the velocity-time relation obtained by differentiating Eq. (*c*) with respect to the time, Thus,

$$\left[v = \frac{dx}{dt}\right] \qquad v = \omega C_1 \cos \omega t - \omega C_2 \sin \omega t$$

Setting $v = v_o$ when $t = 0$, we have

$$v_o = \omega C_1 (1) - \omega C_2(0) \text{ or } \quad C_1 = \frac{v_o}{\omega} \qquad (e)$$

Under the specified conditions, the vibration is described by the equation

$$\boldsymbol{x = \frac{v_o}{\omega} \sin \omega t + x_o \cos \omega t} \qquad (16\text{-}7.1)$$

The period is the time required for the position vector describing the vibration (Section 16-6) and moving with a constant angular velocity ω, to sweep through one revolution, or

$$\tau = \frac{2\pi}{\omega} = 2\pi\sqrt{\frac{W}{kg}} = 2\pi\sqrt{\frac{e_{st}}{g}} \qquad (16\text{-}7.2)$$

where the static elongation is

$$e_{st} = \frac{W}{k} \qquad (f)$$

and represents the elongation produced by the weight when hanging freely from the spring. The frequency is

$$f = \frac{1}{\tau} = \frac{\omega}{2\pi} = \frac{1}{2\pi}\sqrt{\frac{kg}{W}} = \frac{1}{2\pi}\sqrt{\frac{g}{e_{st}}} \qquad (16\text{-}7.3)$$

It will be observed that Eq. (16-7.1) really shows the combined effect of two vibrations of the same frequency. If the position vectors of these vibrations are plotted 90° apart, as in Fig. 16-7.2, the sum o f their projections on the vertical X axis will be

$$x = x_o \cos \omega t + \frac{v_o}{\omega} \sin \omega t = A \cos(\omega t - \beta) \qquad (16\text{-}7.4)$$

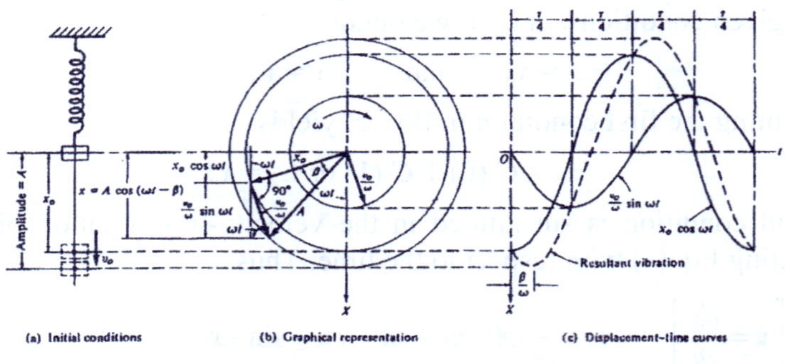

Figure 16-7.2 Combination of vibrations showing phase angle and resultant vibration.

The figure shows that the sum of two simple harmonic motions of the same frequency is another simple harmonic motion of the same frequency. The amplitude of the resultant vibration is

$$A = \sqrt{x_o^2 + \left(\frac{v_o}{\omega}\right)^2} \qquad (g)$$

which lags behind vibration x_o cos ωt by the phase angle β, the value of which is determined from the condition

$$\tan \beta = \frac{v_o}{\omega x_o} \qquad (h)$$

The phase angle β indicates that the resultant vibration reaches its maximum values of displacement, velocity, and acceleration at β/ω see behind vibration x_o cos ωt. The construction shown in Fig. 16-7.2 may be used to represent the displacement curves of each vibration. The resultant vibration is determined by adding these displacements either geometrically or directly from the position vector A.

The following conclusion can be drawn from this discussion: The period and frequency of free vibration of a body are independent of the initial conditions of the motion. The period and frequency of the vibration depend only on the weight W of the body and the scale k of the spring, or on the static elongation e_{st} caused by the weight.

The system of weight and spring shown in Fig. 16-7.1 represents a conventionalized diagram for vibrating bodies in which the supports act like springs, regardless of whether this is their actual purpose. For such cases the period and frequency can also be given by Eqs. (16-7.2) and (16-7.3) by substituting in them the static elongation e_{st} caused by the weight of the body whose vibrations are being considered or the equivalent spring constant of that weight, whichever is more convenient.

ILLUSTRATIVE PROBLEMS

16-7.1 The deflection produced at the free end of the cantilever beam in Fig. 16-7.3 by a static load of 900 N is 5 mm. If a weight W = 1350 N is dropped on the free end of the beam from a height h = 50 mm, compute the frequency of vibration. What will be the maximum deflection? Assume that the weight stays in contact with the beam after striking it.

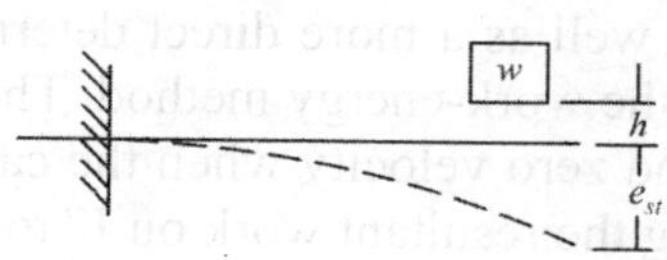

Figure 16-7.3

Solution

Using Eq. (*f*), we find that the spring constant, determined from the deflection produced by the static load, is

$$\left[k = \frac{W}{e_{st}}\right] \qquad k = \frac{900}{5} = 180 \text{ N/mm}$$

The frequency with which the 1335 N weight will vibrate is determined from Eq. (16-7.3).

$$\left[f = \frac{1}{2\pi}\sqrt{\frac{kg}{W}}\right] \qquad f = \frac{1}{2\pi}\sqrt{\frac{(180 \times 1000) \times 9.81}{1350}}$$

$$= 5.76 \text{ vib per sec}$$

The amplitude of vibration is given by Eq. (*g*). Since the initial conditions of the motion are $x_o = - W/k$ and $v_o = \sqrt{2gh}$, and also

since $\omega = \sqrt{kg/W}$, Eq. (*g*) may be rewritten

$$\left[A = \sqrt{x_o^2 + \left(\frac{v_o}{\omega}\right)^2}\right] \qquad A = \sqrt{\left(\frac{W}{k}\right)^2 + \frac{2Wh}{k}}$$

Since the amplitude is measured from the position of static equilibrium, the maximum deflection will be the sum of the static deflection plus the amplitude of vibration. Hence,

$$\left[\begin{aligned} e_{max} &= e_{st} + A \\ &= \frac{W}{k} + \sqrt{\left(\frac{W}{k}\right)^2 + \frac{2Wh}{k}} \end{aligned}\right]$$

$$e_{max} = \frac{1350}{180} + \sqrt{\left(\frac{1350}{180}\right)^2 + \frac{2 \times 1350 \times 50}{180}}$$

$$= 7.5 + 28.4 = 35.9 \text{ mm.} \qquad \textbf{\textit{Ans.}}$$

A check on this result, as well as a more direct determination of the maximum deflection, is afforded by the work-energy method. The weight has zero velocity when initially dropped, and zero velocity when the cantilever beam is deflected through e_{max} mm. Equating the resultant work on W to the zero change in kinetic energy, we have

$$\left[W(h+e_{max})-\frac{1}{2}ke^2_{max}=0\right]$$

$$1350(50+e_{max})-\frac{1}{2}\times 180\times e^2_{max}=0 \qquad e_{max}=35.9\text{ mm} \qquad \textbf{\textit{Ans.}}$$

16-7.2 Consider a body of weight W supported by a system of two springs, as shown in Fig. 16-7.4. Determine the equivalent spring constant for each arrangement.

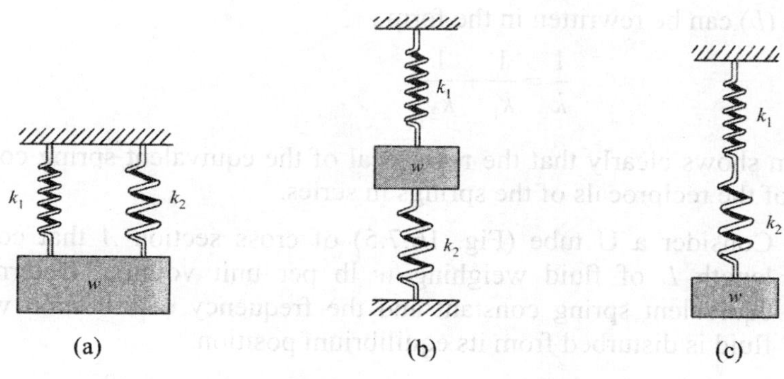

Figure 16-7.4 Springs in parallel and in series.

Solution

In Fig. 16-7.4a, it is evident that a displacement of the body through a unit distance stretches each spring a unit distance. Hence the equivalent spring constant for the system is

$$k = k_1 + k_2 \qquad (a)$$

This arrangement is known as springs in parallel.

A variation of it is shown in Fig. 16-7.4b. At first it is not so evident that the spring constant here is given by Eq. (*a*), but it will be observed that a unit displacement of the weight will stretch one spring a unit amount while compressing the other a unit amount. Hence, the spring force exerted on the weight when displaced a unit distance is also

$$k = k_1 + k_2 \qquad (a)$$

The arrangement in Fig. 16-7.4c is known as springs in series. A unit force applied to W will be transmitted through both springs. Since the displacement of the weight is the sum of the deflection of both springs, a unit force will cause a displacement of the amount

$$s = \frac{1}{k_1} + \frac{1}{k_2}$$

From Hooke's law applied to a spring, $P = ks$, we find $k = P/s$, or for a unit force

$$k = \frac{1}{\dfrac{1}{k_1} + \dfrac{1}{k_2}} \tag{b}$$

Equation (b) can be rewritten in the form

$$\frac{1}{k} = \frac{1}{k_1} + \frac{1}{k_2} \tag{c}$$

This form shows clearly that the reciprocal of the equivalent spring constant is the sum of the reciprocals of the springs in series.

16-7.3 Consider a U tube (Fig. 16:7.5) of cross section A that contains a length L of fluid weighing w lb per unit volume. Determine the equivalent spring constant and the frequency of vibration when the fluid is disturbed from its equilibrium position.

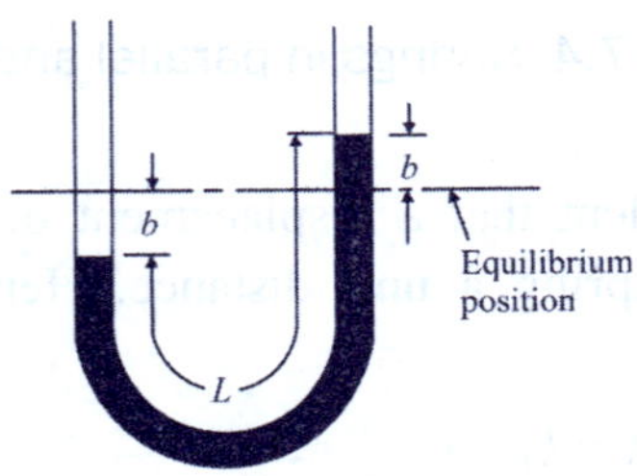

Figure 16-7.5

Solution

If the fluid oscillates back and forth, the weight in motion is $W = wAL$. The unbalanced force acting on the fluid when it is displaced a distance b from its equilibrium position is $2wAb$. This results in a "gravity spring" which tends to restore the fluid to its equilibrium position. The spring constant or force required to displace the fluid through a unit distance is therefore

$$k = 2wA$$

Applying Eq. (16-7.3) gives the frequency

$$\left[f = \frac{1}{2\pi}\sqrt{\frac{kg}{W}}\right] \qquad f = \frac{1}{2\pi}\sqrt{\frac{2\omega Ag}{\omega AL}} = \frac{1}{2\pi}\sqrt{\frac{2g}{L}} \qquad \textit{Ans.}$$

Observe that the frequency of vibration is independent of the fluid. A column of mercury, for example, vibrates with the same frequency as a column of water.

16-7.4 The system shown in Fig. 16-7.6 consists of a stiff rod of length L that has a body of weight W at its free end and is supported by a spring having a constant k_s. Determine the frequency of vibration of the weight W if it is displaced vertically from its position of equilibrium.

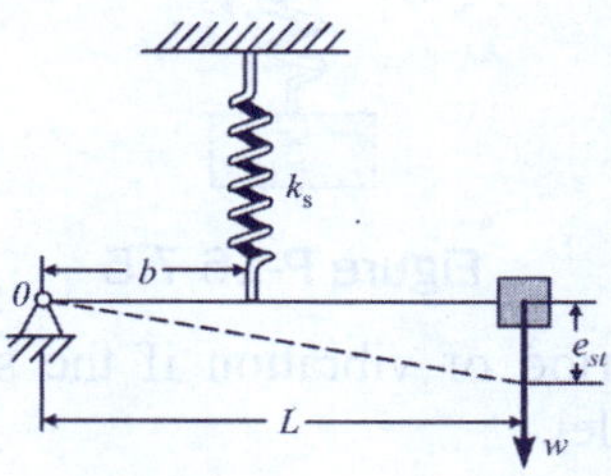

Figure 16-7.6

Solution

It was shown in Eq. (16-7.3) that the frequency of vibration can be expressed in terms of the static deflection produced by the body whose vibrations are being considered.

From the static equilibrium of the rod, moments about O determine the force exerted on the spring to be $P = WL/b$. This stretches the spring an amount

$$e = \frac{P}{k_s} = \frac{WL}{k_s b}$$

From the geometry in Fig. 16-7.6, the corresponding deflection of the weight is

$$e_{st} = e\frac{L}{b} = \frac{W}{k_s}\left(\frac{L}{b}\right)^2$$

Applying Eq. (16-7.3), we find the frequency of vibration to be

$$\left[f = \frac{1}{2\pi}\sqrt{\frac{g}{e_{st}}}\right] \qquad f = \frac{1}{2\pi}\frac{b}{L}\sqrt{\frac{gk_s}{W}} \qquad \textit{Ans.}$$

PROBLEMS

16.7.5 A weight $W = 450$ N is suspended from two springs in series, as shown in Fig. P-16-7.5. If $k_1 = 5$ N/mm and $k_2 = 4$ N/mm, compute the period of free vibration when W is placed vertically from its equilibrium position.

Figure P-16-7.5

16-7.6 Compute the period of vibration if the springs in Prob. 16-7.5 are arranged in parallel.

16-7.7 Assume that the weight in Prob. 16-7.5 has a velocity of 3000 mm per sec as it passes through its equilibrium position. Compute the maximum tension produced in the springs. By how much will this tension be changed if the lower spring is omitted?

16-7.8 A stiff bar of length $L = 1250$ mm and negligible weight is supported by a frictionless hinge at its left end, as in Fig. P-16-7.8. It carries a body of weight $W = 45$ N at a distance $b = 450$ mm from the hinge. The bar is supported at the right end by a spring having a constant $k_s = 10$ N/mm. Determine the frequency of vibration of W if it is displaced vertically from its equilibrium position.

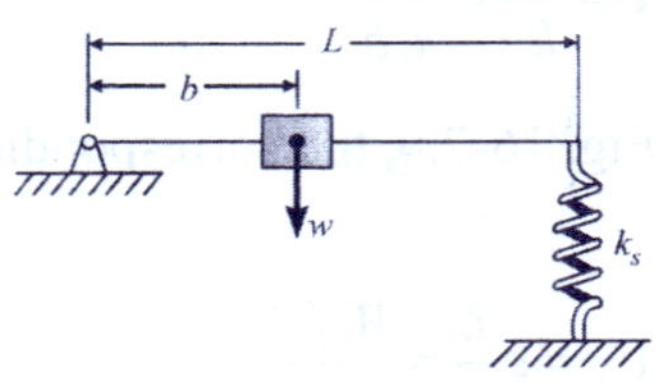

Figure P-16-7.8

16-7.9 As shown in Fig. P-16-7.9, a weight $W = 20$ N rests upon a frictionless surface between two springs having constants $k_1 = 0.3$ N/m and $k_2 = 0.5$ N/mm. The weight is released from rest when it is 20 mm from its equilibrium position. Determine its period of vibration and velocity as it passes through the equilibrium position.

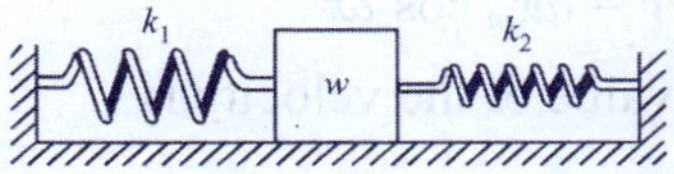

Figure P-16-7.9

16-7-10 An elevator weighing 20 KN is being lowered at the constant rate of 3000 mm per sec when the hoisting drum is stopped suddenly. If the elastic properties of the cable are such that 5 KN stretches it 0.6 mm, compute the frequency of free vibration of the elevator. Determine the maximum tension produced in the cable. Neglect the weight of the cable.

16-7.11 In Prob. 16-7.10, assume that a spring which elongates 25 mm per 10 KN is inserted between the elevator and the end of the cable. Determine the frequency of vibration and the maximum tension produced in the cable.

16-8 FREE VIBRATIONS ANALYZED BY WORK-ENERGY METHOD

The work-energy method provides an alternative technique for determining the frequency and period of vibrating systems. Stated briefly, it consists of determining the elastic energy stored in the system at the instant of maximum displacement and equating this energy to the kinetic energy at the instant the system passes through the position of static equilibrium.

As an example of this method, consider the spring-suspended weight in Fig. 16-7.1. The resultant work is done by the unbalanced spring force, which is zero at the position of static equilibrium and maximum at the position of maximum displacement x_m. The velocity is maximum as the body passes through the position of static equilibrium and at the position of maximum displacement when the body is instantaneously at rest. Expressing the work energy relation between these two positions as an equation, we obtain

$$\left[\text{RW} = \frac{W}{2g}\left(v^2 - v_o^2\right) \right] \qquad \frac{1}{2}kx_m^2 = \frac{W}{2g}(0 + v_m^2) \qquad (a)$$

Since the restoring force kx is always directed toward the origin and is directly proportional to the displacement, we can assume that the motion is harmonic and is defined by the relation

$$x = x_m \sin \omega t$$

The velocity, obtained by differentiating the displacement with respect to the time, is given by

$$v = \omega x_m \cos \omega t$$

Obviously the maximum value of the velocity is

$$v_m = \omega x_m$$

Substituting this value in Eq. (*a*) gives

$$\frac{1}{2}kx_m^2 = \frac{1}{2}\frac{W}{g}\omega^2 x_m^2$$

from which

$$\omega^2 = \frac{kg}{W} \qquad (b)$$

This agrees with the value of natural circular frequency as determined in, Section 16-7. Since x_m^2 cancels out, we conclude as before that time frequency is independent of the amplitude.

ILLUSTRATIVE PROBLEMS

16-8.1 The stiff bar of length L and weight W in Fig. 16-8.1 is hinged freely at one end and supported at the other by a spring for which the constant is k. Determine the period of vibration, using the work-energy method.

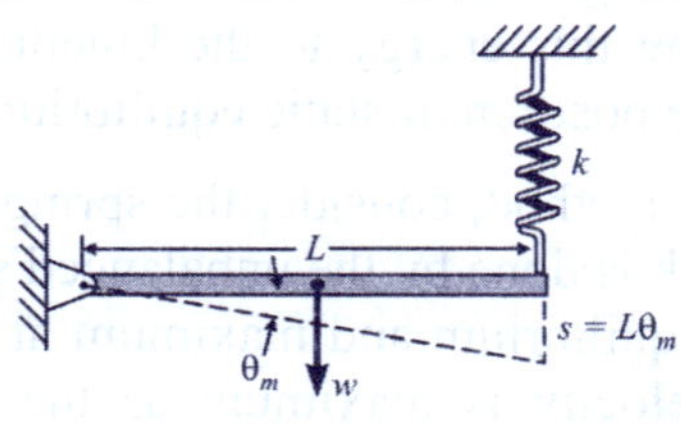

Figure 16-8.1

Solution

Let θ_m be the angle through which the bar rotates from its equilibrium position to reach the position of maximum displacement. Assume that the angular displacement at any instant is defined by the harmonic motion

$$\theta = \theta_m \sin \omega t$$

The angular velocity[3] at any instant, found by differentiation, is

[3] To avoid confusion with natural circular frequency ω, the angular velocity is here denoted by Ω.

$$\Omega = \frac{d\theta}{dt} = \omega\theta_m \cos\omega t$$

from which the maximum angular velocity is seen to be $\Omega_m = \omega\theta_m$.

When the bar is in an extreme position, the spring is stretched an amount $s = L\theta_m$. Equating the elastic energy at an extreme position to the kinetic energy as the bar passes through its equilibrium position, we obtain

$$\frac{1}{2}k(L\theta_m)^2 = \frac{1}{2}I\Omega_m^2 = \frac{1}{2}\left(\frac{1}{3}\frac{W}{g}L^2\right)(\omega\theta_m)^2$$

Therefore,

$$k = \frac{1}{3}\frac{W}{g}\omega^2$$

or

$$\omega = \sqrt{\frac{3kg}{W}}$$

The period is therefore expressed by

$$\tau = \frac{2\pi}{\omega} = 2\pi\sqrt{\frac{W}{3kg}}$$

Comparing this result with that in Prob. 16-2.10, we see that the period is the same as though a body equal to one-third the weigh of the bar were applied directly to the spring.

16-8.2 A uniform rod of weight w and length L is welded to the inside of a hoop of negligible weight and radius L as shown in Fig. 16-8.2. Find the frequency of small oscillations if the assembly is displaced slightly from its equilibrium position and rolls without slipping on the horizontal surface.

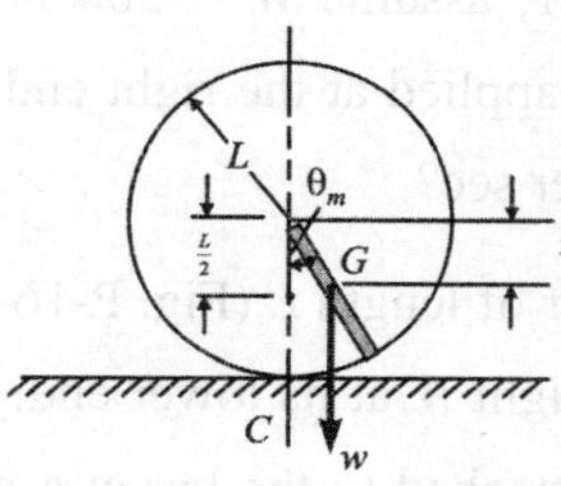

Figure 16-8.2

Solution

As in the previous problem, let θ_m represent the maximum angular displacement from the equilibrium position. Here the system is in plane motion and we use to express the kinetic energy in terms of rotation about the instant center C. Equating the work done in moving from the extreme position to the kinetic energy as the system moves through its equilibrium position, we obtain

$$\left[\text{RW} = \frac{1}{2} I_c \omega^2\right] \qquad W\left(\frac{L}{2} - \frac{L}{2}\cos\theta_m\right) = \frac{1}{2}\left(\frac{1}{3}\frac{W}{g}L^2\right)\Omega_m^2$$

or

$$(1 - \cos\theta_m) = \frac{L}{2g}\Omega_m^2 \qquad (a)$$

Assuming harmonic motion, we have $\Omega_m^2 = \omega^2\theta_m^2$ as in the previous problem. Also, by series expansion for small angles, $\cos\theta_m = 1 - \theta_m^2/2$. Substituting these values in Eq. (a) gives

$$\frac{\theta_m}{2} = \frac{L}{3g}\omega^2\theta_m^2 \qquad \text{or} \qquad \omega^2 = \frac{3g}{2L}$$

whence the frequency is

$$\left[f = \frac{\omega}{2\pi}\right] \qquad f = \frac{1}{2\pi}\sqrt{\frac{3g}{2L}}$$ ***Ans.***

PROBLEMS

16-8.3 In Illus. Prob. 16-8.1, assume W = 300 N and k = 5 N/mm. What additional weight Q applied at the right end of the bar will produce a frequency of 2 vib per sec?

16-8.4 A stiff weightless bar of length L (Fig. P-16-8.4) is hinged at its upper end and carries a weight W at its lower end. Two springs, each having the constant k, are attached to the bar at a distance b from the hinge. Compute the frequency of vibration for small oscillations.

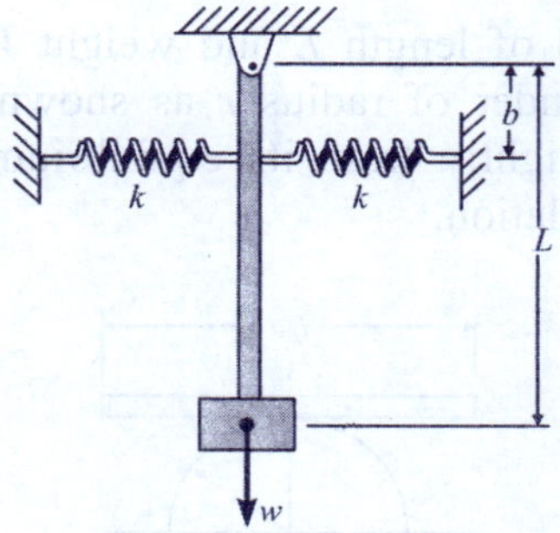

Figure P-16-8.4

16-8.5 If the bar in Fig. P-16-8.4 weighs Q N and the weight W is removed, determine the frequency of vibration, assuming small oscillations.

16-8.6 Let the bar in Fig. P-16-8.4 weigh $Q = 150$ N and the weight $W = 70$ N. If $L = 600$ mm, $b = 150$ mm and $k = 4$ N/mm, compute the period of vibration.

16-8.7 In Fig. P-16-8.7, a homogeneous solid cylinder of radius r and mass m is displaced slightly from its equilibrium position at the bottom of a curved path of radius R. Find the period of small oscillations if the cylinder rolls without slipping.

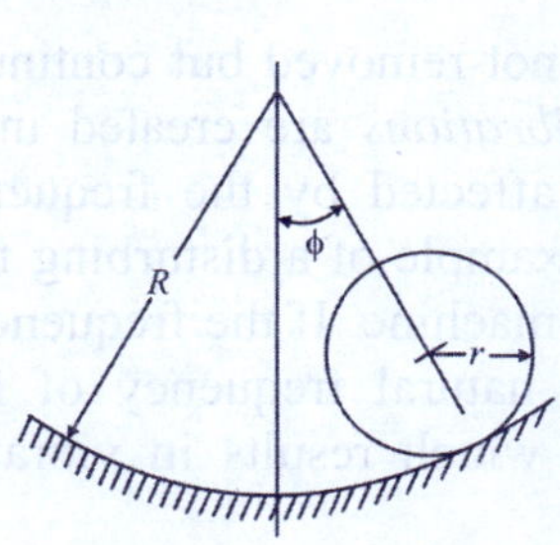

Figure P-16-8.7

16.8.8 A uniform semicircular cylinder of radius r and weight W is displaced through a small angle θ from its equilibrium position as shown in Fig. P-16-8.8. Find the period of its oscillations if it rolls without slipping.

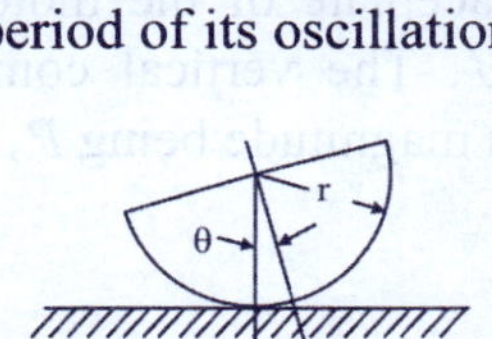

Figure P-16-8.8

16-8.9 A uniform board of length L and weight W is balanced on a fixed semicircular cylinder of radius r as shown in Fig. P-16-8.9. If the plank is tilted slightly from its equilibrium position, determine the period of its oscillation.

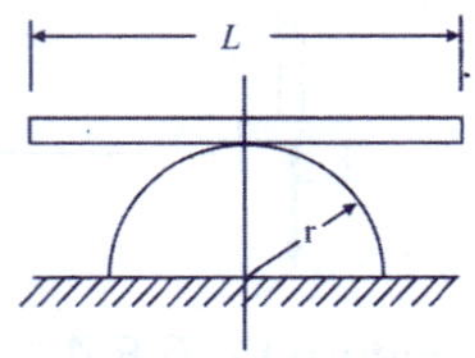

Figure P-16-8.9

16-9 FORCED VIBRATIONS

Thus far we have considered free vibrations in which the amplitude depends only on the initial conditions of the motion. The free vibration is maintained by the action of an unbalanced variable spring force created by the motion itself. The vibration exists only after it is begun by an external disturbing force which, after displacing the body from its position of equilibrium, ceases to act on the vibrating body.

If the disturbing force is not removed but continues to act upon the body at periodic intervals, *forced vibrations* are created in which the frequency and amplitude of vibration are affected by the frequency and magnitude of the disturbing force. A typical example of a disturbing force is the force caused by lack of balance in a rotating machine. If the frequency of the disturbing force is the same as the machine's natural frequency of free vibration, a condition known as *resonance* arises which results in vibrations of dangerously large amplitudes.

In most practical cases the disturbing force varies with the time according to a sine or cosine law. For instance, the motor in Fig. 16-9.1 has a weight W and is mounted on a spring-supported base which permits only vertical movement. Denote the equivalent spring constant of the support by k. A small unbalanced weight W_1 rotates with the faceplate of the motor, and creates a centrifugal inertia force $P_o = (W_1/g)\ r\omega^2$. The vertical component of this force varies harmonically with the time, its magnitude being $P_o \cos \omega t$.

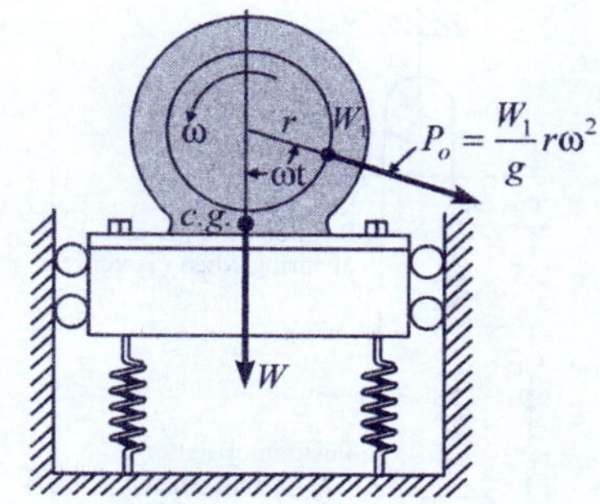

Figure 16-9.1 Forced vibration caused by an unbalanced rotating weight.

If displacements from the position of static equilibrium are taken as positive downward, the equation of motion is

$$\left[\Sigma X = \frac{W}{g} a\right] \qquad W + P_o \cos \omega t - (W + kx) = \frac{W}{g}\frac{d^2x}{dt^2}$$

Multiplying this by $\frac{g}{W}$ gives

$$\frac{d^2x}{dt^2} = -\frac{kg}{W}x + \frac{gP_o}{W}\cos \omega t \tag{a}$$

Before solving this equation, consider another example of forced vibration illustrated by the spring-suspended weight shown in Fig. 16-9.2. The upper end of the spring is connected by an inextensible cord to a disk of radius r that rotates with an angular velocity ω. Assuming the cord to be long compared with r, this creates in the upper end of the spring the simple harmonic motion

$$b = r \cos \omega t$$

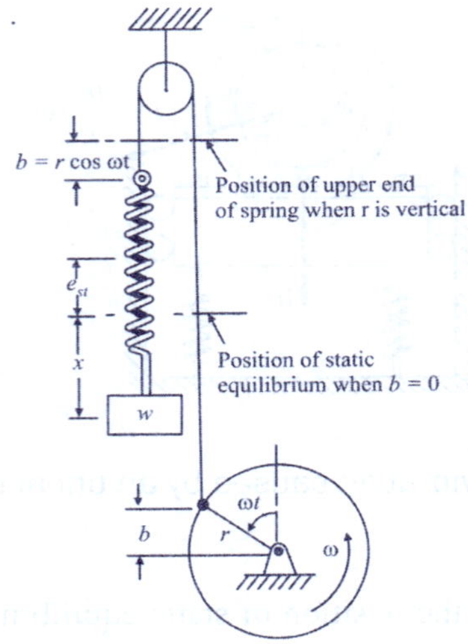

Figure 16-9.2 Forced vibration of a spring-suspended weight due to motion of the point of support.

The time t is measured from the instant radius r is vertically up, and b is measured from the center of the disk. The spring extension is $x - b$ because the downward movement of the top of the spring lends to overcome the downward stretching movement x caused by the vibration of the weight. Hence, at any instant the unbalanced upward spring force is $k(x - b)$.

Forces, displacements, and accelerations being taken as positive downward, the equation of motion is

$$\left[\Sigma X = \frac{W}{g}a\right] \qquad -k(x-b) = \frac{W}{g}\frac{d^2x}{dt^2}$$

If we multiply this by $\frac{g}{W}$ and replace b by $r\cos\omega t$, this becomes

$$\frac{d^2x}{dt^2} = -\frac{kg}{W}x + \frac{krg}{W}\cos\omega t \qquad (b)$$

It will be noticed that Eqs. (a) and (b) are identical except for the form of the constant term preceding the expression cos ωt in each equation. The only difference between the two cases is that in the second case the disturbing force is transmitted to the weight through the spring instead of being applied directly to it. Denoting the coefficient of cos ωt by h, we have in the first case

$$h = \frac{gP_o}{W}$$

and in the second case

$$h = \frac{krg}{W}$$

Also, as in Section 16-7, the expression $\frac{kg}{W}$ can be replaced by ω_n^2.

Note that ω_n is the natural circular frequency of a free vibration, whereas in each of Eqs. (*a*) and (*b*) above, ω is the circular frequency of the disturbing force. Thus, Eqs. (*a*) and (*b*) may be put into the form

$$\frac{d^2x}{dt^2} = -\omega_n^2 x + h\cos\omega t \qquad (c)$$

The forced vibration represented by Eq. (*c*) may be visualized as a combination of a free vibration of the type discussed in Section 16-7 plus a new motion caused by the disturbing force. Thus, if x_1 represents the displacement caused by the free vibration (which is independent of the disturbing force) and x_2 is the additional displacement caused by the disturbing force, the actual displacement is

$$x = x_1 + x_2$$

With this assumption, Eq. (*c*) may be considered to be composed of the following parts:

$$\frac{d^2x_1}{dt^2} = -\omega_n^2 x_1 \qquad (d)$$

$$\frac{d^2x_2}{dt^2} = -\omega_n^2 x_2 + h\cos\omega t \qquad (e)$$

Adding these equations gives the original equation (*c*); Eq. (*d*) is similar to Eq. (*a*) in Section 16-7. Hence the value of x_1 is given by

$$x_1 = C_1 \sin \omega_n t + C_2 \cos \omega_n t \qquad (f)$$

It seems reasonable to suppose that a tentative solution of Eq. (*e*) is $x_2 = A \cos \omega t$, in which the constant A, representing the amplitude of forced vibration, is selected to satisfy the equation. Substituting this solution in Eq. (*e*) yields

$$-A\,\omega^2 \cos \omega_n t = -A\omega_n^2 \cos \omega t + h \cos \omega t$$

from which the value of A is

$$-A\,\omega^2 + A\omega_n^2 = h$$

or

$$A = \frac{h}{\omega_n^2 - \omega^2} \tag{16-9.1}$$

Hence, a solution of Eq. (e) is

$$x_2 = A\cos\omega t = \frac{h}{\omega_n^2 - \omega^2}\cos\omega t \tag{16-9.2}$$

The complete solution of Eq. (c) is the sum of the solutions of its parts, or

$$x = x_1 + x_2 = C_1\sin\omega_n t + C_2\cos\omega_n t + \frac{h}{\omega_n^2 - \omega^2}\cos\omega t \tag{16-9.3}$$

This result is shows graphically in Fig. 16-9.3. Part (a) represents the displacement-time diagram for the free vibration (assuming $C_1 = 0$) and (b) is the diagram of the forced vibration. The summation of these two harmonic motions (which have different periods) is shown in (c).

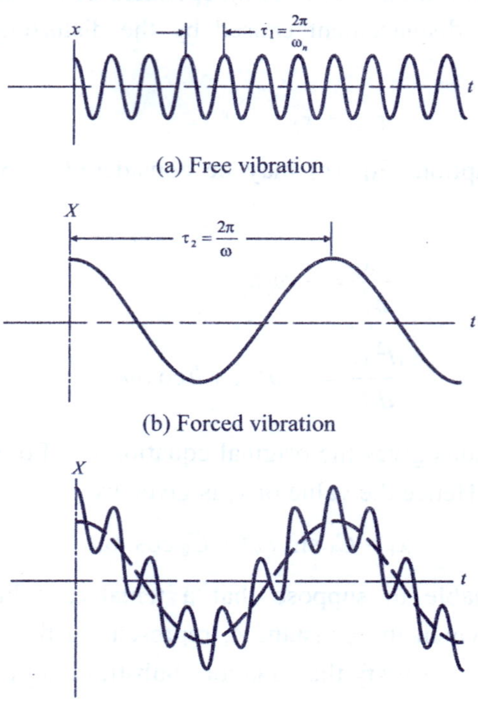

(a) Free vibration

(b) Forced vibration

(c) Resultant vibration

Figure 16-9.3

Since some damping of the free vibration will always exist, the free vibration will die out, as shown in Fig. 16-9.4a. Combining the damped free vibration with the forced vibration results in the steady state of forced vibration shown in Fig. 16-9.4b which is maintained indefinitely by the periodic disturbing force.

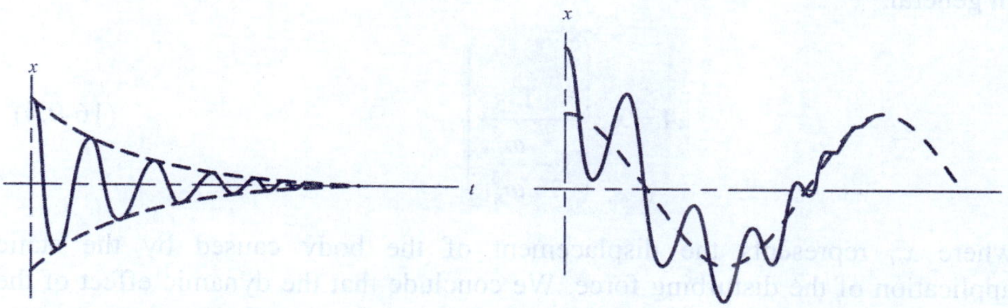

(a) Damped free vibration (b) Resultant steady state of forced vibration

Figure 16-9.4

Magnification Factor

Because of the damping of the free vibration, it is the motion of forced vibration represented in Fig. 16-9.4b that is of practical importance. If the free vibrations which soon die out are neglected, the maximum displacement (or amplitude) of forced vibration will be given by Eq. (16-9.1).

For the spring-mounted motor shown in Fig. 16-9.1,

$\omega_n^2 = \dfrac{kg}{W}$ and $h = \dfrac{gP_o}{W}$. Hence the amplitude of forced vibration reduces to

$$A = \frac{P_o}{k}\left(\frac{1}{1-\dfrac{\omega^2}{\omega_n^2}}\right) \qquad (g)$$

Similarly in the spring-suspended weight in Fig. 16-9.2, if h is replaced by $\dfrac{krg}{W}$, the amplitude of forced vibration becomes

$$A = r\left(\frac{1}{1-\dfrac{\omega^2}{\omega_n^2}}\right) \qquad (h)$$

The factor $\frac{P_o}{k}$ in Eq. (*g*) represents the maximum displacement caused by the static action of the disturbing force P_o. Also r in Eq. (*h*) represents the maximum displacement of W in Fig. 16-9.2 when the disk revolves very slowly. In general,

$$A = x_{st}\left(\frac{1}{1-\frac{\omega^2}{\omega_n^2}}\right) \tag{16-9.4}$$

where x_{st} represents the displacement of the body caused by the static application of the disturbing force. We conclude that the dynamic effect of the disturbing force on the maximum displacement is given by the expression

$$\frac{1}{1-\frac{\omega^2}{\omega_n^2}} \tag{i}$$

This is known as the *magnification factor.*

The effect of this factor is shown in Fig. 16-9.5, in which the factor has been plotted in terms of the ratio ω/ω_n. Thus, when the disturbing force alternates slowly compared with the natural frequency (ω/ω_n less than 1/2), the magnification factor is only slightly larger than unity. Hence, the dynamic deflection is only slightly greater than that produced if the disturbing force were applied statically.

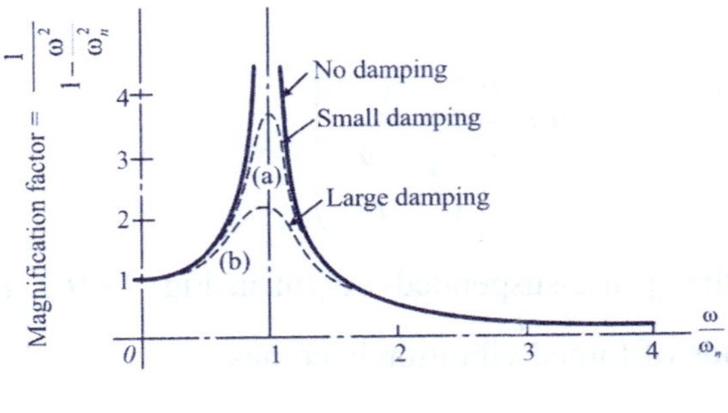

Figure 16-9.5

As the impressed frequency approaches the natural frequency, i.e., as ω becomes equal to ω_n, the magnification factor increases rapidly, reaching infinity when ω equals ω_n. Physically this means that when $\omega/\omega_n = 1$, the frequency of the disturbing force coincides exactly with the natural frequency; i.e., the disturbing force tends to push the body always at the right time in the right direction so that the amplitude increases without limit. The condition existing when the impressed frequency is equal to the natural frequency is known as *resonance*. In a machine, the speed of rotation at which resonance occurs is called the *critical speed*. Obviously a machine should be designed to operate well above or below this speed.

As ω is increased above resonance, the magnification factor becomes negative. Physically this means that the disturbing force is 180° out of phase with the motion; i.e., it is of opposite sense to the displacement. Usually this phase relation is of only little interest and the sign of the magnification factor is neglected. It will be observed that when the impressed frequency is well above the natural frequency the magnification factor approaches zero and the body tends to stand still in space. This is because the disturbing force moves up and down so rapidly that the body cannot follow it because of its inertia.

It can be shown that the effect of a relatively small and a relatively large viscous damping can be represented by curves (*a*) and (*b*) in Fig. 16-9.5. Note that there is practically no change in the magnification factor except near resonance and that the resonant frequency is practically unchanged.

For torsinal forced vibrations, equations similar to those discussed above can be obtained. To do this, we replace force by torque, mass by mass moment of inertia, and linear values of displacement, velocity, and acceleration by corresponding angular values.

PROBLEMS

16-9.1 A horizontal shaft rotates in bearings at its ends. At its midpoint is keyed a disk weighing 200 N, whose center of gravity is 3 mm from the axis of rotation. If a static force of 900 N deflects the shaft and disk through 3 mm, determine the critical speed of rotation of the shaft.

16.9.2 The motor in Fig. 16-9.1 has a weight W = 5 KN and is supported by four springs each having a constant of 45 N/mm. When the motor is rotating at 2000 rpm, the centrifugal force caused by the eccentric weight is P_o = 900 N. Compute the maximum amplitude of forced vibration.

16-9.3 The spring-suspended weight in Fig. 16-9.2 weighs 300 N and the spring constant is k = 20 N/mm. The disk of radius r = 25 mm rotates at 25 rad per sec. Compute the maximum tension produced in the spring.

16-9.4 The bob in Fig P-16-9.4 weighs 35 N and is fastened to the upper end of a flexible rod. The lower end of the rod is embedded in a block that oscillates horizontally with an amplitude of 1.5 mm and a frequency of 1 vib per sec. The elastic properties of the rod are such that a horizontal force of 45 N applied at the upper end deflects the bob through 25 mm. Determine the amplitude of forced vibration of the bob.

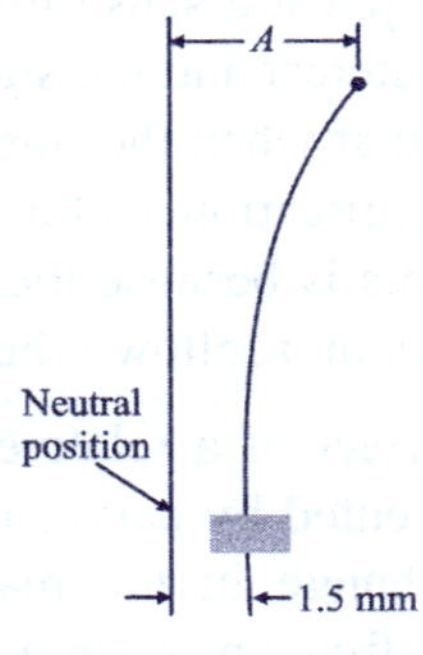

Figure P-16-9.4

SUMMARY

Vibrations are essentially periodic motions which repeat themselves after a definite time interval called the period. The *frequency* is the reciprocal of the period and represents the number of vibrations per second.

Most vibrations have simple harmonic motion. The mathematical description of simple harmonic motion is

$$a = -\omega^2 s \tag{16-2.1}$$

in which ω is the natural circular frequency of the motion. If a motion can be expressed in this form, the period and frequency are respectively

$$\tau = \frac{2\pi}{\omega} \tag{16-2.2}$$

$$f = \frac{\omega}{2\pi} \tag{16-2.3}$$

For purposes of comparison, the period and frequency for various types of pendulums are summarized in the following table:

Pendulum	Period	Frequency	Reference
Simple	$2\pi\sqrt{\frac{L}{g}}$	$\frac{1}{2\pi}\sqrt{\frac{g}{L}}$	Section 16-3
Compound	$2\pi\sqrt{\frac{k_z^2}{g\bar{r}}}$	$\frac{1}{2\pi}\sqrt{\frac{g\bar{r}}{k_z^2}}$	Section 16-4
Torsion	$2\pi\sqrt{\frac{IL}{JG}}$	$\frac{1}{2\pi}\sqrt{\frac{JG}{IL}}$	Section 16-5

The manner in which displacement, velocity, and acceleration vary with time can be visualized by projecting upon a diameter the length of the position, velocity, and acceleration vectors. These vectors rotate about a common axis with the same angular velocity ω, which is the natural circular frequency of the vibration. The magnitude of the position vector is r, the amplitude of the vibration. The magnitude of the velocity vector is $r\omega$ located 90° ahead of the position vector; the magnitude of the acceleration vector is $r\omega^2$ located 90° ahead of the velocity vector.

The general solution of the differential equation of the free vibration of a spring-suspended weight is

$$x = \frac{v_o}{\omega}\sin \omega t + x_o \cos \omega t \tag{16-7.1}$$

in which x_o is the initial displacement from the position of static equlibrium and v_o is the initial velocity at x_o.

The combined effect of the two vibrations of the same frequency expressed in Eq (16-7.1) is equivalent to

$$x = A\cos(\omega t - \beta) \tag{16-7.4}$$

in which A is the amplitude of a resultant vibration that has same frequency as each of the combined vibrations. The magnitude of the amplitude A and of the phase angle β is determined from

$$A = \sqrt{x_o^2 + \left(\frac{v_o}{\omega}\right)^2} \quad \text{and} \quad \tan\beta = \frac{v_o}{\omega x_o}$$

Section 16-7 shows that the period and frequency of a free vibration are independent of the initial conditions of the motion. They depend only on the weight W and the constant k of the spring support or on the static elongation e_{st} caused by the weight as expressed by

$$\tau = 2\pi\sqrt{\frac{W}{kg}} = 2\pi\sqrt{\frac{e_{st}}{g}} \tag{16-7.2}$$

$$f = \frac{1}{2\pi}\sqrt{\frac{kg}{W}} = \frac{1}{2\pi}\sqrt{\frac{g}{e_{st}}} \tag{16-7.3}$$

Hence, the period or frequency of body of weight W which is part of a vibrating system may be found by determining the static elongation caused by that weight or computing the equivalent spring constant of the weight, whichever is more convenient.

An alternative technique for determining the frequency and period of vibrating systems is the work-energy method. In this, the elastic energy in the system at the instant of maximum displacement is equated to its kinetic energy at the instant the system passes through its mid-position. This procedure determines the natural circular frequency of the systems.

In a forced vibration, the maximum displacement (i.e., amplitude) is given by

$$A = x_{st}\left(\frac{1}{1-\frac{\omega^2}{\omega_n^2}}\right) \tag{16-9.5}$$

When x_{st} is the displacement cause by the static application of the disturbing force, ω is the circular frequency of the disturbing force, and ω_n is the natural circular frequency of free vibration of the system.

INDEX

A

Absolute acceleration, in planar motion, 543
Absolute motion, 417
Absolute reference frames, 479
Absolute systems, 4
Absolute systems of units, 5, 481
Acceleration, 419
  angular, 536
  constant, 424, 536
  in curvilinear motion, 445, 454, 465
  from hodograph, 420-421
  instant center of, 569
  of instant center of velocity, 569
  jerk, 442
  normal component of, 421, 493
  polygon, 551
  radial and transverse components of, 465
  tangential component of, 421, 456, 493
  variable, 422-427
Acceleration-time diagram, 432, 539
  displacement from, 433, 539
  velocity from, 432, 539
Accuracy, numerical, 8
Action and reaction, 4, 92, 172, 215, 478
Active-force diagram, 661
Addition of forces, coplanar, 25
  spatial, 30
Addition of vectors, 37
Adhesion, 196
Amplitude of vibrations, 769, 787, 802, 803
Analysis of structures, 137-194
  by graphic methods, 405
  by method of joints, 143
  by method of members, 169
  by method of sections, 154
  by slopes, 144
Angle, of friction, 198
  phase, 787
Angular acceleration, 536
Angular displacement, 534
  from acceleration-time curve, 539
  not a vector, 534
  from velocity-time curve, 539
Angular impulse, 735
Angular momentum, 735
  conservation of, 738
  in plane motion, 589
Angular displacement *(continued)*
  rate of change of, 487, 637
  relative, 487
Angular motion, 533-539
  acceleration in, 536

displacement in, 535
from motion curves, 539
related to linear motion, 536-537
sign convention for, 536
velocity in, 536
Angular velocity, 536
from acceleration-time curve, 539
Apogee, 753
Areal velocity, 753
Areas, centroids of. *See* Centroids
moment of inertia for. *See* Moments of inertia for areas
properties of, 253, 291
Astronomical frame of reference, 439
Auxiliary circle 769, 783
Axes, rotating reference, 608, 612
Axioms of mechanics, 4
Axis, centroidal, 244
of moments, 54
of precession, 810
principal, of inertia, 298
of rotation, 534
of spin, 759
of symmetry, 298

B

Ballistic pendulum, 722
Band brakes, 226
Banking of highway curves, 497
Beams. *See also* Bending moment in beams; Shear in beams
methods of support, 340
statically determinate, 340
statically indeterminate, 341
types of loads, 340
Bearing reactions, 675, 676
Belt friction, 226
Bending moment diagrams,
construction of, 358, 361
shape of, 358, 361
Bending moment in beams, 339-372
defined, 341, 343
diagrams, 348
equations of, 346, 350
relation to shear, 354-378
sign of, 343
Binormal unit vector, 458
Bodily contact, types of, 93
Body, center of gravity of, 242, 270
freely falling, 425
rigid, 478
Bohr, Niels, 479
Bow's notation, 395, 356, 359
Brake, differential band, 228

C

Calculators, pocket, 10-12
Center, instant, of zero acceleration, 569-570
Center, instant, of zero velocity, 566-575
acceleration, 570
kinetic energy about, 686
location of, 566
Center of gravity, 242
of bodies, 243, 270
summary of, 276
Center of moments, 64
Center of oscillation, 737
Center of percussion, 594
Center of suspension, 776
Central force motion. See Satellite motion
Centrifugal inertia force, 493
Centroidal axis, 245
Centroidal rotation, 641

Centroids, 244
    of areas and lines, 244
    of composite figures, 252
    for geometric shapes, 253
    importance of, 244
    by integration, 245
    signs for, 253
    of structural shapes, 252
    summary of, 277
    by tabular computation. 256, 274
    theorems of Pappus, 277
    of volumes, 270
Chasle's theorem, 545
Circle, auxiliary, 769-783
    Mohr's, 309
Circular frequency, 770
Circular orbit, 752
Classification of mechanics, 2
Coefficient of friction, 197
Coefficient of restitution, 728
Coefficient of rolling resistance, 238
Components, 18
    of acceleration. *See* Acceleration
    oblique, 18
    rectangular, 19
    of spatial forces, 30, 33
    of velocity. *See* Velocity
Composite areas, centroids of, 252
    moments of inertia for, 289
Compound pendulum, 776
    center of oscillation, 777
    center of suspension, 776
    equivalent length of, 776
    period of, 776
Compound trusses, 140, 184
Compression in truss members, 141
Concentrated load, 340
Concurrent forces, equilibrium of, 98
    resultant of, 25, 33
Conditions for, stability, 142
    static determinancy, 95
Cone of friction, 200
Conservation of angular momentum, 738
    in satellite motion, 750
Conservation of linear momentum, 719-724
Constant acceleration, in rectilinear motion, 422
    in rotation, 537
Constant of gravitation, 484
Coordinate systems, various right-banded, 40
Coplanar forces, equilibrium of, 93-100, 104-112
    resultant of, 76-78, 81-85
Coriolis acceleration, 615
Coulomb friction. See Friction
Counter diagonals, 169
Couples, 73
    effect of, 74
    independence of moment center, 73
    moment arm of, 73
Couple vector, 74
Criteria for stability, of frames, 171
    of planar trusses, 143
    of space trusses, 185
Critical speed, 855
Cross product, defined, 46
    determinant form of, 48
    evaluation of, 47
    interpretation of, 46
    moment as a, 55
    properties of, 48
    of unit vectors, 47
Curvature, radius of, 458
Curvilinear motion, 444-461
    binormal unit vector, 457
    flight of projectiles, 446
    normal and tangential components of, 454-457

osculating plane, 457
principal normal of, 457
radial and transverse
components in, 463-466
rectangular components of, 444
Curvilinear translation. *See* Translation
Cylindrical coordinates, 463

D

D'Alembert's principle, 482
translation, 518-521
geometrical development, 483
mathematical development, 484
Degrees of freedom, 378
Derivative of vector fixed in a rotating body. *See* Omega - theorem
Determinant, evaluation of, 48
Determinant form, of cross product, 48
for moment of force, 66-68
for scalar triple product, 51
Determinate beams, 340
Diagram, acceleration-time, 432, 539
active-force, 711
displacement-time, 432, 539
force-time, 505, 755
free-body, 6, 96
Maxwell, 405, 408
velocity-time, 432, 539
Diagrams, shear and moment, 348, 357
application of, 357-366
Differential band brake, 228
Differential equations of motion, 422
solution of, 423-424
Differentials and Integrals, 15
Differential work, 641
Dimensional checks, 8
Dimensions, basic, 481
derived, 481
Direct central impact, 727
Direct eccentric impact, 727
Direction cosines, 33
Direction of motion determined, for plane motion, 611
for rotation, 593
for translation, 508
Disk friction, 233
in new surfaces, 234
in worn surfaces, 234
Displacement, 418
from acceleration-time diagram, 432, 539
angular, 534
from force-time diagram, 505, 507
from velocity-time diagram, 432, 539
virtual, 376
Distance, contrasted to displacement, 418
Distributed load, 340
Dot convention, Newton's, 419
Dot product, defined, 40
to determine component by, 40
evaluation of, 41
interpretation of, 40
properties of, 41
of unit vectors, 30
Dynamic equilibrium, defined, 481
explanation of, 518
in plane motion, 625
in rotation, 591, 598
in translation, 518-521
Dynamics, defined, 477
fundamental equations of, 416
general principles of, 477-489
history of, 477
preview of, 415

Dynamics of jet streams, 711
on moving vane, 714
on stationary vane, 712
Dyne, 5

E

Earth satellites. *See* Satellite motion
Eccentric impact, 727
Effective force, on a particle, 481
on a rigid body, in planar motion, 590, 608, 622
Efficiency, 657
Einstein, Albert, 479
Elastic impact. *See* Impact
Elliptic orbit, 753
Energy, kinetic. *See* Kinetic energy
Energy, potential, 646
Engineering mechanics, defined, 1
Equilibrium, of concurrent forces, 98
defined, 91
dynamic. *See* Dynamic equilibrium
equivalence of force and moment summations for, 105
general equations of, 94, 119
by graphic methods, 401
meaning of, 91
of nonconcurrent coplanar forces, 104-112
of spatial systems, 118-127
Equivalence of applied forces and dynamic effects, 590, 595
Escape velocity, 753
Exploratory section, 341

F

Flight of projectiles, 446
maximum height and range; 448
Force, characteristics of, 3
components of, 18, 30, 33
defined, 3
external effects of, 3
impulsive, 706
intercepts of line of action of, 59, 77
resultant, 25, 33
reversed effective. *See* Inertia force
on a spring, 644
Force-couple system, 75
Force-displacement diagram, 540, 643
for a spring, 644
Forced precession, 760
Forced vibrations, 798
amplitude of, 801
critical speed of, 805
magnification factor of, 803
resonance caused by, 799, 804
torsional, 805
Force multiplier, 56
applications of, 33, 120
defined, 34
Force polygon, 394, 397
Forces in space. *See* the Spatial forces
direction cosines, 32
equilibrium of, 119
resultant of, 33, 76
Force-time diagram, 505-507,

Frames, analysis of, 170
Free-body diagrams, defined, 6, 92
  introduction to, 6
  method of constructing, 92, 95, 119
  table of bodily contact, 93
Freedom, degrees of, 377
Freely falling bodies, 425
Free-rolling bodies, kinematics of, 549
  kinetics of, 607-614
Free vibrations, 769, 784
  amplitude of, 771, 787
  frequency of, 771,775, 786
  period of, 768, 775, 777, 781, 786
  phase angle of, 787
  by work-energy method, 793
Frequency, 771, 775, 786
  natural circular, 770
Friction, 195-240
  angle of, 198
  belt, 226
  cone of, 199
  defined, 195
  disk, 233
  kinetic, 197
  laws of, 200
  rolling, 238
  screw, 223
  static, 197
  summary of, 239
  theory of, 195
  in wedges, 214
Friction force on highway curves, 498
Frictionless path, velocity along, 650
Fundamental concepts and axioms, 2
Funicular polygon, 396, 397

G

Galilean reference frame, defined, 479
Galileo, 478
General moment equation, 487
  in a noninertial translating frame, 488
General omega theorem, 585
General plane motion. *See also* Plane motion
  dynamic equilibrium in, 625
  effective forces in, 622
  equivalent sets of dynamic effects, 622
  about instant center of zero acceleration, 622
  kinetics of, 621-629
General principles of dynamics, 477-489
Geometric shapes, centroids for, 253
  moments of inertia for, 290, 328
Graphic methods, 394-410
  acceleration by, 551
  analysis of trusses by, 405
  Bow's notation for, 398, 407
  equilibrium by, 401
  Maxwell diagrams for, 407, 408
  substitute member in, 410
  resultant of coplanar forces by, 25
  simple harmonic motion by, 769, 783, 787
  string polygon for, 396, 397
  summary of, 414
  velocity by, 552
Gravitation, Newton's law of, 479
  universal constant of, 479
Gravitational mass, 480

Gravitational systems, 4
of units, 4, 481
Gravity, acceleration due to, 425
center of, 241, 242, 270
spring, 790
Guldin, theorems of, 264
Gyration, radius of, 281, 317
Gyroscopic action, introduction to, 759-781
precession axis, 760
spin axis, 759
torque axis, 760

H

Hand calculators, 10-12
Harmonic motion, simple, 819
Heisenberg, Werner, 479
Henneberg's method, applied to Wichert truss, 167
Highway curves, banking of. 497
Hodograph, 421
Hooke, Robert, 477
Hooke's law, 643
Horsepower, 567, 657
Huygens, 477, 777

I

Ideal angle of banking, 498
Impact, elastic, 726-730
coefficient of restitution; 727
energy loss in, 728, 740
inelastic or plastic, 728, 743
line of, 726
in plane motion, 734
in rotation, 742
types of, 727
Impending motion, direction of
in rotation, 594
in translation, 508-509
Impulse, angular, 735
linear, 705
Impulse-momentum method, 704-767
advantages of, 738
diagrammed approach to, 737, 739, 742
difference from work-energy, 736
in plans motion, 734-744
principle of, 736
summary of, 764
in translation, 505-507, 704-709
Impulsive motion, 706
Inclined axes, moments, of inertia about, 306
Indeterminate beams, 341
Inertia, 481
moments of. *See* Moments of inertia for areas
principal axes of, 298, 312
products of. See Product of inertia
Inertia components, 598, 628
Inertia force, centrifugal, 493
on a particle, 481
in plane motion, 628
tangential, 492
in translation, 518
Inertial axes, use of, 561
Inertial mass, 481
Inertial reference frame, defined, 479
Inertia tensor, 5
Instant center of zero acceleration, 569-571
Instant center of zero velocity, 566-571
acceleration of, 569
kinetic energy with respect to, 686

location of, 570
use of, 571-575
Instantaneous axis of rotation, 566
Intercepts of resultant force, 60, 77
Internal force, 484
in beams, 341
in structures, 141

J

Jackscrew, 224
Jerk, 442
Jet streams, 711-716
variable mass systems, 711
Joints, method of, 143, 186

K

Kepler, Johann, 754
Kepler's laws, 754
Kilowatt, 657
Kinematic relations in connected systems, 508
for plane motion, 613
for rotation, 539, 593
Kinematics of gyroscope, 596
Kinematics of particles, 415-476
in curvilinear motion, 444, 454, 463
defined, 417
in rectilinear motion, 422, 431
by motion curves, 431-440
sign convention, 421
Kinematics of rigid bodies, 532-634
angular motion, 533-540
in plane motion, 544-540
Chasle's theorem, 545
free-rolling bodies, 549
by geometric analysis, 549-556
by scalar calculus, 543
in rotation, 533-540
by motion curves, 539
Kinetic energy, defined, 640
about instant center, 686
in plane motion, 686
in rotation, 669
of a system of bodies, 661, 669, 688, 695
Kinetic friction, 198
Kinetics of particles, 490-531
in curvilinear motion, 492
effective force on, 481
laws of motion for, 478, 480
in polar coordinates, 493
in rectilinear motion, 492
Kinetics of rigid bodies, 585-638
equivalence of applied, forces to dynamic effects, 588, 592, 595
in fixed-axis rotation, 590
in plane motion, 607, 621

L

Lagrange, J. L., 574
Law of universal gravitation, 479
Laws, of friction, 200
Kepler's, 754
of motion, Newton's, 478
Lead of square-threaded screw, 223
Line, angular motion of, 533, 547
centroid of, 243
Linear displacement, from acceleration-time diagram, 432
from force-time diagram, 505, 507
from velocity-time diagram, 432
Linear Impulse, 705
Linear impulse-momentum equation, 705
component form of, 706
diagram of, 706

Linear momentum, 705
applied to jet streams, 712-714
conservation of, 719-724
principle of, 720
Load-shear relation for beams, 354

M

Machines, analysis of, 182-183
Magnification factor, 803
Marked member, 146
Mass, defined, 3
Mass, gravitational versus inertial, 480
Mass, time-dependent, 712
variable, systems. 711, 716
Mass center, motion of, 485
Mass moments of inertia, 316-337
for composite bodies, 327
determined by compound pendulum, 778
by integration, 319
radius of gyration for, 317
table of, 328
transfer formula for, 317
units of, 317
Maximum and minimum moments of inertia, 312
by Mohr's circle, 309
Maximum bending moment, 341, 358, 368
Maxwell diagrams, 405, 408
Mechanics, defined, 1
Members, marked, 146
multi-force, 170-171
two-force, 141, 193
Method, of joints, 143, 186
of members, 169
of sections, 154, 189
of slopes, 144
of solving problems, 7
of successive approximation, 709
Metric system of units, 482
conversion equivalents, 5
MKS system of units, 482
Modulus, of elasticity, 393, 830
of a spring, 644
Mohr's circle for moments-of inertia, 309
application of, 310
rules for applying, 310
Moment, of areas, 244, 256
bending. *See* Bending moment in beams
coplanar applications of, 57
sign convention, 57
of a couple, 73
diagrams. *See* Bending moment diagrams
of force, defined, 54
of momentum. *See* Angular momentum
spatial applications of, 63
of vector, 54
as vector cross product, 55
Moment effect of external forces, 486
fundamental equation for, 487
Moments of inertia for areas, 278-312
composite areas, 289
defined, 278
by integration, 283
maximum and minimum, 312
Mohr's circle for, 309
polar, 280
principal axes for, 312

radius of gyration for, 281
table of geometric shapes, 290
transfer formula for, 281
units and signs, 279
with respect to inclined axes, 306
Moments of inertia of bodies. *See* Mass moments of inertia
Momentum, angular, 735
conservation of 738
linear, 704
Motion, curvilinear, 394-411
of mass center, 435
plane *See* Plane motion
possible direction of, 458, 593
of projectiles, 446
rectilinear, 422
of rocket, 715
simple harmonic, 769
uniformly accelerated, 424
Motion curves, 431-440
correlation with force-time diagrams, 505
correlation with shear and moment diagrams, 435
geometric significance of, 433
relations among, 431
Multiforce members, 170, 171

N

Natural circular frequency, 770
Neutral axis, 244
Newton, Sir Isaac, 478
Newton as unit of force, 481
Newtonian reference frame, defined, 479
Newton's law of gravitation, 479
universal constant of, 479
Newton's laws of motion, 478
Nodal section, 781
Noncentroidal rotation, dynamic equilibrium in, 591
effective forces in, 591
kinetic energy in, 719
kinetics of, 591
Nonconcurrent forces, equilibrium of, 104-111, 118-126
resultant of, 76-80
Normal acceleration, 454, 456
in terms of rectangular components, 456
unit vector of, 456
Notation, Bow's, 396, 406, 409
dot, 419
vector, 37
Numerical accuracy, 12

O

Oblique central impact, 727, 730
Oblique eccentric impact, 657
Omega theorem, 582-585
applied to time derivatives of unit vectors, 584
in fixed-axis rotation, 584
Orbiting satellites. *See* Satellite motion
Origin for absolute motion, 417
for interplanetary motion, 417
for relative motion, 417
Osculating plane, 458

P

Pappus, theorems of, 264
Parallel, springs in, 789
Parallel-axis theorems. *See* Transfer formulas
Parallelogram law, 18, 394

Particle kinematics. *See* Kinematics of particles
Pendulum, ballistic, 726
compound, 776
simple, 775
torsion, 779
Percussion, center of, 594
Periodic motion. *See* Vibrations
Periods of vibration, 768, 775, 777, 786
Phase angle, 787
Pitch of, square-threaded screw, 223
Pivot bearing, friction in, 235
Planck, Max, 479
Plane motion, kinematics of. *See* Kinematics of rigid bodies
Plane motion, kinetics of, 585-638
angular momentum, 586
in centroidal rotation, 591
dynamic equilibrium in, 625
equations of, 587, 588
physical interpretation of, 588
in general motion, 621
in noncentroidal rotation, 591
of rolling bodies, 607-614
Pocket calculators, 10-15
Point diagram, 111, 727
Polar coordinates. *See* Radial and transverse components
Polar moments of inertia, 283
Polygon, acceleration, 545, 551
force, 395, 398
funicular or string, 396, 398
velocity, 545, 550
Position vector, 418
change in, 418
rate of change of, 583
Potential energy, 646
Potential function, 646
Power, 656
units of, 657
Precession, 760
axis of, 760
Principal normal, 457
Principle, of action end reaction, 4, 92, 215
of conservation of momentum, 720,
of moments, 54
of transmissibility, 3, 56
of virtual work, 376
Principal axes of inertia, 312,
Product of inertia defined, 297
relative to inclined axes, 308
with respect to axis of symmetry, 298
transfer formula for, 298
units and signs, 297
Projectiles, flight of, 446
maximum height and range, 450

Q

Quantum mechanics, 479

R

Radial and transverse components, 463-466
of acceleration, 465
geometric significance, 465
related to normal and tangential components, 465
unit vectors of, 464
rate of change of, 464, 584
of velocity, 414
Radius of curvature, 408, 410, 417, 454

Radius of gyration, 281, 317
Rated speed of highway curve, 448
Rectangular components, of acceleration, 394
of force, 18, 30, 33
of moment, 65
related to normal and tangential components, 406
of velocity, 394
Rectilinear motion, correlation with angular motion, 535
kinematics of, 422-428
motion curves for, 431-440
sign convention of, 423
Rectilinear translation. *See* Translation
Reduction of a system of forces, 75
Redundant members, 167
Reference axes, 417
with fixed origin, 417
inertial, 479, 561
with moving origin, 417
Relative motion, 417
Relative spatial motion,
acceleration equation in, 612
applications of, 618-624
correlation with plane motion, 613, 614, 615
geometric Interpretation of, 609
kinematics of, 608-624
relation between absolute and rotating systems, 610
velocity equation in, 611, 613
Relativistic mechanics, 479
Resonance in forced vibration, 798, 805
Restitution, coefficient of, 726
Resultant of any force system, 74
concurrent, 25, 26
coplanar, 76-78
graphically, 395
reduction to force-couple system, 75
spatial, 79-80
wrench, 76
Resultant work, 643
Retrograde precession, 643
Reversed effective force, 484
See also Inertia force
Reversed normal effective force, 493
Reversed tangential effective force, 493
Right-hand rule, 46
Rigid body, defined, 2
motion of mass center of, 485
types of, motion, 416, 532
Rigid structure, 138
Rocket flight, 715
Rolling bodies, kinematics of, 549, 570, 573
kinetics of, 607-613
Rolling resistance, 238
Rotating reference frames, 558
Rotation, axis of, 533
center of percussion for, 594
centroidal, 591
characteristics of, 534
defined, 533
dynamic equilibrium in, 592, 597
effective forces of, 591
impulse-momentum applied to, 739-744
kinematics of, 533-539
kinetic energy of, 669
noncentroidal, 591
reference axes for, 592
sign convention for, 536, 592
work-energy applied to, 668-670

S

Satellite motion, 750-757
- apogee, 752
- areal velocity, 753
- escape velocity, 753
- Kepler's laws, 754
- perigee, 753
- period of orbit, 753
- various paths of, 753
- work done by gravitational attraction, 750

Scalar product. *See* Dot product
Scalars, defined, 5
Scalar triple product, 51, 68
- evaluation of, 51

Screws, *See* square-threaded screws
Second moment of area, See Moments of inertia for areas
Sections, method of, 154, 185
Series, springs in, 789
Shear diagrams, 348
- construction of, 358, 359, 363
- shape of, 356, 359, 363

Shear in beams, 339-372
- defined, 341, 342
- equations of, 346, 350
- relation to bending moment, 354-366
- section of zero shear, 348, 360, 363,
- sign of, 342

Significant figures, 12
Simple harmonic motion, 769
- amplitude of, 769
- auxiliary circle of, 769
- frequency of, 770
- graphic representation of, 783
- natural circular frequency, 770
- period of, 771

Simple pendulum, 775
Simple trusses, assumptions in, 140
- construction of, 138
- criteria for stability and. determinancy, 143
- defined, 140
- loading of, 140
- method of joints applied to, 143, 145
- method of sections applied to, 154
- redundant members in, 167
- support of, 148
- zero load members in, 143, 149

Slide rule techniques, 13, 18, 22, 32, 99
Slopes, method of, 145
Slug, 6
Solution of problems, 7
Space probe, 452
Space trusses, 184
- compound, 184
- construction of, 184
- criterion for stability, 185
- by method of joints, 185
- by method of sections, 189
- simple, 185

Spatial forces, components of, 30
- direction angles, 32
- direction cosines of, 33
- equilibrium of, 118
- rectangular components, 33
- resultant of, 33, 76

Spatial moment, 63
- component of, in any direction, 66-68
- geometric method, 65
- by vector algebra, 66, 67

Speed, defined, 419
  correlated with velocity, 420
  axis of, 759
Spring, work done by, 644
  from force-displacement diagram, 644
  independent of spring rotation, 645
Spring constant, 644
  equivalent, 788
  torsional, 779
Springs, in parallel, 789
  in series; 790
Square-threaded screws, 223
  lead of, 223
  multiple-threaded, 223
  pitch of, 223
Stability, equation of, in frames, 171
  in trusses, 142, 186
Static friction, 197
Statics, preview of, 1
Step bearing, friction in, 233
String polygon, 396
  pole of, 397
  rays of, 397
Substitute member, 410
Successive approximation, 759
Summary of, analysis of structures, 193
  centroids, 277
  equilibrium of force systems, 135
  friction, 239
  general principles of dynamics, 489
  impulse and momentum, 764
  kinematics of a particle, 474
  kinematics of rigid bodies, 582
  kinetics of particles, 530
  kinetics of rigid bodies, 637
  mechanical vibrations, 806
  moments of inertia, 337
  resultants of force systems, 88
  shear and moment, 412
    work-energy method, 751
Suspension, center of, 776
Symmetry, axis of, 298
Systems of units, 481

T

Table of, centroids, 253
  free-body diagrams, 93
  moments of inertia, 290, 328,
Tabular computation, 256, 274, 293
Tangential acceleration, 454, 457
  unit vector of, 455
  rate of change of, 455, 584
Tangential inertia force, 492
Tension in truss members, 142
Theorems of Pappus, 264
Theory of friction, 195
Theory-solution method, 8
Three-dimensional force systems. *See* Spatial forces
Three-dimensional motion. *See* Spatial dynamics; Spatial motion
Thrust bearing, friction in, 233
Time-dependent mass, 712
Torque axis, 760
Torsional spring constant, 780
Torsional vibrations, forced, 805
Torsion pendulum, 780
  nodal section of, 781
  period of, 780
Torus, 268

Transfer formula
for moments of inertia,
281, 299, 317
for radii of gyration, 281, 319
Translation, defined, 490
analysis as a particle, 491, 504
analysis as a rigid body,
518-522
of connected systems,
504, 507-510
curvilinear, 493, 495-498
dynamic equilibrium in, 519
kinematic characteristic of, 490
possible direction of motion of,
508
rectilinear, 492
use of force-time diagrams in,
505-507
Triangle law, 17
Trigonometric functions, 13
relations, 13
Triple scalar product, 52
Triple vector product, 52
Trusses. *See* Simple trusses
compound, 140
complex, 140
Two-force members, 143, 169
rule for tension or
compression, 143

U

Unbalanced wheel, 659
work-energy applied to, 741
Uniformly accelerated motion, 424
Units, absolute systems of, 481
gravitational systems, 4, 481
Units and dimensions, 4
conversion of, 4
systems of, 4, 481
Unit vector, 38
along arbitrary direction, 39
binormal, 458
along coordinate axes, 38
normal, 454
radial, 463
rate of change of, 584
tangent, 419
transverse, 463
Unit vectors, cross product of, 46
dot product of, 40

V

Variable acceleration, 423, 427
Variable mass systems, 761
Varignon's theorem, 60
V belt, friction in, 231
Vector, defined, 5
notation, 37
representation of a, 5
Vector algebra, 40
cross or vector product, 46
dot or scalar product, 40
Vector calculus, introduction to,
443
Differentiation, rules for, 443
by omega theorem, 582
integration, 443
Vector displacement, 418
Vector notation, 37
various methods of, 37
Vector product. See Cross
product
Vectors, addition of, 16
calculus of. *See* Vector calculus
equality of, 49
free, 17
localized, 17
rate of change of. *See* Omega
theorem
sliding, 37
Vector triple product, 49
reduction formula for, 49

Velocity, absolute
  angular, 535
  areal, 753
  correlated to acceleration,
    420, 421
  defined, 419
  of escape, 753
  in frictionless path, 650
  hodograph derived from, 420
  from instant center, 566
  from motion curve, 566-574
  in plane motion, 546, 550, 553
  radial and transverse
  components of, 463
  rectangular components of, 444
  relative to rotating axes, 559
Velocity-time diagram, 432, 539
  displacement from, 432, 539
Vibrations, 768-808
  amplitude of,
    771, 787, 801, 803
  defined, 768
  forced, 798
  free, 784
  frequency of, 771, 775, 786
  period of,
    768, 775, 777, 780, 786
  torsional, 779
  by work-energy method, 793
Virtual displacement, 375
  geometrically, 382
mathematically, 381
Virtual work, 373-386
  application of, 377-386
  defined, 375
  introductory concepts of, 375
  principle of, 376
  summary of, 412
Volumes, centroids of, 270

W

Watt, 657
Wedges, friction in, 214
Weight, defined, 480
Weight moment of inertia, 327
  See also Mass moments of
  inertia
  conversion to mass moments of
  inertia, 329
Wichert truss, 166
Work, of applied moment, 669
  definition and meaning of, 373
  differential, 641
  from force-displacement
  diagram, 643
  of gravitational attraction, 751
  of internal forces in connected
  systems, 659
  interpretation and calculation
  of, 641
  potential function for, 646
  of a spring, 644
  virtual. *See* Virtual work
Work-energy method, 639-663
  advantages of, 659
  in connected systems, 659-663
  in free vibrations, 793
  general plan of applying, 660
  in plane motion, 685-692
  in rotation, 668-670
  in translation, 640, 646-650
Wrench, as resultant of force
  system, 76, 87